Chapman & Hall's Complete

Fundamentals of Engineering Exam Review Workbook

Chapman & Hall

Professional Engineer Workbook Series

Chapman & Hall and Professional Engineering Review Course (PERC), Inc. have teamed up to provide the best Professional Engineer review material available.

The Chapman & Hall Professional Engineer Workbook Series is designed to prepare candidates to pass the PE exams. Based on the NCEES examination specification, candidates are guided through the preparation process focusing only on the areas presented on the exam. Each book is designed as a teaching reference that contains detailed explanations, worked problems, sample tests, and additional problems and solutions at the end of the book. Every book in the series is the full study kit needed by every candidate and includes an all-important exam strategy to help you focus on passing the exam by preparing in the most efficient and time-tested way.

So, whether you are taking the Fundamentals of Engineering (EIT) exam, or the Principles and Practice exam, Chapman & Hall has all the material you'll need to effectively study and pass.

Chapman & Hall's Complete Civil Engineering Exam Review Workbook includes complete coverage of: transportation systems and facilities; building, bridge and special structures; water and wastewater treatment; geotechnical projects; hydraulics; and foundations.

Hardback (ISBN: 0-412-15011-5)

Chapman & Hall's Complete Electrical Engineering Exam Review Workbook includes complete coverage of: power generation systems; transmission and distribution systems; rotating machines; lightning protection; control systems; electronic devices; instrumentation; digital systems; communication systems; and computer systems.

Hardback (ISBN: 0-412-15031-X)

Chapman & Hall's Complete Mechanical Engineering Exam Review Workbook includes complete coverage of: machine design; stress analysis; kinematics; dynamics; power generation; vibration; heat transfer; thermodynamics; and fire protection.

Hardback (ISBN: 0-412-15021-2)

Chapman & Hall's Complete

Fundamentals of Engineering Exam Review Workbook

PROFESSIONAL ENGINEER REVIEW COURSE, INC.

Northport, New York

CHAPMAN & HALL

INTERNATIONAL THOMSON PUBLISHING
Thomson Science

New York • Albany • Bonn • Boston • Cincinnati • Detroit
London • Madrid • Melbourne • Mexico City • Pacific Grove
Paris • San Francisco • Singapore • Tokyo • Toronto • Washington

Cover design: Curtis Tow Graphics

Copyright © 1998 by PERC, Inc.

Printed in the United States of America

Chapman & Hall
115 Fifth Avenue
New York, NY 10003

Chapman & Hall
2-6 Boundary Row
London SE1 8HN
England

Thomas Nelson Australia
102 Dodds Street
South Melbourne, 3205
Victoria, Australia

Chapman & Hall GmbH
Postfach 100 263
D-69442 Weinheim
Germany

International Thomson Editores
Campos Eliseos 385, Piso 7
Col. Polanco
11560 Mexico D.F
Mexico

International Thomson Publishing–Japan
Hirakawacho-cho Kyowa Building, 3F
1-2-1 Hirakawacho-cho
Chiyoda-ku, 102 Tokyo
Japan

International Thomson Publishing Asia
221 Henderson Road #05-10
Henderson Building
Singapore 0315

Nelson Canada
1120 Birchmount Road
Scarborough, Ontario
Canada M1K 5G4

To request a refund, please send us the following:
  Your name and address
  The Chapman & Hall Exam Review Workbook
  Proof of failure from your state board (this must be dated within two months of requesting
    your refund)
  The original sale receipt or invoice showing the purchase price of the Workbook.
Send the material to:
  Engineering Marketing Manager
  Chapman & Hall
  115 Fifth Avenue
  New York, NY 10003
We'll refund the amount you paid for the Workbook up to the published list price.

All rights reserved. No part of this book covered by the copyright hereon may be reproduced or used in any form or by any means—graphic, electronic, or mechanical, including photocopying, recording, taping, or information storage and retrieval systems—without the written permission of the publisher.

1 2 3 4 5 6 7 8 9 10 XXX 01 00 99 98

Library of Congress Cataloging-in-Publication Data

Chapman &Hall's complete fundamentals of engineering exam review
    workbook / by Professional Engineer Review Course, Inc.
      p.   cm. -- (Professional engineer workbook series)
    Includes bibliographical references and index.
    ISBN 0-412-14961-3 (pb : alk. paper)
    1. Engineering--United States--Examinations--Study guides.
  2. Engineering--Problems, exercises, etc.   I. Professional Engineer
  Review Course, Inc.   II. Series.
  TA159.C48   1998
  620'.0076--dc21   97-43131
                                                                 CIP

British Library Cataloguing in Publication Data available

"Fundamentals of Engineering Exam Review Workbook" is intended to present technically accurate and authoritative information from highly regarded sources. The publisher, editors, authors, advisors, and contributors have made every reasonable effort to ensure the accuracy of the information, but cannot assume responsibility for the accuracy of all information, or for the consequences of its use.

To order this or any other Chapman & Hall book, please contact **International Thomson Publishing, 7625 Empire Drive, Florence, KY 41042.** Phone: (606) 525-6600 or 1-800-842-3636.
Fax: (606) 525-7778. e-mail: order@chaphall.com.

For a complete listing of Chapman & Hall titles, send your request to **Chapman & Hall, Dept. BC, 115 Fifth Avenue, New York, NY 10003.**

TABLE OF CONTENTS

INTRODUCTION xv
EXAM STRATEGY - EIT xvii
PROBLEM SOLVING xxiii

SECTION 1: MATHEMATICS

Selected References 1-iii
Numbers 1-1
  Real Numbers 1-1
  Imaginary Numbers 1-2
Sets 1-3
Algebra 1-5
  Fundamental Laws 1-5
  Exponents 1-5
  Binomial Formula 1-7
  Rules for Determining Terms in a Binomial Expansion 1-8
  Determinants 1-9
  Linear Algebra 1-12
  Vector Analysis 1-21
  Higher Order Algebraic Equations 1-26
  Logarithms 1-28
Trigonometry 1-33
  Angles 1-33
  Trigonometric Functions 1-34
  Polar Coordinates 1-37
  Fundamental Identities 1-38
  Laws for Triangles 1-39
  Hyperbolic Functions 1-41
Analytic Geometry 1-42
  Straight Lines 1-42
  Conic Sections 1-46
Differential Calculus 1-49
  Functions 1-49
  Limits 1-49
  The Derivative 1-52
  Basic Rules of Differentiation 1-53
  Table of Derivatives 1-53
  Implicit Differentiation 1-59
  Higher Order Derivatives 1-59
  Partial Derivatives 1-60
  Applications 1-63
Integral Calculus 1-70
  Indefinite Integral 1-70

v

Definite Integral 1-71
Table of Integrals 1-72
Integration by Parts 1-74
Double Integrals 1-75
Applications 1-87
Differential Equations 1-82
Definitions 1-82
Equations with Variables Separable 1-82
First-Order Linear Differential Equations 1-84
Second-Order Linear Differential Equations Homogeneous
 with Constant Coefficients 1-85
Applications 1-87
Laplace Transformation 1-88
Definition of the Laplace Transform 1-88
Table of Transforms 1-90
Application to the Solution of Differential Equations 1-92
Probability and Statistics 1-94
Probability 1-94
Statistics 1-97
Normal Distribution Functions 1-103

SECTION 2: STATICS

Selected References 2-ii
Notation 2-iii
Coplanar Force Systems 2-1
Parallel Force Systems 2-1
Concurrent Force Systems 2-2
Non-concurrent Force Systems 2-3
Couple 2-3
Varignon's Theorem 2-4
Equations of Equilibrium 2-4
Planar Structure with Parallel System of Forces 2-5
Planar Structure with Concurrent Force Systems 2-6
Planar Structure with Non-concurrent Force System 2-7
Trusses 2-8
Method of Joints 2-10
Short Cuts and Special Cases 2-14
Method of Sections 2-15
Short Cuts and Special Cases 2-17
Illustrative Examples 2-19
Three Force Members 2-21
Illustrated Examples 2-22
Free Body Diagrams of Each Member 2-22
Solved Problems 2-24
Three Dimensional Forces 2-44

Concurrent Forces in Space 2-44
Direction Cosines 2-45
Concurrent Forces Not Mutually Perpendicular 2-46
Non-concurrent Space Forces 2-47
Equations of Equilibrium for 3D Concurrent Forces 2-47
Moment of a Force About an Axis 2-47
Friction 2-55
Summary of Laws of Friction 2-57
Illustrative Problems 2-58
Wedges 2-63
Screw Thread Friction 2-65
Belt Friction 2-67
Solved Problems 2-70
Centroids 2-87
Centroids of Areas 2-89
Centroids of Lines 2-90
Centroids by Integration 2-90
Theorems of Pappus and Guldinus 2-94

SECTION 3: MECHANICS OF MATERIALS

Selected References 3-iii
Notation 3-iv
Tensile and Compressive Strain 3-1
Shearing Strain 3-1
Stress-Strain Diagram 3-2
Definitions 3-2
Proportional Limit 3-2
Yield Point 3-2
Ultimate Strength 3-3
Permanent Set 3-3
Elastic Limit 3-3
Yield Strength 3-3
Hooke's Law 3-4
Poisson's Ratio 3-4
Coefficient of Thermal Expansion 3-4
Statically Indeterminate Structures 3-9
General Solution Procedure 3-9
Illustrated Problems 3-9
Torsion 3-13
Angle of Twist 3-15
Compound Shafts 3-17
Illustrated Examples 3-17
Statically Indeterminate Torsion Problems 3-19
Mechanics of Materials Solved Problems 3-21
Beam Analysis 3-29

TABLE OF CONTENTS vii

FUNDAMENTALS OF ENGINEERING EXAM REVIEW WORKBOOK

Types of Beams 3-29
Types of Loads 3-29
Conditions of Equilibrium 3-30
Sign Convention 3-32
Reactions 3-33
Illustrated Problems 3-33
Shear and Bending Moment Diagrams 3-36
The Shear Forces Diagram 3-36
The Bending Moment Diagram 3-38
Illustrated Example 3-41
The Flexure Formula 3-45
Position of Neutral Axis 3-47
Moment of Inertia of a Rectangle 3-48
Parallel Axis Theorem 3-49
Moments of Inertia of Composite Sections 3-49
General Procedure of Composite I 3-50
Illustrated Example 3-50
Solved Problems 3-51
Combined Stress 3-65
Formulas for Normal and Shear Stress 3-66
Appendix A: Combined Stress (Example) 3-70

SECTION 4: DYNAMICS

Selected References 4-ii
Fundamental Definitions 4-1
Equations of Rectilinear Motion 4-2
Summary for Rectilinear Motion 4-3
Newton's Laws of Motion 4-4
Fundamental Equations of Kinetics for a Particle 4-5
Center of Mass 4-6
Center of Gravity 4-7
Space Motion 4-7
Kinematics of Rectilinear Motion 4-8
Constant Velocity 4-8
Constant Acceleration 4-8
Graphical Representation of Motion 4-13
Displacement as a Function of Time 4-13
Velocity as a Function of Time 4-15
Acceleration as a Function of Velocity 4-19
Constant Force—Constant Acceleration 4-20
d'Alembert's Principle 4-25
Variable Force 4-28
Angular Velocity 4-39
Kinematics of Curvilinear Motion 4-43
Rectangular Coordinates 4-43

TABLE OF CONTENTS

Component Motion 4-46
Flight of Projectiles 4-46
Radial and Tangential Motion 4-48
Kinetics of Curvilinear Motion 4-53
Banking of Curves 4-57
Rotation of a Rigid Body 4-65
Dynamic Equilibrium of Noncentroidal Rotation ... 4-70
Radius of Gyration 4-70
Rotational Dynamic Equilibrium 4-70
Work–Energy 4-75
Work–Energy Method Involving Variable Forces ... 4-77
Equivalent Springs 4-80
Conservative Systems 4-83
Impulse and Momentum 4-84
Dynamic Action of Jets 4-88
Conservation of Momentum 4-91
Elastic Impact 4-94
Table I: Common Examples of Simple Harmonic Motion ... 4-100

SECTION 5: FLUID MECHANICS

Selected References 5-ii
Some Symbols and Constants Used in Fluid Mechanics ... 5-iii
Fluid Pressure 5-1
Hydrostatic Force 5-5
Thin-Walled Cylinders Under Pressure 5-11
Thick-Walled Cylinders 5-13
Buoyancy 5-14
Dams and Gates 5-15
Bernoulli's Theorem for an Incompressible Fluid .. 5-19
Head Loss Due to Pipe Friction 5-24
Orifice Coefficients 5-31
Entrance Losses 5-32
Other Losses 5-35
Viscosity and the Reynolds Number 5-41
Pipe Line Supplied By a Pump 5-47
Branch Pipe Flow 5-50
Flow Measurement 5-56
Nozzles 5-59
Flow Trajectory 5-60
Flow Over Wiers 5-61
Sluice Gate 5-67
Flow in Open Channels 5-67
Dynamic Force and Momentum 5-69
Dimensional Analysis 5-73
Translation and Rotation 5-75

SECTION 6: ELECTRICITY/ELECTRONICS

Selected References 6-iii
ELECTRICITY I 6-1
Electrical Terminology 6-1
Units and Notation 6-2
Electrostatics 6-2
Electrolysis 6-4
Resistance, Capacitance, Inductance 6-5
Direct Current (DC) Circuits 6-11
Alternating Current (AC) Circuits 6-18
Impedance and Admittance 6-21
DC and AC Power 6-28
Complex Circuits 6-31
Resonance 6-40
Simple Filters and Attenuators 6-44
Practice Problems with Solutions 6-47
Appendix 6.1, Selected Unit Relationships 6-56
Appendix 6.2, Review of Complex Numbers 6-57
Appendix 6.3, Algebra of Determinants 6-60
ELECTRICITY II 6-63
Three-Phase circuits 6-63
Magnetics 6-69
Transformers 6-71
Rotating Machines 6-79
DC Generators 6-81
DC Motors 6-88
AC Generators 6-93
AC Motors 6-97
Three-Phase Induction Motor 6-99
Measurements 6-105
Practice Problems with Solutions 6-108
Appendix 6.4, Magnetic Units and Terminology 6-117
Appendix 6.5, DC Motor Characteristics 6-118
Appendix 6.6, AC Motor Characteristics 6-118
ELECTRICITY III 6-119
Time Domain Analysis 6-119
Introduction to Laplace Transforms 6-142
System Engineering 6-149
Examples 6-171
ELECTRICITY IV 6-179
Semiconductors 6-179
Semiconductor Devices 6-185
Field Effect Transistors 6-193
Transistors 6-208
DC Conditions - BJT 6-213

Signal Conditions - BJT 6-219
Appendix - Operational Amplifiers 6-231

SECTION 7: CHEMISTRY

Selected References 7-ii
Periodic Law 7-1
Gas Laws 7-3
Chemical Reactions 7-6
Reaction Rate (Kinetics) and Equilibrium 7-8
Solutions 7-12
Nuclear Chemistry and Radioactivity 7-14
Thermochemistry 7-14
Organic Chemistry—Study of Carbon Compounds 7-15
Valence 7-19
Oxidation-Reduction (Redox) Reactions 7-22
Chemical Mathematics (Stoichiometry) 7-26
Illustrative Problems 7-27
Periodic Chart of the Elements 7-34

SECTION 8: THERMODYNAMICS

Selected References 8-ii
Definition of Symbols 8-iii
Thermodynamic Processes 8-1
General Energy Equation 8-3
Thermodynamic Cycles 8-10
Carnot Cycle 8-10
Ericson Cycle 8-11
Stirling Cycle 8-12
Otto Cycle 8-12
Diesel Cycle 8-13
Illustrative Examples 8-18
Gas Mixtures 8-28
Volumetric Analysis 8-29
Gravimetric Analysis 8-29
Dalton's Law of Partial Pressures 8-30
Combustion 8-31
Heating Value 8-34
Liquids and Vapor 8-35
Throttling Calorimeter 8-38
Rankine Cycle 8-38
Efficiencies 8-39
Solved Problems 8-41
Psychrometrics 8-47
Power Plants 8-50
Reheat Cycle 8-54

Refrigeration 8-57
Heat Transfer 8-61
Radiation 8-66
Solved Problems 8-67
Steam Turbines and Nozzles 8-73

SECTION 9: MATERIALS SCIENCE

MATERIAL SCIENCE AND STRUCTURE OF MATTER 9-1
Crystal structures 9-1
Space lattices 9-2
Phase diagrams 9-5
Heat treating 9-6
Corrosion 9-13
Materials testing 9-13
Impact tests 9-15
Stress–strain relationships 9-18
Physical properties 9-18
Nonmetallic materials 9-23
PERIPHERAL SCIENCES 9-26
Rectilinear motion 9-26
Harmonic motion 9-31
Wave motion 9-39
Sound waves 9-47
Intensity of sound 9-47
Light and optics 9-50
Spherical mirrors 9-51
Refraction 9-54
Prisms 9-56
Lenses 9-57
Keppler's Laws 9-64

SECTION 10: ENGINEERING ECONOMICS

Introduction 10-1
Time Value of Money 10-2
Cash Flow Diagrams 10-3
Financial Calculations 10-4
Equivalency Formulas 10-5
Bond Calculations 10-21
Depreciation Calculations 10-27
Straight Line Depreciation Method (SLD) 10-28
Sinking-Fund Depreciation Method (SFD) 10-28
Declining Balance or Fixed Percentage Depreciation Method (DBD) .. 10-28
Double Declining Balance Depreciation Method (DDBD) 10-29
Sum of the Years Digits Depreciation (SYD) 10-30
Comparative Economic Analysis Methods 10-34

Profit in Business	10-34
Annual Cost (TAC) Method	10-35
Basic Approach	10-37
Approach Including Risk	10-44
Present Worth Method	10-46
Capitalized Cost Method	10-56
Rate of Return Method	10-58
Evaluation of an Investment	10-58
Corporate Rate of Return Calculations	10-65
Incremental Rate of Return (Comparing Alternatives)	10-71
Benefit to Cost Ratio Method	10-76
Other Topics	10-81
Replacement Studies	10-81
Determination of Economic Replacement Interval for the Same Asset	10-81
Determination of Economic Replacement for an Improved Asset	10-82
Break Even Cost Analysis	10-84
Minimum Cost Analysis/Economic Lot Size	10-87
Homework Problems and Solutions	10-87

SECTION 11: PRACTICE PROBLEMS

Mathematics	11-1
Statics	11-5
Mechanics of Materials	11-11
Dynamics	11-18
Fluid Mechanics	11-22
Electricity/Electronics	11-27
Chemistry	11-32
Thermodynamics	11-34
Materials Science	11-39
Engineering Economics	11-41

SECTION 12: SOLUTIONS TO PRACTICE PROBLEMS

Mathematics	12-1
Statics	12-11
Mechanics of Materials	12-24
Dynamics	12-36
Fluid Mechanics	12-47
Electricity/Electronics	12-56
Chemistry	12-68
Thermodynamics	12-73
Materials Science	12-81
Engineering Economics	12-85

SECTION 13: SIMULATED EXAMS

A.M. Section 13-1
P.M. Section 13-16
A.M. Solutions 13-24
P.M. Solutions 13-44

INDEX 14-1

Introduction

I am often asked the question, "Should I get my PE license or not?" Unfortunately the answer is, Probably. First let's take a look at the licensing process and understand why it exists, then take a look at extreme situations for an attempt at a yes/no answer, and finally consider the exams.

All 50 have a constitutionally defined responsibility to protect the public. From an engineering point of view, as well as many other professions, this responsibility is met by the process of licensure and in our case the Professional Engineer License. Though there are different experience requirements for different states, the meaning of the license is common.

The licensee demonstrates academic competency in the Fundamentals of Engineering by examination (Principles and Practices at PE time).

The licensee demonstrates qualifying work experience (at PE time).

The licensee ascribes to the Code of Ethics of the NSPE, and to the laws of the state of registration.

Having presented these qualities the licensee is certified as an Intern Engineer, and the state involved has fulfilled its constitutionally defined responsibility to protect the public.

Thus for you as an individual the EIT certificate states to all concerned that you have demonstrated academic competency in the Fundamentals of Engineering, you live up to a code of ethics, and when you pass the PE exam, you demonstrate academic competence in the Principles of Engineering as well as qualifying work experience, and ascribe to the Code of Ethics of the NSPE. This is understood by any who may inquire, and is evidenced by the letters PE next to your name.

Licensed PEs have an earning potential roughly 14% higher over the course of their career than nonlicensed engineers because they have a broader base of career opportunities available to them. This is primarily of benefit to younger engineers, but those near to the end of their career often find the PE license opens doors to opportunities after retirement.

Many corporations see the licensing of their engineering organizations as a sign of professionalism, and use it as an incentive to gain business.

So, what about your career and the PE license? We look at two extremes:

1. If you are going to go into consulting and work on projects that directly affect the public, you are going to get your PE or you are not going to participate. It is that simple.

2. If you are going to work for a large corporation, and retire at age 65 or whatever, no, don't bother. Take advantage of all the programs of the corporation and even consider a Master's degree in fields of their interest (and obviously yours).

3. If you are going into teaching at the college level, no, get a PhD instead.

Unfortunately, few operate at these extremes, so the choice is not always obvious. In addition to the above, the PE license is a credential that will get a resume read in the job market.

My suggestions is, and you have probably already concluded, having purchased this book, that somewhere along the course of your career, you will find the PE beneficial, and thus worth the time, energy, and money that you invest.

The FIT is passable. People do it all the time. It has been clearly evidenced that if you know the technical material, you will pass this exam, and if you do not, you will not. I have heard every excuse in the book, but almost never do I hear that the reason that a candidate did poorly on the exam is that he or she did not know the Fundamentals of Engineering well enough to meet the pass point of the exam. Objectivity in this area is essential.

Knowledge of the technical material as well as speed are gained by doing practice problems. I know of no other way. Use the NCEES exam specs to make sure you are concentrating on the proper technical areas, then build a strategy.

When your plan is done, start reading the text. Try a practice problem or two, using the NCEES reference handbook. You will get stuck. Consult the solution, then go back to the text. Remain in this learning process until you are confident in the subject. I cannot overemphasize the need to know the subject, and not just how to answer a few practice problems. When you have worked all the practice problem we have provided, go get some more. Use the NCEES Sample Problems as a guide to the degree of difficulty you should expect on the exam, and get the whole program on a realistic emotional level. Too many assign too much fear to this exam. Respect is appropriate, but not fear.

Let me take this opportunity now to thank you for the opportunity to be of assistance to you in this endeavor, and to wish you well in the pursuit of your Professional Engineer License.

Exam Strategy—EIT

The numbers below are not mathematically related in any way, but describe the scores needed to pass the FE exam and what they mean.

$$240 = 100$$
$$110 = 70$$
$$0 = 0$$

A score of 110 points is equal to a passing grade of 70 for the EIT exam. The highest attainable score is 240, equivalent to 100%, and the lowest score is 0.

Though this may seem an oversimplification, to pass this exam, just get 110 points and do not get 109 or less. This exam is not curved. If everyone in the country who takes this exam makes 110 or better, everyone passes. Conversely, if everyone makes 109 or less, everyone fails.

I am often asked, what happens if I do well in the morning, and not in the afternoon? It is unlikely that one will do well in the morning and then suffer a sudden loss of ability, but the answer is—just get 110 points. Conversely, it is not likely that after a poor morning, one will suddenly improve at lunchtime, so the answer is still—just get 110 points.

There is another meaning to 110 points that is important to all 50 State boards and to the NCEES. In short it represents that body of knowledge in which you must demonstrate competency in order to be certified by a state as an Engineer in Training. States, by authority of their constitution, have the responsibility to protect the public. They believe that if you make 110 points you have demonstrated technical competency and can be licensed to practice in public as a professional engineer after you gain the necessary experience.

The NCEES issues lists of subjects and the number of questions for each in the sheet in their FE Sample Examination (also available at www.pe-exam.com). Of the 120 questions available in the morning, you should be prepared to handle enough fields to have access to 100 questions. In the afternoon you will have 60 questions to answer and I suggest you be prepared in enough fields to have access to 50. Attempting all 180 questions will result in undue stress and other undesirable effects.

Each morning question is worth 1 for total available morning points of 120. If you get 110 correct in the morning, it's not necessary to take the afternoon test. This has never happened so it's wise to plan on the afternoon session where each of the 60 questions is worth 2 points.

As you can see on the list of subjects from the NCEES, there are 12 different fields on the FE. If you are to be prepared for 100 questions in the morning and 50 in the afternoon you will find that you must be prepared in at least 7 of those fields. This is where the EIT gets its reputation. Getting all the

xvii

questions correct in your discipline is a sure way to fail. The biggest reason for failing this exam is not being prepared in enough subjects so that you have access to 110 points, rather than missing questions in your own field.

If you know the technical material, in enough fields of engineering you do not have to worry, and if you do not know the technical material, you will not have to worry either, but for very different reasons. *There is no substitute for technical knowledge on this exam.*

Please get the list from the NCEES out now and review the list of morning subjects and prioritize the subjects 1–12 in your order of preference in such a way that you pick enough fields to have access to 100 questions.

We conclude from this:

Everyone's favorite field will be Mathematics.
Then the subjects of the discipline.
Then seriously consider Engineering Economics.
Prepare in enough other fields to have access to enough questions to pass.

Now for the afternoon. Your exam book will contain the questions in all the disciplines that are offered and you must pick the one you want to do. As you can see the list of subjects in any of the technical disciplines very closely resembles the exam spec for the PE exam. If you do questions in your discipline in the afternoon, you will in effect be working on a mini PE exam and and just about double the scope of knowledge required to pass. In, addition you will be obliged to take a PE Review Course in your discipline. However the list of subjects in the General Engineering exam is the same as the list in the morning exam. There is no gain in doing one discipline in the afternoon and your probability of success is much greater by doing both exams in the same subjects for both parts.

Note the list of subtopics on the NCEES exam specs. If a subtopic is not on the list, it is not necessary to prepare for it.

The exam book contains an index of fields and question numbers in the back. Approach the questions by field starting with your first priority and working to your tenth. This ensures that your effort is placed in the fields you are capable in and leaves your guessing to the areas of lesser interest (or even none at all). Using this prioritized list of fields and the question index will maximize the use of time available.

For Electrical Engineers, there is another approach to consider. Everyone is familiar with E = IR. To EE's this is as basic as one can get. Each field on this exam has E = IR level of questions. Just going through the review course should prepare you at this fundamental level in all fields.

EXAM STRATEGY-EIT   xix

Statics: Sums of forces and moments must be zero.

Mechanics of Materials: $\delta = \dfrac{PL}{AE}$

Thermodynamics: $Q = U A \Delta T$

Dynamics: $F = ma$

Economics: There are six financial functions.

Fluid Mechanics: Conservation of mass and energy

There is no reason not to be able to recognize and do problems at this very fundamental level in all fields.

Speed is acquired as a result of doing the practice problems. These practice problems are either representative of old exam problems or included for teaching purposes. There is no substitute for practice as a way of getting both the speed and the accuracy required for this exam.

The EIT is an all multiple choice exam. This type of question tells you how well you are doing, as you have choices to pick from, and must come close to or go back to start. The four choices also show the required decimal accuracy.

The questions will not always be computational. If you are inclined to grab a formula and run, you may have trouble here. Attention to assumptions and methods is important for these noncomputational questions. Some will be simply definitions.

The drawback to multiple choice questions is that a black mark in the right place earns a point, and one in the wrong place earns nothing. To a large extent this can be offset by guessing. As a matter of fact you are expected to guess, and a factor for guessing has been included in the 110 points; so make sure you have 120 black marks on your morning paper and 60 black marks on your afternoon papers before you turn them in. Statistically speaking, single letter guessing is just slightly more efficient than random guessing.

Since the odds of successful guessing are 1 in 4, you will not be able to guess your way to success. There are no predominant letters or sequences on this exam so don't waste time trying to find them.

You must reserve time to guess. The proctors will not announce that it is time to guess now. About 5 min. is adequate, but you must remember. I suggest that you not scramble to get that last question right. Your odds in the scramble mode are nil; your odds of guessing are 1 in 4.

We found psychology to be a factor as well as technical knowledge on both exams, but particularly the EIT because of the time pressure of the exam.

If you have a real phobia when it comes to exams, we cannot help. Other sources must be consulted for situations in which a student cannot do multiple choice questions even though he is very familiar with the material.

The biggest reason for psychological difficulties is not phobias but the fear of failure, and it feeds on itself. The obvious correction to fear of failure is to know the technical material.

Read the question slowly.
Determine what the question wants you to do.
Do it.
Apply a simple order of magnitude test to your answer.

Please remember that this exam is on the Fundamentals of Engineering, not at the PhD level, and that people do pass it regularly. Do not give this exam credit for being more difficult than it is. Most of you already have a degree in engineering and thus have already demonstrated the ability to pass it. NCEES sample exams are a good source of perspective.

The next psychological factor is tension. Normally this is due to uncertainty and is overcome by confidence.

Recognize that you are tense before the panic sets in.
Stop, relax.
Count breaths.
Stare at a large blank area.
Analyze proctors.
Pinch yourself, snap a rubber band.

The tension is probably due to the fact that you are not doing as well as you think you should (knowing the material again). All of the anxiety in the world will not get the questions answered. If you do not know the subject, accept the fact, cancel the anxiety, and go on to what you do know. Take our practice exam to get the jitters out of your system when it doesn't matter. Remember, your goal is to pass with a 70, and a 90 is not required.

Fatigue can be a real problem during this exam. Just sitting for an 8-hour period requires physical energy that one is not normally accustomed to, and the emotional energy of thinking at 100% of capacity for an 8-hour period. The associated stress of being in an exam environment also takes its toll. A good night's sleep the night before is essential. Please do not attempt to cram all night the night before.

The last psychological problem is just hating to have to do it. You might as well have a positive attitude, and just enjoy it. It does cover material that you once enjoyed in college, so why not now?

EXAM STRATEGY-EIT

For EIT students the Practice Exam is as important as any technical session. Please take 3 hours to do it as it is the last chance you have to evaluate how prepared you are while you still have time to do something about it. Time pressure matters on the EIT; thus we insist it be done in the classroom situation. All practice exams are graded and returned at the end of the 3-hour session with solutions.

If you want to contest your result, consult your local board for guidelines. You must show that the machine did something wrong or missed a "gray" vs. "white" but the answers used by the NCEES are correct. Generally speaking, don't bother.

There may or may not be questions that are on the exam for the purpose of evaluation by the NCEES and they will not be scored. Thus it's not necessary to worry about them.

You may have wondered why the results take so long in coming back. The NCEES grades all the papers from the entire country and then evaluates the results of every question to ensure that nothing was overlooked. If something unexpected crops up, all of the papers from the entire country will be regraded against revised standards.

The purpose of this is simple. The NCEES will be pleased to call you and tell you that they made a mistake in grading and you passed after all. They will never be in a situation where they have to call you and say, we are sorry, we made a mistake, please send your certificate back.

Problem Solving

It is necessary to consider the most important variable in the problem solving process, the problem solver, who in this case is you and only you. The state and the NCEES are pleased to provide the academic and testing services, but that is the end of their participation. You are the key element to your success on these exams.

Differences between expert and novice problem solvers can be expressed in four categories: attitude, accuracy, degree of activity, and problem solving methods.

Positive attitude

Believe that problems can be solved through careful, persistent analysis and understanding that at first the problem may be confusing. This is not a situation that "you either know it or you don't", it is one that requires organized analytical thought. Don't fight the problem or the process. This is no place for an ego trip. Not finding any of the given answers means that you have done it wrong, not the NCEES.

Concern for accuracy

Great care must be taken to understand facts and relationships fully and accurately. Reread the problem several times. Don't leap to conclusions, but do pay attention to first impressions. Evaluate conclusions and judgments for accuracy and sensibility.

Breaking the problem into parts

Start at a point which is readily understood and then break the problem into successively smaller steps. Use a few key fundamental concepts as building blocks. These are quantitative formulas, visuals, and verbal descriptions.

Involvement in problem solving

Question the problem and try to create a mental picture of ideas. Redescribe the problem for the purpose of finding key conditions, variables. Be sure what the problem states and requires, as opposed to lying back and hoping the solution will come or that maybe the problem will go away and you can make it up on the next one.

Problem solving methods

When exploring a problem, it is important to get a feel for what the problem and solution are about. Begin by trying to guess the answer, eliminate the remainders then evaluate as stated before. One of the most useful methods for problem solving is the McMaster Five Step Strategy developed by Donald Woods (© Chem. Eng. Educ., Summer 1979, p. 132). The following areas are of particular use in the PE.

FUNDAMENTALS OF ENGINEERING EXAM REVIEW WORKBOOK

Define:
- Understand the stated problem
- Identify the unknown or stated objective
- Isolate the system and identify the knowns and unknowns and constraints stated in the problem
- Identify inferred constraints and criteria
- Identify the states criteria

Explore:
- Identify tentative pertinent relationships between inputs, outputs and unknowns
- Recall past related problems or experiences, pertinent theories and fundamentals
- Hypothesize, visualize, idealize and generalize
- Discover what the real problem and constraints are
- Consider both short-time and long-time implications
- Identify meaningful criteria
- Choose a basis or reference set of conditions
- Collect missing information, resources or data
- Guess the answer or result
- Simplify the problem to obtain an "order-of-magnitude" result
- If you cannot solve the proposed problem, first solve some other related problem or part of the problem

Plan:
- Identify the problem type and select a set of approaches
- Generate alternative ways to achieve the objective
- Map out the solution procedure (algorithm) to be used
- Assemble resources needed

Carry through:
- Follow the procedure developed under the plan phase
- Evaluate and compare alternatives
- Eliminate alternatives that do not meet all objectives and satisfy all constraints
- Select the best alternative of those remaining

Reflect:
- Check that the solution is blunder-free
- Check reasonableness of results
- Check that criteria and constraints are satisfied
- Check procedure and logic of your arguments
- Communicate results (put the black mark in the right circle)

Analyzing Problems

Identify:
- The pieces or elements of the problem
- A suitable classification to group the elements; the elements of problem statements can be classified by:
  - Criteria
  - Function
  - Information
- What's missing from the problem statement
- What's extraneous to the problem statement
- What pieces related to the problems have been solved
- The relationship between the elements
- Omissions from the problem statement

Distinguish:
- Fact from opinion from opinionated fact
- Conclusions from evidence

Detect:
- Fallacies in logic
- Incorrectly defined problems
- Missing information necessary to solve the problem
- Extraneous information

Recognize:
- Unstated assumptions
- What particulars are relevant
- When the problem should no longer be broken down into smaller pieces or subproblems
- What pieces are related to problems that have been solved

Another problem solving strategy by Frazer (© *Chem. Soc. Rev.*, *11* (2), p. 171, 1982):

- Try working backward from the goal, not forward from the information given (i.e., what is the step just before the solution is found?)
- Make a guess at the solution and then work backward to see if the guessed information is consistent with all available information
- Do not try to cope with too much information at one time. Break the problem down into subgoals and work at each separately
- Write down all ideas that come to you no matter how irrelevant or foolish they may seem ("brainstorm")
- Rest to allow time for "incubation of the problem"
- Verbalize the problem-talk to colleagues and experts. (Do this during homework sessions, but make absolutely no conversation during the FiT exam)

1
Mathematics

TABLE OF CONTENTS

Selected References iii
Numbers 1-1
    Real Numbers 1-1
    Imaginary Numbers 1-2
Sets 1-3
Algebra 1-5
    Fundamental Laws 1-5
    Exponents 1-5
    Binomial Formula 1-7
    Rules for Determining Terms in a Binomial Expansion 1-8
    Determinants 1-9
    Linear Algebra 1-12
    Vector Analysis 1-21
    Higher Order Algebraic Equations 1-26
Logarithms 1-28
Trigonometry 1-33
    Angles 1-33
    Trigonometric Functions 1-34
    Polar Coordinates 1-37
    Fundamental Identities 1-38
    Laws for Triangles 1-39
    Hyperbolic Functions 1-41
Analytic Geometry 1-42
    Straight Lines 1-42
    Conic Sections 1-46
Differential Calculus 1-49
    Functions 1-49
    Limits 1-49
    The Derivative 1-52
    Basic Rules of Differentiation 1-53
    Table of Derivatives 1-53
    Implicit Differentiation 1-59

Higher Order Derivatives 1-59
Partial Derivatives 1-60
Applications 1-63
Integral Calculus 1-70
Indefinite Integral 1-70
Definite Integral 1-71
Table of Integrals 1-72
Integration by Parts 1-74
Double Integrals 1-75
Applications 1-75
Differential Equations 1-82
Definitions 1-82
Equations with Variables Separable 1-82
First-Order Linear Differential Equations 1-84
Second-Order Linear Differential Equations Homogeneous
 with Constant Coefficients 1-85
Applications 1-87
Laplace Transformation 1-88
Definition of the Laplace Transform 1-88
Table of Transforms 1-90
Application to the Solution of Differential Equations 1-92
Probability and Statistics 1-94
Probability 1-94
Statistics 1-97
Normal Distribution Functions 1-103

SELECTED REFERENCES

R.S. Burington, "Handbook of Mathematical Tables and Formulas," Handbook Publishers, 1953.

A.E. Taylor, "Advanced Calculus," Ginn and Co., 1955.

R.V. Churchill, "Modern Operational Mathematics in Engineering," McGraw-Hill, 1944.

J.E. Powell and C.P. Wells, "Differential Equations," Ginn and Co., 1950.

C.T. Holmes, "Calculus and Analytic Geometry."

T. Baumeister and L.S. Marks, "Standard Handbook for Mathematical Engineers," 10th edition, McGraw-Hill, 1997.

MATHEMATICS

NUMBERS

Real Numbers

A study of numbers begins with the integers, which are also referred to as whole numbers or natural numbers. A rational number is a number that can be written in the form p/q where p and q are integers and q ≠ 0. Integers are rational numbers because they can be expressed as a ratio of integers.

EXAMPLE:

3/4 is a fraction and a rational number.

√3/2 is a fraction and is not a rational number.

An irrational number is a number that cannot be expressed in the form p/q where p and q are integers.

EXAMPLE:

√3 or 1.732... is irrational, whereas 0.1111... is a rational number because it is equivalent to 1/9.

In very general terms, nonrepeating and nonterminating numbers are irrational.

The real number system is composed of the integers, the rational fractions and the irrational numbers. It can be represented geometrically by the points on a straight line. There is a one-to-one correspondence between the real numbers and the points on the line (Fig. 1.1).

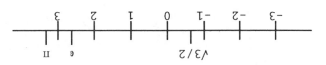

The real number line

Figure 1.1

A prime number is an integer that has no factors except itself and 1.

EXAMPLE:

1, 2, 3, 5, 7, 11, 13...

1-2   FUNDAMENTALS OF ENGINEERING EXAM REVIEW WORKBOOK

The symbol n!, read "n factorial," is an abbreviation for the product of all integers from 1 to n inclusive. This is also represented by the symbol $\underline{n}$.

$$n! = (n)(n-1)(n-2)...(2)(1)$$

EXAMPLE:

$$5! = 5 \times 4 \times 3 \times 2 \times 1 = 120$$

$$\frac{3!}{4!} = \frac{3 \times 2 \times 1}{4 \times 3 \times 2 \times 1} = \frac{1}{4}$$

Note that 0! = 1.

Imaginary Numbers

The square root of a negative number has no real number roots. This mathematical condition introduces imaginary numbers. The determination of the square root of a negative number is handled in the following manner. If P is a positive number, then

$$\sqrt{-P} = \sqrt{-1 \times P} = \sqrt{-1} \times \sqrt{P} = i\sqrt{P}$$

where $i = \sqrt{-1}$. Sometimes j is used in place of i.

Numbers such as i, 3i, $-i\sqrt{2}$ are called pure imaginary numbers. Numbers such as a + bi, where a and b are real numbers, are called complex numbers.

EXAMPLE:

$$\sqrt{-16} = \sqrt{-1} \times \sqrt{16} = 4i$$

(Note that the principal square root of a non-negative number is a non-negative number. For example, $\sqrt{16}$ is generally written as 4, not -4 or ± 4.)

EXAMPLE:

Find the value of $(i)^{23}$.

Note that $i^1 = i$, $i^2 = -1$, $i^3 = -i$, $i^4 = +1$

also $i^{4n+1} = i$, $i^{4n+2} = -1$, $i^{4n+3} = -i$, $i^{4n+4} = +1$

$$i^{23} = i^{(4 \times 5+3)} = -i$$

MATHEMATICS 1-3

EXAMPLE:

$(3 + 2i) + (6 + 4i) = 9 + 6i$

EXAMPLE:

Rationalize $\dfrac{3 + 2i}{3 - i}$

It is generally useful to use the conjugate of the denominator when finding the quotient of two complex numbers. The conjugate of a complex number $a + ib$ (where a and b are real numbers) is $a - ib$. Note that the product of a complex number and its conjugate is a real number. Thus if $A = a + ib$, its conjugate is $\bar{A} = a - ib$ and $A \times \bar{A} = a^2 + b^2$.

Thus, $\dfrac{3 + 2i}{3 - i} = \dfrac{3 + 2i}{3 - i} \cdot \dfrac{(3 + i)}{(3 + i)}$

$= \dfrac{9 + 6i + 3i + 2(i)^2}{9 + 1}$

$= \dfrac{7 + 9i}{10} = \dfrac{7}{10} + \dfrac{9i}{10}$

SETS

A collection of objects or entities is commonly called a *set*. The objects comprising the set are called elements or members of the set. If an element X belongs to a set S, this is indicated by X∈S.

If X does not belong to S, this is indicated by X∉S.

A set can be defined by listing the elements that belong to the set. For example

{1, 2, 8}

is the set composed of the numbers 1, 2, and 8. The set of all the natural numbers may be denoted by

{1, 2, 3, 4, ...}

The null set is the set that has no elements and is generally denoted by the symbol ∅.

1-4 FUNDAMENTALS OF ENGINEERING EXAM REVIEW WORKBOOK

A set may be defined by using a variable to express a common property of all of the elements of the set. For example

$$\{X : X^3 > 27\}$$

is the set of all numbers X such that $X^3 > 27$.

If all the elements of a set A are elements of another set B, A is a subset of B. This is denoted by A⊂B.

The symbol ⊂ is called the inclusion symbol and this relationship is shown in Fig. 1.2.

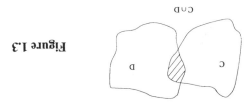

Figure 1.2

The union of two sets consists of all of the elements of either set. This is usually represented by the symbol ∪ so that

$$C \cup D = \{X : X \in C \text{ or } X \in D\}$$

The intersection of two sets consists of all of the elements that belong to both sets. This is usually represented by the symbol ∩ so that

$$C \cap D = \{X : X \in C \text{ and } X \in D\}$$

This relationship is represented by the shaded region in Fig. 1.3.

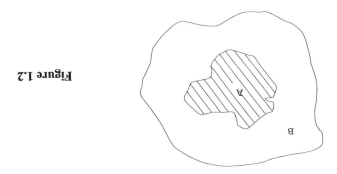

Figure 1.3

MATHEMATICS

EXAMPLE:

If A = {3,5,7,10} and B = {X : X^2 > 26}
then A ∪ B = {7,10}

EXAMPLE:

If A = {2,4,6} and B = {3,5,7} then A ∩ B = ∅.

EXAMPLE:

If A = {1,2,3}, B = {2,4,6,8,10} and C = {1,5,9}
then C ∪ (A ∩ B) = {1,2,5,9}.

ALGEBRA

Fundamental Laws

The fundamental operations of addition and multiplication conform to the following laws:

Law	Addition	Multiplication
Commutative	a+b = b+a	ab = ba
Associative	(a+b)+c = a+(b+c)	(ab)c = a(bc)
Distributive	a(b+c) = ab + ac	

The symbols a, b and c represent any real numbers.

Exponents

The following laws of exponents apply where m and n are integers or fractions:

$$\sqrt[n]{a} = a^{1/n}$$

$$\frac{a^m}{a^n} = a^{m-n}$$

$$(a^n)^m = (a^m)^n = a^{mn}$$

$$(ab)^m = a^m b^m$$

$$\left[\frac{a}{b}\right]^m = \frac{a^m}{b^m}$$

$$a^{-m} = \frac{1}{a^m} \text{ where } a \neq 0$$

$$a^0 = 1 \text{ where } a \neq 0$$

$$a^1 = a$$

When the exponents are fractions, expressions involving radicals are sometimes employed. The meaning of the radical is illustrated by the following expression

$$\sqrt[n]{a} = a^{1/n}$$

where n is the index and a is the radicand. As noted previously, when a is negative and n is even $\sqrt[n]{a}$ represents an imaginary number. Thus, the following can be used, except when a is negative and n is even

$$\sqrt[n]{a^m} = a^{m/n} \quad \text{also} \quad \sqrt[n]{a^m} = a^{m/n}$$

Useful simplifications can be made using the following expressions

$$\sqrt[n]{ab} = \sqrt[n]{a} \cdot \sqrt[n]{b}$$

$$\sqrt[n]{\frac{a}{b}} = \frac{\sqrt[n]{a}}{\sqrt[n]{b}}$$

EXAMPLES:

1) Simplify $(8x^5y)(3x^2y^2)$

Remember, in multiplying we add exponents. Therefore,

$$(8x^5y)(3x^2y^2) = 24 \, x^{5+2} \, y^{1+2} = 24 \, x^7 \, y^3.$$

MATHEMATICS 1-7

2) Perform the following division

$$\frac{a^6}{a^9} = A^{6-9} = A^{-3} = \frac{1}{A^3}$$

3) What is the value of $(.008)^{2/3}$?

$(.008)^{2/3} = (.008^{1/3})^2 = (.2)^2 = .04$

4) Simplify

$$\frac{a^{2/3} b^{1/3}}{a^{-3/2} b^{-1/2}} = a^{2/3} a^{3/2} b^{1/3} b^{1/2}$$

$= a^{(4+9)/6} b^{(2+3)/6} = a^{13/6} b^{5/6}$

5) Simplify

$$\sqrt[3]{\frac{5}{x^3} + 3a} = \frac{\sqrt[3]{3ax^3 + 5}}{\sqrt[3]{x^3}} = \frac{\sqrt[3]{3ax^3 + 5}}{x}$$

6) Simplify

$\sqrt[3]{-2} = \sqrt[3]{-1(2)} = \sqrt[3]{-1}\sqrt[3]{2} = -1\sqrt[3]{2}$
$= -\sqrt[3]{2}$

Binomial Formula

The binomial formula is

$(a+b)^n = a^n + na^{n-1}b + \frac{n(n-1)}{2!}a^{n-2}b^2 + \frac{n(n-1)(n-2)}{3!}a^{n-3}b^3 + \ldots + nab^{n-1} + b^n$

where n is a positive integer. A special case of this formula is obtained by letting a = 1:

$(1 + b)^n = 1 + nb + \frac{n(n-1)}{2!}b^2 + \ldots + nb^{n-1} + b^n$

When n is not a positive integer, an infinite series called the binomial series results.

RULES FOR DETERMINING TERMS IN A BINOMIAL EXPANSION

$$(A+X)^n$$

1. The number of terms in the expansion is one more than the exponent.
2. The first term is A^n. Subsequent exponents of A decrease by 1 for each term.
3. The exponent of X in the second term is 1. Subsequent exponents of X increase by 1 for each term.
4. The sum of the exponents in any term equals the exponent of the binomial.
5. The coefficient of the terms equidistant from the first and last terms are equal.
6. The coefficient of the second term is the same as the exponent of the binomial.
7. The coefficient of successive terms is computed from the previous term by multiplying the coefficient by the exponent of A and dividing by one more than the exponent of X.

Example:

Consider $(A + X)^7$. From the rules, the first two terms are:

$$A^7 + 7 A^6 X$$

where the coefficient of the second term is 7, the exponent of A is 6, and the exponent of X is 1. The coefficient of the third term is computed by Rule 7 as:

$$C_3 = \frac{7 \times 6}{1 + [1]} = 21$$

RULES FOR DETERMINING TERMS IN A BINOMIAL EXPANSION

$$(A + X)^n$$

Shortcut

Pythagoras' Triangle

Start with a triangle of 1s. Always start the next level by placing 1s on either side to form the shape of the triangle (Fig. 1.4).

```
    1
   1 1
  1 x 1
```

Figure 1.4

Compute the interior term(s) by adding the two adjacent terms above (Fig. 1.5).

```
    1
   1 1
  / \ +
   ⇓
  1 2 1
```

Figure 1.5

Thus, coefficients for any binomial can be computed directly (Fig. 1.6).

PYTHAGORAS' TRIANGLE

```
        1
       1 1
      1 2 1
     1 3 3 1
    1 4 6 4 1
   1 5 10 10 5 1
```

Figure 1.6

Determinants

A determinant of order n is a square array (or matrix) of n^2 elements. The evaluation or expansion of the determinant requires the summation of $n!$ terms. Although all determinants can be evaluated by the same basic procedure, the second-order determinant will be discussed first because of its simplicity.

Consider the following second-order determinant:

$$\begin{vmatrix} a_1 & b_1 \\ a_2 & b_2 \end{vmatrix}$$

The element positions in the matrix and determinant are denoted by the subscripts i and j and describe an element a_{ij}. The i indicates the row or vertical position, the top most row being 1, and the j describes the horizontal position, the leftmost column being 1. Therefore element a_{21} is in the second row, first column.

The principal diagonal of the determinant below is a_{11} and b_{22}. Elements a_{21} and b_{12} are the secondary diagonal. The value of the determinant is the product of the principal diagonal minus the product of the secondary diagonal (Fig. 1.7).

1-10 FUNDAMENTALS OF ENGINEERING EXAM REVIEW WORKBOOK

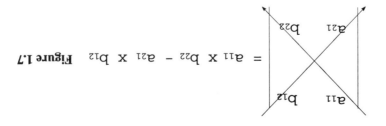

Figure 1.7 $= a_{11} \times b_{22} - a_{21} \times b_{12}$

EXAMPLE:

$$\begin{vmatrix} 5 & -4 \\ 3 & 2 \end{vmatrix} = 5 \times 2 -(-4) \times 3 = 22$$

The expansion of higher order determinants can be accomplished by considering the minor of each element in a column or a row.

In any determinant, if the row and column containing a given element, say e (see example) are blotted out, the determinant formed from the remaining elements is called the minor of e.

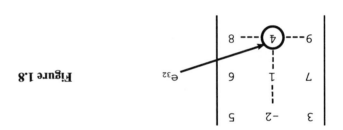

Figure 1.8

In Fig. 1.8 the minor of 4 is $\begin{vmatrix} 3 & 5 \\ 7 & 6 \end{vmatrix}$

The procedure for expanding a determinant by minors is as follows:

1. Multiply each element of a column (or each element of a row) by its minor and give the product a plus or minus sign according as the sum of the position numbers in the row and column containing the element is even (+) or odd (−). See the example that follows.

2. Take the sum of the signed products.

MATHEMATICS 1-11

EXAMPLE:

$$\begin{vmatrix} a_{11} & b_{12} & c_{13} \\ a_{21} & b_{22} & c_{23} \\ a_{31} & b_{32} & c_{33} \end{vmatrix} = (-1^{1+1})a_{11} \begin{vmatrix} b_{22} & c_{23} \\ b_{32} & c_{33} \end{vmatrix} + (-1^{2+1})a_{21} \begin{vmatrix} b_{12} & c_{13} \\ b_{32} & c_{33} \end{vmatrix} + (-1^{3+1})a_{31} \begin{vmatrix} b_{12} & c_{13} \\ b_{22} & c_{23} \end{vmatrix}$$

EXAMPLE:

$$D = \begin{vmatrix} 3 & -1 & 2 \\ 6 & 4 & 1 \\ 2 & -3 & -4 \end{vmatrix} = +3 \begin{vmatrix} 4 & 1 \\ -3 & -4 \end{vmatrix} -6 \begin{vmatrix} -1 & 2 \\ -3 & -4 \end{vmatrix} +2 \begin{vmatrix} -1 & 2 \\ 4 & 1 \end{vmatrix}$$

$$D = 3(-16+3) -6(4+6) +2(-1-8)$$

$$D = -39 -60 -18 = -117$$

Alternatively, D can also be expanded by considering the elements (and associated minors) of any row (row 1 in this illustration).

$$D = 3 \begin{vmatrix} 4 & 1 \\ -3 & -4 \end{vmatrix} = -(-1) \begin{vmatrix} 6 & 1 \\ 2 & -4 \end{vmatrix} +2 \begin{vmatrix} 6 & 4 \\ 2 & -3 \end{vmatrix}$$

$$D = 3(-16+3) +1(-24-2) +2(-18-8)$$

$$D = -39 -26 -52 = -117$$

Higher order determinants may be expanded by repeated application of this process.

EXAMPLE:

$$+a_1 \begin{vmatrix} b_2 & c_2 & d_2 \\ b_3 & c_3 & d_3 \\ b_4 & c_4 & d_4 \end{vmatrix} -a_2 \begin{vmatrix} b_1 & c_1 & d_1 \\ b_3 & c_3 & d_3 \\ b_4 & c_4 & d_4 \end{vmatrix} = \begin{vmatrix} a_1 & b_1 & c_1 & d_1 \\ a_2 & b_2 & c_2 & d_2 \\ a_3 & b_3 & c_3 & d_3 \\ a_4 & b_4 & c_4 & d_4 \end{vmatrix}$$

$$+a_3 \begin{vmatrix} b_1 & c_1 & d_1 \\ b_2 & c_2 & d_2 \\ b_4 & c_4 & d_4 \end{vmatrix} -a_4 \begin{vmatrix} b_1 & c_1 & d_1 \\ b_2 & c_2 & d_2 \\ b_3 & c_3 & d_3 \end{vmatrix}$$

Linear Algebra

Simultaneous linear equations with two unknowns are most easily solved by eliminating one of the variables by subtraction (after obtaining similar terms of opposite sign). Consider the following equations:

$$a_1 x + b_1 y = c_1 \quad (1)$$

$$a_2 x + b_2 y = c_2 \quad (2)$$

Multiply eq. 1 by $-a_2$ and eq. 2 by a_1, and then add the two new equations

$$-a_2 a_1 x - a_2 b_1 y = -a_2 c_1$$
$$a_1 a_2 x + a_1 b_2 y = a_1 c_2$$
$$\overline{0 + a_1 b_2 y - a_2 b_1 y = a_1 c_2 - a_2 c_1} \quad (3)$$

Equation 3 is a linear equation in one unknown, y. This equation can be solved for y and then x can be found from either one of the original equations by substitution of the value of y.

EXAMPLE:

$3x + y = 2$

$x - 2y = 5$

Multiply the first equation by 2 and add the equations to eliminate y

$6x + 2y = 4$
$\underline{x - 2y = 5}$
$7x = 9$ or $x = 9/7$

Now substitute this value of x in either equation and solve for y using the second equation

$9/7 - 2y = 5$
$9 - 14y = (5 \times 7)$
$-14y = 26$
$y = -26/14 = -13/7$

Therefore, $x = 9/7$ and $y = -13/7$

Consider Eqs. (1) and (2) again:

$a_1x + b_1y = c_1$ (1)

$a_2x + b_2y = c_2$ (2)

A general formula for the value of y could have been obtained if Eq. 3 was solved for y.

Writing Eq. 3

$a_1b_2y - a_2b_1y = a_1c_2 - a_2c_1$

Solving for y:

$$y = \frac{a_1c_2 - a_2c_1}{a_1b_2 - a_2b_1}$$

Similarly it can be shown that

1-14 FUNDAMENTALS OF ENGINEERING EXAM REVIEW WORKBOOK

$$x = \frac{c_1b_2 - c_2b_1}{a_1b_2 - a_2b_1}$$

Solutions for x and y can be obtained by substituting directly into these formulas.

EXAMPLE:

$$3x + y = 2$$
$$x - 2y = 5$$

$$x = \frac{(2)(-2) - (5)(1)}{(3)(-2) - (1)(1)} = \frac{-4-5}{-6-1} = \frac{9}{7}$$

$$y = \frac{(3)(5) - (1)(2)}{(3)(-2) - (1)(1)} = \frac{15-2}{-6-1} = \frac{-13}{7}$$

Examination of the general formulas for x and y indicates another approach to the solution of simultaneous linear equations—solution by determinants, "Cramer's Rule."

Notice that the denominators of the x and y formulas are the same and can be obtained by expanding the determinant of the x and y coefficients:

$$\begin{vmatrix} a_1 & b_1 \\ a_2 & b_2 \end{vmatrix} = a_1b_2 - a_2b_1$$

The numerator for the value of x can be obtained expanding

$$\begin{vmatrix} c_1 & b_1 \\ c_2 & b_2 \end{vmatrix} = c_1b_2 - c_2b_1$$

which is the determinant of coefficients with the constants substituted for the x coefficients. Similarly, the determinant of the coefficients with the constants substituted for the y coefficients will give the numerator of y.

Thus

$$x = \frac{\begin{vmatrix} c_1 & b_1 \\ c_2 & b_2 \end{vmatrix}}{\begin{vmatrix} a_1 & b_1 \\ a_2 & b_2 \end{vmatrix}} \quad \text{And } y = \frac{\begin{vmatrix} a_1 & c_1 \\ a_2 & c_2 \end{vmatrix}}{\begin{vmatrix} a_1 & b_1 \\ a_2 & b_2 \end{vmatrix}}$$

MATHEMATICS 1-15

This procedure is applicable to any system of n linear equations in n unknowns and will give a unique solution if, and only if, the determinant of the coefficient matrix is not zero. This procedure is particularly useful for the solution of three (or more) unknowns. For three linear equations in the following form

$$a_1x + b_1y + c_1z = k_1$$

$$a_2x + b_2y + c_2z = k_2$$

$$a_3x + b_3y + c_3z = k_3$$

The determinant of the coefficients, Δ, is

$$\Delta = \begin{vmatrix} a_1 & b_1 & c_1 \\ a_2 & b_2 & c_2 \\ a_3 & b_3 & c_3 \end{vmatrix}$$

If $\Delta \neq 0$, the value of any unknown can be found from the following fraction

1. The denominator is Δ.
2. The numerator is the determinant obtained if in Δ the column of coefficients of the unknown is replaced by the column of constant terms.

Therefore:

$$x = \frac{\begin{vmatrix} k_1 & b_1 & c_1 \\ k_2 & b_2 & c_2 \\ k_3 & b_3 & c_3 \end{vmatrix}}{\Delta} \quad y = \frac{\begin{vmatrix} a_1 & k_1 & c_1 \\ a_2 & k_2 & c_2 \\ a_3 & k_3 & c_3 \end{vmatrix}}{\Delta} \quad z = \frac{\begin{vmatrix} a_1 & b_1 & k_1 \\ a_2 & b_2 & k_2 \\ a_3 & b_3 & k_3 \end{vmatrix}}{\Delta}$$

This system of equations is sometimes written in the following matrix form

$$\begin{bmatrix} a_1 & b_1 & c_1 \\ a_2 & b_2 & c_2 \\ a_3 & b_3 & c_3 \end{bmatrix} \begin{bmatrix} x \\ y \\ z \end{bmatrix} = \begin{bmatrix} k_1 \\ k_2 \\ k_3 \end{bmatrix}$$

EXAMPLE:

Solve for A, B, and C where

$2A + 3B - C = -10$

$-A + 4B + 2C = -4$

$2A - 2B + 5C = 35$

$$\Delta = \begin{vmatrix} 2 & 3 & -1 \\ -1 & 4 & 2 \\ 2 & -2 & 5 \end{vmatrix}$$

$$\Delta = 2 \begin{vmatrix} 4 & 2 \\ -2 & 5 \end{vmatrix} -(-1) \begin{vmatrix} 3 & -1 \\ -2 & 5 \end{vmatrix} +2 \begin{vmatrix} 3 & -1 \\ 4 & 2 \end{vmatrix}$$

$\Delta = 2(20 + 4) + (15 - 2) + 2(6 + 4)$

$\Delta = 48 + 13 + 20 = 81$

A is given by

$$A = \frac{\begin{vmatrix} -10 & 3 & -1 \\ -4 & 4 & 2 \\ 35 & -2 & 5 \end{vmatrix}}{81}$$

$$A = \frac{-10(24) + 4(13) + 35(10)}{81}$$

$$A = \frac{162}{81} = 2$$

$$B = \frac{\begin{vmatrix} 2 & -10 & -1 \\ 2 & -4 & -1 \\ 5 & 35 & 2 \end{vmatrix}}{81}$$

$$B = \frac{2(-90) + 1(-15) + 2(-24)}{81}$$

$$B = \frac{-243}{81} = -3$$

C can be obtained by substituting the values found for A and B into one of the equations:

$$2(2) + 3(-3) - C = -10$$

$$C = 5$$

Thus

$$A = 2, B = -3, C = 5$$

There are certain types of problems in linear algebra that require setting up equations to represent the physical conditions of the problem. The solution to the problem is then obtained by solving the equation(s).

EXAMPLE:

Determine how much alcohol with a purity of 99% must be added to 50 gallons of 90% alcohol to obtain a 95% alcohol solution.

The initial amount of pure alcohol = 0.90 x 50 = 45 gal.

Let x = gal. of 99% alcohol that must be added

Then $0.99 x + 45 = 0.95$
$$\frac{0.99x + 45}{50 + x} = 0.95$$

$$0.99 x + 45 = (0.95)(50) + 0.95 x$$

$$0.04 x = 47.5 - 45$$

$$x = \frac{2.5}{0.04} = 62.5 \text{ gal.}$$

EXAMPLE:

Dan is 20 years older than Andy. Twenty years ago he was twice as old as Andy. How old are both men?

Let   D = Dan's age

A = Andy's age

Then D = A + 20

and D - 20 = 2 (A-20)

A + 20 - 20 = 2 A - 40

A = 40 years old

D = A + 20 = 60 years old

EXAMPLE:

Alan digs 5 holes to every 3 holes dug by Bob. Together Alan and Bob can dig 24 holes a day. How many holes can Bob dig by himself in one day?

Let A = number of holes per day for Alan
     B = number of holes per day for Bob

Then $B = \frac{3}{5} A$ or $A = \frac{5}{3} B$

and A + B = 24

$\frac{5}{3} B + B = 24$

8B = 72

B = 9

Thus, Bob can dig 9 holes per day by himself.

EXAMPLE:

Four men can build 5 cabinets in 3 days. How many men are required to build 15 cabinets in 4 days?

man days/cabinet = $\frac{4 \text{ men} \times 3 \text{ days}}{5 \text{ cabinets}} = \frac{12}{5}$

Let x = number of men required to build 15 cabinets in 4 days.

Then $\frac{4x}{12} = \frac{5}{15}$

$x = \left(\frac{12}{15}\right)\left(\frac{5}{4}\right) = 9$ men

EXAMPLE:

A's swimming pool may be filled with water using his hose, neighbor B's hose, neighbor C's hose, or any combination. Using A's hose and B's hose takes 30 hours. Using A's hose and C's hose takes 24 hours. Using B's hose and C's hose takes 20 hours. How long would it take using A's hose alone?

Let A = hours to fill pool with A's hose
B = hours to fill pool with B's hose
C = hours to fill pool with C's hose

Then the fraction of the pool volume filled hourly under the various combinations is given by

$\frac{1}{A} + \frac{1}{B} = \frac{1}{30}$

$\frac{1}{A} + \frac{1}{C} = \frac{1}{24}$

$\frac{1}{B} + \frac{1}{C} = \frac{1}{20}$

Now we have three equations in three unknowns. In matrix form this is

$\begin{bmatrix} 1 & 1 & 0 \\ 1 & 0 & 1 \\ 0 & 1 & 1 \end{bmatrix} \begin{bmatrix} \frac{1}{A} \\ \frac{1}{B} \\ \frac{1}{C} \end{bmatrix} = \begin{bmatrix} \frac{1}{30} \\ \frac{1}{24} \\ \frac{1}{20} \end{bmatrix}$

$\Delta = 1\,(-1) - 1\,(1) + 0 = -2$

$$\frac{1}{A} = \frac{\begin{vmatrix} 0 & 1 & \frac{1}{30} \\ 1 & 0 & \frac{1}{24} \\ 1 & 1 & \frac{1}{20} \end{vmatrix}}{-2}$$

$$\frac{1}{A} = \frac{1}{30}(-1) - \frac{1}{24}(1) + \frac{1}{20}(1)$$

$$\frac{1}{A} = -\frac{1}{60} - \frac{1}{48} + \frac{1}{40}$$

$$\frac{1}{A} = \frac{-1920 - 2400 + 2880}{115,200}$$

$$A = \frac{115,200}{1440} = 80$$

Thus, it will take 80 hours to fill the pool with A's hose alone.

MATHEMATICS 1-21

Vector Analysis

Definitions

A scaler is a quantity defined by magnitude (mass, length, time, temperature). If to each point (x,y,z) of region R in space, there is a scaler r = OA, then OA is called a scaler point function, and the region R is denoted as a scaler field (Fig. 1.9).

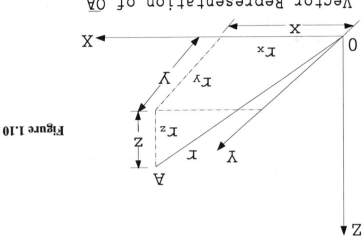

Figure 1.9

Scaler Representation of OA

A vector is a quantity defined by magnitude and direction (force, moment, displacement, velocity, acceleration). If to each point (x,y,z) of a region R in space there is a vector r = OA, then OA is called a vector point function, and region R is denoted as vector field (Fig. 1.10).

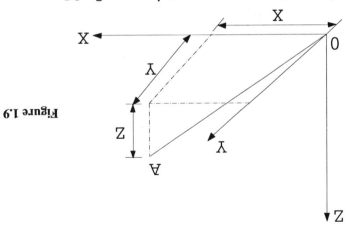

Figure 1.10

Vector Representation of OA

Components and Magnitudes

The vector **r** may be resolved into any number of components.
In the Cartesian coordinate system, **r** is resolved into three mutually perpendicular components, each parallel to the respective coordinate axis.

$$r = r_x + r_y + r_z = r_x i + r_y j + r_z k$$

The magnitude of **r** is given by the magnitudes of its components:

$$r = |r| = \sqrt{r_x^2 + r_y^2 + r_z^2}$$

The unit vector is the ratio of the vector to its magnitude. **i, j, k** are the unit vectors in the x, y, z axis directions respectively. **e** is the unit vector in the r direction.

$$i = \frac{r_x}{r_x}, \quad j = \frac{r_y}{r_y}, \quad k = \frac{r_z}{r_z}, \quad e = \frac{r}{r}$$

The direction cosines of a vector **r** are:

$$\alpha = \frac{r_x}{r}, \quad \beta = \frac{r_y}{r}, \quad \gamma = \frac{r_z}{r}$$

and relate **e** to **i, j, k**.

$$e = \alpha i + \beta j + \gamma k$$

Vector Addition and Subtraction

A sum of vectors **a** and **b** is a vector **c** formed by placing the initial point of **b** on the terminal point of **a** and joining the initial point of **a** to the terminal point of **b**.

A difference of vectors **a** and **b** is a vector **d** formed by placing the initial point of **b** on the initial point of **a** and joining the terminal point of **b** with the terminal point of **a**.

A difference of vectors **b** and **a** is a vector **e** formed by placing initial point of **b** on the initial point of **a** and joining the terminal point of **a** with terminal point of **b**.

Consider the vectors **a** and **b** in Fig. 1.11.

MATHEMATICS 1-23

Scaler–Vector Laws   (**a**,**b** are vectors; m,n are scalars)

m**a** = **a**m   Cumulative law
a(m + n) = m**a** + n**a**   Distributive law
m(**a** + **b**) = m**a** + m**b** Distributive law
m(n**a**) = mn(**a**)   Associative law

Vector Summation Laws

Consider the vectors a, b, c and f in Fig. 1.12.

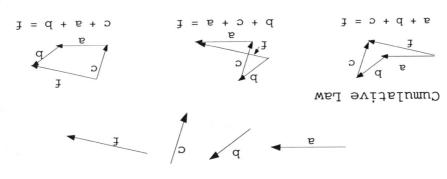

Cumulative Law

a + b + c = f   b + c + a = f   c + a + b = f

Figure 1.12

Associative Law

(a + b) + c = f   a + (b + c) = f

Scaler Product

The scaler product, or a dot product, of two vectors **a** and **b** is defined as the product of their magnitudes and the cosine of the angle between them. The result is a scaler.

$$\mathbf{a} \cdot \mathbf{b} = ab \cos\alpha = a_x b_x + a_y b_y + a_z b_z$$

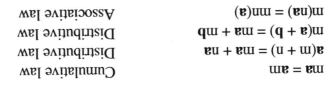

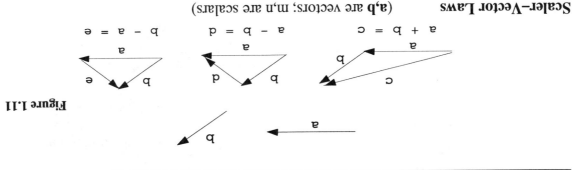

Figure 1.11

a + b = c   a − b = d   b − a = e

Multiplication is non-commutative and distributive.

$$\mathbf{a} \cdot \mathbf{b} = -\mathbf{b} \cdot \mathbf{a}$$

$$\mathbf{a} \cdot (\mathbf{b} + \mathbf{c}) = \mathbf{a} \cdot \mathbf{b} + \mathbf{a} \cdot \mathbf{c}$$

Two vectors $\mathbf{a} \neq \mathbf{0}$, $\mathbf{b} \neq \mathbf{0}$ are normal ($\alpha = 90°$) if $\mathbf{a} \cdot \mathbf{b} = 0$.
Thus,

$$\mathbf{i} \cdot \mathbf{j} = \mathbf{j} \cdot \mathbf{k} = \mathbf{k} \cdot \mathbf{i} = 0$$

Two vectors $\mathbf{a} \neq \mathbf{0}$, $\mathbf{b} \neq \mathbf{0}$ are parallel ($\alpha = 0°$) if

$$\mathbf{a} \cdot \mathbf{b} = ab$$

Thus

$$\mathbf{i} \cdot \mathbf{i} = \mathbf{j} \cdot \mathbf{j} = \mathbf{k} \cdot \mathbf{k} = 1$$

Vector Product

The vector product (or Cross product) of two vectors $\mathbf{a}$ and $\mathbf{b}$ is defined as the product of their magnitudes, the sine of the angle between them, and the unit vector $\mathbf{n}$ normal to their plane. The result is a vector.

$$\mathbf{a} \cdot \mathbf{b} = ab \sin \alpha \; \mathbf{n}$$

$$= (a_y b_z - a_z b_y)\mathbf{i} - (a_x b_z - a_z b_x)\mathbf{j} + (a_x b_y - a_y b_x)\mathbf{k}$$

Multiplication is non cumulative but it is distributive.

$$\mathbf{a} \times \mathbf{b} = -\mathbf{b} \times \mathbf{a} \quad \mathbf{a} \times (\mathbf{b} + \mathbf{c}) = (\mathbf{a} \times \mathbf{b}) + (\mathbf{b} \times \mathbf{c})$$

Two vectors $\mathbf{a} \neq \mathbf{0}$, $\mathbf{b} \neq \mathbf{0}$ are normal ($\alpha = 90°$) if $\mathbf{a} \times \mathbf{b} = ab\mathbf{n}$
Thus

$$\mathbf{i} \times \mathbf{j} = \mathbf{k}$$
$$\mathbf{j} \times \mathbf{k} = \mathbf{i}$$
$$\mathbf{k} \times \mathbf{i} = \mathbf{j}$$

Two vectors $\mathbf{a} \neq \mathbf{0}$, $\mathbf{b} \neq \mathbf{0}$ are parallel ($\alpha = 0°$) if

$$\mathbf{a} \times \mathbf{b} = 0$$

Thus

$$\mathbf{i} \times \mathbf{i} = \mathbf{j} \times \mathbf{j} = \mathbf{k} \times \mathbf{k} = 0$$

MATHEMATICS 1-25

The product of two vectors in determinant form is

$$\mathbf{a} \times \mathbf{b} = \begin{vmatrix} i & j & k \\ a_x & a_y & a_z \\ b_x & b_y & b_z \end{vmatrix} = \begin{vmatrix} i & a_x & b_x \\ j & a_y & b_y \\ k & a_z & b_z \end{vmatrix} = i \begin{vmatrix} a_y & b_y \\ a_z & b_z \end{vmatrix} - j \begin{vmatrix} a_x & b_x \\ a_z & b_z \end{vmatrix} + k \begin{vmatrix} a_x & b_x \\ a_y & b_y \end{vmatrix}$$

Example: Multiply the vectors:

$\mathbf{a} = 4i + 3j - k$
$\mathbf{b} = 2i - 6j + 2k$

Establish and complete the determinant:

$$\mathbf{a} \times \mathbf{b} = \begin{vmatrix} i & j & k \\ 4 & 3 & -1 \\ 2 & -6 & 2 \end{vmatrix} = \begin{vmatrix} i & j & k \\ 2 & 4 & -1 \\ -6 & 3 & 2 \end{vmatrix} = i \begin{vmatrix} 3 & -6 \\ -1 & 2 \end{vmatrix} - j \begin{vmatrix} 4 & 2 \\ -1 & 2 \end{vmatrix} + k \begin{vmatrix} 4 & 2 \\ 3 & -6 \end{vmatrix}$$

$= i(6 - 6) - j(8 - (-2)) + k(-24 - 6)$

$= -10j - 30k$

Example:

The system shown in Fig. 1.13 is to be in equilibrium. What must be the balancing moment?

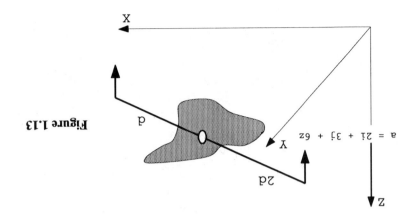

$\mathbf{a} = 2i + 3j + 6z$

Figure 1.13

The given moment is:

$$2d(2i + 3j + 6z) = (4di + 6dj + 12dz)$$

The required moment must be a force acting through a distance d and to balance, the restoring moment must be the given moment divided by the distance through it acts.

The restoring moment is:

$$\frac{(4di + 6dj + 12dz)}{d} = (4i + 6j + 12z)$$

Higher Order Algebraic Equations

The solution of higher order algebraic equations usually involves some procedure for finding the roots of the equation. Roots of an equation in the form f(x) = 0 are values of x that satisfy the equation. The quadratic formula may be used to find the roots of a second degree equation. The roots of the quadratic equation

$$ax^2 + bx + c = 0$$

are

$$x = \frac{-b \pm \sqrt{b^2 - 4ac}}{2a}$$

There are formulas for the roots of cubic and quadratic equations, but these involve lengthy procedures. Roots may be found by factoring. If x = r is a root of the equation f(x) = 0, then (x - r) is a factor of f(x) and dividing f(x) by (x - r) will reduce the equation to one degree less than that of the original equation.

EXAMPLE:

If x = 2 is a root of the equation $x^3 - 7x + 6 = 0$, what are the other roots?

If x = 2 is a root then (x-2) is a factor of $x^3 - 7x + 6$. The reduced equation can be found by division.

$$\begin{array}{r}x^2 + 2x - 3\\ x-2\overline{\smash{\big)}\,x^3 + 0x^2 - 7x + 6}\\ \underline{x^3 - 2x^2}\\ +2x^2 - 7x\\ \underline{+2x^2 - 4x}\\ -3x + 6\\ \underline{-3x + 6}\\ 0\end{array}$$

Now use the quadratic formula to find the roots of the reduced equation:

$$x^2 + 2x - 3 = 0$$

$$x = \frac{-2 \pm \sqrt{(2)^2 - 4(1)(-3)}}{2}$$

$$x = \frac{-2 \pm \sqrt{4 + 12}}{2}$$

$$x = \frac{-2 \pm 4}{2}$$

$x = +1$ and $x = -3$ are roots

Information about the approximate value of a root can be found from the fact that if $y = f(x)$ is plotted on a Cartesian coordinate system, the real roots are values of x where the graph crosses the x-axis (the exact criterion is that $y = 0$; there are certain cases where the graph does not actually cross the x-axis). This indicates that if a function is continuous between two different values of x and $f(x)$ has opposite signs at these two points, then there is at least one root between these two values of x. The approximate value of the root can be found by repeated applications of this principle.

EXAMPLE:

Given the equation $2x^4 + 5x^2 - 4x - 1 = 0$
One root of the equation lies within the range of:

a) 0 to 1   b) 1 to 2   c) 2 to 3   d) -1 to -2; or
e) none of these

1-28 FUNDAMENTALS OF ENGINEERING EXAM REVIEW WORKBOOK

Find f(x) at the limits of the ranges indicated

x	f(x)
-2	+59
-1	+10
0	-1
1	+2
2	+43
3	+194

Based on sign changes, there is at least one root in the range 0 to 1 (there also is at least one root between 0 and -1). Check with a rough plot (Fig. 1.14).

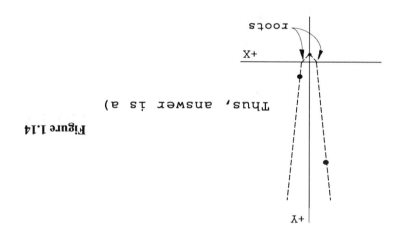

Figure 1.14

Thus, answer is a)

LOGARITHMS

The logarithm of a number to the base b (b>0 and b ≠ 1) is the exponent that must be employed to express the number as a power of b. This definition means that

$$\log_b n = a$$

is equivalent to $b^a = n$

The following can be derived from these two equivalent expressions

$$n = b^{\log_b n}$$

Additional properties of the logarithm are:

$$\log_b 1 = 0$$

$$\log_b b = 1$$

$$\log_b mn = \log_b m + \log_b n$$

$$\log_b \frac{m}{n} = \log_b m - \log_b n$$

$$\log_b \frac{1}{n} = -\log_b n$$

$$\log_b (m)^n = n \log_b m$$

$$\log_b \sqrt[n]{m} = \frac{1}{n} \log_b m$$

When the logarithm to one base is known, the logarithm to any other base may be determined in the following manner. By definition

$$n = b^{\log_b n}$$

Taking the logarithm of each side of the equation to the base a gives

$$\log_a n = \log_a \left[b^{\log_b n} \right]$$

$$\log_a n = (\log_b n)(\log_a b)$$

Solving for $\log_b n$

$$\log_b n = \frac{\log_a n}{\log_a b}$$

Also if we set n = a, the following useful expression is obtained.

$$\log_b a = \frac{1}{\log_a b}$$

A base commonly employed is the base 10 (the Briggsian System). Logarithms in this system are referred to as common Logarithms and $\log_{10} n$ is generally written as log n.

Another base used extensively because it greatly simplifies mathematical expressions in the calculus is the base e. Mathematically e is given by

$$e = \lim_{m \to \infty} (1 + \frac{1}{m})^m = 2.718218...$$

The number e is an irrational number which, therefore, does not repeat or terminate,

$$e = 2.718281828459...$$

Logarithms employing e as the base are referred to as natural logs and $\log_e n$ is generally written as ln n.

Conversion between common logs and natural logs can be made by the following

$$\log n = 0.4343 \ln n$$

$$\ln n = 2.3026 \log n$$

Logarithms can be found directly by using scientific calculators. However, when tables are to be used to find the logarithm, two other terms must be defined: the mantissa and the characteristic. The logarithm is composed of a decimal part called the mantissa which is always positive, and a whole number called the characteristic which may be positive, negative, or zero. Mantissas are found in tables of logarithms.

EXAMPLE (Fig. 1.15):

Log 19.75 = 1.2956

CHARACTERISTIC MANTISSA   Figure 1.15

log 1.975 = 0.2956

Notice that the characteristic changes as the decimal point moves whereas the mantissa is affected only by the sequence of numbers.

Log 0.1975 = -1 + 0.2956 which may be written as -0.7044, however this form can lead to confusion since the mantissa does not appear explicitly. In order to avoid confusion the following manipulation is sometimes employed: add 10 to the characteristic and subtract 10 as follows:

$$\begin{array}{r} -1 + 0.2956 \\ +10 \quad -10 \\ \hline 9 + 0.2956 \quad -10 \end{array} \quad \text{or } 9.2956 - 10 = \log 0.1975$$

MATHEMATICS 1-31

Rules for Characteristics of Common Logs:

Number	Characteristic
1 or greater	The characteristic is positive and is one less than the number of digits to left of the decimal point.
less than 1	The characteristic is negative and one greater than the number of zeros between the decimal point and the first nonzero digit.

Mantissas not found directly in the tables can be determined by linear interpolation.

EXAMPLES:

1) Find 3^4 by logs.

   let $x = 3^4$

   $\log x = \log 3^4$

   $\log x = 4 \log 3$

   $\log x = 4 \times 0.477$

   $\log x = 1.908$

   $x = 81$

2) Which one of the following is correct according to this statement $7 = 10^{0.845}$?

   a) $\log 10 = 0.845$

   b) $\log 7 = 0.845$

   c) $\log 0.845 = 7$

   By inspection we can see that the correct answer is b).

3) What is $\log (0.001)^3$?

   $\log(0.001)^3 = 3 \log 0.001 = 3 (-3) = -9$

   alternate solution: $\log(0.001)^3 = \log (10^{-3})^3 = \log 10^{-9} = -9$

4) Log 0.00069 is:

   a) 6.062 + 10
   b) 5.839 - 10
   c) 4.839 - 10
   d) 6.839 - 10

The correct answer is d).

5) Find x if $x^{-4/5} = 0.0001$. Take the log of both sides

$$\frac{-4}{5} \log x = \log(0.0001)$$

$$\frac{-4}{5} \log x = -4$$

$$\log x = +5$$

$$x = 100,000$$

6) Find $\log_{b^2}(n)$, if $\log_b n = x$.

$$b^x = n$$

$$b = n^{1/x}$$

$$b^2 = n^{2/x}$$

$$\log_{b^2}(n)^{2/x} = 1$$

$$\log_{b^2}(n) = 1$$

$$\log_{b^2}(n) = \frac{x}{2}$$

7) Find $\ln 1/\sqrt{e}$

$$\ln \frac{1}{\sqrt{e}} = \ln e^{-1/2} = -\frac{1}{2} \ln e = -\frac{1}{2}$$

8) Find $\log_7 56$

Since $\log_a n = \dfrac{\log_b n}{\log_a b}$

Then $\log_7 56 = \dfrac{\log 56}{\log 7} = 2.069$

TRIGONOMETRY

Angles

Trigonometry deals with angles. Angles can be expressed in degrees or radians. There are 360° in a complete circle. A radian is the angle that is subtended by an arc on a circle which is equal to the radius of the circle. Since the circumference of a circle is 2π, $360° = 2\pi$ radians. One radian is approximately 57.3°.

Note that the distance along an arc, S, is equal to the radius r, times the included angle measured in radians (Fig. 1.16).

$S = r\theta$

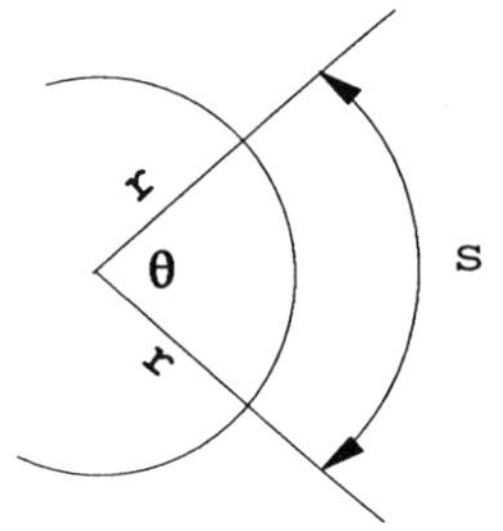

Figure 1.16

Conversions:

$$\text{radians} = \text{degrees} \times \frac{\pi}{180}$$

$$\text{degrees} = \text{radians} \times \frac{180}{\pi}$$

From these conversions, the following values can be found:

$$\begin{aligned}
30° &= \pi/6 \text{ radians} \\
60° &= \pi/3 \text{ radians} \\
45° &= \pi/4 \text{ radians} \\
90° &= \pi/2 \text{ radians} \\
180° &= \pi \text{ radians} \\
360° &= 2\pi \text{ radians}
\end{aligned}$$

1-34 FUNDAMENTALS OF ENGINEERING EXAM REVIEW WORKBOOK

EXAMPLE:

Convert 10 rpm to radians/sec

Since 1 revolution = 360°, 10 rpm is an angular velocity of 3600°/min

Then 10 rpm = 3600 x $\frac{\pi}{180}$ x $\frac{1}{60}$ = $\frac{\pi}{3}$ rad/sec

Trigonometric Functions

Consider a rectangular coordinate system and let a radius vector of length r rotate through an angle θ in a counterclockwise direction from the positive x-axis (Fig. 1.17). (Note that if θ is negative the rotation is in the clockwise direction.)

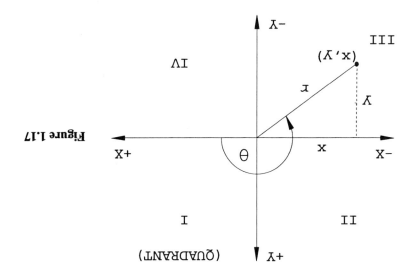

Figure 1.17

If the coordinates of the end point of the vector are (x, y), the trigonometric functions are defined as follows. (note: x and y may be positive or negative depending on the location of the end point, but r is always positive):

$\sin \theta$ = y/r
$\cos \theta$ = x/r
$\tan \theta$ = y/x
$\csc \theta$ = 1/sin θ = r/y
$\sec \theta$ = 1/cos θ = r/x
$\cot \theta$ = 1/tan θ = x/y
$\tan \theta$ = sin θ/cos θ

Note that as θ moves from quadrant to quadrant, the sign of the trig function will change because of the change in sign of x and y. The signs of the trigonometric functions in each of the quadrants is given in the following table:

Quadrant→	I	II	III	IV
sin θ	+	+	-	-
cos θ	+	-	-	+
tan θ	+	-	+	-

EXAMPLE:

Find the sin, cos, and tan of the angle $17\pi/6$ radians.

First we note that $17\pi/6$ is greater than 2π and less than 3π radians.

$$\frac{17\pi}{6} - \frac{12\pi}{6} = \frac{5\pi}{6} \text{ radians}$$

$$\frac{5\pi}{6} = \frac{3\pi}{6} + \frac{2\pi}{6} = 90° + 60°$$

In sketching this out we note that $\theta = 30°$ and $\alpha = 60°$ (Fig. 1.18).

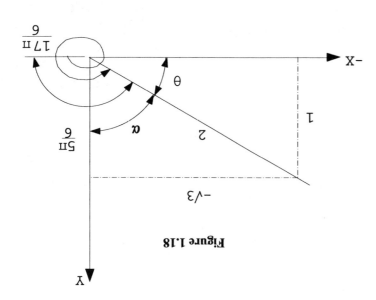

Figure 1.18

Then

$$\sin \frac{17\pi}{6} = \frac{1}{2}$$

$$\cos \frac{17\pi}{6} = -\frac{\sqrt{3}}{2}$$

$$\tan \frac{17\pi}{6} = -\frac{1}{\sqrt{3}}$$

Inverse functions are defined to determine the original angle from the value of one of the trigonometric functions. For example, if

$$\sin \theta = x$$

Then
$$\sin^{-1} x = \theta$$

This is also sometimes written with the prefix "arc." For example

$$\arcsin x = \sin^{-1} x$$

Many problems involve the solution of right triangles. Consider the right triangle in Fig. 1.19.

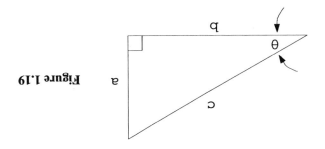

Figure 1.19

By the Pythagorean theorem:

$$a^2 + b^2 = c^2$$

And the trig functions are:

$$\sin \theta = \frac{a}{c} = \frac{\text{opposite}}{\text{hypotenuse}}$$

$$\cos \theta = \frac{b}{c} = \frac{\text{adjacent}}{\text{hypotenuse}}$$

$$\tan \theta = \frac{a}{b} = \frac{\text{opposite}}{\text{adjacent}}$$

EXAMPLE:

Find sin (arctan ½)

Let θ = arctan ½

MATHEMATICS 1-37

Then θ is defined by the triangle in Fig. 1.20.

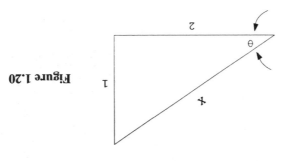

Figure 1.20

By the Pythagorean theorem

$$x^2 = 1^2 + 2^2 = 5$$

$$x = \sqrt{5}$$

Then $\sin \theta = \sin(\arctan \tfrac{1}{2}) = \dfrac{1}{\sqrt{5}}$

Polar Coordinates

Sometimes the locations of points are expressed in polar coordinates. Polar coordinates are in the form (r, θ), where r is a radius vector to the point from the origin and θ is the angle of rotation of the radius vector measured from the x-axis (Fig. 1.21).

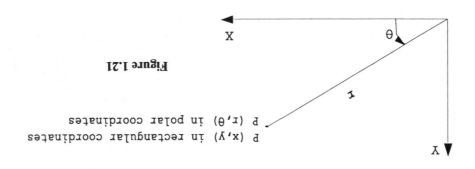

Figure 1.21

If a point is given as (x, y) the polar coordinates can be determined from

$$r = \sqrt{x^2 + y^2}$$

$$\theta = \tan^{-1} y/x = \sin^{-1} y/r = \cos^{-1} x/r$$

If a point is given as (r, θ) the rectangular coordinates can be determined from

$$y = r \sin \theta \qquad x = r \cos \theta$$

EXAMPLE:

Find the polar coordinates of the point (4, 3)

$$r = \sqrt{x^2 + y^2} = \sqrt{4^2 + 3^2} = 5$$

$$\theta = \tan^{-1}(3/4) = 36.9°$$

Fundamental Identities

Pythagorean formulas:

$$\sin^2 A + \cos^2 A = 1$$

$$1 + \cot^2 A = \csc^2 A$$

$$\tan^2 A + 1 = \sec^2 A$$

Double-angle formulas:

$$\sin 2A = 2 \sin A \cos A$$

$$\cos 2A = 2\cos^2 A - 1 = 1 - 2\sin^2 A$$

$$= \cos^2 A - \sin^2 A$$

$$\tan 2A = \frac{2 \tan A}{1 - \tan^2 A}$$

Half-angle formulas:

$$\sin \frac{A}{2} = \pm \sqrt{\frac{1 - \cos A}{2}}$$

$$\cos \frac{A}{2} = \pm \sqrt{\frac{1 + \cos A}{2}}$$

$$\tan \frac{A}{2} = \pm \sqrt{\frac{1 - \cos A}{1 + \cos A}} = \frac{\sin A}{1 + \cos A} = \frac{1 - \cos A}{\sin A}$$

MATHEMATICS 1-39

Two-angle formulas:

$\sin(A \pm B) = \sin A \cos B \pm \cos A \sin B$

$\cos(A \pm B) = \cos A \cos B \mp \sin A \sin B$

$\tan(A \pm B) = \dfrac{\tan A \pm \tan B}{1 \mp \tan A \tan B}$

$\sin A + \sin B = 2 \sin \dfrac{A+B}{2} \cos \dfrac{A-B}{2}$

$\sin A - \sin B = 2 \cos \dfrac{A+B}{2} \sin \dfrac{A-B}{2}$

$\cos A + \cos B = 2 \cos \dfrac{A+B}{2} \cos \dfrac{A-B}{2}$

$\cos A - \cos B = 2 \sin \dfrac{A+B}{2} \sin \dfrac{A-B}{2}$

Laws for Triangles

Given the triangle ABC with sides a, b, and c, (Fig. 1.22), the following laws apply.

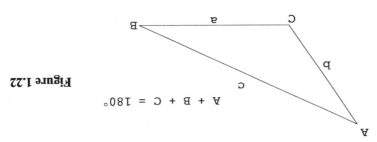

$A + B + C = 180°$

Figure 1.22

Laws of cosines:

$c^2 = a^2 + b^2 - 2ab \cos C$

$b^2 = a^2 + c^2 - 2ac \cos B$

$a^2 = b^2 + c^2 - 2bc \cos A$

1-40 FUNDAMENTALS OF ENGINEERING EXAM REVIEW WORKBOOK

Laws of sines:

$$\frac{a}{\sin A} = \frac{b}{\sin B} = \frac{c}{\sin C}$$

EXAMPLE:

If boat A heads in an easterly direction at 20 knots and boat B heads in a south easterly direction at 30 knots, how far apart are the boats after 2 hours if it is assumed they started from the same location?

If the physical conditions of this problem are sketched (Fig. 1.23), it will be obvious that the solution can be obtained from the law of cosines:

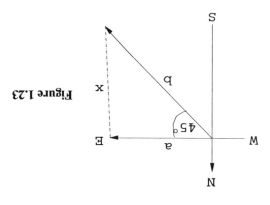

Figure 1.23

let a = distance traveled by boat A in 2 hours.
     b = distance traveled by boat B in 2 hours.
     x = distance of A from B after 2 hours.

a = 2 x 20 = 40 nautical miles
b = 2 x 30 = 60 nautical miles

$x^2 = (40)^2 + (60)^2 - 2(40)(60)\cos 45°$

x = 42.5 nautical miles

EXAMPLE:

If a parallelogram has sides of lengths 3 and 5 and an included angle of 60°, find the length of the diagonals.

MATHEMATICS 1-41

Sketch the parallelogram (Fig. 1.24).

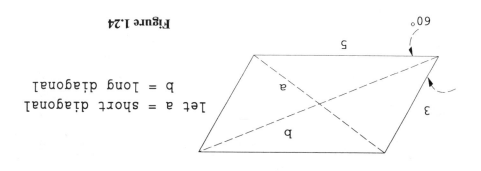

Figure 1.24

Let a = short diagonal
b = long diagonal

$a^2 = (3)^2 + (5)^2 - 2(3)(5) \cos 60°$

$a^2 = 9 + 25 - 30(0.5) = 19$

a = 4.36 (short diagonal)

$b^2 = (3)^2 + (5)^2 - 2(3)(5) \cos(180-60)$

$b^2 = 9 + 25 - 2(3)(5)(-0.5) = 49$

b = 7.0 (long diagonal)

Hyperbolic Functions

Definitions

$$\text{hyperbolic sin } x = \sinh x = \frac{e^x - e^{-x}}{2}$$

$$\text{hyperbolic cos } x = \cosh x = \frac{e^x + e^{-x}}{2}$$

$$\text{hyperbolic tan } x = \tanh x = \frac{e^x - e^{-x}}{e^x + e^{-x}}$$

$$\text{hyperbolic csc } x = \operatorname{csch} x = \frac{1}{\sinh x}$$

$$\text{hyperbolic sec } x = \operatorname{sech} x = \frac{1}{\cosh x}$$

$$\text{hyperbolic cot } x = \coth x = \frac{1}{\tanh x}$$

1-42 FUNDAMENTALS OF ENGINEERING EXAM REVIEW WORKBOOK

Inverse Functions

If $\sinh \theta = x$ then $\theta = \sinh^{-1} x = \text{arcsinh } x$

Identities

$\sinh(-x) = -\sinh x$
$\cosh(-x) = \cosh x$
$\tanh(-x) = -\tanh x$
$\cosh^2 x - \sinh^2 x = 1$
$1 - \tanh^2 x = \text{sech}^2 x$
$\coth^2 x - 1 = \text{csch}^2 x$

ANALYTIC GEOMETRY

Straight Lines

The general equation of a straight line has only first order terms and can be written as

$$Ax + By + C = 0$$

The slope–intercept form of the equation of a straight line is

$$y = mx + b$$

where m is the slope and b is the y intercept (value of y when x = 0) (Figure 1.25).

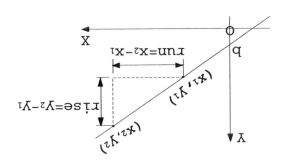

Figure 1.25

$$\text{slope} = m = \frac{\text{rise}}{\text{run}} = \frac{\Delta y}{\Delta x} = \frac{y_2 - y_1}{x_2 - x_1}$$

MATHEMATICS 1-43

In terms of the general equation

$$m = \frac{-A}{B} \text{ and } b = \frac{-C}{B}$$

Another form of the equation for a straight line is

$$y - y_1 = m(x-x_1)$$

where (x_1, y_1) is any point on the line. The two-point form of the equation is

$$\frac{y - y_1}{x - x_1} = \frac{y_2 - y_1}{x_2 - x_1}$$

The slope of a line is the tangent of its inclination. If θ is the inclination, then as shown in Figure 1.26 the slope is positive if y increases as x increases (L_1) and negative if y decreases as x increases (L_2).

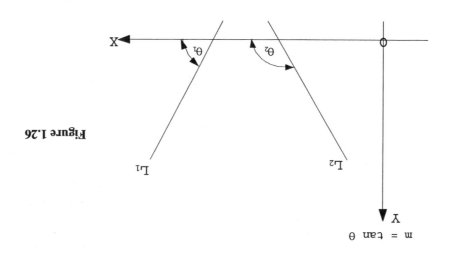

Figure 1.26

$m = \tan \theta$

If two lines are parallel, their slopes are equal.

$$m_1 = m_2$$

If two lines are perpendicular, their slopes must satisfy the following

$$m_1 m_2 = -1$$

or

$$m_1 = \frac{-1}{m_2}$$

1-44 FUNDAMENTALS OF ENGINEERING EXAM REVIEW WORKBOOK

If straight lines L_1 and L_2 (L_1 being the line of greater inclination) have slopes m_1 and m_2, the positive angle, α, from L_2 to L_1 is given by

$$\text{Tan } \alpha = \frac{m_1 - m_2}{1 + m_1 m_2}$$

The distance, d, between any two points can be found using the Pythagorean Theorem

$$d^2 = (y_2 - y_1)^2 + (x_2 - x_1)^2$$

$$d = \sqrt{(y_2 - y_1)^2 + (x_2 - x_1)^2}$$

EXAMPLE:

The point (5,2) is the midpoint of the line connecting (−4, −5) with (x, y). Find x and y (Figure 1.27).

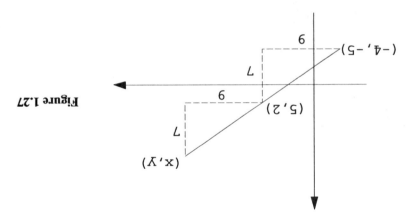

Figure 1.27

The slope from (−4, −5) to (5,2) is found from

$$m = \frac{\Delta y}{\Delta x} = \frac{2 - (-5)}{5 - (-4)} = \frac{7}{9}$$

Since (5,2) is the midpoint between (−4,−5) and (x,y) then

$$x = 5 + \Delta x = 5 + 9 = 14$$
$$y = 2 + \Delta y = 2 + 7 = 9$$

and (x,y) is (14,9)

EXAMPLE:

The angle between line 1 whose slope is –¼ and line 2 is 135°. Find the slope of line 2.

$$\tan \alpha = \frac{m_1 - m_2}{1 + m_1 m_2} \quad \text{but } \tan 135° = -1$$

$$-1 = \frac{-\frac{1}{4} - m_2}{1 - m_2/4}$$

Solving for m_2 yields $m_2 = 3/5$.

EXAMPLE:

Find the equation of the line perpendicular to the line $3x - 2y = 4$ and passing through the point (3, 2).

The equation of the line $3x - 2y = 4$ can be written as

$$y = \frac{3}{2}x - 2$$

The slope of this line is 3/2 thus the slope of the line we are looking for is −2/3. The equation is then

$$y - 2 = \frac{-2}{3}(x-3)$$

$$y = \frac{-2}{3}x + 4$$

EXAMPLE:

Find the equation of the family of lines perpendicular to the line $5x + 7y + 11 = 0$.

The slope of the line $5x + 7y + 11 = 0$ is

$$m = \frac{-A}{B} = \frac{-5}{7}$$

Thus, the slope of the family of curves perpendicular to this line is 7/5 and the equation of the family is

$$7x - 5y + c = 0$$

EXAMPLE:

Find the equation of the line with x and y intercepts of x_o and y_o respectively

Since the y intercept is y_o, we can write

$$y = mx + y_o \qquad\qquad \text{also } 0 = mx_o + y_o$$

$$m = \frac{-y_o}{x_o} \qquad\qquad \text{and } y = \frac{-y_o}{x_o} x + y_o$$

Rearranging

$$x_o y = -y_o x + x_o y_o$$

$$\frac{y}{y_o} + \frac{x}{x_o} = 1$$

Note that this is the equation of a straight line in terms of the x and y intercepts. Similarly it can be shown that the equation of a plane is

$$\frac{x}{x_o} + \frac{y}{y_o} + \frac{z}{z_o} = 1$$

where x_o, y_o, and z_o are the x, y, and z intercepts, respectively.

Conic Sections

An equation with second-order terms will give a conic section. The general equation of a conic section is

$$Ax^2 + Bxy + Cy^2 + Dx + Ey + F = 0$$

The conic section may be identified from the general coefficients as:

$$\text{if} \quad (B^2 - 4AC) < 0$$

a parabola if $(B^2 - 4AC) = 0$

a hyperbola if $(B^2 - 4AC) > 0$

The simplest form of the equation for an ellipse (i.e., with the center at the origin) is

$$\frac{x^2}{a^2} + \frac{y^2}{b^2} = 1$$

If $a^2 = b^2$, the equation reduces to

$$x^2 + y^2 = a^2$$

This is the equation for a circle of radius a (with the center at the origin).

The simple form of the equation for a parabola (vertex at the origin) is

$$y = a x^2 \text{ or } x = a y^2$$

The simple form of the equation for a hyperbola (center at the origin) is

$$\frac{x^2}{a^2} - \frac{y^2}{b^2} = 1 \quad \text{or} \quad \frac{y^2}{a^2} + \frac{x^2}{b^2} = 1$$

When the center (or vertex in the case of the parabola) of the conic section is translated from the origin to a point (h,k), the simple forms of the equations are modified by substituting (x-h) for x and (y-k) for y. For example the equation of a circle of radius r with the center at (h,k) is

$$(x - h)^2 + (y - k)^2 = r^2$$

If the axis of the curve is inclined to the coordinate axes, an x y term is introduced.

EXAMPLE:

Given the following equation: $x^2 - 4x + 2y^2 - 12y = -20$; identify the conic section and determine the location of the center or vertex.

Note that the general coefficients are as follows

$A = 1$
$B = 0$
$C = 2$

Then $B^2 - 4AC = 0 - 4(1)(2) = -8$. $\therefore B^2 - 4AC < 0$ and the curve is an ellipse. To find the center of the ellipse, complete squares

$(x^2 - 4x) + 2(y^2 - 6y) = -20$

$(x^2 - 4x + 4) + 2(y^2 - 6y + 9) = -20 + 4 + 18$

$(x - 2)^2 + 2(y - 3)^2 = 2$

$$\frac{(x-2)^2}{2} + (y - 2)^2 = 1$$

Thus, the center is at (2,3). The simple form of the ellipse can also be recognized.

EXAMPLE:

Find the equation of the tangent to the circle defined by the equation $x^2 + y^2 = 25$ at the point (3,4)

Sketch the figure of the tangent to the circle at (3,4) (Fig. 1.28).

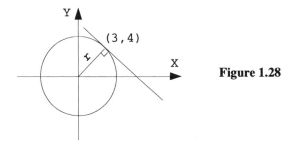

Figure 1.28

Note that the radius will be perpendicular to the tangent at (3,4) and the slope of the radius to this point is

$$m_r = \frac{\Delta Y}{\Delta X} = \frac{4}{3}$$

Slope of tangent $= -\frac{3}{4}$

The equation for the tangent is

$$y = \frac{-3}{4}x + b$$

Now since (3,4) is on the tangent

$$4 = \frac{-3}{4}(3) + b$$

$$b = \frac{25}{4}$$

and the equation of the tangent is

$$y = \frac{-3}{4}x + \frac{25}{4}$$

DIFFERENTIAL CALCULUS

Functions

The symbol f(x) is a shorthand representation of the expression "function of x", where x is any variable. When the expression y = f(x) is written, x is the <u>independent</u> variable and y is the <u>dependent</u> variable (its value depends on the value of x). When we write f(2), this means the value of f(x) when x = 2. For example, if

$$f(x) = 3x^2 + 2x + 3$$
$$f(1) = 2(1)^2 + 2(1) + 2 = 8$$
$$f(2) = 3(2)^2 + 2(2) + 3 = 19$$

If the dependent variable, z, is a function of two independent variables, we write

$$z = f(x,y)$$

When it is required to introduce more than one function into a discussion, additional functional symbols, such as F(x), G(x), etc., are used.

Limits

"Limits" are applied to express the value a function approaches when the independent variable approaches some constant. More precisely, we say that f(x) approaches the limit L as x approaches a if the numerical value of f(x) − L can be made to remain as small as we like by taking x sufficiently close to a. This is written as

$$\lim_{x \to a} f(x) = L$$

Note that the definition does not mean that f(a) = L (although it may) since the function may not actually be defined when x = a (it may not be a continuous function at x = a). A function f(x) is <u>continuous</u> at x = a if f(a) exists and if

$$\lim_{x \to a} f(x) = f(a)$$

Some Properties of Limits:

If $\lim_{x \to a} f(x) = A$ and $\lim_{x \to a} F(x) = B$, then

1) $\lim_{x \to a} \{f(x) \pm F(x)\} = A \pm B$

2) $\lim_{x \to a} \{f(x) F(x)\} = AB$

3) $\lim_{x \to a} \dfrac{f(x)}{F(x)} = \dfrac{A}{B}$, if $B \neq 0$

4) $\lim_{x \to a} \{f(x) + K\} = A + K$, where K = constant

5) $\lim_{x \to a} Kf(x) = KA$

Also $\lim_{x \to 0} \dfrac{1}{x} = \infty$ and $\lim_{x \to \infty} \dfrac{1}{x} = 0$

EXAMPLE:

$$\lim_{x \to 2} (3x^3 - 7)$$

$$\lim_{x \to 2} (3x^3 - 7) = 3(2)^3 - 7 = 17$$

EXAMPLE:

$$\lim_{x \to 2} \left(\frac{5}{2-x} \right) = \infty$$

We can make the function $\left(\frac{5}{2-x} \right)$ exceed any given number by taking x close enough to 2.

EXAMPLE:

Find $\lim_{x \to 3} \frac{(x^2 - 9)}{(x - 3)}$

In this case, the numerator and the denominator approach zero as x approaches 3 and the limit is not apparent. However, if we factor the numerator we have

$$\lim_{x \to 3} \frac{(x^2 - 9)}{(x - 3)} = \lim_{x \to 3} \frac{(x + 3)(x - 3)}{(x - 3)} = \lim_{x \to 3} (x + 3) = 6$$

In the example above, the numerator and denominator were both approaching zero. When it is not possible to factor, this type of problem can be solved by using L'Hospital's Rule:

Suppose that f(x)→0 and F(x)→0 as X→a, then if

$$\lim_{x \to a} \frac{f'(x)}{F'(x)} = A$$

It is also true that

$$\lim_{x \to a} \frac{f(x)}{F(x)} = A$$

Applying this rule to the previous example:

$$\lim_{x \to 3} \frac{(x^2 - 9)}{(x - 3)} = \lim_{x \to 3} \frac{2x}{1} = 6$$

EXAMPLE:

Find $\lim_{x \to 3} \frac{(x^2 - 9)}{(27 - 3x - 2x^2)}$

$$\lim_{x \to 3} \frac{(x^2 - 9)}{(27 - 3x - 2x^2)} = \lim_{x \to 3} \frac{2x}{(-3 - 4x)} = \frac{6}{-15} = \frac{-2}{5}$$

EXAMPLE:

Find $\lim\limits_{x \to 0} \dfrac{e^x + e^{-x} - 2}{5x^2}$

$$\lim_{x \to 0} \frac{e^x + e^{-x} - 2}{5x^2} = \lim_{x \to 0} \frac{e^x - e^{-x}}{10x}$$

The rule must be applied again

$$\lim_{x \to 0} \frac{e^x + e^{-x}}{10x} = \lim_{x \to 0} \frac{e^x - e^{-x}}{10} = \frac{2}{10} = \frac{1}{5}$$

The Derivative

An important limit forms the basis of the differential calculus. By means of this limit, rates of change can be calculated. The term by which all instantaneous rates of change are designated is the <u>derivative</u>. The derivative is defined as:

Let y be a single-valued function of x. Let Δy be the change in y when x changes by an amount Δx. Then the limit of the ration $\Delta y/\Delta x$, as Δx approaches zero, provided that this limit exists, is called the derivative of y with respect to x. In symbols, this is written as

$$\lim_{\Delta x \to 0} \frac{\Delta y}{\Delta x} = \frac{dy}{dx}$$

The derivative is also written in the following forms:

$$D_x y, \frac{df(x)}{dx}, f'(x), y'$$

A function for which the derivative exists is called a differentiable function.

EXAMPLE:

Find $\dfrac{dy}{dx}$, if $y = x^3$.

Let Δy be the change in y when x changes by Δx, then

$$y + \Delta y = (x + \Delta x)^3$$

$$\Delta y = (x + \Delta x)^3 - y$$

$$\Delta y = (x + \Delta x)^3 - x^3$$

$$\Delta y = 3x^2\Delta x + 3x(\Delta x)^2 + (\Delta x)^3$$

$$\frac{\Delta y}{\Delta x} = 3x^2\Delta x + 3x(\Delta x)^2 + (\Delta x)^3$$

$$\lim_{\Delta x \to 0}\frac{\Delta y}{\Delta x} = \lim_{\Delta x \to 0}[3x^2 + 3x\Delta x + \Delta x^2]$$

$$\frac{dy}{dx} = 3x^2$$

Basic Rules of Differentiation

The example above illustrates the use of the definition of the derivative in performing differentiation. However, in practice differentiation is most conveniently performed by use of formulas or basic rules already derived for various common functions. Some of these formulas are provided in the Table of Derivatives.

EXAMPLE:

Find $\frac{dy}{dx}$ if $y = 6x^4$ $\frac{dy}{dx} = 4 \cdot 6x^3 = 24x^3$

EXAMPLE:

Find $\frac{dy}{dx}$ if $y = \sqrt{1-x}$ rewrite $y = (1-x)^{\frac{1}{2}}$

Then

$$\frac{dy}{dx} = \frac{1}{2}(1-x)^{-\frac{1}{2}}(-1) = \frac{-1}{2\sqrt{1-x}}$$

Table of Derivatives

1. $\frac{dc}{dx} = 0$

2. $\frac{d}{dx}(cv) = c\frac{dv}{dx}$

3. $\frac{d}{dx}(u \pm v) = \frac{du}{dx} \pm \frac{dv}{dx}$

4. $\dfrac{d}{dx}(uv) = u\dfrac{dv}{dx} + v\dfrac{du}{dx}$

5. $\dfrac{d}{dx}(x^n) = n\,x^{n-1}$ (power rule)

6. $\dfrac{d}{dx}(u^n) = nu^{n-1}\dfrac{du}{dx}$

7. $\dfrac{d}{dx}(c^u) = c^u \cdot \ln c \cdot \dfrac{du}{dx}$

8. $\dfrac{d}{dx}(v^u) = uv^{u-1}\dfrac{dv}{dx} + v^u \cdot \ln v \cdot \dfrac{du}{dx}$

9. $\dfrac{d}{dx}(e^x) = e^x$

10. $\dfrac{d}{dx}(e^u) = e^u\dfrac{du}{dx}$

11. $\dfrac{d}{dx}\left(\dfrac{u}{v}\right) = \dfrac{v\dfrac{du}{dx} - u\dfrac{dv}{dx}}{v^2}$

12. $\dfrac{d}{dx}(\log_c x) = \dfrac{1}{x}\log_c e$

13. $\dfrac{d}{dx}(\log_c u) = \dfrac{1}{u}\log_c e \cdot \dfrac{du}{dx}$

14. $\dfrac{d}{dx}(\ln x) = \dfrac{1}{x}$

15. $\dfrac{d}{dx}(\ln u) = \dfrac{1}{u}\dfrac{du}{dx}$

16. $\dfrac{d}{dx}(\sin u) = \cos u \dfrac{du}{dx}$

17. $\dfrac{d}{dx}(\cos u) = -\sin u \, \dfrac{du}{dx}$

18. $\dfrac{d}{dx}(\tan u) = \sec^2 u \, \dfrac{du}{dx}$

19. $\dfrac{d}{dx}(\csc u) = -(\csc u)(\cot u) \, \dfrac{du}{dx}$

20. $\dfrac{d}{dx}(\sec u) = (\sec u)(\tan u) \, \dfrac{du}{dx}$

21. $\dfrac{d}{dx}(\cot u) = -\csc^2 u \, \dfrac{du}{dx}$

22. $\dfrac{d}{dx}(\sinh u) = \cosh u \, \dfrac{du}{dx}$

23. $\dfrac{d}{dx}(\cosh u) = \sinh u \, \dfrac{du}{dx}$

24. $\dfrac{d}{dx}(\tanh u) = \operatorname{sech}^2 u \, \dfrac{du}{dx}$

25. $\dfrac{d}{dx}(\operatorname{csch} u) = -(\operatorname{csch} u)(\coth u) \, \dfrac{du}{dx}$

26. $\dfrac{d}{dx}(\operatorname{sech} u) = -(\operatorname{sech} u)(\tanh u) \, \dfrac{du}{dx}$

27. $\dfrac{d}{dx}(\coth u) = -\operatorname{csch}^2 u \, \dfrac{du}{dx}$

Notes: (a) u and v are functions of x.
(b) c and n are constants.
(c) $\lim\limits_{m \to \infty} (1 + \dfrac{1}{m})^m$.
(d) All angles are in radians.

EXAMPLE:

Find $\dfrac{dy}{dx}$ if $y = \dfrac{4x^3 + 3}{2x}$

Let $u = 4x^3 + 3$ and $v = 2x$, then by formula 11 in the Table:

$$\dfrac{dy}{dx} = \dfrac{(2x)(12x^2) - (4x^3 + 3)(2)}{4x^2}$$

$$\dfrac{dy}{dx} = \dfrac{24x^3 - 8x^3 - 6}{4x^2} = \dfrac{8x^3 - 3}{2x^2}$$

EXAMPLE:

Find $\dfrac{dy}{dx}$ if $y = e^{-x} \cos 2x$

$$\dfrac{dy}{dx} = e^{-x}(-\sin 2x)(2) + (-1) e^{-x} \cos 2x$$

$$\dfrac{dy}{dx} = -e^{-x}(2 \sin 2x + \cos 2x)$$

EXAMPLE:

Find $\dfrac{dy}{dx}$ if $y = (\sin ax)(\tan ax)$

$$\dfrac{dy}{dx} = (\sin ax)(\sec^2 ax) + (\tan ax)(\cos ax)$$

$$\dfrac{dy}{dx} = (a \sin ax)(\sec^2 ax) + \dfrac{a \sin ax}{\cos ax} \cos ax$$

$$\dfrac{dy}{dx} = (a \sin ax)(\sec^2 ax + 1)$$

EXAMPLE:

Find $\dfrac{dy}{dx}$ if $y = (3x)^x$

let $u = x$ and $v = 3x$, then by formula 8 in the Table

$$\dfrac{dy}{dx} = x(3x)^{x-1}(3) + (3x)^x (\ln 3x)(1)$$

$$\frac{dy}{dx} = (3x)^x (1 + \ln 3x)$$

EXAMPLE:

Find $\frac{dy}{dx}$ if $y = \cosh^2 3x - \sinh(x^2)$

The solution to this problem requires use of formulas 3, 5, 6, 22, and 23 from the Table.

$$\frac{dy}{dx} = (2 \cosh 3x)(\sinh 3x)(3) - \cosh(x^2)(2x)$$

$$\frac{dy}{dx} = 6 \cosh 3x \sinh 3x - 3x \cosh x^2$$

EXAMPLE:

Find $\frac{dy}{dx}$ if $y = e^{-x} \ln x$

$$\frac{dy}{dx} = e^{-x}\left(\frac{1}{x}\right) + (-1) e^{-x} \ln x$$

$$\frac{dy}{dx} = e^{-x}\left(\frac{1}{x} - \ln x\right)$$

Taking logs prior to differentiation will sometimes simplify the procedure considerably. The following example illustrates this.

EXAMPLE:

Find $\frac{dy}{dx}$ if $y = x^3 \sqrt{\frac{1 - x^2}{1 + x^2}}$.

Taking logs;

$\ln y = \ln x^3 + \ln \sqrt{1 - x^2} - \ln \sqrt{1 + x^2}$

then,

$\ln y = 3 \ln x + \frac{1}{2} \ln (1-x^2) - \frac{1}{2} \ln (1+x^2)$

Differentiating each side:

$$\frac{1}{y}\frac{dy}{dx} = \frac{3}{x} + \frac{1}{2}\left(\frac{1}{1-x^2}\right)(-2x) - \frac{1}{2}\frac{1}{1+x^2}(2x)$$

$$\therefore \frac{dy}{dx} = y\left(\frac{3}{x} - \frac{x}{1-x^2} - \frac{x}{1+x^2}\right)$$

Substituting for y;

$$\frac{dy}{dx} = x^3\sqrt{\frac{1-x^2}{1+x^2}}\left(\frac{3}{x} - \frac{x}{1-x^2} - \frac{x}{1+x^2}\right)$$

Derivatives may be treated as a ratio of differentials that can be manipulated algebraically. This property is most evident when working with the integral calculus, however, it is also useful for illustrating or visualizing various relationships of derivatives. For example,

$$\frac{dy}{dx} = \frac{1}{\left(\frac{dx}{dy}\right)} \quad \text{that is} \quad D_x y = \frac{1}{D_y x}$$

Also, when working with parametric equations such as

$$x = f(t) \qquad y = g(t)$$

the derivative of y with respect to x can be determined without solving explicitly for y as a function of x, by the following

$$\frac{dy}{dx} = \frac{dy/dt}{dx/dt}$$

EXAMPLE:

Find $\frac{dy}{dx}$ if $y = 2t^3 - 6t^2$ and $x = 3t^2$

$$\frac{dy}{dt} = 6t^2 - 12t$$

$$\frac{dx}{dt} = 6t$$

then $\dfrac{dy}{dx} = \dfrac{dy/dt}{dx/dt} = \dfrac{6t^2 - 12t}{6t} = t-2$

Implicit Differentiation

When y is defined as an implicit function of x, $f(x,y) = 0$, $\dfrac{dy}{dx}$ may be determined without solving explicitly for y. The procedure to be used is as follows:

1. Differentiate the terms of each member considering y as a differentiable function of x (use formula 4 in the Table of Derivatives for those terms containing x and y).

2. Solve for $\dfrac{dy}{dx}$

EXAMPLE:

Find $\dfrac{dy}{dx}$ if $x^3 + 3x^2y + y^3 = 8$

Differentiate each term:

$$\dfrac{d}{dx}(x^3) + \dfrac{d}{dx}(3x^2y) + \dfrac{d}{dx}(y^3) = \dfrac{d}{dx}(8)$$

$$3x^2 + 3(x^2 \dfrac{dy}{dx} + y\,2x) + 3y^2 \dfrac{dy}{dx} = 0$$

$$\dfrac{dy}{dx}(3y^2 + 3x^2) = -3x^2 - 6xy$$

$$\dfrac{dy}{dx} = -\dfrac{x^2 + 2xy}{x^2 + y^2}$$

Higher Order Derivatives

The derivatives, f'(x), of a function $y = f(x)$ may be a differentiable function. In this case, the derivative of f'(x) is called the second derivative of y. The second derivative may be written in several forms:

$$\dfrac{d}{dx}\left(\dfrac{dy}{dx}\right),\ \dfrac{d^2y}{dx^2},\ f''(x),\ y'',\ D_x^2 y$$

Similarly, this process may be repeated and derivatives of higher order written in analogous fashion.

EXAMPLE:

Find the Second Derivative of

$$4x^2 + 8x + 16 = y$$

$$\frac{dy}{dx} = 8x + 8$$

$$\frac{d^2y}{dx^2} = 8$$

EXAMPLE:

If $y = 8x^3 - x^2$

$y' = 24x^2 - 2x$

$y'' = 48x - 2$

$y''' = 48$

$y'''' = 0$

EXAMPLE:

Find $\frac{d^3y}{dx^3}$ if $y = 2e^x + \sin 2x$

$$\frac{dy}{dx} = 2e^x + 2\cos 2x$$

$$\frac{d^2y}{dx^2} = 2e^x - 4\sin 2x$$

$$\frac{d^3y}{dx^3} = 2e^x - 8\cos 2x$$

Partial Derivatives

If a function $z = f(x,y)$ is differentiated with respect to x while y is treated as a constant, the resulting derivative is the <u>partial derivative</u> of z with respect to x. The partial derivative treats all independent variables, except for the one indicated, as constants. The partial derivative of z with respect to x is written as

$$\frac{\partial z}{\partial x}$$

Other forms are

$$\frac{\partial f}{\partial x}, \ f_x, \ f'_x(x, y), \ f_1(x,y)$$

The partial derivative is also defined by a limit. If $z = f(x,y)$

$$\frac{\partial f}{\partial x} = f_x = \lim_{\Delta x \to 0} \frac{f(x + \Delta x, y) - f(x,y)}{\Delta x}$$

Likewise the partial derivative of f with respect y is:

$$\frac{\partial f}{\partial y} = f_y = \lim_{\Delta y \to 0} \frac{f(x, y + \Delta y) - f(x,y)}{\Delta y}$$

There are four second-order partial derivatives:

$$\frac{\partial^2 z}{\partial x^2}$$

$$\frac{\partial^2 z}{\partial y^2}$$

$$\frac{\partial^2 z}{\partial x \partial y} = \frac{\partial}{\partial x}\left(\frac{\partial z}{\partial y}\right)$$

$$\frac{\partial^2 z}{\partial y \partial x} = \frac{\partial}{\partial y}\left(\frac{\partial z}{\partial x}\right)$$

The last two derivatives are called <u>cross derivatives</u>. If the cross derivatives are continuous, the order of differentiation is immaterial. That is

$$\frac{\partial^2 z}{\partial x \partial y} = \frac{\partial^2 z}{\partial y \partial x}$$

also

$$\frac{\partial^3 z}{\partial x^2 \partial y} = \frac{\partial^3 z}{\partial x \partial y \partial x} = \frac{\partial^3 z}{\partial y \partial x^2}$$

etc.

EXAMPLE:

Find the first and second partial derivatives of $z = x^4 y^{-2}$.

$$\frac{\partial z}{\partial y} = -2 x^4 y^{-3}$$

$$\frac{\partial z}{\partial x} = 4x^3 y^{-2}$$

$$\frac{\partial^2 z}{\partial y^2} = 6x^4 y^{-4}$$

$$\frac{\partial^2 z}{\partial x^2} = 12x^2 y^{-2}$$

$$\frac{\partial^2 z}{\partial y \partial x} = \frac{\partial}{\partial y}(4x^3 y^{-2}) = -8x^3 y^{-3}$$

check

$$\frac{\partial^2}{\partial x \partial y} = \frac{\partial}{\partial x}(-2x^4 y^{-3}) = -8x^3 y^{-3}$$

In the discussion of partial derivatives thus far, x and y have been treated as independent variables. However, x and y may be dependent on other variables. If $z = f(x,y)$ with continuous first partial derivatives and if x and y are continuous functions of independent variables r, s,... which have first partial derivatives, then

$$\frac{\partial z}{\partial r} = \frac{\partial z}{\partial x}\frac{\partial x}{\partial r} + \frac{\partial z}{\partial y}\frac{\partial y}{\partial r}$$

In the special case when z = f(x,y) and x = g(r) and y = ∅ (r), the derivative of z with respect to r is a total derivative and is given by

$$\frac{dz}{dr} = \frac{\partial z}{\partial x}\frac{dx}{dr} + \frac{\partial z}{\partial y}\frac{\partial y}{\partial r} \quad \text{(Chain Rule)}$$

EXAMPLE:

The altitude of a right circular cone is 15 in. and increases at the rate of 0.2 in./min. The radius of the base is 10 in. and decreases at 0.3 in./min. How fast is the volume changing?

Let x = radius and y = altitude of cone. The volume of the cone, V, is given by

$$V = \frac{\pi x^2}{3} y$$

The change in volume with respect to time can be written as a total derivative because V is a function of one independent variable, t

$$\frac{dv}{dt} = \frac{\partial v}{\partial x}\frac{dx}{dt} + \frac{\partial v}{\partial y}\frac{dy}{dt} = \frac{\pi}{3}(2XY\frac{dx}{dt} + X^2\frac{dy}{dt})$$

$$= \frac{\pi}{3}(2 \times 10 \times 15 (-0.3) + 10^2 (0.2))$$

$$= -\frac{70\pi}{3} \text{ in}^3 \text{ per min. decreasing}$$

Applications

As noted previously, the derivative represents the instantaneous rate of change of the function with respect to x. Various useful physical and geometrical applications can be made using this fact. For example, velocity is the rate of change of position with time. If s is a distance and t is time, the velocity, v, is given by

$$v = \frac{ds}{dt}$$

(Note that when v is positive the motion is in the positive s direction.) In addition, acceleration, a, which is the rate of change of velocity with respect to time, is given by

$$a = \frac{dV}{dt} = \frac{d^2s}{dt^2}$$

Similar applications can be made with regard to other rates of change. A particularly useful application is the relationship of the derivative t the slope of a curve (i.e., the slope of the line tangent to the curve at the point in question). If the equation of a curve is y = f(x), the derivative of y with respect to x is the slope of the curve at any point (x,y) on the curve (Fig. 1.29).

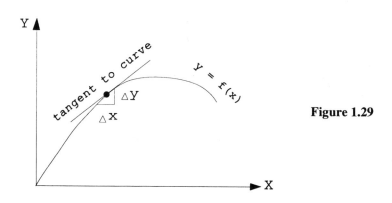

Figure 1.29

$$\text{slope} = \lim_{\Delta x \to 0} \frac{\Delta y}{\Delta x} = \frac{dy}{dx}$$

EXAMPLE:

Find the slope of the parabola $y = x^2 - 1$ when it crosses the X axis.

$$\text{Slope} = \frac{dy}{dx} = 2x$$

When $y = 0$, $x = \pm 1$ therefore when $y = 0$, $\frac{dy}{dx} = \pm 2$ and parabola has a slope of ± 2 when it crosses the X axis (Fig. 1.30).

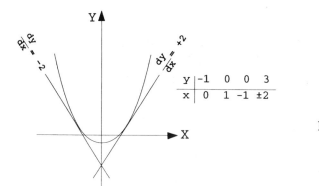

y	-1	0	0	3
x	0	1	-1	±2

Figure 1.30

EXAMPLE:

A point moves along a straight line with its distance S, in feet, from a fixed point given by $S = 8t^3 - 6t^2 + t + 3$ with time t in seconds.

Find the position, velocity and acceleration of the point at t = 10 sec.

At t = 10: $S = 8(1000) - 6(100) + 10 + 3$

$$S = 7413 \text{ ft.}$$

$$V = \frac{ds}{dt} = 24t^2 - 12t + 1$$

$$V = 24(100) - 12(10) + 1$$

$$V = 2281 \text{ ft/sec}$$

$$a = \frac{dv}{dt} = 48t - 12$$

$$a = 48(10) - 12$$

$$a = 468 \text{ ft/sec}^2$$

EXAMPLE:

Find the equation of the normal to the curve $y = 2x^2$ at the point (2,8)

The slope of the tangent, m_T, at the point is

$$m_T = \frac{dy}{dx} = 4x = 4(2) = 8$$

Since the normal is perpendicular to the tangent, the slope of the normal, m_N, is

$$m_N = -\frac{1}{8}$$

The equation of the normal is

$$y = -\frac{x}{8} + b$$

Find b:
$$8 = -\frac{(2)}{8} + b$$

$$b = \frac{33}{4}$$

Thus, the equation of the normal is

$$y = -\frac{x}{8} + \frac{33}{4}$$

EXAMPLE:

The point (1,2,3) is on a mountain whose shape is defined by the equation

$$z = 17 - 2x^2 - 3y^2$$

Is the path drawn from the point (1,2,3) steeper in the positive x direction or in the positive y direction?

The slope in the x direction = $\dfrac{\partial z}{\partial x}$.

The slope in the y direction = $\dfrac{\partial z}{\partial y}$.

$$\frac{\partial z}{\partial x} = -4x = -4 \text{ @ } x = 1$$

$$\frac{\partial z}{\partial y} = -6x = -12 \text{ @ } y = 2$$

∴ The path is steeper in the positive y direction.

EXAMPLE:

The edge of an expanding cube increases at the rate of 3 in. per sec. When its edge is 5 in. long, find the rate of change of its volume and of its total area.

Let x = length of an edge then volume, $V = x^3$, and

total area (6 faces) = $6x^2$.

The change of volume with respect to time is

$$\frac{dV}{dt} = \frac{d}{dt}(x^3) = 3x^2 \frac{dx}{dt} \text{ where } x = 5 \text{ in.}$$

$$\frac{dx}{dt} = 3 \frac{\text{in.}}{\text{sec}} \quad \text{(given)}$$

$$\frac{dv}{dt} = 3(5)^2(3) = 225 \ \frac{in^3}{sec}$$

$$\frac{da}{dt} = \frac{d}{dt} 6x^2 = 12x\left(\frac{dx}{dt}\right)$$

$$12(5)(3) = 180 \ \frac{in^2}{sec}$$

EXAMPLE:

Sand poured on the ground at the rate of 3 ft³/min forms a conical pile whose height is one third the diameter of the base. How fast is the altitude of the pile increasing when the radius of the base is 2 ft?

For a cone, volume = 1/3 area of base x altitude

$$V = \frac{1}{3}\pi r^2 h$$

Since $h = \frac{1}{3}D = \frac{1}{3}(2r)$ is given

$r = \frac{3h}{2}$, Substituting yields

$$V = \frac{\pi}{3}\left(\frac{3h}{2}\right)^2 h = 3/4\pi h^3$$

$$\frac{dV}{dt} = \frac{9}{4}\pi h^2 \frac{dh}{dt}$$

$$\frac{dh}{dt} = \frac{4}{9\pi h^2} \frac{dV}{dt} = \frac{4(3)}{9\pi\left(\frac{4}{3}\right)^2} = \frac{3}{4\pi} \ ft/min$$

Many interesting problems require the determination of the maximum and/or minimum values of a function. A possible method for solving such problems can be seen by inspection of Fig. 1.31 below which shows the relationship between the behavior of a function and the signs of its derivatives (the function is a continuous function with continuous first and second derivatives).

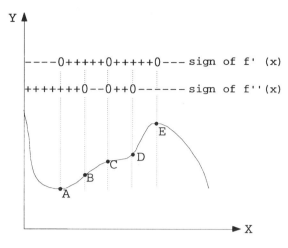

Figure 1.31

Positive values of f'(x) are associated with portions of the curve that rise to the right (positive slope such as from A to E) while negative values of f'(x) are associated with portions of the curve that fall to the right (negative slope). A zero value of f'(x) is associated with points that have a horizontal slope (A, C, and E). Points A and E are "local" minimum and maximum points, respectively. Thus, when the derivative is positive the function increases as x increases and when the derivative is negative the function decreases as x increases. The derivative is zero at maximum or minimum values.

The sign of f''(x) is associated with the rate of change of the slope. A positive f''(x) indicates that the slope is increasing (the curve is concave upward such as that portion up to B or between C and D), while a negative f''(x) indicates a decreasing slope (the curve is concave downward). Points such as B, C, and D where f''(x) is zero and changes sign are points of inflection. Note that at a maximum point (E) f''(x) is negative and at a minimum point (A) f''(x) is positive.

This information is very useful in evaluating a function with respect to possible maxima or minima. The procedure for this evaluation is as follows:

1. Find the first derivative of the function.

2. Set the first derivative equal to zero, and solve the resulting equation for real roots to find the values of the variable. These values are called "critical" values.

3. Find the second derivative.

4. Substitute each critical value of the variable in the second derivative. If the result is negative, then the function is a maximum for that value; if the result is a positive number, the function is a minimum [if f''(x) is zero, the value of f'(x) on either side of the critical value must be evaluated].

EXAMPLE:

Find the maximum value of the function $y = -x^2 + 4x - 3 = 0$

Evaluate critical values:

Y' = − 2x + 4 = 0

∴ x = +2 is a critical value. Check the second derivative: Y" = −2, thus, at x = +2 the function is a maximum. The maximum value of the function $f(2) = -2^2 + 4(2) - 3 = +1$

EXAMPLE:

A farmer has 120 linear feet of fencing with which to enclose a rectangular plot of ground and to divide it into three rectangular parts by partitions parallel to an end (Fig. 1.32). What should the dimensions be so that the area of the plot is a maximum?

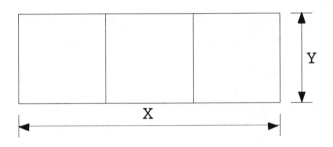

Figure 1.32

Area of plot = A = xy

Length of fence = 4y + 2x = 120

$$\therefore x = \frac{120 - 4y}{2} = 60 - 2y$$

Then $A = (60 - 2y) y = 60y - 2y^2$

$$\frac{dA}{dy} = 60 - 4y$$

Find the critical value: 60 − 4y = 0

$$\therefore y = \frac{60}{4} = 15 \text{ is a critical value}$$

Since $\frac{d^2A}{dy^2} = -4$, the area is a maximum

x = 60 − 2(15) = 30

Thus, the dimensions are 15 ft by 30 ft for maximum area

INTEGRAL CALCULUS

Indefinite Integral

Integration is the process of finding a function whose differential is given. Integration is the inverse of differentiation. The symbol $\int$ means "the function whose differential is." Thus,

$$d \left(\int f(x) \, dx \right) = f(x) \, dx$$

Since the derivative of a constant is zero, all functions whose differentials are equal to the differential of u are in the form u + c, where c is the constant of integration. That is:

$$\int du = u + c$$

Because of the presence of the constant, the resulting function is called an <u>indefinite integral</u>.

EXAMPLE: Find $\int 3x^2 \, dx$

Since the differential of x^3 is $3x^2$, then

$$\int 3x^2 \, dx = x^3 + c$$

Note that this is the opposite of the application of the power rule for differentiation (formula 5 in the Table of Derivatives). The rule for integration of simple algebraic functions in the form ax^n (where a and n are constants) is to increase the exponent by one, divide by the new exponent, and add the constant of integration. That is

$$\int ax^n \, dx = \frac{ax^{n+1}}{n+1} + c$$

Like differentiation, integration is often carried out by use of formulas already derived for various common functions. Some of these formulas are provided in the Table of Integrals, which is a short list of fundamental integrals. A handbook of mathematical tables/formulas should be consulted for a more complete list.

EXAMPLE:

$$\int (5x^3 + 4x^2 + 4) \, dx = \frac{5x^4}{4} + \frac{4x^3}{3} + 4x + c$$

EXAMPLE:

$$\int \frac{(3x-2)}{\sqrt{x}} dx = \int (3x^{1/2} - 2x^{-1/2}) dx$$

$$= \frac{3x^{3/2}}{3/2} - \frac{2x^{1/2}}{1/2} + c$$

$$= 2x^{3/2} - 4x^{1/2} + c$$

EXAMPLE:

$$\int \frac{dx}{\sqrt{9-4x^2}} = \frac{1}{2} \int \frac{2dx}{\sqrt{9-4x^2}} = \text{arc sin } \frac{2x}{3} + c$$

Definite Integral

If F(x) is an indefinite integral of f(x) dx, then

$$\int_a^b f(x) \, dx = F(b) - F(a)$$

When the limits of integration are defined, as in this case, the integral is called a <u>definite integral</u>.

EXAMPLE:

$$\int_2^3 3x^2 \, dx = x^3 \Big|_2^3 = (3)^3 - (2)^3 = 19$$

EXAMPLE:

$$\int_1^2 (2x-4)^2\, dx = \int_1^2 (4x^2 - 16x + 16)\, dx$$

$$= \left. \frac{4x^3}{3} - 8x^2 + 16x \right|_1^2$$

$$= \left(\frac{32}{3} - 32 + 32\right) - \left(\frac{4}{3} - 8 + 16\right) = \frac{4}{3}$$

EXAMPLE:

$$\int_0^{\pi/4} \sin 2x\, dx = \frac{1}{2}\int_0^{\pi/4} \sin 2x (2dx)$$

$$= \left. -\frac{1}{2} \cos 2x \right|_0^{\pi/4} = -\frac{1}{2}(0-1) = \frac{1}{2}$$

TABLE OF INTEGRALS

1. $\int a\, du = a \int du = au + c$

2. $\int (du \pm dv) = \int du \pm \int dv$

3. $\int (u \pm v)\, dx = \int u\, dx \pm \int v\, dx$

4. $\int u\, dv = uv - \int v\, du$

5. $\int u^n\, du = \dfrac{u^{n+1}}{n+1} + c$

6. $\int \dfrac{du}{u} = \ln u + c$

7. $\int a^u\, du = \dfrac{a^u}{\ln a} + c$

8. $\int e^u\, du = e^u + c$

9. $\int \dfrac{du}{u^2 + a^2} = \dfrac{1}{a} \arctan\left(\dfrac{u}{a}\right) + c$

10. $\int \dfrac{du}{u^2 - a^2} = \dfrac{1}{2a} \ln\left(\dfrac{u - a}{u + a}\right) + c$, if $u^2 > a^2$

11. $\int \dfrac{du}{a^2 - u^2} = \dfrac{1}{2a} \ln\left(\dfrac{a + u}{a - u}\right) + c$, if $a^2 > u^2$

12. $\int \dfrac{du}{\sqrt{u^2 \pm a^2}} = \ln(u + \sqrt{u^2 \pm a^2}) + c$

13. $\int \dfrac{du}{\sqrt{a^2 - u^2}} = \arcsin\left(\dfrac{u}{a}\right) + c$

14. $\int \sin u\, du = -\cos u + c$

15. $\int \cos u\, du = \sin u + c$

16. $\int \tan u\, du = -\ln(\cos u) + c$

17. $\int \csc u\, du = \ln(\csc u - \cot u) + c$

18. $\int \sec u\, du = -\ln(\sec u - \tan u) + c$

19. $\int \cot u\, du = \ln(\sin u) + c$

20. $\int \sinh u\, du = \cosh u + c$

21. $\int \cosh u\, du = \sinh u + c$

22. $\int \tanh u\, du = \ln(\cosh u) + c$

23. $\int \operatorname{csch} u\, du = \ln\left(\tanh \dfrac{u}{2}\right) + c$

24. $\int \operatorname{sech} u\, du = \arcsin(\tan u) + c$

25. $\int \coth u\, du = \ln(\sinh u) + c$

Notes:
(a) u and v are functions of x.
(b) a, c, and n are constants.
(c) $e = \lim\limits_{m \to \infty} \left(1 + \dfrac{1}{m}\right)^m$.
(d) all angles are in radians

Integration by Parts

The formula for integration by parts is particularly useful for problems where the integrand is the product of functions of different types. The purpose is to replace one integral with another that is easier to integrate. The formula is based on the differential equation

$$d(uv) = udv + vdu$$

which is integrated and rearranged to give $\int udv = uv - \int vdu$

The selection of the factors u and dv is a matter of trial and error. It is usually best to choose for u a function that is simplified by differentiation. Occasionally it is necessary to repeat the application of the formula.

EXAMPLE:

Find $\int x\ln x\, dx$

Let $u = \ln x$, $dv = xdx$ then $du = \dfrac{dx}{x}$, $v = \dfrac{x^2}{2}$.

$$\int x \ln x\, dx = \frac{x^2 \ln x}{2} - \int \frac{x}{2} dx$$

$$= \frac{x^2}{2} \ln x - \int \frac{x}{2} dx$$

$$= \frac{x^2}{2} \ln x - \frac{x^2}{4} + c$$

EXAMPLE:

Find $\int_a^b \ln x\, dx$

Let $u = \ln x$, $dv = dx$ then $du = \dfrac{dx}{x}$, $v = x$.

$$\int_a^b \ln x\, dx = x\ln x \Big|_a^b - \int_a^b dx$$

$$= b\ln b - a\ln a - b + a$$

$$= b(\ln b - 1) - a(\ln a - 1)$$

Double Integrals

Some problems in integral calculus involve functions of two variables. Such problems may require the evaluation of a double integral such as

$$\int_a^b \int_c^d f(x,y) \, dy \, dx$$

This may be written as

$$\int_a^b \left(\int_c^d f(x,y) \, dy \right) dx$$

The integration is carried out by first performing the integration within the brackets, then performing a second integration employing the outer notations.

EXAMPLE:

$$\int_0^1 \int_1^2 y^2 \, dy \, dx = \int_0^1 \left(\frac{y^3}{3} \Big|_1^2 \right) dx = \int_0^1 \left(\frac{8}{3} - \frac{1}{3} \right) dx = \int_0^1 \frac{7}{3} \, dx$$

$$= \frac{7}{3} x \Big|_0^1 = \frac{7}{3}$$

Applications

Consider the area, A, bounded by the x-axis the vertical line x = a, the vertical line x = b, and the arc of the curve y = f(x) (Fig. 1.33).

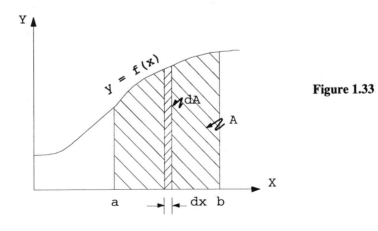

Figure 1.33

The differential of this area A is given by the formula

   $dA = f(x)\, dx$

The area, A, can then be determined by the definite integral

$$A = \int_a^b f(x)\, dx$$

EXAMPLE:

Find the area bounded by the curve $y = 6x^2$, the x-axis, and the vertical lines $x = 2$ and $x = 4$

$$A = \int_2^4 6x^2\, dx = \left.\frac{6x^3}{3}\right|_2^4 = \left. 2x^3 \right|_2^4 = 112 \text{ square units}$$

EXAMPLE:

Find the area bounded by the x-axis, the y-axis and the curve $x + y^2 = 4$ (see Fig. 1.34).

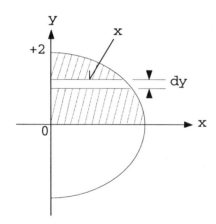

$$A = \int_0^2 x\, dy = \int_0^2 (4 - y^2)\, dy$$

$$A = \left.\left(4y - \frac{y^3}{3}\right)\right|_0^2 = 8 - \frac{8}{3}$$

$$A = \frac{16}{3} \text{ square units}$$

Figure 1.34

EXAMPLE:

Determine the area moment of inertia of the area bounded by $y = x^3$, $y = 0$, and $x = 1$, with respect to the y-axis (Fig. 1.35). All dimensions are in feet.

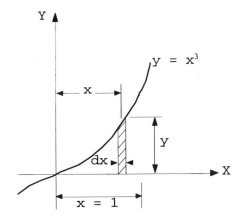

$$I_y = \int x^2 \, dA \text{ where } dA = y \, dx$$

$$I_y = \int_0^1 x^2 \, y \, dx = \int_0^1 x^5 \, dx = \left.\frac{x^6}{6}\right|_0^1$$

$$= 0.167 \text{ ft.}^4$$

Figure 1.35

The area between two curves: $y_1 = f(x)$ and $y_2 = g(x)$ in the region (a,b) where $y_2 > y_1$

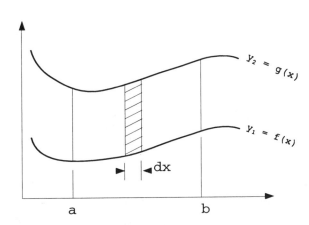

Figure 1.36

is given by $A = \int_a^b (y_2 - y_1) \, dx = \int_a^b [g(x) - f(x)] \, dx$ (Fig. 1.36).

EXAMPLE:

Find the area of the region (Fig. 1.37) bounded by the curves

$$y = \frac{x^2}{4} \text{ and } y = \frac{x}{2} + 2$$

Solve for the points of intersection of the curves by simultaneous solution of the two equations

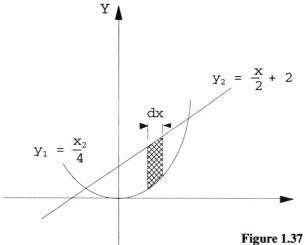

Figure 1.37

At $y_1 = y_2$:

$$\frac{x^2}{4} = \frac{x}{2} + 2$$

$$x^2 - 2x - 8 = 0$$

Factoring:

$$(x + 2)(x - 4) = 0$$

$$x = -2, \quad x = +4$$

These values of x are the limits.

$$\text{Area } A = \int dA = \int_{-2}^{4} (y_2 - y_1)\, dx$$

$$A = \int_{-2}^{4} \left(\frac{x}{2} + 2 - \frac{x^2}{4}\right) dx$$

$$A = \left[\frac{x^2}{4} + 2x - \frac{x^3}{12}\right]_{-2}^{4} = 4 + 8 - 5\frac{1}{3} - 1 + 4 - \frac{2}{3} = 9 \text{ sq. units}$$

When a curve $y = f(x)$ is revolved about the x axis, a surface of revolution is generated. To find the area of this surface, consider the area generated by an element of arc ds (Fig. 1.38). This area is approximately that of a cylinder of radius y and length ds. Summing all such elements of surface yields the surface area A.

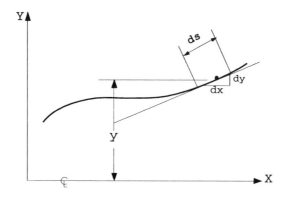

Figure 1.38

Element of Arc = ds

But $ds = \sqrt{(dx)^2 + (dy)^2}$

The area of the cylinder whose radius is y and whose axial length is ds is da where

$$dA = 2\pi y \, ds$$

$$dA = 2\pi y \sqrt{(dx)^2 + (dy)^2}$$

$$dA = 2\pi y \, dx \sqrt{1 + \left(\frac{dy}{dx}\right)^2}$$

Then

$$A = 2\pi \int y \, ds$$

$$A = 2\pi \int y \sqrt{1 + \left(\frac{dy}{dx}\right)^2} \, dx$$

EXAMPLE:

Find the area of the surface of revolution between x = 0 and x = a, generated by revolving the arc of the semicubical parabola $a^2 y = x^3$ about the x axis.

$$y = \frac{x^3}{a^2} \; ; \; \frac{dy}{dx} = 3\frac{x^2}{a^2}$$

$$ds = \sqrt{1 + \left(\frac{dy}{dx}\right)^2} \, dx$$

$$ds = \frac{1}{a^2} (a^4 + 9x^4)^{\frac{1}{2}} \, dx$$

$$dA = 2\pi y \, ds$$

$$dA = \frac{2\pi}{a^4} (a^4 + 9x^4)^{\frac{1}{2}} x^3 \, dx$$

$$A = \frac{2\pi}{a^4} \int_0^a (a^4 + 9x^4)^{\frac{1}{2}} x^3 \, dx$$

$$A = \frac{\pi}{27 a^4} (a^4 + 9x^4)^{3/2} \Big|_0^a$$

$$A = \frac{\pi}{27} (10\sqrt{10} - 1) a^2$$

$$A = 3.6 a^2$$

A fundamental theorem of the integral calculus is the notion of the integral as the limit of a sum. Again consider the area, A, bounded by the x-axis, the lines x = a and x = b, and the curve y = f(x). In this case divide the region (a,b) into n intervals of lengths Δx_1, Δx_2, ..., Δx_n and let x_1, x_2, ..., x_n be values of x, one in each interval (Fig.1.39).

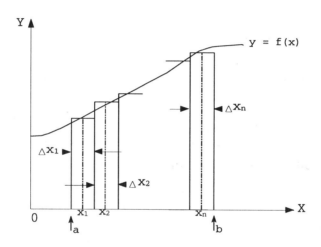

Figure 1.39

Then $\lim_{n \to \infty} \sum_{1}^{n} f(x_k) \Delta x_k = A$

Thus, $\int_a^b f(x) \, dx = \lim_{n \to \infty} \sum_{1}^{n} f(x_k) \Delta x_k$

EXAMPLE:

Find $\lim_{n \to \infty} \sum_{1}^{n} x_k^3 \Delta x_k$ in the interval from x = 0 to x = 4

$$\lim_{n\to\infty} \sum_1^n x_k^3 \Delta x_k = \int_0^4 x^3 dx = \frac{x^4}{4}\Big|_0^4 = 64$$

The relation of the integral to a limit can be used to determine the volume of a solid generated by rotating the curve y = f(x) about the x-axis. Consider the curve y = f(x) in the region (a,b) (Fig. 1.40).

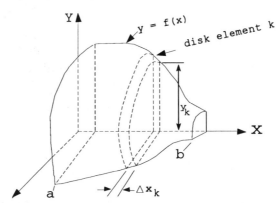

Figure 1.40

The volume of the disk element, V_k, is

$$V_k = \pi Y_k^2 \Delta x_k$$

The total volume is

$$V = \lim_{n\to\infty} \sum_1^n \pi y_k^2 \Delta x_k$$

Then

$$V = \int_a^b \pi y^2 dx$$

EXAMPLE:

Find the volume of the solid generated by rotating about the x-axis the curve $x = 3y^2$ in the region from x = 0 to x = 6.

$$V = \int_0^6 \pi y^2 dx = \int_0^6 \pi \frac{x}{3} dx$$

$$V = \frac{\pi x^2}{6}\Big|_0^6 = 6\pi \text{ cubic units}$$

DIFFERENTIAL EQUATIONS

Definitions

A <u>differential equation</u> is an equation that involves derivatives (or differentials). If there is only one independent variable in the equation, it is called an <u>ordinary</u> differential equation. When there is more than one independent variable, it is called a <u>partial</u> differential equation. The <u>order</u> of the differential equation is the order of the highest derivative, while the <u>degree</u> is the degree of the highest ordered derivative.

The solution to a differential equation is a relationship of the variables that satisfies the equation and contains no derivatives. The <u>general solution</u> of a differential equation of the nth order contains n essential arbitrary constants while a <u>particular solution</u> is a solution obtained from the general solution by assigning values to the constants (usually while evaluating the boundary conditions).

Equations with Variables Separable

Equations with variables separable are first order equations that by simple algebraic manipulation may be written in the following form

$$M(x)\,dx + N(y)\,dy = 0$$

where M(x) is a function of x only (or a constant) and N(y) is a function of y only (or a constant). The solution to such equations can be obtained directly by integration.

EXAMPLE:

Find the solution to $\dfrac{dy}{dx} = \dfrac{x}{y}$.

Separating variables:

$$x\,dx - y\,dy = 0$$

integration yields

$$\frac{x^2}{2} - \frac{y^2}{2} = \frac{C}{2}$$

where the constant of integration is chosen as C/2 to eliminate fractions in the next step.

$$x^2 - y^2 = C \text{ which is the general solution of the equation.}$$

EXAMPLE:

Find the general solution of the equation.

$$y\, dx - y^2\, dy = dy$$

Separate the variables and integrate.

$$y\, dx = (1 + y^2)\, dy$$

$$\int dx = \int \frac{dy}{y} + \int y\, dy$$

$$x = \ln y + \frac{y^2}{2} + c$$

or $\quad 2x = 2 \ln y + y^2 + c_1$

EXAMPLE:

Find the general solution of: $\dfrac{dy}{dt} = ky$.

Separating variables and integrating:

$$\int \frac{dy}{y} = \int k\, dt;\ \ \ln y = kt + C_1 = kt + \ln C$$

$$\ln \frac{y}{c} = kt \text{ recall } \ln_a N = X\ ;\ a^x = N$$

$$\frac{y}{c} = e^{kt} \text{ or } y = C e^{kt}$$

EXAMPLE:

Find the general solution of:

$$x^2\, dy + y^2\, dx = 0$$

Separating the variables yields

$$\frac{dy}{y^2} + \frac{dx}{x^2} = 0$$

Using $\int u^n \, du = \dfrac{u^{n+1}}{n+1}$ yields

$$-y^{-1} - x^{-1} = C$$

now multiplying through by $-xy$

$$x + y = c_1 \, xy$$

First-Order Linear Differential Equations

The first-order linear equation has the form

$$\frac{dy}{dx} + Py = Q$$

where P and Q are constants or functions of x alone. This equation has the following solution.

$$y = Ce^{-\emptyset} + e^{-\emptyset} \int Q e^{\emptyset} \, dx$$

where $\emptyset = \int P \, dx$

EXAMPLE:

Solve $\dfrac{dy}{dx} = \dfrac{2y}{x+1} + (x+1)^{5/2}$.

rearranging:

$$\frac{dy}{dx} - \frac{2y}{x+1} = (x+1)^{5/2}$$

shows the equation to be Linear where

$$P = -\frac{2}{x+1} \quad \text{and} \quad Q = (x+1)^{5/2}$$

$$\emptyset = \int P \, dx = -\int \frac{2}{x+1} \, dx = -2 \ln(x+1) = -\ln(x+1)^2$$

$$y = Ce^{\ln(x+1)^2} + e^{\ln(x+1)^2} \int (x+1)^{5/2} \, e^{-\ln(x+1)^2} \, dx$$

$$y = C(x+1)^2 + (x+1)^2 \int \frac{(x+1)^{5/2}}{(x+1)^2} \, dx$$

$$y = C(x+1)^2 + (x+1)^2 \int (x+1)^{1/2} \, dx$$

$$y = C(x+1)^2 + (x+1)^2 \frac{(x+1)^{3/2}}{3/2}$$

$$y = C(x+1)^2 + \frac{2}{3}(x+1)^{7/2}$$

Second-Order Linear Differential Equations Homogeneous With Constant Coefficients

This type of equation has the following form (P and Q are constants):

$$\frac{d^2y}{dx^2} + P\frac{dy}{dx} + Qy = 0$$

Note that this equation is satisfied by the substitution of y and its successive derivatives, therefore, its solution will be in a form where the derivatives repeat themselves upon differentiation. Thus, assume the solution is in the form:

$$y = e^{ax}$$

and determine a so that the differential equation is satisfied.

$$\frac{dy}{dx} = a e^{ax} \quad ; \quad \frac{d^2y}{dx^2} = a^2 e^{ax}$$

Substituting the derivatives into the differential equation and cancelling, e^{ax} yields:

$$a^2 + Pa + Q = 0.$$

This is called the <u>auxiliary equation</u>. If the auxiliary equation has real roots a_1 and a_2, then $y = e^{a_1 x}$ and $y = e^{a_2 x}$ are particular solutions of the differential equation.

The general solution is:

$$y = C_1 e^{a_1 x} + C_2 e^{a_2 x}$$

The constants in the general solution can be determined from the boundary conditions. Note that if the roots of the auxiliary equation are complex the solution will contain trigonometric functions. For example, assume the two roots are a + bi and a - bi. Then the solution takes the following form

$$y = C_1 e^{(a+bi)x} + C_2 e^{(a-bi)x}$$

now since $e^{a+bi} = e^a(\cos b + i \sin b)$ and $e^{a-bi} = e^a(\cos b - i \sin b)$

$$y = C_1 e^{ax}(\cos bx + i \sin bx) + C_2 e^{ax}(\cos bx - i \sin bx)$$

$$y = e^{ax}\left[(C_1 + C_2)\cos bx + i(C_1 - C_2)\sin bx\right]$$

now let $A = C_1 + C_2$ and $B = i(C_1 - C_2)$, then the final form of the solution is

$$y = e^{ax}(A\cos bx + B\sin bx)$$

If the roots are pure imaginary roots such as $\pm bi$, the solution will have the form

$$y = A\cos bx + B\sin bx$$

If the roots are two real <u>identical</u> roots, the solution is

$$y = C_1 x e^{ax} + C_2 e^{ax}$$

where a is the multiple root.

EXAMPLE:

Solve $\dfrac{d^2x}{dt^2} + \dfrac{5dx}{dt} + 6x = 0$.

For boundary conditions:

$$\dfrac{dx}{dt} = 12 \text{ at } t = 0$$

$$\dfrac{d^2x}{dt^2} = 0 \text{ at } t = 0$$

This is a homogeneous equation of order 2 and is linear with constant coefficients. The solution is in the form of $x = e^{at}$; where $x' = ae^{at}$; $x'' = a^2 e^{at}$

Substituting into the differential equation

$$a^2 e^{at} + 5a\, e^{at} + 6 e^{at} = 0$$

The auxiliary equation is

$$a^2 + 5a + 6 = 0$$

Hence;

$$(a + 3)(a + 2) = 0 \therefore a = -2, -3 \text{ and } x = C_1 e^{-2t} + C_2 e^{-3t}$$

The <u>particular</u> solution is found as follows:

$$\frac{dx}{dt} = -2C_1 e^{-2t} - 3C_2 e^{-3t}$$

at $t = 0$:

$$\frac{dx}{dt} = -2C_1 - 3C_2 = 12 \tag{1}$$

$$\frac{d^2x}{dt^2} = 4C_1 e^{-2t} + 9C_2 e^{-3t}$$

at $t = 0$:

$$\frac{d^2x}{dt^2} = 4C_1 + 9C_2 = 0 \tag{2}$$

Multiplying (1) times (2) yields

$$-4C_1 - 6C_2 = 24 \tag{3}$$

Adding (2) and (3): $3C_2 = 24$. Therefore $C_2 = 8$. Substitute into (1):

$$-2 C_1 - 3(8) = 12$$

$$-2 C_1 = 36$$

Therefore $C_1 = -18$

$$X = -18 e^{-2t} + 8e^{-3t}$$

Applications

Many physical problems involve the solution to differential equations.

EXAMPLE:

A 100 gal tank is filled with brine containing 60 lbs of dissolved salt. Water runs into the tank at the rate of 2 gal/per min and the mixture, kept uniform by stirring, runs out at the same rate. How much salt is in the tank after 1 hour?

Let S = lbs of salt in tank after t minutes.

$\dfrac{S}{100}$ = lbs/gal concentration after t minutes

During interval dt, 2 dt gal of water flows in and 2 dt gal of brine containing $\dfrac{2S}{100}$ dt = $\dfrac{S}{50}$ dt lbs of salt flows out.

Change in amount of salt in the tank is dS:

$$dS = -\dfrac{S}{50} dt \quad \text{or} \quad \dfrac{dS}{S} = -\dfrac{dt}{50}$$

Integrating:

$$\ln S = -\dfrac{t}{50} \qquad \therefore S = Ce^{-t/50}$$

Determine C:

At t = 0   S = 60

Therefore C = 60 and $S = 60e^{-t/50}$

At t = 60, $S = 60e^{-6/5} = 60(0.301) = 18$ lbs. Answer

LAPLACE TRANSFORMATION

A modern form of the operational calculus consists of the use of the Laplace transformation. The theory of the Laplace transformation introduces many rules that are important in the analysis of problems in engineering. One of the most important of these rules deals with the transformation of derivatives of functions. By the use of this rule, significant simplifications in certain types of problems in differential equations can be made.

Definition of the Laplace Transformation

The Laplace transformation of F(t) is given by:

$$\int_0^\infty e^{-st} F(t)\, dt = f(s)$$

This operation is called the Laplace transform of F(t) and is written in the following forms:

$$f(s) = L\{F(t)\} = L\{F\} = F(s)$$

The Laplace transform of F(t) exists (i.e., the integral converges) if F(t) is continuous and provided some constant α exists such that

$$e^{-\alpha t} |F(t)|$$

is bounded as t approaches infinity. The Laplace integral converges when $s > \alpha$.

EXAMPLE:

Find the Laplace transform of F(t) = 3.

$$L\{F\} = \int_0^\infty 3e^{-st}\, dt = -\frac{3}{S} e^{-st} \Big|_0^\infty$$

when $s > 0$ the integral converges and $L\{F\} = 3/s$.

EXAMPLE:

Find the Laplace transform of $F(t) = e^{at}$ when $t > 0$.

$$L\{F\} = \int_0^\infty e^{at} e^{-st}\, dt = \frac{1}{a-s} e^{-(s-a)t} \Big|_0^\infty$$

when $s > a$ the integral converges and

$$L\{e^{at}\} = \frac{1}{s-a}$$

Some useful transforms are shown in the following table.

TABLE OF TRANSFORMS

F(t)	f(s)	α (s > α)
1	$\dfrac{1}{s}$	0
C (constant)	$\dfrac{c}{s}$	0
e^{at}	$\dfrac{1}{s-a}$	a
t^n (n = 1,2,...)	$\dfrac{n!}{s^{n+1}}$	0
sin kt	$\dfrac{k}{s^2+k^2}$	0
cos kt	$\dfrac{s}{s^2+k^2}$	0
sinh kt	$\dfrac{k}{s^2-k^2}$	$|k|$
cosh kt	$\dfrac{s}{s^2-k^2}$	$|k|$
e^{-at} sin kt	$\dfrac{k}{(s+a)^2+k^2}$	-a
e^{-at} cos kt	$\dfrac{s+a}{(s+a)^2+k^2}$	-a
$\sqrt{t}$	$\dfrac{\sqrt{\pi}}{2\sqrt{s^3}}$	0
$\dfrac{1}{\sqrt{t}}$	$\sqrt{\dfrac{\pi}{s}}$	0

The Laplace transformation is linear so that if A and B are constants then:

$$L\{AF(t) + BG(t)\} = AL\{F(t)\} + BL\{G(t)\}$$

Note that if: $L\{F(t)\} = f(s)$

Then: $F(t) = L^{-1}\{f(s)\}$

where $L^{-1}\{f(s)\}$ denotes the function whose Laplace transformation is f(s). F(t) is the inverse transform of f(s). The inverse transformation is linear so that if A and B are constants,

$$L^{-1}\{Af(s) + Bg(s)\} = AL^{-1}\{f(s)\} + BL^{-1}\{g(s)\}$$

This relationship and algebraic manipulation of fractions can be used to find the inverse transforms of quotients of polynomials of s. The procedure is to rearrange the expression (by considering factors) in the form of the sum of functions with known inverses.

EXAMPLE: Find $L^{-1}\left(\dfrac{s+1}{s^2+2s}\right)$.

Consider the function $\dfrac{s+1}{s^2+2s} = \dfrac{s+1}{s(s+2)}$

assume $\dfrac{s+1}{s(s+2)} = \dfrac{A}{s} + \dfrac{B}{s+2}$

clear fractions:

$$\dfrac{s+1}{s(s+2)} = \dfrac{A(s+2) + Bs}{s(s+2)} = \dfrac{(A+B)s + 2A}{s(s+2)}$$

equating like powers of s gives

$A + B = 1$ and $2A = 1$

then $A = \frac{1}{2}$ and $B = \frac{1}{2}$

and $L^{-1}\left\{\dfrac{s+1}{s^2+2s}\right\} = \dfrac{1}{2}L^{-1}\left\{\dfrac{1}{s}\right\} + \dfrac{1}{2}L^{-1}\left\{\dfrac{1}{s+2}\right\} = \dfrac{1}{2} + \dfrac{1}{2}e^{-2t}$

EXAMPLE:

Find $L^{-1}\left\{\dfrac{k^2}{s(s^2+k^2)}\right\}$.

note that $\dfrac{k^2}{s(s^2+k^2)} = \dfrac{1}{s} - \dfrac{s}{s^2+k^2}$

Then $L^{-1}\left\{\dfrac{k^2}{s(s^2 = k^2)}\right\} = L^{-1}\left\{\dfrac{1}{s}\right\} - L^{-1}\left\{\dfrac{s}{(s^2+k^2)}\right\} = 1 - \cos kt$

Application to the Solution of Differential Equations

Consider the transformation of $F'(t)$.

By the formula for integration by parts

$$L\{F'(t)\} = \int_0^\infty e^{-st} F'(t)\, dt$$

$$= e^{-st} f(t)\Big|_0^\infty + s\int_0^\infty e^{-st} F(t)\, dt$$

If $e^{-\alpha t}|F(t)|$ is bounded as t approaches infinity, then for every $s > \alpha$ the first term on the right is $-F(o)$ and

$$L\{F'(t)\} = sL\{F(t)\} - F(o)$$

This formula is the fundamental operational property of the Laplace transformation and makes it possible to substitute an algebraic expression of the transform for the operation of differentiation. This property is very useful in the solution of certain differential equations.

Also it can be shown that

$$L\{F''(t)\} = s[sL\{F(t)\} - F(+0)] - F'(0)$$

or

$$L\{F''(t)\} = s^2 f(s) - sF(0) - F'(0)$$

and in general:

$$L\{\{F^{(n)}(t)\} = s^n f(s) - s^{n-1} F(0) - s^{n-2} F'(0)$$
$$- s^{n-3} F''(0) - \ldots - F^{(n-1)}(0)$$

The Laplace transformation is useful in solving some linear ordinary differential equations.

EXAMPLE:

Find the general solution of the differential equation

$$Y''(t) - k^2 Y(t) = 0$$

with $Y(o) = A, Y'(o) = B$

Now

$$L\{y'' - k^2 y\} = L\{0\}$$

Since the Laplace transform is linear

$$L\{y''\} - k^2 L\{y\} = 0$$

Now by the rule for transformation of derivatives

$$L\{Y''\} = s^2 y(s) - sA - B$$

Then (substituting into the previous equation)

$$s^2 y(s) - sA - B - k^2 y(s) = 0$$

This is a simple algebraic equation that can be solved for y(s):

$$y(s) = A \frac{s}{s^2 - k^2} + B \frac{1}{s^2 - k^2}$$

or

$$y(s) = A \frac{s}{s^2 - k^2} + \frac{B}{K} \frac{K}{s^2 - k^2}$$

Now by the inverse transforms of the functions on the right:

$$Y(t) = A \cosh Kt + \frac{B}{k} \sinh Kt$$

If

$$C = \frac{B}{k}$$

then

$$Y(t) = A \cosh kt + C \sinh kt$$

where A and C are arbitrary constants

PROBABILITY AND STATISTICS

Probability

If an event E can happen in h ways out of a total of t possible, equally likely ways, then the probability of E occurring (success) is denoted

$$p = \text{probability of E} = \frac{h}{t} = P(E)$$

The probability that the event E will not occur, designated $\overline{E}$ (failure), is denoted

$$q = \text{probability of "not E"} = \frac{t-h}{t} = P(\overline{E})$$

Then

$$q = 1 - \frac{h}{t} = 1 - p$$

Therefore

$$p + q = 1 \quad \text{or} \quad P(E) + P(\overline{E}) = 1$$

EXAMPLE:

Let E be the event that the number 4 or 5 occurs in one toss of a die. If the die is fair (not loaded), 6 numbers are equally likely to occur. Since E can happen in two ways

$$P(E) = \frac{h}{t} = \frac{2}{6} = \frac{1}{3}$$

The probability of not tossing a 4 or 5 is

$$P(\overline{E}) = 1 - P(E) = 1 - \frac{1}{3} = \frac{2}{3}$$

If an event cannot occur its probability is 0. If it must occur, its probability is 1 and its occurrence is certain.

If p is the probability that an event will occur and q is the probability that it will not occur, the odds in favor of its happening are p to q. The odds against its happening are q to p.

Thus, in the above example the odds in favor of tossing a 4 or 5 are $P(E)$ to $P(\overline{E})$ which is 1/3 to 2/3, or 1 to 2. The odds of tossing a number other than 4 or 5 are 2/3 to 1/3, or 2 to 1.

The following are some <u>basic laws of probability</u>:

1. Events of a set are <u>independent</u> if they can happen without affecting the occurrence of others, otherwise they are <u>dependent</u>.

   The probability of tossing a head on the second toss of a coin is independent of the probability of tossing a head on the first toss.

   In a box containing 3 red and 3 blue balls, the probability of drawing a red ball on the first draw is E_1. The event of drawing a blue ball on the second draw is E_2. If the balls are not replaced after being drawn, then E_1 and E_2 are <u>dependent</u> events.

2. Events of a set are <u>mutually exclusive</u> if the occurrence of any one event at a particular time excludes the occurrence of any other event at that time.

   If E_1 is the event of drawing an ace from a deck of cards and E_2 is the event of drawing a king, the probabilities of drawing either an ace or a king in a single draw are <u>mutually exclusive</u> events. The probability of E_1 or E_2 is written as $P(E_1 + E_2)$

   $$P(E_1 + E_2) = P(E_1) + P(E_2) = \frac{1}{13} + \frac{1}{13} = \frac{2}{13}$$

   If E_1 is the event of drawing an ace from a deck and E_2 is the event of drawing a spade then E_1 and E_2 are not mutually exclusive events since the ace of spades can be drawn. If $P(E_1 E_2)$ is the probability of the simultaneous occurrence of E_1 and E_2

   $$P(E_1 + E_2) = P(E_1) + P(E_2) - P(E_1 E_2) = \frac{4}{52} + \frac{13}{52} - \frac{1}{52} = \frac{16}{52} = \frac{4}{13}$$

   which is the probability of drawing either an ace or a spade or both.

3. If two events, E_1 and E_2 are independent events whose probabilities are $P(E_1)$ and $P(E_2)$, the probability of simultaneous occurrence is $P(E_1) \times P(E_2)$.

   The probability of drawing the ace of spades from a full deck is the probability of drawing an ace times the probability of drawing a spade.

   $$P(E_1) = \frac{1}{13}, \quad P(E_2) = \frac{1}{4}$$

   $$P(E_1 E_2) = P(E_1) \times P(E_2) = \frac{1}{13} \times \frac{1}{4} = \frac{1}{52}$$

Two important concepts, <u>permutations</u> and <u>combinations,</u> are illustrated in the following example.

EXAMPLE:

How many different 5 card hands may be dealt from a deck of 52 cards?

The first card may be selected in 52 ways. Since there are now only 51 cards left in the deck, the second card may be selected in 51 ways. With only 50 cards left in the deck, the third card may be selected in only 50 ways, and so forth. The 5 cards may be selected in 52 x 51 x 50 x 49 x 48 ways.

This total represents the number of possible permutations of the 5 cards, that is, the number of different arrangements of the 52 cards taking 5 at a time. In general, the number of permutations can be determined from

$$_nP_x = \frac{n!}{(n-x)!}$$

where n is the total number of items and x is the number of items selected. Note for this example

$$_{52}P_5 = 52 \times 51 \times 50 \times 49 \times 48$$

as derived above. In determining the number of different hands the arrangement of the cards in a hand is not important (i.e., the order of the cards doesn't change the hand). Thus, the cards selected may be arranged in 5 x 4 x 3 x 2 x 1 sequences. The total number of different hands is then

$$\frac{52 \times 51 \times 50 \times 49 \times 48}{5 \times 4 \times 3 \times 2 \times 1}$$

This total represents the number of combinations for the 5 cards. The number of combinations does not consider arrangement. In general, the number of combinations of n things taken x at a time is given by

$$_nC_x = \frac{n!}{(n-x)!x!}$$

For this example

$$_{52}C_5 = \frac{52!}{(52-5)!5!} = \frac{52 \times 51 \times 50 \times 49 \times 48}{5 \times 4 \times 3 \times 2 \times 1}$$

as derived above.

EXAMPLE:

What is the probability of being dealt a royal flush in poker (i.e., A, K, Q, J, 10 of the same suit)

$$\text{Probability} = \frac{h}{t} = \frac{4}{_{52}C_5} = \frac{(52-5)!5!4}{52!}$$

$$= \frac{5 \times 4 \times 3 \times 2 \times 4}{52 \times 51 \times 50 \times 49 \times 48} = 1.5 \times 10^{-6}$$

(1.5 in 1 million hands)

EXAMPLE:

What is the probability of being dealt a regular flush (any 5 cards in the same suit)

In each suit there is $_{13}C_5$ combinations of flushes

∴ Total flush combinations = $4 \times {}_{13}C_5$

then $$\frac{h}{t} = \frac{4 \times {}_{13}C_5}{{}_{52}C_5}$$

$$= \frac{4 \times 13 \times 12 \times 11 \times 10 \times 9 \times 5 \times 4 \times 3 \times 2}{5 \times 4 \times 3 \times 2 \times 52 \times 51 \times 50 \times 49 \times 48}$$

$$= 0.002$$

(1 in 500 hands)

EXAMPLE:

A box contains 5 white balls and 3 black balls. What is the probability of drawing 3 white balls in succession from the box (without returning balls selected)?

Since there are 8 balls in the box to begin with, the probability of drawing a white ball on the first selection is 5/8. If a white ball is drawn on the first draw the probability of drawing a white ball on the second selection is 4/7.

After two successful draws, the probability of drawing a white ball on the third selection is 3/6. Thus, the total probability is

$$\frac{5}{8} \times \frac{4}{7} \times \frac{3}{6} = 0.179$$

Statistics

Sample statistics are employed in engineering analyses for the purpose of inferring certain properties of a "population" from a very small portion of the population (i.e., the sample). From the data, sample statistics are computed as estimates of the population properties. The statistical methods and definitions discussed in this section require that the sample be a random one (i.e., each item of the sample has an equal chance of being included in the sample).

The mean (arithmetic average) of a sample is given by

$$\bar{X} = \frac{\sum x_i}{n}$$

where $\bar{X}$ = mean

x_i = individual data values

n = sample size

The median is the "middle" value of a sample (50th percentile). The mode is the most frequently occurring value. The geometric mean is a specialized measure of central tendency; the nth root of the product of the n values. The geometric mean is given by

$$\log \bar{x}_g = \frac{\sum \log x_i}{n}$$

where $\log \bar{x}_g$ = logarithm of the geometric mean

$\log x_i$ = logarithm of the individual values

n = sample size

The geometric mean can be shown to be

$$\bar{x}_g = \sqrt[n]{\text{(product of the n values)}}$$

EXAMPLE:

Determine the mean, median, mode, and geometric mean of the following sample:

23, 26, 26, 27, 30, 31, 40

$$\bar{x} = \frac{\sum x_i}{n} = \frac{23 + 26 + 26 + 27 + 30 + 31 + 40}{7}$$

$\bar{x} = 29$

median = 27 ("middle value")

mode = 26 (most frequently occurring value)

$$\bar{x}_g = \sqrt[7]{23 \times 26 \times 26 \times 27 \times 30 \times 31 \times 40}$$

$$\bar{x}_g = 28.6$$

The <u>range</u> of a sample is the difference between the greatest and least of a set of values. The <u>standard deviation</u> is a measure of dispersion and is given by

$$S = \sqrt{\frac{\sum(x_i - \bar{x})^2}{n - 1}}$$

where   s = standard deviation

x_i = individual data values

$\bar{x}$ = sample mean

n = sample size

It can be shown that standard deviation is also given by

$$S = \sqrt{\frac{\sum x_i^2 - \frac{(\sum x_i)^2}{n}}{n - 1}}$$

The <u>variance</u>, a measure of dispersion, is the square of the standard deviation (s^2). The <u>standard geometric deviation</u>, a specialized measure of dispersion is given by

$$\log s_g = \sqrt{\frac{\sum (\log x_i)^2 - \frac{(\sum \log x_i)^2}{n}}{n - 1}}$$

where $\log s_g$ = logarithm of the standard geometric deviation

$\log x_i$ = logarithm of the individual values

n = sample size

EXAMPLE:

Determine the range, standard deviation, variance, and standard geometric deviation of the following sample:

23, 26, 26, 27, 30, 31, 40

range = 40 - 23 = 17

x_i	x_i^2
23	529
26	676
26	676
27	729
30	900
31	961
40	1600
$\sum x_i = 203$	$\sum x_i^2 = 6071$

$$\frac{(\sum x_i)^2}{n} = \frac{(203)^2}{7} = 5887$$

$$s = \sqrt{\frac{6071 - 5887}{6}} = 5.54$$

variance $= s^2 = 30.67$

Standard geometric deviation:

x_i	$\log x_i$	$(\log x_i)^2$
23	1.3617	1.8543
26	1.4150	2.0021
26	1.4150	2.0021
27	1.4314	2.0488
30	1.4771	2.1819
31	1.4914	2.2242
40	1.6021	2.5666
$\sum \log x_i = 10.1937$		$\sum (\log x_i)^2 = 14.8800$

$(\sum \log x_i)^2 = 103.911$

$$\log s_g = \sqrt{\frac{14.8800 - \frac{103.911}{7}}{6}} = 0.0769$$

$s_g = 1.19$

Histograms (constructed on a rectangular coordinate system with the measured values on the abscissa and the frequencies on the ordinate) are graphical representations of the frequency distribution. The histogram is a rectilinear approximation of the distribution curve such that the area under the curve between two values of the variable is proportional to the frequency of observations in that interval. A distribution curve that is very useful and that approximates many empirical frequencies is the <u>normal</u> or <u>gaussian</u> distribution.

The standard form for the normal distribution is given by

$$\emptyset = \frac{1}{2\Pi} \exp\left(-\frac{z^2}{2}\right)$$

where
$\emptyset$ = normal density function
$Z = (x - \mu)/\sigma$
μ = mean value of x
σ = standard deviation
x = the random variable

Based on general practice, the mean and the standard deviation are represented by μ and σ, respectively, in this discussion (distribution analysis), while x and s were used for these parameters in the discussion on statistical inference.

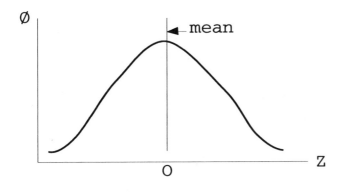

Figure 1.41

The graphical representation of the normal density function is the familiar bell-shaped curve (Fig. 1.41). This curve has the following characteristics:
(1) The maximum occurs at $Z = 0$ ($x = \mu$)
(2) It is symmetrical about a vertical through the maximum
(3) There are inflection points at $Z = \pm 1.0$ ($x = \mu \pm \sigma$)
(4) It is asymptotic to the horizontal axis.

Integration of the normal density function gives the normal distribution function, Φ:

$$\Phi(Z') = \int_{-\infty}^{Z'} \emptyset \, dZ$$

The normal distribution function at a point Z', $\Phi(Z')$, represents the probability of the variable being equal to or less than Z'. The graph of Φ is shown below in Fig. 1.42.

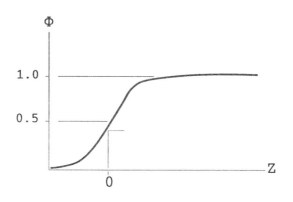

Figure 1.42

Several important characteristics of the normal distribution and density functions are:

(1) the density function $\theta(Z)$ is symmetrical about $Z = 0$, thus

$$\theta(Z) = \theta(-Z)$$

(2) the symmetry of $\theta(Z)$ implies that

$$\int_{-\infty}^{a} \theta(Z)\, dZ = \int_{a}^{\infty} \theta(Z)\, dZ = 1 - \int_{-\infty}^{a} \theta(Z)\, dZ$$

or

$$\Phi(-a) = 1 - \Phi(a)$$

(3) The probability of regions or intervals on the random variables are calculated by the use of the distribution function values

$$P(a \leq z \leq b) = \Phi(b) - \Phi(a)$$

Values of the normal distribution and related functions are generally tabulated in Statistical tests. A simplified table is provided below for reference. Note that the table gives Φ for the values of $Z \geq 0$ only. Values for $Z < 0$ can be determined by the equation

$$\Phi(-Z) = 1 - \Phi(Z)$$

NORMAL DISTRIBUTION FUNCTIONS

Z	Φ(Z)	1 - Φ(Z)	θ(Z)
0.0	.5000	.5000	.3989
0.1	.5398	.4602	.3970
0.2	.5793	.4207	.3910
0.3	.6179	.3821	.3814
0.4	.6554	.3446	.3683
0.5	.6915	.3085	.3521
0.6	.7257	.2743	.3332
0.7	.7580	.2420	.3123
0.8	.7881	.2119	.2897
0.9	.8159	.1841	.2661
1.0	.8413	.1587	.2420
1.5	.9332	.0668	.1295
2.0	.9773	.0227	.0540
2.5	.9938	.0062	.0175
3.0	.9987	.0013	.0044
3.5	.9998	.0002	.0009

EXAMPLE:

If x is a normally distributed random variable, what is the probability of x ≤ 11 with μ = 10.5 and σ = 1.0?

$$\text{at } x = 11, Z = \frac{x - \mu}{\sigma} = \frac{11 - 10.5}{1.0} = 0.5$$

from the Table, Φ(0.5) = 0.6915

EXAMPLE:

For the distribution described in the last example:

(1) What is the probability of x ≤ 10?

(2) What is the probability of 10 ≤ x ≤ 11?

$$\text{at } x = 10, Z = \frac{10 - 10.5}{1.0} = -0.5$$

from the Table, Φ(-0.5) = 1 - Φ(0.5) = 0.3085

$$P(10 \le x \le 11) = \Phi(0.5) - \Phi(-0.5)$$
$$= 0.6915 - 03085$$
$$= 0.383$$

2
Statics

TABLE OF CONTENTS

Selected References	2-ii
Notation	2-iii
Coplanar Force Systems	2-1
Parallel Force Systems	2-1
Concurrent Force Systems	2-2
Non-concurrent Force Systems	2-3
Couple	2-3
Varignon's Theorem	2-4
Equations of Equilibrium	2-4
Planar Structure with Parallel System of Forces	2-5
Planar Structure with Concurrent Force Systems	2-6
Planar Structure with Non-concurrent Force System	2-7
Trusses	2-8
Method of Joints	2-10
Short Cuts and Special Cases	2-14
Method of Sections	2-15
Short Cuts and Special Cases	2-17
Illustrative Examples	2-19
Three Force Members	2-21
Illustrated Examples	2-22
Free Body Diagrams of Each Member	2-22
Solved Problems	2-24
Three Dimensional Forces	2-44
Concurrent Forces in Space	2-44
Direction Cosines	2-45
Concurrent Forces Not Mutually Perpendicular	2-46
Non-concurrent Space Forces	2-47
Equations of Equilibrium for 3D Concurrent Forces	2-47
Moment of a Force About an Axis	2-47
Friction	2-55
Summary of Laws of Friction	2-57
Illustrative Problems	2-58
Wedges	2-63
Screw Thread Friction	2-65
Belt Friction	2-67
Solved Problems	2-70
Centroids	2-87
Centroids of Areas	2-89
Centroids of Lines	2-90
Centroids by Integration	2-90
Theorems of Pappus and Guldinus	2-94

SELECTED REFERENCES

The primary intent of these notes and solved problems are to aid your recall of the fundamentals of *statics*. These notes will concentrate valuable reference material for use in your state examination. They, however, are not meant to be an exhaustive treatment of the subject. Additional study is recommended from the following references:

F. Singer, "Engineering Mechanics," Harper & Row, New York

A. Higdon and W. Stiles, "Engineering Mechanics," Prentice Hall, Englewood Cliffs, NJ

W.G. McLean and E.W. Wilson, "Theory and Problems of Engineering Mechanics," Schaum's Outline Series, McGraw-Hill, New York

R.H. Perry, "Engineering Manual," McGraw-Hill, New York

F. Seely, N. Ensign, and P. Jones, "Analytical Mechanics for Engineers," John Wiley & Sons, New York

I. Levinson, "Statics and Strength of Materials," Prentice Hall, Englewood Cliffs, NJ

E.S. Sherry, "Elements of Structural Engineering," International Textbook Company, Scranton, PA

C.W. Hudson and E.J. Squire, "Elements of Graphic Statics," McGraw-Hill, New York

I.H. Shames, "Engineering Mechanics—Statics," Prentice Hall, Englewood Cliffs, NJ

E.C. Harris, "Elements of Structural Engineering," The Ronald Press, New York

NOTATION

↦	Positive sense of applied forces	NA	Neutral axis, generally the centroidal axis
↻	Positive sense of applied moments	O	Origin of coordinate system or reference to centroidal axis when used as subscript
AB	Force in member AB, lbs, kips		
A_x	Force at point A in x direction, lbs, kips	psi	Pounds per square inch
CG	Center of gravity	P	Force, lbs, kips
f	Coefficient of friction, dimensionless	Q	Force, lbs, kips
F	Force, lbs, kips	r	Mean radius, inches
F_{AX}	Force at point A in x direction, lbs, kips	R	Resultant force or reaction, lbs, kips
		R_{AX}	Reaction at point A in x direction, lbs, kips
FBD	Free body diagram showing all forces left by body	R_{123}	Resultant of force vectors F_1, F_2, and F_3; lbs, kips
F_X	Force in x direction	t	Thickness of member, in.
k	kilopound = kip = 1000 lbs	T	Torque, in-lbs, ft-lbs
ksi	kips per square inch	T_s	Tension in belt on slack side, lbs, kips
L	Length of member	T_t	Tension in belt on taut side, lbs, kips
ln	Natural logarithm, log to base e	w	Intensity of uniformly distributed load, lbs/ft, k/ft
M	Applied moment, in lbs, ft-lbs, kip-ft	W	Total load or force, lbs, kips
M_{AX}	Applied moment at point A about the x axis, in-lbs, ft-lbs	x,y,z	Orthogonal coordinate axes
M_X	Applied moment about the x axis, in-lbs, ft-lbs	α	Angle, radians or degrees
N	Normal force, lbs, kips		

β   Angle of wrap of belt around pulley radians, degrees

θ_x
θ_y Orientation of angles of force vector with respect to coordinate axes, radians, degrees
θ_z

ρ   Radius vector, in., ft or density of material, lbs/ft^3, lbs/in^3

ø   Angle of friction, radians, degrees

Statics is a branch of Engineering Mechanics that deals with the effects and distribution of forces applied to rigid bodies that are at rest. A rigid body, whether a beam, truss, bolt, or shaft, is composed of elements of mass that are fixed in position relative to one another. All bodies will deform under load, to be sure, however, the study of Statics neglects these small deflections. Deflections of members are treated in the study of Mechanics of Materials.

We begin the review of Statics with the fundamentals of FORCE SYSTEMS. It is essential to master the techniques of resolving and combining forces to derive the equations of equilibrium in the two-dimensional domain.

COPLANAR FORCE SYSTEMS

Parallel Force System

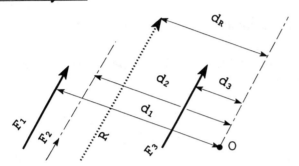

Figure 2.1

All forces are parallel and may be extended along their line of action without changing their effect (Fig. 2.1).

$$\text{Resultant} = R = \Sigma F = F_1 + F_2 + F_3$$

Locate R by taking moments about an arbitrary point 0.

$$\left\{ \begin{array}{l} \text{Moment of resultant} \\ \text{about 0} \end{array} \right\} = \left\{ \begin{array}{l} \text{Sum of moments of each force,} \\ F_1, F_2 \text{ and } F_3 \text{ about 0.} \end{array} \right\}$$

$$R(d_R) = F_1(d_1) + F_2(d_2) + F_3(d_3)$$

$$d_R = \frac{F_1 d_1 + F_2 d_2 + F_3 d_3}{R}$$

The system of forces F_1, F_2 ..., may be put into equilibrium by the addition of a force that is equal, opposite, and colinear with R. This force, which creates equilibrium, is called the equilibrant.

Concurrent Force Systems

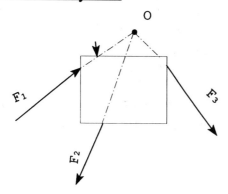

Figure 2.2

The lines of action of all the forces intersect in a common point (Fig. 2.2).

The resultant force is obtained by tip-to-tail vector addition, or by the parallelogram method of addition illustrated in Fig. 2.3.

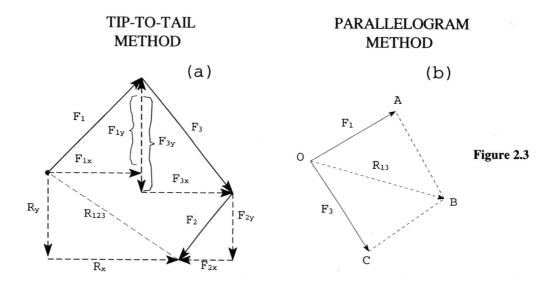

Figure 2.3

$\Sigma F_x = F_{1x} + F_{3x} - F_{2x} = R_x$ Diagonal OB = resultant of F_1 and F_3

$\Sigma F_y = F_{1y} - F_{3y} - F_{2y} = R_y$ Diagonal OE = resultant of R_{13} and F_2

$$R_{123} = \sqrt{R_x^2 + R_y^2}$$

The subscript x denotes the component of force in the x direction, etc.

Non-Concurrent Force System

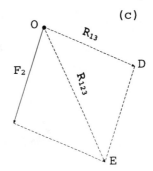

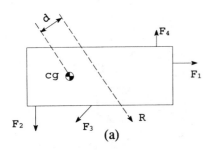

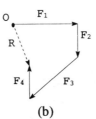

Figure 2.4

Lines of action of force are in various directions and do not intersect in a common point (Fig. 2.4).

An unbalanced non-concurrent force system gives body motion consisting of translation and rotation.

Translation is proportional to the magnitude of the resultant.

Rotation is proportional to the moment of the resultant force.

$$R_x = \Sigma F_x \qquad R_y = \Sigma F_y$$

(Moment due to R) = R x d, where d is the perpendicular distance between R and moment center. Calculate "d" by the procedure illustrated for parallel force system.

Couple

A couple is a torque produced by two equal and opposite forces separated by distance "d". The magnitude of a couple is a measure of its ability to produce twisting (Fig. 2.5).

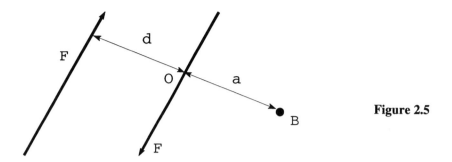

Figure 2.5

The moment of a couple is independent of moment center. (Arrow indicates positive direction.)

$$\circlearrowleft^{+} \Sigma M_O = F \times d \qquad \circlearrowleft^{+} \Sigma M_B = F(d + a) - F \times a = F \times d$$

Varignon's Theorem

The moment of a force = the sum of moments of the components of the force.

In Fig. 2.6 the moment of force F about the origin is F(d) which is identical to the moments about O produced by the x and y components of F.

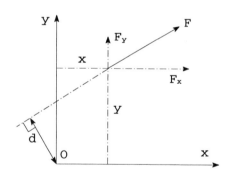

Figure 2.6

$$\circlearrowleft^{+} \Sigma M_O = F \times d = F_x(y) - F_y(x)$$

Equations of Equilibrium

Considering the equilibrium of a body produces one of the most powerful means to analyze the forces acting on it. If a body is in equilibrium it will neither rotate nor translate in space.

The resultant force and moment acting on a body must be zero for equilibrium to exist.

$$\left. \begin{array}{l} \Sigma F_x = 0 \\ \Sigma F_y = 0 \\ \Sigma M = 0 \end{array} \right\} \text{equations of equilibrium}$$

Planar Structure with Parallel System of Forces (Fig. 2.7)

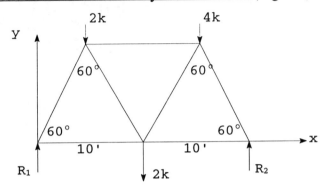

Figure 2.7

(k = kip = kilopound = 1000 lbs)

Because equilibrium means zero resultant, $\Sigma F_Y = 0$ and $\Sigma M = 0$ are the available equations of equilibrium.

$\Sigma F_x = 0$ is trivial because there are no forces with components in the X direction:

R_1 and R_2 are forces produced by the supports of the truss. The five forces shown constitute a parallel system of forces felt by the truss.

↑ $\Sigma F_Y = 0$ (arrow indicates positive direction)

↑ $\Sigma F_Y = -2 -4 -2 + R_1 + R_2 = 0$ $R_1 = 8 - R_2$

↷ $\Sigma M_1 = 0$ (arrow indicates clockwise moment is considered positive)

↷ $\Sigma M_1 = 2 \times 5 + 2 \times 10 + 4 \times 15 - 20 R_2 = 0$

$$R_2 = \frac{10 + 20 + 60}{20} = 4.5^k$$

and $R_1 = 8 - 4.5 = +3.5^k$

Planar Structure with Concurrent Force Systems

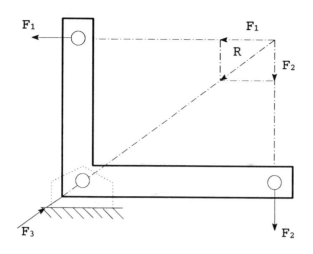

Figure 2.8

If crank is in equilibrium, forces F_1 and F_2 produce a resultant which passes through the fulcrum; that is, the resultant has no moment arm (Fig. 2.8).

This is another way of stating that the moments produced by F_1 and F_2 exactly balance.

If the reaction F_3, at the fulcrum, is to hold the crank in equilibrium it must be equal, colinear, and opposite to the resultant of F_1 and F_2.

Solving for F_3: $\rightarrow \Sigma F_X = 0$

yields $-F_1 + F_{3x} = 0$ $F_{3x} = F_1$

$$+ \Sigma F_Y = 0$$

yields $-F_2 + F_{3y} = 0$ $F_{3y} = F_2$

Planar Structure with Non-concurrent Force Systems

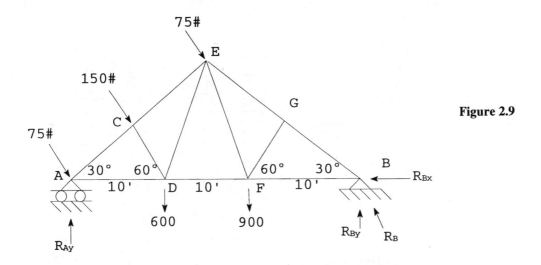

Figure 2.9

A state of equilibrium exists when the resultant effect of a system of forces is neither a force nor a couple.

The resultant of force system = 0 when

$$\Sigma F_x = 0$$

$$\Sigma F_y = 0$$

$$\Sigma M = 0$$

Consider truss as a rigid body and solve for reactions: Reaction at A must be vertical due to roller while reaction at B is unknown both in magnitude and direction. R_B can be replaced by R_{Bx} and R_{By} (as shown in Fig. 2.9). Note there are three equations of equilibrium available and 3 unknowns, R_{Ay} and the magnitude and direction of R_B or R_{Bx} and R_{By}. Simplify the solution by combining loads along AE into their resultant, (300 lbs) acting at joint C and normal to chord AE. Next, extend the resultant R to line AB and resolve it into its components, R_x and R_y.

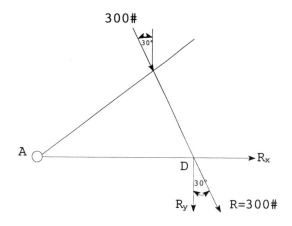

Figure 2.10

$R_x = 300 \sin 30 = 150$ lbs $R_y = 300 \cos 30 = 260$ lbs

$\rightarrow \Sigma F_X = 0, R_x - R_{Bx} = 0$

$R_{Bx} = R_x = 150$ lbs

$\circlearrowright \Sigma M_A = 0, -30 R_{By} + 900 \times 20 + 600 \times 10 + R_y(10) = 0$

$R_{By} = 877$ lbs

$\uparrow \Sigma F_Y = 0, -R_y - 600 - 900 + R_{By} + R_{Ay} = 0$

$\qquad R_{Ay} = 883$ lbs

Trusses

An ideal or simple truss is composed of members joined together by pins or bolts in a manner that prevents motion between members. The center lines at the ends of the members intersect in a common point so that there is no secondary bending (Fig. 2.11).

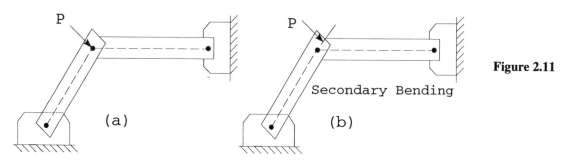

Figure 2.11

Generally, the weight of the members of a truss is small compared to the loads and can be neglected or considered included in the applied load.

All applied loads act at the ends, or pinned joint, of the members producing tension or compression. There is no bending of the members, as there would be in the members shown in Figs. 2.11b and 2.12b.

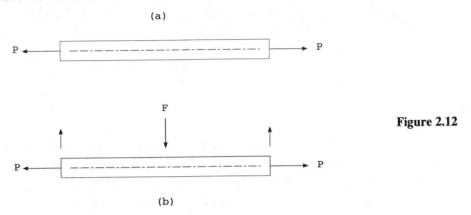

Figure 2.12

Any member of a truss, or rigid frame, may deform elastically under load, but these deformations will be small compared to the dimensions of the truss and do not alter its geometry.

Note that the underlying unit of construction of all trusses is three members arranged in the shape of a triangle (Fig. 2.13). A triangle is rigid and can sustain its shape under load, whereas a 4-bar rectangle is unstable.

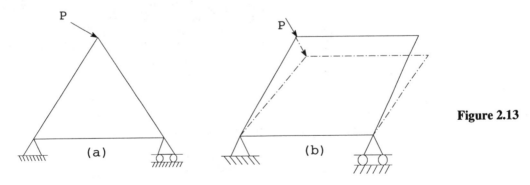

Figure 2.13

Simple trusses are built up from the basic triangular unit as shown in Fig. 2.14.

2-10 FUNDAMENTALS OF ENGINEERING EXAM REVIEW WORKBOOK

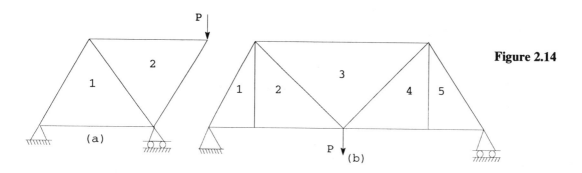

Figure 2.14

Method of Joints

If a truss is in equilibrium, then every portion of the truss must also be in equilibrium. Thus a portion of the truss can be isolated and considered to be in equilibrium under the action of all the forces that it experiences. A convenient portion of the truss to isolate is the pin at each joint. Isolate a pin by drawing a free body diagram of it and apply all the forces acting on it, including the internal forces produced by the bars and any external forces acting on the pin. The forces acting on the pin constitute a concurrent force system for which only two equations of equilibrium are available. Therefore only two unknowns can be considered at each joint.

For convenience, let compressive forces act toward the pin and tensile forces away from the pin (Fig. 2.15).

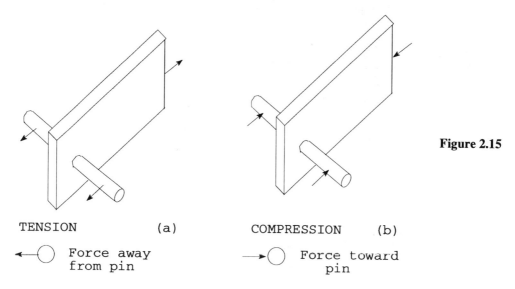

Figure 2.15

Summarizing, the procedure of the <u>method of joints</u> is:

1. Calculate the reactions.

2. Select a pin (joint) with only two unknowns.

STATICS 2-11

3. Draw the free body diagram of the pin. Show <u>all</u> forces experienced by the pin. Assume unknown forces to be <u>tensile</u>. Forces toward the pin are compressive, those away from the pin are tensile.

4. Solve equilibrium equations of the concurrent force system. Positive values of force indicate a correct assumption of direction of forces.

5. Mark the original truss diagram with correct directions of force.

6. Select the next pin where there are only two unknowns.

7. Repeat steps 3 through 6 until forces in all members are known.

<u>Illustrated Example</u>

Determine the forces in the members of the truss, loaded as shown in Fig. 2.16.

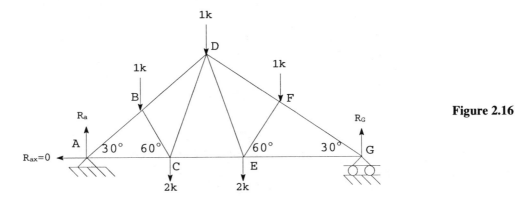

Figure 2.16

Reactions:

$\Sigma M_A = 0$ Yields R_G

$\Sigma F_y = 0$ Yields R_A

Or, note from symmetry of loading and symmetry of truss, that 1/2 of the total load must act at each support.

$$R_A = R_G = \frac{F_y}{2} = 3.5^k$$

Select Pin A, draw its Free Body Diagram.

AB and AC are the forces in members AB and AC. They are assumed to be in tension (Fig. 2.17).

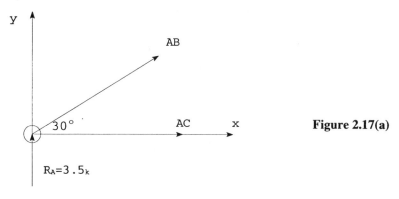

Figure 2.17(a)

Solve Equilibrium Equations:

$\uparrow \Sigma F_Y = 0 : 3.5 + AB \sin 30 = 0$

$$AB = -\frac{3.5}{1/2} = -7.0^k$$

A negative result indicates the assumed direction of force was incorrect. The force should be toward the pin, indicating compression.

Now, correcting Fig. 2.17 to show force AB acting toward the pin, solve for AC.

Figure 2.17(b)

$\rightarrow \Sigma F_X = 0 : AC - AB \cos 30° = 0$

$$AC = 7(.8667) = 6.06^k$$

The positive result obtained from member AC means the direction of the force in AC was chosen correctly. Force AC is tensile.

Mark the original truss with the correct direction of the forces AB and AC as shown in Fig. 2.18.

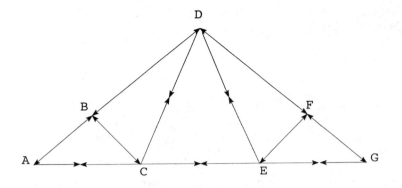

Figure 2.18

Because member AB is in compression we know its direction. When considering pin B, the compressive force in member AB must be shown acting toward pin B.

Pin B

Select the x-axis along chord AD for convenience (Fig. 2.19).

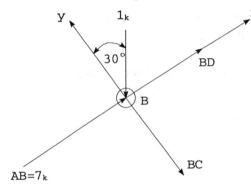

Figure 2.19

$\nwarrow \Sigma F_y = 0 : -BC - 1 \{\cos 30\} = 0$
$BC = -0.867^k$

$\nearrow \Sigma F_x = 0 : 7 - 1 \sin 30 + BD = 0$
$BD = -6.5^k$

Pin C

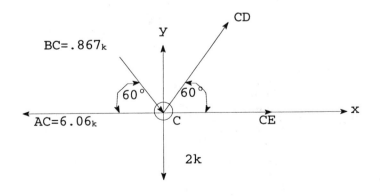

Figure 2.20

↑ $\Sigma F_Y = 0$: CD sin 60 - BC sin 60 - 2 = 0

$$CD = 3.18^k$$

→ $\Sigma F_X = 0$: CE + 3.18 cos 60 + .867 cos 60 - 6.06 = 0

$$CE = 4.04^k$$

The forces in the remaining members can be determined by considering the equilibrium of the remaining pins. The complete results are indicated in Fig. 2.18. Note that the upper chords, ABDFG, are all in compression and the lower chords, ACEG, are all in tension. Generally, this is true for trusses of this type with downward loads.

Short Cuts and Special Cases for Trusses in Equilibrium

1. Where a truss is symmetrical with respect to <u>loading</u> and geometry only half the truss need be calculated. Corresponding members of the other half will have corresponding forces.

2. If the force in a particular member is required, the method of joints can be used on any pin involving that member as long as only two unknowns act at that joint.

3. For the joint with applied loads and more than two unknowns, the Method of Joints can be used when all the unknowns, except one, have the same line of action.

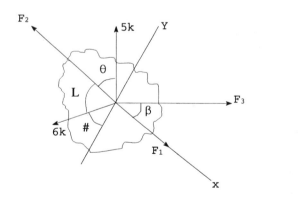

Figure 2.21

↗ $\Sigma F_Y = 0$

$$F_3 \sin \beta + 5 \sin \theta - 6 \sin \alpha = 0 \quad \text{yields} \quad F_3 = \frac{6 \sin \alpha - 5 \sin \theta}{\sin \beta}$$

4. Where three unknowns and no applied loads exist at a pin and two unknowns are colinear, the third unknown force must be zero (Fig. 2.22).

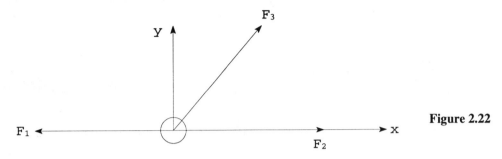

Figure 2.22

F_3 must be zero to satisfy ↑ $\Sigma F_Y = 0$

5. Where only two unknowns exist at a pin and they are not colinear, each must be zero (Fig. 2.23).

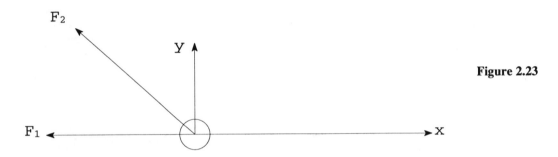

Figure 2.23

F_1 and F_2 must be zero to satisfy $\Sigma F_x = 0$ and $\Sigma F_Y = 0$.

Method of Sections

The method of sections is a powerful tool for analyzing the distribution of forces in a loaded truss, particularly if the force in only one or two members is desired. Essentially, the method consists of dividing a truss into two parts, discarding one part and applying the forces that the discarded part would have produced on the other part to keep it in equilibrium. Generally, the method requires application of the conditions of equilibrium to a nonconcurrent force system. The procedure is illustrated by the following example.

Procedure for the Method of Sections

Determine the force in the member EG of the truss shown in Fig. 2.24.

2-16 FUNDAMENTALS OF ENGINEERING EXAM REVIEW WORKBOOK

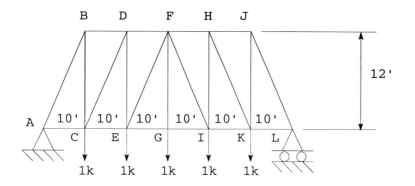

Figure 2.24

1. Calculate the reactions

    $\circlearrowright \Sigma M_A = 0$ Yields $R_L = 2500$ lbs

    From Symmetry $R_A = 2500$ lbs

2. Pass a cutting plane through not more than three members (Fig. 2.25).

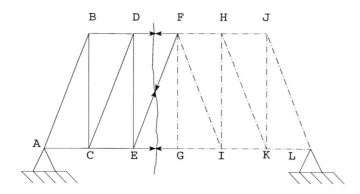

Figure 2.25

3. Apply an external force to each cut member of the section of truss containing the member to be solved.

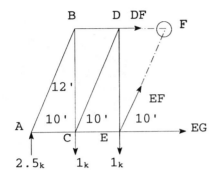

Figure 2.26

These forces must hold this section of truss in equilibrium if they equal forces in members before the members were cut.

The stresses in the uncut members are unaffected by the hypothetical cutting plane. Note that the section constitutes a system of nonconcurrent forces for which there are three equations of equilibrium.

4. Calculate unknown forces, considering forces in cut members and applied loads as external loads. Truss ABDECA is considered a rigid body.

Choose the intersection of members of DF and EF as a moment center, then taking moments about this point we have one equation in the unknown we seek.

$\circlearrowright \Sigma M_F = 0$

$2500(30) - 1000(20) - 1000(10) - EG(12) = 0$

EG = 3750 lbs

The main advantage of the method of sections is that the force can be found in any desired member without the necessity of calculating forces in other members.

Short Cuts and Special Cases for Trusses in Equilibrium

1. In special cases some unknown bar forces can be found when more than three unknowns exist. Here the lines of action of all but one unknown force, F_4, intersect in a common point (Fig. 2.27).

$\circlearrowright \Sigma M_B = 0$ yields F_4

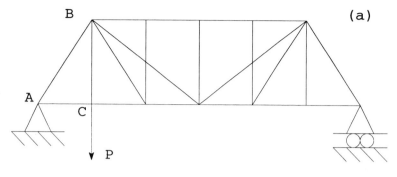

Figure 2.27

The other unknowns cannot be solved by the remaining two equations of equilibrium.

2. After isolating a section of the truss a $\Sigma M = 0$ can be taken about any convenient moment center. Consider the truss in Fig. 2.28.

$\circlearrowleft^+ \Sigma M_o = 0$ Yield

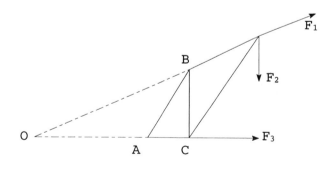

Figure 2.28

3. The cutting plane need not be a straight line, so long as it does not cut more than three members with unknown forces; excluding special cases.

The method of joints is a special case of the method of sections (Fig. 2.29).

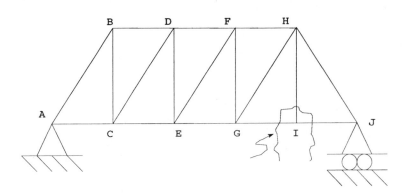

Figure 2.29

The cutting plane isolates joint 1.

Illustrated Examples

1. The A frame shown in Fig. 2.30 has two members 12 ft long and a tie rod 6 ft long connecting the members at points 8 ft from the hinged apex. Determine the tension in the tie rod when a load of 4000 lb is suspended from the apex. Neglect the weight of the frame and assume that it rests on a smooth surface.

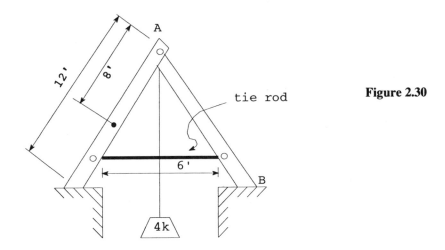

Figure 2.30

With symmetrical loading and smooth surfaces, reactions at the supports must be equal and vertical; 2^K each.

<u>Approach</u>: Isolate one leg of the truss, and draw its free body diagram showing all the forces <u>experienced</u> by the leg (Fig. 2.31).

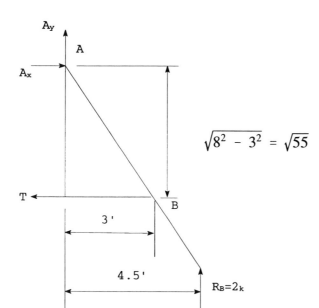

Figure 2.31

Solving for the distance from the ℄ to point B; $3/8 = x/12$

$X = 3/8 \times 12 = 4.5$

$\circlearrowright \Sigma M_A = 0$ Yields

$T \times \sqrt{55} - R_B \times 4.5 = 0$

$T = \dfrac{2000 \times 4.5}{7.42} = 1210$ lbs   Answer

2. The L-shaped crank shown is pivoted at point A and is subjected to the action of the two forces indicated. Find the thrust at the pivot and the upward force exerted by the wheel at C. Neglect the weight of the crank itself (Fig. 2.32).

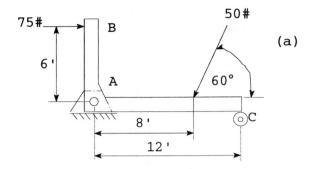

(a)

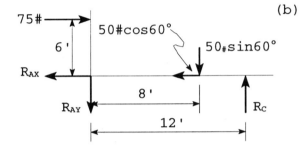

(b)

Figure 2.32

$\circlearrowleft \Sigma M_A = 0 \quad 75 \times 6 + (50 \sin 60°) \, 8 - R_c \times 12 = 0$

$12 \, R_c = 450 + 400 \times 3/2$

$R_c = 66.4$ lbs   Answer

$\rightarrow \Sigma F_X = 0 \quad 75 - R_{AX} - 50 \cos 60° = 0$

$R_{AX} = 75 - 50 \times 1/2 = 50$ lbs

$+ \Sigma F_Y = 0 \quad R_C - R_{AY} - 50 \sin 60° = 0$

$R_{AY} = 23.1$ lbs

$R_A = \sqrt{(R_{AX})^2 + (R_{AY})^2} = \sqrt{50^2 + 23.1^2} = 55.1$ lbs   Answer

Three Force Members

Up to this point we have been analyzing members of trusses subjected to axial loads. Frames and trusses can be constructed in a manner that imparts transverse loads in addition to axial loads. These are called <u>three force members</u> (See Fig. 2.12b). Such members are always subjected to bending as well as axial loads. The end forces are not directed along the axis of the member; therefore it is not possible to analyze a structure containing three force members by the method of sections.

Three force members generally are statically indeterminate, but they can be solved if they are members of larger frameworks. Often, each member is acted upon by more unknowns than there are equations of equilibrium. The solution almost always lies with the member causing these unknown forces. Logically, the causing member experiences the equal and opposite reactions of these unknown forces. Examining a free body diagram of each interacting member may yield up to two sets of three simultaneous equations in the common unknowns. These equations result from considering the equilibrium of each isolated member.

As a general rule, it is best to solve first for the unknown common to two free body diagrams.

Illustrated Example

Determine the interacting forces between the members of the frame shown in Fig. 2.33.

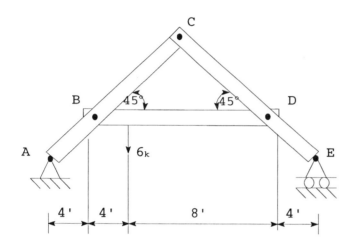

Figure 2.33

Solving for reactions at A and E.

$\circlearrowright \Sigma M_A = 0$ yields

$6(8) - R_E(20) = 0 \quad R_E = 2.4^k$

$\uparrow \Sigma F_y = 0$ yields

$R_A - 6 + 2.4 = 0 \quad R_A = 3.6^k$

Free Body Diagrams of Each Member

Isolate each member of the truss as a free body diagram showing all forces felt by the member. Assume the direction of forces acting on one member and reverse the direction of reaction to this force that acts on interacting member (Fig. 2.34).

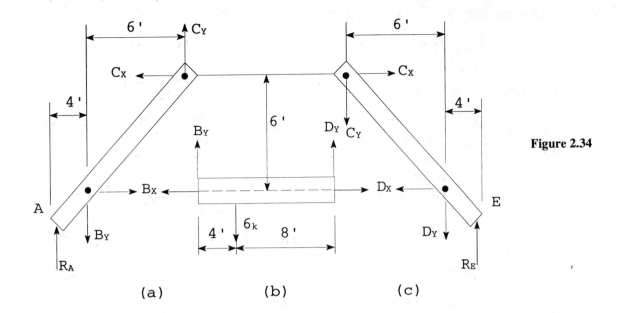

Figure 2.34

Note there is a nonconcurrent force system acting on each member, which means there are three equations of equilibrium available for each member.

$$\circlearrowright \Sigma M = 0 \qquad \rightarrow \Sigma F_X = 0 \qquad \uparrow \Sigma F_Y = 0$$

For the member in Figure 2.34a:

$$\circlearrowright \Sigma M_B = 0 : R_A \times 4 - 6C_X - C_Y \times 6 = 0$$

(recall $R_A = 3.6^k$)

For the member in Figure 2.34 c:

$$\circlearrowright \Sigma M_D = 0 : -R_E \times 4 + C_X \times 6 - C_Y \times 6 = 0$$

(recall $R_E = 2.4^k$)

Solving these two simultaneous equations yields:

$C_X = 2000$ lbs $\qquad C_Y = 400$ lbs

If either C_X or C_Y were negative it would mean that the chosen direction was incorrect. Before proceeding the incorrect direction should be reversed.

Now using the remaining equations of equilibrium for the member in Figure 2.34a:

$\rightarrow \Sigma F_X = 0$: $B_X - C_X = 0$ $B_X = 2000$ lbs

$\uparrow \Sigma F_Y = 0$: $+C_Y - B_Y + R_A = 0$ $B_Y = 3600 + 400 = 4000$ lbs

For the Member in Figure 2.34c:

$\rightarrow \Sigma F_X = 0$: $C_X - D_X = 0$ $D_X = 2000$ lbs

$\uparrow \Sigma F_Y = 0$: $R_E - D_Y - C_Y = 0$ $D_Y = 2400 - 400 = 2000$ lbs

All unknowns have been solved. However, the equations of equilibrium applying to member BD have not been used, they can be used to check the solution.

$\rightarrow \Sigma F_X = 0$: $-B_X + D_X = 0$ $-2000 + 2000 = 0$ checks

$\uparrow \Sigma F_Y = 0$: $B_Y - 6000 + D_Y = 0$

$4000 + 2000 = 6000 = 0$ checks

$\circlearrowright \Sigma M_B = 0$: $6000 \times 4 - 12 \times D_Y = 0$

$24000 - 2000 \times 12 = 0$ checks

Solved Problems

1. Two smooth cylinders rest (as shown in Fig. 2.35) on an inclined plane and a vertical wall. Compute the value of the force, F, exerted by the wall.

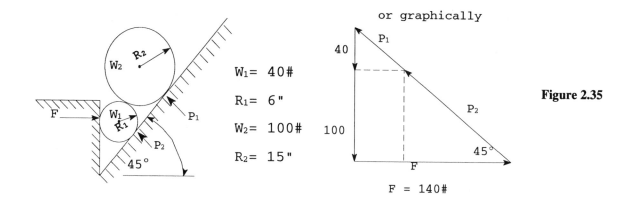

Figure 2.35

$W_1 = 40\#$
$R_1 = 6"$
$W_2 = 100\#$
$R_2 = 15"$

F = 140#

Approach

Establish a 45° incline as the x axis, show all forces produced by constraints, and resolve all forces into x and y components. Examine equilibrium in the x direction (Fig. 2.36).

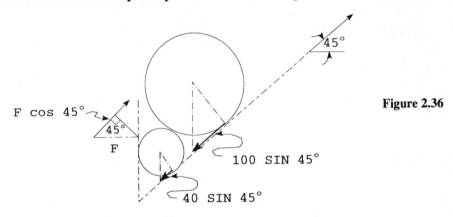

Figure 2.36

$\nearrow \Sigma F_Y = 0$: $F \cos 45° - 40 \sin 45° - 100 \times \sin 45° = 0$

$F = 100 + 40 = 140$ lbs

2. Determine the forces in each member of the truss, loaded as shown in Fig. 2.37, and check your result for member CD by solving for it with the method of joints.

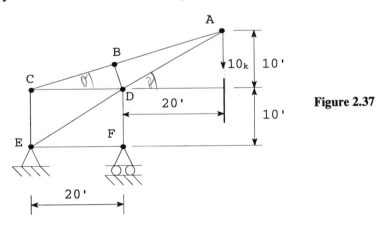

Figure 2.37

Approach

Solve for the reactions, treating the entire truss as a rigid body. Use method of joints to determine the forces in each member.

$\circlearrowright \Sigma M_E = 0 \quad 10 \times 40 - R_F \times 20 = 0$

$$R_F = \frac{400}{20} = 20^k \uparrow$$

$\uparrow \Sigma F_Y = 0 \qquad -R_E + R_F - 10 = 0$

$$R_E = R_F - 10 = 10^k \downarrow$$

Joint F (Fig. 2.38)

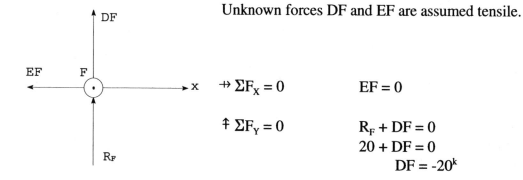

Unknown forces DF and EF are assumed tensile.

$\rightarrow \Sigma F_X = 0 \qquad EF = 0$

$\uparrow \Sigma F_Y = 0 \qquad R_F + DF = 0$
$\qquad\qquad\qquad\quad 20 + DF = 0$
$\qquad\qquad\qquad\qquad\quad DF = -20^k$

Figure 2.38

Joint E (Fig. 2.39)

CE and DE are assumed tensile

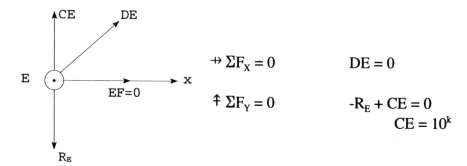

$\rightarrow \Sigma F_X = 0 \qquad DE = 0$

$\uparrow \Sigma F_Y = 0 \qquad -R_E + CE = 0$
$\qquad\qquad\qquad\qquad CE = 10^k$

Figure 2.39

Joint C (Fig. 2.40)

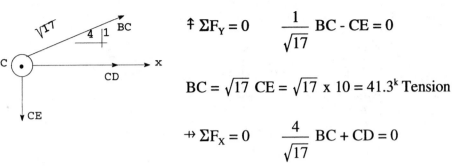

Figure 2.40

$\uparrow \Sigma F_Y = 0 \qquad \dfrac{1}{\sqrt{17}} BC - CE = 0$

$BC = \sqrt{17} \; CE = \sqrt{17} \times 10 = 41.3^k$ Tension

$\rightarrow \Sigma F_X = 0 \qquad \dfrac{4}{\sqrt{17}} BC + CD = 0$

$CD = \dfrac{4}{\sqrt{17}} \times 41.3 = -40^k$ compression

Joint D (Fig. 2.41)

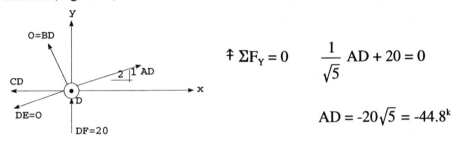

Figure 2.41

$\uparrow \Sigma F_Y = 0 \qquad \dfrac{1}{\sqrt{5}} AD + 20 = 0$

$AD = -20\sqrt{5} = -44.8^k$

Summary of Answers

Member	Force
AB	41.3^{k}T
AD	44.8 C
BC	41.3 T
BD	0
CE	10 T
CD	40 C
DE	0
EF	0
DF	20 C

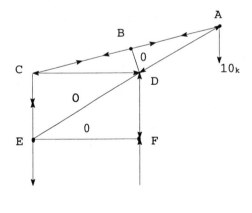

Figure 2.42

3. Using the method of sections, find that force in member FI (Fig. 2.43).

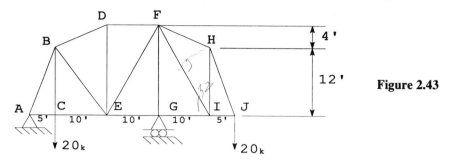

Figure 2.43

Approach

Pass a cutting plane as shown in Fig. 2.44, draw a free body diagram of the right hand section, then extend the line of action of FH until it intersects IJ extended; call this point K.

$\Sigma M_H = 0$ yields one equation in FI.

$$\circlearrowright \Sigma M_H = 0 \ : \ FI \ \frac{16}{\sqrt{356}} (30) - 20(25) = 0$$

$$FI = \frac{20 \times 25 \times \sqrt{356}}{16 \times 30} = 19.7^k \text{ tension}$$

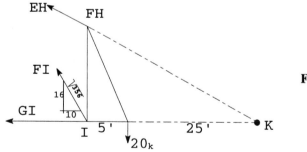

Figure 2.44

4. Determine the forces in members CD and DG (Fig. 2.45).

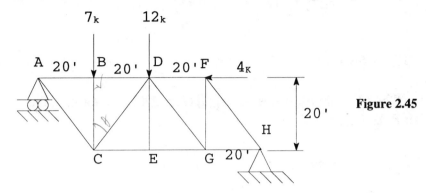

Figure 2.45

Approach

Find the reaction at A, then pass the cutting plane through BD, CD, CE, draw the FBD (free body diagram) of left hand section (Fig. 2.46). Take $+\Sigma F_Y = 0$ to get one equation in unknown CD.

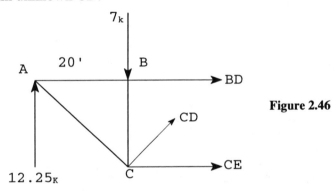

Figure 2.46

$\circlearrowright \Sigma M_H = 0 : R_A(80) - 7(60) - 12(40) - 4(20) = 0$

$$R_A = \frac{30 + 480 + 420}{80} = 12.25^k$$

From the FBD $+\Sigma F_Y = 0 \quad 12.25 - 7 + \dfrac{CD}{\sqrt{2}} = 0$

$$CD = -5.25\sqrt{2} = -7.43^k \text{ compression}$$

Now consider the FBD of section isolated by cutting plane through DF, DG, EG.

$⤉ \Sigma F_Y = 0 : 12.25 - 7 - 12 - \dfrac{DG}{\sqrt{2}} = 0$

$DG = -6.75\sqrt{2} = -9.55^k$ compression

5. The base angles at E and F of this A frame are equal (Fig. 2.47). Determine the reaction applied to member BD.

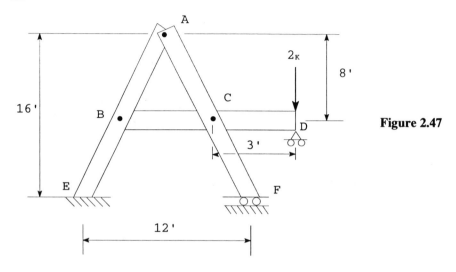

Figure 2.47

Approach

Treating the frame as a rigid body, solve for the reaction at F. Unlock all joints and draw the FBD of all members. Show all forces felt by members. Take $↻ \Sigma M_A = 0$ for member AF, this yields one equation in two unknowns, C_x and C_y. Take $↺ \Sigma M_B = 0$ for member BD to get another equation in C_x and C_y. Take $⤉ \Sigma F_Y = 0$ for member BD to solve for B_y. Take $⇀ \Sigma F_X = 0$ to solve B_x.

$↺ \Sigma M_E = 0 : 2(12) - 12 R_F = 0 \qquad R_F = 2^k$

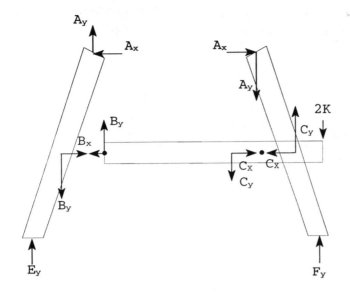

Figure 2.48

Member AF

$\circlearrowright \Sigma M_A = 0 : -C_Y(3) + C_X(8) - 2(6) = 0$

Member BD

$\circlearrowright \Sigma M_B = 0 : C_Y(6) + 2(9) = 0$

$$C_Y = -\frac{18}{6} = -3^k$$

Substituting C_Y in equilibrium equation for member AF:

$-(-3)\,3 + 8\,C_X - 12 = 0 \qquad C_X = 0.375^k$

For Member BD

$\uparrow \Sigma F_Y = 0 : B_Y - C_Y - 2 = 0, \quad B_Y = 2 + C_Y = -1^k$

$\rightarrow \Sigma F_X = 0 : -B_x + C_X = 0, \qquad B_x = C_X = 0.375^k$

6. The frame shown in Fig. 2.49 has hinged supports at A and E. Determine the components of the hinge forces at A and E and the load in members BC and BD

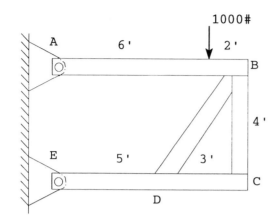

Figure 2.49

Approach

Using the entire truss as a FBD solve for horizontal reactions at A and E (Fig 2.50). From the FBD of member AB solve for vertical reactions at A and E. Now draw the FBDs of members BD and BC, and solve for the required forces.

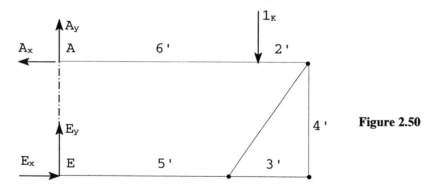

Figure 2.50

$\circlearrowright \Sigma M_E = 0 : - A_x(4) + 1(6) = 0$

$$A_x = 1.5^k$$

$\rightarrow \Sigma F_X = 0 \quad - A_x + E_x = 0$

$$E_x = 1.5^k$$

From the FBD of AB (Fig. 2.51)

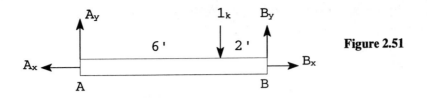

Figure 2.51

$\circlearrowright \Sigma M_B = 0$, $A_y(8) - 1(2) = 0$

$$A_y = 0.25^k$$

$\rightarrow \Sigma F_X = 0: -A_x + B_x = 0 \qquad B_x = A_x = 1.5^k$

$\uparrow \Sigma F_Y = 0: A_y - 1 + B_Y = 0 \quad B_Y = 1 - A_y = 0.75^k$

Now for the entire truss:

$\uparrow \Sigma F_Y = 0: A_y + E_y - 1 = 0 \quad E_y = 1 - A_y = 0.75^k$

Now draw FBDs of BC, DB, and EC (Fig. 2.52).

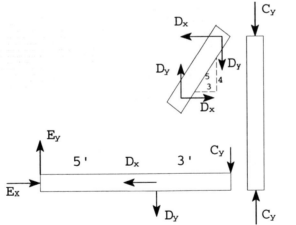

Figure 2.52

For member CE

$\circlearrowright \Sigma M_D = 0 : \dfrac{3}{4}(5) + C_Y(3) = 0$

$$C_Y = -\dfrac{15}{4 \times 3} = -1.25^k = \text{load in BC, tension}$$

↑ $\Sigma F_Y = 0 : \frac{3}{4} - D_y - (-1.25) = 0 \quad D_y = +2^k$

→ $\Sigma F_X = 0 : E_x - D_x = 0 \quad D_x = 1.5^k$

From the geometry of member BD, load in BD is $\frac{5}{3} D_x = \frac{5}{3} (1.50) = 2.5^k$ compression.

Check of calculations at joint B.

↑ $\Sigma F_Y = 0 : B_Y + C_Y - D_y = 0 \quad 0.75 + 1.25 - 2 = 0$, check

↳ direction has been reversed.

7. A disc weighs 100 lbs and is supported on a 30° incline by bar AB whose top is supported by cable AC (Fig. 2.53). Determine the reactions at B and the tension in the cable.

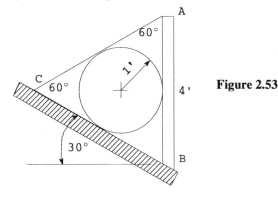

Figure 2.53

Approach (Fig. 2.54)

From the geometry of the disc, determine the horizontal component of force W_x and the line of action of force. From the FBD of member AB determine the reactions. Using geometry at point A find the tension in AC.

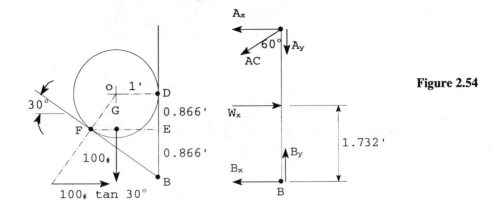

Figure 2.54

From the geometry of disc, GE = 1', FG = 1 sin 30° = 0.5', DE = 0.866.

In triangle FEB, EB = FE tan 30° = 1.5 x (0.577) = 0.866.

Then W_x acts on AB a distance of 1.732' from B.

For member AB

$\circlearrowright \Sigma M_B = 0$: - A_x (4) + W_x (1.732) = 0, A_x = 25.0 lbs

From the geometry of point A

AC sin 60° = A_x,   AC = $\dfrac{25}{0.866}$ = 28.9 lbs

$\rightarrow \Sigma F_X = 0$: - A_x + W_x - B_x = 0

B_x = 32.7 lbs

$+ \Sigma F_Y = 0$: AC cos 60° - B_Y = 0

B_Y = 14.45 lbs

8. A 2 ft diameter pulley is mounted on beam AB and supports a load of 1000 lbs The beam is simply supported at A and C (Fig. 2.55). Neglect the weight of the beam and determine the reactions at A and C.

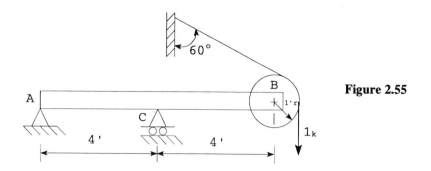

Figure 2.55

Approach (Fig. 2.56)

Draw the polygon of forces acting at point B, and solve for the reactions acting on the beam. Use a free body diagram of beam to solve for the reactions.

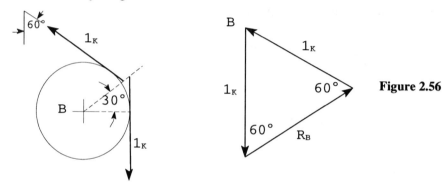

Figure 2.56

$R_{Bx} = 1 \cos 30° = 0.866^k \qquad R_{By} = 1 \sin 30° = 0.5^k$

From the FBD of beam:

$\circlearrowleft \Sigma M_A = 0 : - R_c (4) + R_{By} (8) = 0$

$R_c = 2 R_{By} = 1^k$

$\circlearrowleft \Sigma M_C = 0 : R_{Ay} (4) + R_{By} (4) = 0$

$- R_{Ay} = R_{By}' ; R_{Ay} = - \dfrac{1^k}{2}$

$\circlearrowleft \Sigma F_x = 0 : - R_{Bx} + R_{Ax} = 0 \quad R_{Ax} = 0.866^k$

9. Determine the reactions at the supports A and B of the truss loaded as shown in Fig. 2.57.

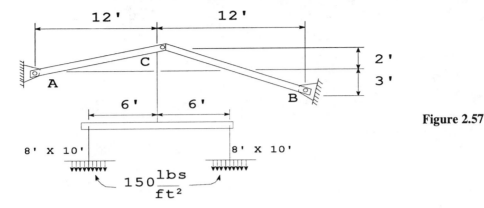

Figure 2.57

Approach

Using the force per unit area times the total area, determine the total load acting on the junction C. Resolve the force at C into the components along the members AC and CB.

Force acting on the joint C is 2(8)(10)(150) = 24,000 lbs

From the polygon of forces acting at junction C (Fig. 2.58):

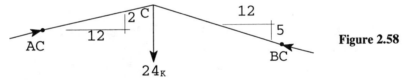

Figure 2.58

$$\uparrow \Sigma F_Y = 0 : \frac{2}{\sqrt{148}} \, AC + \frac{5}{13} \, BC - 24{,}000 = 0$$

$$\rightarrow \Sigma F_X = 0 : \frac{12}{\sqrt{148}} \, AC - \frac{12}{13} \, BC = 0$$

$$AC = \frac{12}{13} \, BC \, \frac{\sqrt{148}}{12} = \frac{\sqrt{148} \, BC}{13}$$

Substituting in the first equation:

$$\frac{2}{\sqrt{148}} \frac{\sqrt{148}}{13} BC + \frac{5}{13} BC = 24,000$$

$$BC = \frac{24,000 \times 13}{7} = 44571 \text{ lbs}$$

$$AC = \frac{\sqrt{148}}{13}(44,700) = 41710 \text{ lbs}$$

10. Find the total hinge force at point B of the three hinged arch (Fig. 2.59).

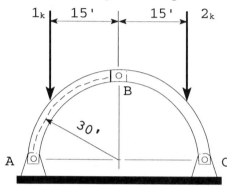

Figure 2.59

An arch is a structure that develops a horizontal thrust reactions under the action of vertical loads.

Approach (Fig. 2.60)

Draw a free body diagram of members AB and BC, and use equations of equilibrium to solve for reaction at B.

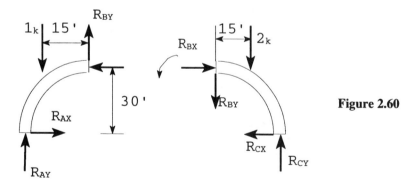

Figure 2.60

From member AB:

$$\circlearrowleft \Sigma M_A = 0 : 1(15) - R_{Bx}(30) - R_{By}(30) = 0 \qquad (1)$$

From member BC:

$$\circlearrowright \Sigma M_C = 0 : -2(15) + R_{Bx}(30) - R_{By}(30) = 0 \qquad (2)$$

From Eq. (1):

$$R_{Bx} = -\frac{R_{By}(30) + 15}{30} = -R_{By} + \frac{1}{2}$$

Substituting in Eq. (2):

$$-2(15) + (-R_{By} + \frac{1}{2})\,30 - R_{By}(30) = 0$$

$$-1 - R_{By} + \frac{1}{2} - R_{By} = 0 \quad R_{By} = -0.25^k$$

$$R_{Bx} = 0.25 + 0.5 = 0.75^k$$

11. Find the tension in cable EF (Fig. 2.61).

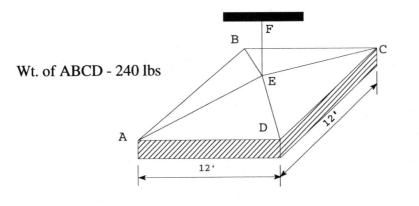

Wt. of ABCD - 240 lbs

Figure 2.61

(A) 104 lbs (B) 416 lbs

(C) 240 lbs (D) 60 lbs

By inspection, the answer is (C).

12. What is the reaction at point C of the beam? (See Fig. 2.62.)

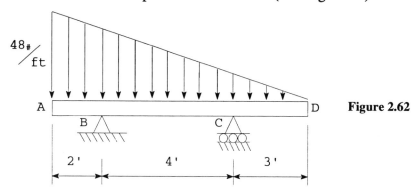

Figure 2.62

(A) 54 lbs   (B) 108 lbs

(C) 36 lbs   (D) zero lbs

Find the center of gravity (CG) of the loading diagram 3 ft from A

$\circlearrowright \Sigma M_B = 0 \quad 48(9)(0.5)(1) - R_c(4) = 0$

$R_c = \dfrac{24\,(9)}{4} = 54$ lbs

The answer is (A).

13. Determine the reaction at joint A.

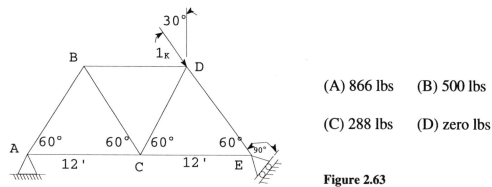

(A) 866 lbs   (B) 500 lbs

(C) 288 lbs   (D) zero lbs

Figure 2.63

The applied load is normal to the support at E and is co-linear with DE, hence there is no reaction at A.

The answer is (D).

14. Determine the reaction at point C of beam BD (Fig. 2.64).

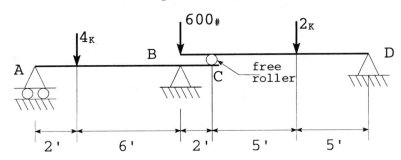

Figure 2.64

(A) 600 lbs (B) 3150 lbs

(C) 1720 lbs (D) 1000 lbs

From member BD, $\curvearrowleft^+ \Sigma M_D = 0$:

$- 600(12) + R_c(10) - 2000(5) = 0 \quad R_c = 1720$ lbs

The answer is (C).

15. What is the force in member DE of this pin jointed truss (Fig. 2.65)?

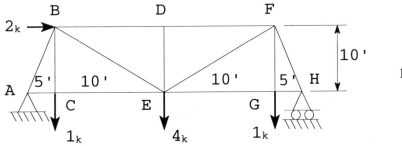

Figure 2.65

(A) 2^k (B) 4^k

(C) 6^k (D) zero lbs

$\uparrow \Sigma F_Y = 0$ at point D yields DE = 0

The answer is (D)

16. Find the load in member BD (Fig. 2.66).

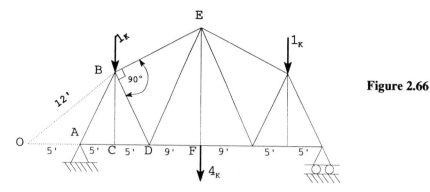

Figure 2.66

(A) compression (B) zero lbs (C) tension

$\circlearrowright \Sigma M_o = 0$ yields $1\,(OC) + BD\,(OB) - 3\,(OA) = 0$

$$BD = \frac{3\,(OA) - OC}{OB}$$

The answer is (C).

17. Under what conditions is a body in equilibrium (Fig. 2.67)?

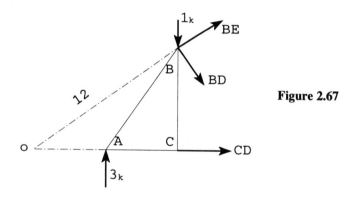

Figure 2.67

(A) Summation of all forces equals zero.
(B) Sum of all horizontal forces equals zero.
(C) Sum of moments is zero.
(D) Sum of all forces and moments are zero.

The answer is (D): it satisfies three equations of equilibrium.

18. If members AC and BD are cables, which member can be removed from the truss without affecting its stability when the truss is loaded as shown in Fig. 2.68?

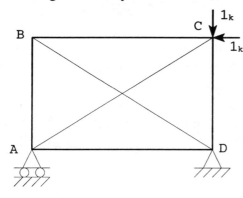

Figure 2.68

(A) AB   (B) AC

(C) CD   (D) BD

From the given choices, member AC can be removed because the load is compressive and a cable (or rope) cannot take compression.

The answer is (B).

19. What type of load cannot be transmitted through a pinned joint?

   (A) shear   (B) double shear

   (C) moment   (D) axial

   The answer is (C).

20. Under what circumstances is the method of sections more practical than the method of joints for analyzing the forces in a member of a truss?

   (A) When three members meet at a joint
   (B) When the truss has many members
   (C) When the truss is unstable
   (D) When four members meet at a joint

   The answer is (B).

THREE-DIMENSIONAL FORCES

Concurrent Forces in Space

If several forces in space have the same point of application they can be reduced to a single resultant force by using the principle of addition by the parallelogram of forces as was done for the coplanar case.

Consider three mutually perpendicular forces intersecting at point O (Fig. 2.69). Establish a set of mutually perpendicular axes, with origin at O and their point of intersection, and let each force lie along a coordinate axis.

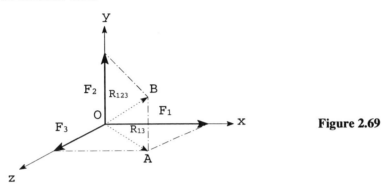

Figure 2.69

Use the parallelogram method to combine forces F_1 and F_3, this yields vector OA which is designated R_{13}. Similarly add F_2 to R_{13} to get vector OB = R_{123}.

Concurrent forces can be added in any order to obtain their resultant. The resultant always acts through the point of concurrence.

Concurrent forces can also be combined by the tip-to-tail method of vector addition. Operating again on the three forces in Fig. 2.69, draw the vector F_1 colinear with the X axis, as show in Fig. 2.70, draw vector F_3 from the tip of F_1 parallel to the Z axis, and from the tip of F_3, draw F_2 parallel to the Y axis.

The resultant is the vector from point O to the tip of F_2. It must be identical in magnitude and direction to the resultant found in Fig. 2.69.

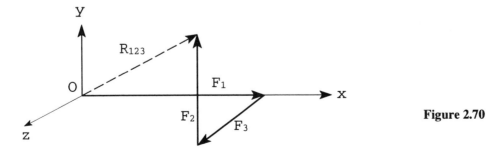

Figure 2.70

Direction Cosines

Very often it is convenient to investigate the effect of a force by dealing with its components. To find the components of a force R in the x direction, simply multiply R by the cosine of the angle between the force and the x axis. Similarly, the components of R in any direction can be determined.

Direction Cosines:

$$F_x = R \cos \theta_x$$

$$F_y = R \cos \theta_y$$

$$F_z = R \cos \theta_z$$

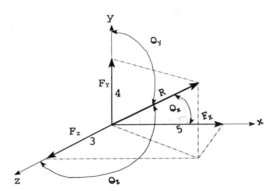

Figure 2.71

The functions $\cos \theta_x$, $\cos \theta_y$ and $\cos \theta_z$ are called the <u>direction cosines</u> of the force R. They can be found by considering any two convenient points defining the line of action of the force R. Select two convenient points on the vector R, say, the origin (0,0,0), and the point x = 5, y = 4, and z = 3 designated (5, 4, 3) (Fig. 2.71).

Distance d:

$$d = \sqrt{(x_2 - x_1)^2 + (y_2 - y_1)^2 + (z_2 - z_1)^2}$$

$$= \sqrt{(5 - 0)^2 + (4 - 0)^2 + (3 - 0)^2}$$

$$= \sqrt{50} = 7.07$$

and

$$\cos \theta_x = \frac{x}{d} = \frac{5}{7.07} = .707$$

$$\cos \theta_y = \frac{y}{d} = \frac{4}{7.07} = .567$$

$$\cos \theta_z = \frac{z}{d} = \frac{3}{7.07} = .424$$

Concurrent Forces Not Mutually Perpendicular

This is a system of concurrent forces that do not lie in three orthogonal planes. They can also be combined by the tip-to-tail method of vector addition (Fig. 2.72).

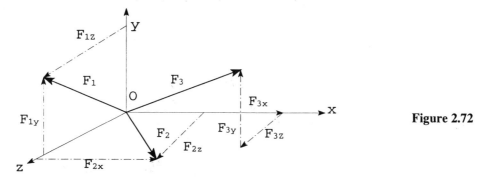

Figure 2.72

Force F_1 in plane yz has two components.

Force F_2 in plane xz has two components.

Force F_3 not in any orthogonal plane has three components.

To combine the three forces, add all the x components; all the y components, and all the z components to get the components of the resultant (Fig. 2.73). Using the tip-to-tail method of vector addition, it is clear that the components of the resultant are:

$$R_x = F_{3x} + F_{2x}$$

$$R_y = F_{1y} + F_{3y}$$

$$R_z = F_{1z} + F_{2z} + F_{3z}$$

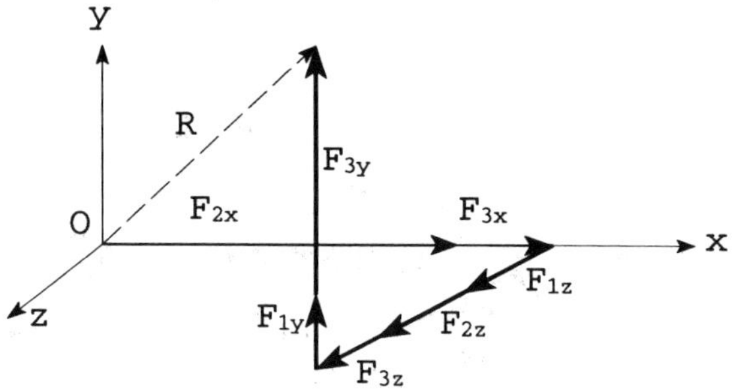

Figure 2.73

Nonconcurrent Space Forces

The powerful concept of equilibrium applies equally well to space forces as it does to coplanar forces.

> EQUILIBRIUM IS A CONDITION IN WHICH THE RESULTANT OF A SYSTEM OF FORCES MUST EQUAL ZERO.

If a system of concurrent forces is in equilibrium in space, the resultant R must be zero. This can occur only when there are no net x, y, or z components. Stated in another manner, concurrent forces in space are in equilibrium when their orthogonal projections on any plane represent a coplanar system of forces in equilibrium.

> WHERE A SYSTEM OF CONCURRENT SPACE FORCES IS IN EQUILIBRIUM THE ALGEBRAIC SUMS OF THE PROJECTIONS OF THESE FORCES ON MUTUALLY ORTHOGONAL AXES, X, Y, AND Z, MUST VANISH.

Equations of Equilibrium for Three-Dimensional Concurrent Forces

$$\Sigma F_x = 0 \; ; \; \Sigma F_y = 0 \; ; \; \Sigma F_z = 0$$

On the basis of these equilibrium equations, the following rules can be stated.

1. Three concurrent forces <u>not</u> in one plane cannot be in equilibrium unless all three forces are zero.

   <u>Proof</u>: Refer to Fig. 2.69. Equate to zero the algebraic sum of the projections of F_1, F_2 and F_3 on axes <u>perpendicular</u> to plane xz. F_2 is the only force that has a nonzero projection, therefore F_2 must equal zero if equilibrium is to exist. Extend the same reasoning to the other axes.

2. If two of four concurrent forces, not all in one plane, are colinear, equilibrium can exist only if the other two forces are zero. The two colinear forces must be equal and opposite.

3. If all but one of many concurrent forces are co-planar, equilibrium can exist only if the odd force is zero.

Moment of a Force About an Axis

The moment of a force about an axis is a measure of its ability to produce rotation about the axis.

Consider the force F in space and find the moment it produces about the y axis.

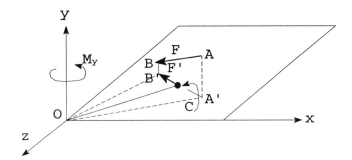

Figure 2.74

THE MOMENT OF A FORCE ABOUT ANY AXIS IS DUE TO THE COMPONENTS OF THE FORCE LYING IN THE PLANE PERPENDICULAR TO THE AXIS OF MOMENTS.

Plane xz is $\perp$ y

F' = projection of F on xz

$\overline{M_F}$ about y = F'(OC) = 2 (area of triangle OA'B')

The moment of force about an axis is represented geometrically by a vector directed along the axis in such a manner that it points in the direction of the extended thumb of the right hand when the fingers of that hand are curled about the axis of the moment, in the sense of the moment. With the thumb pointing in the positive direction of the coordinate axis, a positive moment corresponds to the curling fingers of that hand. This is the right-hand rule for moments.

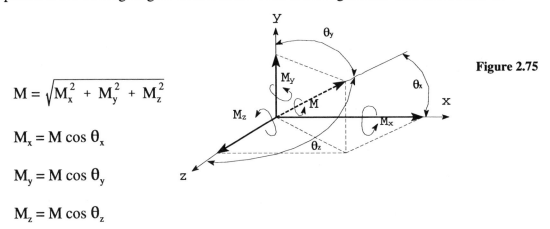

Figure 2.75

$$M = \sqrt{M_x^2 + M_y^2 + M_z^2}$$

$M_x = M \cos \theta_x$

$M_y = M \cos \theta_y$

$M_z = M \cos \theta_z$

A FORCE, OR ITS COMPONENT, THAT IS PARALLEL TO AN AXIS PRODUCES NO MOMENT ABOUT THAT AXIS.

A FORCE THAT INTERSECTS AN AXIS PRODUCES NO MOMENT ABOUT THAT AXIS.

Equations of Equilibrium (For Concurrent Moment Vectors)

$$\Sigma M_x = 0 \qquad \Sigma M_y = 0 \qquad \Sigma M_z = 0$$

Either the three moment equations above or the three force equations of equilibrium may be used to solve the unknown quantities required to maintain equilibrium of a concurrent system of space forces. Usually, any combination of these equations (not more than three) can be used to solve the three unknowns.

Illustrated Example:

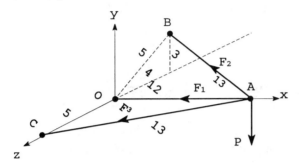

Figure 2.76

Determine the forces in the members of this truss, loaded by a vertical force P at point A.

To solve this problem, isolate the point A which is the intersection of a system of concurrent space forces and examine its equilibrium.

Use the same sign convention as for the coplanar case; forces away from the joint designate tension, those toward the joint designate compression. Assume the unknown forces F_1, F_2, and F_3 are tensile axial forces in the bars. Equate to zero the algebraic sum of all projections of the forces at A parallel to each orthogonal axis.

$$\rightarrow \Sigma F_x = 0: \quad -F_1 - \frac{12}{13}F_2 - \frac{12}{13}F_3 = 0$$

$$\uparrow \Sigma F_y = 0: \quad -P + \frac{3}{13}F_2 = 0$$

(where the component of F_2 is 5/13 F_2 x 3/5 = 3/13 F_2).

$$\swarrow \Sigma F_z = 0: \quad \frac{5}{13}F_3 - \frac{4}{13}F_2 = 0$$

Solving the set of simultaneous equations yields:

$$F_1 = 7.2P, \quad F_2 = 4.33P, \quad F_3 = 3.47P$$

This problem can also be solved by using two moment equations of equilibrium.

$\Sigma M_x = 0$ (This equation is trivial since all forces intersect the X axis).

$$\Sigma M_y = 0: +\frac{4}{13}F_2 \cdot 12 - \frac{5}{13}F_3 \cdot 12 = 0$$

$$\Sigma M_z = 0: -P \cdot 12 + \frac{3}{13}F_2 \cdot 12 = 0$$

Solve these equations for F_2 and F_3 then use $\Sigma F_x = 0$ to determine F_1

For a body to be in equilibrium, it must neither translate nor rotate. Therefore, the equations of equilibrium of a system of forces in space must be based upon the fact that the resultant of all forces and the resultant of all moments vanish.

$\Sigma F_X = 0$ $\qquad$ $\Sigma M_X = 0$

$\Sigma F_Y = 0$ $\qquad$ $\Sigma M_Y = 0$

$\Sigma F_Z = 0$ $\qquad$ $\Sigma M_Z = 0$

Six independent equations are necessary and sufficient to maintain equilibrium in space.

<u>Illustrated Problem</u>

A derrick, shown in Fig. 2.77 lifts a load of 4000 lbs. The mast is supported at point G in a ball and socket joint. The boom, member BC is perpendicular to the mast and lies in a plane at 30° to the plane XY. Guy wires AD and AE are fixed by ball joints. Determine the force in the guy wires and the reaction at G.

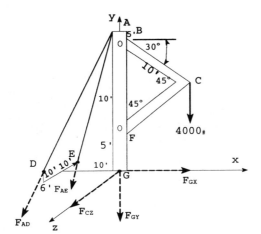

Figure 2.77

Solution

Since members AD and AE are pin jointed at their ends they can sustain only axial forces. Each axial force, however, will have three unknown components. Further, since the directions of the forces in members AD and AE are known from the geometry of the structure, the direction cosines of these forces can be determined. This implies that when any one component at D is known, the others can be solved through geometrical relations.

The mast, AG, is a three force member that will have three unknown reaction components at G. Now draw all unknown reactions, assuming for the moment that they are tensile and act away from the joint. (The unknown reactions are shown dotted in Fig. 2.77. There are five unknown quantities to be determined: F_{AD}, F_{AE}, F_{GX}, F_{GY} and F_{GZ}).

By inspection of the overall structure and the reactions, note that a summation of moments about the X axis taken at joint A eleminates the forces F_{AD}, F_{AE} and F_{GY} because they intersect joint A; force F_{GX} is eliminated because it is parallel to the X axis. Recall the rules following Fig. 2.75.

$$\sum M_{AX} = 0; \ -F_{GZ}(20) + 4000\,(10 \sin 30) = 0$$

$$F_{GZ} = \frac{20{,}000}{20} = 1000 \text{ lbs}$$

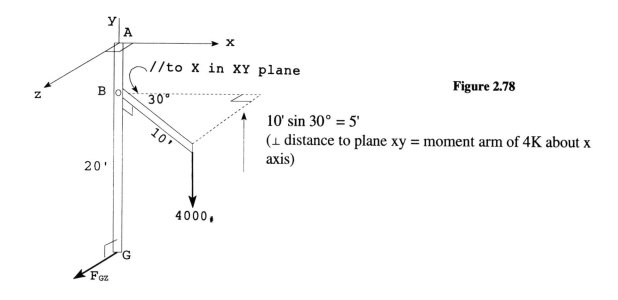

Figure 2.78

10' sin 30° = 5'
(⊥ distance to plane xy = moment arm of 4K about x axis)

Similarly $\circlearrowright M_{AZ} = 0$; $F_{GX}(20) - 4000(10 \cos 30) = 0$

$$\therefore F_{GX} = 1732 \text{ lbs}$$

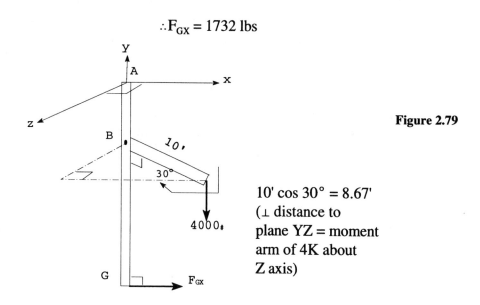

Figure 2.79

10' cos 30° = 8.67'
(⊥ distance to plane YZ = moment arm of 4K about Z axis)

Now take $\circlearrowright M_{DY} = 0$; $-F_{GZ}(10) + F_{EX}(20) - F_{GX}(10) = 0$

$$F_{EX} = \frac{1000(10) + 1732(10)}{20} = 1366 \text{ lbs}$$

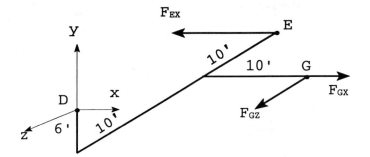

Figure 2.80

Note

The 4000 lb load is parallel to the y-axis, hence it does not produce any moment about the y-axis.

Using $\rightarrow \Sigma F_x = 0$, solve for F_{DX} which is assumed acting in the positive X direction.

$$-F_{EX} + F_{GX} + F_{DX} = 0$$

$F_{DX} = 1366 - 1732 = -366$ lbs or 366 lbs in the negative X direction.

To find the force in AE we must use the geometry of the derrick and the force F_{EX}. From Fig. 2.81, note that the length $EG = \sqrt{10^2 + 10^2}$ then the length of AE is:

$$AE = \sqrt{10^2 + 10^2 + 20^2}$$

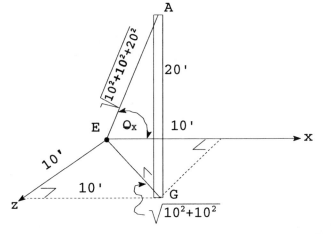

Figure 2.81

The direction cosine of AE with respect to the x-axis is:

$$\cos \theta_x = \frac{10}{\sqrt{10^2 + 10^2 + 20^2}}$$

Now $F_{EX} = F_{AE} \cos \theta_x$

$$F_{AE} = \frac{F_{EX}}{\cos \theta_x} = \frac{1366 \sqrt{10^2 + 10^2 + 20^2}}{10} = \frac{1366 \times 24.5}{10} = 3346 \text{ lb}$$

F_{AD} is found in a like manner. Using Fig. 2.82 the length of AD is found as follows:

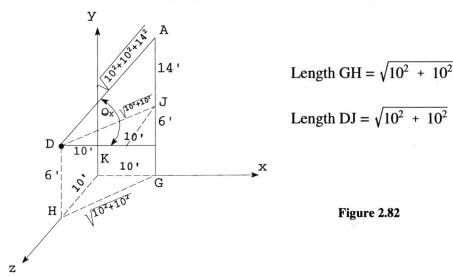

Length GH $= \sqrt{10^2 + 10^2}$

Length DJ $= \sqrt{10^2 + 10^2}$

Figure 2.82

Length AD $= \sqrt{10^2 + 10^2 + 14^2}$

Direction cosine of AD with respect to the x-axis is:

$$\cos \theta_x = \frac{10}{\sqrt{10^2 + 10^2 + 14^2}}$$

Therefore $F_{DX} = F_{AD} \cos \theta_x$ or $F_{AD} = \dfrac{F_{DX}}{\cos \theta_x}$

$$F_{AD} = \frac{366 \sqrt{10^2 + 10^2 + 14^2}}{10}$$

$$F_{AD} = \frac{366 \times 19.9}{10} = 728 \text{ lbs tension}$$

Length JK is parallel to the z-axis

Length DK = 10 ft

Friction

Friction is defined as contact resistance exerted by one body upon another when one body moves, or tends to move, relative to the other. Friction is always a retarding force opposing motion or the tendency to move. Friction arises by virtue of the roughness of the contact surfaces. It can be both an asset and a liability.

Consider a block of weight W resting on a clean, dry horizontal surface and subjected to a horizontal force P as in Fig. 2.83.

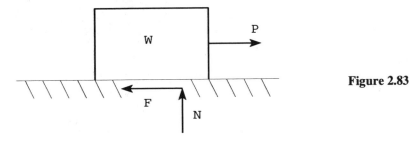

Figure 2.83

As P increases, so does the friction force F, up to the moment where motion occurs. Then the friction force suddenly decreases approaching a kinetic value which remains fairly constant. See Fig. 2.84.

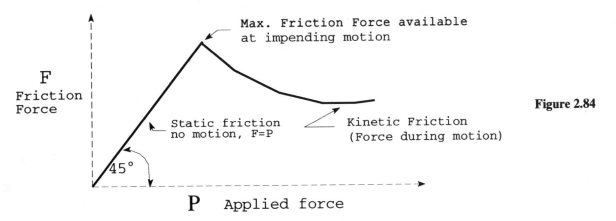

Figure 2.84

The maximum frictional resistance is proportional to the normal force N according to $F = \mu N$ where μ is the coefficient of friction. This expression is meaningful only when motion is impending, i.e. when F is maximum.

At the point of impending motion the static coefficient of friction μ_s, must be used in the above equation. When motion occurs, the kinetic coefficient of friction, μ_k, must be used.

The value of μ varies with the materials in contact, whether there is static friction or kinetic friction and on the condition of the surfaces in contact.

The total reaction to the block resting on the surface, is R, as shown in Fig. 2.85. R is the resultant of the friction force F and the normal force N.

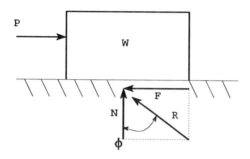

Figure 2.85

The angle between R and N is a function of the magnitude of F. As F increases, so does the angle. When the maximum friction force is developed (point of impending motion), the value of this angle, ϕ is called the angle of friction. From Fig. 2.85:

$$\tan \phi = \frac{F}{N} \qquad \tan \phi = \mu$$

For coplanar forces, the static reaction must lie within the angle BAC in Fig. 2.86 in order for no motion to occur.

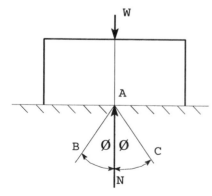

Figure 2.86

In the three-dimensional case, the reaction must lie within the cone generated by rotating AB about N, in order for no motion to occur. The cone is called the <u>cone of friction</u>.

Summary of the Laws of Friction

1. <u>Static friction</u> is the tangential force between two bodies which opposes the sliding of one body relative to the other.

2. <u>Limiting friction</u> is the maximum value of static friction which occurs when motion is pending.

3. <u>Kinetic friction</u> is the frictional force between two bodies after relative motion begins.

4. <u>Angle of friction</u> is the angle between the total reaction of one body on another and the normal to the common tangent when motion is impending.

5. <u>Coefficient of static friction</u> is the ratio of the limiting friction force to the normal force.
   $\mu_s = F/N$

6. <u>Coefficient of kinetic friction</u> is the ratio of the kinetic friction force to the normal force.
   $\mu_k = F'/N$

7. The coefficient of friction is independent of the normal force, however, the friction force is proportional to the normal force.

8. The coefficient of friction is independent of the area of contact.

9. The coefficient of kinetic friction is always less than that of static friction.

10. At low speeds kinetic friction is independent of the speed. At higher speeds generally there is a decrease in friction.

11. The static friction force is never greater than that required to hold the body in equilibrium.

Illustrated Problems

1. A 100 lb block is to be pushed up a 30° inclined plane by pushing horizontally (Fig. 2.87). If the coefficient of friction is 0.20, what is the force required to produce impending motion up the plane?

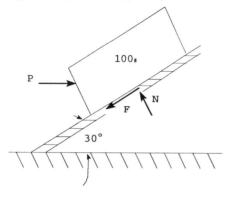

Figure 2.87 (a)

Approach

Draw a free body diagram of the block, showing all forces that it feels. Use the equations of equilibrium and the friction equation $F = \mu N$, to solve for the unknown forces. Select the x-axis parallel to and positive in the direction of impending motion.

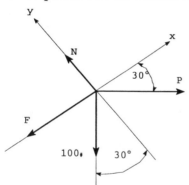

Figure 2.87 (b)

$\nwarrow \Sigma F_y = 0$: $N - 100 \cos 30 - P \sin 30 = 0$

$$N = 86.6 + \frac{P}{2}$$

$F = \mu N$: $F = 0.2 \left(86.6 + \frac{P}{2}\right) = 17.32 + 0.1 P$

↛ $\Sigma F_x = 0$: P cos 30 - F - 100 sin 30 = 0

0.866 P - (17.32 + 0.1 P) - 50 = 0

$$\therefore P = \frac{50 + 17.32}{0.766} = 88.0 \text{ lbs}$$

2. A person interested in pulling the block of problem 1 up the incline naturally wants to expend the least amount of force. Determine the least value of P and its direction, necessary to cause motion up the incline.

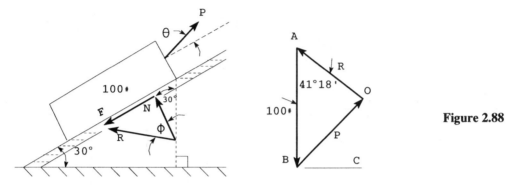

Figure 2.88

Approach

Draw a free body diagram of the block. Since motion is impending up the plane, the friction force F is directed down the plane. Use the equation tan ∅ = f to find the direction of the resultant reaction. Using the tip-to-tail method of vector addition, draw the polygon of forces acting on the block and solve for P.

$\tan \phi = \mu = 0.2$ $\phi = 11° 18'$

The angle of R with respect to the vertical is:

30° + (11° 18') = 41° 18'

The force polygon must close because equilibrium exists. Starting at A in Fig. 2.88b, draw the weight of the block (100 lbs) parallel to its line of action, then draw the total reaction R at 41° 18' to the vertical intersecting A. The force that closes the polygon is P. It should be perpendicular to the vector R, in order for it to be a minimum.

P = 100 sin (41° 18') = 66 lbs

Angle ABO = 90 - (41° 18')

∴ Angle OBC = 41° 18'

Hence P makes 11° 18' with the incline.

NOTE THIS ANGLE IS EXACTLY THE ANGLE OF FRICTION.

3. Two blocks separated by a rigid link (pivoted at the attachments) are to be pushed to the right. What is the force P required at impending motion? f = 0.2

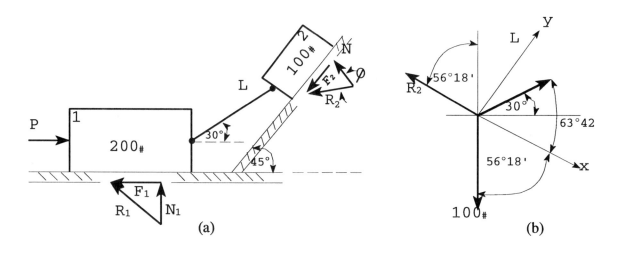

Figure 2.89

Approach

Draw FBD of each block, use the equations of equilibrium and the friction equation to solve for the unknowns.

The direction of R_2 can be found as follows:

$\tan \emptyset = 0.2$ $\quad \emptyset = 11° 18'$

Therefore R_2 is (11° 18') + 45° = 56° 18' from the vertical.

Solving for L can best be accomplished by selecting the x-axis co-linear with R_2 (Figure 2.55 b) and equating to zero all forces perpendicular to X.

$\nearrow \Sigma F_y = 0$: L sin 63° 42′ − 100 sin 56° 18′ = 0

$\qquad\qquad$ L = 93 lbs.

Now consider the 200-lb block (Fig. 2.90). It is more convenient here to work with the components of R_1.

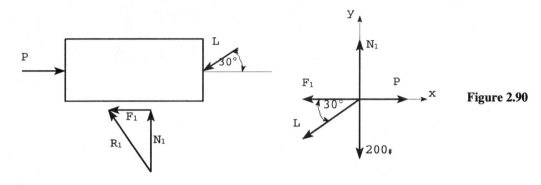

Figure 2.90

$\uparrow \Sigma F_y = 0$: N_1 − 200 − L sin 30° = 0

$\qquad\qquad N_1 = 246.5$ lbs.

Since motion is impending $F_1 = \mu N_1 = 0.2 \times 246.5 = 49.3$ lbs

Then $\rightarrow \Sigma F_x = 0$ yields: $-F_1 - L \cos 30° + P = 0$

$\qquad\qquad \therefore P = 129.8$ lbs

4. A man weighing 200 lbs walks across a plank from left to right. The plank is 10 ft long, weighs 100 lbs, and rests on the rough side walls of a gulley as shown in Fig. 2.91. The plank is at 15° to the horizontal. Assume the angle of friction is 15°. At what point will the man be when the plank begins to slip?

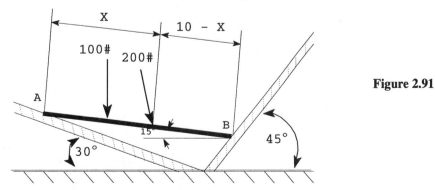

Figure 2.91

When the motion of the plank is impending assume that point B moves down the incline. This identifies the direction of the friction force.

Draw the force polygon of the loads on the plank.

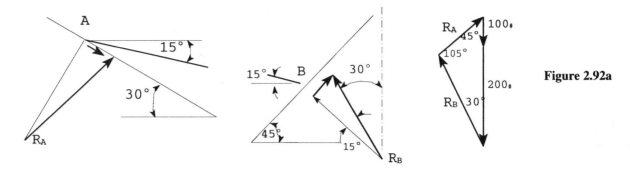

Figure 2.92a

Using the law of Sines

$$\frac{R_B}{\sin 45} = \frac{300}{\sin 105} \qquad R_B = \frac{300 \times 0.707}{\sin(90-15)} = 219.5 \text{ lbs}$$

$\circlearrowright \Sigma M_A = 0$ will locate the man.

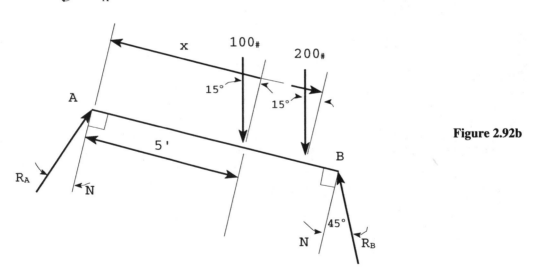

Figure 2.92b

$\circlearrowright \Sigma M_A \quad R_B \cos 45° \times 10' - (200 \cos 15°)X - (100 \cos 15°) 5' = 0$

$$X = \frac{219.5 \times 0.707 \times 10 - 100 \times 0.966 \times 5}{200 \times 0.966} = 5.5 \text{ ft. to the right of A.}$$

STATICS 2-63

Wedges

All of the foregoing laws of friction apply to wedges at the moment of impending motion.

5. The 20 degree wedge shown in Fig. 2.93 is used to raise a 10 kip load. Assume the angle of friction to be 15° for both the wedge and the block. What force P is required to start moving the wedge? Neglect all inertial effects.

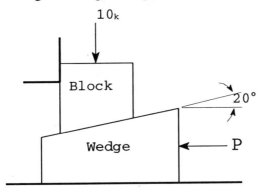

Figure 2.93

Approach

Draw the free body diagram of the wedge and the block, showing the applied forces and the friction forces opposing motion. Use the equations of equilibrium to solve for the unknown forces.

The free body diagram and the point diagram of forces for the wedge are shown in Fig. 2.94.

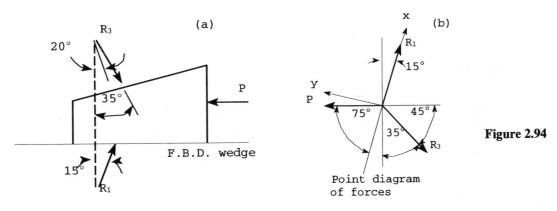

Figure 2.94

Note from the point diagram of forces on the wedge that there are three unknowns. These cannot be solved until R_3, which is common to both the wedge and the block, is determined from the FBD of the block.

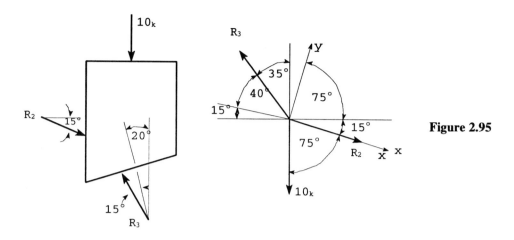

Figure 2.95

$\nearrow \Sigma F_y = 0$ yields R_3

$R_3 \sin 40° - 10^k \sin 75° = 0$

$\therefore R_3 = \dfrac{10^k \times 0.966}{0.643} = 15^k$

R_2 need not be solved since the equations of equilibrium applied to the wedge will yield P.

From Fig. 2.94 select the x-axis colinear with R_1.

$\nwarrow \Sigma F_y = 0$: $P \sin 75° - R_3 \sin 50° = 0$

$\therefore P = \dfrac{15^k \times 0.766}{0.966} = 11.92^k$

Self-Locking Wedges

In Fig. 2.93 if the wedge angle was great enough, the wedge could be forced from under the block by the 10^k load. The wedge angle that produces R_3 on the wedge co-linear with R_1 assures that the wedge is self-locking.

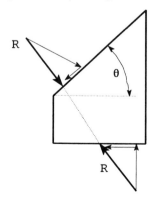

Figure 2.96

Screw Thread Friction

Power screws, such as screw jacks, generally use a square, acme, or buttress thread. A square threaded screw is basically an inclined plane wrapped around a cylinder in the form of a helix. Modifications of the square thread are the Acme and Buttress Thread.

Consider an inclined plane, ABC, wrapped around a cylinder. The student can demonstrate this by folding a piece of paper into a triangle and curling it into a cylinder.

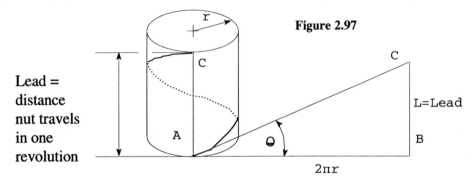

Figure 2.97

The base length of the equivalent inclined plane is $2\pi r$, where r is the mean radius of the thread. The pitch angle of the equivalent inclined plane is defined by:

$$\tan \theta = \frac{L}{2\pi r}$$

Tremendous mechanical advantages can be obtained by inclined planes, which is why practical applications of this basic machine are found in screw jacks and power screws. Generally, the force analysis of the square thread can be accomplished with sufficient accuracy by considering the load to be acting on one thread, even though the nut may span several threads.

A simplified screw jack, shown in Fig. 2.98 is to be used to raise a load W by applying force P to the handle.

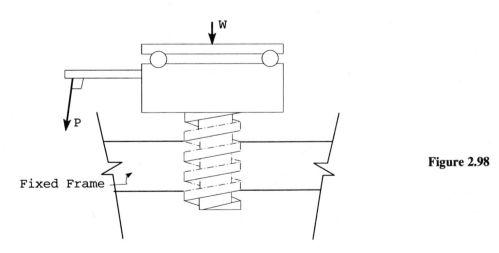

Figure 2.98

The forces on the thread (inclined plane) can be visualized as shown in Fig. 2.99.

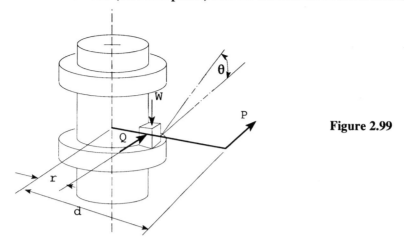

Figure 2.99

Raising the load implies impending motion UP the inclined plane. Lowering the load implies motion impending DOWN the incline. A free body diagram of the loads acting on the inclined plane appears in Fig. 2.100.

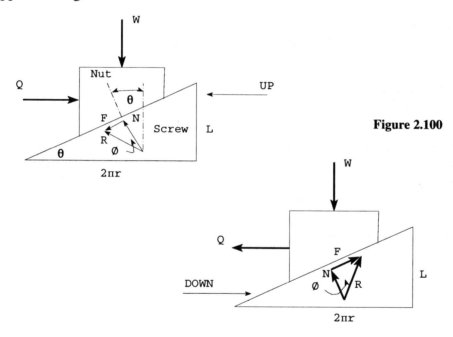

Figure 2.100

From the force polygon for raising the load (Fig. 2.101), the force Q can be expressed in terms of W.

$$\tan(\emptyset + \theta) = \frac{Q}{W}$$

$Q_{up} = W \tan(\emptyset + \theta)$

$Q_{down} = W \tan(\emptyset - \theta)$

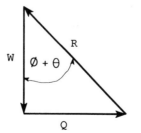 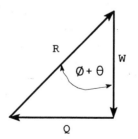

Figure 2.101

For the thread to be self locking the angle of friction, $\emptyset$, must be greater than the pitch angle, θ.

Now using $\circlearrowleft$ ΣM axis of screw = 0 yields $P \times d - Q \times r = 0$

$$P = \frac{Qr}{d} = \frac{W\, r \tan \emptyset \pm \theta}{d}$$

(UP / DOWN)

Belt Friction

A belt or band passing over a rough pulley often is used to transmit power or is used as a brake. In either machine, the principles of friction previously developed apply equally well.

Figure 2.102 represents a pulley being driven by a belt where the tension T_t on the tight side exceeds the tension T_s on the slack side. The belt has a wrap angle of β which defines the area of contact on which the friction force acts. The difference in the tensions is caused by the friction force F.

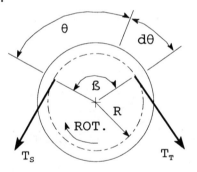

Figure 2.102

Neglecting the thickness of the belt, examine the equilibrium of the portion of the belt of length $d\theta$, as shown in Fig. 2.103.

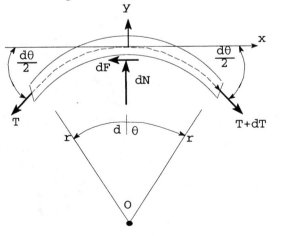

Figure 2.103

$\circlearrowright \Sigma M_0 = 0 \quad -Tr - (dF)r + (T + dT)r = 0 \quad dF = dT$

$\uparrow \Sigma F_y = 0 \quad -T \sin \frac{(d\theta)}{2} + dN - (T + dT) \sin \frac{(d\theta)}{2} = 0$

Simplify this expression by neglecting products of the differentials and let $\sin d\theta = d\theta$ since we are dealing with small angles. Then $d\theta$ must be expressed in radians.

$$dN = T \sin \frac{(d\theta)}{2} + T \sin \frac{(d\theta)}{2} + \sin \frac{(d\theta)}{2} dt$$

$$= 2T \frac{(d\theta)}{2} = T\, d\theta$$

At the instant of slipping $F = \mu N \quad dF = \mu\, dN = \mu T\, d\theta$

Solving for dT from the above equation, where $dF = dT$

$$\therefore dT = \mu T\, d\theta$$

This differential equation can be solved by separating variables and integrating between limits.

$$\int_0^\beta \frac{dT}{T} = \mu \int_0^\beta d\theta = \mu\beta$$

Note β must be expressed in radians. This belt friction equation can be solved with the use of common logarithms.

$$\log T_t - \log T_s = \mu\beta\,[\log_{10} e] = 0.434\,\mu\beta$$

Illustrated Example

A differential band brake is used to measure the torque output of an engine by attaching a drum to its output shaft. The device is shown in Fig. 2.104a. Find the torque M when a force P of 100 lbs is applied to the brake handle. Use the coefficient of kinetic friction $f_k = 0.1$.

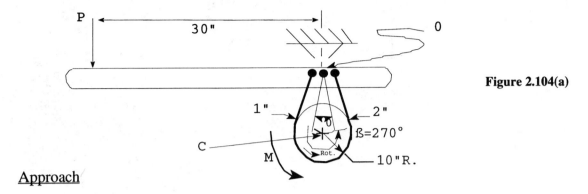

Figure 2.104(a)

Approach

Take a summation of moments about the fulcrum of the brake, 0, to derive an expression in terms of T_t and T_s, then use the belt friction equation as a second simultaneous equation to solve for the tensions. Use the values of T in a moment equation about the center of the drum to solve for M.

For the brake handle:

$$\circlearrowright \Sigma M_0 = 0 \text{ yields } - T_t \times 1 \; T_s \times 2 - P \times 30 = 0$$

$$2T_s - T_t = 3000 \text{ lbs}$$

Using the belt friction equation

$$\log \frac{T_t}{T_s} = 0.434 \, f \, \beta \quad \text{where } \beta = 270° = \frac{3}{2}\pi \text{ radians}$$

$$\log \frac{T_t}{T_s} = 0.434 \times 0.1 \times \frac{3}{2}\pi = 0.2045$$

Hence $\frac{T_t}{T_s} = 1.60$ and $2T_s - 1.60T_s = 3000$

$$T_s = 7500 \text{ lbs}$$

$$T_t = 12{,}000 \text{ lbs}$$

$\circlearrowright \Sigma M_c = 0$

$-M + T_t \times 10 - T_s \times 10 = 0$

$M = [12000 - 7500] \, 10 = 45000 \text{ in-lbs}$

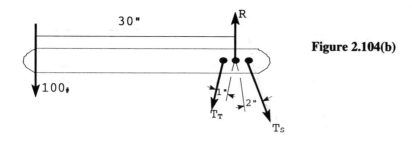

Figure 2.104(b)

Solved Problems

1. What angle does P make with the horizontal for the system shown, in order for it to have a minimum value? Assume that the pulley is frictionless and the coefficient of static friction is 0.2 for all surfaces.

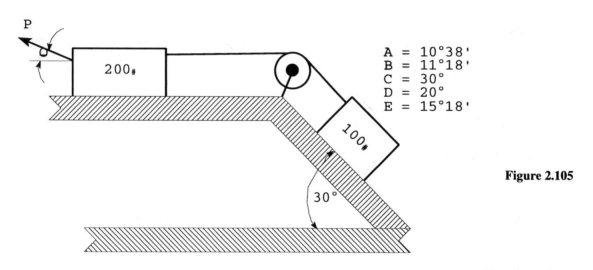

A = 10°38'
B = 11°18'
C = 30°
D = 20°
E = 15°18'

Figure 2.105

Approach

Draw a free body diagram of the 200 lb. block, take summation of forces normal to the frictional resultant.

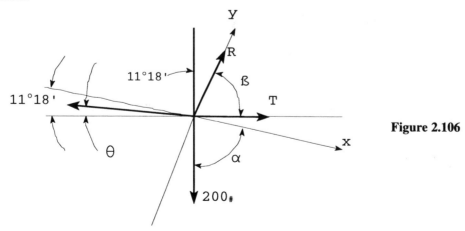

Figure 2.106

$\rightarrow \Sigma F_x = 0$; $- P \cos (11° \ 18' - [\theta]) + T \cos\beta + 200 \cos\alpha = 0$

$$P = \frac{T \cos \beta + 200 \cos \alpha}{\cos (11° \ 18' - [\theta])}$$

which is a mimimum when $\theta = 11° \ 18'$ (denominator is maximized)

Answer is 11° 18', (B)

2. What is the least value of P in problem 1 required to cause impending motion to the left?

(A) 115 lbs   (B) 140 lbs

(C) 125 lbs   (D) 150 lbs

(E) None of these

Approach

From summation of forces acting on the 100 lb block, find T. In diagram above, $\beta = 11° \ 18'$. $\alpha = 78° \ 42'$. The equation for P in problem 1 can be solved.

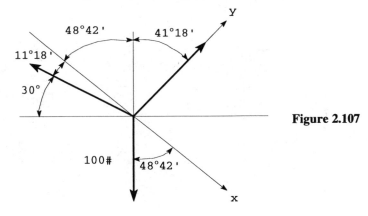

Figure 2.107

R makes an angle 30° with the vertical. $- T \cos (11° \ 18')$

$+ 100 \cos (48° \ 42') = 0$ $T = \dfrac{66}{.98} = 68$ lbs

Now using the diagram in problem 1, where $\theta = 11° \ 18'$,

$- P + 68 \cos (11° \ 18') + 200 \cos (78° \ 42') = 0$

$P = 68 \times 0.98 + 200 \times 0.196 = 66.6 + 39.2 = 105.8$ lbs

Answer is (E), none of the given answers.

3. A drum of water weighing 3000 lbs rests on an incline, as shown. A crow bar is used to start rolling the drum up the hill. If the coefficient of friction is 0.25 for all contact surfaces, what force P is required to start the drum up the hill?

Approach

At the point of impending motion assume that the drum is about to rotate and roll uphill. Draw a FBD of the drum, solve for normal force acting on the crow bar. Draw a FBD of the crow bar, take moments about pivot, and solve for P.

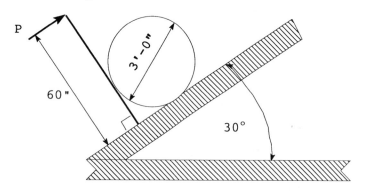

Figure 2.108

$\nearrow \Sigma F_x = 0$: $N_1 - F_2 - 3000 \sin 30 = 0$

$\nwarrow \Sigma F_y = 0$: $-F_1 + N_2 - 3000 \cos 30 = 0$

$$F_1 = \mu N_1, \quad F_2 = \mu N_2$$

Then $N_1 - \mu N_2 - 1500 = 0$

$-\mu N_1 + N_2 - 2600 = 0$

$N_1 = 1500 + 0.25 N_2 - 0.25 (1500 + 0.25 N_2) + N_2 = 2600$

$-0.0625 N_2 + N_2 = 2600 + 375$

$N_2 = \dfrac{2975}{0.9375} = 3180$ lbs

$N_1 = 1500 + 0.25 (3180) = 2295$ lbs

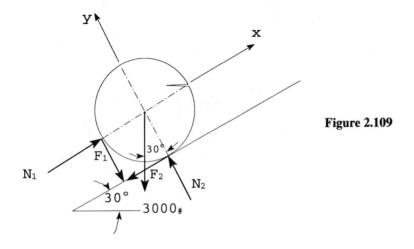

Figure 2.109

$\circlearrowright \Sigma M_R = 0 \quad P \times 60 - N_1 \times 18 = 0$

$$P = \frac{18\,N_1}{60} = \frac{18 \times 2295}{60} = 688.5 \text{ lbs}$$

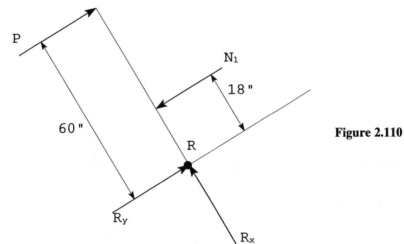

Figure 2.110

4. A single thread screw with square thread has 2.5 threads per inch. The root diameter is 2.6 in. and the O.D. is 3.0 in. Assuming a coefficient of friction of 0.1, determine the turning moment necessary to start lifting an axial load of 20 tons. What is the moment required to lower the load?

Approach

Using data of screw, determine lead and helix angle. Solve formula for screw jack.

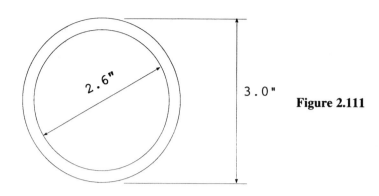

Figure 2.111

Mean diameter = $\frac{3 + 2.6}{2}$ = 2.8 in. r_{mean} = 1.4 in.

Pitch = lead for single thread = distance from a point on one thread to a corresponding point on adjacent thread.

Lead = L = $\frac{1}{\text{no. of threads per in.}}$ = $\frac{1}{2.5}$ = 0.4 in. thd. or in. rev.

$\tan \theta = \frac{L}{2 \pi r_m} = \frac{0.4}{2 \pi \times 1.4} = 0.0455$ $\theta = 2° - 36'$

$\tan \theta = 0.1$ $\phi = 5° - 43'$

Moment required to raise load = M_R = Wr tan($\phi + \theta$)

= 40,000 x 1.4 x tan (5° 43' + 2° 36')

= 40,000 x 1.4 x 0.146 = 8180 lbs in. = 682 lb. ft.

Moment required to lower load = M_L = W r tan ($\phi + \theta$)

= 40,000 x 1.4 x tan 3° 07' = 3040 lb. in. = 253.3 lb. ft.

5. A boat hauling on a hauser uses a capstan on the dock to moor. The tension in the hauser is 3000 lbs If the coefficient of friction is 0.25, how many turns must the hauser make around the capstan so that the pull on the capstan end of the hauser does not exceed 35 lbs?

Approach

Draw a FBD of capstan, then use belt friction equation to solve for (angle of wrap around capstan).

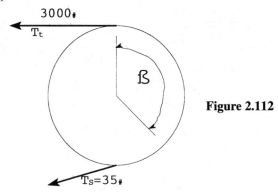

Figure 2.112

$\mu = 0.25$ then $\log T_t - \log T_s = 0.434\, \mu\beta$

$$\beta = \frac{\log 3000 - \log 35}{0.434 \times 0.25} = \frac{3.4771 - 1.5441}{0.1082} = 17.85 \text{ radians}$$

$$= 17.85 \times \frac{1 \text{ rev}}{2\pi \text{ rad.}}$$

$$= 2.85 \text{ revolutions or turns}$$

6. A man lowers a bucket of cement down the side of a building by wrapping a rope around a drum, as shown. Using 1.25 turns around the drum and exerting 100 lbs, what weight of cement can be lowered? Use a coefficient of friction of 0.3.

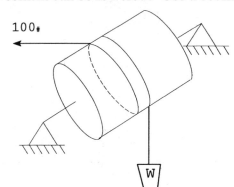

Figure 2.113

2-76   FUNDAMENTALS OF ENGINEERING EXAM REVIEW WORKBOOK

Approach

Draw a FBD of drum and use same method as problem 5. (See diagram next page).

$$f = 0.3 \therefore \log T_t - \log T_s = 0.434 \times 0.3 \times 1.25 \text{ rev.} \times \frac{2\pi \text{ rad.}}{\text{rev.}}$$

$$\log T_t = 1.023 + 2.000 = 3.023$$

$$\therefore T_t = 1055 \text{ lbs}$$

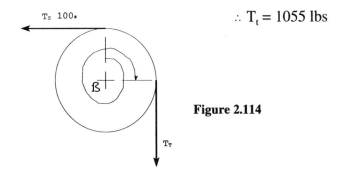

Figure 2.114

7. The tripod shown has hinged joint at A, B, C, and D. Determine the forces in each leg when 600 lbs is applied at A parallel to the x-axis.

Point A is on the y-axis

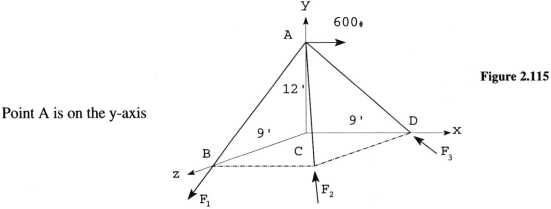

Figure 2.115

Approach

Assume forces exist in each member and indicate their direction on the diagram. Resolve these assumed forces into their components along the coordinate axes, use the equations of equilibrium.

Length of required members are shown on the sketch.

Length $AC = \sqrt{12^2 + (9\sqrt{2})^2} = \sqrt{306} = 17.5'$

$\rightarrow \Sigma F_x = 0$

$$600 - F_3 \times \frac{9}{15} - F_2 \times \frac{9\sqrt{2}}{17.5} \times \frac{9}{9\sqrt{2}} = 0$$

$\uparrow \Sigma F_y = 0$

$$F_3 \times \frac{12}{15} + F_2 \times \frac{12}{17.5} - F_1 \times \frac{12}{15} = 0$$

$\swarrow \Sigma F_z = 0$

$$F_1 \times \frac{9}{15} - F_2 \times \frac{9\sqrt{2}}{17.5} \times \frac{9}{9\sqrt{2}} = 0$$

From ΣF_z $F_1 = F_2 \times \frac{9}{17.5} \times \frac{15}{9} = F_2 \times \frac{15}{17.5}$

From ΣF_x $F_3 = \frac{15}{9}(-F_2 \times \frac{9}{17.5} + 600) = -F_2 \times \frac{15}{17.5} + \frac{600 \times 15}{9}$

Substituting in F_y:

$$\frac{12}{15}(-F_2 \times \frac{15}{17.5} + \frac{600 \times 15}{9}) + F_2 \times \frac{12}{17.5} - \frac{12}{15}(F_2 \times \frac{15}{17.5}) = 0$$

$$-F_2 \times \frac{12}{17.5} + F_2 \times \frac{12}{17.5} - F_2 \times \frac{12}{17.5} = -\frac{600 \times 15}{9} \times \frac{12}{15}$$

$$-F_2 \times \frac{12}{17.5} = -\frac{600 \times 12}{9} \qquad F_2 = 1168 \text{ lbs}$$

$$F_1 = \frac{15}{17.5}(1168) = 1000 \text{ lbs}$$

$$F_3 = -1168 \times \frac{15}{17.5} + \frac{600 \times 15}{9} = 0$$

8. Determine the force P necessary to drive the two blocks A apart. Assume the angle of friction for all contact surfaces is 10°.

<u>Approach</u>

Draw a FBD of one block and the wedge, showing all forces experienced by the members. Draw the polygon of forces for each FBD, use equations of equilibrium to solve the unknowns.

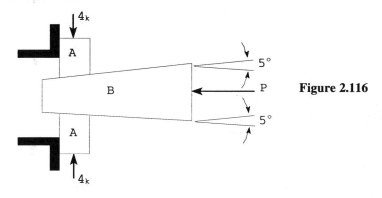

Figure 2.116

$\Sigma F_y = 0 \quad 4 \cos 10 - R_B \cos 25 = 0$

$\therefore R_B = \dfrac{4 \times .9848}{.9063} = 4.34^k$

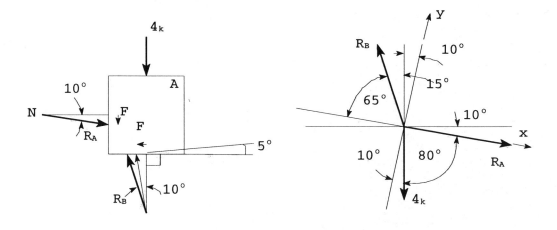

Figure 2.117

$\Sigma F_x = 0 \quad P - 2 R_B \sin 15 = 0$

$\therefore P = 2 \times 4.34 \times 0.2588 = 2.24^k$

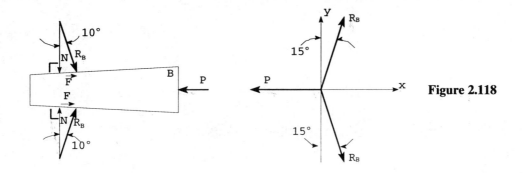

Figure 2.118

9. What is the break-free torque M when a 200 lb braking force is applied as shown?

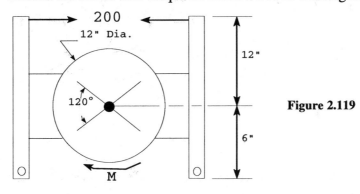

Figure 2.119

Approach

To break free implies impending motion. Take moments about the pivot of the brake lever to determine the normal force acting on the brake drum. Then take $M_{drum} = 0$ to solve for torque. If no coefficient of friction is given, assume a value.

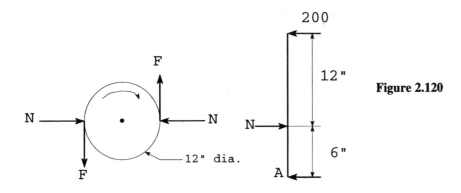

Figure 2.120

$\circlearrowright \Sigma M_A = 0$ N x 6 - 200 x 18 = 0   N = 600 lbs

F = f N   F = 0.20 x 600 = 120 lbs

$\circlearrowright \Sigma M_{drum} = 0$ 120 x 12 in. = 1440 in. lbs

10. Find the reactions at B, C, and D. Assume all joints are pinned.

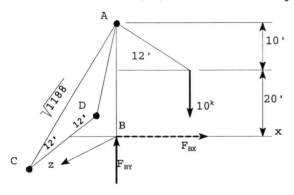

Figure 2.121

Approach

Indicate the true lengths of each member on the diagram. Draw the assumed forces in each member and then the assumed directions of the unknown reactions. Use equations of equilibrium to solve the unknowns.

$\circlearrowright \Sigma M_{CD} = 0$ $-F_{BY} \times 12 + 24(10) = 0$

$F_{BY} = 20^k$

$F_{AC} = F_{AD}$ From symmetry of loading and symmetry of structure.

$\uparrow \Sigma F_y = 0; -10^k - F_{AC} \times 2 \times \dfrac{30}{\sqrt{1188}} + 20 = 0$

$F_{AC} = F_{AD} = \dfrac{10 \times \sqrt{1188}}{2 \times 30} = 5.75^k$

11. Determine the force P required to move block B to the right. μ = 0.15

(A) 3 lbs (B) 6 lbs

(C) 9 lbs (D) 12 lbs

(E) none of these

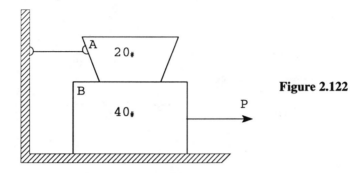

Figure 2.122

Approach

Draw a FBD of the lower block. Note that the normal force acting upward on the lower block is 60 lbs

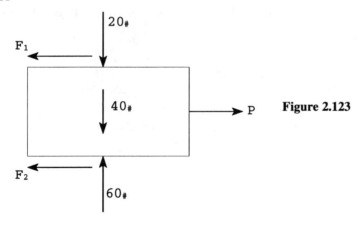

Figure 2.123

$\Sigma F_x = 0 \qquad -F_1 - F_2 + P = 0 \quad$ but

$F_1 = 0.15 \times 20 = 3$ lbs, $F_2 = 0.15 \times 60 = 9$ lbs

∴ $P = 12$ lbs

Answer is (D)

12. For an applied moment M = 2000 in. lbs what is the force P required to stop the drum? Assume the kinetic coefficient of friction is 0.2; neglect the inertial effect of the drum.

(A) 132 lbs   (B) 160 lbs

(C) 264 lbs   (D) 465 lbs

(E) None of these

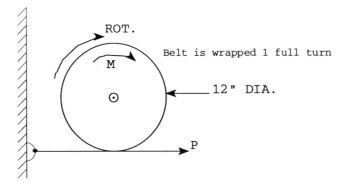

Figure 2.124

Approach

Use the belt friction equations with β = 2π radians. Find relationship between T_t and T_s, then take moments about the center of the drum.

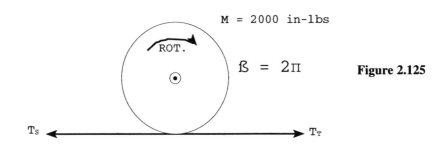

Figure 2.125

$$\log \frac{T_t}{T_s} = 0.434 \times 2\pi \times 0.20 = 0.545$$

Taking logs $\frac{T_t}{T_s} = 3.508$

$+\circlearrowright \Sigma M_c = 0 \quad 2000 - 6 T_t + 6 T_s = 0$

$$6\,T_t - \frac{6\,T_t}{3.508} = 2000$$

$$T_t\,(6 - 1.71) = 2000$$

$$\therefore T_t = P = 467 \text{ lbs}$$

13. What is the angle θ required to force the block B out from under A? Take $\mu = 0.1$ and assume the blocks to be weightless.

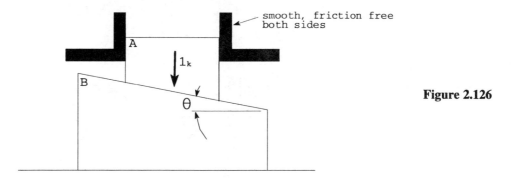

Figure 2.126

Approach

Draw a FBD of both blocks in order to visualize the forces experienced by the blocks. From the polygon of forces find the direction of forces.

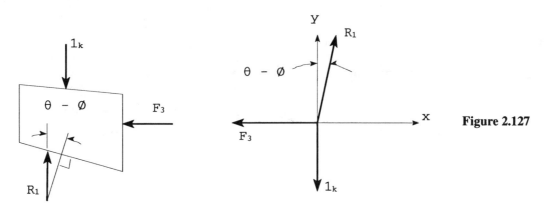

Figure 2.127

$\uparrow \Sigma F_y = 0 \quad \therefore R_1 \cos(\theta - \emptyset) - 1 = 0$

$$R_1 = \frac{1}{\cos(\theta - \emptyset)}$$

From polygon of forces, motion can be impending only when R_1 and R_2 are colinear and equal and opposite.

$\theta - \phi = 5° 43'$ $\theta = 5° 43' + 5° 43' = 11° 26'$

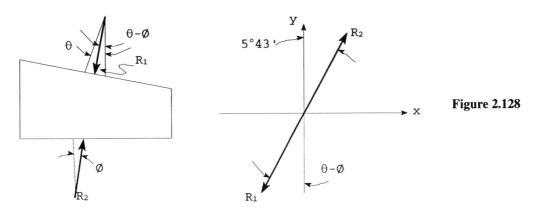

Figure 2.128

14. Determine the force in member AC.

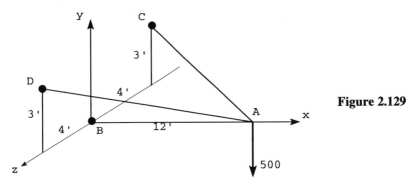

Figure 2.129

Approach

Use $\Sigma F_y = 0$ and symmetry of structure, $(F_{AC} = F_{AD})$ $F_{AC} = 1083$ lbs

15. For a right handed, square thread, screw jack, determine the force P required to raise a load of 5000 lbs. Use the data: $\mu = 0.20$, $\theta = 8° 42'$, $r = 1.0$ in., and $d = 30$ in. See Figs. 2.99 and 2.100 for definitions of data.

Approach

Use $P = \dfrac{W \times r \tan(\theta + \phi)}{d} = 60.5$ lbs

16. Two blocks, equal in weight, rest on bases of unequal area. Each block has the same force P applied to it, in the same direction. If the force P is gradually increasing from zero, which block will move first?

   (A) small base   (B) large base

(C) both at same time

(D) not enough information to conclude

The friction force is independent of the contact area, therefore the answer is (C).

17. Under what conditions in space is the moment about a given axis due to a force eliminated?

   (A) Force is parallel to axis

   (B) Force is skewed to axis

   (C) Force intersects axis

   (D) Force has an equal and opposite component equidistant on the other side of axis

   A force parallel to an axis cannot produce twisting of the axis; therefore the answer is (A).

18. Which force or moment applied to the wheel yields the maximum torque?

   (A) $M_1 = 120$ in. - lbs

   (B) $M_2 = 60$ in. - lbs

   (C) $M_3 = 140$ in. - lbs

   (D) $P = 12$ lbs

   P x r = 144 in. lbs; therefore the answer is (D)

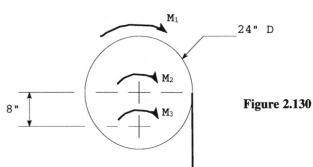

Figure 2.130

19. The block shown in Fig. 2.131 is to remain in equilibrium. If the coefficient of friction is 0.40 and the block weighs 80 lbs, what is the maximum value of the force P?

(A) 20 lbs

(B) 32 lbs

(C) 40 lbs

(D) 80 lbs

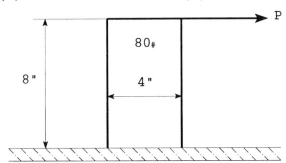

Figure 2.131

Approach

Draw a FBD of the block, and take moments about the overtuning point.

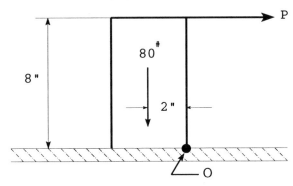

Figure 2.132

$\circlearrowleft \Sigma M_o = 0 \ -80 \times 2 + P \times 8 = 0$

$\qquad P = 20 \text{ lbs}$

Check to see if P exceeds the friction force, if it does the answer is the lesser of the two forces.

The answer is (A).

STATICS 2-87

20. Find the force in the member DH of this rigid truss. Assume that all joints are pinned and that all bays are stable.

(A) 2000 lbs   (B) 3000 lbs

(C) 4000 lbs   (D) 6000 lbs

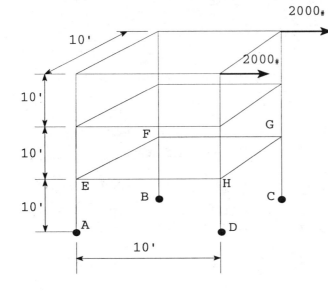

Figure 2.133

Approach

Take a summation of moments about axis AB. $R_D = R_C$

$$\circlearrowleft \Sigma M_{AB} = 0 - 4000 \times 30 + 2R_D \times 10 = 0$$

$R_D = 6000$ lbs

The answer is (D).

CENTROIDS

Consider the body in Fig. 2.134, suspended from point 1 by a flexible wire. The only forces acting on the body are the tension T in the wire and the weight W. Equilibrium can exist only when T and W are equal, opposite and colinear. That is to say the line of action of T must pass through the point where the resultant weight of the body acts. This point is called the center of gravity of the body.

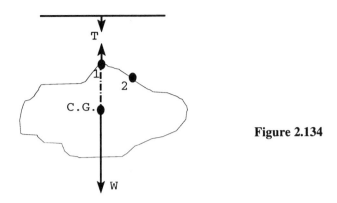

Figure 2.134

Should the body be suspended from point 2, the line of action of T would again pass through the center of gravity. Experimentally, this is how the CG can be determined.

Now consider the flat plate shown in Fig. 2.135. Divide the plate into small square elements, each having weight w_i. These elemental weights constitute a parallel system of forces whose resultant, W, is the sum of all the elemental weights w_i.

$$W = \Sigma w_i = w_1 + w_2 + w_3 + \ldots\ldots + w_n$$

THE MOMENT OF A RESULTANT FORCE ABOUT AN AXIS IS EQUAL TO THE SUM OF THE MOMENTS OF ITS CONSTITUENT FORCES ABOUT THAT AXIS.

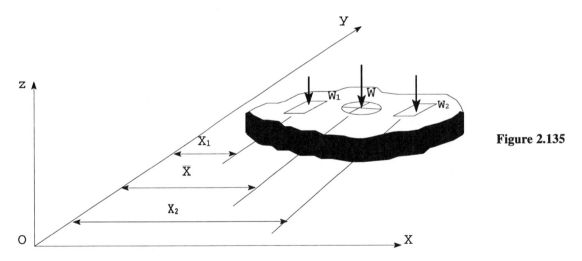

Figure 2.135

Select coordinate axes X and Y and let the coordinates for w_1 be (X_1, Y_1); the coordinates for w_2 be (X_2, Y_2) and so on. Let the coordinates of the total weight, W, acting at the center of gravity, be $(\overline{X}, \overline{Y})$.

Taking moments about the y-axis yields:

$$W \overline{X} = w_1 X_1 + w_2 X_2 + \ldots + w_n X_n = \sum_{i=1}^{i=n} w_i X_i$$

Similarly, Moments about the x axis yields:

$$W \overline{Y} = w_1 Y_1 + w_2 Y_2 + \ldots + w_n Y_n = \sum_{i=1}^{i=n} w_i Y_i$$

solving for the coordinates of the C.G.,

$$\overline{X} = \frac{\Sigma wx}{W} \qquad \overline{Y} = \frac{\Sigma wY}{W} \quad \text{where } W = \Sigma w.$$

Centroids of Areas

The term centroid, rather than center of gravity, is used when referring to areas, lines, and volumes because these figures do not have weight.

If the material of the plate in Fig. 2.135 is homogeneous, the total weight can be expressed as the thickness, t, times the total area, A, times the density ρ Therefore

$$W = t \rho A$$

and the moment of this force about the y axis is:

$$W(X)_{c.g.} = t \rho (X_{c.g.}) A \text{ where } X_{c.g.} = \overline{X}$$

Similarly, for an elemental area a_i, the moment about the y-axis is:

$$w_i X_i = t \rho X_i a_i$$

Equating the moment of the resultant force about an axis, to the sum of the moments of its elements, we have:

$$t \rho \overline{X} A = t \rho X_1 a_1 + t \rho X_2 a_2 + \ldots t \rho X_n a_n$$

Canceling the constants t and ρ we have:

$$A \overline{X} = \Sigma a_i X_i \quad \text{or} \quad \overline{X} = \frac{\Sigma aX}{A} \quad \text{where} \quad A = \Sigma a.$$

Similarly, for moments about the y-axis:

$$\overline{Y} = \frac{\Sigma aY}{A}$$

Centroids of Lines

A line may be assumed to be the axis of a slender filament of constant cross sectional area a. With a density of ρ the total weight of the length L of the filament is:

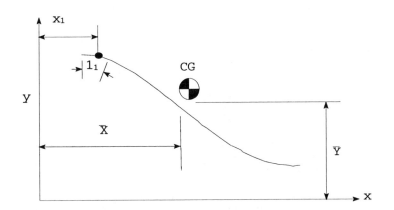

Figure 2.136

$$W = \rho\, a\, L$$

and by analogy with the above:

$$\rho\, a\, L\, \overline{X} = \rho\, a\, l_1 X_1 + \rho\, a\, l_2 X_2 + ... + \rho\, a\, l_n X_n$$

canceling constants yields;

$$L\, \overline{X} = l_1 X_1 + l_2 X_2 + ... + l_n X_n = \Sigma l\, X \quad \text{or} \quad \overline{X} = \frac{\Sigma l\, X}{L}$$

similarly, $\overline{Y} = \dfrac{\Sigma l\, Y}{L}$

Centroids by Integration

While the above formulas serve the purpose of illustrating the underlying theory used to determine the centroids, it is more convenient to determine the centroids by integration. For example, the length L in the formula for the centroid of a line is not easily determined, except by the use of calculus.

Replacing the elemental area in the above formulas by the differential dA, the equations for the centroid of an area become

$$A \overline{X} = \int X \, dA \quad \text{and} \quad A \overline{Y} = \int Y \, dA$$

similarly for the line,

$$L \overline{X} = \int X \, dL \quad \text{and} \quad L \overline{Y} = \int Y \, dL$$

Illustrated Example:

1. Determine the centroid of the arc of a circle, subtending an angle α that is symmetrical about the vertical axis and has a radius of r (Fig. 2.137).

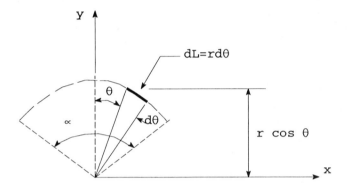

Figure 2.137

The element of arc, dL, is $dL = r \, d\theta$

and its distance from the x-axis is

$$Y = r \cos \theta \quad \text{where } \theta \text{ is measured from the y-axis}$$

$$\overline{Y} = \frac{\int Y \, dL}{L} = \frac{\int_{-\alpha/2}^{+\alpha/2} r \cos \theta \, (r \, d\theta)}{r\alpha} = \frac{r^2 \int_{-\alpha/2}^{+\alpha/2} \cos \theta \, d\theta}{r\alpha} = \frac{r}{\alpha} \sin \theta \Big]_{-\alpha/2}^{\alpha/2}$$

$$= \frac{r}{\alpha} \left[\sin \frac{\alpha}{2} - \left(-\sin \frac{\alpha}{2} \right) \right] = \frac{2r \sin \frac{\alpha}{2}}{\alpha}$$

Since the arc straddles the y-axis and is symmetrical about it, $\overline{x}$ must lie on the y-axis.

If the arc is a semi-circle, $\alpha = 180°$ and $\sin \frac{\alpha}{2} = 1$, then

$$\overline{Y} = \frac{2r(1)}{\pi} = \frac{2r}{\pi}$$

2. Find the centroid of a circular sector bounded by the arc of problem 1 and the radii (Fig. 2.138).

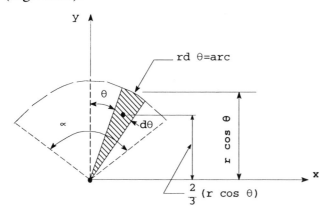

Figure 2.138

Use the expression $\overline{Y} = \frac{\int Y\, dA}{A}$, select as the element of area the shaded sector which can be approximately by a triangle.

$$dA = \frac{1}{2}(r)\, rd\theta \quad \text{and} \quad A = \frac{1}{2} r^2 \alpha$$

The distance of its centroid from the axis is found as follows. Recall that the centroid of a triangle is 2/3 the length of the altitude, measured from the vertex. Therefore, the distance from the origin in this problem is 2/3r, and its distance from the x-axis is:

$$Y = 2/3\, r \cos\theta$$

and

$$\overline{Y} = \frac{\int 2/3\, r \cos\theta\, [1/2\, r^2\, d\theta]}{1/2\, r^2\, \alpha}$$

$$\overline{Y} = \frac{2r^3}{3r^2 \alpha} \int \cos\theta\, d\theta = \frac{2r}{3\alpha} \sin\theta \Big]_{-\alpha/2}^{\alpha/2}$$

$$= \frac{2}{3}\frac{r}{\alpha}(\sin\frac{\alpha}{2} - \sin\frac{-\alpha}{2}) = \frac{2r}{3\alpha}(2\sin\frac{\alpha}{2}) \text{ and as in problem 1, } X = 0.$$

If the sector is a semi-circle, then $\alpha = 180° = \pi$, $\sin \alpha/2 = \sin 90 = 1$.

$$\overline{Y} = \frac{2r}{3\pi}(2) = \frac{4r}{3\pi} = 0.424r$$

3. Locate the centroid of the area bounded by the parabola $Y = kX^3$, the x-axis and the line $X = a$ (Fig. 2.139).

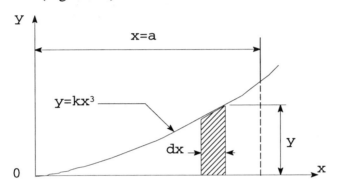

Figure 2.139

The element of area, dA (shaded), = Y dX

$$A_{total} = \int_0^a dA = \int_0^a Y\, dX = \int_0^a k\, X^3 dX = \frac{KX^4}{4}\Big|_0^a = \frac{Ka^4}{4}$$

To determine $\overline{X}$ use $A\overline{X} = \int X\, dA$

$$\frac{ka^4}{4}\overline{X} = \int_0^a X\, dA = \int_0^a XY\, dx = k\int_0^a X^4\, dX$$

$$\overline{X} = \frac{\dfrac{ka^5}{5}}{\dfrac{ka^4}{4}} = \frac{4}{5}\, a.$$

The centroids of composite figures can be handled by superposition of the individual parts. Holes and voids enter the calculations as negative entries.

THEOREMS OF PAPPUS AND GULDINUS

The <u>Theorem of Pappus</u> states that the surface area is the product of the length of the path generated by the centroid of the line x.

As an illustration of this theorem, rotate line 1–2 of Fig. 2.140 about the y-axis to form the spitoon shaped vessel. Find the surface area of the figure.

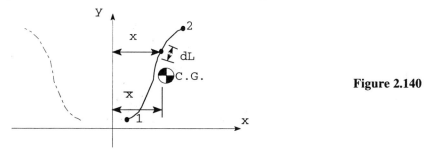

Figure 2.140

Revolving dL about the y-axis generates the surface area dA, where

$$dA = dL\,[2\pi(X)]$$

and for the total area

$$A = \int 2\pi(X)\,dL = 2\pi \int X\,dL = 2\pi \overline{X}\,L$$

The <u>Theorem of Guldinus</u> states that the volume is the product of the area of Fig. 2.14 multiplied by the length of the path described by the centroid of the area.

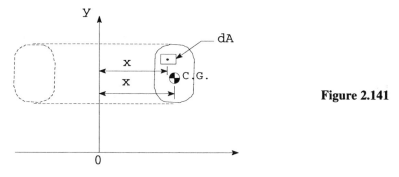

Figure 2.141

Rotate area A about the y-axis, note each elemental area dA generates a circular ring whose volume is:

$$dV = 2\pi(X)\,dA,$$

therefore the total volume is:

$$V = \int dV = 2\pi \int X\,dA = 2\pi \overline{x}\,A$$

3
Mechanics of Materials

MECHANICS OF MATERIALS

TABLE OF CONTENTS

Selected References 3-iii
Notation 3-iv
    Tensile and Compressive Strain 3-1
    Shearing Strain 3-1
    Stress-Strain Diagram 3-2
    Definitions 3-2
        Proportional Limit 3-2
        Yield Point 3-2
        Ultimate Strength 3-3
        Permanent Set 3-3
        Elastic Limit 3-3
        Yield Strength 3-3
        Hooke's Law 3-4
        Poisson's Ratio 3-4
        Coefficient of Thermal Expansion 3-4
    Statically Indeterminate Structures 3-9
        General Solution Procedure 3-9
        Illustrated Problems 3-9
    Torsion 3-13
        Angle of Twist 3-15
        Compound Shafts 3-17
        Illustrated Examples 3-17
    Statically Indeterminate Torsion Problems 3-19
    Mechanics of Materials Solved Problems 3-21
Beam Analysis 3-29
    Types of Beams 3-29
    Types of Loads 3-29
    Conditions of Equilibrium 3-30
    Sign Convention 3-32
    Reactions 3-33
    Illustrated Problems 3-33
    Shear and Bending Moment Diagrams 3-36
        The Shear Forces Diagram 3-36
        The Bending Moment Diagram 3-38
        Illustrated Example 3-41
    The Flexure Formula 3-45
    Position of Neutral Axis 3-47
    Moment of Inertia of a Rectangle 3-48
    Parallel Axis Theorem 3-49
    Moments of Inertia of Composite Sections 3-49
    General Procedure of Composite I 3-50
    Illustrated Example 3-50
    Solved Problems 3-51

Combined Stress 3-65
Formulas for Normal and Shear Stress 3-66
Appendix A: Combined Stress (Example) 3-70

SELECTED REFERENCES

The primary intent of these notes and solved problems are to aid your recall of the fundamentals of Mechanics of Materials. These notes will concentrate valuable reference material for use in your state examination. They, however, are not meant to be an exhaustive treatment of the subject. Additional study is recommended from the following references.

F. Singer, "Strength of Materials," Harper & Row, New York

F. Seely, "Resistance of Materials," John Wiley & Sons, New York

S. Timoshenko, "Strength of Materials" (two volumes), D. Van Nostrand Co., New York

R.J. Roark, "Formulas for Stress and Strain," McGraw-Hill, New York

S.A. Nash, "Theory and Problems of Strength of Materials," Schaum's Outline Series, McGraw-Hill, New York

NOTATION

↦	Positive sense of applied forces
↻	Positive sense of applied moments
a	Distance; in., ft
A	Area, in.^{2}
b	Distance or width of beam; in., ft
c	Outside radius of shaft or distance from NA to remotest stressed fiber
d	Distance from point of investigation to NA or moment arm; in., ft
e	Total elongation of a specified length of material
E	Young's modulus of elasticity, psi
F	Applied force, lbs, kips
G	Shear modulus of rigidity, psi
h	Depth of beam, inc.
i	When used as a subscript or superscript, denotes i^{th} particle
I	Moment of inertia, in.^{4}
J	Polar moment of inertia; in.^{4}
k	kip = kilopound = 1000 lbs or distance, in., ft
L	Length, ft
M	Bending moment, in.-lbs, or ft-lbs
NA	Neutral axis
O	Origin of coordinate system or reference to centroidal axis when used as a subscript
P	Applied force, lbs, kips

P_{xy}	Product of inertia with respect to x- and y-axes
r	Radius of shaft; in.
t	Thickness of material; in.
T	Torque; in.-lbs or ft-lbs
V	Transverse shear force; lbs
w	Applied unit load or uniformly distributed load; lbs/ft, k/ft
W	Total applied load; lbs, kips
x	Distance or subscript denoting x-axis; in. or ft
$\bar{x}$	Centroidal distance in x direction from datum; in. or ft
y	Distance or subscript denoting y-axis; in. or ft
$\bar{y}$	Centroidal distance in y direction from datum; in. or ft
z	Distance or subscript denoting z-axis; in. or ft
$\bar{z}$	Centroidal distance in z direction from datum; in. or ft
α	Coefficient of thermal expansion; in./in. - °F
γ	Unit shear strain; in./in.
δ	Total deformation; in.
ϵ	Unit tensile strain; in./in.
θ	Angle or angle of twist in plane of cross-section about twist axis; radians, degrees
μ	Poisson's ratio
σ	Tensile stress; psi
τ	Shear stress; psi

In contrast to the study of <u>statics</u> where the dimensions of a member are assumed to remain constant under load, the study of Mechanics of Materials does consider changes in the dimensions of a member. These changes, called deformations or deflections, are caused by the loading and are a function of the properties of the material.

<u>Tensile and Compressive Strain</u> (Fig. 3.1)

In any member the external loads must be in equilibrium with the internal forces, or rupture will occur. These actions or loads are called stresses and the deformations produced by these stresses are called strains.

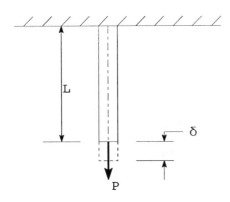

Figure 3.1

Under an axial tensile load, the bar elongates an amount δ (the total elongation). Normal tensile unit strain = ϵ = δ/L. ϵ = unit elongation in inches per inch. For compressive load $\epsilon = -\delta/L$.

<u>Shearing Strain</u> (Fig. 3.2)

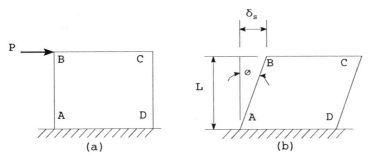

Figure 3.2

Imagine a rectangle ABCD built up of layers, as a deck of cards, where one layer slides relative to adjacent layers.

The total deformation due to sliding layers in length L is δ_s; the straining per unit length = γ, where $\gamma = \delta_s/L = \tan \emptyset = \emptyset$ (radians) for small angles.

Stress–Strain Diagrams (Figs. 3.3 and 3.4)

Using a test specimen of known dimensions in a tensile testing machine, the load and deformation can be measured simultaneously. Determine the unit stress from:

$$\sigma = \frac{P}{A} = \frac{\text{Load}}{\text{Initial Cross Section Area}}$$

Plot unit stress vs. unit strain for the material under test.

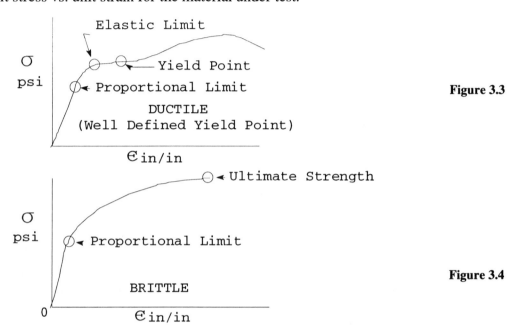

Figure 3.3

Figure 3.4

DEFINITIONS

Proportional Limit

The proportional limit is the maximum unit stress that can be developed in a material without causing the unit strain to increase disproportionately.

Yield Point

The yield point is the unit stress at which the material continues to deform without an increase in load.

Ultimate Strength

The ultimate strength is the maximum unit stress that can be developed in a material, as determined from the original cross-sectional area of the specimen.

Permanent Set

The permanent set is the unit deformation retained by the test specimen after the load has been reduced to zero.

Elastic Limit

The elastic limit is the maximum unit stress that can be developed in a material without permanent set.

Yield Strength

The yield strength is the maximum unit stress that can be developed in a material without causing more than a specified permanent set.

Method of Determining Yield Strength of Brittle Material (Fig. 3.5)

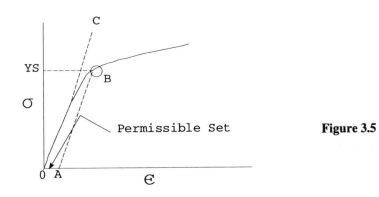

Figure 3.5

1. Plot permissible set $\overline{OA}$.

2. Draw $\overline{AB}$ // to $\overline{OC}$. $\overline{OC}$ is the line tangent to the stress–strain curve through the point O.

3. Where $\overline{AB}$ intersects the stress–strain diagram the ordinate is the yield strength.

Hooke's Law

For stresses within the proportional limit, all elastic samples of a material behave according to the same law.

Ratio of $\dfrac{\text{Unit stress}}{\text{Unit strain}}$ = Constant

The modulus of elasticity, E, is the ratio of unit stress to the accompanying unit strain within the proportional limit of a material.

$$E = \dfrac{\sigma}{\epsilon} = \text{slope of straight line portion of stress–strain curve}$$

Substituting $\dfrac{P}{A}$ for σ and $\dfrac{\delta}{L}$ for ϵ

$$E = \dfrac{P/A}{\delta/L}, \text{ solving for deformation } \delta = \dfrac{PL}{AE}$$

where δ = in., A = in.²

$\quad\quad\quad\quad$ P = lbs., E = p.s.i

$\quad\quad\quad\quad$ L = in.

Poisson's Ratio

Poisson's ratio, μ, is the ratio of lateral unit strain to the longitudinal unit strain. It is usually negative because one quantity of the ratio must reduce as the other increases.

The Coefficient of Thermal Expansion

The coefficient of thermal expansion, α, is the unit elongation of material caused by a 1°F change in temperature.

The deformation due to temperature change is

$\quad\quad \delta_T = L\alpha(\Delta T) \quad\quad\quad \Delta T$ = change in temperature in °F

Illustrated Problems

1. A steel and an aluminum bar are coupled together end to end and loaded axially at the extreme ends (Fig. 3.6). Both bars are 2 in. in diameter; the steel bar is 6 ft long, and the

aluminum bar is 4 ft long. When the load is applied, it is found that the steel bar elongates 0.0052 in. in a gage length of 8 in. Poisson's ratio for this steel is 1/4 and the modulus of elasticity of the aluminum is 10.6 x (10^6) psi and 30.0 (10^6) for steel. Determine:

(A) The load
(B) The total change in length from A to C
(C) The change in diameter of the steel bar

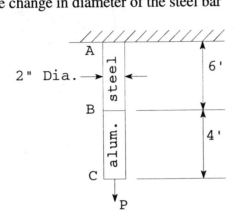

Figure 3.6

a) Let the subscript "s" denote steel, and "a" denote aluminum.

$$\epsilon_s = \frac{.0052}{8} = .00065$$

$$E_s = \frac{\sigma_s}{\epsilon_s}; \quad \sigma_s = E_s \epsilon_s$$

$$\sigma_s = 30\,(10^6)(.00065) = 19{,}500 \text{ psi}$$

The load in each member must be identical and $= P = \sigma A$

From the free body diagram (FBD) of cut through steel member

$$P = \sigma_s(A_s) = 19{,}500 \times \frac{\pi(2)^2}{4} = 61{,}230 \text{ lbs} \quad \text{Answer}$$

b) Total change in length is $\delta_s + \delta_a = \epsilon_s L_s + \epsilon_a L_a$

where $\epsilon_a = \dfrac{\sigma_a}{E_a} = \dfrac{19{,}500}{10.6\,(10^6)} = .00184 \dfrac{\text{in}}{\text{in}}$

$\epsilon_s = .00065$

$\delta_s + \delta_a = .00065 (72) + .00184 \times (48) = .135$ in.

c) Using

$$\mu = \frac{\epsilon \text{ lateral}}{\epsilon \text{ longitudinal}} = \frac{1}{4}$$

$\epsilon_{lat.} = \frac{1}{4} \times .00065 = .00016 \quad \Delta D = D \times \epsilon_{lat.} = 2 \times .00016 = .00033$ in.

2. A bar of uniform cross section and homogenous material hangs vertically while suspended from one end (Fig. 3.7). Determine, in terms of W, L, A and E, the extension of the bar due to its own weight.

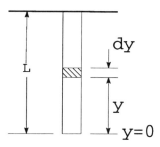

Figure 3.7

Let W = total weight of bar, lbs

w - weight per in., lbs/in.

L = length of bar, in.

A = cross sectional area, in.^{2}

E = Young's modulus

Equate forces at distance y

$$\sigma_y (A) = w (y)$$

$$\sigma_y = \frac{w(y)}{A}$$

Unit strain at section = $\epsilon_y = \frac{\sigma_y}{E_y} = \frac{w(y)}{A(E_y)}$

Elongation of length dy is de_y, a constant equal to: $de_y = \epsilon_y d_y$

Total elongation $= \int de_y$

$$e_{tot.} = \int_0^L \epsilon_y dy = \int_0^L \frac{w\,y}{A\,E_y} dy = \frac{w}{AE} \int_0^L y\,dy = \frac{wL^2}{2\,AE} \quad \text{Answer}$$

3. The A frame shown in Fig. 3.8 has two members 12 ft long and a tierod 6 ft long connecting the members at point 8 ft from the hinged apex. Determine the tension in the tierod when a load of 4000 lbs is placed at the apex. Neglect the weight of the frame and assume that it rests on a smooth surface. Determine the elongation of the tierod if it has a cross-sectional area of 0.25 in.² and is made of steel.

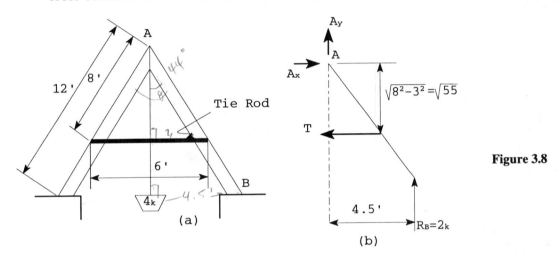

Figure 3.8

With symmetrical loading and smooth surfaces, reactions at supports must be equal and vertical, 2^k each.

Solving from distance from ℄ to point B:

$$\frac{3}{8} = \frac{X}{12}$$

$$X = \frac{3}{8}(12) = 4.5'$$

$$\circlearrowright \Sigma M_A = 0$$

$$T \times \sqrt{55} - R_B(4.5) = 0$$

$$T = \frac{2000\ (4.5)}{7.42} = 1210 \text{ lbs.}$$

$$\delta_{rod} = \frac{1210(6)(12)}{0.25(30)(10^6)} = 0.0013 \text{ in. Answer}$$

4. A pendulum 30 in. long from pivot to bob is built of vertical brass and steel rods so arranged that the differential expansion of the two metals is compensating and keeps the length of the pendulum constant under temperature changes (Fig. 3.9). The expansion of the brass rods raises the pendulum bob while the expansion of the steel rods lowers it. What total effective length of each material must be used? Coefficients of expansion per degree Fahrenheit are 0.0000103 in./in. and 0.0000065 in./in. for brass and steel, respectively.

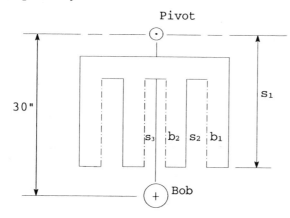

Figure 3.9

Let ——— = steel and ·–·–·–· brass.

L_s = total effective length of steel rods

$L_s = S_1 + S_2 + S_3$

L_b = total effective length of brass rods

$L_b = b_1 + b_2$

Since expansions are compensating $L_s(.0000065) - L_b(.0000103) = 0$

$$L_s = \frac{103}{65} L_b$$

Note that there are two unknowns that require two equations for solution. The following statement yields the second equation.

Total length of steel minus total length of brass = 30 in.

$L_s - L_b = 30"$

$\therefore \dfrac{103}{65} L_b - L_b = 30, \; L_b \left(\dfrac{103}{65} - 1\right) = 30$

$\therefore L_b = 51.4$ in., $L_s = 30 + 51.4 = 81.4$ in. Answer

Statically Indeterminate Structures

A statically indeterminate structure is one in which the number of unknown reactions or applied forces exceeds the available equations of equilibrium. Solution of this type of problem requires a knowledge of the elastic properties of the body to obtain the necessary additional equations.

General Solution Procedure

1. Draw a free body diagram of selected member.

2. Note number, magnitude, and positions of unknowns.

3. Note type of force system on FBD to determine available equations of equilibrium.

4. Write deformation equation for each unknown in excess of the available equations of equilibrium.

5. Solve the set of equilibrium and deformation equations simultaneously. Deformations and forces must be related through Hooke's Law. The theory is invalid for stresses above the proportional limit of the material.

Illustrated Problems

1. In Fig. 3.10, S is a steel bar 0.5 by 0.8 in. and B is a brass bar 1 by 0.5 in. Bar C may be regarded as rigid. A load of 3850 lbs is hung from C at such a place that the bar C remains horizontal. The modulus of elasticity of brass is $15(10^6)$ psi. Determine

(a) The axial stress in each material.

(b) The vertical displacement of bar C.

(c) The position of the line of action of the 3850-lb force.

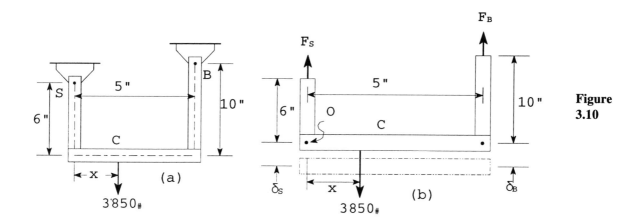

Figure 3.10

The force system is parallel; ∴ two equations of equilibrium are available, but there are three unknowns, F_s, F_b, X.

The equations of equilibrium are ↻ $\Sigma M_O = 0$ ↑ $\Sigma F_Y = 0$

The deformation equation is $\delta_s = \delta_B$

↻ $\Sigma M_O = 0$ yields $3850X - F_B(5) = 0$

↑ $\Sigma F_Y = 0$ yields $F_s - 3850 + F_B = 0$

using $\delta = \dfrac{PL}{AE}$

$$\delta_s = \dfrac{F_s(6)}{.5(.8)(30)(10^6)} = \dfrac{1}{2} F_s (10^{-6})$$

$$\delta_B = \dfrac{F_B(10)}{1(.5)(15)(10^6)} = \dfrac{4}{3} F_B (10^{-6})$$

∴ $\dfrac{1}{2} F_s (10^{-6}) = \dfrac{4}{3} F_B (10^{-6})$

$F_s = \dfrac{8}{3} F_B$

substituting in $\Sigma F_Y = 0$

$\dfrac{8}{3}F_B - 3850 + F_B = 0$, $\dfrac{11}{3}F_B = 3850$, $F_B = 1050$ lbs

$F_s = 2800$ lbs.

$\sigma_B = \dfrac{P}{A} = \dfrac{1050}{.5} = 2100$ psi $\sigma_s = \dfrac{2800}{.4} = 7000$ psi

From $\Sigma M = 0$, $X = \dfrac{5(1050)}{3850} = 1.36$ in. to right of point 0.

2. The pin connected structure in Fig. 3.11 occupies the position shown in Fig. 3.11(a) under no load. When the loads D (= 2000 lbs) and P are applied, the rigid bar C becomes horizontal. Determine the load P. The following data apply.

Bar	Material	Area (in.²)	E (psi)
A	Aluminum alloy	2	$10(10^6)$
B	Monel	4	$24(10^6)$

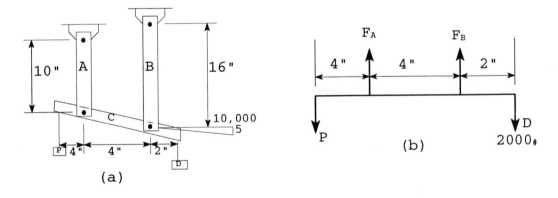

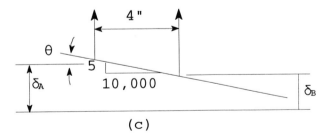

Figure 3.11

The available equations of equilibrium = 2.

The unknowns are F_A, F_B, and P relating deflections of links A and B.

$$\delta_A = \delta_B + 4 \sin \theta$$

For small angles $\sin \theta = \tan \theta$

$$\delta_A = \delta_B + 4 \tan \theta$$

↑ $\Sigma F_Y = 0$: $-P + F_A + F_B - 2000 = 0$

↻ $\Sigma M_D = 0$: $-P(10) + F_A(6) + F_B(2) = 0$

Using $\delta = \dfrac{PL}{AE}$

$$\frac{F_A(10)}{2(10)(10^6)} = \frac{F_B(16)}{4(24)(10^6)} + \frac{4(5)}{1(10^4)}$$

$$\frac{F_A}{2} = \frac{F_B}{6} + 2000 \qquad F_A = \frac{F_B}{3} + 4000$$

substitute in $\Sigma M_D = 0$

$$-P(10) + (\frac{F_B}{3} + 4000) 6 + F_B(2) = 0$$

$2 F_B + 24{,}000 + 2 F_B = 10 P$

$4 F_B = 10 P - 24{,}000 \therefore F_B = \dfrac{5}{2}P - 6000$

$$-P + \frac{\frac{5}{2}P - 6000}{3} + 4000 + (\frac{5}{2}P - 6000) - 2000 = 0$$

$\therefore P = \dfrac{36000}{14} = 2571$ lbs.

3. A rectangular steel beam between two rigid walls, 96 in. apart, is heated to a uniform temperature 80°F above ambient (Fig. 3.12). What is the resultant stress developed in the beam? Assume E - 30 (10⁶) psi.

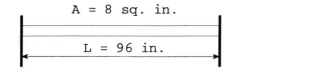

Figure 3.12

Approach

Remove one wall constraint, allow the beam to expand due to temperature rise, then apply force to return beam to original length, and calculate stress (Fig. 3.13).

$$\delta = L \alpha \Delta T = 96 \ (6.5) \ (10^{-6}) \ (80) = 0.499 \text{ in.}$$

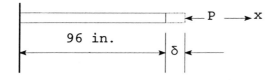

Figure 3.13

Apply a compressive load to produce deformation equal to δ.

$$-\delta = \frac{+PL}{AE} \qquad P = -\frac{\delta AE}{L} = -\frac{.0499(8)(30)(10^6)}{96.0499} = -124{,}690 \text{ lbs. compression}$$

$$\sigma = \frac{P}{A} = -\frac{124690}{8} = -15{,}586 \text{ psi compression}$$

Torsion

Torsional loads are forces that cause a bar to twist about its central axis (Fig. 3.14). They are applied in a plane perpendicular to the axis of twist and are resisted by internal shear stresses.

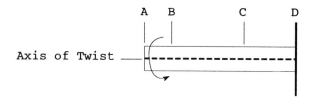

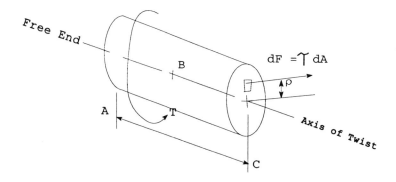

Figure 3.14

For equilibrium:

<u>Moment of External Forces = Moment of Internal Forces</u>

Moment of external forces = T_1

τ = shear stress, psi, F = shear force

Internal force on element of cross-sectional area = $dF = \tau \, dA$

Element of internal resisting moment = $\tau \, dA \rho = dM_r$

ρ = distance to center of area dA.

Total internal resisting moment = $\int_{Area} \tau \, \rho \, dA = M_r$

The shear stress τ varies directly as the distance from the center of the twist (Fig. 3.15).

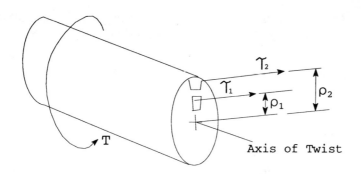

Figure 3.15

$$\frac{\tau}{\delta_1} = \frac{\tau}{\delta_2} = \text{constant}$$

Using $T_1 = \int \tau \rho \, dA$ and multiplying inside the integral sign by ρ/ρ

$$T_1 = \int \frac{\tau}{\rho} \rho^2 \, dA$$

Since $\tau/\rho = $ constant, expression becomes $T_1 = \int \frac{\tau}{\rho} \rho^2 \, dA$

where $\int \rho^2 \, dA = $ Polar Moment of Inertia of the cross-sectional area, which is denoted by J.

$$\left.\begin{array}{l} T = \dfrac{\tau J}{\rho} \\[1em] \tau = \dfrac{T\rho}{J} \end{array}\right\} \text{Formula for shear stress in a shaft subjected to external torque T}$$

$$\tau_{max} = \frac{TC}{J}$$

where c = the outside radius of shaft. For a solid circular shaft $J = \dfrac{\pi d^4}{32} = \dfrac{\pi r^4}{2}$

For a hollow circular shaft $J = \dfrac{\pi}{32} \left[d_{outside}^4 - d_{inside}^4 \right]$

(essentially J for the solid shaft minus J for the removal shaft).

Angle of Twist

For optical and precision mechanical equipment, stiffness rather than strength may be the controlling factor in design. It then becomes important to know the angle of twist. The angle of twist is regarded as a torsional deformation which can be expressed, as in the case of tension, by Hooke's Law.

To develop these torsional deformation relations, consider the stress–strain diagram for a sample of ductile material in torsion. The curve is similar to that for pure tension (Fig. 3.3).

In Fig. 3.16 Hooke's Law applies; that is, stress is proportional to strain up to the proportional limit.

$$\frac{\tau}{\gamma} = G \qquad \text{Where } \tau = \text{unit shear stress psi}$$

$$\gamma = \text{shear strain in./in.}$$

G = Modulus of elasticity in shear or torsion

Figure 3.16

Fig. 3.17 illustrates a section of a twisted shaft, where the left end is fixed and line AB deforms to AB', after twisting. The shearing unit strain at the surface of the shaft is:

$$\gamma_c = \frac{\widehat{BB'}}{L} = \frac{C\theta}{L}$$

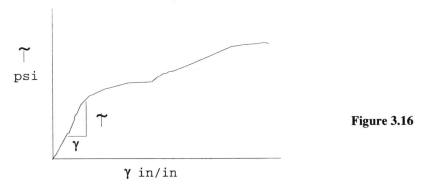

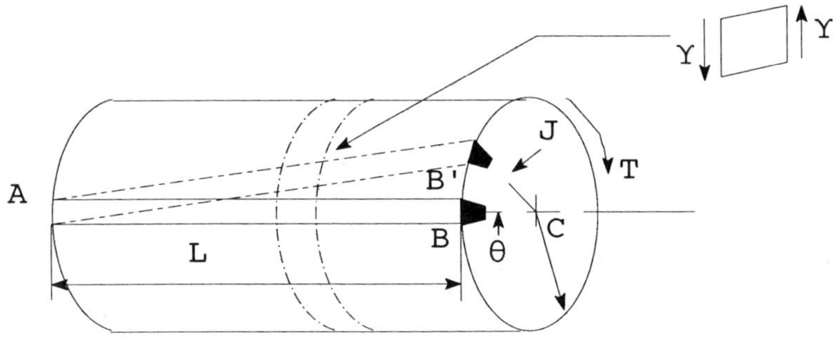

Figure 3.17

But $\tau = \dfrac{TC}{J}$, substituting $\dfrac{TC}{J}$ for τ and $\dfrac{C\theta}{L}$ for γ in the expression for G.

$$G = \frac{\frac{TC}{J}}{\frac{c\theta}{L}} = \frac{TL}{\theta J} \qquad \therefore \theta = \frac{TL}{GJ}$$

the angle of twist for a shaft twisted by torque T (θ is in radians).

Compound Shafts

Note that the shear stress in a shaft, $\tau = \frac{T\rho}{J}$, is a function of the applied torque (T) and the geometry (J) of the shaft. It does not depend upon the material of the shaft.

The angle of twist of each portion of shaft, however, depends on the applied torque, the shaft geometry and the rigidity of the material of that portion of shaft.

To find the maximum shear stress due to torsion of a compound shaft, it is generally necessary to determine the shear stress wherever the torque and/or shaft geometry changes.

THE ANGLE OF TWIST OF A COMPOUND SHAFT IS THE ALGEBRAIC SUM OF THE TWISTS OF EACH SECTION WHERE THE GEOMETRY, APPLIED TORQUE, OR MATERIAL CHANGES.

Illustrated Examples

1. A shaft is composed of two 6-in. diameter sections coupled together (Fig. 3.18). Part AB is bronze and part BC is steel. A torque is applied at the free end which causes a maximum shear stress in the bronze shaft of 10,000 psi. Find the angle of twist of the free end. Assume the modulus of rigidity of bronze is 6 (10^6) psi and that of steel is 12(10^6) psi.

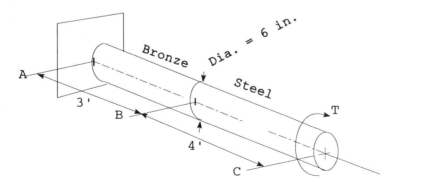

Figure 3.18

$$\theta_{total} = \theta_B + \theta_S = \frac{TL_B}{G_B J} = \frac{TL_S}{G_S J} = \frac{T}{J}\left[\frac{L_B}{G_B} + \frac{L_S}{G_S}\right]$$

But $\tau = \frac{T\rho}{J} = 10{,}000$ psi

$$\therefore T = \frac{10{,}000\, J}{\rho}\,;\; \rho = 3 \text{ in.}$$

$$\theta_{total} = \frac{10{,}000\, J}{3\, J}\left[\frac{36}{6(10^6)} + \frac{48}{12(10^6)}\right] = .033 \text{ radians}$$

$$= 1.90° \text{ clockwise}$$

2. Rod AB is a rigid pointer fastened to the end of the 4-in. diameter solid circular shaft (Fig. 3.19). Assume the shaft is steel where $G = 12\,(10^6)$ psi. Determine:

   a) The maximum shearing stress in the shaft.

   b) The angular distance through which point A moves, measured from the no load position.

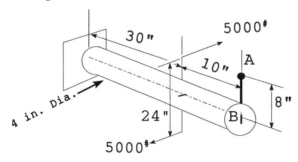

Figure 3.19

$G(\text{bronze}) = 6.5 \times 10^6$
$G(\text{aluminimum}) = 4 \times 10^6$

a) $T = 5000$ lbs $(24$ in.$) = 120{,}000$ in.-lbs

$$\tau = \frac{T\rho}{J} = \frac{120{,}000(2)}{\frac{\pi}{32} \times (4)^4} = 9540 \text{ psi} \underline{\text{ Answer}}$$

b) $$\theta = \frac{TL}{GJ} = \frac{120{,}000(30)}{12(10^6) \times \frac{\pi}{32} \times 4^4} = 0.0119 \text{ radians}$$

$$.0119 \text{ radians} \times \frac{360°}{2\pi \text{ radians}} = 0.683° \qquad \underline{\text{Answer}}$$

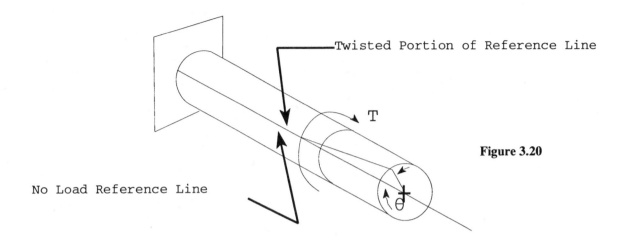

Figure 3.20

Statically Indeterminate Torsion Problems

The statically indeterminate torsion problem requires the use of distortion equations in addition to the available equilibrium equations in order to solve for the unknowns. Whenever there are more unknowns than there are equilibrium conditions, the problem is identical to that for statically indeterminate tension problems.

Solution Procedure

1. Draw a free body diagram.

2. Note the number of unknowns, which may be magnitude, position, and direction of forces and moments.

3. Examine the type of force system to determine the available equations of equilibrium.

4. Make a simplified distortion diagram and write a distortion (deformation) equation for each extra unknown.

5. Solve the set of equilibrium and distortion equations simultaneously.

Illustrated Example

1. The circular shaft AC is fixed to rigid walls at A and C. The section AB is solid annealed bronze and the hollow section BC is made of aluminum alloy 2024 T-4. There is no stress in the shaft before the 20,000 ft-lb torque is applied. Determine the maximum shearing stress in each shaft after T is applied.

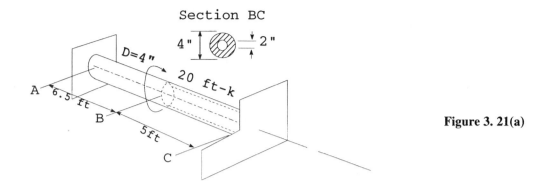

Figure 3. 21(a)

From the FBD, (Fig. 3.21b), we have replaced the rigid walls by unknown torques T_A and T_C.

↻ $\Sigma M_{axis} = 0$ yields

$20,000 (12) - T_A - T_C = 0$ equilibrium equation

$\theta_{AB} = \theta_{BC}$

The angle of twist at B must be the same for both shaft sections, otherwise the shaft has been broken.

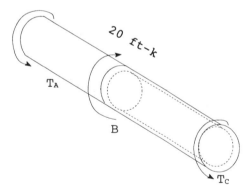

Figure 3.21(b)

$G(bronze) = 6.5 \times 10^6$
$G(aluminimum) = 4 \times 10^6$

$T = \dfrac{\tau J}{\rho}$

$$\dfrac{t_{AB} \pi \dfrac{4^4}{32}}{2} + \dfrac{t_{BC} \pi \left(\dfrac{4^4 - 2^4}{32}\right)}{2} = 240,000 \ (1)$$

Since $\theta = \dfrac{\gamma L}{c} = \dfrac{\tau L}{Gc}$

The deformation equation becomes

$$\frac{\tau_{AB}(6.5)(12)}{6.5(10^6)(2)} = \frac{\tau_{BC}(5)(12)}{4(10^6)(2)}$$

$$\tau_{AB} = \frac{5}{4}\tau_{BC}$$

Solving (1) and (2) simultaneously

$$\tau_{AB} = 10{,}900 \text{ psi}$$

$$\tau_{BC} = 8{,}730 \text{ psi}$$

Mechanics of Materials Solved Problems

1. The truss shown in Fig. 3.22 supports the single load of 120,000 lbs. If the working stress of the material in tension is taken to be 20,000 psi, determine the required cross-sectional area of the bars DE and AC. Find the elongation of bar DE. Assume all joints of the truss to be pinned and the modulus of elasticity, $E = 30 \times 10^6$ psi.

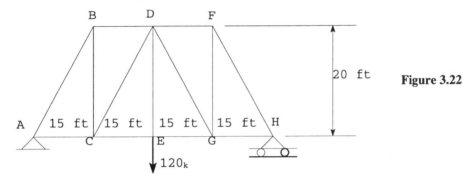

Figure 3.22

Approach

Using equations of equilibrium, solve for the reactions at A and H. Use equilibrium of pins at A and E ($\Sigma F_x = 0$, $\Sigma F_y = 0$), to determine forces in required members. Use $\sigma = P/A$ to find the area of each bar, then use Hooke's Law to find the elongation of DE.

$\circlearrowright \Sigma M_A = 0$ and $\uparrow \Sigma F_Y = 0$ can be used to determine the reactions

However, notice the symmetry of loading and geometry of the truss; conclude that each reaction must be 1/2 of the 120^k load.

$$R_A = R_H = 60^k$$

Equilibrium of Joint A (Fig. 3.23)

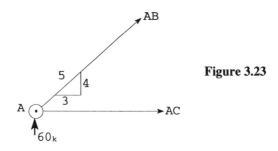

Figure 3.23

$\uparrow \Sigma F_Y = 0; \ 60 + \dfrac{4}{5} AB = 0 \quad AB = -60 \times \dfrac{5}{4} = -75^k$

The force in AB was chosen in the wrong direction. It is acting toward the pin which means the bar AB is in compression.

Correct the direction of the force in AB before proceeding.

From Fig. 3.23 above: $\rightarrow \Sigma F_X = 0$ yields $-\dfrac{3}{5} AB - AC = 0$

$AC = \dfrac{3}{5} AB = \dfrac{3}{5}(75) = 45^k$ tension. From $\sigma = P/A$

$A = \dfrac{P}{\sigma} = \dfrac{45^k}{20 \ \text{ksi}} = 2.25 \ \text{in}^2 = $ area of AC <u>Answer</u>

Equilibrium of Joint E (Fig. 3.24)

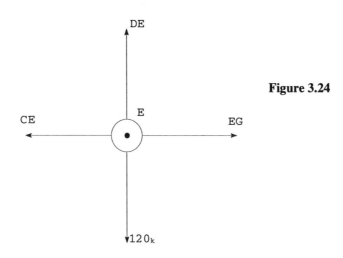

Figure 3.24

↟ $\Sigma F_Y = 0$; DE - $120^k = 0$ DE = 120^k tension

$$A = \frac{P}{\sigma} = \frac{120^k}{20 \text{ ksi}} = 6.0 \text{ in}^2 = \text{area of DE} \quad \underline{\text{Answer}}$$

Elongation of DE

$$\delta = \frac{PL}{AE} = \frac{120,000(20 \text{ ft})(12) \text{ in/ft}}{6.0 \text{ in}^2(30)(10^6) \text{ psi}} = 160(10^{-3}) = 0.16 \text{ in.} \quad \underline{\text{Answer}}$$

2. A bolt with 0.7 in.² cross-sectional area extends through a bronze tube and has 20 threads per inch (Fig. 3.25). The bronze tube is 27 in. long and just makes contact with the washer so that there are no initial stresses. Calculate the unit stress in the unthreaded length of the bolt and in the bronze tube that would result from a quarter turn of the nut. Assume the modulus of elasticity of bronze is 14 x 10⁶ psi, the cross-sectional area of the bronze is 1.9 in.², and the bolt is steel.

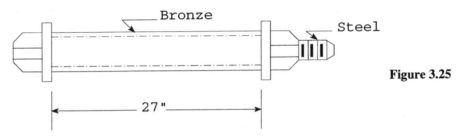

Figure 3.25

Approach (Fig. 2.26)

Assume the bolt to be absolutely rigid, and advance the nut axially a distance equivalent to 1/4 turn, thus compressing the tube. Now permit the bolt to elongate under the force produced by the compressed tube. The actual elongation of bolt and compression of tube must add up to the axial distance equivalent to 1/4 turn of the nut.

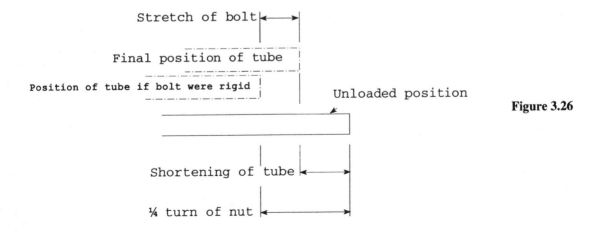

Figure 3.26

Let P = force in the bolt which must be the same force compressing the tube.

$$\delta_{Bolt} + \delta_{Tube} = \frac{1}{4} \times \frac{1}{20} = \frac{1}{80} \text{ in.}$$

$$\frac{PL_b}{A_b E_b} + \frac{PL_t}{A_t E_t} = \frac{1''}{80} \qquad P\left[1 + \frac{L_t A_b E_b}{L_b A_t E_t}\right] = \frac{A_b E_b}{80 L_b}$$

$$P = \frac{0.7(30)(10^6)}{80(27)\left[1 + \frac{27(0.7)(30)(10^6)}{27(1.9)(14)(10^6)}\right]} = \frac{21(10^6)}{80(27)(1.79)} = 5430 \text{ lbs}$$

$$\sigma_{Bolt} = \frac{5430}{0.7} = 7750 \text{ psi} \qquad \underline{\text{Answer}}$$

$$\sigma_{Tube} = \frac{5430}{1.9} = 2850 \text{ psi} \qquad \underline{\text{Answer}}$$

3. Due to a misalignment of the pin holes at A, B and C, after the pins at A and B are in place, a force P = 6 kips upward at D is necessary to permit insertion of the pin at C (Fig. 3.27). Determine the axial stress in the bar CE when the force P is removed with all pins in place.

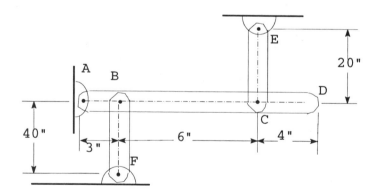

Figure 3.27

$\overline{CE}$ is aluminum; $\overline{BF}$ is steel $\overline{AD}$ is rigid

$E = 10\,(10^6)$ psi $E = 30\,(10^6)$ psi

$A = 4 \text{ in.}^2$ $A = 2 \text{ in.}^2$

Approach

Apply 6^k at point D to raise AD, thus stretching BF. Determine the stretch of BF and rise of AD along the line of link EC. After inserting pin at C, 6^k is released and the force in BF will pull AD down but is resisted by the force in CE. When moments produced by forces in BF and CE cancel, AD will come to rest (Fig. 3.28).

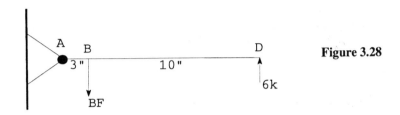

Figure 3.28

↻ $\Sigma M_A = 0$ yields BF (3) = 6^k(13) = 0 ; BF = 78/3 = 26^k tension

$$\delta_{BF} = \frac{PL}{AE} = \frac{26000(40)}{2(30)(10^6)} = 0.017 \text{ in.}$$

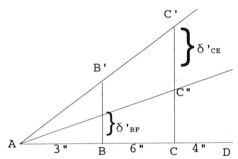

Figure 3.29

In Fig. 3.29, above, BB' is the deflection of member BF. Triangle AB'B ≃ AC'C.

BB' = 0.017 in.

$$\frac{BB'}{3} = \frac{CC'}{9}, \quad CC' = 3 (B') = 0.051 \text{ in.}$$

$$\frac{CC''}{9} = \frac{\delta'_{BF}}{3}, \quad CC \text{ in.} = 3 \, \delta'_{BF}$$

$3\delta'_{BF} + \delta'_{CE} = 0.051$ in.

$\delta'_{CE} = 0.051 - 3 \, \delta'_{BF}$

Final deformation of BF is δ'_{BF}; the final deformation of CE is δ'_{CE}.

where:
$$\delta'_{BF} = \frac{(BF)'(40)}{2(30)(10^6)}$$

Then:
$$(BF)' = \frac{\delta'_{BF}(2)(30)(10^6)}{40} = \frac{3}{2} \times \delta'_{BF} \times 10^6$$

$$\delta'_{CE} = 0.051 - 3\delta'_{BF} = \frac{(CE)'(20)}{4(10)(10^6)}$$

$$(CE)' = 2(10^6)(0.051 - 3\delta'_{BF})$$

$\circlearrowleft \Sigma M = 0$ yields ; $(BF)'(3) - (CE)' \times (9) = 0$

Substituting:

$$\frac{3}{2}\delta'_{BF}(10^6)(3) - (0.051 - 3\delta'_{BF}) \times 2(9)(10^6) = 0$$

$\delta'_{BF}(0.5 + 6) = 0.102$ or $\delta'_{BF} = 0.051$ in.

$(CE)' = [0.051 - 3(0.0157)]\, 2(10^6) = 7800$ lbs,

Then

$$\sigma_{CE} = \frac{7800}{4} = 1950 \text{ psi, tension} \qquad \underline{\text{Answer}}$$

4. A composite bar mounted between two rigid walls is free from stress at ambient temperature (Fig. 3.30). If the temperature of the structure drops by T degrees Fahrenheit, determine the minimum temperature to which the assembly may be subjected in order that the stress in the aluminum does not exceed 24,000 psi.

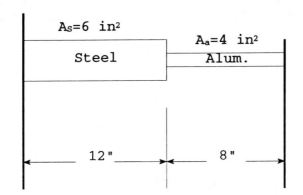

Figure 3.30

Approach

Remove the right hand constraint, allow the bar to contract, then apply force at the end to return the bar to its original position (Fig. 3.31).

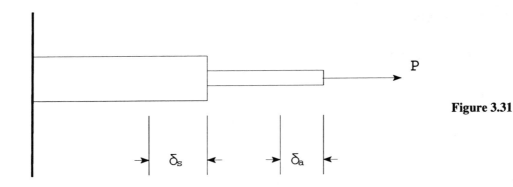

Figure 3.31

$\Delta T = T°F$, $\alpha_s = 6.5(10^{-6}) \dfrac{\text{in.}}{\text{in.} - °F}$

$\alpha_a = 12.5(10^{-6}) \dfrac{\text{in.}}{\text{in.} - °F}$

$\delta_{total} = L_s \alpha_s \Delta T + L_a \alpha_a \Delta T = \Delta T [12(6.5)(10^{-6}) + 8(12.5)((10^{-6})]$

$\qquad = 178 (10^{-6}) \times \Delta T$

Applying force P to the free end:

$\delta_{total} = \dfrac{P L_s}{A_s E_s} + \dfrac{P L_a}{A_a E_a} = P \left[\dfrac{L_s}{A_s E_s} + \dfrac{L_a}{A_a E_a} \right]$

solving for P, we have:

$$P = \frac{178 \times 10^{-6} \Delta T}{12/(6 \times 30 \times 10^6) + 8/(4 \times 10.5 \times 10^6)} = 692 \Delta T$$

$$\sigma_a = 24,000 = \frac{P}{A_a} = \frac{692 \Delta T}{4} = 173 \Delta T \quad \text{therefore,} \quad \Delta T = \frac{24000}{173} = 138.5°F$$

Since the bar was subjected to a drop in temperature of 138.5°F, the actual temperature is:

$$T = 70°F - 138.5°F = -68.5°F$$

5. The ratio of stress to strain, within the proportional limit, is called:

    a) the modulus of rigidity
    b) Hooke's constant
    c) Young's modulus
    d) Poisson's ratio
    e) the reversible region

    The answer is (c), Young's modulus.

6. This stress–strain diagram for a certain metal (Fig. 3.32) indicates that

    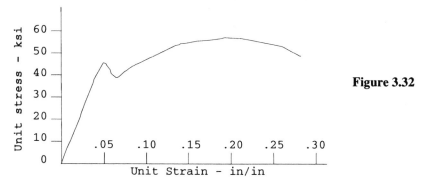

    Figure 3.32

    a) it is probably steel (mild)
    b) it is probably cold rolled Type 304 stainless steel
    c) it is probably monel
    d) it is probably cast iron
    e) it is not a metal at all

    The answer is (a). It is probably mild steel; it has a well defined yield point.

7. Is the material in the stress strain diagram of problem 6:

   a) brittle
   b) rigid
   c) strong
   d) ductile
   e) annealed

   The answer is (d), because it has a well defined yield point.

BEAM ANALYSIS

Types of Beams

Generally beams fall into two classifications: statically determinate and statically indeterminate, where the classification indicates the design approach to be used. A statically determinate beam is one where the number of unknowns (which can be the reactions or loads), are exactly equal to the number of available equations of equilibrium. These unknowns can be found by using the equations of equilibrium.

A statically indeterminate beam is one in which the number of unknowns exceeds the available equations of equilibrium. Solution of this type of problem requires a knowledge of the elastic properties of the beam in order to obtain the necessary additional equations.

Some examples of beam problems that appeared in past examinations are shown in Figs. 3.33 and 3.34.

Types of Loads

Beams often are subjected to many types and combinations of loads as shown in Figs. 3.33 and 3.34.

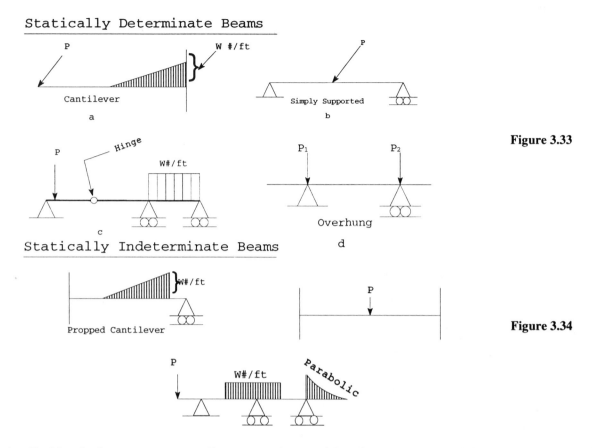

Figure 3.33

Figure 3.34

Applied loads that act over a small area may be considered as <u>concentrated</u>, i.e., acting at a single mathematical point, (Fig. 3.33c). A true single point application, though rarely achieved in practice, is a valid assumption for beam analyses except where local bearing stresses are to be analyzed.

Bending moments generally are also considered as acting at a single point.

The overwhelming majority of beams are analyzed as two-dimensional members with a coplanar loading system.

Characteristics of the more common types of loads are summarized in Fig. 3.35 where the origin of the coordinate system is taken at the left end. In each case the axis of the beam coincides with the positive x-axis while the loads generally act normal to beam and parallel to the y-coordinate axis.

Conditions of Equilibrium

The fundamental assumption in the design of beams (as well as most problems in Statics) is that the beam is in equilibrium under the action of the applied load and the reactions provided at the supports. A beam in equilibrium subjected to loads will deform, to be sure, but as a rigid body it can neither translate nor rotate. If such is the case, we may say that the summation of forces in

MECHANICS OF MATERIALS   3-31

Figure 3.35

BEAM LOAD CHARACTERISTICS

TYPE OF LOAD	LOAD DIAGRAM	RESULTANT OF LOAD, R	LOCATION OF RESULTANT
CONCENTRATED	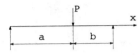	$R = P$	$x = a$
UNIFORMLY DISTRIBUTED	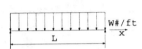	$R = w \times L$	$x = \dfrac{L}{2}$
TRIANGULAR DISTRIBUTION	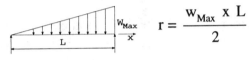	$r = \dfrac{w_{Max} \times L}{2}$	$x = \dfrac{2}{3} L$
TRUNCATED TRIANGULAR	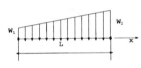	$R = \left[\dfrac{w_1 + w_2}{2}\right] \times L$	$x = \left[1 + \dfrac{w_2}{w_1 + w_2}\right]\dfrac{L}{3}$
TRIANGULAR DISTRIBUTION	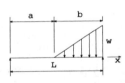	$R = \dfrac{w_b}{2}$	$x = a + \dfrac{2}{3} b$
PARABOLIC DISTRIBUTION $Y = f(X^2)$	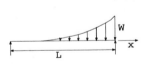	$R = \dfrac{w L}{3}$	$x = \dfrac{3}{4} L$
PARABOLIC DISTRIBUTION $Y = F(x^2)$	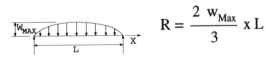	$R = \dfrac{2 w_{Max}}{3} \times L$	$x = \dfrac{L}{2}$
CONCENTRATED BENDING MOMENT	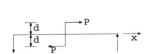	$R = M = 2 P d$	M = Constant for all values of x.

each coordinate direction must equal zero and the algebraic sum of moments about any point must also equal zero.

These equations may be used for the solution of both statically determinate and statically indeterminate beams. However, the indeterminate case as previously mentioned, requires additional equations.

Sign Convention

Before proceeding with the actual calculations, it is necessary to establish a sign convention for forces and moments. Whatever convention you establish is immaterial so long as you remain constant throughout the solution of the problem.

In this text, we shall consider the following convention for axial force, shear, and bending moment at any transverse cross-section of a beam.

<u>Axial force</u> is the load applied at the centroid of the cross-section parallel to the long axis of the beam. It is positive when it tends to pull the beam apart, i.e., when it produces a tensile stress (Fig. 3.36).

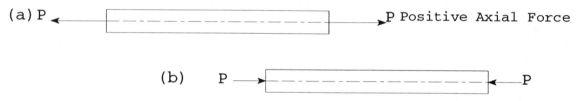

Figure 3.36

<u>Shear force</u> is the transverse force tending to cut the beam as would be done by a large pair of shears. Shear is positive when it tends to push the left portion upward with respect to the right portion (Fig. 3.37).

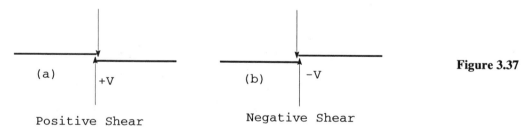

Figure 3.37

<u>Bending moment</u> is the moment experienced by the beam which tends to curve or bend the axis of the beam. The moment is about an axis normal to the plane of loading and is positive when it tends to produce tension in the lower fibers of the beam and compression in the upper fibers (Fig. 3.38).

POSITIVE BENDING MOMENT   NEGATIVE BENDING MOMENT

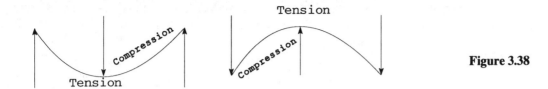

Figure 3.38

Standardizing your approach in setting up equations will help minimize errors. For example, when summing forces, always take those acting to the right as positive and those acting upward as positive. When summing moments, always treat clockwise moments as positive when acting on the portion of the beam to the right of the moment center. Therefore, the equilibrium equations indicating the chosen positive directions are:

$$\rightarrow \Sigma F_X = 0 \qquad \uparrow \Sigma F_Y = 0 \qquad \curvearrowright \Sigma M = 0$$

Reactions

Reactions are the forces exerted on a beam by its supports. The applied loads and the reactions constitute the load system required to hold the beam in static equilibrium.

A hinge support, generally portrayed by the left support of Fig. 3.33c, supplies a reaction passing through the center of the hinge. The magnitude and direction of the reaction are unknown. These two unknowns are most easily handled by resolving them into vertical and horizontal components both acting through the center of the hinge pin. The roller support, on the right of Fig. 3.33b, can move horizontally, therefore, the only reaction must be normal to the rolling surface through the center of the pin.

Fixed supports (Fig. 3.33a) prevent translation <u>and</u> rotation, thus the reaction at each fixed support is composed of three unknown quantities, a force of unknown magnitude and direction plus a bending moment. As before, the reaction is most easily manipulated by resolving it into its components.

Illustrated Problems

1. Find the reactions at the fixed end, B, of the cantilever beam in Fig. 3.39.

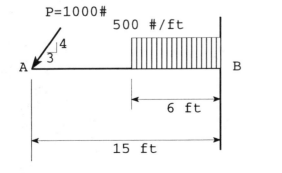

Figure 3.39

Approach

Resolve the force P into its components, apply the reactions at B, and sketch a free body diagram of the entire beam (Fig. 3.40).

Solution

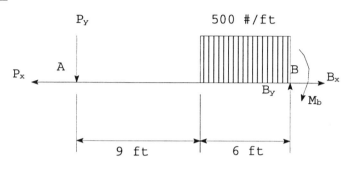

Figure 3.40

Note the components of P are:

$$P_x = \frac{3}{5}P = 600 \text{ lbs}, \quad P_y = \frac{4}{5}P = 800 \text{ lbs}$$

$\circlearrowleft \Sigma M_B = 0: -800(15) - 500(6)(3) + M_B = 0$

$M_B = 12,000 + 9,000 = 21,000$ Ft. lbs <u>Answer</u>

$\rightarrow \Sigma F_X = 0: -600 + B_x = 0$

$B_x = 600$ lbs <u>Answer</u>

$\uparrow \Sigma F_Y = 0 ; -800 - 500(6) + B_y = 0$

$B_y = 3800$ lbs <u>Answer</u>

2. Determine the reaction of the beams at A, B, and C (Fig. 3.41). Neglect the weights of the members.

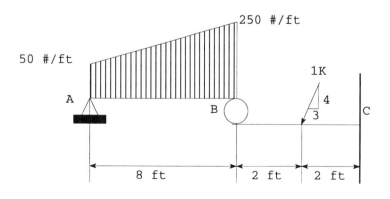

Figure 3.41

Approach

Note that there are two separate beams, where the reaction of one is an applied load on the other. Consider beam $\overline{AB}$ alone. Using equations of equilibrium, find the reactions at A and B. Consider beam $\overline{BC}$ alone with reaction at B as an applied load. Use equations of equilibrium to solve for reactions at C.

Solution

At point A, the intensity of the load is 50 lbs and the intensity at point B is 250 lbs Resolve the distributed load into a rectangle and a triangle and sketch a free body diagram of beam $\overline{AB}$ (Fig. 3.42).

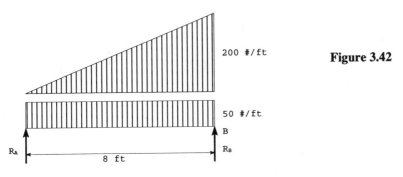

Figure 3.42

Note there are only two equations of equilibrium available here, since there are no horizontal forces $\Sigma F_x = 0$.

$\circlearrowright \Sigma M_A = 0$ (See Fig. 3.35 for Loading Characteristics)

$$50(8)(4) + 200(8)\frac{(1)}{2}(8)\frac{(2)}{3} - 8R_B = 0$$

$$R_B = \frac{50(8)(4) + 200(8)\frac{(1)}{2}\frac{(16)}{3}}{8} = 733.3 \text{ lbs}$$

$\uparrow \Sigma F_Y = 0 \quad -50(8) - 200(8)\frac{(1)}{2} + 733.3 + R_A = 0$

$R_A = 400 + 800 - 733.3 = 466.7 \text{ lbs}$

Now from a free diagram of $\overline{BC}$ (Fig. 3.43):

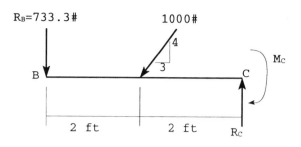

Figure 3.43

$$\circlearrowright \Sigma M_C = 0 \; ; \; -733.3(4) - \frac{4}{5}(1000)(2) + M_c = 0$$

$$M_c = 2933.2 + 1600 = 4533 \text{ ft.-lbs}$$

$$\rightarrow \Sigma F_X = 0 \; ; \; -\frac{3}{5} \times 1000 + R_{c_x} = 0$$

$$R_{c_x} = 600 \text{ lbs}$$

$$\uparrow \Sigma F_Y = 0 \; ; \; -733.3 - \frac{4}{5}(1000) + R_{c_y} = 0$$

$$R_{c_y} = 733.3 + 800 = 1533.3 \text{ lbs}$$

Shear and Bending Moment Diagrams

Shear and bending moment diagrams are graphical representations of the values of shear and bending moment at any point along the length of a loaded beam. These diagrams offer convenient means to determine the maximum values of shear and bending moment as well as their location.

The Shear Force Diagram

The shear diagram is simply a curve whose ordinate at any point on the beam represents, in magnitude and sign, the shear at that point on the beam.

When a section of a loaded beam in equilibrium is isolated, that section must remain in equilibrium under the action of the forces applied to it. The forces experienced by the section may be reactions, loads and internal forces. Consider as an example the equilibrium of section $\overline{AB}$ cut from the simply supported beam in Fig. 3.44a.

(a)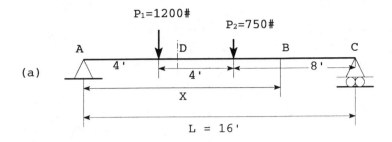

(b)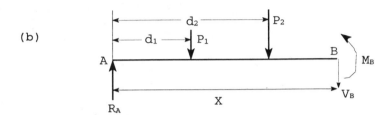

Figure 3.44

A free body diagram of section $\overline{AB}$ is shown in Fig. 3.44b, where the forces experienced by the beam section are the reaction at A, the applied loads P_1 and P_2 and the internal forces at B which are V_B and M_B. The forces V_B and M_B represent the action of the removed portion $\overline{BC}$ on segment $\overline{AB}$. Equilibrium of $\overline{AB}$ exists when the internal forces equilibrate the external forces.

The shearing force at any cross-section equals the algebraic sum of the external forces acting from the left of that section.

$$V_B = R_A - P_1 - P_2$$

Similarly, the shear at section D is:

$$V_D = R_A - P_1$$

Clearly the magnitude of the shear force varies with the coordinate X. At point A, the shear force is zero and just to the right of A (in a distance approaching zero), the magnitude jumps discontinuously to a value of R_A. The shear remains constant for increasing values of X up to the point of application of P_1 where "just to the right" of P_1 the shear force is $(R_A - P_1)$. At $X = L$ the shear curve returns abruptly to zero.

THE SHEAR DIAGRAM ALWAYS STARTS AT ZERO AND TERMINATES WITH ZERO AT THE BEAM ENDS.

Once you have determined the reactions, the shear diagram can be drawn.

For the beam in Fig. 3.44a:

$$\circlearrowright \Sigma M_C = 0 \quad \text{yields} \quad R_A = 1275 \text{ lbs}$$

$$\updownarrow \Sigma F_Y = 0 \quad \text{yields} \quad R_C = 675 \text{ lbs}$$

Then $V_A = R_A = 1275$ lbs

$V_{P_1} = R_A - P_1 = 1275 - 1200 = 75$ lbs

$V_{P_2} = R_A - P_1 - P_2 = 75 - 750 = -675$ lbs

$V_c = R_c = 675$ lbs

The shear diagram is most easily plotted below a sketch of the beam showing all loads, as in Fig. 3.45.

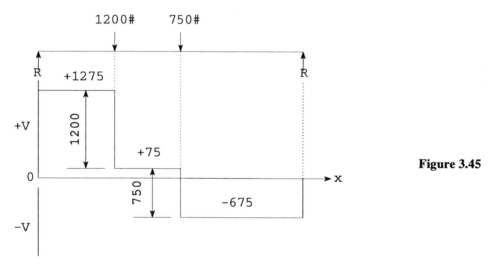

Figure 3.45

From the shear diagram, the value of the shear can be determined by inspection. Note the abrupt change in the value of the shear at the point of application of the loads and that the magnitude of each change exactly equals the magnitude of the applied load.

The Bending Moment Diagram

The bending moment diagram is a curve whose ordinate at any point on the beam represents, in magnitude and sign, the bending moment at that point on the beam.

From Fig. 3.44b, the sum of the bending moments, about point B, caused by the <u>external</u> forces is:

$$M_B = R_A X - P_1 (x - d_1) - P_2 (x - d_2)$$

Clearly, this is not equal to zero and can equal zero only when balanced by the resisting internal bending moment M_B.

> THE BENDING MOMENT AT ANY Cross-section EQUALS THE ALGEBRAIC SUM OF THE BENDING MOMENTS CAUSED BY THE EXTERNAL FORCES ACTING TO THE LEFT OF THAT SECTION.

$$M = R_A X - P_1 (x - d_1) - P_2 (x - d_2)$$

Isolating a segment of the beam, dx in length, between the loads and applying the internal forces acting on the segment, you can draw its free body diagram (Fig. 3.46).

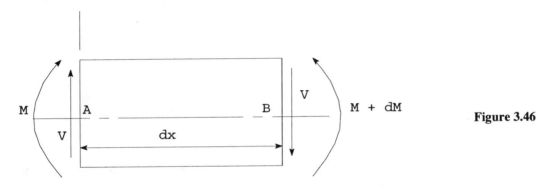

Figure 3.46

Since there is no external load acting on this segment and the weight of the beam is neglected, the shearing forces at the two cross-sections must be equal. The bending moments, however, are not equal.

$$\circlearrowright \Sigma M_B = 0 \quad \text{yields} \quad M - (M + dM) + V\,dx = 0$$

$$dM = V\,dx$$

The increase in the bending moment equals the moment caused by the shear force V. Integrating the above equation:

$$\int_A^B dM = \int_A^B V\,dX$$

$$M_o - M_A = \int_A^B V\,dX$$

$$\left.\begin{array}{l}\text{Change in Bending}\\ \text{Moment between}\\ \text{Cross sections A and B}\end{array}\right\} = \left\{\begin{array}{l}\text{Area under shear curve}\\ \text{between cross sections A and B}\end{array}\right.$$

THE CHANGE IN BENDING MOMENT BETWEEN TWO CROSS-SECTIONS OF A BEAM EQUALS THE AREA UNDER THE SHEAR CURVE BETWEEN THOSE TWO CROSS-SECTIONS.

Returning to the beam in Fig. 3.44a, we proceed to draw the bending moment diagram by using the shear diagram (Fig. 3.47).

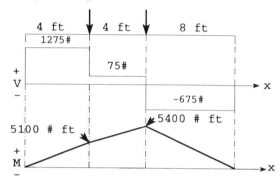

Figure 3.47

The value of the bending moment at x = 4 ft was determined by the area under the shear curve between x = 0 and x = 4 ft, or the product of 1275 x 4. The value of the bending moment at x = 8 ft, was arrived at by adding the area under the shear curve between x = 4 ft and x = 8 ft to the value of the bending moment at x = 4 ft.

M_x @ 8 ft = 5100 + 75 x 4 = 5400 ft-lbs

Since the shear has a negative value when x exceeds 8 ft, consider the area under the shear curve as "negative area" and subtract from the previous value of the bending moment.

M_x @ 16 ft - 5400 - 8 x 675 = 0

The bending moment at the right end returns to zero, as it should since there is a hinged joint there.

Once the bending moment diagram has been drawn, the maximum bending moment and its location are easily determined by inspection.

There is one exception where the bending moment diagram cannot be drawn directly from the shear diagram and that is <u>whenever a concentrated bending moment is applied to the beam</u>. In this case, you merely add algebraically the value of the concentrated moment, at its point of application, to the bending moment found from the shear diagram.

The shear and bending moment diagrams are not always linear; a distributed load produces a parabolic moment curve.

The technique for handling a concentrated bending moment and non-linear moment diagram is shown in the following illustrated example:

Illustrated Example

1. For the beam loaded as shown in Fig. 3.48, determine the shear and bending moment diagrams.

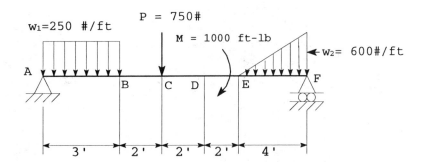

Figure 3.48

Approach

Use equations of equilibrium to solve for reactions at A and F; start at A where shear equals the reaction R_A and draw the shear diagram for the entire beam. From the area under the shear curve draw the moment curve including the concentrated moment M at point D.

Solution

$\circlearrowright \Sigma M_F = 0$

$R_A \times 13 - (W_1 \times 3) \times 11.5 - (P \times 8) + M - (W_2 \times 4) \times \dfrac{1}{2} \times \dfrac{1}{3} \times 4 = 0$

$R_A = \dfrac{34.5\, W_1 + 8P - M + \dfrac{8}{3} W_2}{13}$

$= \dfrac{34.5 \times 250 + 8 \times 750 - 1000 + \dfrac{8}{3} \times 600}{13} = 1171.15 \text{ lbs}$

↑ $\Sigma F_Y = 0$

$(R_A - W_1) \times (3 - P - W_2) \times 4 \times (\frac{1}{2} + R_F) = 0$

$R_F = -R_A + 3 W_1 + P + 2 W_2$

$= -1171.15 + (3 \times 250) + 750 + (2 \times 600) = 1528.85$

Check

↻ $\Sigma M_A = 0$

$250 \times 3 \times \frac{3}{2} + 750 \times 5 + 1000 + 600 \times 4 \times \frac{1}{2} (9 + \frac{8}{3}) - (13 \times 1528.85) = 0$

$1125 + 3750 + 1000 + 14000 = 19875 = 0$; Check

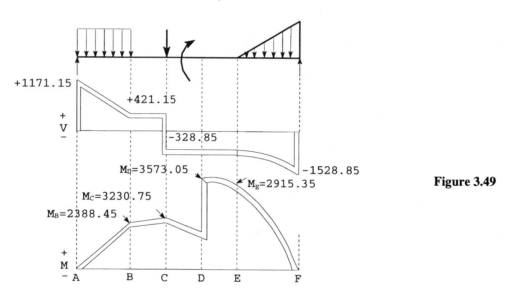

Figure 3.49

To determine the value of the shear at B consider the vertical loads to the left of B.

$V_B = R_A - W_1 \times 3 = 1171.15 - 3 \times 250 = 421.15$ lbs

Note the slope of the straight line between A and B is 250 lbs/ft, the rate at which the uniformly distributed load is applied.

There are no applied loads between B and C, therefore, the shear remains constant. At C, the vertical force P produces a negative value of shear.

$V_c = B_B - P = 421.15 - 750 = -328.85$ lbs

Again, there are no applied shear loads between C and E, hence the shear of -328.85 lbs remains constant between C and E.

To the right of point E the triangularly distributed load does create additional negative shear force within the beam. (Review the sign convention established on p. 3-33.)

Now recall that the shearing force at any cross-section equals the algebraic sum of the external forces acting to the left of that section. Therefore to find the shear at F, add algebraically, the external loads between E and F to the value of shear at E. The load between E and F is the area under the load diagram or - 600 x 4 x ½ = -1200 lbs. Adding this to - 328.85 lbs, the value of shear at E, yields - 1528.85 lbs. The reaction at F is exactly this value and since it is in a positive direction, it closes the shear diagram.

The shape of the shear curve between E and F is a second degree parabola because load (or area under the loading diagram) is being added at a varying rate. Whether the shape of the parabola is convex or concave is easily determined by examining the load between the two adjacent unit lengths of the beam. If the load increases as x increases, then the slope of the shear curve increases as x increases.

Now utilizing the fact that the change in bending moment between two cross-sections of a beam equals the area under the shear curve between those two cross-sections, proceed to draw the bending moment diagram. The area under the shear curve between A and B is (Fig. 3.50):

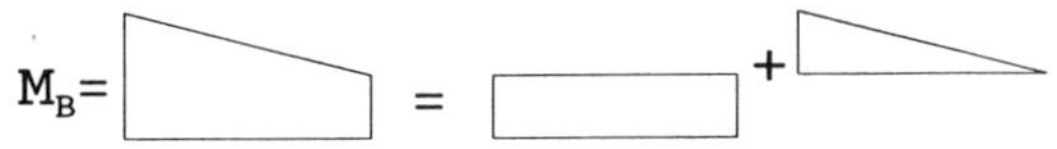

Figure 3.50

$= (1171.15 - 421.15) \times 3 \times ½ + 421.15 \times 3 = 2388.45$ ft.-lbs

This is the magnitude of the moment at B. The question is, how does the moment curve vary between A and B? Note that the shear is decreasing as x increases, therefore the slope of the moment curve must be decreasing as x increases. In other words shear area is added at a decreasing rate. Between B and C, shear area is added at a constant rate and the moment curve results in a straight line. Between C and D, shear area is constant, but negative, and must be subtracted at a constant rate.

Since only forces to the left of the point of investigation are considered, the concentrated moment does not influence the moment diagram until the point of investigation is to the right of point D, the point where the concentrated moment is applied. At the right of point D the clockwise concentrated moment is seen as causing tensile stresses in the lower fibers, and compression in the upper fibers; a positive moment. This moment must be added to the magnitude of the moment at D; see Fig. 3.49.

The moment between E and F is determined by adding the (negative) shear area between point E and the point under investigation.

3-44 FUNDAMENTALS OF ENGINEERING EXAM REVIEW WORKBOOK

The moment at F is:

$M_F = M_E$ minus

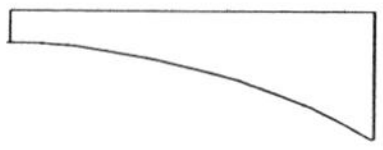

Figure 3.51

$= 2915.35 -$

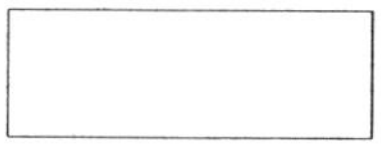

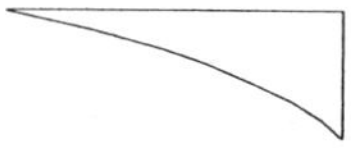

$= 2915.35 - 328.85 \times 4 - 1/3 \times 4 \, (1528.85 - 328.85)$

$= 0$

Which it must be, since there is a hinged joint at F that cannot transmit a moment.

The shear and bending moment can also be found for any point along the beam through the equations for shear and moment. Let point A denote the origin of the x-axis and consider only forces to the left of the point of investigation.

$$V_{x_0 < x < 3} = R_A - w_1(x)$$

$$V_{x_0 < x < 3} = R_A x - w_1 \left(\frac{x^2}{2}\right)$$

$$V_{x_3 < x < 5} = R_A - w_1(3)$$

$$M_{x_3 < x < 5} = R_A x - 3w_1(x - 1.5)$$

$$V_{x_5 < x < 7} = R_A - 3w_1 - P$$

$$M_{x_5 < x < 7} = R_A x - 3w_1(x - 1.5) - P(x - 5)$$

$$V_{x_7 < x < 9} = R_A - 3w_1 - P$$

$$M_{x_7 < x < 9} = R_A x - 3w_1(x - 1.5) - P(x - 5) + M$$

$$V_{x_9 < x < 13} = R_A - 3w_1 - P - \frac{w_2}{4}(x - 9)\frac{(x - 9)}{2}$$

$$V_{x_9 < x < 13} = R_A x - 3w_1(x - 1.5) - P(x - 5)$$

$$+ M - \frac{W_2}{2(4)} (x-9)^2 \frac{(x-9)}{3}$$

Note W_2 is the maximum intensity of the triangularly distributed load in lbs/ft.

M in each moment equation is obtained by evaluating the constant of integration when each shear expression is integrated.

The Flexure Formula

Consider the FBD of the beam in Fig. 3.44b and examine the internal stresses produced by the bending moment at point B. Our theory is based upon the assumption that plane sections before bending remain plane after bending and therefore produce a linear stress distribution (Fig. 3.52).

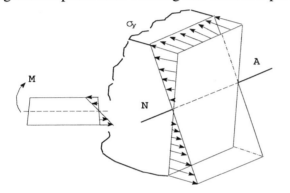

Figure 3.52

For equilibrium, the external bending moment at the point under investigation must equal the resisting moment produced by the internal stresses (Fig. 3.53).

Figure 3.53

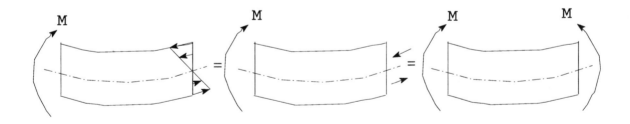

The tensile or compressive stress on a fiber is proportional to its distance from the neutral axis. The neutral axis is the line in the plane of the cross-section where the fiber stress is zero (Fig. 3.54).

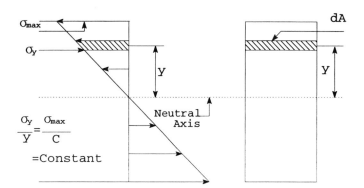

Figure 3.54

Imagine the beam to be composed of layers and that each layer produces a force that resists the applied bending moment.

Force on area of one layer = $\sigma_y dA$

Moment of force on layer = $\sigma_y(y dA)$

Sum of moments of all forces on all layers is:

$$M_r = \int \sigma_y (y\, dA) = \int \sigma_y\, y\, \frac{(y)}{y}\, dA = \frac{\sigma y}{y} \int y^2 dA$$

but

$$\frac{\sigma y}{y} = \frac{\sigma_{max}}{c}$$

therefore

$$M_r = \frac{\sigma_{max}}{c} \int y^2\, dA$$

but $\int y^2\, dA = I =$ moment of inertia of cross-sectional area.

$$M_r = \frac{\sigma_{max} I}{c} = \text{resisting moment due to internal stresses.}$$

For equilibrium $M = M_r$ and $M = \dfrac{\sigma_{max} I}{c}$

$$\sigma_{max} = \frac{Mc}{I} \quad \text{where M is in in.-lbs}$$

$$c = \text{in., and } I = \text{in}^4.$$

Position of Neutral Axis

The value of I in the flexure formula cannot be found unless the position of the neutral axis in the cross-sectional area is known. Recall the condition of equilibrium $\Sigma F_x = 0$. In Fig. 3.54 note that

$$\int \sigma_y \, dA = 0 \quad \text{or} \quad \frac{\sigma}{y} \int y \, dA = 0$$

Since $\frac{\sigma}{y}$ = constant but $\frac{\sigma y}{y}$ cannot be zero.

$$\int y \, dA = 0$$

where y is measured from the neutral axis.

$\int y \, dA$ is often called the Moment of Area of the cross-section or $A \bar{y}$.

$$\int y \, dA = A \bar{y}$$

where A = total cross-sectional area, and $\bar{y}$ is the distance from the neutral axis to the centroidal axis.

But A is not zero, therefore $\bar{y} = 0$ and the centroidal axis is coincident with the neutral axis.

The centroidal axis (or neutral axis) may easily be found by taking the moment of the area about some convenient reference axis (Fig. 3.55).

$$\Sigma M_{ref} = A_1 y_1 + A_2 y_2 = \Sigma A_i y_i$$

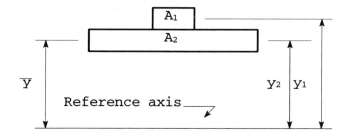

Figure 3.55

If the entire area is assumed to be at $\bar{y}$, the distance from the reference axis to the centroidal axis, then the moment about the reference axis would be (Fig. 3.56):

$$M_{ref} = (A_1 + A_2) \bar{y} = \bar{y} \Sigma A_i$$

equating moments

$$\Sigma A_i y_i = \bar{y}\Sigma A_i \text{ therefore } \bar{y} = \frac{\Sigma A_i y_i}{\Sigma A_i}$$

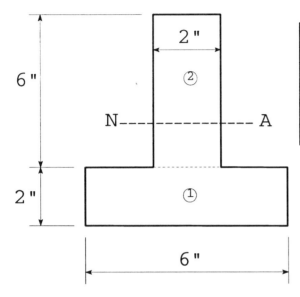

Part	Area	y	Ay
①	2 x 6 = 12	1	12
②	2 x 6 = 12	5	60
	$\Sigma A = 24$		$\Sigma Ay = 72$

$$\bar{y} = \frac{72}{24} = 3 \text{ in.}$$

Figure 3.56

Moment of Inertia of a Rectangle

Moment of inertia = I = second moment of area = $\int y^2 \, dA$

Consider the rectangle section shown in Fig. 3.57, where the height is h and the neutral axis is at h/2, the width is b, and y is measured from the neutral axis.

$$I_{NA} = \int y^2 \, dA \quad \text{where} \quad dA = b \, dy$$

$$I_{NA} = \int_{-h/2}^{+h/2} y^2 \cdot b \cdot dy = b \int y^2 dy = \left. \frac{b y^3}{3} \right|_{-h/2}^{+h/2}$$

$$= \frac{b\left[\frac{h}{2}\right]^3}{3} - \frac{b\left[-\frac{h}{2}\right]^3}{3} = 2\frac{bh^3}{24} = \frac{bh^3}{12}$$

$$I_{base} = b \int_0^h y^2 dy = \left. \frac{by^3}{3} \right|_0^h = \frac{bh^3}{3}$$

Parallel Axis Theorem

The moment of inertia about any axis parallel to the neutral axis can be found by considering the second moment of the area about the axis required.

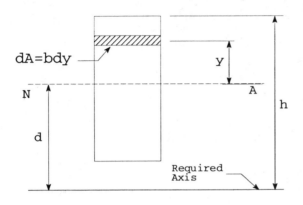

Figure 3.57

$$I_b = \int (y + d)^2 dA = \int y^2 dA + 2d\int y \, dA + d^2 \int dA$$

where the $\int y^2 \, dA = I_{NA}$

$d^2 \int dA = d^2 A$ since $\int y dA = A . \bar{y}$

$2d \int y . da = 2dA \bar{y}$

but $\bar{y}$ is the distance from the centroidal axis to the neutral axis which was shown to be zero.

$$2d \int y dA = 0$$

and $\underline{I_b = I_{NA} + AD^2}$

The values of the moment of inertia for various cross-section appear in most engineering handbooks.

Moments of Inertia of Composite Sections

The moments of inertia of composite sections are found by dividing the composite section into several simple parts. The required moment of inertia can be determined by adding, or subtracting, the effects of component areas (Fig. 3.58).

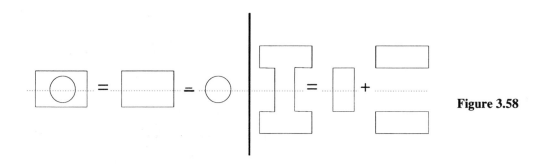

Figure 3.58

General Procedure of Composite I

1. Divide cross-section into several simple parts with known I_{cg}.

2. Locate the neutral axis (NA).

3. Again divide composite section into convenient areas with respect to the NA.

4. Calculate I of each area about its own centroid.

5. Use the parallel axis theorem to obtain I about the composite NA.

Illustrated Example

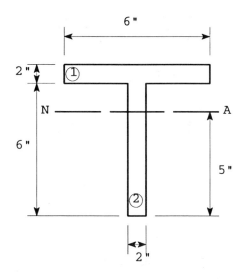

Part	Area	y	Ay
①	12	7	84
②	12	3	36
	ΣA = 24		ΣAy = 120

$\bar{y} = 120/24 = 5$ in.

Figure 3.59

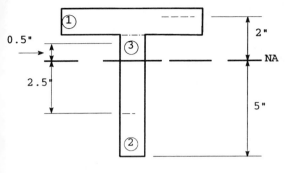

Part	Area	$I_o = \dfrac{bh^3}{12}$	d	d^2	Ad^2
①	12	$\dfrac{48}{12} = 4$	2	4	48
②	10	20.83	2.5	6.25	62.5
③	2	.167	.5	.25	.5

$$\Sigma I_o = 25 \qquad \Sigma Ad^2 = 111$$

$$I_{NA} = 25 + 111 = 136 \text{ in.}^4$$

Figure 3.60

SOLVED PROBLEMS

1. a) Draw the shear and bending moment diagrams for the cantilever beam (Fig. 3.61).

   b) Locate the maximum bending moment.

   c) Determine the maximum bending stress.

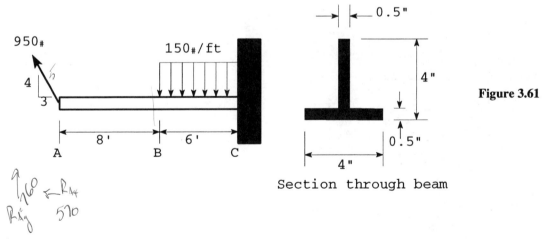

Figure 3.61

Section through beam

Approach

Resolve the 950-lb load into horizontal and vertical components, find the reactions, and draw the shear and bending moment diagrams (Fig. 3.61). The coordinate where the shear is zero locates the maximum bending moment. Use the area under shear diagram to determine maximum moment. Locate the neutral axis of the section, and determine I of section. Use the flexure formula to find maximum stress (Fig. 3.62).

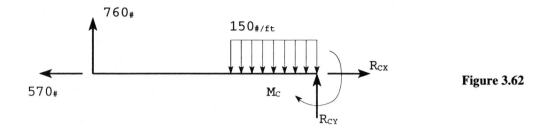

Figure 3.62

↑ $\Sigma F_y = 0$; $760 - 150 \times 6 + R_{cy} = 0$

$$R_{cy} = 900 - 760 = 140 \text{ lbs}$$

→ $\Sigma F_x = 0$; $-570 + R_{cx} = 0$; $R_{cx} = 570 \text{ lbs}$

↺ $\Sigma M_c = 0$; $760 \times 14 - 150 \times 6 \times 3 + M_c = 0$

$$M_c = -10640 + 2700 = -7940$$

Before proceeding, correct for direction of M_c (Fig. 3.63).

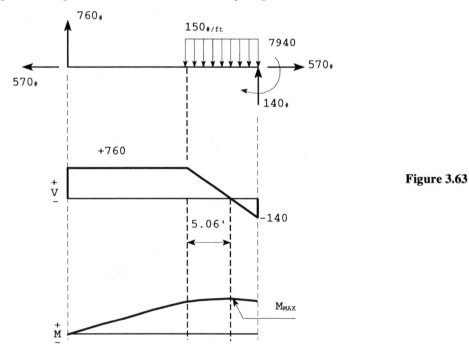

Figure 3.63

$V_D = 760 - 150X = 0$

$X = \dfrac{760}{150} = 5.06$ ft to the right of B

$M_{max} = 760 \times 8 + 5.06 \times \dfrac{760}{2} = 8000$ ft-lbs.

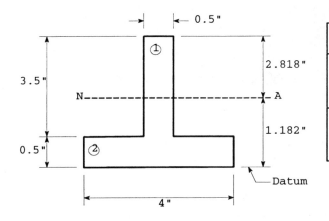

Part	Area	Y	Ay
①	$\dfrac{1}{2} \times 3.5 = 1.75$	2.25	3.94
②	$\dfrac{1}{2} \times 4 = 2$	.25	.5

$\Sigma A = 3.75$ in.^{2} $\Sigma A_y = 4.44$ in.^{3}

$\bar{y} = \dfrac{4.44}{3.75} = 1.182$ in.

Figure 3.64

Calculation for I

Part	A	I_o	d	d^2	Ad^2
①	1.75	$\dfrac{\dfrac{1}{2} \times \overline{3.5^3}}{12} = 1.785$	1.068	1.14	2.00
②	2	$\dfrac{4 \times \dfrac{1}{8}}{12} = .042$	.932	.870	1.74

$\Sigma I_o = 1.827$ $\Sigma Ad^2 = 3.74$

$I_{NA} = I_o + Ad^2 = 5.567$ in.^{4}

$\sigma_{max} = \dfrac{Mc}{I} = \dfrac{8000 \text{ ft-lbs} \times 12 \frac{\text{in}}{\text{ft}} \times 2.818}{5.567} = 48,600$ psi

2. a) Draw the shear and bending moment diagram for this simply supported beam, loaded as shown in Fig. 3.65.

b) Locate and determine the maximum bending stress. (Hint: check the maximum stress in each section.)

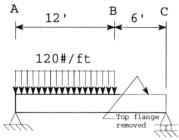

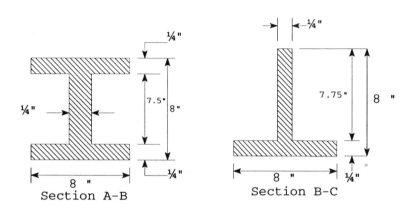

Figure 3.65

Approach (Figs. 3.66–3.69)

Use the same general approach as that of problem 1.

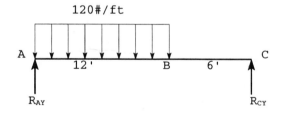

Figure 3.66

$\circlearrowright \Sigma M_A = 0$: $120 \times 12 \times 6 - R_{cy} \times 18 = 0$

$R_{cy} = \dfrac{120 \cdot 12 \cdot 6}{18} = 480$ lbs

$\rightarrow \Sigma F_y = 0$: $R_{ay} - 120 \times 12 + 480 = 0$; $R_{ay} = 960$ lbs

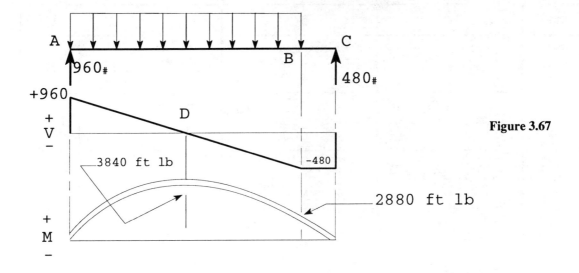

Figure 3.67

$V_D = 960 = 120X = 0 \qquad X = \dfrac{960}{120} = 8 \text{ ft}$

$M_D = 960 \cdot 8 \cdot 1/2 = 3840 \text{ ft-lbs}; \; M_B = 480 \cdot 6 = 2880 \text{ ft-lbs}$

Due to symmetry the NA is 4 in. from base.

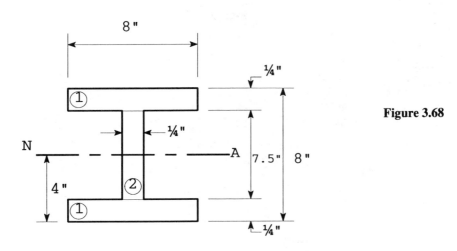

Figure 3.68

Part	A	I_o	d	d^2	Ad^2
①	2	$\frac{8}{12}\frac{1}{64} = .01$	3.875	15	30
①	2	.01			30
②	$\frac{15}{2}\frac{1}{4} = 1.875$	$\frac{1}{4}\frac{\overline{7.5^3}}{12} = 8.80$	0	0	0

$\Sigma I_o = 8.82$ $\Sigma Ad^2 = 60$

$I_{N.A.})_{A-B} = 68.82$ in⁴ for Section A-B
Now locate NA for beam Section B-C.

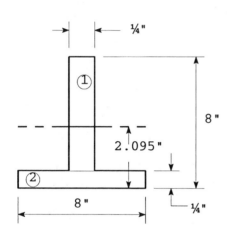

Figure 3.69

Part	A	y	Ay
①	$7.75 \times \frac{1}{4} = 1.94$	4.125	8
②	2	$\frac{1}{8}$	.25

$\Sigma A = 3.94$ $\Sigma A_y = 8.25$

$\bar{y}\ \frac{8.25}{3.94} = 2.09$ in.

Part	A	I_o	d	d^2	Ad^2
①	1.94	$\dfrac{1}{4}\dfrac{\overline{7.75^3}}{12} = 9.72$	2.095	4.4	8.5
②	2	$\dfrac{8\left(\frac{1}{4}\right)^3}{12} = .01$	1.905	3.625	7.25

$$\Sigma I_o = 9.73 \qquad \Sigma Ad^2 = 15.75$$

$I_{N.A.})_{BC} = 9.73 + 15.75 = 25.48 \text{ in.}^4$

$\sigma_D = \dfrac{3840 \times 12 \times 4}{68.82} = 2680 \text{ psi} = $ Max. stress in section A-B of beam

$\sigma_B = \dfrac{2880 \times 12 \times 5.97}{25.48} = 8100 \text{ psi} = $ Max. stress in section B-C of beam

3. Determine the moment of inertia about the horizontal neutral axis (Fig. 3.70).

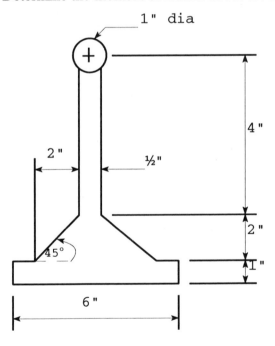

Figure 3.70

Approach (Fig. 3.71)

Break up the composite shape into elemental areas whose data can be taken from available tables. Determine the location of the neutral axis and then use the Parallel Axis Theorem.

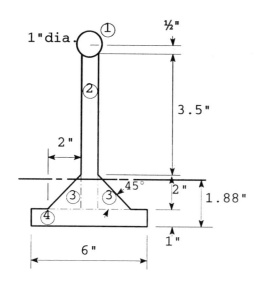

Figure 3.71

Part	A	y	A y
①	$\frac{\Pi}{4} = .785$	7	5.50
②	$\frac{1}{2} \times 5.5 = 2.75$	3.75	10.30
③	$2 \times 2 \times \frac{1}{2} = 2$	1.67	3.33
③	2	1.67	3.33
④	6 x 1 = 6	0.5	3

$\Sigma A = 13.535$ $\qquad$ $\Sigma A_y = 25.46$

$$\bar{y} = \frac{\Sigma A\, y}{\Sigma A} = \frac{25.46}{13.535} = 1.88 \text{ in.}$$

Part	A	I_o	d	d^2	Ad^2
①	.785	.0491	5.12	26.2	20.55
②	2.75	$\dfrac{.5\,(5.5)^3}{12} = 6.932$	1.87	3.50	9.63
③	2	.444	.21	.04	.08
③	2	.444	.21	.04	.08
④	6	$\dfrac{6(1)^3}{12} = .5$	1.38	1.90	11.40

$$\Sigma I_o = 8.369 \qquad \Sigma Ad^2 = 41.74$$

$I_{NA} = \Sigma I_o + \Sigma Ad^2 = 8.369 + 41.74 = 50.11 \text{ in}^4$

4. A simply supported beam, 18 ft long, carries a uniformly distributed load of 7^k/ft from the left support to a point 8 ft to the right and 4^k/ft for the rest of the beam. The vertical shear 7 ft from the left end is

   a) 15.33 kips   d) 56.0 kips
   b) 5.67 kips   e) None of these
   c) 49.0 kips

Approach (Fig. 3.72)

Use equations of equilibrium. Take a summation of moments about the right end, evaluate the left end reaction, and sum all loads to the left of the 7ft mark.

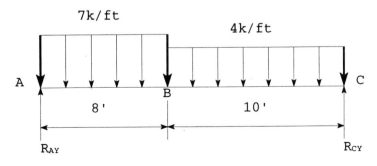

Figure 3.72

$\curvearrowright \Sigma M_C = 0$: $R_{ay} \times 18 - 7 \times 8 \times 14 - 4 \times 10 \times 5 = 0$

$$R_{ay} = \frac{7 \times 8 \times 14 + 4 \times 10 \times 5}{18} = 54.67^k$$

$V_7 = 54.67^k - 7 \text{ k/ft} \times 7 \text{ ft} = 5.67^k = $ shear @ $X = 7$ ft

5. Calculate the reactions at the supports, neglect the weight of the beam, and assume the connection at B is a rigid right angle (Fig. 3.73). Hint: There are two components to the reaction at C.

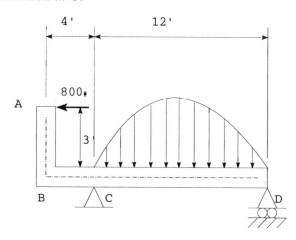

Figure 3.73

Symmetrical/parabolic distribution maximum intensity = 600 lbs at center.

Approach (Fig. 3.74)

Use $\Sigma M_C = 0$ to solve for the reaction at D, $\Sigma F_y = 0$ to solve for the reaction C, and $\Sigma F_x = 0$ to find the reaction in the horizontal direction at C.

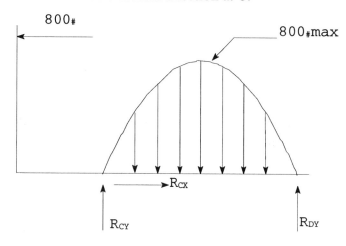

Figure 3.74

$\circlearrowleft^+ \Sigma M_C = 0: \quad \dfrac{2}{3} \times 600 \times 12 \times \dfrac{12}{2} - 800 \times 3 \times R_{Dy} \times 12 = 0$

$R_{DY} = 2200$ lbs

$\updownarrow \Sigma F_y = 0$: $-\dfrac{2}{3} \times 600 \times 12 + R_{Dy} + R_{Cy} = 0$; $-4800 + 2200 = -R_{Cy}$

$$R_{Cy} = 2600 \text{ lb.}$$

$\rightarrow \Sigma F_x = 0$: $-800 + R_{Cx} = 0$; $R_{Cx} = 800$ lbs

6. Determine the reactions at A and B (Fig. 3.75).

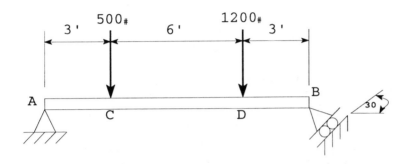

Figure 3.75

Approach

Use equations of equilibrium to solve for the reactions:

$\circlearrowleft \Sigma M_B = 0$ yields R_{Ay} $\circlearrowleft \Sigma M_B = 0$: $R_{Ay} \times 12 - 500 \times 9 - 1200 \times 3 = 0$

$$R_{Ay} = \dfrac{4500 + 3600}{12} = 675 \text{ lbs}$$

$\updownarrow \Sigma F_y = 0$ yields R_{BY} $\updownarrow \Sigma F_y = 0$: $R_{By} - 500 - 1200 + 677 = 0$

$$R_{By} = 1023 \text{ lbs}$$

$$R_B = \dfrac{1023}{\cos 30} = 1182 \text{ lbs}$$

$\rightarrow \Sigma F_x = 0$ yields R_{Bx} $\rightarrow \Sigma F_x = 0$: $R_{Bx} = 1182 \sin 30 = 591$ lbs

7. What are the reactions at the built in end of this cantilever beam (Fig. 3.76)?

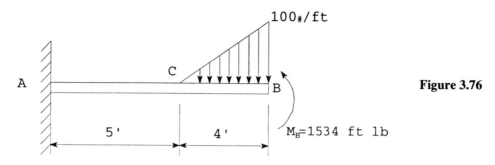

Figure 3.76

Approach (Fig. 3.77)

Use equilibrium equations to determine the reactions

$\uparrow \Sigma F_y = 0$ yields R_{Ay}

$\circlearrowleft^+ \Sigma M_A = 0$ yields the moment at the wall.

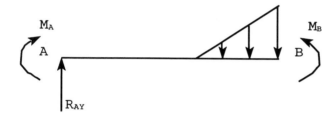

Figure 3.77

$\uparrow \Sigma F_y = 0$: $R_{Ay} - \dfrac{1}{2} \times 100 \times 4 = 0$ $R_{AY} = \frac{1}{2} \times 100 \times 4 = 200$ lbs

$\circlearrowleft^+ \Sigma M_A = 0$: $M_A + 200 \times (5 + \dfrac{8}{3}) - 1534 = 0$ $M_A = 1534 - 200 (7.67) = 0$

8. Neglect the weight of the beams, assume rigid connections at C and D and determine the reactions at A and B (Fig. 3.78).

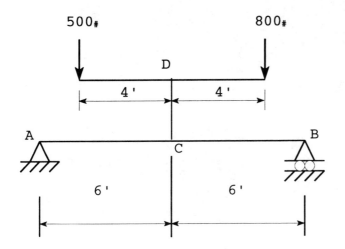

Figure 3.78

Approach

Use equations of equilibrium. Assume that the forces act on beam AB where their lines of action intersect the beam (Fig. 3.79).

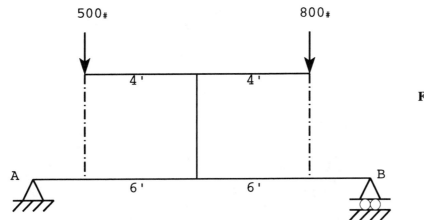

Figure 3.79

$\circlearrowright \Sigma M_A = 0 \quad 500 \times 2 + 800 \times 10 - 12 R_{By} = 0$

$\qquad R_{By} = 750$ lbs acting upward

$\uparrow \Sigma F_y = 0 \quad R_{Ay} - 500 - 800 + 750 = 0$

$\qquad R_{Ay} = 550$ lbs acting upward

9. If the supporting spring at A transmits a load of 192 lbs, what is the value of the uniformly distributed load W (Fig. 3.80)?

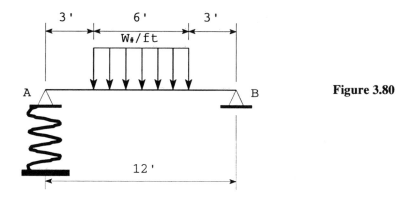

Figure 3.80

Approach

The spring force is the reaction at A; therefore the unknowns are the load intensity W and the reaction at B. Use the equations of equilibrium to solve for the reactions.

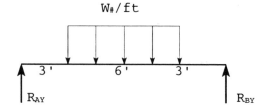

Figure 3.81

$\circlearrowleft \Sigma M_B = 0$: $192 \times 12 - W \times 6 \times 6 = 0$

$$W = \frac{192 \times 12}{36} = 64 \text{ lbs/ft}$$

10. Find the reactions at point B of this idealized version of the lift acting upon the wing beam of an airplane (Fig. 3.82).

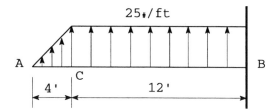

Figure 3.82

Approach

Use summation of vertical forces = 0 to solve for R_{By} and summation of moments about B to solve for the moment at the wall.

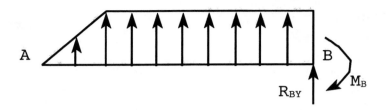

Figure 3.83

$+\uparrow \Sigma F_y = 0: \quad \dfrac{1}{2} \times 25 \times 4 + 25 \times 12 + R_{By} = 0$

$R_{By} = -350$ lbs or 350 lbs acting downward

$\circlearrowright^+ \Sigma M_B = 0: \quad 25 \times 12 \times 6 + \dfrac{1}{2} \times 25 \times 4 \,(12 + \dfrac{4}{3}) + M_B = 0$

$M_B = -2466.5$ ft-lbs or

2466.5 ft-lbs counterclockwise

Combined Stress

There are times when an axial stress will add to or subtract from a bending stress. In such cases the <u>maximum</u> <u>stress</u> is usually called for.

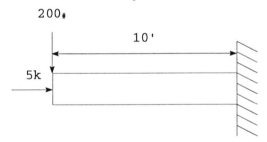

Figure 3.84

Beam is 2 in. x 2 in. cross-section.

What is the maximum stress in the beam?

a. +16750 psi   b. -19250 psi
c. +36000 psi   d. -37250 psi
e. +1250 psi

$$\sigma_{bend} = \dfrac{Mc}{I} = \dfrac{200 \left(10 \text{ in} \times \dfrac{12 \text{ in}}{\text{ft}}\right)}{\dfrac{2 \times 2^3}{12}1} = 18000 \text{ psi}$$

$$\sigma_{axial} = \frac{P}{A} = \frac{-5000}{2 \times 2} = -1250 \text{ psi}$$

σ_{bend} is + on the upper surface (tension) and - on the lower surface (compression). σ_{axial} is in compression.

On the upper surface, $\sigma_{max} = +18000 - 1250 = 16750$ psi
On the lower surface, $\sigma_{max} = -18000 - 1250 = -19250$ psi
$\therefore \sigma_{max} = -19250$ psi

The answer is b).

Formulas for Normal and Shear Stress

Figure 3.85 shows an elastic material under edge loading consisting of stresses S_x and S_y and S_{xy} ($S_{xy} = S_{yx}$). Note these stresses act on surfaces described by the x- and y-axes. That is, S_x acts perpendicular to (normal to) the surface having the x-axis as its normal direction and S_{xy} acts in the same plane but parallel to the plane.

S_x, S_y and S_{xy} may be considered reference stress values. S_n and S_s are stresses acting on an inclined plane ϕ degrees counterclockwise from the x plane as shown in Figure 3.85. Summation of forces in the n and s direction lead to formulas for S_n and S_s.

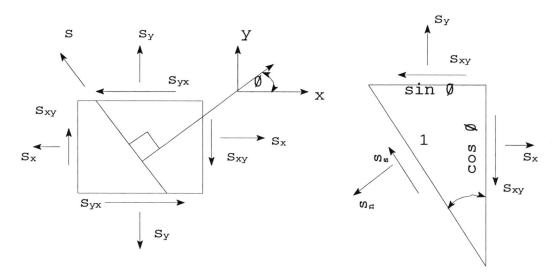

Figure 3.85

$$S_n = S_x \cos^2 \phi + S_y \sin^2 \phi - S_{xy} \sin \phi \cos \phi \qquad (1)$$

$$S_s = (S_x - S_y) \sin \phi \cos \phi + S_{xy} (\cos^2 \phi - \sin^2 \phi) \qquad (2)$$

Using formulas from trigonometry we have

$$S_n = \frac{S_x + S_y}{2} \pm \frac{S_x - S_y}{2} \cos 2\phi \pm S_{xy} \sin 2\phi \qquad (1a)$$

$$S_s = -\frac{S_x - S_y}{2} \sin 2\phi + S_{xy} \cos 2\phi \qquad (2a)$$

Planes of Maximum or Minimum Normal Stress (Principal Stresses–Principal Planes)

Using calculus and Eq. (1a) obtain the maximum (or minimum) of S_n [S_x, S_y, S_{xy} - constants; ϕ variable]

$$\frac{dS_n}{d\phi} = 0 = -\frac{S_x - S_y}{2}(\sin \overline{2\phi}) - S_{xy} \cos \overline{2\phi}$$

$$\tan(\overline{2\phi}) = -\frac{2 S_{xy}}{S_x - S_y} \qquad (3)$$

ϕ is the counterclockwise angle to plane of max(min) S_n from reference plane (x plane)

Note: Two values of ϕ satisfy this equation $0 \leq \overline{2\phi} \leq 360°$. One value is the plane of $(S_n)_{max}$. The other is the plane of $(S_n)_{min}$. Since the two values of 2ϕ are 180° apart, the values of ϕ are 90° apart.

Planes of Zero Shear, $S_s = 0$

Using Eq. (2a)

$$S_s = 0 = \frac{S_x - S_y}{2} \sin 2\phi + S_{xy} \cos 2\phi$$

This is exactly the same equation used to obtain $\tan \overline{2\phi}$. Hence the principal planes are not only planes of maximum or minimum normal stress. They are also planes of zero shear stress.

Planes of Maximum or Minimum Shear Stress

Using $\dfrac{dS_n}{d\phi} = 0$ with Eq. (2a)

$$0 = \frac{S_x - S_y}{2} \cos \overline{\overline{2\emptyset}} - S_{xy} \sin \overline{\overline{2\emptyset}}$$

$$\tan \overline{\overline{2\emptyset}} = \frac{S_x - S_y}{2 S_{xy}} \tag{4}$$

Note: $\tan \overline{\overline{2\emptyset}} = -\dfrac{1}{\tan \overline{2\emptyset}}$ The angle $2\emptyset$ is perpendicular to $\overline{\overline{2\emptyset}}$. This means that the plane described by $\overline{\overline{\emptyset}}$ is 45° away from the plane described by $\overline{\emptyset}$.

The Following Statements Can Be Made

1. The principal stresses occur on planes that are 90° apart.

2. The planes of maximum shearing stress are 90° apart.

3. The maximum and minimum normal stresses (principal stresses) occur on those planes where the shear stress is zero.

4. The planes of maximum shearing stress make angles of 45° with the planes of principal stress.

Mohr's Circle–Graphical Representation

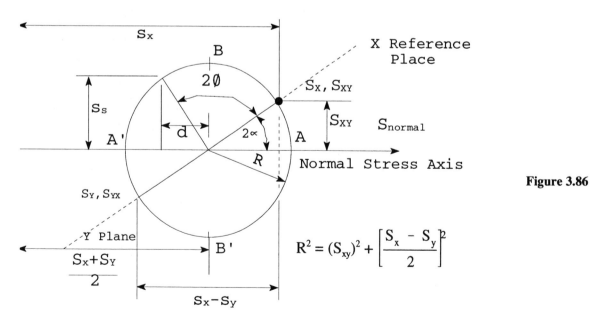

Figure 3.86

Reference planes (x and y) are 90° apart in material, 180° apart in plot.

Verification of Drawing

$$\sin(2\alpha + 2\theta) = \frac{S_s}{R} = \sin 2\alpha \cos 2\theta + \cos 2\alpha \sin 2\theta$$

$$\frac{S_s}{R} = \frac{S_{xy}}{R} \cos 2\theta + \frac{S_x - S_y}{2R} \sin 2\theta$$

$$S_s = \frac{S_x - S_y}{2} \sin 2\theta + S_{xy} \cos 2\theta \quad \text{Same as (2a)}$$

$$\cos(2\alpha + 2\theta) = \frac{d}{R} = \frac{\frac{S_x + S_y}{2} - S_n}{R} = \sin 2\alpha \sin 2\theta - \cos 2\alpha \cos 2\theta$$

$$\frac{\frac{S_x + S_y}{2} - S_n}{R} = \frac{S_{xy}}{R} \sin 2\theta - \frac{S_x - S_y}{2R} \cos 2\theta$$

$$S_n = \frac{S_x + S_y}{2} - S_{xy} \sin 2\theta + \frac{S_x - S_y}{2} \cos 2\theta \quad \text{Same as (1a)}$$

Rules for Plotting Mohr's Circle

1. Normal stress—horizontal axis tension positive to right

2. Shear stress—vertical axis positive shear couples are clockwise (Note Shear couples on x faces Fig. 3.85)

3. Radius value = $\sqrt{(S_{xy})^2 + \left(\frac{S_x - S_y}{2}\right)^2}$

   Center location always on horizontal axis = $\left(\frac{S_x + S_y}{2}\right)^2$

4. Points A and A' (Fig. 3.86) principal stresses ($\overline{2\theta}$ from x reference stress, angle $\overline{2\theta}$ not indicated in figure)

5. Points B and B' (Fig. 3.86) maximum shear stress ($\overline{\overline{2\theta}}$ from x reference stress, angle $\overline{\overline{2\theta}}$ not indicated in figure)

APPENDIX A
COMBINED STRESS - (EXAMPLE)

The following stresses act at a point in a stressed body: $S_x = -600$ psi; $S_y = 1600$ psi; $S_{xy} = -S_{yx} = 1000$ psi. Find the magnitudes of the normal, shearing, and resultant stresses acting on (1) principal planes, (2) planes of maximum shearing stress, (3) the plane whose normal makes an angle of 30° with the x-axis (counterclockwise).

1. Refer to P.XXX, points A and A', principal stress points.

$$S_n = \frac{S_x + S_y}{2} \pm \sqrt{\left[\frac{S_x - S_y}{2}\right]^2 + (S_{xy})^2}$$

S_n = Abscissa of center ± radius of circle

$$= \frac{-600 + 1600}{2} \pm \sqrt{\left(\frac{-600-1600}{2}\right)^2 + (1000)^2} = 500 \pm 1486.6$$

$S_1 = 1986.6$ psi ; $S_2 = -986.6$ psi

$$\text{Resultant stress} = \sqrt{(S_{normal})^2 + (S_{shear})^2}$$

Hence resultant stresses are the same as S_1 and S_2 as the shear stress is zero on the principal planes

Location of principal planes from reference planes ($\overline{\emptyset}$ counterclockwise angle in material, $\overline{2\emptyset}$ counterclockwise angle on Mohr's Circle)

$$\text{Tan } \overline{\overline{2\emptyset}} = -\frac{2 S_{xy}}{S_x - S_y} = \frac{-2(1000)}{-600-1600} = \frac{1}{1.1} = .9090$$

$\overline{\overline{2\emptyset}} = 42.27°, -137.73°$ (negative angle clockwise)

2. Planes of maximum shearing stress (see P.XXX)

$$\text{Tan } \overline{\overline{2\emptyset}} = -\frac{S_x - S_y}{2 S_{xy}} = \frac{-600-1600}{2(1000)} = -1.1$$

$\overline{\overline{2\emptyset}} = -47.73°, 132.27°$

Note: $\tan \overline{\overline{2\phi}} = -\dfrac{1}{\tan \overline{2\phi}}$ Hence perpendicular (see P.XXX) to principal planes

$-47.73 + 90° = 42.27°$

$132.27° + 90° = 222.27° - 360° = -137.73°$

Maximum shearing stress—radius of Mohr's Circle

$$S_{s\,max} = \sqrt{\left[\dfrac{S_x - S_y}{2}\right]^2 + (S_{xy})^2} = 1486.6 \text{ psi}$$

The normal stress acting on these planes has the value corresponding to the center of Mohr's Circle. Points C and C'

$$S_n = \dfrac{S_x + S_y}{2} = \dfrac{-600 + 1600}{2} = 500 \text{ psi}$$

Resultant stress $= \sqrt{(1486.6)^2 + (500)^2} = 1568.4$ psi

3. $\phi = 30°$ $2\phi = 60°$ $\sin 2\phi = \dfrac{\sqrt{3}}{2} = .866$

$\cos 2\phi = .5$

$$S_n = \dfrac{S_x + S_y}{2} + \dfrac{S_x - S_y}{2} \cos 2\phi - S_{xy} \sin 2\phi$$

$$= \dfrac{-600 + 1600}{2} + \dfrac{-600 - 1600}{2}(.5) - 1000(.866) = -916 \text{ psi}$$

$$S_s = \dfrac{S_x - S_y}{2} \sin 2\phi + S_{xy} \sin 2\phi$$

$$= \dfrac{-600 - 1600}{2}(.866) + 1000(.5) = -452.6 \text{ psi}$$

Resultant stress $= \sqrt{(916)^2 + (452.6)^2} = 1021.7$ psi

See point D. $S_n = -916$ psi $2\phi = 60°$
                $S_s = -452.6$ psi   on Mohr's Circle

Mohr's Circle Construction (Fig. 3.87)

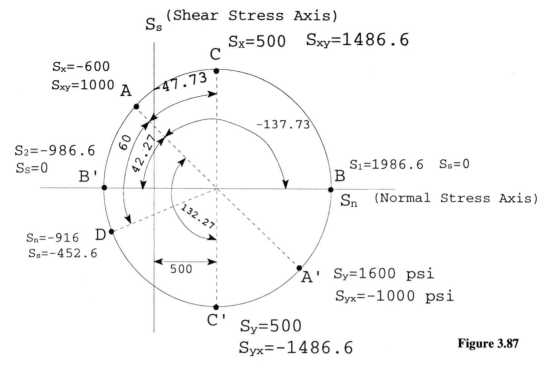

Figure 3.87

1. Draw S_n axis (normal stress axis horizontal).

   Draw S_s axis (shear stress axis vertical).

2. Locate center of Mohr's Circle (+500 psi) Point 0 is the center at $\left(\dfrac{S_x + S_y}{2}\right)$ along the horizontal axis.

3. Calculate radius $R = \sqrt{\left[\dfrac{S_x - S_y}{2}\right]^2 + (S_{xy})^2} = (1486.6 \text{ psi})$. Draw circle.

4. Note reference stresses

   A $\quad \begin{aligned} S_x &= -600 \\ S_{xy} &= 1000 \end{aligned}$

   A' $\quad \begin{aligned} S_y &= 1600 \\ S_{yx} &= -1000 \end{aligned}$

5. Note principal stresses

   B $\quad \begin{aligned} S_1 &= 1986.6 \text{ psi} \\ S_{shear} &= 0 \end{aligned}$

   B' $\quad \begin{aligned} S_2 &= -986.6 \\ S_{shear} &= 0 \end{aligned}$

6. Note maximum shear stress

$$C \begin{matrix} S_x = 500 \\ S_{xy} = 1486.6 \end{matrix} \qquad C' \begin{matrix} S_y = 500 \\ S_{xy} = 1486.6 \end{matrix}$$

7. Note point D ($2\phi = 60°$)

$$S_n = -916 \text{ psi}$$
$$S_s = -452.6 \text{ psi}$$

4
Dynamics

TABLE OF CONTENTS

Selected References 4-ii
Fundamental Definitions 4-1
Equations of Rectilinear Motion 4-2
    Summary for Rectilinear Motion 4-3
Newton's Laws of Motion 4-4
    Fundamental Equations of Kinetics for a Particle 4-5
    Center of Mass 4-6
    Center of Gravity 4-7
    Space Motion 4-7
    Kinematics of Rectilinear Motion 4-8
    Constant Velocity 4-8
    Constant Acceleration 4-8
    Graphical Representation of Motion 4-13
        Displacement as a Function of Time 4-13
        Velocity as a Function of Time 4-15
        Acceleration as a Function of Velocity 4-19
    Constant Force–Constant Acceleration 4-20
    d'Alembert's Principle 4-25
    Variable Force 4-28
    Angular Velocity 4-39
Kinematics of Curvilinear Motion 4-43
    Rectangular Coordinates 4-43
Component Motion 4-46
    Flight of Projectiles 4-46
    Radial and Tangential Motion 4-48
Kinetics of Curvilinear Motion 4-53
Banking of Curves 4-57
Rotation of a Rigid Body 4-65
Dynamic Equilibrium of Noncentroidal Rotation 4-70
    Radius of Gyration 4-70
    Rotational Dynamic Equilibrium 4-70
Work–Energy 4-75
    Work–Energy Method Involving Variable Forces 4-77
Equivalent Springs 4-80
Conservative Systems 4-83
Impulse and Momentum 4-84
Dynamic Action of Jets 4-88
Conservation of Momentum 4-91
Elastic Impact 4-94
Table I: Common Examples of Simple Harmonic Motion 4-100

SELECTED REFERENCES

S. Timoskenko and D.H. Young, "Engineering Mechanics," McGraw-Hill, New York

J.P. Den Hartog, "Mechanics," McGraw-Hill, New York

A. Sloane, "Engineering Kinematics," Macmillan, New York

G. Housner and D. Hudson, "Applied Mechanics," Van Nostrand, New York

S. Ballard., E.P. Slack, and E. Hausmann, "Physics Principles," Van Nostrand, New York

"Applied Mechanics: More Dynamics," John Wiley & Sons, New York

D.H. Pletta and D. Frederick, "Engineering Mechanics: Statics and Dynamics," Ronald Press

FUNDAMENTAL DEFINITIONS

Dynamics is the study of the motion of particles and bodies. In describing particles and bodies, the criterion of relative size determines whether a given object is considered a particle or a body. For example, in astronomical calculations, the Earth may be considered as a particle because its dimensions are negligible compared to the size of its path.

A particle denotes an object of point size, so small that differences in the motions of its parts can be neglected.

A body denotes a system of particles that form an object of appreciable size, wherein the relative motion of the particles may require consideration.

Dynamics is comprised of two branches, kinematics and kinetics. The study of kinematics involves the geometry of motion without regard to the forces causing the motion. Kinetics is the branch that relates the forces acting on a body, to its motion.

The motion of a particle is defined in terms of four fundamental quantities: force, displacement, velocity, and acceleration. These quantities are vectors, that is, represented by direction and magnitude as opposed to scalars, which are represented by magnitude alone. Vector quantities are designated by means of a bar over the quantity. Thus $\overline{F}$ represents a force of magnitude F in a specified direction.

Displacement ($\overline{S}$) is the vector distance of a particle from its origin to the position occupied by the particle on its path of travel.

If the path of a particle is a straight line, the displacement and the distance traveled are numerically equal. However, if the path is curved, the distance traveled along its path differs from the displacement.

Example (Fig. 4.1)

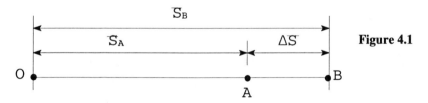

Figure 4.1

Rectilinear displacement (See Fig. 4.2) from A to B, $\Delta \overline{S} = \overline{S}_B - \overline{S}_A$

Rectilinear distance from A to B, $\Delta S = S_B - S_A$; $\Delta S \neq \Delta \overline{S}$

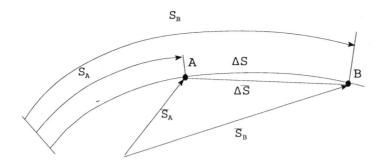

Figure 4.2

Curvilinear displacement from A to B, $\Delta S = S_B - S_A$

Curvilinear distance from A to B, $\Delta S = S_B - S_A$; $\Delta S = \Delta S$

The most common units of displacement in engineering are ft (ft) and miles (mi), soon to be changed to meters and kilometers.

<u>Velocity (V)</u> of a particle is the time rate of change of its displacement.

<u>Speed</u>, because it is being a scalar quantity, represents the magnitude of the velocity vector only and not its direction. The most common units of velocity in engineering are ft per second (fps) and miles per hour (mph).

<u>Acceleration ($\bar{a}$)</u> is the time rate of change of velocity. The most common unit of acceleration in engineering is ft per second (fps^2).

EQUATIONS OF RECTILINEAR MOTION

If a particle traverses a straight line distance ΔS during any time interval Δt, its average speed is

$$V_{avg} = \frac{\Delta S}{\Delta t}$$

The instantaneous speed, V, is obtained as the time Δt approaches zero.

$$V = \lim_{\Delta t \to 0} \frac{\Delta S}{\Delta t} = \frac{dS}{dt}$$

Differentials <u>with respect to time</u> are often indicated by a dot over the symbol, or

$$V = \frac{dS}{dt} = \dot{S}$$

If a particle undergoes a change of speed ΔV during a time interval Δt the average acceleration of the particle is

$$a_{avg} = \frac{\Delta V}{\Delta t}$$

The instantaneous acceleration, a, is obtained as the time Δt approaches zero

$$a = \lim_{\Delta t \to 0} \frac{\Delta V}{\Delta t} = \frac{dV}{dt} = \frac{d}{dt}\left(\frac{dS}{dt}\right) = \frac{d^2S}{dt^2}$$

Second differentials <u>with respect to time</u> are often indicated by two dots over the symbol, or

$$a = \dot{V} = \ddot{S}$$

Another useful differential equation for the acceleration of a particle is obtained as follows:

$$a = \frac{dV}{dt}\frac{dS}{dS} = \frac{dS}{dt}\frac{dV}{dS} = V\frac{dV}{dS} \quad \text{or} \quad V\,dV = a\,dS$$

In rectilinear motion the resultant acceleration is in the direction of the straight line path. However, the acceleration may be either in the sense in which the particle is moving indicating <u>increasing</u> velocity, or in the opposite sense indicating <u>decreasing</u> velocity called <u>deceleration</u> or <u>retardation</u>. This difference is accounted for by the sign of the acceleration.

<u>Summary for Rectilinear Motion</u>

$$V = \frac{dS}{dt} \qquad V = \dot{S}$$

$$a = \frac{dV}{dt} = \frac{d^2S}{dt^2} \qquad a = \dot{V} = \ddot{S}$$

$$V\,dV = a\,ds$$

Bear in mind that the above equations are algebraic equations, derived from vector quantities and hence may be written for each coordinate direction.

NEWTON'S LAWS OF MOTION

For our purposes we shall rephrase Newton's Laws to be more pertinent to the present discussion.

1. A PARTICLE ACTED UPON BY A BALANCED FORCE SYSTEM HAS NO ACCELERATION.

Under this condition the particle either has zero velocity or moves with constant velocity.

2. A PARTICLE ACTED UPON BY AN UNBALANCED SYSTEM OF FORCES MUST ACCELERATE IN LINE AND DIRECTLY PROPORTIONAL TO THE RESULTANT OF THE FORCE SYSTEM.

If the unbalanced force F acts in line with the velocity vector V to increase the velocity's magnitude, the particle will accelerate (Fig. 4.3).

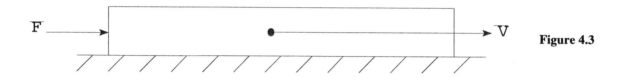

Figure 4.3

If the unbalanced force F acts in line with the velocity vector V to decrease the magnitude of the velocity, the particle will decelerate (Fig. 4.4).

Figure 4.4

If the unbalanced force F acts normal to the velocity vector V the particle will experience a change in direction only (velocity magnitude constant - acceleration directed inward) (Fig. 4.5).

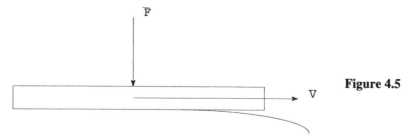

Figure 4.5

If the unbalanced force F does not act normal to the velocity vector V the particle will experience both a change in magnitude and direction (Fig. 4.6).

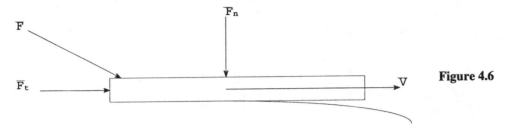

Figure 4.6

3. **ACTION AND REACTION FORCES BETWEEN PARTICLES ARE ALWAYS EQUAL AND OPPOSITELY DIRECTED.**

Note that the action forces are exerted <u>on</u> the particles, the reaction forces <u>by</u> the particles. Free body diagrams in dynamics show either all forces exerted <u>by</u> a particle or all forces exerted <u>on</u> a particle but not both or some combination of both.

<u>Fundamental Equations of Kinetics for a Particle</u>

Consider particle P acted upon by forces F_1, F_2, F_3. Since forces are vector quantities their resultant F may be determined by vector addition. By Newton's second law particle P experiences an acceleration $\bar{a}$ in line with and proportional to F, thus one obtains Fig. 4.7

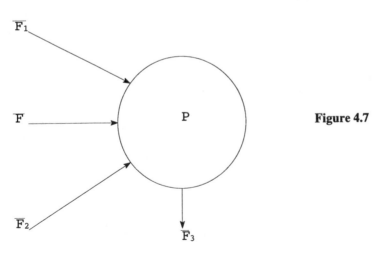

Figure 4.7

or $F = k\bar{a}$

where k is a constant of proportionality when F and $\bar{a}$ are parallel to the coordinate axis. This equation reduces to the algebraic expression $F = ka$.

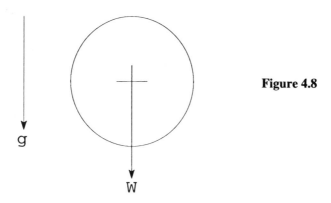

Figure 4.8

Consider the same particle free to accelerate in a vacuum under the action of gravity, where W, its weight, is the resultant unbalanced force acting and g is the resultant gravitational acceleration (Fig. 4.8). We then have

$$W = kg$$

(assuming the same constant of proportionality k since particle remains unchanged)

or $k = \dfrac{W}{g}$ which is defined as the <u>Mass</u>, m, of the particle Newton's second law may then be expressed

$$\Sigma F = m\bar{a}$$
$$\Sigma F = \dfrac{W}{g}\bar{a}$$

When ΣF and $\bar{a}$ are parallel to the coordinate axis this equation becomes

$$\Sigma F = \dfrac{W}{g} a$$

<u>Center of Mass</u>

In translational motion each point on a body experiences the same displacement as any other point so that the motion of one particle represents the motion of the whole body. But even when a body rotates or vibrates as it moves, there is one point on the body, called the <u>center of mass</u>, that moves in the same way that a single particle subject to the same external forces would move.

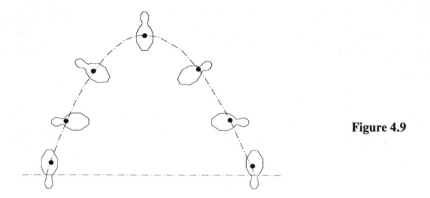

Figure 4.9

Figure 4.9 shows the simple parabolic motion of the center of mass of a ten pin as it travels through the air. No other point in the club moves in such a simple way.

Note that if the ten pin were moving in pure translation every point in it would experience the same displacement as does the center of mass. The motion of the center of a mass of a body is therefore called the translational motion of the body.

Center of Gravity

If a body of mass m is imagined to be divided into a large number of particles, n, a gravitational force will be exerted on each particle. The sum of these individual forces may be shown to be equivalent to a single resultant gravitational force of magnitude mg, where g represents the gravitational acceleration of the field in which the mass m is located. The point at which this force acts is called the center of gravity of the mass.

A single force equal in magnitude and oppositely directed to the equivalent resultant gravitational force applied at the center of gravity will keep the body in static equilibrium.

For most engineering problems the center of mass and center of gravity may be assumed coincident and will be used interchangeably in these notes.

Space Motion

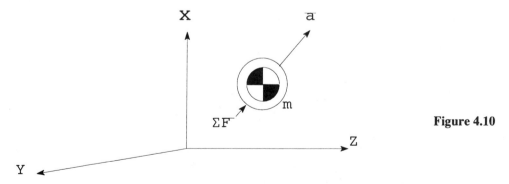

Figure 4.10

Assume that mass m, (Fig. 4.10) is subjected to a resultant unbalanced force F oriented with respect to a Cartesian coordinate system as shown. The body will experience an acceleration $\bar{a}$ in accordance with Newton's Law whose motion along the x-, y- and z-axes may be represented by the following algebraic equation, respectively

$$\Sigma F_x = \frac{W}{g} a_x \quad (m = \frac{W}{g})$$

$$\Sigma F_y = \frac{W}{g} a_y$$

$$\Sigma F_z = \frac{W}{g} a_z$$

where

$$F = \sqrt{(\Sigma F_x)^2 + (\Sigma F_y)^2 + (\Sigma F_z)^2}$$

$$a = \sqrt{a_x^2 + a_y^2 + a_z^2}$$

KINEMATICS OF RECTILINEAR MOTION

Constant Velocity

Consider a particle of mass m moving with constant velocity V. Then
V = constant

$$\text{Acceleration} = a = \frac{dV}{dt} = \frac{d}{dt}(\text{constant}) = 0 \text{ and } F = ma = 0$$

which is in agreement with Newton's first law of motion.

Constant Acceleration

Consider a particle of mass m subject to a constant force resulting in a constant acceleration, a.

$$\frac{dV}{dt} = a = \text{constant} \qquad dv = a\, dt$$

DYNAMICS

Assigning initial conditions for the velocity, namely that at

$$t = 0 \quad V = V_o$$

$$t = t \quad V = V_f$$

The equation of motion becomes

$$\int_{V_o}^{V_f} dV = a \int_0^t dt$$

Integration yields

$$V \Big]_{V_o}^{V_F} = a\, t \Big]_0^t$$

or

$$V_f - V_o = a\,t \quad \text{and} \quad V_f = V_o + a\,t \qquad a = \frac{V_f - V_o}{t}$$

Recall that $\dfrac{dS}{dt} = V \quad dS = V\,dt = V_f\,dt \quad dS = (V_o + a\,t)\,dt$

Assigning initial conditions for the displacement, namely that at

$$t = 0 \quad S = 0$$

$$t = t \quad S = S$$

$$\int_{0+}^{S} dS = \int_0^t (V_o + a\,t)\,dt$$

$$S \Big]_0^S = V_o t + \tfrac{1}{2} a\,t^2 \Big]_0^t$$

$$\therefore S = V_o t + \frac{a\,t^2}{2}$$

Recall that $V\,dV = a\,ds$

then $\int_{V_o}^{V_f} V\,dV = a \int_0^S dS$

$$\left.\frac{V^2}{V}\right]_{V_o}^{V_F} = a\ \left. S\right]_0^S$$

$$\frac{V_f^2 - V_o^2}{2} = aS\ ;\ V_f^2 = V_o^2 + 2as$$

Summary: When a = constant

$$V_f = V_o + at$$

$$S = V_o t + \tfrac{1}{2} a t^2$$

$$V_f^2 = V_o^2 + 2as$$

The above equations may also be derived as follows:

$$\frac{dV}{dt} = a$$

$$V = a \int dt$$

$$V = at + C_1$$

$$\frac{dS}{dt} = V$$

$$\frac{dS}{dt} = at + C_1$$

$$dS = at\,dt + C_1\,dt$$

$$S = a \int t\,dt + C_1 \int dt$$

$$S = \frac{at^2}{2} + C_1 t + C_2$$

Initial Conditions

At $t = 0$; $S = 0$, $V = V_o$

$$V = at + C_1$$

$$V_o = a(0) + C_1$$

$$C_1 = V_o$$

$$V = at + V_o$$

$$S = \frac{at^2}{2} + V_o t + C_2$$

$$(0) = \frac{a}{2}(0)^2 + V_o(0) + C_2$$

$$C_2 = 0$$

$$S = \frac{at^2}{2} + V_o t$$

Example

A stone is dropped down a well and $5\frac{25}{70}$ sec later the sound of the splash is heard. If the velocity of sound is 1120 ft per second, how deep is the well?

Approach:

Use the basic kinetic equation $S = V_o t + \frac{1}{2} g t^2$ to derive an expression for the time to drop to the water. Then use $S = Vt$ to derive another expression for the time to hear the sound. The total time is given, solve for the distance (Fig. 4.11).

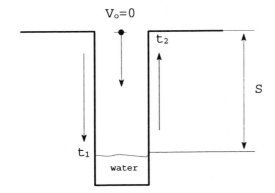

Figure 4.11

$S = V_o t + \tfrac{1}{2} g t^2$, but $V_o = 0$

solving for t: $t = \sqrt{\dfrac{2S}{g}} = t_1$

now using $S = V t$ we have $t_2 = \dfrac{S}{V}$ but $t_1 + t_2 = 5\dfrac{25}{70}$ sec

$\sqrt{\dfrac{2S}{g}} + \dfrac{S}{V} = 5\dfrac{25}{70}$

Rearranging terms, squaring, substituting V - 1120 ft/sec, find S = 400 ft.

Example

A stone is thrown vertically upward from the top of a 100 ft high building with an initial velocity of 50 ft per second and at the same time a second stone is thrown vertically upward from the ground at 75 ft per second. When and where will the stones be at the same height from the ground? Will this occur when the stones are moving up or down?

Approach:

Use the distance equation $S = V_o t + \tfrac{1}{2} a t^2$ with the proper signs for the terms.

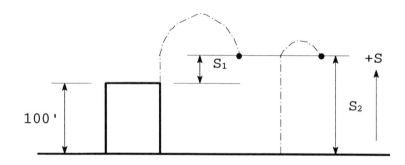

Figure 4.12

Assume the upward direction to be positive for all terms of the equations.

Stone 1: $S_1 = 50 t - \tfrac{1}{2}(32.2 t^2)$

Stone 2: $S_2 = 75 t - \tfrac{1}{2}(32.2 t^2)$

When the stones reach the same height, stone 2 will have traveled 100 ft further than stone 1, hence $S_2 - S_1 = 100$ (Fig. 4.12). Subtracting the original equations:

$$S_2 - S_1 = 25\,t = 100 \therefore t = 4 \text{ sec}$$

Substituting $t = 4$ in the original equations yields:

$$S_1 = -57.6 \text{ ft and } S_2 = 42.4 \text{ ft}$$

Stone 1 passes stone 2 below the top of the building.

To determine if stone 2 passes stone 1 on the way up or down solve for the time required for stone 2 to reach its maximum height from

$$V_f = V_o + a\,t \text{ when } V_f = 0, V_o = 75 \text{ fps}, a = -32.2 \text{ fps}^2$$

$$t = \frac{V_f - V_o}{a} = \frac{0 - 75}{-32.2}$$

$$t = 2.33 \text{ sec}$$

4 sec > 2.33 sec hence stone 2 passes stone 1 on the way down.

Graphical Representation of Motion

The variables that appear in the kinematic equations of motion, i.e., S, V, a and t may be plotted in terms of each other to provide curves that are useful in understanding the motion. Three of these relations, namely, displacement as a function of time, $S = S(t)$; velocity as a function of time, $V = V(t)$; and acceleration as a function of time $a = a(t)$ are discussed below.

Displacement as a Function of Time

Displacement diagrams (S-t curves)

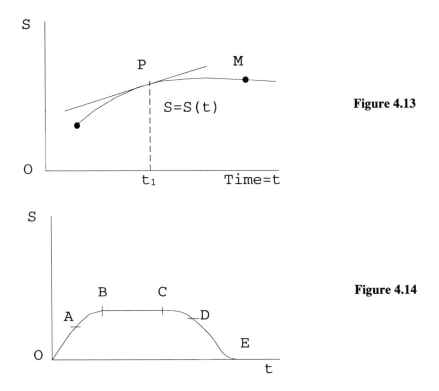

Figure 4.13

Figure 4.14

Consider the curve S = S(t) shown in Fig. 4.13. The slope of the curve at any point is dS/dt or

$$S = S(t)$$

$$dS = S'(t)\, dt$$

$$\frac{dS}{dt} = S'(t)$$

but $\frac{dS}{dt}$ is the instantaneous velocity, V. Hence at time t, the instantaneous velocity is represented by the slope at point P as shown. Note that the steeper the curve, the greater is the speed so that at point M the speed is zero (zero slope corresponding to no change in S).

Based on these characteristics of displacement diagrams Fig. 4.14 may be interpreted to represent a motion as follows:

1. Constant speed from 0 to A (constant slope)

2. Decreasing speed from A to B (decreasing slope)

3. Zero speed from B to C (no motion or displacement—zero slope)

4. Increasing speed from C to D (increasing slope)

5. Maximum speed at point D = ∞

6. Decreasing speed from D to E (decreasing slope)

The negative slope from C to E indicates that the velocity vector is pointing in the negative sense, that is, the body has changed its direction of motion.

Velocity as a Function of Time

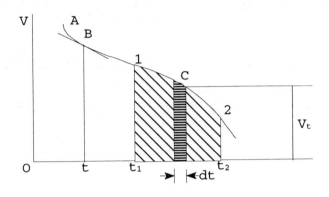

Figure 4.15

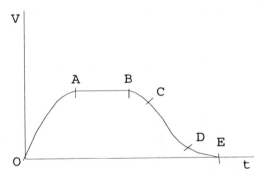

Figure 4.16

Consider the curve V = V(t) shown in Fig. 4.15. The slope of the curve at any point is $\dfrac{dV}{dt}$ or $V = V(t); dV = V'(t) dt; \dfrac{dV}{dt} = V'(t).$

But $\dfrac{dV}{dt}$ is the instantaneous acceleration, $a = \dot{v}$. Hence at the time t the instantaneous acceleration is represented by the slope at point B as shown.

Another characteristic of the V-t curve may be derived from a re-examination of the equation relating velocity and displacement, namely

$$V = V(t) = \frac{dS}{dt}$$

Introducing limits, for velocity and displacement

$$\int_{S_1}^{S_2} dS = \int_{t_1}^{t_2} V(t)\, dt$$

Figure 4.15 shows that the quantity (V(t) dt represents the area of an elemental strip centered about point C of width dt and height V(t) and that the quantity

$$\int_{t_1}^{t_2} V(t)\, dt$$

represents the total area under the V-t curve from point 2, corresponding to times t_1 and t_2 respectively. But this quantity is the total displacement from point 1 to point 2 given by $\int_{S_1}^{S_2} dS$.

Hence, <u>the area under the V-t curve represents to scale the distance traveled in the time interval in question</u>.

Based on the above characteristics of velocity diagrams Fig. 4.16 may be interpreted to represent a motion as follows:

1. Decreasing acceleration from 0 to A (decreasing slope)

2. Zero acceleration - constant velocity from A to B (zero slope)

3. Increasing deceleration from B to C (increasing negative slope)

4. Constant deceleration from C to D (constant slope)

5. Decreasing deceleration from D to E (decreasing negative slope)

Example

For motion defined by $S = 6t + 4t^3$, determine the velocity and acceleration after 3 sec.

$$V = \frac{dS}{dt} = \frac{d}{dt}(6t + 4t^3) = 6 + 12t^2$$

$$a = \frac{dV}{dt} = \frac{d}{dt}(6 + 12t^2) = 24t$$

After 3 sec:

$$V = 6 + 12t^2 = 6 + 12(3)^3 = 114 \text{ ft/sec}$$

$$a = 24t = 24(3) = 72 \text{ ft/sec}^2$$

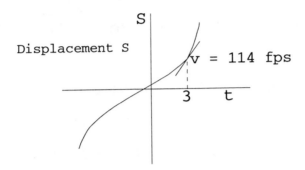

Figure 4.17

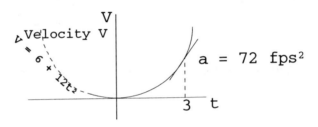

Figure 4.18

Acceleration as a function of time,

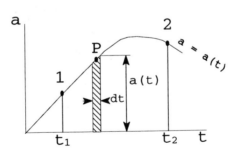

Figure 4.19

Figure 4.19 is an acceleration diagram whose equation is

$$a = a(t) = \frac{dV}{dt}$$

$$dV = a(t)\, dt$$

Introducing limits for velocity and acceleration

$$\int_{V_1}^{V_2} dV = \int_{t_1}^{t_2} a(t)\, dt$$

Figure 4.19 shows that the quantity a(t) dt represents the area of an elemental strip centered about point P of width dt and height a(t) and that the quantity

$$\int_{t_1}^{t_2} a(t)\, dt$$

represents the total area under the a-t curve from point 1 to point 2 which is equal to the change in speed $\int_{V_1}^{V_2} dV$ for the time period between t_2 and t_1.

Hence, <u>the area under the a-t curve represents to scale the change in speed for the time interval in question</u>.

Example

Determine the velocity and displacement of a particle after 3 sec if the motion is defined by a = 4t and if it is known that S = 6 ft and V = 3 ft per sec after 1 sec.

$$dV = a\, dt$$

$$\int_{3}^{V} dV = \int_{1}^{t} 4t\, dt$$

Integrating

$$V \Big]_{3}^{V} = 2t^2 \Big]_{1}^{t}$$

or

$$V - 3 = 2t^2 - 2(1) = 2t^2 - 2 \quad \therefore V = 2t^2 + 1$$

Substituting V above in dS = V dt yields

$$\int_{6}^{S} dS = \int_{1}^{t} V\, dt = \int_{1}^{t} (2t^2 + 1)\, dt$$

$$S\Big]_6^S = \frac{2t^3}{3} + t\Big]_1^t$$

$$S - 6 = \frac{2t^3}{3} + t - \frac{2}{3} - 1$$

$$S = \frac{2t^3}{3} + t + \frac{13}{3}$$

Hence if $t = 3$ sec, substitution yields

$$V = 2t^2 + 1 = 2(9) + 1 = 19 \text{ ft/sec}$$

$$S = \frac{2t^3}{3} + t + \frac{13}{3} = \frac{2}{3}(9) + 3 + \frac{13}{3} = 25\frac{1}{3} \text{ ft.}$$

Acceleration as a function of Velocity

Assume that the acceleration of a particle is defined in terms of its velocity, i.e., $a = f(V)$.

From our earlier work we showed that $a = f(v) = \dfrac{dV}{dt}$ and
$VdV = a\, ds = f(V)\, dS$

Separating variables in each equation yields

$$dt = \frac{dV}{f(V)} \qquad ds = \frac{VdV}{f(V)}$$

$$\int_{t_o}^{t} dt = \int_{V_o}^{V} \frac{dV}{f(V)} \quad ; \quad \int_{S_o}^{S} dS = \int_{V_o}^{V} \frac{VdV}{f(V)}$$

AND

$$t - t_o = F_1(V) \qquad S - S_o = F_2(V)$$

Example:

Determine the time and displacement of a particle having velocity of 2 ft/sec where the motion is defined by $a = \dfrac{1}{V^2}$ and it is known that t and S are zero when V = 0.

$$\int_{t_o}^{t} dt = \int_{V_o}^{V} \frac{dV}{f(V)} = \int_{0}^{V} \frac{dV}{\dfrac{1}{V^2}} = \int_{0}^{V} V^2\, dV$$

$$t = \left. \frac{V^3}{3} \right|_{0}^{V} = \frac{V^3}{3}$$

$$\int_{0}^{S} ds = \int_{V_o}^{V} \frac{V\,dV}{f(V)} = \int_{0}^{V} \frac{V\,dV}{\dfrac{1}{V^2}} = \int_{0}^{V} V^3\, dV$$

$$S = \frac{V^4}{4}$$

Substituting V = 2 ft/sec yields

$$t = \frac{V^3}{3} = \frac{(2)^3}{3} = \frac{8}{3}\ \text{sec}$$

$$S = \frac{V^4}{4} = \frac{(2)^4}{4} = 4\ \text{ft}$$

Constant Force—Constant Acceleration

Example (Figs. 4.20 and 4.21)

Determine the acceleration of the system, the tension in the cable connecting the two blocks, and the distance traveled at the moment the blocks attain a velocity of 10 ft/sec. Assume that the system is at rest initially and that the kinetic coefficient of friction, μ, is 0.2.

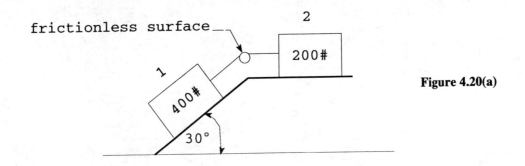

Figure 4.20(a)

Assume all forces acting on the blcks act through their center of mass.

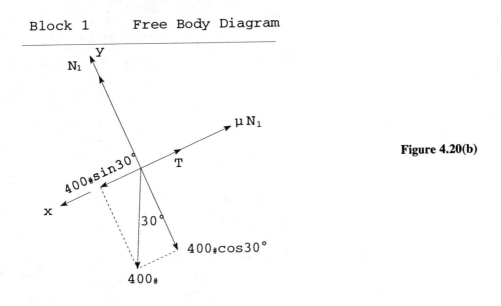

Figure 4.20(b)

$\nwarrow \Sigma F_y = 0$

$N_1 - 400 \cos 30° = 0$

$N_1 = 346.4 \text{ lbs}$

$\swarrow \Sigma F_x = \dfrac{W_1}{g} a \ ;$

$400 \sin 30° - T - \mu N_1 = \dfrac{400}{g} a$

$200 - T - 0.2 (346.4) = \dfrac{400}{32.2} a$

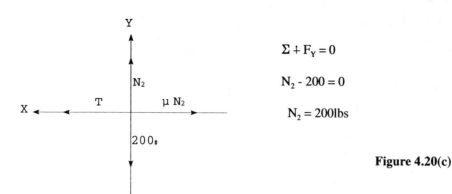

Figure 4.20(c)

$\Sigma + F_Y = 0$

$N_2 - 200 = 0$

$N_2 = 200 \text{ lbs}$

$\Sigma \leftarrow F_x = 0$

$$T - \mu N_2 = \frac{W_2}{g} a$$

$$T - (.2)(200) = \frac{200}{32.2} a$$

$$T - 40 = \frac{200}{32.2} a$$

Solution of the equations for T and a yield a = 4.88 ft/sec² T = 70.25 lbs

From earlier work on constant acceleration

$$V^2 = V_0^2 + 2aS$$

$$S = \frac{V^2}{2a} = \frac{(10 \text{ fps})}{2(4.88 \text{ fps}^2)} = 10.25 \text{ ft}$$

In general the result for the acceleration may be obtained directly from the equation $\Sigma F = ma$ if ΣF is considered for all the forces affecting the motion of the system and m is considered to be the mass of the entire system, thus:

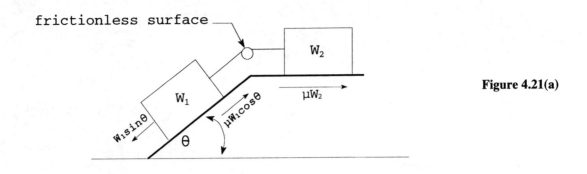

Figure 4.21(a)

$$\Sigma f = m\,a = \frac{W}{g}\,a \text{ yields}$$

$$W_1 \sin\theta - \mu W_1 \cos\theta - \mu W_2 = \frac{W_1 + W_2}{g} a$$

$$a = \frac{w_1 \sin\theta - \mu(w_2 + w_1 \cos\theta)}{w_1 + w_2} g$$

$$a = \frac{400(.5) - 0.2\,(200 + 400\,(.866))}{400 + 200}\,(32.2)$$

$$= 4.88 \text{ ft/sec}^2$$

If the pulley surface is assumed to have a coefficient of friction f then free body diagrams of each weight plus the pulley must be employed taking into account the difference in tension on either side of the pulley.

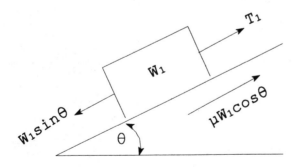

Figure 4.21(b)

$\swarrow \Sigma F = m\,a = \dfrac{W}{g} a$ yields

$$W_1 \sin\theta - \mu W_1 \cos\theta - T_1 = \dfrac{W_1}{g}$$

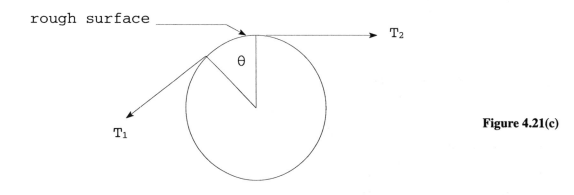

Figure 4.21(c)

$T_1 = T_2 e^{f\theta}$

where

T_1 is the tight tension

T_2 is the slack tension

f is the cable friction coefficient

and θ is the angle of contact in radians (in this case the angle of the slope)

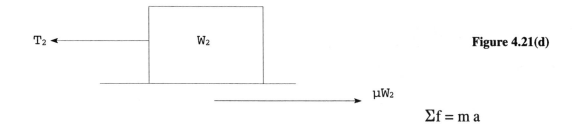

Figure 4.21(d)

$\Sigma f = m\,a$

yields $T_2 - \mu W_2 = \dfrac{W_2}{g} a$

Solving for a

$$a = \frac{W_1 \sin\theta - \mu(w_1 \cos\theta + w_2 e^{f\theta})}{w_1 + w_2 e^{f\theta}} g$$

D'Alembert's Principle

Consider the basic kinetic equation of motion of dynamics, namely

$$\Sigma F = m\bar{a}$$

If the right-hand term in the equation is transposed to the left-hand side, we have

$$\Sigma F - m\bar{a} = 0$$

From this equation we see that if a force equal to $m\bar{a}$, collinear and opposite in sense to the resultant ΣF is added to a free body, the result is a system of forces in equilibrium. Under this condition, the laws of static equilibrium apply, and the ΣF in any direction would be zero. This statement applies when the free body is undergoing linear motion only and is not experiencing rotary motion. See Fig. 4.22.

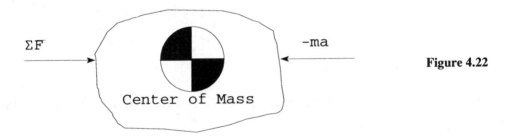

Figure 4.22

The resultant force, ΣF, is called the effective force: hence this opposite force, $-m\bar{a}$, is called the reversed effective force (REF), inertia force or D'Alembert force. The line of action of the reversed effective force for a body in translation passes through its center of mass.

From the above discussion we conclude that either of two methods may be used in solving problems in dynamics. One method involves making a free body diagram showing all external forces acting. The conditions that then apply for plane rectilinear motion are

$$\Sigma F_x = ma_x \qquad \Sigma F_y = ma_y \qquad \Sigma M_{cg} = 0$$

(REF not included in free body diagram) where the sum of the moments ($\Sigma M_{cg} = 0$) is taken about the center of mass.

The other method involves adding an inertia force $m\bar{a}$ to the free body diagram such that it is opposite to the body's acceleration and passes through its center of gravity (CG). The conditions of static equilibrium then apply, or

$$\Sigma F_x = 0 \qquad \Sigma F_y = 0 \qquad \Sigma M = 0$$

(REF included in free body diagram) where $\Sigma M = 0$ applies for all points in the body.

Example

If bar AB weighs 96.6 lbs and bar BC weighs 161.0 lbs and the truck has an acceleration of 6 ft/sec² to the right, determine the components of the reaction at B (Fig. 4.23).

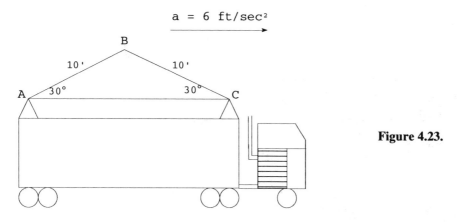

Figure 4.23.

Assume center of mass is halfway between points.

Method 1: Apply a D'Alembert force at the center of mass of each bar and apply the second condition of static equilibrium as shown.

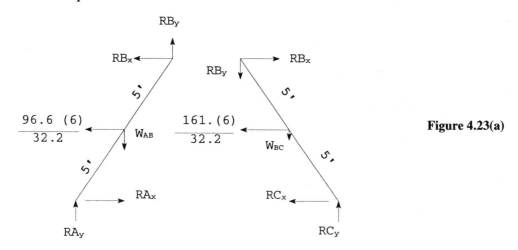

Figure 4.23(a)

$\circlearrowleft \Sigma M_{A_{(AB)}} = 0; \quad -\dfrac{96.6}{32.2}(6)5\sin 30 + 96.6(5\cos 30) - R_{Bx}(10\sin 30) - R_{By}(10\cos 30) = 0$

$8.66\, R_{B_y} + 5\, R_{B_x} = 374$

$\circlearrowleft \Sigma M_{C_{(CB)}} = 0; \quad -\dfrac{161}{32.2}(6)5\sin 30 - 161(5\cos 30) - R_{BY}(10\cos 30) + R_{Bx}(10\sin 30) = 0$

$5\, R_{Bx} - 8.66\, R_{By} = 770 \text{ lbs}$

Solving these equations yields:

$R_{By} = 23 \text{ lbs and } R_{Bx} = 115 \text{ lbs}$

<u>Method 2</u>: Apply the dynamic equations of motion (including $\Sigma M_{C_g} = 0$)

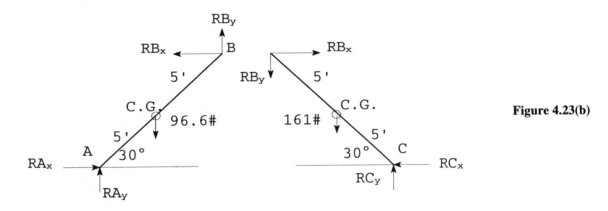

Figure 4.23(b)

<u>Rod AB</u>

$\uparrow \Sigma F_Y = 0: \quad R_{A_y} + R_{B_y} = 96.6 \text{ lbs}$

$\rightarrow \Sigma F_X = \dfrac{W}{g} a_x: \quad R_{Ax} - R_{Bx} = \dfrac{96.6}{32.2}(6 \text{ ft/sec}^2) = 18 \text{ lbs}$

$\circlearrowleft \Sigma M_{C_g} = 0$

$-R_{B_y}(5\cos 30°) - R_{Bx}(5\sin 30°) - R_{Ax}(5\sin 30°) + R_{Ay}(5\cos 30°) = 0$

$-R_{By}(4.33) - R_{Bx}(2.5) - R_{Ax}(2.5) + R_{Ay}(4.33) = 0$

Substituting

$R_{Ay} = 96.6 - R_{By}$; $R_{Ax} = 18 + R_{Bx}$

$-4.33 R_{By} - 2.5 R_{Bx} - (18 + R_{Bx})(2.5) + (96.6 - R_{By})(4.33) = 0$

$-4.33 R_{By} - 2.5 R_{Bx} - 45 - 2.5 R_{Bx} + 418.3 - 4.33 R_{By} = 0$

$-8.66 R_{By} - 5 R_{Bx} + 373.3 = 0$

$8.66 R_{By} + 5 R_{Bx} = 373.3$ lbs

Rod BC (Check of results)

$\Sigma F_y = 0 : R_{C_y} - R_{B_y} = 161$ lbs

$\Sigma F_x = \dfrac{W}{g} a_x : R_{B_x} - R_{C_x} = \dfrac{161}{32.2}(6) = 30$ lbs

$\circlearrowright \Sigma M_{C_g} = 0$

$R_{Bx}(5 \sin 30°) - R_{By}(5 \cos 30°) + R_{CX}(5 \sin 30°) - R_{Cy}(5 \cos 30°) = 0$

$2.5 R_{Bx} - 4.33 R_{B_y} + 2.5 R_{CX} - 4.33 R_{C_y} = 0$

Substituting

$R_{C_y} = 161 + R_{B_y}$; $R_{CX} = R_{Bx} - 30$

$2.5 R_{Bx} - 4.33 R_{B_y} + 2.5(R_{Bx} - 30) - 4.33(161 + R_{By}) = 0$

$2.5 R_{Bx} - 4.33 R_{B_y} + 2.5 R_{Bx} - 75 - 695 - 4.33 R_{By} = 0$

$5 R_{Bx} - 8.66 R_{By} = 770$ or $R_{B_y} = 23$ lbs and $R_{Bx} = 115$ lbs

Use of the D'Alembert force is seen to result in a more direct and less complex solution to the above problems.

Variable Force

In the following discussion, force will be taken to mean the resultant force.

Force as a Function of Time, F = f(t)

Let x = displacement of a body then $\frac{dx}{dt} = \dot{x} = V$ and $\frac{d^2x}{dt^2} = \ddot{x} = a$.

If F is a function of time, F = f(t) the equation of motion is

$m\ddot{x} = f(t)$ or $\ddot{x} = \frac{1}{m} f(t)$

Integrating twice with respect to time yields

$$\dot{x} = \frac{1}{m} \int f(t) \, d + C_1$$

$$x = \frac{1}{m} \iint f(t) \, dt \, dt + C_1 t + C_2$$

Thus the velocity and position of the body are determined in terms of time when sufficient data concerning the initial conditions are available to determine the constants of integration, C_1 and C_2.

Example

A body starts from rest and moves along a frictionless horizontal plane under the action of a horizontal force F which varies with time according to the relationship $F = 5 + t + \frac{t^2}{4}$. Determine the motion given the conditions that at t = 0, x = 0, $\dot{x}$ = 0.

$$m\ddot{x} = f(t) = \frac{t^2}{4} + t + 5, \text{ integration yields}$$

$$m\dot{x} = \frac{t^3}{12} + \frac{t^2}{2} + 5t + C_1, \text{ integration yields}$$

$$mx = \frac{t^4}{48} + \frac{t^3}{6} + \frac{5t^2}{2} + C_1 t + C_2 : \text{at } t = 0, \dot{x} = 0, x = 0$$

hence $C_1 = 0$, $C_2 = 0$, then $\dot{x} = \frac{gt}{W} \left(\frac{t^2}{12} + \frac{t}{2} + 5 \right)$

$$x = \frac{gt^2}{W} \left(\frac{t^2}{48} + \frac{t}{6} + \frac{5}{2} \right)$$

Force as a Function of Distance, F = f(x)

Examples of this type of problem are those investigating the forces between charged particles and those involving gravitational attraction. If $F = f(x)$ then the equation of motion is

$$m\ddot{x} = f(x) \quad \text{or} \quad m\frac{d^2x}{dt^2} = f(x)$$

expressing $\dfrac{d^2x}{dt^2}$ as $\dfrac{d}{dt}\dfrac{(dx)}{dt} = \dfrac{d}{dt}(\dot{x})$ and transporting,

$$m\frac{d}{dt}(\dot{x}) = f(x) \quad ; \quad d\dot{x} = \frac{1}{m} f(x)\, dt$$

Multiplying both sides of the equation by $2\dot{x}$ and integrating over $\dot{x}$.

$$2\dot{x}\, d\dot{x} = \frac{2}{m} f(x)\, dt \quad ; \quad \dot{x} = \frac{2}{m} f(x)\, dt\, \frac{dx}{dt}$$

$$2\dot{x}\, d\dot{x} = \frac{2}{m} f(x)\, dx. \text{ integrating, } (\dot{x})^2 = \frac{2}{m} \int f(x)\, dx + C_1$$

$$V = \dot{x} = \frac{dx}{dt} = \pm\sqrt{\frac{2}{m}} \sqrt{\int f(x)\, dx + C_1}$$

Solving for dt

$$dt = \pm\sqrt{\frac{m}{2}} \frac{dx}{\sqrt{\int f(x)\, dx + C_1}}$$

integration yields:

$$t = \pm\sqrt{\frac{m}{2}} \int \frac{dx}{\sqrt{\int f(x)\, dx + C_1}} + C_2$$

With sufficient initial conditions the constant of integration C_1 and C_2 and the algebraic sign preceding the roots may be determined.

Example

A body of weight W falls from rest at a height above the earth's surface equal to the radius of the Earth. Find the velocity with which it strikes the Earth, assuming the Earth to be a fixed body and that the force of attraction varies inversely as the square of the distance from the center of the Earth (Fig. 4.24).

Solution

Let the origin be taken at the center of the Earth and let the x axis be coincident with the line joining the weight and the center of the Earth.

The force F acting on the weight when it is at distance x is:

$$F = -\frac{k}{x^2}$$

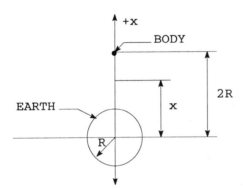

Figure 4.24

where k is a constant of proportionality. The force is negative because it is directed toward the earth in a negative x direction.

When $x = R$, $F = -W$ Thus $K = -Fx^2 = WR^2$

and $F = f(x) = -\dfrac{WR^2}{x^2}$

The equation of motion then is

$$\frac{W}{g}\ddot{x} = -\frac{WR^2}{x^2} \quad ; \quad \ddot{x} = -\frac{gR^2}{x^2}$$

$$\frac{d}{dt}\frac{dx}{dt} = -\frac{gR^2}{x^2} \qquad d(\dot{x}) = -\frac{gR^2}{x^2} dt$$

Multiplying the left side by $2\dot{x}$ and the right side by its equivalent, $2\frac{dx}{dt}$, yields an exact differential on the left.

$$2\dot{x}\, d\dot{x} = -\frac{2g R^2}{x^2} dt \frac{dx}{dt}, \text{ integration yields;}$$

$$\dot{x}^2 = -2g R^2 \int \frac{dx}{x^2} = \frac{2gR^2}{x} + C_1$$

Initial, or boundary conditions, are:

$$x = 2R,\ \dot{x} = 0,\ C_1 = -\frac{2g R^2}{2R} = -gR$$

$$\dot{x}^2 = \frac{2g R^2}{x} - gR$$

$$\dot{x}^2 = gR\left(2\frac{R}{x} - 1\right)$$

$$\dot{x} = \pm\sqrt{gR}\sqrt{2\frac{R}{x} - 1}$$

$$V = -\sqrt{gR}\sqrt{2\frac{R}{x} - 1}$$

where the negative sign indicates that the velocity is negative since it is directed toward the origin. When the weight strikes the earth $x = R$, hence

$$V = -\sqrt{gR}\sqrt{2\frac{R}{R} - 1} = -\sqrt{gR}$$

$$V = \sqrt{(32.2 \text{ fps}^2)(20.9 \times 10^6 \text{ ft})} = 25,900 \text{ fps}$$

Force as a Function of Velocity, F = f(V)

The equation of motion is: $ma = m\dfrac{dV}{dt} = f(V)$

Solving for dt and integrating yields $dt = \dfrac{mdV}{f(V)}$

$$t = m \int \dfrac{dV}{f(V)} + C_1$$

Since $V = \dfrac{dx}{dt}$, $dt = \dfrac{dx}{V} = \dfrac{mdV}{f(V)}$

Solving for dx and integrating yields $dx = mV\dfrac{dV}{f(V)}$

$$x = m \int \dfrac{VdV}{f(V)} + C_2$$

The constants of integration, C_1 and C_2 are evaluated from the initial conditions.

Example

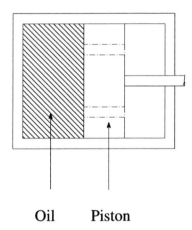

Figure 4.25(a)

Oil   Piston

Figure 4.25a shows a hydraulic damping device which provides a retardation force proportional to the velocity of the input. It consists essentially of a piston which is free to move in a fixed cylinder filled with oil. As an initial velocity V_o is imparted to the piston rod, the piston moves and oil is forced through orifices in the piston, causing a decelerating force proportional to its velocity, i.e., $F = -kV$. For k = 2 lb-sec/ft, piston weight of 3.22 lbs, and an initial velocity of 20 ft/sec determine: (a) The velocity and displacement after 0.1 sec; (b) the distance the piston will travel before coming to rest.

$$m \frac{dV}{dt} = -kV$$

(a)
$$\frac{dV}{V} = -\frac{k}{m} dt$$

For initial conditions we have that at $t = 0$, $V = V_o$ at $t = t$, $V = V$

Thus
$$\int_{V_o}^{V} \frac{dV}{V} = -\frac{k}{m} \int_0^t dt$$

$$\ln V \Big]_{V_o}^{V} = -\frac{kt}{m}$$

Recall that $\ln A - \ln B = \ln \frac{A}{B}$

$$\ln A + \ln B = \ln AB$$

thus
$$\ln \frac{V}{V_o} = -\frac{kt}{m}$$

or
$$\frac{V}{V_o} = e^{\frac{-kt}{m}} \quad \text{and} \quad V = V_o \left[e^{\frac{-kt}{m}} \right]$$

(recall that if $y = \ln u$ then $e^y = u$)

This result is plotted in Fig. 4.25b.

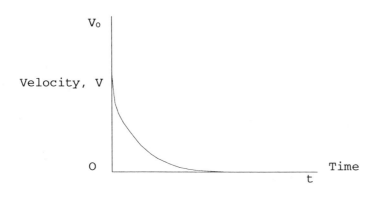

Figure 4.25(b)

Note: at $t = 0$ $V = V_o$ as $t \to \infty$, $V \to 0$

$$V = \frac{dS}{dt} = V_o e^{\frac{-kt}{m}} \qquad dS = V_o e^{\frac{-kt}{m}} dt$$

For initial conditions we have at $t = 0$, $S = 0$; at $t = t$, $S = S$

$$\int_0^S dS = V_o \int_0^t e^{\frac{-kt}{m}} dt$$

$$S = -\frac{m}{k} V_o e^{\frac{-kt}{m}} \Big]_0^t = \frac{m V_o}{k} - \frac{m V_o}{k} e^{\frac{-kt}{m}}$$

$$S = \frac{m V_o}{k} (1 - e^{\frac{-kt}{m}})$$

This result is plotted in Fig. 4.25c.

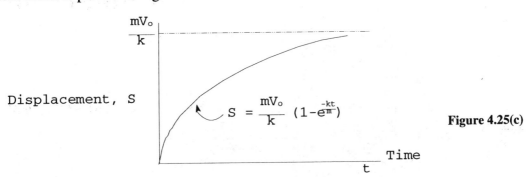

Figure 4.25(c)

Note: In Figure 25b at $t = 0$, $S = 0$.

$$\text{as } t \to \infty \quad S \to \frac{m V_o}{k}$$

$$V dV = a\, dS = -\frac{kV}{m} dS$$

$$dV = -\frac{k}{m} dS$$

$$\int_{V_o}^{V} dV = -\frac{k}{m} \int_0^S dS$$

$$V - V_o = -\frac{k}{m} S$$

$$V = V_o - \frac{k}{m} S$$

This result is plotted in Fig. 4.25d.

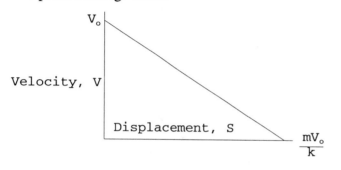

Figure 4.25(d)

(a) $\quad V = V_o e^{\frac{-kt}{m}} = 20 \, e^{-\frac{2 \times 0.1}{3.22/32.2}} = \frac{20}{7.389} = 2.71 \text{ fps}$

(b) $\quad S = \frac{m V_o}{k} (1 - e^{\frac{-kt}{m}}) = \frac{3.22(20)}{32.2(2)} (1 - \frac{1}{7.389}) = \frac{6.389}{7.389} = 0.865 \text{ ft.}$

$$V = V_o - \frac{k}{m} S = 0, \quad V = V_o - \frac{g}{w} kS = 0$$

$$S = \frac{mV_o}{k} = \frac{(3.22)(20)}{(32.2)(2)} = 1 \text{ ft}$$

Example

An object falling under the influence of gravity in air initially encounters a resistance proportional to the square of its velocity. With time the object undergoes a diminishing acceleration until a point is reached at which the acceleration is zero. The velocity of the object at this time is said to be its <u>terminal velocity</u>. Given the above conditions, derive an expression for the object's velocity as a function of time and as a function of the distance fallen.

Solution

If the body is at the origin when $t = 0$ and it moves in the positive direction along the x-axis, the equation of motion is:

$$m\ddot{x} = F - kV^2 \quad \text{or} \quad \frac{W}{g}\frac{dV}{dt} = W - kV^2$$

The accelerating force, $W - kV^2$, is approximately W for small velocities. The critical velocity for which $W - kV^2 = 0$ is called the terminal velocity and will be designated by V_T.

$$V_T = \sqrt{\frac{W}{k}} = \text{terminal velocity}$$

It is evident that V_T depends upon the weight and the proportionality factor.

The proportionality factor depends upon the resistance of the medium to the moving body. The velocity of the object always approaches its terminal velocity.

Substituting the value of k from the expression for terminal velocity in the original equation of motion and reducing terms yields:

$$\frac{dV}{V_T^2 - V^2} = \left[\frac{g}{V_T^2}\right] dt$$

Integrating,
$$\int \frac{dV}{V_T^2 - V^2} = g \int \frac{dt}{V_T^2}$$

yields (determining the constant of integration from the initial condition that $V = V_o$ when $t = 0$),

$$\ln\left[\frac{V_T + V}{V_T - V} \cdot \frac{V_T - V_o}{V_T + V_o}\right] = 2gt/V_T$$

from which

$$\frac{V_T + V}{V_T - V} = \left[\frac{V_T + V_o}{V_T - V_o}\right](e^{2gt/V_T}) \quad \text{or}$$

$$\frac{V_T - V}{V_T + V} = \left[\frac{V_T - V_o}{V_T + V_o}\right](e^{-2gt/V_T})$$

$$-V = V_T \left[\frac{V_T - V_o}{V_T + V_o}\right](e^{-2gt/V_T}) - V_T$$

$$+V \left[\frac{V_T - V_o}{V_T + V_o}\right](e^{-2gt/V_T})$$

$$V = V_T \left[\frac{(V_T + V_o) - (V_T - V_o) e^{-2gt/V_T}}{(V_T + V_o) + (V_T - V_o) e^{-2gt/V_T}}\right]$$

This equation expresses the velocity after any time in terms of the initial velocity V_o, and the time t. As t increases, the second term in both the numerator and the denominator approach zero. The bracketed expression approaches unity and the velocity approaches the terminal velocity V_T.

The relationship between the distance fallen and the velocity are obtained as follows:

$$\frac{dV}{V_T^2 - V^2} = g/V_T^2 \, dt \text{ or}$$

$$\left[\frac{1}{V_T^2 - V^2}\right]\frac{dV}{dt} = \frac{g}{V_T^2}$$

Multiplying each side by dX

$$\frac{a \, dX}{V_T^2 - V^2} = \frac{g}{V_T^2} dX$$

Since $a = \frac{dV}{dt}$, $V = \frac{dX}{dt}$ then $a \, dX = V \, dV$

Substituting:

$$\frac{dV}{V_T^2 - V^2} = \frac{g}{V_T^2} dX.$$ Integrating and determining the constant of integration from the initial condition where $X = 0$ when $V = V_o$.

$$\ln\left[\frac{V_T^2 - V^2}{V_T^2 - V_o^2}\right] = -\frac{2g}{V_T^2} X$$

Solving for V^2: $V^2 = V_T^2 \left[1 - \left(1 - \dfrac{V_0^2}{V_T^2}\right) e^{\dfrac{-2qx}{V_T^2}} \right]$

This equation expresses the velocity in terms of the initial velocity, the terminal velocity and the distance fallen. As the distance fallen increases, the velocity approaches the terminal velocity.

Angular Velocity

Assume that body B rotates about axis 0 in such a way that line OA moves through angle $\Delta\theta$ to position OA'. The average <u>angular velocity</u>, ω, of body B is said to be the time rate of change of angular displacement of line OA, or $\omega = \dfrac{\Delta\theta}{\Delta t}$

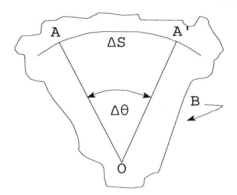

Figure 4.26

The instantaneous angular velocity of B is obtained by the limiting process

$$\omega = \lim_{\Delta t \to 0} \dfrac{\Delta\theta}{\Delta t} = \dfrac{d\theta}{dt} = \dot\theta$$

ω is usually measured in radians per second, radians per minute, revolutions per minute (rpm) or revolutions per second (rps).

Relation Between Angular and Linear Speeds

In Fig. 4.26, point A is seen to travel distance ΔS in moving to point A' in time interval Δt. Thus we may state that

$$\Delta S = r \Delta \theta$$

$$\dfrac{\Delta S}{\Delta t} = r \dfrac{\Delta\theta}{\Delta t}$$

$$\lim_{\Delta t \to 0} \frac{\Delta S}{\Delta t} = r \frac{\Delta \theta}{\Delta t}$$

$$\frac{dS}{dt} = r \frac{d\theta}{dt}$$

But $\frac{dS}{dt} = V$ and $\frac{d\theta}{dt} = \omega$ hence $V = r\omega$

When r is in ft and ω in radians/sec, V will be in ft per second.

<u>Angular Acceleration</u>

If during the time interval Δt the angular velocity of B in Fig. 4.26 changes by $\Delta \omega$, then its average <u>angular acceleration</u>, α, is

$$\alpha = \frac{\Delta \omega}{\Delta t}$$

and its instantaneous angular acceleration is

$$\alpha = \lim_{\Delta t \to 0} \frac{\Delta \omega}{\Delta t} = \frac{d\omega}{dt} = \frac{d}{dt}\frac{d\theta}{dt} = \frac{d^2\theta}{dt^2} = \ddot{\theta}$$

Multiplying α by $\frac{d\theta}{d\theta}$ we obtain

$$\alpha = \frac{d\omega}{dt}\frac{d\theta}{d\theta} = \frac{d\theta}{dt}\frac{d\omega}{d\theta}$$

$$\alpha = \omega \frac{d\omega}{d\theta} \quad \text{or} \quad \alpha\, d\theta = \omega\, d\omega$$

The equations for angular acceleration apply both for constant or variable acceleration. The positive sense of an angular velocity or an angular acceleration may be chosen arbitrarily to be either clockwise or counterclockwise. When the angular acceleration α is in the same sense as the angular velocity ω, the angular velocity is increasing. When the sense of α is opposite ω, then ω is decreasing. Angular acceleration is usually expressed in radians per second, per second (rad per sec^2).

Relation Between Linear and Angular Motion

The relation between linear and angular motion may be summarized as follows:

Linear Motion	Angular Motion
$V = \dfrac{dS}{dt}$	$\omega = \dfrac{d\theta}{dt}$
$a = \dfrac{dV}{dt}$	$\alpha = \dfrac{d\omega}{dt}$
$a\,ds = V\,dV$	$\alpha\,d\theta = \omega\,d\omega$

Conversion from Linear to Angular Motion

$$S = r\,\theta$$

$$V = r\,\omega$$

For Constant Linear Acceleration a	For Constant Angular Acceleration
$V_f = V_o + a\,t$	$\omega_f = \omega_o + \alpha\,t$
$S = V_o t + \frac{1}{2} a\,t^2$	$\theta = \omega_o t + \frac{1}{2}\alpha\,t^2$
$V_f^2 = V_o^2 + 2\,a\,s$	$\omega_f^2 = \omega_o^2 + 2\,\alpha\,\theta$

Example

A wheel that is rotating 330 rpm is slowing down at the rate of 3 rad per sec² (Fig. 4.27).

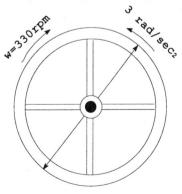

Figure 4.27

(a) What time will elapse before the wheel stops?

(b) At what rate in rpm is the wheel revolving after 5 sec?

(c) Through how many revolutions has it turned during the first 5 sec? Until the wheel stops?

Answers

(a) $\omega_o = 330 \dfrac{\text{rev}}{\text{min}} \times 2\pi \dfrac{\text{rad}}{\text{rev}} \times \dfrac{\text{min}}{60 \text{ sec}} = 11\pi \dfrac{\text{rad}}{\text{sec}}$

$\omega_f = \omega_o + \alpha t$ where $\omega_f = 0$

$\omega_o = 11\pi \dfrac{\text{rad}}{\text{sec}}$

$\alpha = -3 \dfrac{\text{rad}}{\text{sec}^2}$

$t = -\dfrac{\omega_o}{\alpha} = \dfrac{-11\pi}{-3} = 11.5$ secs

(b) $\omega_f = \omega_o + \alpha t$

$\omega_f = 11\pi - 3(5) = 34.6 - 15 = 19.6$ rad/sec

$\omega_f = 19.6 \dfrac{\text{rad}}{\text{sec}} \times \dfrac{\text{rev}}{2\pi \text{ rad}} \times \dfrac{60 \text{ sec}}{\text{min}} = \dfrac{19.6 \times 30}{\pi} = 187$ rpm

(c) $\theta = \omega_o t + \frac{1}{2} \alpha t^2$

$\theta = 11\pi (5) - 3/2 (5)^2$

$\theta = 173 - 37.5 = 135.5$ radians

$N = \dfrac{135.5 \text{ radians}}{2\pi \text{ radians/rev}} = 21.5$ revolutions

$\omega_f^2 = \omega_o^2 + 2\alpha \theta$ where $\omega_f = 0$

$\theta = \dfrac{-\omega_o^2}{2L} = \dfrac{-(11\pi)^2}{-2 (3)} = \dfrac{121\pi^2}{6} = 199.5$ rad $= 31.7$ revolutions

Kinematics Of Curvilinear Motion

Rectangular Coordinates

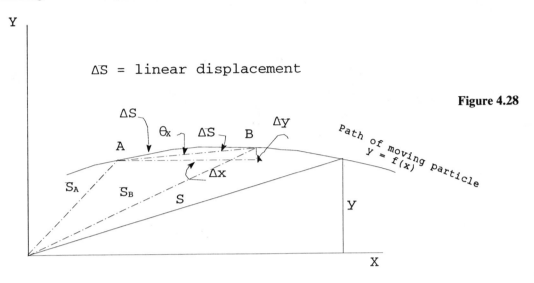

Figure 4.28

The motion of a particle on a curved path is called curvilinear motion. Although the most general case involves a space curve defined by three coordinates, discussion of the two-dimensional case will present the essential ideas. The path of the particle may be represented by the curve $y = f(x)$ and its position A at any time t by a position vector $\overline{S}_A$ measured from the origin of the coordinate system. Some time later, at $t + \Delta t$, the particle will be at position B defined by vector $\overline{S}_B$ (Fig. 4.28). The displacement of the particle will then be given by the vector $\Delta \overline{S}$ which, in vector notation is seen to be

$$\Delta \overline{S} = \overline{S}_B - \overline{S}_A$$

(Note that the displacement $\Delta \overline{S}$ in curvilinear motion is different than the distance traveled ΔS. This was not the case in rectilinear motion.)

Vector $\Delta \overline{S}$ may be expressed in terms of rectilinear motion parallel to the coordinate axes as

$$\Delta \overline{S} = \Delta \overline{x} + \Delta \overline{y}$$

The average velocity between point A and B is found from

$$\frac{\Delta \overline{S}}{\Delta t} = \frac{\Delta \overline{x}}{\Delta t} + \frac{\Delta \overline{y}}{\Delta t}$$

and the instantaneous velocity at point A as

$$\lim_{\Delta t \to 0} \frac{\Delta \overline{S}}{\Delta t} = \frac{\Delta \overline{x}}{\Delta t} + \frac{\Delta \overline{y}}{\Delta t} \quad ; \quad \therefore \quad \frac{d\overline{S}}{dt} = \frac{d\overline{x}}{dt} + \frac{d\overline{y}}{dt}$$

or $\quad \overline{V} = \overline{V}_x + \overline{V}_y$

These vectors are plotted in Fig. 4.29.

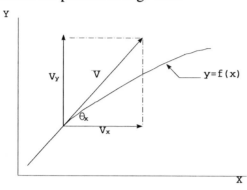

Figure 4.29

The magnitude of $\overline{V}$ is determined from $V = \sqrt{V_x^2 + V_y^2}$

and its direction from

$$\tan \theta_x = \frac{V_y}{V_x} = \frac{\frac{dy}{dt}}{\frac{dx}{dt}} = \frac{dy}{dx}$$

But $\frac{dy}{dx}$ is the slope of the curve $y = f(x)$ at point A hence the velocity $\overline{V}$ is the directed tangent to the curve at every point.

Consider next the velocities of the particle at point A and B as shown in Fig. 4.30.

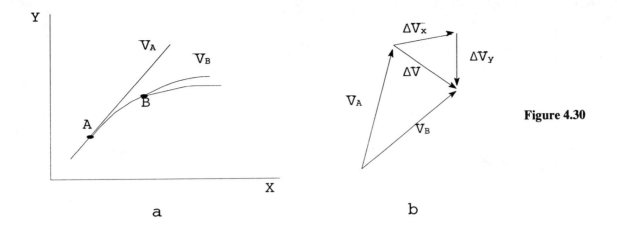

Figure 4.30

The change in velocity between points A and B is determined as shown in Fig. 4.30b as:

$$\Delta \overline{V} = \overline{V}_B - \overline{V}_A$$

Expressed in terms of rectilinear motion parallel to the coordinate axis we have

$$\Delta \overline{V} = \Delta \overline{V}_x + \Delta \overline{V}_y$$

The average acceleration between point A and B is found from

$$\frac{\Delta \overline{V}}{\Delta t} = \frac{\Delta \overline{V}_x}{\Delta t} + \frac{\Delta \overline{V}_y}{\Delta t}$$

And the instantaneous acceleration at point A as:

$$\lim_{\Delta t \to 0} \frac{\Delta \overline{V}}{\Delta t} = \frac{\Delta \overline{V}_x}{\Delta t} + \frac{\Delta \overline{V}_y}{\Delta t} \ ; \ \frac{d\overline{V}}{dt} = \frac{d\overline{V}_x}{dt} + \frac{d\overline{V}_y}{dt}$$

or

$$\overline{a} = \overline{a}_x + \overline{a}_y$$

These vectors are plotted in Fig. 4.31.

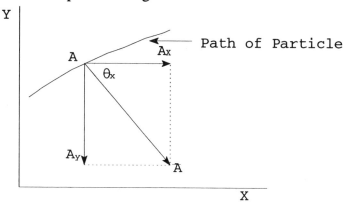

Figure 4.31

The magnitude of $\bar{a}$ is determined from $a = \sqrt{a_x^2 + a_y^2}$

and its direction from

$$\tan \theta = \frac{a_y}{a_x} = \frac{\dfrac{d^2y}{dt^2}}{\dfrac{d^2x}{dt^2}} = \frac{d^2y}{d^2x}$$

$\dfrac{d^2y}{d^2x}$ in general is not tangent to the curve (except for motion along a straight line or at a point of inflection in the curved path) as indicated in Fig. 4.31.

From the foregoing discussion we conclude that plane curvilinear motion is merely the superposition of the coordinates of two simultaneous rectilinear motions in the x and y directions.

Component Motion

Flight of Projectiles

The flight of projectiles is most easily analyzed by treating the projectile as a particle of constant mass with constant acceleration. This approach neglects the effect of air resistance.

DYNAMICS 4-47

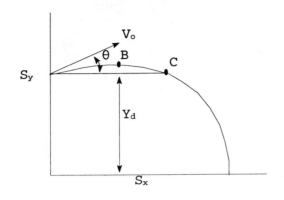

Figure 4.32

Displacement upward is considered positive. Displacement to the right is considered positive (Fig. 4.32).

Direction	Acceleration	Velocity	Displacement
y	$a_y = -g$	$(V = V_{initial} + a\,t)$	$(S = V_{initial}\,t + \tfrac{1}{2}\,a\,t^2)$
		$V_y = V_o \sin\theta - g\,t$	$S_y = V_o\,t \sin\theta - \tfrac{1}{2}\,g\,t^2 + y_o$
x	$a_x = 0$	$V_x = V_o \cos\theta$	$S_x = V_o\,t \cos\theta$

Example

A projectile is fired from the top of a cliff with a velocity of 1414 ft/sec at 45 degrees to the horizontal (Fig. 4.33). Find the range on a horizontal plane through the bottom of the cliff.

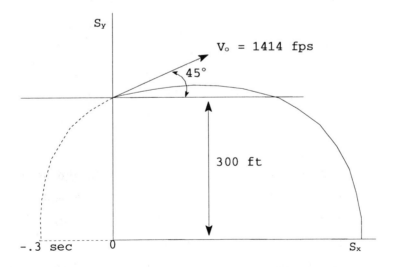

Figure 4.33

Solution

Use $S_y = (V_o \sin \theta) t - \frac{1}{2} g t^2 + y_o$ and find t when $S_y = 0$

$0 = 1414 \sin 45 (t) - 16.1 t^2 + 300$

$t^2 - 62.1 t - 18.63 = 0$

$t = 62.4$ sec or $t = -0.3$ sec

Using the positive value of t,

$S_x = V_o \cos \theta (t) = 1414 \times 0.707 \times 62.4 = 62,400$ ft

Radial and Tangential Coordinates

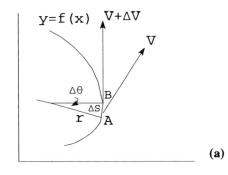

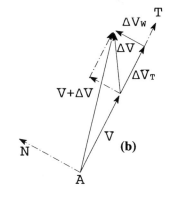

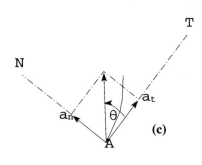

Figure 4.34

The acceleration of a particle is the time rate of change of its velocity. For curvilinear motion the derivative must account for the change in both magnitude and direction of the velocity vector $\overline{V}$. In Fig. 4.34 the velocity of the particle at A is $\overline{V}$ and at B is $\overline{V} + \Delta \overline{V}$, a time Δt later. The vector change $\Delta \overline{V}$ has two components, one $\Delta \overline{V}_t$ due to the change in the magnitude of $\overline{V}$ and directed along the tangent T to the path and the other $\Delta \overline{V}_N$ due to the change in the direction of $\overline{V}$ and directed along the normal N to the path (Fig. 4.34b). The acceleration equation may then be derived as

$$\Delta \overline{V} = \Delta \overline{V}_N + \Delta \overline{V}_T$$

$$\lim_{\Delta t \to 0} \frac{\Delta \overline{V}}{\Delta t} = \frac{\Delta \overline{V}_n}{\Delta t} + \frac{\Delta \overline{V}_T}{\Delta t}$$

$$\frac{d\overline{V}}{\Delta t} = \frac{d\overline{V}_n}{\Delta t} + \frac{d\overline{V}_T}{\Delta t}$$

$$\overline{a} = \overline{a}_N + \overline{a}_T$$

From the vector diagram (Fi. 4.34b) we observe that as $\Delta t \to 0$

$$\Delta \overline{V}_T \to 0 \text{ and } \Delta \overline{V}_N \to \overline{V}[\Delta \theta]$$

Thus

$$\lim_{\Delta t \to 0} \frac{\Delta \overline{V}_N}{\Delta t} = \overline{V}\left[\frac{\Delta \theta}{\Delta t}\right]$$

$$\frac{d\overline{V}_N}{dt} = \overline{V}\left[\frac{d\theta}{\Delta t}\right] = \overline{V}\omega$$

and the magnitude of the normal component of acceleration becomes

$$a_n = v\omega$$

$$a_n = (\omega r)\omega = r\omega^2$$

$$a_n = V\left[\frac{V}{r}\right] = \frac{V^2}{r}$$

where the direction of the normal component of acceleration is always <u>toward the center of curvature</u>.

From the vector diagram, we observe that

as $\quad \Delta t \to 0, \Delta \overline{V}_N \to 0$

and $\quad \Delta \overline{V}_T \to \Delta \overline{V}$

and the magnitude of the tangential component of acceleration becomes

$$a_t = \frac{dV_T}{dt} = \frac{dV}{dt} \quad \text{but } V = r\frac{d\theta}{dt}$$

Hence

$$a_t = \frac{d}{dt}\left(r\frac{d\theta}{dt}\right) = r\frac{d^2\theta}{dt^2} + \frac{dr}{dt}\frac{d\theta}{dt}$$

$$a_t = r\alpha + 2\omega\frac{dr}{dt}$$

The second term in the expression for a_t will be zero when $\frac{dr}{dt} = 0$ which occurs when r is constant, as for a circle or at a point that represents a maximum or minimum radius of curvature. The magnitude of the total acceleration, a (Fig. 4.34c) is then

$$a = \sqrt{a_n^2 + a_t^2}$$

and its direction is given by $\tan\theta = \dfrac{a_n}{a_T}$

where $a_n = V\omega = r\omega^2 = \dfrac{V^2}{r}$

$$a_t = \frac{dV}{dt} = r\alpha + 2\omega\frac{dr}{dt}$$

a_n is known as the <u>centripetal</u> acceleration and is always directed inward along r. $(-a_n)$ is known as the <u>centrifugal</u> acceleration and is a reversed effective acceleration or D'Alembert acceleration which is added to problems in curvilinear motion to convert the dynamic problem to one in static equilibrium. The centrifugal acceleration is added so as to act outward from the curve.

Example

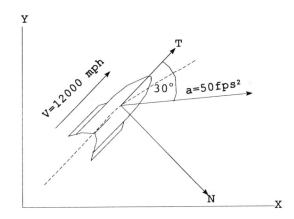

Figure 4.35

A rocket has a total acceleration of 50 ft/sec² in the direction shown at a certain point in its trajectory. Its speed at that time is 12,000 miles/hr (Fig. 4.35).

(a) Calculate the radius of curvature r of the path at this position.

(b) If r is increasing at the rate of 20 miles/sec at the instant considered, calculate the instantaneous angular acceleration α of the radius of curvature to the rocket.

$$V = 12{,}000 \text{ mi/hr} \times \frac{44}{30} = 1.76 \times 10^4 \text{ ft/sec}$$

$$\frac{dr}{dt} = 20 \frac{\text{mi}}{\text{sec}} \times 5280 = 10.56 \times 10^4 \text{ ft/sec}$$

$$a_n = a \sin 30° = 50 (.5) = 25 \text{ ft/sec}^2$$

$$a_t = a \cos 30° = 50 (.866) = 43.3 \text{ ft/sec}^2$$

(a) $a_n = \dfrac{V^2}{r}$ $\therefore r = \dfrac{V^2}{a_n} = \dfrac{(1.76 \times 10^4 \text{ ft/sec})^2}{25 \text{ ft/sec}} = 1.24 \times 10^7 \text{ ft}$

$r = 2350$ miles

(b) $a_t = r\alpha + 2\omega \dfrac{dr}{dt}$

$$\alpha = \dfrac{1}{r}\left[a_t - 2\left(\dfrac{V}{r}\dfrac{dr}{dt}\right)\right] = \dfrac{1}{1.24 \times 10^7 \text{ ft}}\left[43.3 \text{ fps}^2 - 2\left(\dfrac{1.76 \times 10^4 \text{ fps}}{1.24 \times 10^7 \text{ ft}}\right)\left(10.56 \times 10^4 \dfrac{\text{ft}}{\text{sec}}\right)\right]$$

$$\alpha = -86 \times 10^{-7} \dfrac{\text{rad}}{\text{sec}^2}$$

The minus sign shows that the angular acceleration is in the opposite sense to the angular velocity of the radius of curvature.

Example

A particle moves in a circular path of 20 ft radius so that its arc length from an initial position A is given by the relation $S = 6t^3 - 4t$ (Fig. 4.36). Determine the normal and tangential components of acceleration of the particle when $t = 2$ sec and also its angular velocity and acceleration.

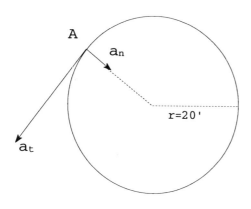

Figure 4.36

$S = 6t^3 - 4t$

$V = \dfrac{dS}{dt} = 18t^2 - 4$; $a_t = \dfrac{dV}{dt} = 36t$

at $t = 2$ sec ; $V = 18t^2 - 4 = 18(2)^2 - 4 = 68$ fps

$a_t = 36t = 36(2) = 72$ ft/sec^2

$$a_n = \frac{V^2}{r} = \frac{(68 \text{ fps})^2}{20 \text{ ft}} = 231 \text{ ft/sec}^2$$

$$\omega = \frac{V}{r} = \frac{68 \text{ fps}}{20 \text{ ft}} = 3.4 \text{ rad/sec}$$

$$\frac{dr}{dt} = 0$$

$$\therefore \alpha = \frac{a_t}{r} = \frac{72 \text{ fps}^2}{20 \text{ ft}} = 3.6 \text{ rad/sec}^2$$

Kinetics of Curvilinear Motion

The dynamic forces required to provide the normal and tangential accelerations associated with curvilinear motion are obtained from application of Newton's Law, namely

Normal to Path
$$\Sigma F_n = \frac{W}{g} a_n = \frac{W V^2}{gr}$$

Tangent to Path
$$\Sigma F_t = \frac{W}{g} a_t = \frac{W}{g} \frac{dv}{dt}$$

Where all forces act through the center of gravity of the body.

Problem

The bob of a conical pendulum weighs W lbs and is supported by a string of length L as shown in Fig. 4.37. The bob moves in a circle with constant velocity V. Determine the tension in the string, the angle the string makes with the vertical and the period or time to complete one revolution.

Solution

(1)   Drawing a free body diagram and writing the dynamic equations of motion we have:

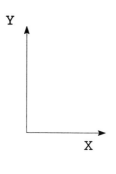
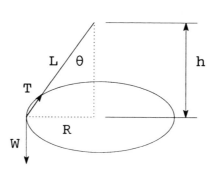

Figure 4.37

↑ $\Sigma F_y = 0$

$T \cos \theta - W = 0$

$W = T \cos \theta$ or $T = \dfrac{W}{\cos \theta}$

→ $\Sigma F_x = ma_n$ $T \sin \theta = \dfrac{W}{g} \dfrac{V^2}{R}$

Combining these equations results in $\tan \theta = \dfrac{V^2}{gR}$

or $\sin \theta = \dfrac{V^2}{gR} \cos \theta$

$\sin \theta = \dfrac{V^2}{g \, L \, \sin \theta} \cos \theta$

$\sin^2 \theta = \dfrac{V^2}{gL} \cos \theta$

$\dfrac{V^2}{gL} \cos \theta - \sin^2 \theta = 0$

$\dfrac{V^2}{gL} \cos \theta - (1 - \cos^2 \theta) = 0$

$\cos^2 \theta + \dfrac{V^2}{gL} \cos \theta - 1 = 0$

Using the quadratic formula yields:

$$\cos\theta = -\frac{V^2}{2gL} + \frac{1}{2}\sqrt{\left(\frac{V^2}{gL}\right)^2 + 4}$$

$S = Vt$ and $t = \dfrac{S}{V}$

For one revolution $S = 2\pi R$

$\tan\theta = \dfrac{V^2}{gR} \qquad \therefore V = \sqrt{gR\tan\theta}$

$$t = \frac{2\pi R}{\sqrt{gR\tan\theta}} = 2\pi\sqrt{\frac{R^2}{gR\tan\theta}} = 2\pi\sqrt{\frac{R}{g\tan\theta}}$$

for small θ, $\tan\theta \approx \sin\theta = \dfrac{R}{L}$

$$t = 2\pi\sqrt{\frac{R}{\frac{gR}{L}}} = 2\pi\sqrt{\frac{L}{g}}$$

Solution 2

If a reversed effective force (REF) or D'Alembert force is added to the system, the laws of static equilibrium may be applied. Since the normal acceleration is directed inward, the REF must be directed outward as shown in Fig. 4.38a and the force diagram closes upon itself as in Fig. 4.38b.

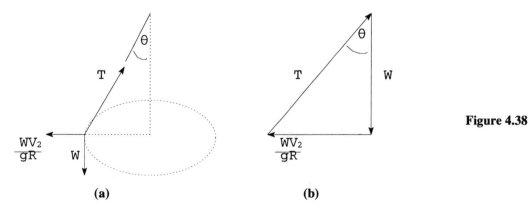

Figure 4.38

From Fig. 4.38b we have

$$\tan \theta = \frac{\dfrac{W}{g} \dfrac{V^2}{R}}{W} = \frac{V^2}{gR}$$

This result is seen to be the same as that obtained in Solution 1.

Example

A Porter governor shown in Fig. 4.39 is used to control the speed of a steam engine. Each governor ball weighs 16.1 lbs and the central weight is 40 lbs Determine the speed of rotation in rpm, at which the weight begins to rise.

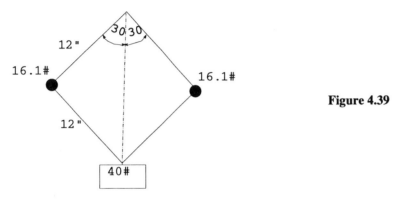

Figure 4.39

Solution

Draw a free body diagram of the 40 lb weight at the moment at which it is about to rise and a free body diagram of the governor ball, employing a reversed effective force.

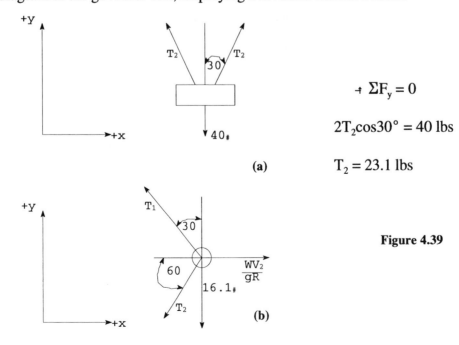

$+ \Sigma F_y = 0$

$2T_2 \cos 30° = 40$ lbs

$T_2 = 23.1$ lbs

Figure 4.39

$\updownarrow \Sigma F_y = 0$

$T_1 \cos 30° - T_2 \cos 30° - 16.1 = 0$

$(T_1 - T_2) \cos 30° = 16.1$

$T_1 = \dfrac{16.1}{\cos 30°} + T_2$

$T_1 = 41.7$ lbs

$\leftrightarrow \Sigma F_x = 0$

$\dfrac{W}{g} \dfrac{V^2}{R} - (T_1 + T_2) \cos 60° = 0$

$V^2 = \dfrac{gR}{W}(T_1 + T_2) \cos 60° = \dfrac{(32.2)(.5)}{16.1}(41.7 + 23.1)(.5)$

$V = 5.6$ fps

$\omega = \dfrac{V}{R} = \dfrac{5.69 \text{ fps}}{.5 \text{ ft}} \times \dfrac{60}{2\pi} = 108.55$ rev/min

Banking of Curves

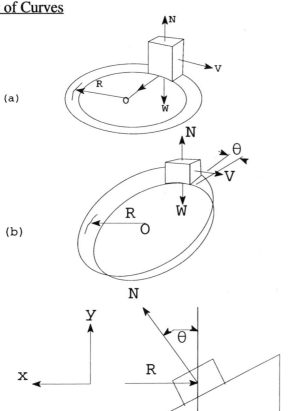

Figure 4.40

Figure 4.40a represents a block moving at constant speed V on a level surface around a curve having a radius of curvature R. In addition to the vertical forces, namely the force of gravity W and a normal force N, a horizontal centripetal force C must be present to maintain the block on the curve. This centripetal force is supplied by a sidewise frictional force exerted by the surface on the tires in the case of a wheeled vehicle and by the rails in the case of a railway car. If the curve is banked on an angle θ as shown in Figs. 4.40b and 4.40c, the normal force N has not only a vertical component, as before, but also a horizontal component that supplies the centripetal force necessary for uniform circular motion; no additional sidewise forces are needed for the vehicle to negotiate the curve. The equations of motion from Fig. 4.40c are:

$\updownarrow \Sigma F_y = 0 : \quad N \cos \theta = W \qquad N = \dfrac{W}{\cos \theta}$

$\leftrightarrow \Sigma F_x = \dfrac{W}{g} \dfrac{V^2}{R} : \quad N \sin \theta = \dfrac{W}{g} \dfrac{V^2}{R} \qquad \tan \theta = \dfrac{V^2}{gR}$

θ is called the <u>ideal bank angle</u> and depends upon the speed of the vehicle (called the rated speed of the curve) and the curvature of the road only.

If a vehicle enters a curve at a speed less than that of the rated speed it tends to <u>slip down the curve</u> unless prevented by friction. The relation between the bank angle and the velocity is determined from Fig. 4.41 as:

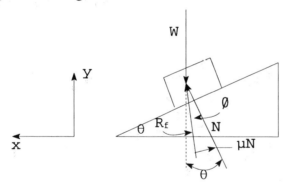

Figure 4.41

W = vehicle weight

R_f = road reaction between vehicle and roadbed also the resultant between the friction force and the normal force.

N = normal force

μ = coefficient of friction

μN = friction force (assumed acting through center of mass to oppose impending motion)

$\emptyset$ = angle between R_f and N

Note that N and μN are components of R_f.

The equations of motion are then

$$\Sigma F_y = 0$$

$$R_f \cos(\theta - \emptyset) = W$$

$$\Sigma F_x = \frac{W}{g} \frac{V^2}{R}$$

$$R_f \sin(\theta - \emptyset) = \frac{W}{g} \frac{V^2}{R}$$

Dividing the latter equation by the former yields:

$$\tan(\theta - \emptyset) = \frac{V^2}{gR}$$

Expanding $\tan(\theta - \emptyset)$

from the Trigonometric Identities

$$\frac{\tan\theta - \tan\emptyset}{1 + \tan\theta \tan\emptyset} = \frac{V^2}{gR}$$

for θ and $\emptyset$ small angles, $\tan\theta \tan\emptyset \approx 0$

$$\tan\theta - \tan\emptyset = \frac{V^2}{gR}$$

$$\tan\theta = \frac{V^2}{gR} + \tan\emptyset$$

Recall from Statics, a stationary block of weight W resting on a plane inclined at an angle $\emptyset$. The angle $\emptyset$ has a value such that there is impending motion down the incline. Remember that the force of friction always opposes motion. The free body diagram of the block is shown in Fig. 4.42.

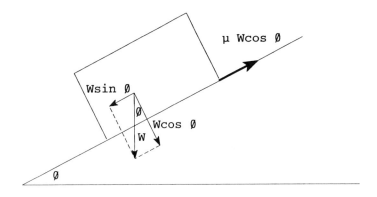

Figure 4.42

$W \sin \emptyset = \mu W \cos \emptyset$

$\tan \emptyset = \mu$

where μ = coefficient of friction and $\tan \theta = \dfrac{V^2}{gR} + \mu$

for motion impending <u>down</u> the curve.

If a vehicle enters a curve at a speed greater than that of the rated speed, it tends to <u>slip up the curve</u>, unless prevented by friction. The relation between bank angle and velocity is determined from Fig. 4.43 where:

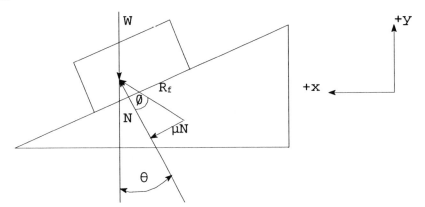

Figure 4.43

$\Sigma F_y = 0 \quad R_f \cos(\theta + \emptyset) = W$

$$R_f = \dfrac{W}{\cos(\theta + \emptyset)}$$

$\Sigma F_x = \dfrac{W}{g} \dfrac{V^2}{R} : R_f \sin(\theta + \emptyset) = \dfrac{W}{g} \dfrac{V^2}{R}$

Substituting R_f found from $\Sigma F_y = 0$, yields $\tan(\theta + \emptyset) = \dfrac{V^2}{gR}$

Expanding $\tan(\theta + \emptyset)$

$$\dfrac{\tan\theta - \tan\emptyset}{1 + \tan\theta \tan\emptyset} = \dfrac{V^2}{gR}$$

for θ and $\emptyset$ small angles, $\tan\theta \tan\emptyset \approx 0$ $\qquad \tan\theta + \tan\emptyset = \dfrac{V^2}{gR}$

$$\tan\theta = \dfrac{V^2}{gR} - \mu$$

for motion impending up the curve.

Example

In a 24-ft roadway on a curve of 1000 ft radius what variation in height is permissible between the outside edge of the roadway and the inside edge to prevent slipping for cars traveling at 50 mph. Assume a value of 0.1 for the coefficient of friction between the tires and the pavement.

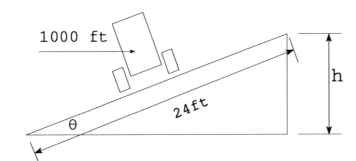

$R = 1000$ ft

$F = 50$ mph $= 73.3$ ft/sec

$\mu = 0.1$

Figure 4.44

(a) <u>Ideal bank angle</u>

$$\tan\theta_i = \dfrac{V^2}{gR} = \dfrac{(73.3 \text{ ft/sec})^2}{(32.2 \text{ ft/sec}^2)(1000 \text{ ft})} = 0.167$$

$\theta_i = 9°\ 30'$, $\quad \sin\theta_i = 0.165$

$h = 24 \sin\theta_i = 24(.165) = 3.96$ ft

(b) <u>Motion pending up</u>

$$\tan \theta_u = \frac{V^2}{gR} - \mu = 0.167 - 0.1 = 0.067$$

for small θ $\tan \theta \approx \sin \theta \approx \theta$

$$h = 24 \sin \theta_u = 24 (0.067) = 1.6 \text{ ft}$$

(c) <u>Motion pending down</u>

$$\tan \theta_d = \frac{V^2}{gR} + \mu = 0.167 + 0.1 = 0.267$$

$$\theta_d = 15°, \quad \sin 15° = 0.259$$

$$h = 24 \sin \theta_d = 24 (0.259) = 6.2 \text{ ft}$$

Permissible variation = 1.6 to 6.2 ft

For a vehicle entering a curve at a speed greater than that of the rated speed of the curve there is a tendency not only for the vehicle to slip upward but also for the vehicle to turn over about the outer wheels. This condition may be studied as shown in Fig. 4.45 by introducing a reversed effective force directed outward from the center of mass and including the individual normal and frictional forces acting at each wheel.

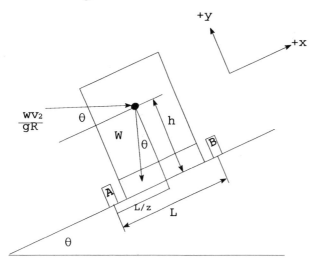

Figure 4.45

N_1 = normal reaction at Wheel A

N_2 = normal reaction at Wheel B

μN_1 = friction force at Wheel A

μN_2 = friction force at Wheel B

h, L = dimensions as shown

N_1 is determined as follows:

$$\circlearrowright \Sigma M_B = 0 : N_1 L - W \cos\theta \left(\frac{L}{2}\right) - W \sin\theta\,(h) + \frac{W}{g}\frac{V^2}{R}\cos\theta(h)$$

$$-\frac{W}{g}\frac{V^2}{R}\frac{L}{2}\sin\theta = 0$$

$$LN_1 = W\left(\frac{L}{2}\cos\theta + h\sin\theta\right) + \frac{W}{g}\frac{V^2}{R}\left(\frac{L}{2}\sin\theta - h\cos\theta\right)$$

$$N_1 = W\left[\frac{\cos\theta}{2} + \frac{h}{L}\sin\theta + \frac{V^2}{gR}\left(\frac{\sin\theta}{2} - \frac{h}{L}\cos\theta\right)\right]$$

$$N_1 = W\left[\left(\tfrac{1}{2} - \frac{V^2 h}{gRL}\right)\cos\theta + \left(\frac{h}{L} + \frac{V^2}{2gR}\right)\sin\theta\right]$$

The speed at which the vehicle tends to overturn may be determined by setting N1 = 0 in the above equation and solving for V. N_2 is determined from

$$\Sigma M_A = 0 \text{ or } \Sigma F_y = 0$$

Example

A roadbed, 30 ft wide, is curved to a radius of 400 ft and elevated 3 ft (Fig. 4.46). If the coefficient of friction is 0.23 what is the maximum speed of the car if it does not skid? What are the normal and frictional forces on the wheels at this speed if the center of mass is 2 ft above the ground, the wheel tread, is 6 ft and the weight of the car is 3220 lbs? At what speed will the car overturn?

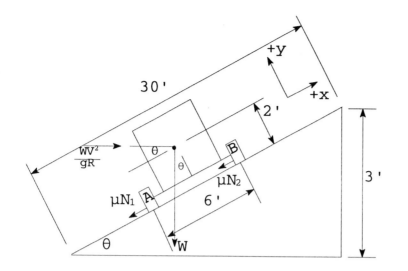

Figure 4.46

(a) $\tan(\theta + \emptyset) = \dfrac{V^2}{gR}$

$\emptyset = \tan^{-1} 0.23 \,;\, \emptyset = 13°$ $\qquad \theta = \sin^{-1} \dfrac{3}{30} = .1, \theta = 6°$

$\theta + \emptyset = 19°, \tan(\theta + \emptyset) = 0.3443$

$V = \sqrt{gR \tan(\theta + \emptyset)} = \sqrt{(32.2)(400)(0.3443)}$

$V = 66.5$ fps $= 45.03$ mph

(b) $\dfrac{W}{g}\dfrac{V^2}{R} = \dfrac{3220 \,(66.5)^2}{32.2 \,(400)} = 1105$ lbs

$\circlearrowleft \Sigma M_B = 0$

$6N_1 + 1105 \cos 6°(2) - 1105 \sin 6°(3) - 3220 \cos 6°(3)(- 3220 \sin 6°(2) = 0$

$6N_1 + (2210 - 9660)(.995) - (3315 + 6440)(.1) = 0$

$6N_1 = 7450(.995) + (9755)(.1) = 7410 + 975.5 = 8385.5$

$N_1 = 1398$ lbs, $\mu N_1 = (.23)(1398) = 321$ lbs

$\Sigma F_y = 0$

$$N_1 + N_2 - 3220 \cos 6° - 1105 \sin 6° = 0$$

$$N_2 = 3220 (.995) + 1105 (.1) - 1398$$

$$N_2 = 3204 + 111 - 1398 = 1917 \text{ lbs}$$

$$\mu N_2 = (.23)(1902) = 441 \text{ lbs}$$

(c) $\circlearrowleft \Sigma M_B = 0$ where $N_1 = 0$

$$\frac{3220 \, V^2}{32.2 \, (400)} (2) \cos 6° - \frac{3220 \, V^2}{32.2 \, (400)} (3) \sin 6° - 322(3) \cos 6° - 322(2) \sin 6° = 0$$

$$\frac{V^2}{4} (1.99 - .3) = 3220 (2.985 + .2)$$

$$V = \sqrt{\frac{(3220)(3.185)(4)}{1.69}} = 156 \text{ fps} = 106 \text{ mph}$$

Rotation of a Rigid Body

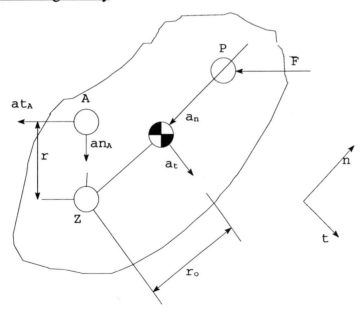

Figure 4.47

Consider a rigid body of mass m constrained to rotate about axis Z subjected to an unbalanced force $\bar{F}$ acting at point P as shown in Fig. 4.47. Based on the action of this force an instantaneous angular acceleration of α and angular velocity of ω are imparted to the body resulting in an acceleration $\bar{a}$ at the center of mass where

$$\bar{a} = \bar{a}_n + \bar{a}_t$$

$\bar{a}_n$ and $\bar{a}_t$ represent the normal and tangential acceleration of the center of mass about Z directed along n and t respectively as shown. The magnitude of these accelerations are:

$$a_n = r_0 \omega^2 \quad \text{and} \quad a_t = r_0 \alpha$$

where r_0 is the displacement of the center of mass from the axis of rotation Z. If ΣF_n and ΣF_t represent the magnitude of the sum of all the forces in the normal and tangential directions, respectively, the instantaneous equations of motion of the center of mass are

$$\Sigma F_n = ma_n = \frac{W}{g} r_0 \omega^2$$

$$\Sigma F_t = ma_t = \frac{W}{g} r_0 \alpha$$

Next consider particle of mass dm at point A. The incremental forces acting on this particle are

$$dF_{n_A} = a_{n_A} dm = r\omega^2 dm$$

$$dF_{t_A} = a_{t_A} dm = r\alpha dm$$

Here r is the distance of point A from the axis of rotation Z as shown.

If we sum the moments acting about Z due to the forces at A we have

$$\Sigma dM_z = (dF_{tA})(r) + (dF_{nA})(\alpha)$$

or $\quad \Sigma dM_z = dm\, r\alpha r = \frac{dW}{g} r^2 \alpha$

$$\Sigma M_z = \alpha \int \frac{dW}{g} r^2 \,;\, \int \frac{dW}{g} r^2 = I_z = \text{mass moment of inertia}$$

Hence: $\quad \Sigma M_z = I_z \alpha$

In summary the equations of motion of a body rotating about a point other than its center of mass, are

$$\Sigma F_n = \frac{W}{g} r_o \omega^2$$

$$\Sigma F_t = \frac{W}{g} r_o \alpha$$

$$\Sigma M_z = I_z \alpha$$

For rotation about the center of mass, $r_o = 0$ and $\Sigma F t = 0$

$$\Sigma F_n = 0 \quad \Sigma M = I_o \alpha$$

From the transfer moment of inertia formula $I_z = I_o + m\, r_0^2$

where I_o is the moment of inertia about the center of mass hence

$$\Sigma M = (I_o + m\, r_0^2)\, \alpha = I_z \alpha$$

In summary, the equations of motion of a body rotating about its center of mass are

$$\Sigma F_n = 0$$

$$\Sigma F_t = 0$$

$$\Sigma M = I_o \alpha$$

Example

A rotating drum has a centroidal mass moment of inertia = 10 ft-lb sec². The coefficient of friction is 0.25. At the instant the brake is applied, the weight has a downward velocity of 20 ft/sec. (See Fig. 4.48). What is the constant brake force P required to stop the block in a distance of 10 ft?

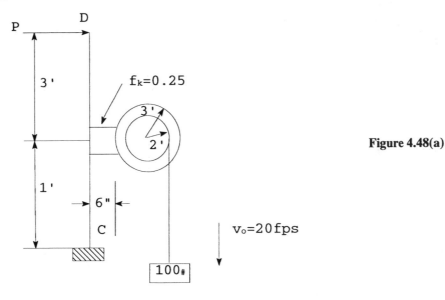

Figure 4.48(a)

4-68 FUNDAMENTALS OF ENGINEERING EXAM REVIEW WORKBOOK

Solution

The linear acceleration of block B is determined from

$$V_f^2 = V_o^2 + 2aS$$

Where $V_f = 0$, a = tangential deceleration of inner drum = a_t

$$V_o = 20 \text{ fps}, S = 10 \text{ ft}$$

$$a_t = -\frac{V_o^2}{2S} = -\frac{(20 \text{ fps})^2}{2 \text{ ft } (10 \text{ ft})} = -20 \frac{\text{ft}}{\text{sec}^2} \text{ (deceleration)}$$

The angular acceleration of the drum is determined from $a_t = r\alpha$

$$\alpha = \frac{a_t}{r} = -\frac{20 \text{ ft/sec}^2}{2 \text{ ft}} = -10 \text{ rad/sec}^2 \text{ (deceleration)}$$

The tension in the cable decelerating block B is determined from a free body diagram

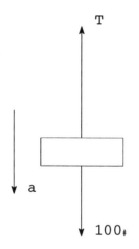

Figure 4.48(b)

$$\downarrow \Sigma F = \frac{W}{g} a$$

$$100 - T = \frac{100}{32.2} (-20)$$

$$T = 100 + \frac{2000}{32.2} = 162.1 \text{ lbs}$$

where T is the force applied by the drum on the weight. The force required to decelerate the drum is obtained from a FBD.

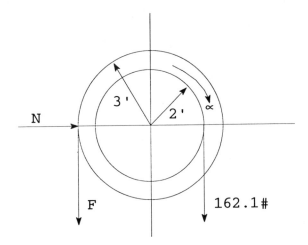

Figure 4.48(c)

$\circlearrowright +) \ \Sigma M_0 = I_0 \alpha$

$(162.1)(2 \text{ ft}) - F(3 \text{ ft}) = (10 \text{ ft-lb-sec}^2)(-10 \frac{\text{rad}}{\text{sec}^2})$

$3F = 424.2 \ ; \ F = 141.4 \text{ lbs} \ ; \ N = \dfrac{F}{\mu} = \dfrac{141.4 \text{ lbs}}{0.25} = 565.6 \text{ lbs}$

The brake force P is determined from a FBD of the brake.

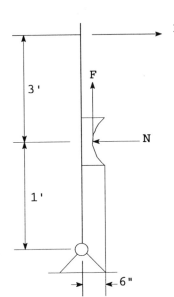

Figure 4.48(d)

$\circlearrowright \Sigma M_c = 0$

P (4 ft) - N (1 ft) - F (½ ft) = 0

4P = (565.6) (1) + (141.4) (½)

4P = 636.3 ; P = 159.1 lbs

Dynamic Equilibrium of Noncentroidal Rotation

Radius of Gyration

The mass moment of inertia is sometimes expressed in terms of the product of its mass and the square of a distance k such that

$$I = mk^2 \text{ or } k = \sqrt{\frac{I}{m}}$$

where k is called the radius of gyration. The distance k may be regarded as the distance from the axis of inertia to the point at which the mass of the body may be assumed to be concentrated and still have the same moment of inertia of the actual distributed mass of the body. This is shown in Fig. 4.49a and 49b.

I_z = mass moment of intertia of m about z

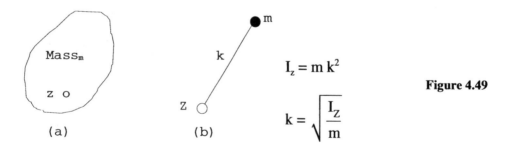

$I_z = m k^2$

$k = \sqrt{\frac{I_z}{m}}$

Figure 4.49

Rotational Dynamic Equilibrium

Before dynamic equilibrium can be created, it is necessary to determine the magnitude, direction, and position of the resultant effective force. Having done this, we can apply an equilibrant in accordance with d'Alembert's Principle.

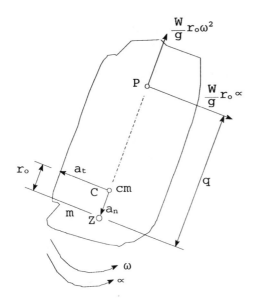

Figure 4.50

Consider a rigid body m whose instantaneous angular velocity and acceleration are ω and α as shown in Fig. 4.50. As a result of this motion an acceleration is established at the center of mass with components a_n and a_t. To create dynamic equilibrium of m we anticipate that two inertia forces must be added equal and opposite to those associated with a_n and a_t. We introduce these forces at a point P located on a radial line through C a distance q from axis of rotation Z. It then remains to determine the relation between r_o, the distance of the center of mass from the axis of rotation, and q. If we assume that the introduction of the reversed normal force $\frac{W}{g}[r_o\omega^2]$ and the reversed tangential force $\frac{W}{g}[r_o\alpha]$ place the body in equilibrium, we may write for the moments about Z

$$\circlearrowright^+ \Sigma M_z = I_z \alpha$$

$$(\frac{W}{g} r_o \alpha) q - I_z \alpha = 0$$

Expressing the moment of inertia I_z in terms of the radius of gyration

$$\frac{W}{g} r_o \alpha q = \frac{W}{g} k_z^2 \alpha$$

$$\therefore r_o q = k_z^2 \quad \text{or} \quad q = \frac{k_z^2}{r_o}$$

Hence we may state that dynamic equilibrium is achieved when a normal force $\frac{W}{g} r_o \omega^2$ and a tangential force $\frac{W}{g} r_o \alpha$ are introduced at a distance q from the axis of rotation where $q = \frac{k_z^2}{r_o}$ is located on a radial line through C as shown. The resulting equations are then

$$\Sigma F_t = 0; \; \Sigma F_N = 0; \; \Sigma M_z = 0$$

Point P is called the <u>center of percussion</u>. It may be thought of as equivalent to the "center of gravity" of the inertia forces. We next show that dynamic equilibrium may also be achieved by adding reversed normal and tangential inertial forces through the center of mass if we also add a couple of magnitude $I_o \alpha$ about the body. Consider Fig. 4.51a which shows body m in dynamic equilibrium.

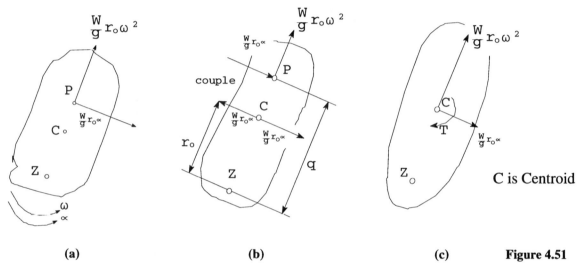

(a) (b) (c) Figure 4.51

C is Centroid

If at C two forces are introduced of equal magnitude $\frac{W}{g} r_o \alpha$, but oppositely directed as shown in Figure 51b the equilibrium of body m is unaltered. These four forces are seen to constitute a single force $\frac{W}{g} r_o \omega^2$ directed radially through C, a single force $\frac{W}{g} r_o \alpha$ directed tangentially through C and a Couple M of magnitude

$$M = \frac{W}{g} r_o \alpha (q - r_o)$$

Substituting for q

$$M = \frac{W}{g} r_o \alpha \left(\frac{k_z^2}{r_o} - r_o \right)$$

$$M = \frac{W}{g}\alpha(k_z^2 - r_o^2) = \alpha(mk_z^2 - mr_o^2)$$

recalling that $I_z = mk_z^2$, $I_o = I_z - mr_o^2$

we have $m = \alpha[I_z - mr_o^2] = I_o\alpha$

In summary, in order to establish dynamic equilibrium for a body rotating about a point other than its center of mass (1) add forces $\frac{W}{g}r_o\alpha$ and $\frac{W}{g}r_o\omega^2$ so as to oppose the tangential and normal accelerations of the body, at a point P located distance $q = \frac{k_z^2}{r_o}$ from axis of rotation. Point P is known as the <u>center of percussion</u>.

or (2) add forces $\frac{W}{g}r_o\alpha$ and $\frac{W}{g}r_o\omega^2$ so as to oppose the tangential and normal acceleration of the body at the center of mass plus a moment of magnitude $I_o\alpha$ (about any point in the body) directed so as to oppose the angular acceleration.

Example

In the rotating linkage shown in Fig. 4.52a the member AB is a thin homogeneous rod 10 ft long, weighing 120 lbs At the instant shown, member AB has an angular acceleration of 1 rad/sec² and an angular velocity of 2 rad/sec. All joints are assumed to have frictionless bearings. Determine the bearing reactions at A and B.

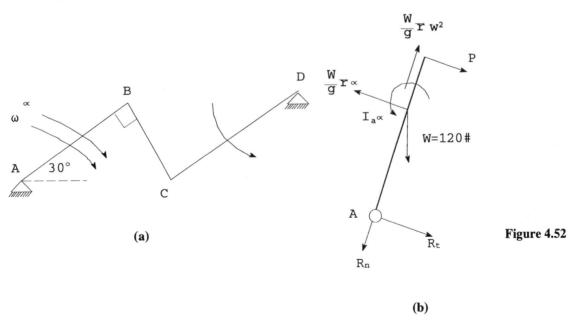

Figure 4.52

Solution

Draw a free body diagram of member AB, (Fig. 4.52b) and apply the inertia forces at the center of gravity and the inertia couple $I_o \alpha$ about the CG to create dynamic equilibrium.

$$\frac{W}{g} \bar{r} \omega^2 = \frac{120}{32.2} \times 5 \times (2)^2 = 74.53 \text{ lbs}$$

$$\frac{W}{g} \bar{r} \alpha = \frac{120}{32.2} \times 5 \times 1 = 18.63 \text{ lbs}$$

$$I_a \alpha = (\frac{1}{3} \frac{W}{g} L^2)\alpha = \frac{1}{3} \times \frac{120}{32.2} (10)^2 (1) = 124.22 \text{ ft-lbs}$$

The value of P is determined from $\Sigma M_A = I_a \alpha$

$10 P + 120 (5 \cos 30) - (18.63 \times 5) - 124.22 = 0$

$P = - 30.22$ lbs (the bearing reaction at B)

From Fig. 4.52c, after resolving the reactions at A into normal and tangential components we can apply the equation of equilibrium.

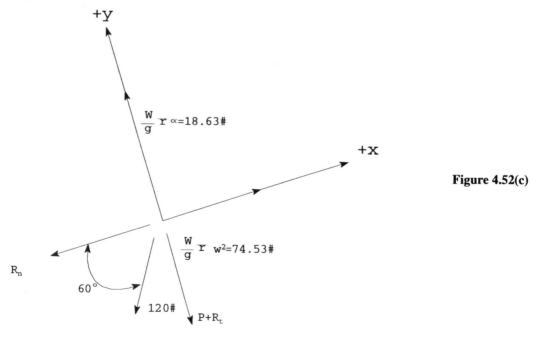

Figure 4.52(c)

$\nearrow \Sigma F_x = 0 : 74.53 - 120 \cos 60 - R_n = 0$

$\therefore R_n = 14.53$ lbs

$\Sigma F_y = 0 : 18.63 - (P + R_t) - 120 \sin 60° = 0$

$\therefore R_t = -55.07 \text{ lbs}$

$R = \sqrt{(14.53)^2 + (55.07)^2} = 57.0 \text{ lbs}$

Work–Energy

Consider a translating body where the resultant force is defined by

$$\Sigma F_x = \frac{W}{g} a$$

Recall a dS = V dV, eliminate "a" from both expressions as follows:

$$\Sigma F_x \frac{g}{W} = a \text{ and } \frac{V \, dV}{dS} = a$$

$$\Sigma F_x \frac{g}{W} = \frac{V \, dV}{dS} \text{ or } \underbrace{\frac{\Sigma F_x \, dS}{\text{work}}}_{} = \frac{W}{g} V \, dV$$

If the initial velocity is V_o when S = 0, and the final velocity is V_f when the displacement is S, we have by integration:

$$\int_0^S \Sigma F_x \, dS = \frac{W}{g} \int_{V_o}^{V_f} V \, dV$$

$$\Sigma F_x \times S = \frac{W}{g} \left[\frac{V_f^2}{2} - \frac{V_o^2}{2} \right] = \frac{W}{2g} (V_f^2 - V_o^2)$$

This is the fundamental work-energy equation. THE RESULTANT WORK ON A TRANSLATING BODY IS EQUAL TO THE CORRESPONDING CHANGE IN KINETIC ENERGY. (Note: The force <u>must</u> be in the direction of the motion.)

Example

A truck with rolling resistance (caused by wheel friction) of 12 lbs per ton starts from rest at A and rolls without power added down a 1000 ft incline with a 4% grade to B. Then, without power added, it goes 800 ft up an incline with a 3% grade to C and without added power goes down an incline with a 2% grade to D. Finally, it travels a distance S along a level track where it comes to rest at E (Fig. 4.53). Find the distance S.

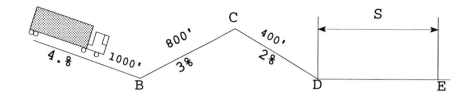

Figure 4.53

Resultant work is:

$$\Sigma F_{x_1} S_1 + \Sigma F_{x_2} S_2 + \Sigma F_{x_3} S_3 + \Sigma F_{x_4} S_4$$

where, on a per ton basis:

$$\Sigma F_{x_1} S_1 = (2000 \sin \theta_1 - 12) 1000 = 68{,}000 \text{ ft-lbs}$$

($\frac{4}{100}$ for small angles, $\sin \theta = \theta$)

$$\Sigma F_{x_2} S_2 = (-2000 \times \frac{3}{100} - 12) 800 = -57{,}600 \text{ ft-lbs}$$

$$\Sigma F_{x_3} S_3 = (2000 \times \frac{2}{100} - 12) 400 = 11200 \text{ ft-lbs}$$

$$\Sigma F_{x_4} S = (-12) S$$

Adding all the terms yields: 21,600 - 12S

Which must be equal to the change in kinetic energy. But, the car started from rest and ended at rest, therefore the change in kinetic energy is zero.

$$21{,}600 - 12 S = 0 \text{ and } S = 1800 \text{ ft}$$

Caution must be used in this method to make certain that the truck would traverse the uphill grade. If BC had been 1500 ft long, the work expended in going up hill would be - 72,000 ft-lbs, indicating that the truck would not reach C.

Work-Energy Method Involving Variable Forces

In a spring the force is proportional to the deformation, as shown in Fig. 4.54.

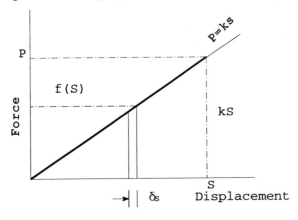

Figure 4.54

$$dW = \Sigma F_s \, dS$$

where dW = incremental work

ΣF_s = forces in S direction = $f(s)$

dS = incremental distance

$$dW = f(s) \, dS$$

$$W = \int f(s) \, dS = \text{area of element } dS$$

The spring force $f(s)$ is directly proportional to the spring deformation or

$$P = f(s) = kS$$

$$W = \int_0^k kS \, dS = \frac{kS^2}{2} \quad \text{which is the area under the curve}$$

Examples:

(1) A weight of 50 lbs is dropped from a height h on to a spring. If the spring has a modulus of 125 lbs per in., what is the value of h which causes a 4 in. deformation of the spring? Assume that the weight remains in contact with the spring after contact.

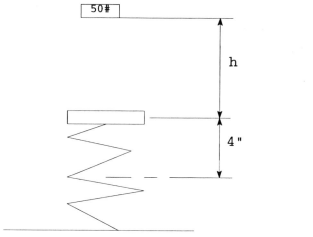

Figure 4.55

Work done = Change in kinetic energy

Solution (1)

$$F_x \, dS = \frac{W}{2g} (V_f^2 - V_o^2)$$

Construct a force displacement curve as shown in Fig. 4.55a.

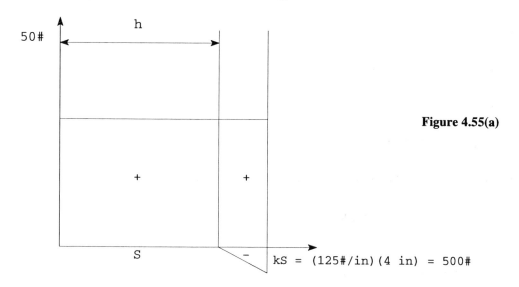

Figure 4.55(a)

kS = (125#/in)(4 in) = 500#

$$\Sigma F_x \, dS = 50 (h + 4) - \tfrac{1}{2} \times 500 \times 4$$

$$= 50h + 200 - 1000 = 50h - 800$$

But the change in kinetic energy is zero because $V_f = V_o = 0$.

∴ 50h − 800 = 0 ; h = 16 in.

Solution (2)

$$\int \Sigma F_x \, dS = \frac{W}{2g}(V_f^2 - V_o)^2$$

$$V_f = V_o = 0$$

$$\Sigma F_x = W - kS$$

$$\int (W - kS) \, dS = 0$$

$$\int_0^{h+4} W \, dS - k \int_0^4 S \, dS = 0$$

$$W \, S \Big]_0^{h+4} - k \, \frac{S^2}{2}\Big]_0^4 = 0$$

$$W(h+4) - \frac{k}{2}(4)^2 = 0$$

$$W = 50 \text{ lbs}$$

$$k = 125 \text{ lbs/in.}$$

$$50(h+4) = \frac{125}{2}(4)^2 = 1000$$

$$h + 4 = 20$$

$$h = 16 \text{ in.}$$

(2) The 322 lbs is solid homogeneous disc is released from rest on the inclined plane as shown in Fig. 4.56. If there is no initial tension in the spring, find the maximum distance that the disc will roll down the plane. The constant of the spring is k = 120 lbs/ft.

The potential energy of position of the disc will be transformed into potential energy stored in the spring.

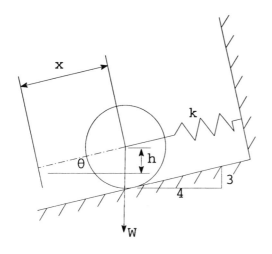

Figure 4.56

$$\text{Work} = \int_0^x F\,dx = \int_0^x kx\,dx = \frac{k x^2}{2}$$

$$\Delta P.E. = Wh = \frac{k x^2}{2} \;;\; h = x \sin \theta$$

$$\therefore x = \frac{2W}{k}(\sin \theta) = \frac{2(322)}{120} \cdot \frac{3}{5} = 3.22 \text{ ft along plane}$$

Equivalent Springs

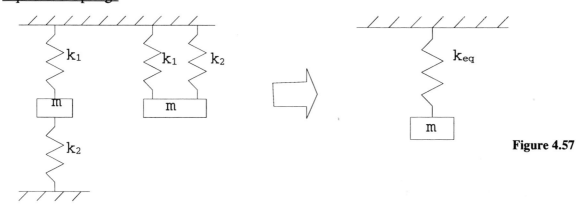

Figure 4.57

Figure 4.57 shows springs of modulus k_1 and k_2 arranged in <u>parallel</u>. The action of these springs may be duplicated by a single spring of modulus k_{eq} which is related to k_1 by $k_{eq} = k_1 + k_2$.

In general "n" number of springs in parallel may be replaced by a single equivalent spring k_{eq} whose modulus is determined from

$$k_{eq} = k_1 + k_2 + \ldots + k_n$$

Springs in parallel are characterized by experiencing equal displacements under loading.

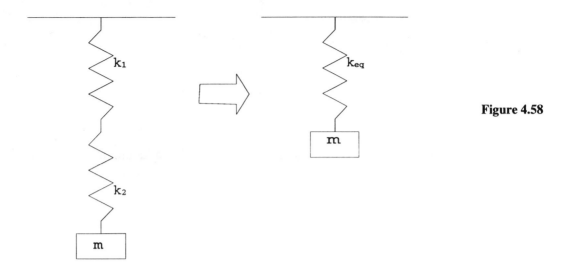

Figure 4.58

Figure 4.58 shows springs of modulus k_1 and k_2 arranged in <u>series</u>. The action of these springs may be duplicated by a single spring of modulus k_{eq} which is related to k_1 and k_2 by

$$\frac{1}{k_{eq}} = \frac{1}{k_1} + \frac{1}{k_2}$$

In general "n" number of springs in series may be replaced by a single equivalent spring k_{eq} whose modulus is determined from

$$\frac{1}{k_{eq}} = \frac{1}{k_1} + \frac{1}{k_2} + \ldots + \frac{1}{k_n}$$

Springs in series are characterized by experiencing the same force under loading.

<u>Example</u>

Determine the frequency of vibration of the 4 lb cart shown in Fig. 4.59. The springs are unstretched when the cart is in its equilibrium position.

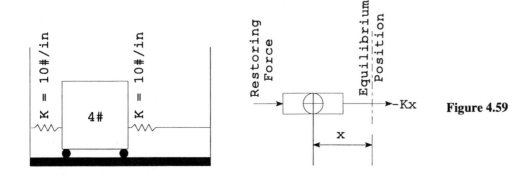

Figure 4.59

$$\Sigma F_x = m\ddot{x}$$

$$-kx - kx = m\ddot{x}$$

$$m\ddot{x} + 2kx = 0 \; ; \; \ddot{x} + \frac{2k}{m} x = 0$$

Assume a solution $x = A \cos \theta = A \cos \omega t$ that relates the linear displacement x to an angular displacement θ where

ω = angular velocity associated with θ

$$x = A \cos \omega t$$

$$\dot{x} = -A \omega \sin \omega t$$

$$\ddot{x} = -A\omega^2 \cos \omega t$$

$$(-A\omega^2 \cos \omega t) + \left(\frac{2k}{m}\right)(A \cos \omega t) = 0$$

$$(A \cos \omega t)\left(-\omega^2 + \frac{2k}{m}\right) = 0$$

Rejecting the solution $A \cos \omega t = 0$ we have

$$-\omega^2 + \frac{2k}{m} = 0$$

$$\omega = \sqrt{\frac{2k}{m}} \quad \text{or} \quad \sqrt{\frac{K_{eq}}{m}}$$

where ω = angular frequency in rad/sec

The linear frequency f_n in cycles/sec or Hertz is given by

$$f_n = \frac{\omega}{2\pi} = \frac{1}{2\pi} \sqrt{\frac{2k}{m}}$$

Note the period of oscillation in seconds is given by:

$$T = \frac{1}{f_n} = 2\pi \sqrt{\frac{m}{2k}}$$

$$f_n = \frac{1}{2\pi} \sqrt{\frac{2k}{m}} = \frac{1}{2\pi} \sqrt{\frac{2 \times 10 \text{ lbs/in.} \times 12 \text{ in./ft}}{4 \text{ lbs/32.2 ft/sec}^2}}$$

$$f_n = 7 \text{ Hertz}$$

Conservative Systems

Conservative systems are characterized by the absence of friction or dissipative type forces to expend the energy of the systems. At any instant the potential and kinetic energy of such systems is a constant or

$$KE + PE = K + U = \text{constant}$$

By the laws of calculus we may also state that

$$\frac{d}{dt}(K + U) = 0$$

Example

A small one pound weight is suspended from a 3 ft rod of negligible mass. A spring is fastened as shown in Fig. 4.60a and the system oscillates through a small arc. Determine the frequency of the resulting vibration.

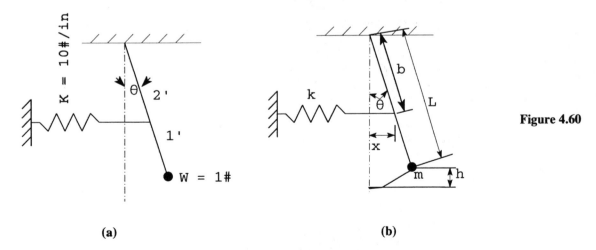

(a)   (b)

Figure 4.60

Solution

Assume the system to be conservative, then

$$PE = U = Wh + \frac{k x^2}{2}$$

$$= mg(L - L\cos\theta) + \frac{k}{2}(b\sin\theta)^2$$

$$= mgL(1 - \cos\theta) + \frac{k}{2}b^2(\sin^2\theta)$$

$$KE = K = \tfrac{1}{2} m V^2, \quad V = \dot{\theta}L \quad \therefore K = \tfrac{1}{2} m \dot{\theta}^2 L^2$$

$$PE + KE = mgL(1 - \cos\theta) + \frac{k}{2}b^2\sin^2\theta + \tfrac{1}{2} m \dot{\theta}^2 L^2$$

$$\frac{d}{dt}(PE + KE) = mgL\sin\theta\,[\dot{\theta}] + kb^2\cos\theta\sin\theta\,[\dot{\theta}] + m\dot{\theta}(\ddot{\theta})L^2$$

for θ small, $\sin\theta \approx \theta$, $\cos\theta \approx 1$, $mgL\,\theta + kb^2\theta + mL^2\ddot{\theta} = 0$

$$\ddot{\theta} + \left(\frac{mgL + kb^2}{mL^2}\right)\theta = 0$$

$$f_n = \frac{1}{2\pi}\sqrt{\frac{WL + kb^2}{\frac{W}{g}L^2}} = \frac{1}{2\pi}\sqrt{\frac{(1)(3) + (10 \times 12)(4)}{\frac{(1)(9)}{32.2}}}$$

$f_n = 6.61$ Hertz

Impulse and Momentum

The momentum $\bar{p}$ of a mass m moving with a velocity $\bar{V}$ was defined by Newton as:

$$\bar{p} = m\bar{V}$$

Since m is a scalar quantity and V a vector, the quantity $\bar{p}$ must be a vector quantity. To effect a change of momentum, Newton postulated that a force $\bar{F}$ was required such that:

$$\bar{F} = \frac{d}{dt}(m\bar{V})$$

For a constant mass, m, the equation reduces to

$$\bar{F} = m\frac{d\bar{V}}{dt}$$

where $\frac{d\bar{V}}{dt}$ may be

1. A change in the magnitude of $\bar{V}$
2. A change in the direction of $\bar{V}$
3. A change in both the magnitude and direction of $\bar{V}$

If we define an <u>impulse</u> as a force acting a short duration of time such as a blow from a sledge hammer, the impact of a bullet, or the dynamic action of a stream of water, we have

$$\bar{F}\,dt = m\,d\bar{V} \text{ where } \bar{F} = \bar{F}(t)$$

and $\bar{F}(t)\,dt$ = the impulse force

$m\,d\bar{V}$ = the change in momentum over dt for a constant m

If we assign the boundary conditions at

$$t = t_o \qquad t = t$$
$$\bar{V} = V_o \qquad \bar{V} = \bar{V}$$

we have

$$\int_{t_o}^{t} \bar{F}(t)\,dt = m \int_{V_o}^{V} d\bar{V}$$

$$\int_{t_o}^{t} \bar{F}(t)\,dt = m(\bar{V} - \bar{V}_o)$$

If the vectors $\overline{F}(t)$ and $\overline{V} = \overline{V}_o$ are parallel to the chosen coordinate axis we have the algebraic equation

$$\int_{t_o}^{t} F(t)\, dt = m(V - V_o)$$

If the vectors $\overline{F}(t)$ and $\overline{V} - \overline{V}_o$ are not parallel to the chosen coordinate axis we have the algebraic equations

$$\int_{t_o}^{t} \Sigma F_x(t)\, dt = \frac{W}{g}(V_x - V_{ox})$$

$$\int_{t_o}^{t} \Sigma F_y(t)\, dt = \frac{W}{g}(V_y - V_{oy})$$

$$\int_{t_o}^{t} \Sigma F_z(t)\, dt = \frac{W}{g}(V_z - V_{oz})$$

Example

A 450 lb steel ingot resting on a flat, level surface is pulled by a force P. P varies with time according to the relationship $P = 20t$ (Fig. 4.61). What is the velocity of the ingot as time varies from 0 to 10 sec? Assume the static coefficient of friction is 0.3 and the kinetic coefficient is 0.2.

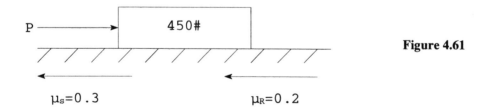

Figure 4.61

Solution (1)

Friction force to overcome before motion occurs = $450\mu_s = 450(0.3) = 135$ lb

Time to build up to 135 lbs; $P = 20t = 135$ lb $\therefore t = 6.75$ sec

Dynamic friction force $F = 450\mu_k = 450(0.2) = 90$ lbs

V_f after 10 sec

$$\int_{t_o}^{t} \Sigma F(t)\, dt = \frac{W}{g} \int_{V_o}^{V} dV$$

The initial conditions are at $t = 6.75$ sec $\quad t = 10$ sec
$$V_o = 0 \quad\quad V = V_f$$

$$\int_{6.25}^{10} (20t - 90)\, dt = \frac{450}{32.2} \int_{0}^{V_f} dV$$

$$10 t^2 - 90 t \Big]_{6.75}^{10} = \frac{450}{32.2} V_f$$

$$1000 - 900 - 455.6 + 607.5 = \frac{450}{32.2} V_f$$

$$V_f = \frac{251.9 \times 32.2}{450} = 18.0 \text{ fps}$$

Solution (2) Force-Time Curve

Draw a diagram of the force-time relationship (Fig. 4.62).

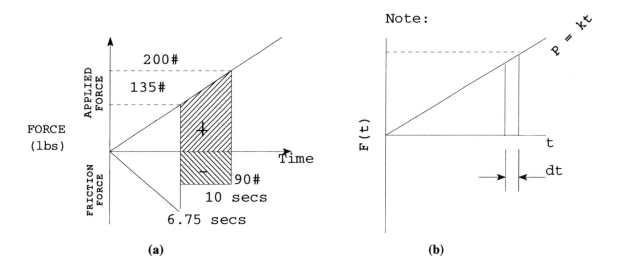

Figure 4.62

4-88   FUNDAMENTALS OF ENGINEERING EXAM REVIEW WORKBOOK

$$\int_0^t F(t)\, dt = \text{area under curve} = \frac{W}{g} V$$

$$\int_0^t F(t)\, dt = \frac{(200 + 135)}{2}(10 - 6.75) - 90(10 - 6.75) = \frac{W}{g} V_f$$

$$544.38 - 292.5 = \frac{450}{32.2} V_f \text{ or } V_f = 18.0 \text{ fps}$$

Dynamic Action of Jets

In these problems the water is treated as a continuous stream of moving masses. It is convenient to treat the jet of water as a free body diagram.

Example

A high pressure hose is used to direct a 2 in. diameter stream of water tangentially against a deflector. The velocity of the water is constant at 100 ft/sec. If the deflector diverts the stream 45°, what force does the water exert on the deflector and what is the angle of this force with respect to the initial stream direction?

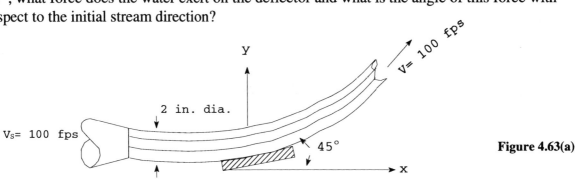

Figure 4.63(a)

$V_s = 100$ fps

Solution (1)

Draw a vector diagram of the change in direction of the jet stream (Fig. 4.63b).

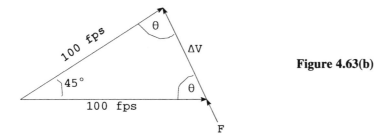

Figure 4.63(b)

From the law of cosines:

$$\Delta V^2 = 2(100)^2 - 2(100)^2 \cos 45°$$

$$\Delta V^2 = 5858 \quad \Delta V^2 = 76.54 \text{ fps}$$

$$F \Delta t = m \Delta V = \frac{\Delta W}{g} \Delta V$$

$\Delta W = \rho A V \Delta t$ = total pounds of water impinging in Δt sec

$$= 62.5 \frac{\text{lb}}{\text{ft}^3} \frac{\pi}{4} \left(\frac{2 \text{ in.}}{12 \text{ in./ft}}\right)^2 \left(100 \frac{\text{ft}}{\text{sec}}\right) \Delta t$$

$$\Delta W = 136.5 \Delta t$$

$$F \Delta t = \frac{136.5 \Delta t}{32.2} \Delta V$$

$$F = \frac{136.5}{32.2}(76.54) = 320.22 \text{ lbs} \; ; \; 2\theta = 180 - 45 = 135°; \; \theta = 67.5°$$

Solution (2)

Resolve jet velocity into components.

$$F_x \Delta t = m \Delta V_x \; ; \; F_y \Delta t = m \Delta V_y$$

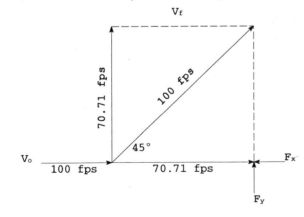

Figure 4.63(c)

Figure 4.63(d)

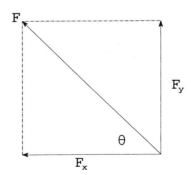

$$F_x \Delta t = \frac{\Delta W}{g}(V_{f_x} - V_{o_x})$$

$$F_x \Delta t = \frac{136.5 \Delta t}{32.2}(70.71 - 100)$$

$F_x = -124.16$ lbs (to the left)

$$F_y \Delta t = \frac{\Delta W}{g}(V_{f_y} - V_{o_y})$$

$$F_y \Delta t = \frac{136.5 \, t}{32.2}(70.71 - 0)$$

$F_y = 299.75$ lbs (up)

$$F = \sqrt{F_x^2 + F_y^2}$$

$$F = \sqrt{(-124.16)^2 + (299.75)^2} = 324.45 \text{ lbs}$$

$$\tan \theta = \frac{|F_y|}{|F_x|} \frac{299.75}{124.16} = 2.4142$$

$\theta = 67.5°$

Consider next the case when the change in jet direction is 90°.

$V_{fy} = 100$ fps

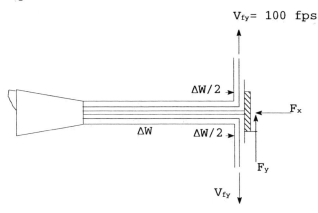

Figure 4.64

Upper half flow

$V_{o_x} = 100$ fps ; $V_{f_x} = 0$; $V_{o_y} = 0$; $V_{f_y} = 100$ fps

$$\Sigma F_x \Delta t = \frac{\Delta W/2}{g}(V_{f_x} - V_{o_x})$$

$$F_x \Delta t = \frac{136.5 \Delta t}{2(32.2)} (0 - 100); \quad F_x = -212 \text{ lb.}$$

$$\Sigma F_y \Delta t = \frac{\Delta W/2}{g} (V_{f_y} - V_{o_y})$$

$$F_y \Delta t = \frac{136.5 \Delta t}{2(32.2)} (100 - 0)$$

$$F_y = +212 \text{ lbs}$$

Lower half flow

$V_{o_x} = 100 \text{ fps} \qquad V_{f_x} = 0$

$V_{o_y} = 0 \qquad V_{f_y} = -100$

By symmetry

$$F_x = -212 \text{ lbs}$$
$$F_y = -212 \text{ lbs}$$

$\Sigma F_x = -212 - 212 = -424 \text{ lbs}$

$\Sigma F_y = 212 - 212 = 0$

Conservation of Momentum

The relation between force and momentum was postulated as

$$\Sigma \overline{F} = \frac{d}{dt}(m\overline{V})$$

where $\Sigma \overline{F}$ = resultant of all external forces acting on a system

$m\overline{V}$ = linear momentum of the system

If the resultant force acting on the system is zero

$$\frac{d}{dt}(m\overline{V}) = 0 \quad \text{and} \quad m\overline{V} = \text{constant}$$

The <u>total momentum</u> of a system is defined as the vector sum of the momenta of its individual parts or

$$m\overline{V} = m_A \overline{V}_{A1} + m_B \overline{V}_{B1} + ... + m_N \overline{V}_{N1}$$

If there is no net external force acting, the momenta of the individual parts may change but their sum remains constant. Under this condition the internal forces being equal and opposite produce equal and opposite changes in momentum which cancel one another. We may then write

$$m_A \overline{V}_{A1} + m_B \overline{V}_{B1} + ... + m_N \overline{V}_{N1} =$$
$$m_A \overline{V}_{A2} + m_B \overline{V}_{B2} + ... + m_N \overline{V}_{N2}$$

or

$$W_A \overline{V}_{A1} + W_B \overline{V}_{B1} + ... + W_N \overline{V}_{N1} = W_A \overline{V}_{A2} + W_B \overline{V}_{B2} + ... + W_N \overline{V}_{N2}$$

where the subscript 1 refers to the initial state of the system and the subscript 2 to its final state.

The condition that the momentum be a constant applies only to the system as a whole.

Note that the velocities in the momenta equation are vector quantities whose direction must be taken into account in its application.

Example

The muzzle velocity of bullets can be determined by the ballistic pendulum portrayed in Figure 4.65. A 4 oz bullet is fired into a 15.75-lb sand box, and remains imbedded. The impulse swings the box through an angle of 24°. What is the velocity of the bullet?

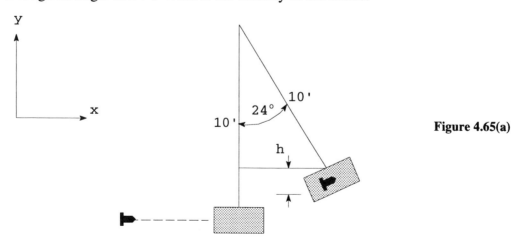

Figure 4.65(a)

Solution

Use the work-energy method to find the common velocity of the shell and the box at the moment of impact. Use the conservation of momentum principle for the velocity of the shell before impact.

$$\Delta \text{Work} = \Delta \text{KE}$$

but $\Delta \text{Work} = \Delta \text{PE} = -Wh = -WR(1 - \cos 24°)$

$$\Delta \text{KE} = \frac{1}{2} \frac{W}{g} (V_f^2 - V_o^2) = \frac{W}{64.4}(0 - V_o^2)$$

where V_o = common velocity of bullet and box at impact

$$-W(10)(1 - .914) = \frac{W}{64.4}(-V_o^2)$$

$$V_o = 7.46 \text{ ft/sec}$$

From conservation of momentum [where subscript S is for the bullet (<u>S</u>hell) and subscript B is for the <u>B</u>ox of sand]

$$W_S \overline{V}_S + W_B \overline{V}_B = (W_S + W_B) \overline{V}_o$$

Where $W_S \overline{V}_S$ = momentum of bullet before impact

$W_B \overline{V}_B$ = momentum of box before impact

$(W_S + W_B) \overline{V}_o$ = momentum of bullet and box after impact

Note that the velocities $\overline{V}_S$, $\overline{V}_B$ and $\overline{V}_o$ are collinear and in the same direction hence we may write

$$W_S V_S + \overbrace{W_B V_B}^{0} = (W_S + W_B) V_o$$

$$V_S = V_o \left(\frac{W_S + W_B}{W_S} \right) = 7.46 \left(\frac{15.75 + \frac{1}{4}}{\frac{1}{4}} \right) = 477.4 \, \frac{\text{ft}}{\text{sec}}$$

The work done in moving the sand box may also be obtained from the definition of work, namely,
dWork = f dS

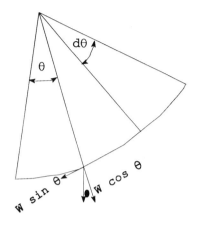

Figure 4.65(b)

dWork = F dS

dS = R dθ   F = -W sin θ

dWork = (-W sin θ) R dθ

$$\text{Work} = -WR \int_0^\theta \sin\theta \, d\theta$$

$$\text{Work} = WR\,[\cos\theta]\,\Big|_0^\theta = WR\cos\theta - WR$$

Work = -WR (1 - cos θ)

Elastic Impact

The largest class of problems in which the principles of momentum apply are those involving sudden impact. These problems range from destructive collisions between fast moving vehicles to encounters between atomic particles.

There are two basic types of collisions: inelastic and elastic. Those that remain in contact after impact are illustrative of inelastic collision. Those where colliding particles separate after impact are called elastic collisions. In both types of collision problems the time interval during which impact occurs is considered infinitesimal. This enables these problems to be handled without considering the effects which would be appreciable if the time interval were longer.

Consideration of pre-impact and post-impact conditions permits the application of the Law of Conservation of Momentum.

When two bodies in motion collide, the resulting impact is subject to precisely the same laws as when one is stationary and the other is moving. It is the relative velocity with which the two meet that is important.

When the perfectly elastic bodies collide, they are deformed at first then separate due to the action of restoring elastic forces. Throughout elastic impact mutual action and reaction forces exist.

Consider two bodies W_A and W_B in "head-on" collision (without rotation). By conservation of momentum we may write

$$W_A \overline{V}_{A_1} + W_B \overline{V}_{B_1} = W_A \overline{V}_{A_2} + W_B \overline{V}_{B_2}$$

The second condition regulating elastic impact follows a law stated by Isaac Newton, namely that, <u>The Velocity of Separation After Collision Is Always Proportional to the Velocity of Approach</u>. The constant of proportionality depends upon the elastic properties of the colliding bodies and is called the <u>coefficient of restitution</u>, e. It is defined as the ratio of the separation velocity to the approach velocity, or

$$e = \frac{\text{relative velocity after impact}}{\text{relative velocity before impact}} = \frac{\text{separation velocity}}{\text{approach velocity}}$$

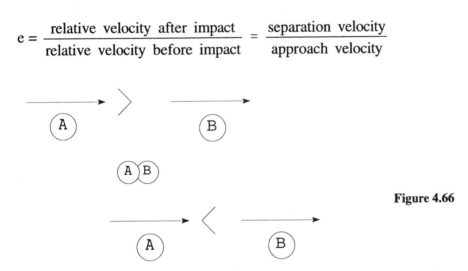

Figure 4.66

If objects A and B have initial velocities $\overline{V}_{A_1}$ and $\overline{V}_{B_1}$ respectively which are collinear and in the same direction as shown in Figure 4.66 but where $\overline{V}_{A_1} > \overline{V}_{B_1}$; object A will overtake object B and impact will occur. After impact $\overline{V}_{B_2}$ will increase while $\overline{V}_{A_2}$ decreases such that $\overline{V}_{B_2} > \overline{V}_{A_2}$ and object B will separate from object A.

Under these conditions the momentum and coefficient of restitution equations become respectively

$$W_A V_{A_1} + W_B V_{B_1} = W_A V_{A_2} + W_B V_{B_2}$$

$$e = \frac{V_{B_2} - V_{A_2}}{V_{A_1} - V_{B_1}}$$

If the bodies are perfectly elastic, e = 1, and the velocities of approach and separation are the same. However, since no material is perfectly elastic e is generally less than unity. Values of e for various materials may be found in most engineering handbooks. The loss in kinetic energy of colliding bodies manifests itself by a small rise in temperature due to internal friction forces.

When there is direct central impact, the velocities of both bodies are directed along the same straight line. In oblique impact, the velocities of the colliding bodies to be used in the above equations are the components along the common normal.

Example

Two smooth spheres, A and B with velocities of $\overline{V}_{A_1}$ = 10 fps and $\overline{V}_{B_1}$ = 12 fps, collide when these velocities are directed at angles of 47° and 32° with the line of centers as shown. The spheres are of equal size but A weighs W_A = 6 lbs and B weighs W_B = 3 lbs The coefficient of restitution is e = 0.8 (a) What are the absolute velocities of these spheres immediately after impact? (b) What is the loss of kinetic energy?

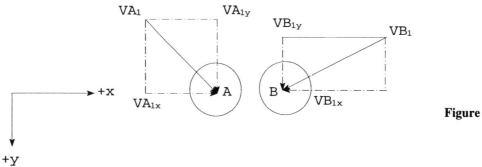

Figure 4.67(a)

Imagine that the spheres are on a horizontal surface and that the view is from above

V_{A_1} = 10 fps W_A = 6 lb V_{B_1} = 12 fps W_B = 3 lb   e = 0.8

1. A line normal to the surfaces of contact is chosen as the x reference axis (as shown in Fig. 4.67a).

$V_{A_{1x}} = V_{A_1} \cos 47° = (10)(.682) = + 6.82$ fps

$V_{B_{1x}} = -V_{B_1} \cos 32° = -(12)(.848) = -10.19$ fps

$V_{A_{1y}} = V_{A_1} \sin 47° = (10)(.731) = + 7.31$ fps

$$V_{B_{1y}} = V_{B_1} \sin 32° = (12)(.530) = +6.36 \text{ fps}$$

2. For smooth spheres, the y components of the velocity <u>will not be changed</u> by the impact because there is no force to cause acceleration, hence

$$V_{A_{2y}} = V_{A_{1y}} = +7.31 \text{ fps} \qquad V_{B_{2y}} = V_{B_{1y}} = +6.36 \text{ fps}$$

3. From Conservation of Momentum

$$W_A V_{A_{1x}} + W_B V_{B_{1x}} = W_A V_{A_{2x}} + W_B V_{B_{2x}}$$

$$(6)(6.82) + (3)(-10.19) = (6) V_{A_{2x}} + (3) V_{B_{2x}}$$

$$6 V_{A_{2x}} + 3 V_{B_{2x}} = 11.5$$

4. From coefficient of restitution

$$e = \frac{V_{B2} - V_{A2}}{V_{A1} - V_{B1}}$$

$$V_{B_{2x}} - V_{A_{2x}} = e(V_{A_{1x}} - V_{B_{1x}})$$

$$V_{B_{2x}} - V_{A_{2x}} = (.8)(6.82 + 10.19)$$

$$V_{B_{2x}} - V_{A_{2x}} = 13.61$$

$$3 V_{B_{2x}} + 6 V_{A_{2x}} = 11.5$$

Solution yields $V_{A_{2x}} = -3.26 \text{ fps} \qquad V_{B_{2x}} = +10.35 \text{ fps}$

5. Absolute Velocities of A and B (Fig. 4.67b)

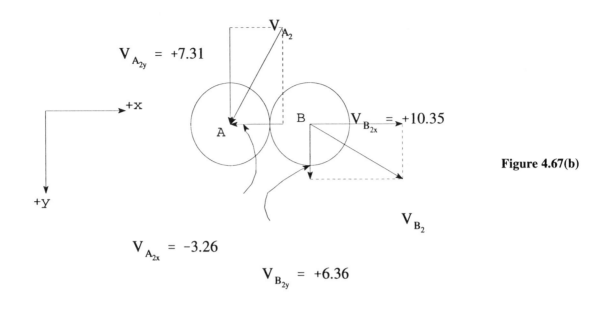

Figure 4.67(b)

$$V_{A_2} = \sqrt{(V_{A_{2x}})^2 + (V_{A_{2y}})^2} = \sqrt{(-3.26)^2 + (+7.31)^2}$$

$$V_{A_2} = 8.00 \text{ fps}$$

$$\tan \alpha = \frac{|V_{A_{2y}}|}{|V_{A_{2x}}|} = \frac{7.31}{3.26} = 2.2423$$

$$\alpha = 65° 58'$$

$$V_{B_2} = \sqrt{(V_{B_{2x}})^2 + (V_{B_{2y}})^2} = \sqrt{(10.35)^2 + (6.36)^2}$$

$$V_{B_2} = 12.1 \text{ fps}$$

$$\tan \beta = \frac{|V_{B_{2y}}|}{|V_{B_{2x}}|} = \frac{6.36}{10.35} = 0.616$$

$$\beta = 31° 40'$$

6. ΔKE (Although momentum is conserved kinetic energy <u>always</u> decreases except when e = 1)

$$KE_1 = \frac{W_A V_{A1}^2}{2g} + \frac{W_B V_{B1}^2}{2g} = \frac{(6)(10)^2 + (3)(12)^2}{64.4} = 16.05 \text{ ft-lbs}$$

$$KE_2 = \frac{W_A V_{A2}^2}{2g} + \frac{W_B V_{B2}^2}{2g} = \frac{(6)(8.00)^2 + (3)(12.1)^2}{64.4} = 12.78 \text{ ft-lbs}$$

ΔKE = loss of KE due to impact = 16.05 - 12.78 = 3.27 ft-lbs

TABLE 4.1 COMMON EXAMPLES OF SIMPLE HARMONIC MOTION

System		Assumptions	Remarks	Period T	Frequency f	Angular Frequency
Spring Mass		• Linear Spring • No Damping	The results are valid independent of gravitation field. Results apply to system in horizontal position.	$T = 2\pi \sqrt{\dfrac{m}{kg_c}}$ $T = \dfrac{1}{f}$ $T = \dfrac{2\pi}{\omega}$	$f = \dfrac{L}{2\pi}\sqrt{\dfrac{kg_c}{m}}$ $f = \dfrac{1}{T}$ $f = \dfrac{\omega}{2\pi}$	$\omega = \sqrt{\dfrac{kg_c}{m}}$ $\omega = 2\pi f$ $\omega = \dfrac{2\pi}{T}$
Simple Pendulum		• Point Mass • No Damping • Small Angles	Results are independent of amount of mass. Results depend on local value of g.	$T = 2\pi \sqrt{\dfrac{L}{g}}$	$f = \dfrac{1}{2\pi}\sqrt{\dfrac{g}{L}}$	$\omega = \sqrt{\dfrac{g}{L}}$
Compound or Physical Pendulum		• No Damping • Small Angles	The simple pendulum with equal period has length $L_e = \dfrac{I}{mL}$	$T = 2\pi \sqrt{\dfrac{I}{mgL}}$	$F = \dfrac{1}{2\pi}\sqrt{\dfrac{mgL}{I}}$	$\omega = \sqrt{\dfrac{mgL}{I}}$
Torsional Spring		• Linear Spring • $\tau = -K\theta$ • No Damping	The results are valid independent of gravitation field or orientation of system	$T = 2\pi \sqrt{\dfrac{I}{K}}$	$f = \dfrac{1}{2\pi}\sqrt{\dfrac{K}{I}}$	$\omega = \sqrt{\dfrac{K}{I}}$
Conical Pendulum		• Point Mass • No Damping	Not Harmonic Motion	$T = 2\pi \sqrt{\dfrac{L\cos\Psi}{g}}$	$F = \dfrac{1}{2\pi}\sqrt{\dfrac{g}{L\cos\Psi}}$	$\omega = \sqrt{\dfrac{g}{L\cos\Psi}}$

General Notes:
- The constants A (amplitude) and ∅ (phase angle) are established from the initial conditions of the problems, i.e., initial displacement and velocity.
- V_{max} occurs when X or θ = 0, which is when a = 0.
- a_{max} occurs when X or θ = ± A, which is when V = 0.
- Kg_c Spring Constant in $\dfrac{\#}{Ft}$.

5
Fluid Mechanics

FLUID MECHANICS 5-i

TABLE OF CONTENTS

Selected References 5-ii
Some Symbols and Constants Used in Fluid Mechanics 5-iii
Fluid Pressure 5-1
Hydrostatic Force 5-5
Thin-Walled Cylinders Under Pressure 5-11
Thick-Walled Cylinders 5-13
Buoyancy 5-14
Dams and Gates 5-15
Bernoulli's Theorem for an Incompressible Fluid 5-19
Head Loss Due to Pipe Friction 5-24
Orifice Coefficients 5-31
Entrance Losses 5-32
Other Losses 5-35
Viscosity and the Reynolds Number 5-41
Pipe Line Supplied By a Pump 5-47
Branch Pipe Flow 5-50
Flow Measurement 5-56
Nozzles 5-59
Flow Trajectory 5-60
Flow Over Wiers 5-61
Sluice Gate 5-67
Flow in Open Channels 5-67
Dynamic Force and Momentum 5-69
Dimensional Analysis 5-73
Translation and Rotation 5-75

SELECTED REFERENCES

V.L. Streeter, "Fluid Mechanics," McGraw-Hill, New York

G.E. Russell, "Hydraulics," Henry Holt & Co., New York

R.L. Daugherty and A.C. Ingersol, "Fluid Mechanics With Engineering Applications," McGraw-Hill, New York

R.W. Henke, "Introduction to Fluid Mechanics," Addison-Wesley, Reading, MA

SOME SYMBOLS AND CONSTANTS USED IN FLUID MECHANICS

A   Area

a   Acceleration

C   Constant in Hazen–Williams Formula

CG   Center of gravity

CP   Center of pressure

d   Diameter

e   Surface roughness (sometimes shown as ϵ)

F   Force

f   Friction factor

g   Gravitational constant

h   Static head

h_d Discharge loss

h_e Entrance loss

h_f Friction loss

H   Same as h except usually used for total

H_L Total head losses

h_o Pressure head at center of gravity

I   Moment of inertia

I_o Moment of inertia about axis through center of gravity

L   Length

m   Mass

N_R Reynolds Number

P	Pressure
R	Hydraulic radius
S	Specific gravity
V	Volume
v	Velocity (sometimes shown as V)
w	Specific weight
X_c	Distance from free surface to CP
X_o	Translation distance between axis through CG and new axis
X_o	Also used for distance from free surface to CG
Z	Elevation from a datum
ρ	Density
μ	Dynamic viscosity
γ	Kinematic viscosity
Σ	Summation

w for water = 62.4 lbs/ft^3

w for sea water = 64.0 lbs/ft^3

S for mercury = 13.6

Atmospheric pressure = 29.92 in. Hg = 14.7 psi

g = 32.2 ft/sec^2

1 Horsepower = 550 ft lbs/sec

1 Gallon = 0.1337 ft^3

1 Inch of water column = 0.0361 psi

FLUID PRESSURE

Pressure is defined as force per unit area (P = F/A). It is generally expressed in pounds per square inch (psi) or pounds per square foot (psf). The pressure on the bottom of a tank having a base area of one square foot and filled to a depth of one foot with water having a specific weight of 62.4 lb/ft³ is

$$P = \frac{F}{A} = \frac{62.4 \text{ lb}}{1 \text{ ft}^2} = 62.4 \text{ psf}$$

In this case the force F was the weight of the one cubic foot of water and this weight rested on the one square foot area.

Our atmosphere exerts a pressure on the surface of the earth. The air is a fluid of a certain specific weight that "rests" on the surface area of the earth. This atmospheric pressure is generally taken as 14.7 psi. This is known as the atmospheric or barometric pressure. Pressures measured with respect to atmospheric pressure are called gage pressures. The units for gage pressure have a g added such as psig, psfg, etc. Total pressure is the sum of the gage pressure plus the atmospheric pressure. Total pressure is also known as absolute pressure and is designated by an a added to the units such as psia, psfa, etc.

Consider the prismatic volume of fluid shown in Figure 5.1. The element under consideration has a length L and a cross-sectional area dA. One end of this element is at a depth h_1 below the free surface of a fluid while the other end is located at a depth h_2. The specific weight is w lb/ft³

The weight of the prismatic volume is W = w L dA. W acts vertically downward. The angle between the centerline of the volume and the line of action of W is shown as α. The force acting on the end of the volume which is at depth h_2 is P_2 dA and acts perpendicular to this face. Similarly the force acting on the opposite face is P_1dA.

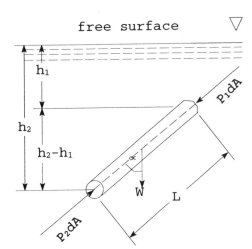

Figure 5.1

Prism length = L
Cross-sectional area = dA
Weight of prismatic volume = W

Since the element under study is in equilibrium, the summation of the forces acting on the element must be zero. $\Sigma F = 0$. The component of the weight W acting along the axis of the element is W cos α. Substituting the value of W yields (w LdA) cos α.

Then: $P_2 \, dA = P_1 \, dA + w \, L \, dA \cos\alpha$

Rearranging: $P_2 \, dA - P_1 \, dA = (w \, L \cos\alpha) \, dA$

Cancel dA: $P_2 - P_1 = w \, L \cos\alpha$

But from the geometry of Fig. 5.1, we see that $h_2 - h_1 = L \cos\alpha$.

Substituting this value gives

$$P_2 - P_1 = w(h_2 - h_1)$$

or

$$P_2 - P_1 = w \, h_2 - w \, h_1$$

from which we can see that $P_2 = w \, h_2$ and $P_1 = w \, h_1$ or generally, $P = w \, h$. That is, for an incompressible fluid (one with a constant w) the pressure varies directly as the depth. (NOTE: Generally, water is taken as incompressible fluid.) h is often called the pressure head or static head.

We can now write $P_{absolute} = (P \text{ ambient} + wh)$ where h is known as the static or pressure head. The units of h are linear units of measure such as inches, ft, meters, etc.

Example

Two sources of pressure M and N, are connected by a water-mercury differential gage as shown in Fig. 5.2. What is the difference in pressure between M and N in psi.

w = 62.4 lb/ft³

w' = (13.6 x 62.4) lb/ft³

(Specific gravity of mercury is 13.6)

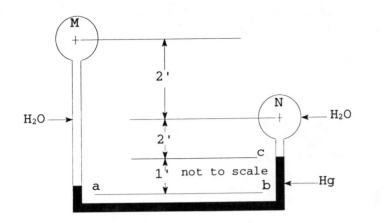

Figure 5.2

In solving this type of problem, it is best to start at one end point in the system and follow round the system to the other end point. Keep in mind that as we move downward from M the pressure increases.

$$P_m + (2+2+1)\,w = P_a \text{ (pressure at point a)}$$

$$P_m + 5w = P_a$$

As we continue downward from point a and across the bottom of the mercury tube and up to point b, the level, or datum, of the system has not changed. Thus the pressure at point a is the same as the pressure at point b.

$$P_b - 1\,w' = P_c$$

Substituting P_a for P_b and then $(P_m + 5w)$ for P_a, we get

$$P_m + 5w - 1w' = P_c$$

But $P_c - 2w = P_n$

$$P_m + 5w - 1w' - 2w = P_n$$

$$P_m + 3w - 1w' = P_n$$

But the problem asks for $P_m - P_n$

$$P_m - P_n = 1w' - 3w$$

$$P_m - P_n = 1(13.6 \times 62.4) - 3(62.4)$$

$$P_m - P_n = 661 \text{ psf.}$$

Is this the correct answer? NO, the problem asked for $P_m - P_n$ in units of psi.

$$P_m - P_n = 661 \frac{lb}{ft^2} \times \frac{1 \ ft^2}{144 \ in^2}$$

$$P_m - P_n = 4.59 \frac{lb}{in^2} \qquad \underline{Answer}$$

If the problem asked for $P_m - P_n$ in units of ft of water:

$$P = wh \quad h = P/w$$

$$h = \frac{661 \frac{lb}{ft^2}}{62.4 \frac{lb}{ft^3}} = 10.59 \ ft \ of \ water$$

Example

Two pipelines containing water under pressure connected by a mercury manometer as shown in Fig. 5.3. Find $P_m - P_n$ in terms of differential of manometer fluid. The specific weight of water is w, the specific weight of the manometer fluid is w' and the specific gravity of the manometer fluid is s.

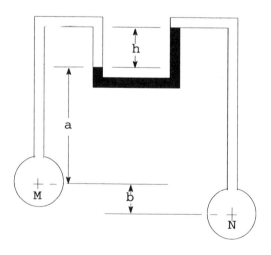

Figure 5.3

As before, we follow the system from end to end.

$$P_m - w \ a - w' \ h + w \ (h+a+b) = P_n$$

$$P_m - P_n = w' \ h + w \ (a-h-a-b) = w' \ h + w \ (-h-b)$$

$P_m - P_n = w'h - wh - wb$

But $w' = sw$

$P_m - P_n = swh - wh - wb = wh(s-1) - wb$ Answer

If M and N were at the same level or datum, $b = 0$. Then $P_m - P_n = wh(s-1)$.

HYDROSTATIC FORCE

A hydrostatic force has magnitude, direction, and line of action. The total force F is equal to the pressure times the immersed area.

$F = P \times A = whA$

For a plane surface in any position, $F = wh_oA$ where h_o is the pressure head at the centroid of the surface area.

Example

Find the total force, F, acting on a rectangular plate measuring 6 ft x 4 ft if the plate is located vertically with the 4 ft edge 10 ft below the surface of the water (Fig. 5.4).

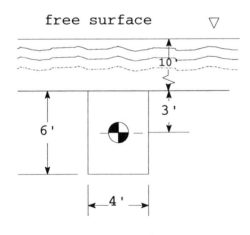

Figure 5.4

$F = w h_o A$

$F = 62.4 \dfrac{\text{lb}}{\text{ft}^3} \times (10+3) \text{ ft} \times (6 \times 4) \text{ ft}^2$

$F = 19{,}469 \text{ lbs.}$

We know that the total force F of 19,469 lbs is acting on the 6 ft x 4 ft area. We know it is acting perpendicular to the plate since pressure can act only perpendicular to a surface. In other words, we know the magnitude and direction of the force F. We do not yet know the location of the line of action.

The force F acts not at the center of gravity (CG) but at the center of pressure (CP), a point on a plane surface where a single resultant force can be assumed to act. The line of action of the hydrostatic force is through the center of pressure.

Take a rectangular plate b wide by d high and submerge it in fluid so that CG is located at a depth h_o below the free surface as shown in Fig. 5.5.

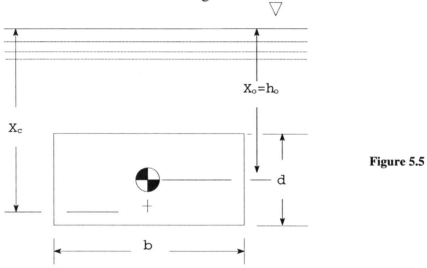

Figure 5.5

Let $h_o = X_o$ and X_c be the depth to the center of pressure, CP We recall that

$$X_c = \frac{I}{S} = \frac{\text{Moment of inertia of the plate}}{\text{Static moment}}$$

$$X_c = \int \frac{X^2 dA}{X\, dA}$$

But $I = I_o + AX_o^2$ where I_o is the moment of inertia of the plane body about its axis through the C.G. The term AX_o^2 is the translation term for referring the moment of inertia to another axis not through the C.G. In this case, the new axis is at the free surface of the fluid.

$$X_c = \frac{I_o + AX_o^2}{AX_o} = \frac{I_o}{AX_o} + X_o$$

$$X_c - X_o = \frac{I_o}{AX_o}$$

For a rectangle, $I_o = \dfrac{bd^3}{12}$ and $A = bd$

$$X_c - X_o = \frac{\dfrac{bd^3}{12}}{bdX_o} = \frac{bd^3}{12\ dbX_o} = \frac{d^2}{12\ X_o}$$

$$X_c - X_o = \frac{d^2}{12\ X_o} \quad \underline{\text{for a rectangle}}$$

This expression for $X_c - X_o$ defines the distance between the CG and the CP. Since the right side of the equation is always greater than zero, the CP will always be located <u>below</u> the CG

Example

A dam is placed across a stream of parabolic cross-section described by the equation $x^2 = y$ where x is the width of the stream and y is its depth. Water of density 1.94 slugs/ft^3 flows in the stream at a depth of 9 ft (Fig. 5.6). Determine the total force against the dam and locate the center of pressure.

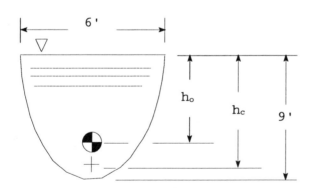

Figure 5.6

$1.94\ \dfrac{\text{slugs}}{\text{ft}^3} = \dfrac{w}{g}$ $\qquad\qquad x^2 = y$

$w = 32.2 \times 1.94 = 62.468$ lb/ft^3 $\qquad x^2 = 9$

Width of stream = 6 ft $\qquad\qquad x = \pm 3$ ft

from an engineering handbook we find for a parabolic section that

$$A = \frac{4}{3} y \left[\frac{x}{2}\right]$$

$$h_o = \frac{2}{5} y$$

$$A = \frac{2}{3} y (2x) = \frac{2}{3} \times 9 \times 6 = 36 \text{ ft}^2$$

$$h_o = \frac{2}{5} y = \frac{2}{5} [9] = 3.6 \text{ ft}$$

$$F = w h_o A = 62.468 \times 3.6 \times 36 = 8096 \text{ lbs.}$$

$$h_c = \frac{I_o + A h_o^2}{A h_o} \quad \text{where} \quad I_o = \left(\frac{16}{175}\right)(y^3)\left(\frac{x}{2}\right)$$

$$I_o = \frac{16}{175} \times 729 \times 3 = 199.95 \text{ ft}^4$$

$$h_c = \frac{200 + 36 (3.6)^2}{36 (3.6)} = 5.14 \text{ ft}$$

Total force = 8096 lbs .   <u>Answer</u>

Center of pressure = 5.14 ft below the surface .   <u>Answer</u>

Now let us consider a plane surface that is not vertical as shown in Fig. 5.7. In this figure the plane under consideration is at an angle with the free surface and extends into the plane of the paper.

Extend the plane of the plate until it intersects the free surface at S. The head, h, at any point along the plate at a slant distance, L, from the surface is $h = L \sin\alpha$.

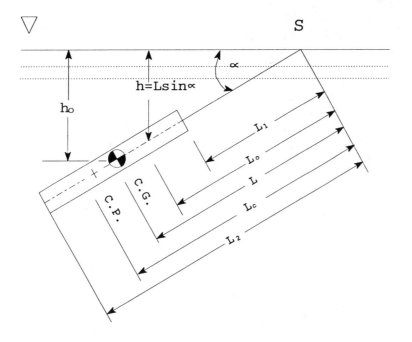

Figure 5.7

The force F on the plate can now be expressed as

$$dF = w\, h\, dA$$

$$F = \int_{L_1}^{L_2} w\, L\, \sin\alpha\, dA$$

$$F = w\, h_o\, A$$

$$F = w\, L_o\, [\sin\alpha]\, A.$$

Now that we have defined the force, we need to know its point of application. Let us take moments about the free surface at S.

$$M_s = F\, L_c \text{ where } L_c = \text{dimension from the surface to the center of pressure}$$

$$M_s = F\, L_c = \int L\, dF \quad \text{but } dF = w\, L\, \sin\alpha\, dA$$

$$M_s = F\, L_c = \int w\, L^2\, \sin\alpha\, dA$$

$$L_c = \int \frac{w\, L^2\, \sin\alpha\, dA}{F}$$

$$L_c = \frac{\int wL^2 \sin\alpha \, dA}{\int wL \sin\alpha \, dA}$$

But $\int L^2 \, dA = I$ and $\int L \, dA = AL_o^2$

$$L_c = \frac{I}{AL_o} \text{ where } I = I_o + AL_o^2$$

$$L_c = \frac{I_o + AL_o^2}{AL_o} = \frac{I_o}{AL_o} + L_o$$

$$L_c - L_o = \frac{I_o}{AL_o} \text{ which reduces to } \frac{d^2}{12L_o}$$

where $d = L_2 - L_1$

Note that the dimensions L are along the slant.

Example

A flat (ft) plate 10 ft x 6 ft lies below the surface of a body of water. The 6 ft edge lies parallel to and 10 ft below the surface and the plane of the plate is inclined at an angle of 45° with the horizontal (Fig. 5.8). Locate the center of pressure.

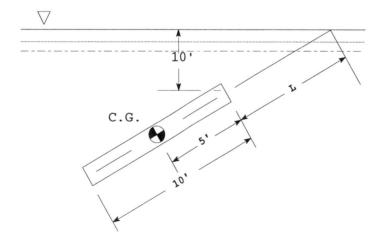

Figure 5.8

$$L_c - L_o = \frac{d^2}{12L_o} \text{ where } L_o = 5 + L$$

$$L = \frac{10}{\sin 45°} = \frac{10}{.707} = 14.14 \text{ ft}$$

$$L_o = 5 + 14.14 = 19.14 \text{ ft}$$

$$L_c - L_o = \frac{(10)^2}{12\ (19.14)} = 0.435 \text{ ft}$$

Center of pressure lies 0.435 ft below the CG of the plate along the plane of the plate.   <u>Answer</u>

THIN-WALLED CYLINDERS UNDER PRESSURE

A thin-walled cylinder can be defined as a structure in which the wall thickness is less than 1/10 the diameter of the section. T = tension force

A pipe is horizontal and of unit length (L) into the plane of the paper.

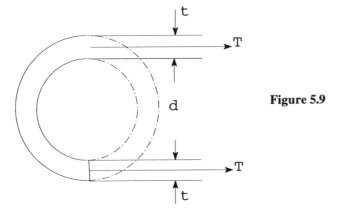

Figure 5.9

The pipe shown in Fig. 5.9 is subjected to an internal pressure, P. This pressure acts normal to the inside wall of the pipe. Remove one half of the pipe as shown by the dotted lines. To maintain the system in equilibrium, we replace the half of the pipe removed by the tension force, T. The force due to the pressure on the remaining half pipe is equal to the pressure times the cross-section area of the pipe or P x d x L. This force is balanced by the two tension forces as shown.

$$2T = P\ d\ L = P\ (2r)\ L = 2\ P\ r\ L$$

But T = stress in the material (S) x cross-sectional area (tL)

$$T = S_t L$$

Substituting in the previous equation 2 (StL) = 2 p r L

Cancelling out 2 L ; St = pr

S is known as the "hoop stress."

Example

Given a 48 in. diameter steel pipe with a wall thickness of 0.25 in. The pipe carries water at an internal pressure of 100 psi.

(A) Find the hoop stress in the steel.

(B) If the internal pressure is raised to 250 psi and the allowable stress in the steel is 15 ksi, what pipe wall thickness is required?

(A) St = pr

$$s = \frac{pr}{t} = 100 \frac{lb}{in.^2} \times \frac{24 \text{ in.}}{.25 \text{ in.}} = 9600 \text{ psi}$$

Check to verify that the cylinder has a "thin wall."

(B) $t = \frac{pr}{s} = \frac{250 \times 24}{15 \times 10^3} = 0.4$ in.

t < .1d or .4 < .1(48) or .4 < 4.8, OK

(A) Hoop stress, s, = 9600 psi   <u>Answer</u>

(B) Pipe wall thickness, t = 0.4 in.   <u>Answer</u>

An application of hoop stress is in the area of stave pipes and tanks.

Example

A wood stave pipeline (Fig. 5.10) is designed for a pressure head of 50 ft of water. The internal diameter of the pipe is 8.5 ft and the wood staves are 2½ in. thick. Circular steel bands 1 in. in diameter and 10 in. center to center hold the staves together from the outside. What is the stress in each band in psi? If the spacing of the supporting piers is 10 ft center to center, what is the load on each pier? Neglect the weight of the staves and bands.

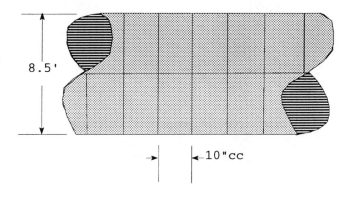

Figure 5.10

$$A = \frac{\pi d^2}{4} = \frac{3.14 \times 1^2}{4} = .7854 \text{ in}^2 \quad P = \frac{w\,h}{144} = \frac{62.4 \times 50}{144} = 21.67 \text{ psi}$$

For 10 in. of pipeline, $T = p \times r \times 10$

$$T = 21.67 \times (4.25 \times 12) \times 10$$

$$T = 11{,}050 \text{ lbs}$$

$$S = \frac{T}{A} = \frac{11{,}050}{.7854} = \underline{14{,}069 \text{ psi}} \quad \underline{\text{Answer}}$$

Wt of water for 10 ft length of pipe = $\dfrac{\pi \times (8.5)^2 \times 10 \times 62.4}{4}$

Wt of water = 35,409 lb. <u>Answer</u>

THICK-WALLED CYLINDERS

For thick-walled cylinders such as high pressure pipes or gun barrels, the stress in the outside diameter is significantly less than the ID stress. The formula most generally used for thick-walled cylinders is from "Hydraulics and Its Applications" by A. H. Gibson.

$$f = \left[\frac{r_i^2 + r_o^2}{r_o^2 - r_i^2}\right] = \frac{\text{Inside Hoop Stress}}{\text{Internal Pressure}} = \frac{S}{P}$$

Where r_o and r_i are outside and inside radii respectively.

Example

Given a 2-in. ID steel pipe with an internal pressure of 8000 psi. The hoop tension at the bore should not exceed the allowable stress of 24,000 psi. Find the OD.

Assume this to be a thick walled cylinder.

$$S = P \left[\frac{r_i^2 + r_o^2}{r_o^2 - r_i^2} \right]$$

$$Sr_o^2 - Sr_i^2 = P(r_i^2 + r_o^2)$$

$$r_o^2 (S - P) = r_i^2 (S + P)$$

$$r_o^2 = r_i^2 \left[\frac{S + P}{S - P} \right] \qquad r_o = r_i \sqrt{\left[\frac{S + P}{S - P} \right]}$$

Substituting values

$$r_o = 1 \sqrt{\frac{24000 + 8000}{24000 - 8000}} = \sqrt{\frac{32000}{16000}} = \sqrt{2} = 1.414 \text{ in.}$$

OD = 2 r_o = 2.828 in.

Check: t = (1.414 - 1.000) = 0.414 in.

Id = .1 x 2 = .2 in.

This is thick-walled as assumed.

Outside diameter is 2.828 in.   **Answer**

Note: In this problem the numerical values were not assigned until the equation was in its simplest form. This not only simplifies the arithmetic with its resultant saving in time but makes it easier to check the work for errors.

BUOYANCY

The Archimedes Principle states that the buoyant force is equal to the weight of the displaced fluid. In other words, the weight of a body immersed in a liquid is apparently decreased by an amount that is equal to the weight of the displaced liquid.

Example

A stone weighs 100 lbs in air and 60 lbs when immersed in water. Calculate its volume and specific gravity.

Wt of water displaced = Wt in air − Wt in water

Wt of water displaced = 100 lbs − 60 lbs = 40 lbs

Volume of stone = volume of water displaced

$$\text{Volume of stone} = \frac{40 \text{ lbs}}{62.4 \frac{\text{lbs}}{\text{ft}^3}} = 0.64 \text{ ft}^3$$

$$\text{Specific Gravity} = \frac{\text{Wt of stone in air}}{\text{Wt of water displaced}} = \frac{100}{40} = 2.5$$

Volume of stone = 0.64 ft³   <u>Answer</u>

Specific gravity = 2.5   <u>Answer</u>

The specific gravity of 2.5 found in the example above indicates that the stone weighs 2.5 times as much as water for the same volume. The specific weight of the stone is equal to 2.5 x 62.4 lbs/ ft = 156 lbs/ft³ This can also be calculated by 100 lbs/0.64 ft³ = 156.3 lbs/ft³. The difference in the answers is caused by rounding off.

The line of action of the buoyant force is through the CG of the submerged volume.

DAMS AND GATES

Dams and gates are devices to stop or contain water in a reservoir, channel, basin, or the like. The forces acting on a dam or gate are hydrostatic forces and gravity forces. These combine to form a system of overturning and restoring moments as well as forces tending to slide a dam along its base.

Example

Figure 5.11 shows a cross-section of a masonry bulk-head dam that is 1 ft long perpendicular to the paper. The masonry is assume to weigh 150 lbs/ft³. The dam is located in a moderate climate upon a hard rock formation so that no trouble is expected from either ice forces or uplift water pressure under the foundation. At what point along the base AB will the total resultant force act?

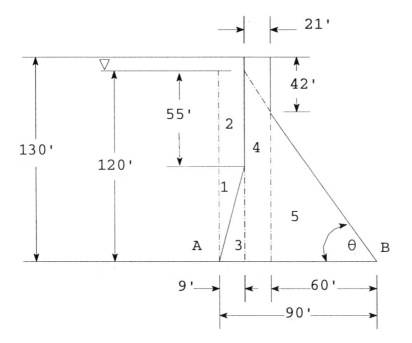

Figure 5.11

Break up the cross-section area of the dam into simple geometric shapes for calculation weight and CG.

Section	Weight (unit width)		Location of CG to left of B
1	½ x 9 (120-55) x 62.4	= 18,252 lbs	$60 + 21 + \frac{2}{3} \times 9 = 87$ ft
2	9 x 55 x 62.4	= 30,888 lbs	$60 + 21 + \frac{9}{2} = 85.5$ ft
3	½ x 9 (120-55) x 150	= 43,875 lbs	$60 + 21 + \frac{9}{3} = 84$ ft
4	(120+10)(21) x 150	= 409,500 lbs	$60 + \frac{21}{2} = 70.5$ ft
5	½ (60)(130-42) x 150	= 396,000 lbs	$\frac{2}{3} \times 60 = 40$ ft
		Σ = 898,515 lbs	

Note that 1 and 2 are volumes of water, not masonry. The weight of water above section 3 is acting down on section 3, causing a clockwise movement about A.

For equilibrium,

$$\Sigma \text{Forces (H and V)} = 0 \text{ and } M = 0$$

The only horizontal force is the hydrostatic force acting horizontally on the face of the dam.

$$F_H = w\, h_o A = 62.4 \times \frac{120}{2} \times (120 \times 1) = 449{,}280 \text{ lbs}$$

$$X_c - X_o = \frac{d^2}{12\, x_o} \text{ but } X_o = d/2$$

$$X_c - \frac{d}{2} = \frac{d^2}{12 \times \frac{d}{2}} = \frac{d}{6}$$

$$X_c = \frac{d}{6} + \frac{d}{2} = \frac{2}{3} d.$$

$$X_c = \frac{2}{3} \times 120 = 80 \text{ ft from the surface or } 120 - 80 = 40 \text{ ft above point A.}$$

$$F_v = \Sigma \text{weights} = W_1 + W_2 + W_3 + W_4 + W_5$$

$$F_v = 18{,}252 + 30{,}888 + 43{,}875 + 409{,}500 + 396{,}000$$

$$F_v = 898{,}515 \text{ lbs}$$

$$\Sigma M_b = 0$$

$$18252(87) + 30{,}888(85.5) + 43875(84) + 409500(70.5) + 396{,}000(40) - R_v(\bar{x}) - 449{,}280(40) = 0$$

$$\bar{x} = \frac{34{,}652{,}898}{898{,}515} = 38.57 \text{ ft to the left of B}$$

The resultant force will act 38.57 ft from point B.   <u>Answer</u>

<u>Example</u>

In order to control the water surface in a reservoir, a hinged leaf gate is provided that has the dimensions as shown in Fig. 5.12. The gate is shown in its upper limiting position and presses on the seat at the end of the bottom leaf. If the weight of the gate itself is neglected, what would be the amount of force per linear foot of crest that is exerted at the seat in a direction normal to the gate?

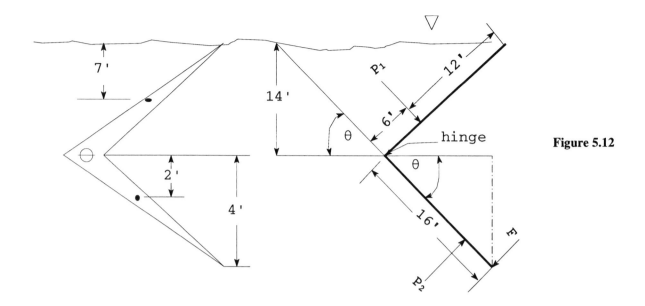

Figure 5.12

The approach to this problem is to find the hydrostatic forces on each leaf (P_1 and P_2) and their point of application (CP) so we can take moments about the hinge in order to find the force F. The crest is the width of the water passage and is directed into the page.

$P_1 = A \, w \, h_o = (18 \times 1) \times 62.4 \times 7 = 7862$ lbs

Note that 7 ft (½ of 14 ft dimension) is the depth to the CG of the upper leaf.

$P_2 = A \, w \, h_o = (16 \times 1) \times 62.4 \times (14 + 2) = 15{,}974$ lbs

For the upper leaf

$$X_c - X_o = \frac{d^2}{12 \, X_o} = \frac{(18)^2}{12 \times 9} = 3 \text{ ft}$$

This puts the CP for the upper left at 6 ft from the hinge.

For the lower leaf, $X_o = \dfrac{16}{2} + d$ where $d = \dfrac{14}{\sin\theta}$ and $\sin\theta = \dfrac{4}{16} = \dfrac{1}{4}$

$d = \dfrac{14}{\frac{1}{4}} = 14 \times 4 = 56$ ft

$X_o = 56 + 8 = 64$ ft

$$X_c - X_o = \frac{d^2}{12 X_o} = \frac{(16)^2}{12(64)} = .33 \text{ ft}$$

P_2 acts $8 + .33 = 8.33$ ft from the hinge.

$\Sigma M_{Hinge} = 0$

$6 P_1 + 16F = 8.33 P_2$

$$F = \frac{(8.33 \times 15974) - 6(7862)}{16}$$

$F = 5368$

Force per linear foot of crest is 5368   <u>Answer</u>

BERNOULLI'S THEOREM FOR AN INCOMPRESSIBLE FLUID

Let us derive Bernoulli's equation for an incompressible, frictionless liquid. An elementary mass of fluid with a cross-section dA. The length ds is the distance the mass moves in dt sec as shown in Fig. 5.13.

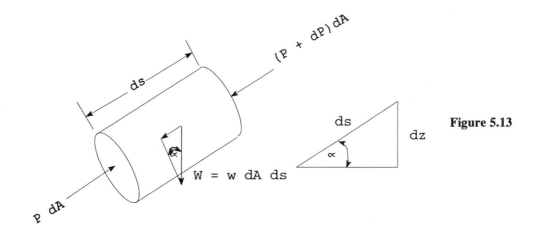

Figure 5.13

The instantaneous velocity $v = \frac{dS}{dt}$. The pressure forces on the ends of the element are PdA and (P + dP) dA respectively.

The weight, $W = w\, dA\, ds$; $W = g \rho\, dAds$ where the density, ρ is equal to $\frac{w}{g}$ slugs/ft^3. The component of the weight in the direction of motion is $(dA\, ds) \times (g \rho \sin\alpha)$. But, from Fig. 5.13 ds

$\sin\alpha = dZ$ where dZ is the change in potential height. The component of the weight in the direction of motion is therefore equal to $dA\,dZ\,g\rho$.

Since $F = ma$

$$PdA - (P+dP)dA - g\rho dZ dA = [\rho\,ds\,dA]\,a$$

The dA terms can be cancelled leaving

$$P - (P+dP) - g\rho dZ = \rho\,ds\,a$$

but $v = \dfrac{dS}{dt}$; $v\,dv = \dfrac{dS}{dt}dv = a\,s$

$$a = \frac{v\,dv}{ds}$$

Substituting this value for a yields:

$$P - (P+dP) - g\rho dZ = \rho ds\,\frac{v\,dv}{ds}$$

Simplifying $P - P - dP - g\rho dZ = \rho v\,dv$

Rearranging: $\rho v\,dv + g\rho dZ + dP = 0$

Dividing then by ρ and rearranging terms

$$\frac{dP}{\rho} + g\,dZ + v\,dv = 0$$

Substituting $\dfrac{w}{g}$ for ρ and cancelling g terms

$$\frac{dP}{w} + dZ + \frac{v}{g}\,dv = 0$$

Integrating this expression gives:

$$\frac{1}{w}\int_1^2 dP + \int_1^2 dZ + \frac{1}{g}\int_1^2 v\,dv = 0$$

which equals $\dfrac{P_2}{w} - \dfrac{P_1}{w} + Z_2 - Z_1 + \dfrac{\overline{V_2^2}}{2g} - \dfrac{\overline{V_1^2}}{2g} = 0$

At this point we introduce the notation $\overline{V}$ which will be used for velocity.

Rearranging gives $\dfrac{P_1}{w} + Z_1 + \dfrac{\overline{V_1^2}}{2g} = \dfrac{P_2}{w} + Z_2 + \dfrac{\overline{V_2^2}}{2g}$

which is the more recognizable form for Bernoulli's equation for an incompressible liquid and no friction.

Simply stated, $\dfrac{P}{w} + Z + \dfrac{\overline{V^2}}{2g} = $ Constant.

Where

$\dfrac{P}{w}$ = Pressure head

Z = Static or elevation head, height measured from a datum.

$\dfrac{\overline{V^2}}{2g}$ = Velocity head

Figure 5.14 illustrates Bernoulli's equation.

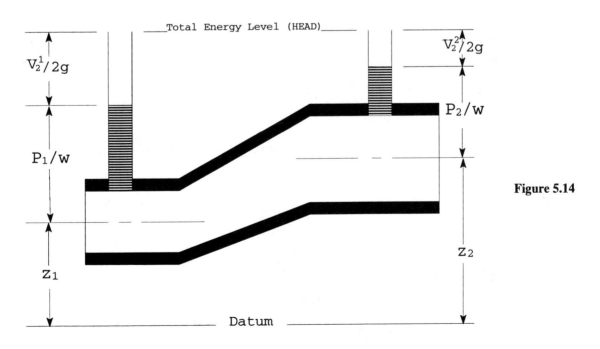

Figure 5.14

5-22 FUNDAMENTALS OF ENGINEERING EXAM REVIEW WORKBOOK

Since we do have friction forces, Bernoulli's equation must be modified as follows:

$$\frac{P_1}{w} + Z_1 + \frac{\overline{V_1}^2}{2g} = \frac{P_2}{w} + Z_2 + \frac{\overline{V_2}^2}{2g} + \text{lost head.}$$

The lost head (H_L) contains the friction losses. Other losses, as may be defined, can be included in the H_L term.

The value for $\overline{V}$ will vary from point to point across a pipe or conduit. The <u>average</u> value is used in computing the term $\overline{V}^2/2g$ by taking the velocity from the continuity equation $\overline{V} = Q/A$. The <u>actual</u> mean velocity can be computed but since the value of the velocity head $\overline{V}^2/2g$ is usually small as compared with the values for P/w and Z, the errors in the value for $\overline{V}$ can be neglected.

Example

A pipe line gradually enlarges from 24-in. diameter at A to 36 in. diameter at B. $\overline{V}_a = 5$ fps, $P_a = 50$ psi.

Assume 2 ft of head lost between A and B. Find P_b if B is 15 ft lower than A.

1. Write Bernoulli's equation.

$$\frac{P_a}{w} + Z_a + \frac{\overline{V_a}^2}{2g} = \frac{P_b}{w} + Z_b + \frac{\overline{V_b}^2}{2g} + H_L$$

2. Calculate $\overline{V}_b$:

$$Q = A_a(\overline{V}_a) = A_b(\overline{V}_b)$$

$$\overline{V}_b = \frac{A_a}{A_b}(\overline{V}_a) = \frac{\pi \times (24)^2 \times 4 \times 5 \text{ fps}}{4 \times \pi \times (36)^2} = 2.22 \text{ fps.}$$

3. Assume the datum to be at the level of B. This makes $Z_b = 0$ and $Z_a = 15$ ft since it was given that B was 15 ft below A.

4. Substitute numerical values in Bernoulli's equation. (Assume w - 62.4 lbs/ft³ since a pipe line was given).

$$\left(\frac{50 \times 144}{62.4}\right) \text{ft} + 15 \text{ ft} + \frac{(5)^2 \text{ ft}}{2(32.2)} = \frac{P_b}{w} + \frac{(2.22)^2}{2(32.2)} + 0 + 2$$

The solution yields, $\dfrac{P_b}{w} = 128.7$ ft $= 8030$ psf $= 55.77$ psi

$P_b = 55.77$ psi.   Answer

Example

A vertical line of pipe (Fig. 5.15) carrying water includes an 18 in. pipe above a 15 in. pipe connected by a reducer. A pressure gage in the 18-in. pipe, 5 ft above the center of the reducer, registers 7 psi. Another gage set in the 15-in. pipe, 5 ft below the center of the reducer, registers 10 psi. Flow in the pipe equals 5 cfs. In which direction is the flow and what is the friction loss in head between the two gages?

Here, we must assume a direction of flow and then justify our choice. Let us assume the flow is from the 18-in. pipe through the reducer to the 15-in. pipe. We must also choose a datum. If we select our datum at the level of the 10 psi gage, one of our Z terms in Bernoulli's equation will be zero. Let us call the location of the 7 psi gage in the 18-in. pipe point 1 and the location of the 10 psi gage in the 15-in. pipe point 2. Writing Bernoulli's equation between state points 1 and 2:

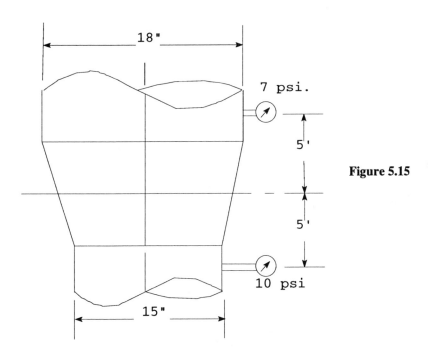

Figure 5.15

$$\dfrac{\overline{V_1^2}}{2g} + \dfrac{P_1}{w} + Z_1 = \dfrac{\overline{V_2^2}}{2g} + \dfrac{P_2}{w} + Z_2 + H_L$$

$$\frac{\overline{V_1^2}}{64.4} + \frac{7 \times 144}{62.4} + 10 = \frac{\overline{V_2^2}}{64.4} + \frac{10 \times 144}{62.4} + 0 + H_L$$

Remember to use the proper conversion in the P/w terms. Note the use of 62.4 lbs/ft³ for w in the term for pressure head.

To solve for H_L, we first need to find $\overline{V}_1$ and $\overline{V}_2$. From the continuity equation, $Q = A\overline{V}$, we find:

$$\overline{V}_1 = \frac{Q}{A_1} = \frac{5}{1.77} \frac{\text{ft}^3/\text{sec}}{\text{ft}^2} = 2.82 \text{ fps}$$

NOTE: $A_1 = \dfrac{\pi d^2}{4} = \dfrac{3.14 \times \left(\dfrac{18}{12}\right)^2}{4} = 1.77 \text{ ft}^2$

$$\overline{V}_2 = \frac{Q}{A_2} = \frac{5}{1.23} \frac{\text{ft}^3/\text{sec}}{\text{ft}^2} = 4.07 \text{ fps}$$

Now we can solve for H_L

$$\frac{(2.82)^2}{64.4} + \frac{7 \times 144}{62.4} + 10 - \frac{(4.07)^2}{64.4} - \frac{10 \times 144}{62.4} - 0 = H_L$$

Solving we find $H_L = +2.94$ ft

Friction head is +2.94 ft   Flow is downward.   **Answer**

If we had assumed the flow upward,

$$H_L = \frac{(4.07)^2}{64.4} + \frac{10 \times 144}{62.4} + 0 - \frac{(2.82)^2}{64.4} - \frac{7 \times 144}{62.4} - 10 = -2.94 \text{ ft}$$

Since the friction loss can never be negative, the flow must be reversed.

HEAD LOSS DUE TO PIPE FRICTION

Resistance to flow is:

1. Directly proportional to the length of pipe. The longer the pipe, the greater the resistance to flow.

2. Inversely proportional to some power of the diameter. The smaller the diameter, the greater the resistance to flow.

3. Variable with roughness for turbulent flow. The rougher the pipe the greater the resistance to flow.

4. Variable with some power of the velocity.

5. Independent of pressure.

This can be expressed mathematically as $h_f = f \dfrac{L}{d} \dfrac{\overline{V^2}}{2g}$ where

h_f = friction head loss, ft

f = friction factor, dimensionless

L = length of pipe, ft

d = ID of pipe, ft

$\dfrac{\overline{V^2}}{2g}$ = Velocity head, ft

This expression is sometimes called the Weisbach formula, or the Darcy equation.

The value for f is given in textbooks and handbooks for different material pipes and at different velocities of flow. The Moody Chart gives the friction factor f as a function of the Reynolds number for pipes of different roughness as seen in Fig. 5.48.

The Hazen–Williams formula for the velocity of flow and the continuity equation can be used to compute volummetric flow. The effects of the friction factor f are taken into account in the constant S in the formula:

$V = 1.318\ C\ R^{0.63}\ S^{0.54}$ where

C = Constant depending on the pipe (roughness)

R = Hydraulic radius (area of flow/wetted perimeter)

S = Hydraulic gradient = h_f ft/1000 ft for the nomograph and ft/ft when used in the equation.

The Hazen-Williams nomograph and values for C are given in Fig. 5.16.

Example

A water supply line consisting of new CI pipe, 12 in. diameter, runs from a pumping station to a town 4000 ft away (Fig. 5.17). The pressure at the town end of the pipe is 50 psi. At a point along the pipe 500 ft from the station, there is an elevated storage tank 130 ft high to the base of the tank. The water in the tank is 20 ft deep and open to the atmosphere. If the level in the tank is maintained by the pump, how much water is flowing in the pipe?

Sketch the layout as described in the problem. If the water level in the tank is kept constant, the flow of water is the same anywhere along the pipe. We know the conditions at the storage tank and at the town, so we can write Bernoulli's equation and find the friction loss between the tank and the town. We will assume the pipe to be level giving $Z_1 = Z_2$ and since Q is constant and the cross-section area of the pipe is constant, $A \overline{V}_1 = A \overline{V}_2$ or $\overline{V}_1 = \overline{V}_2$.

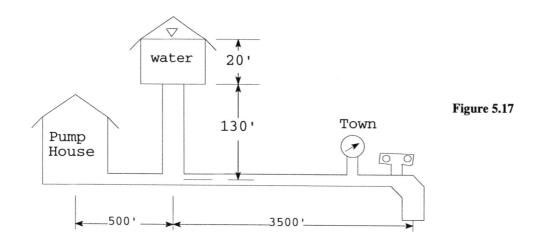

Figure 5.17

$$\frac{P_1}{w} + Z_1 + \frac{\overline{V}_1^2}{2g} = \frac{P_2}{w} + Z_2 + \frac{\overline{V}_2^2}{2g} + h_L$$

Assume all loss due to friction ($h_L = h_f$)

Since $Z_1 = Z_2$ and $\overline{V}_1 = \overline{V}_2$; $\dfrac{P_1}{w} - \dfrac{P_2}{w} = h_f$

but $\dfrac{P_1}{w} = 130 + 20 = 150$ ft

$\dfrac{P_2}{w} = \dfrac{50 \text{ psi} \times 144}{62.4} = 115.4$ ft

$h_f = 150 - 115.4 = 34.6$ ft

Figure 5.16

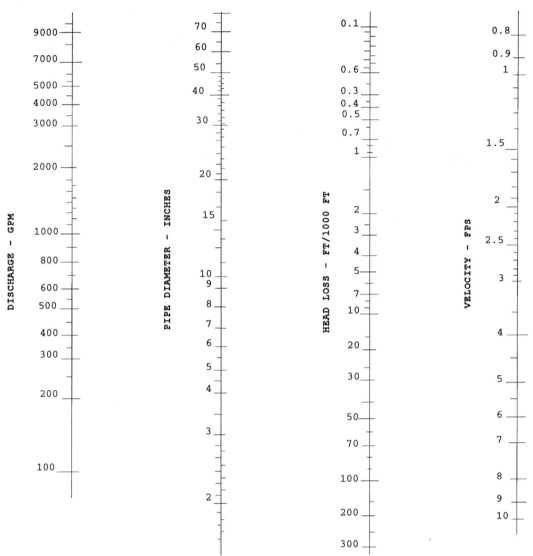

Nomograph for Solution of Hazen-Williams Equation
(C = 100)

To use the nomograph for values of C other than 100:

C=	40	60	80	90	100	110	120	130	140
Divide H_L by	5.45	2.58	1.5	1.22	1	0.84	0.7	0.62	0.54

We now write the Hazen-Williams formula

$$\overline{V} = 1.318 \times C \times R^{0.63} \times S^{0.54}$$

From our reference handbook we find the C for the new CI pipe to be 130.

$$R = \frac{\text{Area of pipe}}{\text{Wetted perimeter}} = \frac{\frac{\pi d^2}{4}}{\pi d} = \frac{d}{4} = \frac{1}{4} = .25 \text{ ft}$$

$$S = \frac{h_f}{\text{length}} \approx \frac{35 \text{ ft}}{3500 \text{ ft}} = 0.01$$

$$\overline{V} = 1.318 \times 130 \times (.25)^{.63} \times (.01)^{.54}$$

$$\overline{V} = 5.95 \text{ fps}$$

However, we can use the Hazen-Williams nomograph to solve the equation. Our chart is for C = 100. If we take the handbook correction for H_L, we may use the chart with C = 100.

$$H_L = \frac{10}{0.62} = 16.1 \text{ ft}$$ Note that H_L in the nomograph is the same as S, loss of head in ft per 1000 ft of pipe length.

If we draw a straight line across the chart passing through diameter = 12 in. and loss of head per 1000 ft = 16.1 ft we find the line intersects at velocity = 5.9 fps and Q = 4.6 cfs.

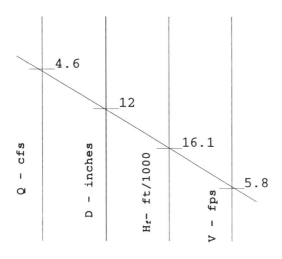

Figure 5.18

Q = 4.6 cfs.   <u>Answer</u>

This problem can also be solved by using $h_f = f \dfrac{L}{d} \dfrac{\overline{V^2}}{2g}$. We found h_f previously (34.6 ft)

$$V = \sqrt{\dfrac{2gh_f\, d}{fL}} \text{ by rearranging terms.}$$

$$V = \sqrt{\dfrac{2 \times 32.2 \times 34.6 \times 1}{f \times 3500}}$$

Since we don't know f, we will assume its value. The best assumption is f - 0.02.

$$V = \sqrt{\dfrac{2 \times 32.2 \times 34.6 \times 1}{.02 \times 3500}} = 5.64 \text{ fps}$$

From the friction factor table in our handbook for d = 12 in. new CI pipe, V = 5.64 fps, f - 0.19.

$$\text{New } V = \sqrt{\dfrac{2 \times 32.2 \times 34.6 \times 1}{.019 \times 3500}} = 5.79 \text{ fps}$$

We find f = 0.19 in the same manner as before (by interpolation).

$$Q = AV = \dfrac{\pi \times 1^2}{4} \times 5.79 = 4.55 \text{ cfs}$$

Q = 4.55 cfs.   **Answer**

Refer to the section "Viscosity and the Reynolds Number," for a review of the Reynolds Number. An example of its use follows.

$$\text{Reynolds Number} = N_R = \dfrac{vd}{\gamma} = \dfrac{vd}{\dfrac{\mu}{\rho}}$$

where γ is the kinematic viscosity, ρ is density, and μ is the absolute viscosity.

By experiment, we have laminar flow at N_R = 2000. For laminar flow, $f = \dfrac{64}{N_R}$.

Example

A fuel oil tank is drained by a welded steel pipe 4 in. in diameter that extends 100 ft along a pier to be used in filling barges. The connection to the tank is at right angles to the surface of the tank and the pipe is level. What is the head loss in the system when the flow is 200 gpm? Assume μ = .0109 lbs sec/ft^2, the specific gravity, s = 0.94, and γ = 0.006 ft^2/sec.

One topic not yet covered in this course is entrance and exit (discharge) losses. This is covered later but for now simply assume that the entrance loss for the pipe at right angles to the tank is one half a velocity head and the exit loss is one velocity head.

Now, apply Bernoulli's equation. Let state point 1 be at the surface of the oil in the tank and state point 2 at the discharge.

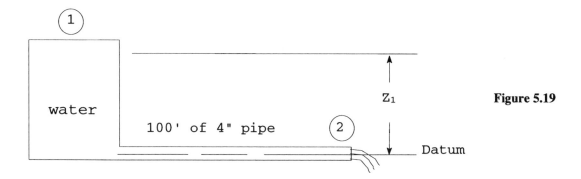

Figure 5.19

$$\frac{P_1}{w} + \frac{\overline{V_1}^2}{2g} + Z_1 = \frac{P_2}{w} + \frac{\overline{V_2}^2}{2g} + Z_2 + H_L$$

Where Z_1 is to be found, $Z_2 = 0$ (at the chosen datum).

$\overline{V}_1 = 0$ (what is the velocity in such a large volume?)

$$\overline{V}_2 = \frac{Q}{A} = \frac{200 \; \frac{\text{gal}}{\text{min}} \times \frac{1 \text{ min}}{60 \text{ sec}} \times \frac{1 \text{ ft}^3}{7.48 \text{ gal}}}{\frac{\pi}{4} \left[\frac{4}{12}\right]^2} = 5.11 \text{ ft/sec}$$

$$\frac{\overline{V}_2^2}{2g} = \frac{(5.11)^2}{2 \times 32.2} = 0.41$$

Reynolds Number = $N_R = \dfrac{vd}{\gamma} = \dfrac{5.11 \times \dfrac{4}{12}}{0.006} = 283.89$

Since $N_R < 2000$, the flow is laminar and $f = \dfrac{64}{N_R}$

$h_f = f \dfrac{L}{d} \dfrac{\overline{V}_2}{2g} = \dfrac{64}{283.89} \times \dfrac{100}{\dfrac{4}{12}} \times 0.41 = 27.73 \text{ ft}$

$h_e = \text{entrance loss} = \tfrac{1}{2} \dfrac{\overline{V}_2}{2g} = 0.205 \text{ ft}$

$H_L = h_f + h_e + h_{exit} = 27.73 + 0.21 + 0.41 = 28.35 \text{ ft}$

Substituting these values found in Bernoulli's equation:

$0 + 0 + Z_1 = 0 + 0.41 + 0 + 28.35$

$Z_1 = 28.76$

Surface of oil is 28.76 ft above the discharge.   **Answer**

ORIFICE COEFFICIENTS

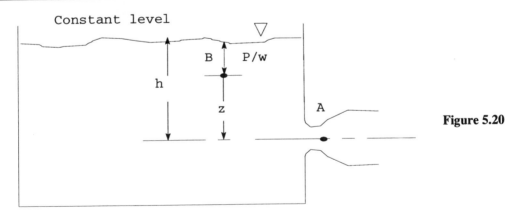

Figure 5.20

Given the tank and orifice shown in Fig. 5.20. Note that as the fluid discharges from the orifice, the diameter of the jet decreases and then increases. The point where the diameter is smallest is known as the <u>vena contracta</u>. The ratio of the area of the jet to the area of the orifice at the vena contracta is known as the coefficient of contraction.

$$C_c = \frac{\text{Area of jet}}{\text{Area of orifice}}$$

Write Bernoulli's equation between some point B in the tank and point A at the vena contracta.

$$\frac{\overline{V}_B^2}{2g} + \frac{P_B}{w} + Z_B = \frac{\overline{V}_A^2}{2g} + \frac{P_A}{w} + Z_a \text{ with no friction loss.}$$

$$0 + h = \frac{\overline{V}_A^2}{2g} + 0 + 0 \text{ where } h = \frac{P_B}{w} + Z_B$$

This reduces in general terms to $\overline{V} = \sqrt{2gh}$ (Toricelli's equation).

Here, $\overline{V}$ is the ideal velocity. Actually, the real velocity is slightly less than ideal.

Actual velocity = $\overline{V}_a = C_v \sqrt{2gh}$

Where C_v is a coefficient of velocity.

From the continuity equation, $Q = A\,\overline{V}_a$ where A is the actual area of the jet and $\overline{V}_a$ is the actual velocity of the fluid.

A = area of orifice x $C_c = A_o \times C_c$

$\overline{V}_a$ = ideal velocity x $C_v = \sqrt{2gh}\ C_v$

$Q = A\,\overline{V}_a = C_c C_v A_o \sqrt{2gh}$

$Q = C_d A_o \sqrt{2gh}$

where $C_d = C_c C_v$ is the coefficient of discharge.

ENTRANCE LOSSES

$\overline{V}_a = = C_v \sqrt{2gh}$

$\overline{V}_a^2 = C_v^2 (2gh)$

$$\frac{\overline{V_a^2}}{2g} = C_v^2 h$$

If there was no head loss due to friction the velocity head would be h.

$$\text{Lost Head} = h - C_v^2 h = h(1 - C_v^2) = \left[\frac{1}{C_v^2} - 1\right]\frac{\overline{V^2}}{2g}$$

For an orifice, a general value for $C_v = 0.98$

$$\text{Lost Head} = \left[\frac{1}{(.98)^2} - 1\right]\frac{\overline{V_2}}{2g} = .041\,\frac{\overline{V_2}}{2g}$$

For a short cylindrical tube, 2½ to 3 diameters long connected at right angles to the supply, a general value of C_v is 0.82.

$$\text{Lost Head} = \left[\frac{1}{(.82)^2} - 1\right]\frac{\overline{V_2}}{2g} \approx .50\,\frac{\overline{V_2}}{2g}$$

Similarly, for a short pipe projecting into a tank or reservoir, $C_v = 0.75$.

$$\text{Lost Head} = \left[\frac{1}{(.75)^2} - 1\right]\frac{\overline{V_2}}{2g} \approx .80\,\frac{\overline{V_2}}{2g}$$

Example

A rectangular steel box floats with a draft of 4 ft. If the box is 20 ft long, 10 ft wide, 6 ft deep, compute the time necessary to sink it to its top edge by opening an orifice 6 in. in diameter in its bottom. Neglect the thickness of the vertical sides and assume $C = 0.60$.

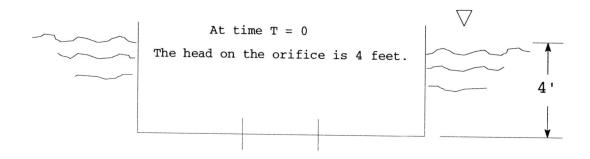

Figure 5.21

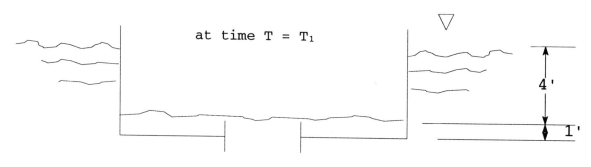

Figure 5.22

The head on the orifice is still 4 ft even though the draft is now 5 ft Similarly, when the draft is 6 ft because there is 2 ft of water in the barge, the head on the orifice remains 4 ft

$$V = \sqrt{2gh} = \sqrt{2 \times 32.2 \times 4} = 16.05 \text{ fps}$$

$$Q = CAV = 0.60 \times .196 \times 16.05 = 1.89 \text{ cfs}$$

Since the problem didn't state which C was given, C_v, C_c, or C_d and the value was 0.60, we can assume it to be C_d. The area of a 6-in, diameter orifice is 0.196 ft².

The time needed to sink the barge is then equal to the volume to be filled divided by the rate of filling. Since we want to sink the barge to its top edge and it floats with a draft of 4 ft when empty, we have to put 2 ft of water into the barge.

$$\text{Time to sink barge} = \frac{(20 \times 10 \times 2) \text{ ft}^3}{1.89 \text{ ft.}^3/\text{sec}}$$

Time to sink barge = 211.6 sec   **Answer**

OTHER LOSSES

On the discharge side, if we discharge into the atmosphere or into a large reservoir, we must disperse one velocity head of energy.

Some of the other losses encountered in piping systems are summarized below.

From pipe to a tank or reservoir $H_D = \dfrac{\overline{V}_2}{2g}$

Sudden Enlargement

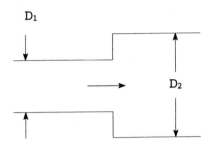

$$H_X = \dfrac{(\overline{V}_1 - \overline{V}_2)^2}{2g}$$

Figure 5.23

Gradual Enlargement

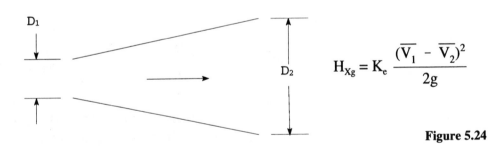

$$H_{Xg} = K_e \dfrac{(\overline{V}_1 - \overline{V}_2)^2}{2g}$$

Figure 5.24

K_e depends on $\dfrac{D_2}{D_1}$ and the cone angle.

Sudden Contraction

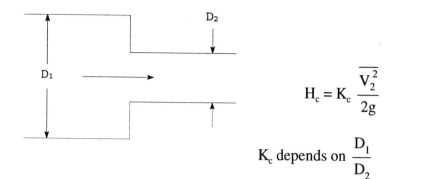

Figure 5.25

$$H_c = K_c \frac{\overline{V_2}^2}{2g}$$

K_c depends on $\dfrac{D_1}{D_2}$

Nozzles, Orifices, Venturis

$$H_L = \left[\frac{1}{C_v^2} - 1\right] \frac{\overline{V_2}^2}{2g}$$

Elbows, Fittings, Valves

$$H_L = k \frac{\overline{V_2}^2}{2g}$$

k is generally supplied by the fitting manufacturer. The entrance from a reservoir to a pipeline is taken as $H_e = \frac{1}{2} \dfrac{\overline{V_2}^2}{2g}$.

Example

A reservoir discharges water into a pool through a 3000 ft horizontal pipe line consisting of 2000 ft of 18-in. And 1000 ft of 12-in. clean CI pipe. If the surface of the reservoir is 25 ft above the surface of the water in the pool, how many cubic ft/sec of water are flowing? $k_c = .28$.

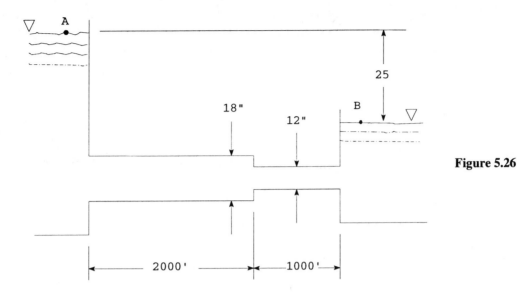

Figure 5.26

Take the datum at a known location. The surface of the pool is known with respect to the surface of the reservoir so it is a good choice.

Write Bernoulli's equation

$$\frac{P_a}{w} + \frac{V_a^2}{2g} + Z_a + = \frac{P_b}{w} + \frac{\overline{V_b^2}}{2g} + Z_b + H_L$$

Since P_a and P_b are atmospheric, and $\overline{V}_a$ and $\overline{V}_b$ are zero since they are in large bodies of water, and $Z_b = 0$ since it is at the datum, we find $H_L = Z_a = 25$ ft

But H_L = loss for entry from reservoir to pipe

+ friction loss in the 18 in. pipe

+ loss by contracting from 18 in. to 12 in. pipe

+ friction loss in the 12 in. pipe

+ loss for discharge into the pool

$$H_L = h_e + h_{f_1} + h_c + h_{f_2} + h_d = 25 \text{ ft}$$

But $h_e = \dfrac{1}{2} \dfrac{\overline{V_1^2}}{2g}$

$$h_{f_1} = f_1 \frac{L_1}{d_1} \frac{\overline{V_1}^2}{2g} = f_1 \times \frac{2000}{1.5} \frac{\overline{V_1}^2}{2g}$$

$$h_c = k_c \frac{\overline{V_2}^2}{2g} \quad \text{where } k_c \text{ was given as .28}$$

Rather than have terms in $\overline{V}_1$ and $\overline{V}_2$, let us convert all terms of $\overline{V}_1$.

$$Q = A_1 \overline{V}_1 = A_2 \overline{V}_2$$

$$\overline{V}_2 = \frac{A_1}{A_2} \overline{V}_1 = \frac{4\pi d_1^2}{4\pi d_2^2} \overline{V}_1 = \frac{d_1^2}{d_2^2} \overline{V}_1 \quad \text{or}$$

$$\overline{V}_2 = \frac{(1.5)^2}{(1)} \overline{V}_1 = 2.25 \overline{V}_1$$

$$\overline{V}_2^2 = 5.06 \overline{V}_1^2$$

$$h_c = .28 \frac{\overline{V}_2^2}{2g} = .28 \times 5.06 \frac{\overline{V_1}^2}{2g} = 1.42 \frac{\overline{V_1}^2}{2g}$$

$$h_{f_2} = f_2 \frac{L_2}{d_2} \frac{\overline{V_2}^2}{2g} = f_2 \times \frac{1000}{1} \times 5.06 \frac{\overline{V_1}^2}{2g} = f_2 \times 5600 \frac{\overline{V_1}^2}{2g}$$

$$h_d = \frac{\overline{V_2}^2}{2g} = 5.06 \frac{\overline{V_1}^2}{2g}$$

We now have to assume an f_1 and an f_2, solve for $\overline{V}_1$ and $\overline{V}_2$, check the values for f_1 and f_2 and continue to do this until there is little or no change in $\overline{V}_1$ and $\overline{V}_2$.

Assume f_1 and $f_2 = .02$. This is a good value when an assumption is required.

$$h_e = 0.50 \frac{\overline{V_1}^2}{2g} \qquad h_{f_2} = 112 \frac{\overline{V_1}^2}{2g}$$

$$h_{f_1} = 26.70 \frac{\overline{V}_1^2}{2g} \qquad h_d = 5.06 \frac{\overline{V}_1^2}{2g}$$

$$h_c = 1.42 \frac{\overline{V}_1^2}{2g} \qquad H_L = 145.68 \frac{\overline{V}_1^2}{2g} = 25$$

Solving for $\overline{V}_1$ yields $\overline{V}_1 = 3.32$ fps

Substituting back yields $\overline{V}_2 = 2.25$ $\overline{V}_1 = 7.48$ fps

Using these values in Table 6.1, we find $f_1 = .0203$ and $f_2 = .0191$.

Substituting these values in the expressions for h_{f_1} and h_{f_2} and solving for $\overline{V}_1$ yields

$$\overline{V}_1 = 3.51 \text{ fps and } \overline{V}_2 = 7.89 \text{ fps}$$

This is close enough so that no further iteration is required.

$$Q = A_1 \overline{V}_1 = A_2 \overline{V}_2 = \frac{\pi \times (1.5)^2}{4} \times 3.51 = 6.2 \text{ cfs}$$

Flow is 6.2 ft³/sec   **Answer**

Table 5.1 is based on <u>average</u> values. The values will vary from table to table, and from reference to reference. The values given in table compare favorably with the Hazen-Williams formula for C = 130.

If one were doing this problem and the friction factor table were not available assume a temperature of the water, check a handbook for the kinematic viscosity, and calculate the Reynolds Number. Then check a handbook for roughness, ϵ, and calculate relative roughness ϵ/D. These values could be entered into the Moody Chart and the value for f read off the chart. This method is time consuming and prone to error so unless absolutely necessary, it is not a recommended method of solution to be used on the examination.

Table 5.1 AVERAGE FRICTION FACTOR (f) FOR NEW CI, WROUGHT IRON, STEEL, AND SMOOTH CONCRETE PIPE

Velocity ft/sec	1	2	3	5	8	10	20	30
Pipe Diameter (inches)								
4	300	270	250	235	225	215	195	190
6	280	255	240	225	210	205	185	180
8	270	245	230	215	200	195	180	175
10	265	240	220	210	195	190	175	170
12	260	230	215	205	190	185	170	165
14	255	225	210	200	190	180	170	160
16	250	220	210	200	185	180	165	160
18	245	215	205	195	185	175	165	160
20	240	215	205	195	180	175	165	155
24	235	210	200	190	180	170	160	155
30	230	205	195	185	175	170	160	155
36	225	200	190	180	170	165	155	150
42	220	200	190	175	170	165	155	145
48	215	200	185	170	165	160	150	145

Multiply values by 10^{-4}.

Some Values of C for the Hazen-Williams Equation:

Cement lined pipe	140	Vitrified Clay pipe	110
New Cast Iron pipe	130	Brick pipe	100
Welded Steel pipe	120	15-20 yr old CI pipe	100
Wrought Iron pipe	110	Over 25 yr old CI pipe	80

VISCOSITY AND THE REYNOLDS NUMBER

The viscosity of a fluid is a measure of its resistance to flow. <u>Absolute</u> viscosity is the ratio of the applied shear stress to the rate of deformation (or the velocity gradient normal to the velocity) in laminar flow. Referring to Fig. 5.27 is expressed as

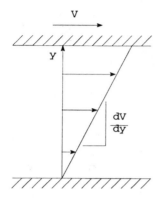

$$\mu = \frac{\tau}{\frac{dv}{dy}}$$

Figure 5.27

The absolute viscosity is also referred to as the <u>dynamic</u> viscosity.

The <u>kinematic</u> viscosity is the ratio of the absolute viscosity to the density. This is expressed as

$$\gamma = \frac{\mu}{\rho} = \frac{\mu g}{w}$$

The units of viscosity in the English system and the metric system are given in Table 5.2

Table 5.2 VISCOSITY UNITS

System	Absolute Viscosity (μ)	Kinematic Viscosity (v)
English	lb-sec/ft^2	ft^2/sec
Metric (cgs)	dyne-sec/cm^2 (poise)	cm^2/sec (stokes)

Although most problems will be worked in the English system, metric units are often given for viscosity. In the metric system, it is customary to use centipoise (0.01 poise) and centistokes (0.01 stokes) for the units of absolute viscosity and kinematic viscosity, respectively.

Example

Determine the dynamic viscosity of an oil that has a kinematic viscosity of 82 centistokes and a specific gravity of 0.83.

English system:

Convert 82 centistokes to English units

$$\nu = 82 \times 1.076 \times 10^{-5}$$

$$\nu = 8.8 \times 10^{-4} \text{ ft}^2/\text{sec}$$

The specific weight of the oil - $(0.83)(62.4) = 51.8 \text{ lb/ft}^3$

Using (1-8), $\mu = \dfrac{w\nu}{g}$

$$\mu = \dfrac{(51.8)(8.8 \times 10^{-4})}{32.2}$$

$$\mu = 0.0014 \text{ lb-sec/ft}^2$$

or $\mu = (0.0014)(47,900) = \underline{68 \text{ cp}}$

Metric system:

Note: $g = 981 \text{ cm/sec}^2$

1 dyne = 1.02×10^{-3} grams

Then $\mu = \dfrac{(0.83)(82)}{(981)(1.02 \times 10^{-3})} = \underline{68 \text{ cp}}$

Industrially, viscosity is measured by determining the time for the liquid to flow through an opening at the bottom of a standard container. The instrument used for petroleum products and lubricating oils is called the Saybolt Universal Viscometer. Thus, the Saybolt viscosity is given in seconds. For heavy oils and viscous liquids, the Saybolt Furol viscometer is used. ASTM tables are available to convert the fluid efflux time to kinematic viscosity. Empirical equations are also available to convert the efflux time to kinematic viscosity. The approximate value of the kinematic viscosity, in stokes, can be obtained from the following equations where t is the time in seconds:

Saybolt Universal

When $32 < t < 100$

$$v = 0.00226t - 1.95/t$$

When $t > 100$

$$v = 0.0022t - 1.35/t$$

Saybolt Furol,

When $25 < t < 40$

$$v = 0.0224t - 1.84/t$$

When $t > 40$

$$v = 0.0216t - 0.60/t$$

The viscosity of liquids decreases as temperature increases. The effect of pressure on the viscosity of liquids can usually be ignored.

Fluids such as water and mineral oil are called Newtonian fluids. Generally, raw sewage can also be considered to behave like a Newtonian fluid. The viscosity of Newtonian fluids is independent of the rate of shear. Non-Newtonian fluids exhibit viscosities that are dependent on the rate of shear. Fluids such as sewage, sludge, grease, syrups, and certain gels show a reduction in viscosity as the rate of shear is increased. For sludge, when the stress is removed the floc structure reforms and the viscosity increases. Such fluids are called thixotropic. For thixotropic fluids, piping is usually designed to keep the viscosity above a certain value (e.g., the velocity of sewage sludge ≥ 5 fps).

Certain liquids such as clay slurries, some starches, and paint show an increase in viscosity as the rate of shear is increased. Such fluids are called dilatant. For these fluids, piping is usually designed to keep the velocity as low as possible.

The effect of the rate of shear on viscosity is illustrated in Fig. 5.28 for the types of fluids discussed above.

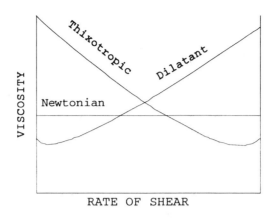

Figure 5.28

Reynolds Number

In general the state of flow may be characterized as laminar or turbulent. The type of flow is governed by the magnitude of the various forces (i.e., inertial, viscous, or gravity) acting on the fluid. Note that surface tension forces are usually not important except in capillary flow.

In laminar flow, also called streamline or viscous flow, the individual particles of the fluid flow in lines parallel to the axis of the conduit without appreciable radial component. The velocity distribution in a pipe in laminar flow is parabolic as shown in Fig. 5.29.

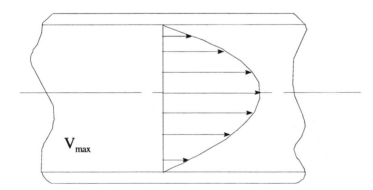

Figure 5.29

The predominant force in laminar flow is the viscous force.

For turbulent flow, the fluid particles move in irregular paths and eddies or vortexes are present in the major portion of the conduit. The velocity distribution in turbulent flow is shown in Fig. 5.30.

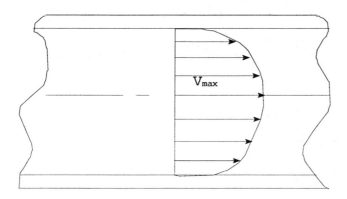

Figure 5.30

The predominant force in turbulent flow is the inertial force.

A dimensionless parameter used to characterize the nature of the flow is called the Reynold's Number. The Reynold's Number is the ratio of the inertial forces of the viscous forces within the fluid. The Reynold's Number in pipe flow is expressed as

$$R = \frac{\rho v d}{\mu} \tag{3-8}$$

where d is the pipe diameter or equivalent diameter as described later. In terms of the kinematic viscosity

$$R = \frac{Vd}{\nu}$$

In pipe flow, the flow is characterized as laminar when the Reynold's Number is less than approximately 2000. Above a Reynold's Number of 4000 the flow is generally turbulent (although under certain conditions this value may be considerably higher). In the transition region, the flow may be laminar or turbulent.

In open channel flow, the Reynold's Number is represented by

$$R = \frac{vL}{\nu}$$

L is referred to as the characteristic length in this equation and is equivalent to the hydraulic radius (described below). When the Reynold's Number in open channel flow is less than approximately 500 the flow is said to be laminar. For practical purposes, the transitional range for the Reynold's Number may be assumed to be 500 to 2000 although there is no definite limit for all flow conditions. It should be noted that in most open channels laminar flow rarely occurs.

Example

Water is flowing at 5 fps in a 2-in. pipe. The temperature of the water is 100°F. Is the flow laminar or turbulent?

At 100°F the kinematic viscosity is 0.74×10^{-5} ft^2/sec (Table 5.3).

Then $$R = \frac{(5)(\frac{2}{12})}{0.74 \times 10^{-5}} = 1.13 \times 10^5$$

Thus, the flow is turbulent.

Table 5.3 PROPERTIES OF WATER

Temperature (°F)	Specific Weight w (lb/ft³)	Density ρ (slug/ft³)	Absolute Viscosity μ × 10⁻⁵ (lb-s/ft²)	Kinematic Viscosity ν × 10⁻⁵ (ft²/s)	Vapor Pressure P_{vp} (lb/in²)
32	62.42	1.940	3.75	1.93	0.09
40	62.43	1.940	3.23	1.66	0.12
50	62.41	1.940	2.74	1.41	0.18
60	62.37	1.938	2.36	1.22	0.26
70	62.30	1.936	2.05	1.06	0.36
80	62.22	1.934	1.80	0.93	0.51
90	62.11	1.931	1.60	0.83	0.70
100	62.00	1.927	1.42	0.74	0.95
110	61.86	1.923	1.28	0.67	1.27
120	61.71	1.918	1.17	0.61	1.69
130	61.55	1.913	1.07	0.56	2.22
140	61.38	1.908	0.98	0.51	2.89
150	61.20	1.902	0.91	0.48	3.72
160	61.00	1.896	0.84	0.44	4.74
170	60.80	1.890	0.78	0.41	5.99
180	60.58	1.883	0.73	0.39	7.51
190	60.36	1.876	0.68	0.36	9.34
200	60.12	1.868	0.64	0.34	11.52
212	59.83	1.860	0.59	0.32	14.70

PIPE LINE SUPPLIED BY A PUMP

Pipe lines are more often supplied by a pump than a gravity reservoir or tank. The point of discharge is generally at a higher elevation than the supply. The pump must provide the necessary head to maintain the required discharge flow.

On leaving a pump, the head (or energy per lb) is:

$$\frac{\overline{V}^2}{2g} + \frac{p}{w} + h$$

Energy delivered = $Qw \left[\dfrac{\overline{V}^2}{2g} + \dfrac{p}{w} + h \right]$

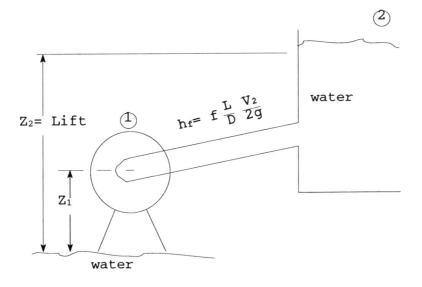

Figure 5.31

$$\frac{\overline{V}_1^2}{2g} + \frac{p_1}{w} + Z_1 = \text{Lift} + \text{friction loss} + \text{discharge loss.}$$

$$\frac{\overline{V}^2}{2g} + \frac{p_1}{w} + Z_1 = \text{Lift} + f \frac{L}{d} \frac{\overline{V}^2}{2g} + \frac{\overline{V}^2}{2g}$$

Energy delivered = $Qw \left[Z_2 + f \dfrac{L}{d} \dfrac{\overline{V}^2}{2g} + \dfrac{\overline{V}^2}{2g} \right]$

The pump 1) raised the water level (Z_2)

2) supplied head lost by friction $\left[f \dfrac{L}{d} \dfrac{\overline{V}^2}{2g}\right]$

3) supplied a velocity head ($\overline{V}^2/2g$)

Example

In a portion of a water supply system, a centrifugal pump delivers 4000 gpm to a reservoir 120 ft higher than the pump sump, through a new CI pipe 12 in. in diameter and 400 ft long. The pump efficiency is 80%. What total input in HP is required at the pump?

First convert 4000 gpm to cfs.

$$Q = 4000 \; \dfrac{\text{gal}}{\text{min}} \times .1337 \; \dfrac{\text{ft}^3}{\text{gal}} \times \dfrac{1 \; \text{min}}{60 \; \text{sec}} = 8.9 \; \dfrac{\text{ft}^3}{\text{sec}}$$

Writing Bernoulli's equation between entrance and outlet yields:

$$\dfrac{\overline{V}_1^2}{2g} + \dfrac{P_1}{w} + Z_1 + H_P = \dfrac{\overline{V}_2^2}{2g} + \dfrac{P_2}{w} + Z_2 + H_L$$

where H_P is head supplied by the pump. In other words, the input must equal the output.

$H_f + H_d$. We must assume that H_e is accounted for in the pump efficiency.

$$\overline{V}_{\text{pipe}} = \dfrac{Q}{A} = \dfrac{8.9 \; \text{cfs}}{.785 \; \text{ft}^2} = 11.3 \; \text{fps}$$

$$H_L = f \dfrac{L}{d} \dfrac{\overline{V}^2}{2g} + \dfrac{\overline{V}^2}{2g} = \left[\dfrac{fL}{d} + 1\right] \dfrac{\overline{V}^2}{2g}$$

From Table 6.1 for a 12-in. pipe with a $\overline{V} = 11.3$ fps, we find $f = 0.01857$

$$H_L = \left[.01857 \left(\dfrac{400}{1}\right) + 1\right] \left[\dfrac{11.3^2}{2 \times 32.2}\right] = 16.71 \; \text{ft}$$

If we take our datum at Z_1, $Z_1 = 0$

P_1 and P_2 are atmospheric pressure or 0 psi gage so $P_1/w = P_2/w = 0$

We can also assume the pump sump is large so that $\overline{V}_1$ is extremely small so $\overline{V}_1^2/2g \approx 0$. The same reasoning for $\overline{V}_2$ in the reservoir gives $\overline{V}_2^2/2g \approx 0$.

Bernoulli's equation now reduces to $0 + 0 + 0 + H_P = 0 + 0 + 120 + 16.7$ ft; $H_P = 136.7$ ft

Energy delivered = Q w h ft lb/sec.

Using 550 ft lb/sec per HP yields

$$HP = \frac{Q \, w \, h}{550} = \frac{8.9 \times 62.4 \times 136.7}{550} = 138 \text{ HP}$$

But this assumes 100% efficiency. HP actual = $\frac{HP}{v} = \frac{138}{.80} = 172.5 HP$

Input HP = 172.5 HP   <u>Answer</u>

<u>Example</u>

In making a test of a centrifugal pump, Hg gages were attached as shown in Fig. 5.32. Q = 5 cfs of cold water and requires 40 HP. Compute the head and pump efficiency.

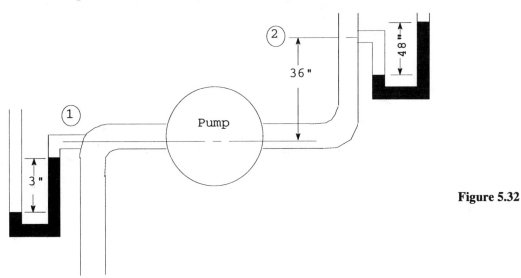

Figure 5.32

Suction and discharge lines are 10 in.

Take datum through ① making $Z_1 = 0$ and $Z_2 = 3$ ft Since we are testing the pump, we can assume the line lengths are short and the friction loss in the lines between ① and ② can be neglected.

$P_1 = -3$ in. Hg $= -3 \times .491 = -1.473$ psi

$P_2 = 48$ in. Hg $= 48 \times .491 = 23.568$ psi

Writing Bernoulli's equation between ① and ② and simplifying yields

$$\frac{P_1}{w} + H_p = \frac{P_2}{w} + Z_2$$

$$\frac{-1.473 \times 144}{62.4} + H_p = \frac{23.568 \times 144}{62.4} + 3$$

Solving for H_p, $H_p = \dfrac{23.568 \times 144}{62.4} + 3 + \dfrac{1.473 \times 144}{62.4}$

$H_p = 60.79$ ft of head supplied by the pump.

$$HP = \frac{Q\,w\,h}{550} = \frac{5.0 \times 62.4 \times 60.79}{550} = 34.5 \text{ HP output.}$$

$$\text{Efficiency} = v = \frac{HP \text{ out}}{HP \text{ supplied}} = \frac{34.5}{40.0} = 86\%$$

Pump head = 60.79 ft   <u>Answer</u>

Pump efficiency = 86%   <u>Answer</u>

<u>BRANCH PIPE FLOW</u>

Water supply systems or sewage disposal systems often have branches from the main. They present a special kind of problem. A classic in this field is known as the three-reservoir problem diagrammed in Fig. 5.33.

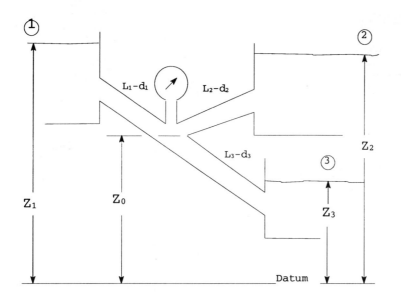

Figure 5.33

The problem is to find the flow between reservoirs.

As we have previously noted, the friction in a long pipe overshadows the minor losses. Since the roughness and friction factors are empirical, we can neglect the minor losses for this problem.

We can then write the following for pipe 1

$$Z_1 - (Z_o + \frac{P}{w}) = f_1 \frac{L_1}{d_1} \frac{\overline{V}_1^2}{2g} \quad (1)$$

and for pipe 2

$$(Z_o + \frac{P}{w}) - Z_2 = f_2 \frac{L_2}{d_2} \frac{\overline{V}_2^2}{2g} \quad (2)$$

and for pipe 3

$$(Z_o + \frac{P}{w}) - Z_3 = f_3 \frac{L_3}{d_3} \frac{\overline{V}_3^2}{2g} \quad (3)$$

From the continuity equation we find

$$Q = A_1 \overline{V}_1 = A_2 \overline{V}_2 + A_3 \overline{V}_3 \quad (4)$$

We now have four equations in four unknowns $\overline{V}_1$, $\overline{V}_2$, $\overline{V}_3$, and P/w.

Combining Equations (1) and (2) gives

$$Z_1 - Z_2 = f_1 \frac{L_1}{d_1} \frac{\overline{V}_1^2}{2g} + f_2 \frac{L_2}{d_2} \frac{\overline{V}_2^2}{2g}$$

And combining Equations (1) and (3) gives

$$Z_1 - Z_3 = f_1 \frac{L_1}{d_1} \frac{\overline{V}_1^2}{2g} + f_3 \frac{L_3}{d_3} \frac{\overline{V}_3^2}{2g}$$

The foregoing assumes flow from reservoir ① to reservoir ② and ③.

To calculate the actual direction of flow, note the energy levels. For example, if $(Z_o + \frac{P}{w}) > Z_2$ the flow is into reservoir ②.

However, if $Z_2 > (Z_o + \frac{P}{w})$, the flow is from reservoir ②.

It is also possible to have a divided flow as shown in Fig. 5.34.

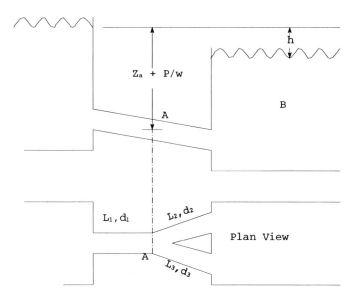

Figure 5.34

$$h_1 = f_1 \frac{L_1}{d_1} \frac{\overline{V}_1^2}{2g} \ ; \ h - h_1 = f_2 \frac{L_2}{d_2} \frac{\overline{V}_2^2}{2g} \ \text{which also equals} \ f_3 \frac{L_3}{d_3} \frac{\overline{V}_3^2}{2g}$$

since the loss in head from Ⓐ to Ⓑ must be the same no matter which pipe we choose to consider.

Therefore $h = f_1 \dfrac{L_1}{d_1} \dfrac{\overline{V}_1^2}{2g} + f_2 \dfrac{L_2}{d_2} \dfrac{\overline{V}_2^2}{2g}$ and $h = f_1 \dfrac{L_1}{d_1} \dfrac{\overline{V}_1^2}{2g} + f_3 \dfrac{L_3}{d_3} \dfrac{\overline{V}_3^2}{2g}$

We need a third equation which is

$$Q = A_1 \overline{V}_1 = A_2 \overline{V}_2 + A_3 \overline{V}_3$$

Example

A pipe system of welded steel is set up from A to C as shown in Fig. 5.35. Q = 6 cfs. What is the drop from A to C in psi?

$\gamma = 1.2 \times 10^{-5}$ ft²/sec.

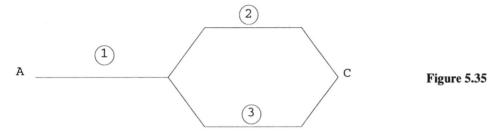

Figure 5.35

2000 ft of 12 in. pipe is ① $A_1 = .785$ ft

3000 ft of 8 in. pipe is ② $A_2 = .349$ ft

4000 ft of 10 in. pipe is ③ $A_3 = .545$ ft

Since we know nothing about how the pipe is split, we can assume minor losses to be negligible.

We can look up values for C for the Hazen-Williams formula and either solve the equation or use the nomograph to find the friction drop in the pipes.

In a table we can look up values for f vs $\overline{V}$ and d for welded steel pipe, assume an f, and find the actual flows by iteration and therefore find the drop due to friction.

We can use the Moody Chart for our solution. Since v is given, it is obvious that this is the solution the examiner is looking for.

For the parallel run of pipe, we have already shown that the drop in pipe ② is the same as the drop in pipe ③.

$$f_2 \frac{L_2}{d_2} \frac{\overline{V}_2^2}{2g} = f_3 \frac{L_3}{d_3} \frac{\overline{V}_3^2}{2g}$$

But $Q_1 = Q_2 + Q_3$

$$\overline{V}_1 = \frac{Q}{A_1} = \frac{6 \text{ ft}^3/\text{sec}}{.785 \text{ ft}^2} = 7.64 \text{ fps}$$

$$N_{r_1} = \frac{d_1 \overline{V}_1}{\nu} = \frac{1 \times 7.64}{1.2 \times 10^{-5}} = 6.37 \times 10^{+5}$$

Since $N_R > 2000$, we must use the Moody Chart itself and cannot use $f = 64/N_R$ which is good <u>only</u> for laminar flow.

From the Moody Chart for smooth pipe, at a $N_R = 6.37 \times 10^5$ we find $f = .0123$

$$N_{f_1} = f_1 \frac{L_1}{d_1} \frac{\overline{V}_1^2}{2g} = .0123 \times \frac{2000}{1} \times \frac{(7.64)^2}{2 \times 32.2} = 22.3 \text{ ft}$$

From the continuity equation 6 cfs = .349 $\overline{V}_2$ + .545 $\overline{V}_3$

Dividing through by .349 to find an expression for $\overline{V}_2$

$17.2 = \overline{V}_2 + 1.56 \overline{V}_3$

Now assume $f_2 = f_3$. They will be slightly different because of a difference in N_R but for a first cut we can assume the equality.

$$f\frac{L_2}{d_2} \frac{\overline{V}_2^2}{2g} = f\frac{L_3}{d_3} \frac{\overline{V}_3^2}{2g} \quad \text{or} \quad \frac{3000}{8/12} \frac{\overline{V}_2^2}{64.4} = \frac{4000}{5/6} \frac{\overline{V}_3^2}{2g}$$

Solving for $\overline{V}_2$ yields $\overline{V}_2 = 1.03 \overline{V}_3$

Then substituting this value of $\overline{V}_2$ in the equation

$17.2 = \overline{V}_2 + 1.56 \overline{V}_3$

and $17.2 = (1.03 + 1.56) V_3$

we obtain $V_3 = \dfrac{17.2}{2.59} = 6.64$ fps

and $\overline{V}_2 = 1.03 \times 6.64 = 6.84$ fps

$$N_{R_2} = \dfrac{d_2 \overline{V}_2}{\nu} = \dfrac{8/12 \times 6.84}{1.2 \times 10^{-5}} = 3.80 \times 10^5$$

From the Moody Chart $f_2 = .0136$

$$N_{R_3} = \dfrac{d_3 \overline{V}_3}{\nu} = \dfrac{5/6 \times 6.84}{1.2 \times 10^{-5}} = 4.75 \times 10^5$$

From the Moody Chart $f_3 = .0131$

Substituting these values for f_2 and f_3 gives

$$.0136 \times \dfrac{3000}{8/12} \times \dfrac{\overline{V}_2^2}{64.4} = .0131 \times \dfrac{4000}{5/6} \times \dfrac{\overline{V}_3^2}{64.4}$$

Solving this gives $\overline{V}_2^2 = 1.03\ \overline{V}_3^2$ or $\overline{V}_2 = 1.01\ \overline{V}_3$ or $\overline{V}_2 \approx \overline{V}_3$

Substituting back into the continuity equation yields

$\overline{V}_2 = \overline{V}_3 = 6.7$ fps

$$h_2 = h_3 = .0136 \times \dfrac{4000}{10/12} \times \dfrac{(6.7)^2}{64.4} = 45.5 \text{ ft}$$

Therefore the total friction loss between Ⓐ and Ⓒ is $22.3 + 45.5 = 67.8$ ft

$$P = wh = \dfrac{62.4 \times 67.8}{144} = 29.38 \text{ psi}$$

Pressure drop Ⓐ from Ⓒ to is 29.34 psi   **Answer**

FLOW MEASUREMENT

Flow through orifices has already been covered. The other most common simple flow measuring devices are the pitot tube, the Venturi tube, and the nozzle.

Pitot Tube

As can be seen in Fig. 5.36, the difference between the stagnation pressure head and the static pressure head is the dynamic pressure head, h. For a flow without any turbulence the stagnation tube pointed directly upstream, the velocity is $V = \sqrt{2g\,h}$

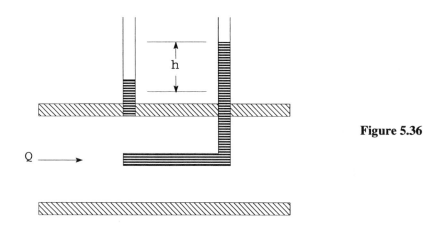

Figure 5.36

Generally, conditions are less than ideal, causing local velocities to fluctuate in magnitude and direction which results in the pitot tube giving a high reading. To compensate for this a coefficient, C_p, which is slightly less than unity (0.995 to 0.98) is used to correct the velocity.

$$V = C_p \sqrt{2g\,h}$$

Venturi Tube

Figure 5.37 showzs a Venturi tube. This discussion will show the derivation of the Venturi formula. Point ① is located several diameters upstream from the constriction. Point ② is at the throat of the Venturi.

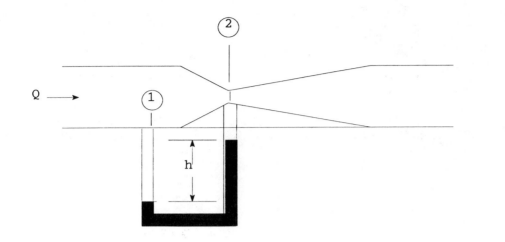

Figure 5.37

Write Bernoulli's equations between ① and ②

$$\frac{\overline{V}_1^2}{2g} + \frac{P_1}{w} + Z_1 = \frac{\overline{V}_2^2}{2g} + \frac{P_2}{w} + Z_2 + H_L$$

The loss between ① and ② is small enough to be neglected. Venturis are usually positioned so they are horizontal so that $Z_1 = Z_2$.

This leaves $\dfrac{\overline{V}_1^2}{2g} + \dfrac{P_1}{w} = \dfrac{\overline{V}_2^2}{2g} + \dfrac{P_2}{w}$

Rearranging terms

$$\frac{\overline{V}_1^2}{2g} - \frac{\overline{V}_2^2}{2g} + \frac{P_1}{w} - \frac{P_2}{w} = 0$$

or

$$(\overline{V}_1^2 - \overline{V}_2^2) + 2g\frac{(P_1 - P_2)}{w} = 0$$

but

$$A_1^2 \, \overline{V}_1^2 = A_2^2 \, \overline{V}_2^2 \quad \text{or} \quad \overline{V}_2^2 = \frac{A_1^2 \, \overline{V}_1^2}{A_2^2}$$

5-58 FUNDAMENTALS OF ENGINEERING EXAM REVIEW WORKBOOK

Substituting, $\overline{V}_1^2 \left[1 - \dfrac{A_1^2}{A_2^2}\right] + 2g \dfrac{(P_1 - P_2)}{w} = 0$

Solving for $\overline{V}_1 = A_2 \sqrt{\dfrac{2g(P_1 - P_2)/w}{A_1^2 - A_2^2}}$

Q Ideal = $A_1 \overline{V}_1$

Q Actual = $C_d A_1 \overline{V}_1 = C_d A_1 A_2 \sqrt{\dfrac{2g(P_1 - P_2)/w}{A_1^2 - A_2^2}}$

Example

In a water supply pipeline 10 ft in diameter there is a level Venturi meter with a throat 5 ft in diameter (Fig. 5.38). At the upper end of the meter the internal pressure is measured by an open water manometer which rises to a height of 30 ft above the center of the pipe. The meter discharge coefficient is 0.98. If 300 ft³/sec of water are flowing through the meter, what would be the height of the water in the manometer at the throat?

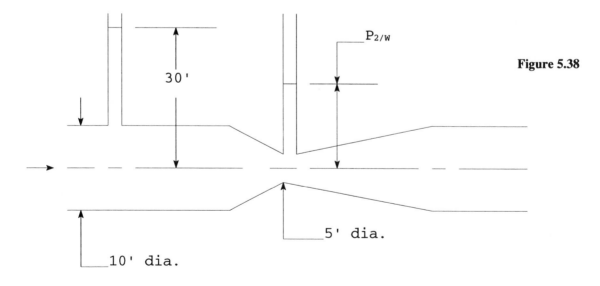

Figure 5.38

$A_1 = \dfrac{\pi \times 10^2}{4} = 78.54 \text{ ft}^2 \qquad A_2 = \dfrac{\pi \times 5^2}{4} = 19.63 \text{ ft}^2$

$$Q = C_d A_1 A_2 \sqrt{\frac{2g\left[\frac{P_1}{w} - \frac{P_2}{w}\right]}{A_1^2 - A_2^2}}$$

$$300 = .98 \times 78.54 \times 19.65 \sqrt{\frac{2 \times 32 \left[30 - \frac{P_2}{w}\right]}{(78.54)^2 - (19.63)^2}}$$

$$300 = 1512.4 \sqrt{\frac{1932 - 64.4 \left[\frac{P_2}{w}\right]}{5783}}$$

$$.1984 = \sqrt{.334 - .01114 \left[\frac{P_2}{w}\right]}$$

Squaring both sides yields

$$.0394 = .334 - .01114 \frac{P_2}{w}$$

$$-.2946 = -.01114 \frac{P_2}{w}$$

or $\dfrac{P_2}{w} = \dfrac{.2946}{.01114} = 26.44$ ft

Height of water in the manometer at the throat is 26.44 ft above ₵ of Venturi.   **Answer**

NOZZLES

Previously we discussed head losses for short tubes and pipes as well as in a general form.

$$H_L = \left[\frac{1}{C_v^2} - 1\right] \frac{\overline{V}^2}{2g}$$

This and Bernoulli's equation can be used to solve nozzle problems such as appear on the examination.

FLOW TRAJECTORY

In this type of problem we are usually given the jet trajectory or points thereon and asked to find conditions at the orifice or nozzle, or are given these conditions and asked to find a point or points on the jet trajectory.

Example: A jet discharges from a circular sharp edged orifice, 2 in. in diameter and under a head of 9 ft $Q = 18.9$ cfm. Coordinates of the ℄ of the jet are $x = 8.14$ ft and $Y = 2$ ft measured from the orifice. Find C_v, C_d, C_c, and the diameter of the jet at the vena contracta.

In the horizontal direction, distance $X = \overline{V} t$

In the vertical direction, the equation is distance $Y = 1/2 \, gt^2$

The jet stream moves horizontally by virtue of its velocity but it also moves downward under the influence of gravity.

From $X = \overline{V} t$ we obtain $\overline{V} = \dfrac{x}{t}$

From $Y = 1/2 \, gt^2$ we obtain $t = \sqrt{\dfrac{2y}{g}}$

Combining these equations yields $\overline{V} = \dfrac{x}{t} = \dfrac{x}{\sqrt{\dfrac{2y}{g}}} = x \sqrt{\dfrac{g}{2y}}$

$\overline{V}$ Actual $= 8.14 \text{ ft} \sqrt{\dfrac{32.2}{2 \times 2}} = 23.1$ fps

$\overline{V}$ Ideal $= \sqrt{2gh} = \sqrt{2 \times 32.2 \times 9} = 24.1$ fps

$C_v = \dfrac{\overline{V}_a}{\overline{V}_i} = \dfrac{23.1}{24.1} = 0.958$

$Q_a = 18.9 \, \dfrac{\text{ft}^3}{\text{min}} \times \dfrac{1 \text{ min}}{60 \text{ sec}} = 0.315$ cfs

$Q_i = A \overline{V} = A \text{ (orifice)} \times \sqrt{2gh} = .02182 \times 24.1$ fps $= .526$ cfs

$$C_d = \frac{Q_a}{Q_i} = \frac{.315}{.526} = 0.60$$

$$C_d = C_v \, C_c$$

$$C_c = \frac{C_d}{C_v} = \frac{0.60}{0.96} = 0.625$$

$$C_c = \frac{A \text{ Jet}}{A \text{ Orifice}} \text{ or } A_j = A_o \, C_c$$

$$A_j = .02182 \text{ ft}^2 \times .625 = 0.0136 \text{ ft}^2$$

$$D_j = \sqrt{\frac{A_j \times 4}{\pi}} \text{ Where } A_j = \frac{\pi D_j^2}{4}$$

$$D_j = \sqrt{\frac{0.136 \times 4}{3.14}} = 0.1316 \text{ ft} = 1.58 \text{ in.}$$

$C_v = 0.958$

$C_d = 0.60$

$C_c = 0.625$ Answer

$D_j = 1.58$ in.

Flow Over Weirs

Weirs are notched openings in vertical walls through which a liquid flows and the flow can be measured. Weirs are used in overflow dams and spillways, for flood control, and for irrigation control. The three most common weirs are the rectangular, triangular, and trapezoidal.

Rectangular Weir

The horizontal side of a rectangular weir is known as the crest and is generally designated by the letter b. The head, H, is the vertical distance from the crest to the level upstream surface. If the crest and sides of the weir are sufficiently removed from the bottom and sides of the upstream reservoir to permit free lateral approach of the water along the face of the weir, the stream will be contracted on these three sides as it emerges from the notch. This is a contracted weir. It has contractions as shown in Fig. 5.39. If the crest of the weir is extended so far as to make the ends of the notch coincident with the boundary or walls of the reservoir or channel, the end contractions will be suppressed. The number of end contractions, designated n, can be either 2, 1 or 0.

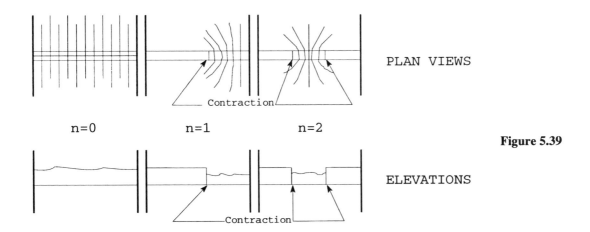

Figure 5.39

If the upstream side of the weir is a large tank or reservoir, the velocity of approach of the water is negligible or zero. In sluices, channels, rivers, etc, the velocity of approach sometimes causes a significant effect on the discharge.

The Francis formula is an empirical formula developed for rectangular weirs.

Q = flow in cfs   n = number of end contractions

b = width of crest $h = \overline{V}_o^2/2g$ where $\overline{V}_o$ = approach velocity.

H = head on the weir

Francis Formula (H > 0.5)

$$Q = 0.622 \times 2/3 \left[b - \frac{nH}{10}\right] \sqrt{2g} \left[(H+h)^{3/2} - h^{3/2}\right]$$

For a contracted weir, we have:

$$Q = 3.33 \left[b - \frac{2H}{10}\right] \left[(H+h)^{3/2} - h^{3/2}\right] ; (\overline{V}_o \neq 0)$$

$$Q = 3.33 \left[b - \frac{2H}{10}\right] H^{3/2} ; (\overline{V}_o = 0)$$

For a suppressed weir, we have:

$$Q = 3.33\, b \left[(H + h)^{3/2} - h^{3/2}\right]; (\overline{V}_o \neq 0)$$

$$Q = 3.33\, b\, H^{3/2}; (\overline{V}_o = 0)$$

If H is the vertical distance from the crest to the free surface and Z is the vertical distance from the crest to the bottom of the reservoir or channel, then H + Z is the depth of approach.

$$Q = 3.33\, b\, H^{3/2} \left[1 + 0.259 \left[\frac{H}{H + Z}\right]^2\right]$$

for a suppressed weir with $V_0 \neq 0$

Other formulae for rectangular weirs are:

Fetley and Stearns Formula

$$Q = 3.31\, b\, (H + \alpha h)^{3/2} + .007\, b\, ; (V_o \neq 0)$$

where α is a function of the head on the weir and the depth of the approach channel. Values of α can be found in handbooks and Hydraulics texts.

$$Q = 3.31\, b\, H^{3/2} + .007\, b; (V_o = 0)$$

Bazins formula is as follows:

$$Q = C \times 2/3\, \sqrt{2g} \left[b\, H^{3/2}\right]$$

Where $C = \left[0.6075 + \frac{.01476}{H}\right]\left[1 + 0.55 \left(\frac{H}{H + Z}\right)^2\right]$ $(V_o \neq 0)$

and $C = (0.6075 + \frac{.01476}{H}); (V_o = 0)$

Use Fetley & Stearns for H < 0.50 ft

Use Francis for H > 0.50 ft

Use Bazin for rough or corroded weirs

Example

A suppressed weir having a crest of 10.58 ft discharges under a head of 0.682 ft If the depth of the channel of approach is 2.2 ft, find the discharge per second.

Since we have a suppressed weir, the width of the channel must be the same as the width of the crest or 10.58 ft

We can get a first approximation by assuming $\overline{V}_o = 0$. Since H $>$ 0.5, use the Francis formula.

$Q = 3.33\ b\ H^{3/2} = 3.33 \times 10.58 \times (0.682)^{3/2} = 19.84$ cfs

Under these conditions,

$$\overline{V}_o = \frac{Q}{A} = \frac{19.84}{10.58 \times 2.2} = 0.85 \text{ fps}$$

then

$$h = \frac{\overline{V}_o^2}{2g} = \frac{(0.85)^2}{2 \times 32.2} = 0.011 \text{ ft}$$

Now we can calculate Q with $\overline{V}_o \neq 0$.

$$Q = 3.33\ b\ \left[(H + h)^{3/2} - h^{3/2}\right]$$

$$Q = 3.33 \times 10.58 \left[(0.693)^{3/2} - (0.011)^{3/2}\right]$$

$Q = 20.3$ cfs

$$\overline{V}_o = \frac{20.3}{10.58 \times 2.2} = 0.872 \text{ fps}$$

$$h = \frac{\overline{V}_o^2}{2g} = \frac{(0.872)^2}{2 \times 32.2} = .0118 \text{ ft}$$

The change in h is so small that no other calculation is necessary.

$Q = 20.3$ ft^3/sec.   **Answer**

Triangular Weirs

Triangular weirs are usually used for small flows. They are very sensitive to roughness of the sides or edges of the notch. The contraction is hard to define exactly. Pour a glass of milk from the V spout of a cardboard milk container and observe the contractions. To account for friction, roughness, and contractions, we use the empirical constant C in the equation.

$$Q = C \times \frac{4}{15} \, b \, \sqrt{2g} \, H^{3/2} \qquad \text{where } b = 2H \tan \alpha$$

Substituting, $Q = C \times \frac{8}{15} \tan \alpha \, \sqrt{2g} \, H^{5/2}$

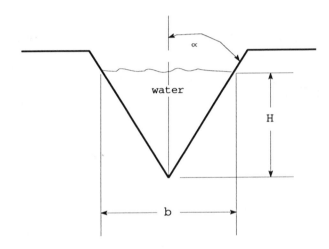

Figure 5.40

$$\tan \alpha = \frac{b/2}{H} = \frac{b}{2H} \qquad \alpha = 1/2 \text{ the notch angle.}$$

For the special case where $\alpha = 45°$ $\quad Q = C \times 4.28 \, H^{5/2}$

An accepted value for C is given by Prof. James Thomson (1861) as 0.593. Therefore, $Q = 2.54 \, H^{5/2}$ for a 90° opening.

Remember that $H^{5/2} = H^{4/2} \times H^{1/2} = H^2 \sqrt{H}$

Example

At an activated sludge plant, return sludge is controlled by flow over a sharp crested triangular weir. If the weir has a total angular opening of 90° and the sludge is 10 in. deep over the notch, how much sludge is flowing over the weir? The sludge may be assumed to flow like water at a temperature of 65°F.

Simply stated, we have water flowing in a triangular weir where $\alpha = 45°$ and $H = \dfrac{(10)}{12}$ ft

$$Q = 2.54 \, H^{5/2} = 2.54 \, \dfrac{(10)}{12}^{5/2} = 1.61 \text{ ft}^3/\text{sec} \qquad \underline{\text{Answer}}$$

C. Trapezoidal Weir

The trapezoidal weir is generally used in irrigation work. The design is to make the trapezoidal weir equivalent to a rectangular suppressed weir of crest b.

A sketch of a trapezoidal weir is shown in Fig. 5.41.

$$Q = 3.367 \, b \, H^{3/2}; \quad (\overline{V}_o = 0)$$

$$Q = 3.367 \, b \left[(H + h)^{3/2} - h^{3/2} \right] \quad (\overline{V}_o \ne 0)$$

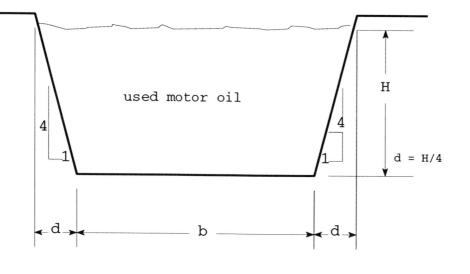

Figure 5.41

SLUICE GATE

Sometimes the discharge from a reservoir or lake occurs through a gate at the bottom of a dam where the discharge is along the bottom of the downstream channel.

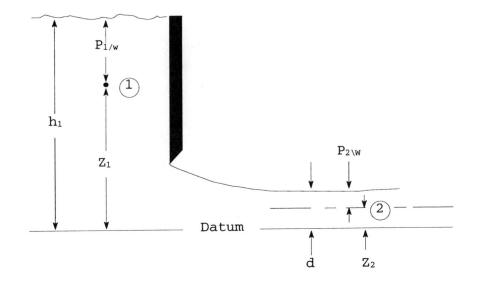

Figure 5.42

The state point ② is in the cross-section of the stream where contraction is complete.

Bernoulli's equation yields:

$$h_1 = \frac{\overline{V}_2^2}{2g} + \frac{P_2}{w} + Z_2 = \frac{\overline{V}_2^2}{2g} + d$$

$$\overline{V}_i = \sqrt{2g(h-d)} \qquad \overline{V}_a = C_v \sqrt{2g(h-d)} \qquad Q = CA\sqrt{2g(h-d)}$$

Where A is the area of the opening and C is the discharge coefficient.

FLOW IN OPEN CHANNELS

The slope of the channel is $S = \dfrac{h}{L}$, where h is the vertical fall in the length L ft.

The hydraulic radius, R, is equal to the area of flow divided by the wetted perimeter as before.

$$\text{For uniform flow } \overline{V} = \sqrt{\frac{8g}{f}} \; RS$$

The Chezy Formula is $\bar{V} = C\sqrt{RS}$,

where $C = \dfrac{157.6}{1 + \dfrac{m}{\sqrt{R}}}$ and m is a measure of absolute roughness.

Values of m can be found in handbooks and Hydraulics text books.

The Manning Formula is $\bar{V} = \dfrac{1.486}{n} R^{2/3} S^{1/2}$,

Where n is a roughness factor, and V is in fps

Example

In an irrigation project it is required to deliver 50 ft³ of water per second through a rectangular (half square) concrete lined flume from one canal to another that is 30 ft below and 2000 ft away. What size should the flume be?

$$A = b \times \frac{b}{2} = \frac{b^2}{2} \qquad Q = 50 \text{ cfs}$$

$$\bar{V} = \frac{Q}{A} = \frac{50}{\frac{b^2}{2}} = \frac{100}{b^2} \qquad R = \frac{A}{W.P.} = \frac{\frac{b^2}{2}}{2b} = \frac{b}{4}$$

$$S = \frac{30}{2000} = .015$$

From a handbook, n = 0.015 for cement lining.

Manning's Formula then yields:

$$\bar{V} = \frac{1.486}{n} R^{2/3} S^{1/2}$$

$$\frac{100}{b^2} = \frac{1.486}{.015} \left[\frac{b}{4}\right]^{2/3} (.015)^{1/2}$$

Solving for <u>b</u> gives 3.12 ft as follows:

$$\frac{100}{b^2} = \frac{1.486}{.015} \left[\frac{b}{4}\right]^{2/3} (.015)^{1/2}$$

$$\frac{100}{b^2} = 99 \left[\frac{b}{4}\right]^{2/3} (.122) = 12.12 \left[\frac{b}{4}\right]^{2/3}$$

$$\frac{100}{12.12} = b^2 \left[\frac{(b)}{4}\right]^{2/3}$$

$$8.25 = b^2 \left[\frac{(b)}{(4)}\right]^{2/3} = \frac{(b)^{8/3}}{(4)^{2/3}}$$

$$(8.25 \times 2.52)^3 = b^8 = 8988$$

$$3.12 = b$$

DYNAMIC FORCE AND MOMENTUM

The change in either magnitude or direction of the velocity of a fluid requires a force that depends on the change in momentum of the fluid.

$$F = M \Delta \overline{V}$$

where M is the mass of the fluids flowing past a normal section, ΔV is the change in velocity, and F is the resultant force acting upon the fluid.

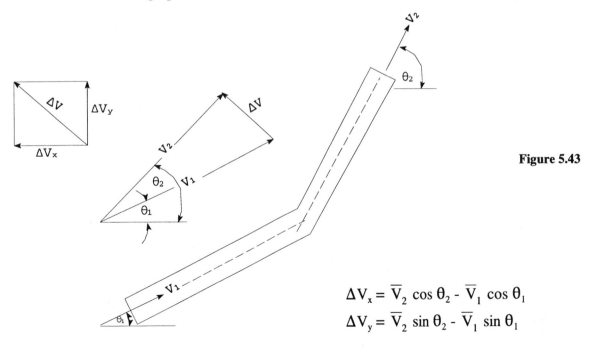

Figure 5.43

$$\Delta V_x = \overline{V}_2 \cos\theta_2 - \overline{V}_1 \cos\theta_1$$
$$\Delta V_y = \overline{V}_2 \sin\theta_2 - \overline{V}_1 \sin\theta_1$$

Figure 5.43 shows a fluid entering a pipe bend with velocity $\overline{V}_1$ making an angle θ_1 with the horizontal. The fluid leaves the pipe bend with a velocity $\overline{V}_2$ at an angle θ_2 with the horizontal. Figure 5.43 shows $\overline{V}_1$ and $\overline{V}_2$ (the input and output velocities, equal in magnitude) assuming negligible loss in the flow through the pipe bend and the value of ΔV must be the closing side of the vector diagram.

In steady flow, the mass of fluid entering must be equal to the mass of fluid leaving the pipe bend.

$$\rho_1 A_1 \overline{V}_1 = \rho_2 A_2 \overline{V}_2 = \frac{Qw}{g} \text{ or } \frac{Qw}{g} = \rho A \overline{V}$$

The X component of the change of momentum is:

$$\frac{Qw}{g} dt \ (\overline{V}_2 \cos \theta_2 - \overline{V}_1 \cos \theta_1)$$

Equating this to the X component of the impulse, F_x dt:

$$F_x = \frac{w}{g} (\overline{V}_2 \cos \theta_2 - V_1 \cos \theta_1) = \frac{w}{g} \Delta V_x$$

Similarly for the y component

$$F_y = \frac{w}{g} (\overline{V}_2 \sin \theta_2 - V_1 \sin \theta_1) = \frac{w}{g} \Delta V_y$$

But $F^2 = F_x^2 + F_y^2$, $F = \sqrt{F_x^2 + F_y^2}$ and $\overline{V} = \sqrt{\Delta \overline{V}_x^2 + \Delta \overline{V}_y^2}$

The resultant dynamic force exerted by the body on the fluid is

$$F = \frac{Qw}{g} \Delta \overline{V} \quad \text{The direction of F will be the same as that of } \Delta \overline{V}.$$

Example

A horizontal jet of water 1 in. in diameter impinges on a plate that changes the direction of the stream 30° upward. The initial velocity of the jet is 100 fps. Friction losses reduce the velocity to 75 fps. Find the force on the deflector.

$\Delta \overline{V}_x = 100 - 75 \cos 30° = 100 - 75 (.866) = 35.0$ fps

$\Delta \overline{V}_y = 75 \sin 30° = 75 (.50) = 37.5$ fps

$\Delta \overline{V} = \sqrt{V_x^2 + V_y^2} = \sqrt{(35.0)^2 + (37.5)^2} = \sqrt{2631} = 51.3$ fps

$\dfrac{Qw}{g} = \rho A \overline{V} = 62.4 \, \dfrac{lb}{ft^3} \times .00545 \, ft^2 \times \dfrac{100^{ft/sec}}{32.2 \, \frac{ft}{sec^2}} = 1.06 \, \dfrac{lb-sec}{ft.}$

$F = \dfrac{w}{g} \Delta \overline{V} = 1.06 \times 51.3 = 54.38$ lb.

By using components of $\Delta \overline{V}$ we find:

$F_x = \dfrac{w}{g} \Delta \overline{V}_x = 1.06 \times 35.0 = 37.1$ lbs

$F_y = \dfrac{w}{g} \Delta \overline{V}_y = 1.06 \times 37.5 = 39.75$ lbs

Drawing a vector diagram and remembering we want the force <u>on</u> the deflector.

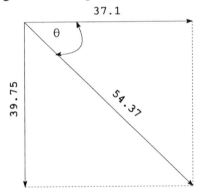

$R = \sqrt{(37.1)^2 + (39.75)^2} = \sqrt{2956.47}$

$R = 54.37$ lbs

$\theta = \sin^{-1} \dfrac{39.75}{54.37} = \sin^{-1} .73$

$\theta = 46.97°$

Figure 5.44

<u>The resultant force on the deflector is 54.37 lbs. to the right and 46.97° below the horizontal.</u>
<u>Answer</u>

<u>Example</u>

Oil having a specific gravity of 0.80 flows through a horizontal pipe at the rate of 25.0 cfs. At a 30° bend the pipe diameter diminishes from 16 to 12 in. The fluid pressure at the entrance to the bend is 15 psi. Calculate the horizontal force at the bend. Assume friction and secondary losses to be negligible.

First we tabulate some of our known values.

$w = 0.80 \times 62.4 = 49.92 \text{ lbs/ft}^3$

$A_1 = 1.396 \text{ ft}^2$ (16in. diameter pipe)

$A_2 = 0.785 \text{ ft}^2$ (12in. diameter pipe)

$\overline{V}_1 = \dfrac{Q}{A_1} = \dfrac{25.0}{1.396} = 17.91 \text{ fps}$

$\overline{V}_2 = \dfrac{Q}{A_2} = \dfrac{25.0}{.785} = 31.85 \text{ fps}$

$\sin 30° = 0.500$

$\cos 30° = 0.866$

$P_1 = 15 \text{ psia} \times 144 = 2160 \dfrac{\text{lb}}{\text{ft}^2}$

Note here the wording ".... at a 30° bend the pipe diminishes from 16 to 12 in. ...." Note, also, that we do not have to account for a loss in head due to the reduction since we are instructed to neglect secondary losses.

Writing Bernoulli's equation for the reduction yields:

$$\dfrac{P_1}{W} + Z_1 + \dfrac{\overline{V}_1^2}{2g} = \dfrac{P_2}{W} + Z_2 + \dfrac{\overline{V}_2^2}{2g}$$

Since $Z_1 = Z_2$, this reduces to:

$P_2 = P_1 + \dfrac{W}{2g}(\overline{V}_1^2 - \overline{V}_2^2) = 2160 + \dfrac{49.92}{64.4}\left[(17.91)^2 - (31.85)^2\right]$

$P_2 = 2160 + .775(320.77 - 1014.42) = 1622 \text{ psf}$

$R_x = \dfrac{QW}{g}(\overline{V}_1 - \overline{V}_2 \cos 30°) + P_1 A_1 - P_2 A_2 \cos 30°$

where the latter two terms account for the force due to the pressure in the system.

$R_x = \dfrac{25.0 \times 49.8}{32.2}(17.91 - 31.85 \times .866) + 2160(1.396)$

$-(1622 \times .785 \times .866) = 1539$ lbs

Horizontal force at the bend is 1539 lbs   <u>Answer</u>

Although not asked for:

$$R_y = \frac{QW}{g}(\overline{V}_2 \sin\theta) + A_2P_2 \sin\theta$$

$$R_y = \frac{25 \times 49.92}{32.2}(31.85 \times .5) + (.785 \times 1622 \times .50)$$

$$R_y = 617.2 + 636.6 = 1253.8 \text{ lb.}$$

$$R = \sqrt{(1539)^2 + (1254)^2} = 1985$$

DIMENSIONAL ANALYSIS

The various physical quantities in fluid mechanics are expressed in certain units. The two most common systems are the MLT or mass, length, time, and the FLT or force, length, time.

Dimensional analysis provides us with a method to express the relationship of various physical quantities. The method of use can best be described by its application to a problem. First, a very abbreviated table of units in both systems.

QUANTITY	MLT	FLT
Mass	M	$FL^{-1}T^2$
Force	MLT^{-2}	F
Density	ML^{-3}	$FL^{-4}T^2$
Pressure	$ML^{-1}T^{-2}$	FL^{-2}

Pressure in the FLT system is $\dfrac{F}{L^2}$ or $\dfrac{\text{pounds}}{\text{square linear unit}}$

Mass in the FLT system is $\dfrac{w}{g}$ or $\dfrac{\text{pounds}}{\frac{\text{ft}}{\text{sec}^2}}$ or $\dfrac{\text{pounds sec}^2}{\text{ft}}$

Example

The flow of an ideal liquid through an orifice is assumed to be a function of the orifice diameter, the fluid density, and the pressure drop through the orifice. By dimensional analysis establish the relationship.

Q is a function of D, ρ, and p.

$$Q = f(D, \rho, P) = D^A \rho^B P^C$$

which means that Q is a function of the diameter to some power, A, a function of the density to some power B, and a function of the pressure to some power C.

We shall use the FLT system, pound, foot, second system.

Dimensionally

$$\rho = FL^{-4}T^2 \quad P = FL^{-2} \quad D = L \quad Q = L^3 T^{-1}$$

We use the equation $Q = D^A \rho^B P^C$ and substitute the dimensional units. Remember that any unit raised to the zero power equals 1.

$$Q = D^A \rho^B P^C$$

$$F^0 L^3 T^{-1} = (L)^A (FL^{-4}T^2)^B (FL^{-2})^C$$

$$F^0 L^3 T^{-1} = L^A F^B L^{-4B} T^{2B} F^C L^{-2C}$$

For this equation to be correct, the exponents for F must be equal on the left and right sides of the equation. Similarly for L and T.

For F $0 = B + C$ For T $-1 = 2B$

For L $3 = A - 4B - 2C$

This gives us three equations in three unknowns.

For the equation for the exponents of T we find $B = -1/2$.

Substituting this value in the equation for the exponents of F yields:

$$0 = -\frac{1}{2} + C \quad \text{From which we find } C = 1/2$$

Substituting the values for B and C into the expression for the exponents of L:

$3 = A - 4(-1/2) - 2(1/2) = A + 2 - 1$

$2 = A$

Taking equation $Q = D^A \rho^B P^C$ and substituting the values of the exponents and adding a constant of proportionality, k:

$Q = k D^2 \rho^{-1/2} P^{1/2}$

$Q = k D^2 \sqrt{\dfrac{P}{\rho}}$ <u>Answer</u>

TRANSLATION AND ROTATION

Occasionally this type of problem appears on the examination. In translation we have the force diagram as shown in Fig. 5.45.

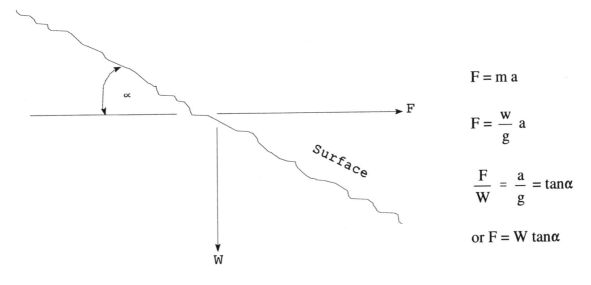

$F = ma$

$F = \dfrac{W}{g} a$

$\dfrac{F}{W} = \dfrac{a}{g} = \tan\alpha$

or $F = W \tan\alpha$

Figure 5.45

<u>Example</u>

A rectangular open container 8 ft long, 6 ft wide, and 4 ft high is mounted on a truck and half filled with water (Fig. 5.46). What is the maximum horizontal acceleration of the truck without spilling any water?

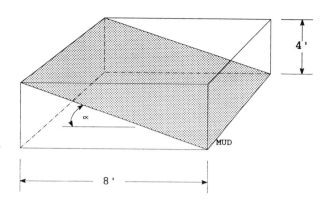

Figure 5.46

$$\tan\alpha = \frac{4}{8} = \frac{a}{g}$$

$$a = \frac{1g}{2} = \frac{1}{2} \times 32.2$$

$$a = 16.1 \text{ ft/sec}^2$$

Maximum horizontal acceleration = 16.1 ft/sec.² <u>Answer</u>

In rotation, we have the force diagram shown in Fig. 5.47.

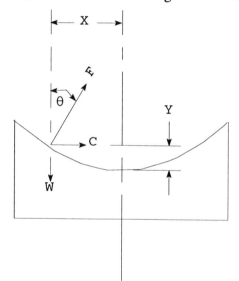

$$C = \frac{W}{g}\omega^2 x$$

$$\tan\theta = \frac{C}{W} = \frac{\omega^2}{g}x$$

$$Y = \frac{\omega^2 x^2}{2g}$$

Figure 5.47

Example

A cylindrical vessel is filled to a depth of 8 ft with water. The vessel is 10 ft deep and 4 ft wide. Determine the angular velocity required to raise the level of the water to the brim.

$$y = \frac{\omega^2 x^2}{2g} = \frac{\omega^2 \times 2^2}{2 \times 32.2} = 4 \text{ ft}$$

$$\omega^2 = \frac{4 \times 2 \times 32.2}{4} = 64.4 \frac{\text{rad}^2}{\text{sec}^2} \qquad \omega = 8.02 \frac{\text{radians}}{\text{sec}}$$

Figure 5.48

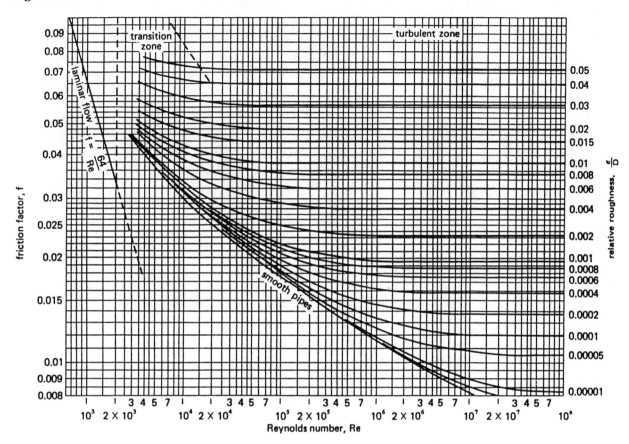

Reproduced from *Principles of Engineering Heat Transfer*, Giedt, published by D. Van Nostrand, Inc., 1957 with permission from Wadsworth Publishing Co., Inc., Belmont, CA.

6

Electricity/Electronics

ELECTRICITY/ELECTRONICS 6-i

TABLE OF CONTENTS

Selected References 6-iii

ELECTRICITY I 6-1
    Electrical Terminology 6-1
    Units and Notation 6-2
    Electrostatics 6-2
    Electrolysis 6-4
    Resistance, Capacitance, Inductance 6-5
    Direct Current (DC) Circuits 6-11
    Alternating Current (AC) Circuits 6-18
    Impedance and Admittance 6-21
    DC and AC Power 6-28
    Complex Circuits 6-31
    Resonance 6-40
    Simple Filters and Attenuators 6-44
    Practice Problems with Solutions 6-47
Appendix 6.1, Selected Unit Relationships 6-56
Appendix 6.2, Review of Complex Numbers 6-57
Appendix 6.3, Algebra of Determinants 6-60

ELECTRICITY II 6-63
    Three-Phase circuits 6-63
    Magnetics 6-69
    Transformers 6-71
    Rotating Machines 6-79
    DC Generators 6-81
    DC-Motors 6-88
    AC Generators 6-93
    AC Motors 6-97
    Three-Phase Induction Motor 6-99
    Measurements 6-105
    Practice Problems with Solutions 6-108
Appendix 6.4, Magnetic Units and Terminology 6-117
Appendix 6.5, DC Motor Characteristics 6-118
Appendix 6.6, AC Motor Characteristics 6-118

ELECTRICITY III 6-119
    Time Domain Analysis 6-119
    Introduction to Laplace Transforms 6-142
    System Engineering 6-149
    Examples 6-171

ELECTRICITY IV 6-179
    Semiconductors 6-179
    Semiconductor Devices 6-185
    Field Effect Transistors 6-193
    Transistors 6-208
    DC Conditions - BJT 6-213
    Signal Conditions - BJT 6-219
Appendix - Operational Amplifiers 6-231

SELECTED REFERENCES

F.W. Sears and M.W. Zemansky, "University Physics," Addison-Wesley, Reading, MA

V. Del Toro, "Principles of Electrical Engineering," Prentice Hall, Englewood Cliffs, NJ

Stein and Hunt, "Electric Power System Components," Van Nostrand Reinhold, New York

H. Milcaf, "Electricity One-Seven," Hayden Publications.

W.R. Wellman, "Elementary Electricity," D. Van Nostrand, New York

E. Norman Lurch, "Electric Circuits," John Wiley & Sons, New York

E.K. Kraybill, "Electric Circuits for Engineers," Macmillan, New York

D.F. Shaw, "An Introduction to Electronics," John Wiley & Sons, New York

E. Malmstadt and W.A. Toren, "Electronics for Scientists," Benjamin

ELECTRICITY I

ELECTRICAL TERMINOLOGY

Electricity problems involve the behavior of voltage and current in an electrical system consisting of resistor, capacitor, and inductive components. Practical definitions of the more frequently used terms are:

Current, I: A measure of the flow of electrical charges in a conductor, such as a wire, in ampere units.

Charge, Q: The quantity of current over a 1-sec period in Coulomb units ($Q = I \times t$).

Electromotive Force, EMF: A potential difference in the system that causes a current to flow in volt units (V).

Resistance, R: A discrete element in an electrical system that opposes the flow of current in Ohm units.

Capacitance, C: Property of a pair of conductors separated by an insulator (dielectric) in Farad units.

Inductance, L: Property of a device, usually a coil of wire, that acts to oppose the flow of current by inducing in the device an opposing voltage, in Henry units.

Energy, W: Considered as the amount of work a system is capable of doing measured in Joules (1 Joule = 1 Newton-meter; N-m).

Power, P: Rate of doing work in watts ($P = w/t$) 1 watt = 1 Joule/sec.

Horsepower, hp: One horsepower (hp) = 746 watts = 2546 Btu/hour.

Efficiency: Defined as $\dfrac{\text{Power delivered to load}}{\text{Power generated}} = \dfrac{P_o}{P_i} \times 100$

Regulation: Defined as

$$\dfrac{\text{No load voltage} - \text{Full load voltage}}{\text{Full load voltage}} \times 100$$

Gram-Calorie: Energy required to raise the temperature of 1 gram of water 1 degree centigrade.

Mechanical Force: F (newtons) = mass (kg) x acceleration (m/sec^2)

<u>Mechanical Work</u>: W (Joules) = F (Newtons) x d (meters)

<u>Vector</u>: Parameter having direction and magnitude, force, for example = $\vec{F}, \overleftarrow{F}$

<u>Scaler</u>: Parameter having magnitude only, for example, temperature, T°

UNITS AND NOTATION

Electricity units are International Units (SI) or the meter, kilogram, second (MKS). However, some electrical terminologies use the centimeter, gram, second (CGS) and the so-called Engineering or English Units (FPS).

Appendix 6.1 contains a selected listing of units and conversions. In the PERC review the units used reflect what seems to be the prevailing system for the topic discussed. Fortunately most electrical terms are in SI units which are also the same in MKS.

It should also be noted that scientific notation is frequently utilized. A few of the popular notations are indicated below.

$$\text{Milliampere, mA} = 10^{-3} \text{ Ampere}$$
$$\text{Microfarad, } \mu\text{F} = 10^{-6} \text{ Farad}$$
$$\text{Picofarad, pF} = 10^{-12} \text{ Farad}$$
$$\text{Kilovolt, kV} = 10^{+3} \text{ Volt}$$
$$\text{Megavolt, mV} = 10^{6} \text{ Volt}$$
$$\text{Millihenry, mH} = 10^{-3} \text{ Henry}$$
$$\text{Kilowatt, kW} = 10^{3} \text{ Watts}$$

ELECTROSTATICS

<u>Coulomb's Law</u>

This establishes the relationship between two statically charged bodies by defining the electrostatic force, F, in Newtons.

$$F = \frac{kq_1 q_2}{d^2}$$

$k \approx 9 \times 10^9$ N-M²/Coulomb², q_1, q_2 = charge in Coulombs; like charges repel, unlike charges attract; d = separation between charges in meters.

Electric Field Intensity

A charged body has a field associated with it. A unit positive test charge placed in the field is used to measure the intensity. Thus

$$\overline{E} = \frac{F}{q_1} \frac{\text{Newton}}{\text{Coulomb}};$$

a vector whose magnitude is measured by a test charge, q_1 placed in the field.

Example

Two similarly charged points, each charged to 10×10^{-12} coulombs, are separated by 3.0 cm. What is the force between the charges? What is the electric field intensity at a point midway between the charges?

Solution

The conditions of the problem are illustrated in Fig. 6.1.

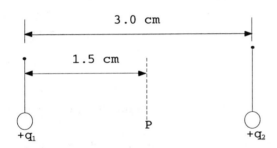

Figure 6.1

a. By Coulomb's Law:

$$F = \frac{k\, q_1\, q_2}{d^2} = \frac{(9 \times 10^9)(10 \times 10^{-12})(10 \times 10^{-12})}{(.03)^2}$$

$$F = \frac{900 \times 10^{-15}}{9 \times 10^{-4}} = 100 \times 10^{-11} = 1000 \text{pN}$$

If charges are assumed to be positive then the force is one of repulsion. This is also true if both charges are assumed to be negative.

b. Electric field intensity $\overline{E} = \dfrac{F}{q}$ and is measured by placing a unit positive test charge q_p at a point midway between q_1 and q_2.

$$\vec{E}_T = \vec{E}_1 + \overleftarrow{E}_2 \qquad = \overline{E}_T = \frac{k\,q_1}{d_1^2} - \frac{k\,q_2}{d_2^2} = 0$$

The test charge at P is repelled to the right due to $\vec{E}_1$ and to the left due to $\overleftarrow{E}_2$. If q_2 were − then

$$\overline{E}_T = \vec{E}_1 + \overleftarrow{E}_2$$

$$\overline{E}_T = \frac{k\,(q_1 + q_2)}{d^2}$$

If the test charge were located at an angle with respect to q_1 and/or q_2, then the field $\overline{E}$ is resolved into vertical and horizontal vectors and $\overline{E}_T$ = Vector resultant of horizontal and vertical components.

Conductors and Insulators: Electricity can be considered as the motion of charges. Most metals, such as copper and aluminum, are good conductors of electricity with an atomic structure such that there are valence electrons relatively free to move as opposed to their other embedded electrons which are more tightly attracted to the atoms nucleus.

Insulators: Usually non-metals, such as rubber and ceramic, are poor conductors characterized in the atomic micro world as having very few mobile valence electrons.

Semiconductors: Germanium and silicon have characteristics somewhere between a conductor and an insulator.

ELECTROLYSIS

Faraday's Law: In electrolysis the mass of a substance liberated (or deposited) at an electrode immersed in an electrolytic solution is proportional to the quantity of electricity passed through the system expressed as:

m_{grams} = gram equivalent weight (g-eqwt) times number of Faradays transferred.

and $\dfrac{m_1}{m_2} \propto \dfrac{\text{eq. wt 1}}{\text{eq. wt 2}}$

One Faraday is required for each gram equivalant of a substance.

also 1 Faraday ≡ 96,500 Coulombs

with 1 electrochemical eq. wt ≡ atomic weight/valence

so that m_{grams} = electrochemical equivalent weight times number of Coulombs transferred.

Example

A current is applied for 1 hour to plate nickel onto an electrode immersed in a nickel sulfate bath. Determine the current required to deposit 10 grams on the electrode. Assume an atomic weight of Ni = 58.71 and a valence of 2.

Solution

Using Faraday's Law relationships,

$$\text{eq wt of Ni} = \frac{\text{Atomic wt}}{\text{valence}} = \frac{58.71}{2} = 29.4$$

a. m_{grams} = 29.4 = 1 gram eq wt x 96,500 coulombs

b. m_{grams} = 10 = 1 gram eq wt x N coulombs

c. $Q = It$

$$N \text{ coulombs} = \frac{10}{29.4} \times 96.500 = 32823$$

$$I = \frac{32823}{3600} = 9.1A$$

RESISTANCE, CAPACITANCE, AND INDUCTANCE

Calculation of Resistance: Resistance of a circular conductor is calculated using the expression:

$$R = \frac{\rho L}{A} \; ; \; \rho = \text{resistivity in } \Omega\text{-CM per ft,}$$

L = length in feet, and A = cross-sectional area of the conductor in circular mils. (1 mil = .001 in.) Area A, in cm = d^2, where d is the diameter of the conductor in mils.

$$1 \text{ cm} = .7854 \text{ mil}^2 : \frac{1 \text{ sq mil}}{.7845} = \text{cm} : \frac{\text{area (mils}^2)}{.7854} = \text{area(cm)}$$

If the material is a rectangular bar then A = cross-sectional area of the bar; i.e. length x width and resistivity is defined as Ω-A per L. Units may be in MKS.

Example

What is the resistance of 625 ft of wire of .013 in² cross-section and resistivity of 10.37 Ω-cm/ft?

Solution

$$A = (.001)^2 (.7854); .013 \text{ sq. in. has } \frac{.013}{(.001)^2 \times .7854} \text{ cm}$$

$$R = \frac{\rho L}{A} = \frac{(10.37)(625)}{.0165 \times 10^{+6}} = .392 \Omega$$

The effect of temperature on resistance is expressed as:

$$R_2 = R_1 (1 + \alpha t)$$

α = temperature coefficient of resistance
R_1 = resistance at a reference temperature
R_2 = resistance at a second temperature
t = elevation of the second temperature above the reference temperature

Example

A certain type of thermometer is constructed using wire with $\alpha = 0.0038$ over the temperature range of operation. The resistance of the wire is 12 ohms at 20°C and stabilizes to 15 ohms operating in the oven. What is the temperature of the oven?

Solution

$$R_2 = R_1 + R_1 \alpha t$$

$$\Delta T = \frac{R_2 - R_1}{R_1 \alpha} = \frac{15 - 12}{12(.0038)}$$

$$\Delta T = 65.7°C$$

Temperature of oven is 65.7°C + 20°C = 85.7°C

There are four systems for temperature measurement.

$$\text{Rankin, } °R = °F + 459.67$$
$$\text{Kelvin, } °K = °C + 273.15$$
$$\text{Fahrenheit, } °F = (9/5 \times °C) + 32$$
$$\text{Celcius, } °C = 5/9 \times (°F - 32)$$

At two different temperatures the resistance of copper can be expressed as

$$\frac{|T| + t_1}{R_1} = \frac{|T| + t_2}{R_2}$$

For copper $T = |234.5| \equiv$ inferred absolute temperature. Each metal has its own inferred absolute temperature T. The T is the extension of a linear decrease of R with decreasing temperature rather than the non-linear decrease to absolute zero, -273°C.

<u>Equivalent Resistance</u>

R series equivalent of Fig. 6.2a

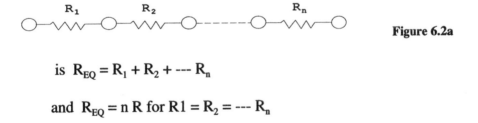

Figure 6.2a

is $R_{EQ} = R_1 + R_2 + \text{---} R_n$

and $R_{EQ} = nR$ for $R_1 = R_2 = \text{---} R_n$

R parallel equivalent of Fig. 6.2b

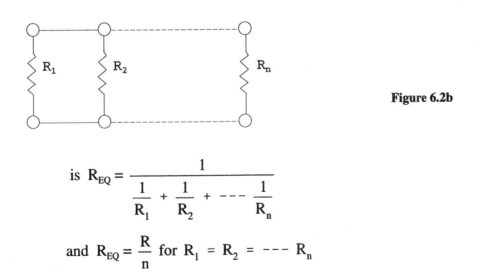

Figure 6.2b

is $R_{EQ} = \dfrac{1}{\dfrac{1}{R_1} + \dfrac{1}{R_2} + \text{---} \dfrac{1}{R_n}}$

and $R_{EQ} = \dfrac{R}{n}$ for $R_1 = R_2 = \text{---} R_n$

The reciprocal of resistance is conductance G, Siemens in (SI) of mhos in (MKS); $G = \dfrac{1}{R}$

$$G_{Series} = \dfrac{1}{\dfrac{1}{G_1} + \dfrac{1}{G_2} + \cdots \dfrac{1}{G_n}}$$

$$G_{Parallel} = G_1 + G_2 = \cdots G_n$$

Example

Determine the equivalent resistance in ohms of the circuit shown in Fig. 6.3.

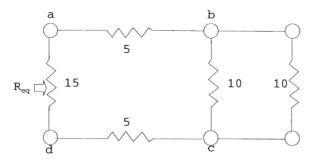

Figure 6.3

Solution

$$R_{bc} = \dfrac{1}{\dfrac{1}{10} + \dfrac{1}{10}} = \dfrac{1}{\dfrac{1+1}{10}} = \dfrac{10}{2} = 5$$

Note: For any two resistors in parallel, R_1 and R_2

$$R_{EQ} = \dfrac{R_1 R_2}{R_1 + R_2}$$

$$R_{bc} = \dfrac{10 \times 10}{10 + 10} = \dfrac{100}{20} = 5$$

$$R_{eq} = (R_{ab} + R_{bc} + R_{cd}) \parallel R_{ad}$$

$$R_{eq} = 15 \parallel 15 = \dfrac{15}{2} = 7.5$$

Calculation of Capacitance

The capacitance of a pair of parallel plates separated by a dielectric is given by:

$$C = \frac{K \epsilon A}{d} (n - 1) \text{ FARADS}$$

where K = dielectric constant
ϵ = 8.85 x 10^{-12} Coulombs^2 per N-M^2
A = Cross-sectional area of one plate in sq. meters
d = plate separation in meters
n = no. of plates

Example

What is the capacitance of 10 plates, each 10 cm x 10cm and separated by sheets of mica 1 mm thick? The K of mica is 6. If the capacitor is connected to a 10-volt battery source, what is the stored charge and how much stored energy does this charge represent?

Solution

A = .1 x .1 = .01 m^2
d = 1MM = .001 m
ϵ = 8.85 x 10^{-12} Coulombs^2 per N-M^2
K = 6
n = 10

$$C = \frac{6 \times 8.85 \times 10^{-12} \times .01 \times 9}{.001}$$

C = 4780 x 10^{-12} = 4780 micro micro farads

or C = 4780 uuf

Charge stored Q = CV = 4780 X 10^{-12} x 10

Q = 478 x 10^{-10} Coulombs

Average Energy W ≡ ½ CV^2 = ½ x 4780 x 10^{-12} x 100

W = 24 x 10^{-8} Joules

Equivalent Capacitance

C series equivalent of Fig.6.4b.

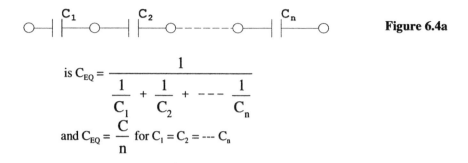

Figure 6.4a

is $C_{EQ} = \dfrac{1}{\dfrac{1}{C_1} + \dfrac{1}{C_2} + \cdots \dfrac{1}{C_n}}$

and $C_{EQ} = \dfrac{C}{n}$ for $C_1 = C_2 = \cdots C_n$

C parallel equivalent of Fig. 6.4b

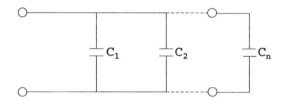

Figure 6.4b

is $C_{EQ} = C_1 + C_2 + \cdots C_n$

and $C_{EQ} = N\,C$ for $C_1 = C_2 = \cdots C_n$

Calculation of Inductance:

The induction of a coil of wire is calculated as:

$$L = \frac{N^2 \mu A}{l} \text{ in Henry's (MKS)}$$

where N = Number of turns

        A = Cross-sectional area of a cylinder form upon which the coil is formed

        l = Average length of form

        μ = Constant of magnetic permeability

Equivalent Inductance

L series equivalent of Fig. 6.5a

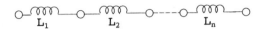

Figure 6.5a

is $L_{EQ} = L_1 + L_2 + \text{---} L_n$

and $L_{EQ} = nL$ for $L_1 = L_2 = \text{---} L_n$

L parallel equivalent of Fig. 6.5b

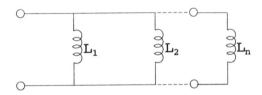

Figure 6.5B

is $L_{EQ} = \dfrac{1}{\dfrac{1}{L_1} + \dfrac{1}{L_2} + \text{---} \dfrac{1}{L_n}}$

and $L_{EQ} = \dfrac{L}{n}$ for $L_1 = l_2 = \text{---} L_n$

DIRECT CURRENT (DC) CIRCUITS

A closed electrical system with a voltage source, such as a battery, and an equivalent resistive load is a basic DC circuit.

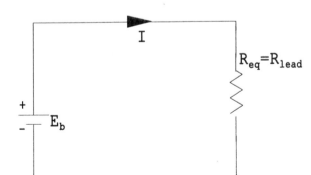

Figure 6.6

By OHMS Law $I = \dfrac{E_b}{R_{EQ}}$ amperes

Also $E_b = I R_{EQ}$ volts

And $R_{EQ} = \dfrac{E_b}{I}$ ohms

Example 7

In a closed electrical system an emf of 10 kV results in a current flow of 100 amperes. What is the equivalent resistance of the system?

Solution

A sketch of the problem is shown in Fig.6.7.

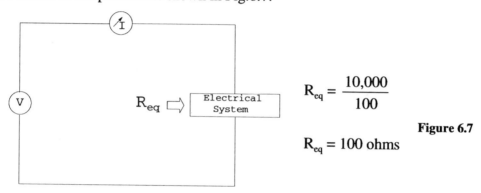

$$R_{eq} = \dfrac{10{,}000}{100}$$

$$R_{eq} = 100 \text{ ohms}$$

Figure 6.7

1. **Kirchoff's Rules.** In any closed electrical system the current flow into any point always equals the current flow away from the same point or

$$\sum_P I_{in} = \sum_P I_{out}$$

2. In any system the total potential around a closed loop is always equal to zero or

$$\sum_{circuit} V_{rise} = \sum_{circuit} V_{drop}$$

Where $V_{rise} = -$ to $+$ and $V_{drop} = +$ to $-$

Example

For the circuit shown in Fig. 6.8 determine R_1, I_1, I_2, E_{bd}, E_{cd}.

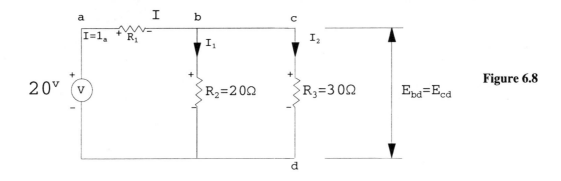

Figure 6.8

Solution

To demonstrate the applicability of Kirchoff's Rules and Ohms's Law the problem is solved as follows:

a. At b: $I = I_1 + I_2$ by Kirchoff's Current Rule

b. $(20 \text{ V}) = (E_{ab}) + (E_{bd})$ by Kirchoff's Voltage Rule

c. $(20 \text{ V}) = (E_{ab}) + (E_{cd})$ by Kirchoff's Voltage Rule

d. $(E_{bd}) = (E_{cd})$ Since the voltage drops across parallel (//) elements are equal.

e. $I_1 R_2 = I_2 R_3$ by Ohm's Law

f. $I_1 = I_2 \dfrac{R_3}{R_2} = I_2 \dfrac{30}{20} = 1.5 \, I_2$

g. Substitute (f) into (a) $I = 1.5 \, I_2 + I_2 = 1\text{A}$

h. $I_2 (1.5 + 1) = 1$; $I_2 = \dfrac{1}{2.5} = .4\text{A}$

i. Substitute (h) into (a) $I_1 = 1 - .4 = .6\text{A}$

j. $E_{bd} = I_1R_2 = .6 \times 20 = 12V$

k. $E_{cd} = I_2R_3 = .4 \times 30 = 12V$

l. $E_{ab} = 20 - E_{bd} = 20 - E_{cd} = 20 - 12V = 8V$

m. $R_1 = E_{ab}/I = 8/1 = 8\Omega$

Answers: $R_1 = 8\Omega$

$I_2 = 0.4A$ $I_1 = 0.6A$

$E_{bd} = E_{cd} = 12V$

DC Voltage Source

In most direct current electrical system problems the voltage source is a battery such as shown in Fig. 6.9.

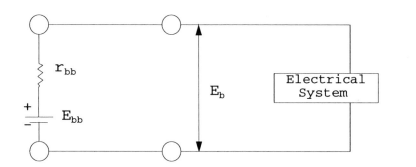

Figure 6.9

E_{bb} = Battery Voltage; r_{bb} = internal battery resistance;
E_b = Battery terminal voltage where connected to a load.
Often V_{bb} is considered negligible and $E_{bb} = E_b$.

Example

The battery system shown in Fig. 6.10a is connected to a DC load of 1Ω. Determine the current flow and voltage across the load, R_L.

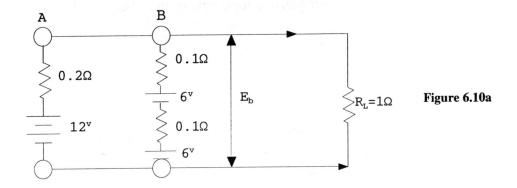

Figure 6.10a

At B, the equivalent of two identical batteries in series aiding (+-+-) is $2\,r_{bb}$ and $2\,E_{bb}$. For n batteries, the equivalent is nr_{bb} and nE_{bb}.

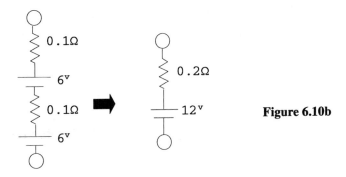

Figure 6.10b

At A, the equivalent of two identical batteries in parallel is $\dfrac{r_{bb}}{2}$ and E_{bb}. For n identical batteries the equivalent is $\dfrac{r_{bb}}{n}$ and E_{bb}.

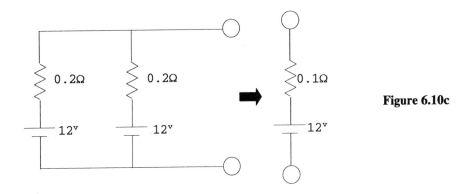

Figure 6.10c

The circuit of Fig. 6.10a is therefore reduced to the circuit of Fig. 10b.

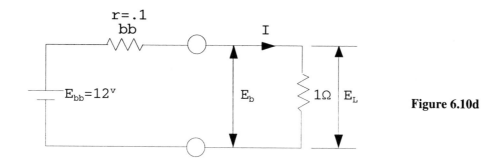

Figure 6.10d

$$I = \frac{12}{1.1} = 10.9 A$$

$$E_b = 12 - (10.9)(.1) = 12 - 1.09$$

$$E_b = 10.9 = E_L$$

Example

For the DC circuit shown in Fig. 6.11 determine the current through the 15 ohm resistor.

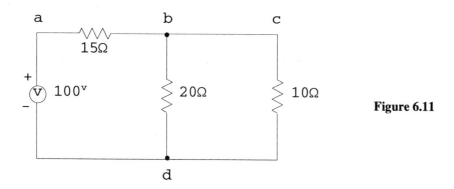

Figure 6.11

Solution

$$R_{EQ} = R_{ab} + R_{bd} \mathbin{//} R_{cd}$$

$$R_{EQ} = 15 + \frac{20 \times 10}{20 + 10}$$

$$R_{EQ} = 15 + \frac{200}{30} = 15 + 6.67$$

$$R_{EQ} = 21.67\Omega$$

$$I_{ab} = I_{circuit} = \frac{100}{21.67} = 4.61A$$

A general rule for resolving the current flow between two parallel resistors is the <u>Rule of Current Division</u> (see Fig. 6.12).

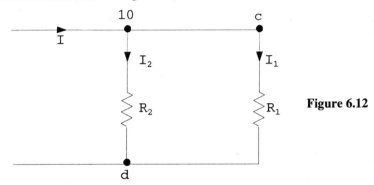

Figure 6.12

$$I_1 = \frac{R_2}{R_1 + R_2} \times I \qquad I_2 = \frac{R_1}{R_1 + R_2} \times I$$

In the preceding example

$$I_{cd} = \frac{20}{20 + 10} \times 4.61 = 3.073A$$

$$I_{bd} = \frac{10}{20 + 10} \times 4.61 = 1.54A$$

$$I = I_{cd} + I_{bd} = 3.073 + 1.54 = 4.61A$$

The dual of the rule for current division is the <u>Rule for Voltage Division</u> (see Fig. 6.13)

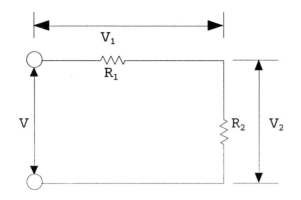

Figure 6.13

$$V_2 = \frac{R_2}{R_1 + R_2} \times V$$

$$V_1 = \frac{R_1}{R_1 + R_2} \times V$$

ALTERNATING CURRENT (AC) CIRCUITS

In electricity the AC voltage source is usually a sine wave. The terminology for this source is shown in Fig. 6.14.

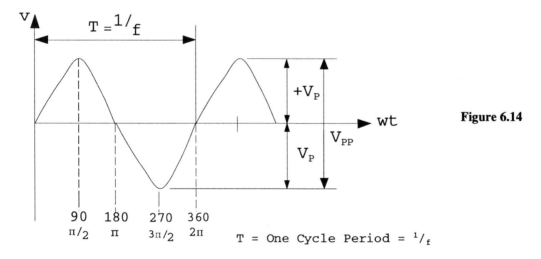

Figure 6.14

T = One Cycle Period = $1/f$

(1) $v = V_p \sin wt$ where v = instantaneous voltage at time, t.

(2) $wt = 2\pi f t = 377t$ at $f = 60$ Hz, $\pi = 3.14$

(3) $V_{R.M.S.} = \left[\dfrac{1}{T}\int_0^t v^2\, dt\right]^{1/2} = \dfrac{V_p}{\sqrt{2}} = .707\, V_p$

(4) $V_{d.c.} = \dfrac{1}{T}\int_0^t v\, dt = \dfrac{2V_p}{\pi} = .637\, V_p$ (full wave rectified)

($V_{d.c.}$ integration performed over ½ cycle; result over one cycle is twice the ½ cycle)

Example

Determine the effective value of the current in the circuit shown in Fig. 6.15.

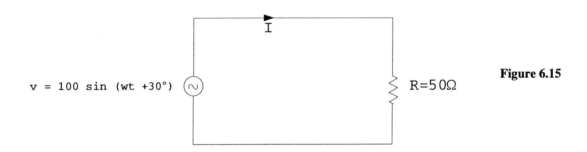

Figure 6.15

Solution

Using Ohm's Law applied to an AC voltage source.

$$i = \dfrac{v}{r} = \dfrac{100\,\mathrm{Sin}\,(wt + 30°)}{50}$$

$i = 2\,\mathrm{Sin}\,(wt + 30°)$ where $I_p = 2$

$I_{R.M.S.} = .707\, I_p = .707\,(2) = 1.414\,A.$

$\qquad = 1.414\,\mathrm{Sin}\,(wt + 30°)$

$I_{R.M.S.} = I_{effective}$, the effective value of current is equivalent to $I_{d.c.}$ from a battery source (not $.637\ I_p$) which will cause the equivalent heat in watts across the resistor, R.

The instantaneous value of v and the calculated i have the same form, $\sin(wt + 30°)$, which means that the voltage and current are in phase with each other but shifted $+30°$ from an arbitrary reference as shown in Fig. 6.16a. The phase angle, $30°$, is a phase lead if it has a + sign and is a phase lag if it has a - sign.

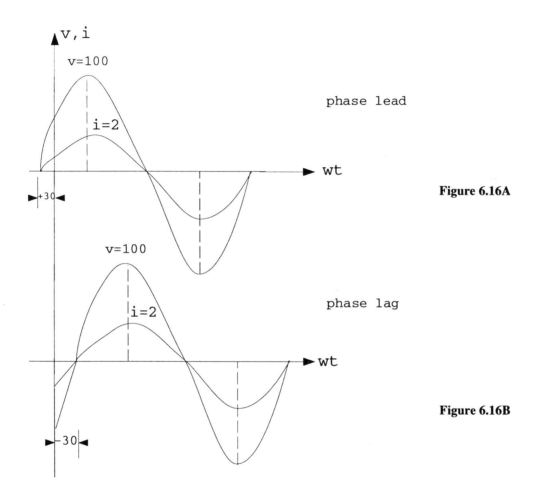

Figure 6.16A

Figure 6.16B

The previous example illustrated that the phase angle must be known to completely define the instantaneous voltage or current. If $v = V_p \sin wt$ then it is understood that $\theta = 0°$. Most AC problems use the RMS voltage for calculations. In the preceding example $v = V_p \sin(wt + \theta)$, $V_p = 100$ and $\theta = 30°$, from this data $V_{R.M.S.} = 70.7$ volts. The RMS voltage can be represented in Polar Form as $70.7\ \underline{|+30°}$. This can be converted to Rectangular Form as follows.

$70.7 \angle +30° = 70.7[\cos 30° + j \sin 30°]$

$= 61.2 + j\, 35.5$

The Rectangular Form is displayed as a vector diagram in Fig. 6.17.

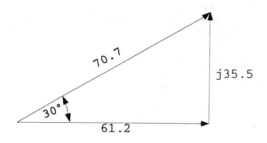

Figure 6.17

$\tan 30° = \dfrac{35.5}{61.2}$ and $70.7 = (61.2^2 + 35.5^2)^{½}$

See Appendix 6.2, Complex Numbers

IMPEDANCE AND ADMITTANCE

In the AC circuit the opposition to the flow of current is the impedance, Z. Using Ohm's Law applied to the AC circuit, we obtain Fig. 6.18.

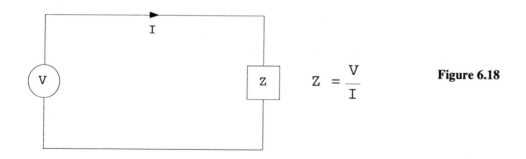

Figure 6.18

The dual of impedance is admittance, Y

$$Y = \dfrac{1}{Z} = \dfrac{I}{V} \, \mho, \text{ mhos or Siemens.}$$

(a) $Z = R$

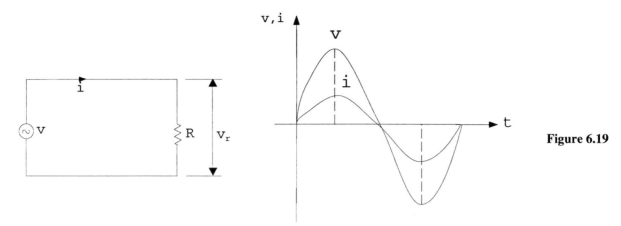

Figure 6.19

$$i = \frac{v}{R} \text{ and } I = \frac{V}{R} \text{ where } V = .707\, V_p \text{ and } V = v_p \sin wt$$

The current, i is in phase with the voltage, v but scaled in amplitude by $\frac{1}{R}$ (Fig. 6.19).

(b) Z = pure capacitance, C (see Fig. 6.20).

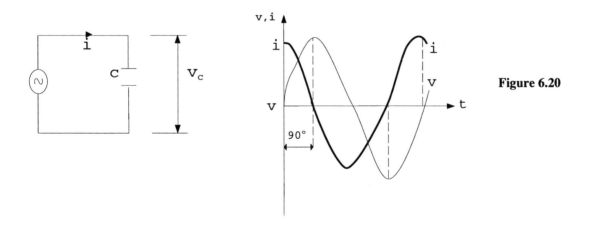

Figure 6.20

$Q = CV = It$

$$i = \frac{\Delta Q}{\Delta t} = \frac{d(Q)}{d(t)} = C\frac{d(v)}{d(t)}$$

$$d(v) = \frac{i}{c}d(t) \; ; \; d(V_p \sin wt) = \frac{i}{c}d(t)$$

$$\int d(V_p \sin wt) = \int \frac{i}{c}d(t)$$

$$w V_p \cos wt = \frac{i}{c}$$

$$i = wC V_p \cos wt = \frac{V_p}{\frac{1}{wC}} \sin(wt + 90°)$$

The current, i, leads the voltage, v by 90° or the effect of a pure capacitance is to cause the voltage across the capacitance to lag the current. The term $\frac{1}{wC} \equiv X_c$ the steady-state reactance in ohms, (w = 2πf).

Also since $Z = \frac{V}{I}$ then $Z = \frac{V \lfloor 0°}{I \lfloor 90°} = X_c \lfloor -90° \equiv -JX_c = -J\frac{1}{2\pi fC}\Omega$

$$Y = \frac{1}{-JX_c} = \frac{1}{-JX_c} \times \frac{J}{J} = JB \; (J^2 = -1)$$

$X_c \equiv$ reactance in ohms Ω

$B_c =$ susceptance in mhos ℧ or Siemens

(c) Z = pure inductance, L (see Fig. 6.21).

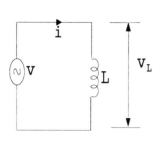
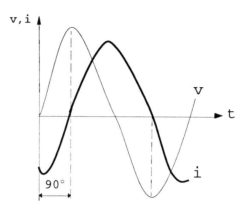

Figure 6.21

$$v = -L \frac{d(i)}{d(t)} \text{, Lenz's Law}$$

$$d(i) = -\frac{v}{L} d(t) = -\frac{V_p \sin wt \, d(t)}{L}$$

$$d(i) = -\int \frac{V_p \sin wt \, d(t)}{L}$$

$$i = -\frac{V_p}{WL} \cos wt = \frac{V_p}{X_L} \sin(wt - 90°)$$

The current, i lags the voltage v by 90° or the effect of a pure inductance is to cause the voltage across the coil to LEAD the current. The term $wL \equiv X_L$, the steady state reactance in ohms, ($w = 2\pi F$).

Also since $Z = \frac{V}{I}$ then $\frac{V \angle 0°}{I \angle -90°} = X_L \angle 90° \equiv JX_L \Omega \equiv J \, 2\pi fL \Omega$

$$Y = \frac{1}{Z} = \frac{1}{JX_L} = -JB$$

X_L = reactance in ohms

B = susceptance in mhos ℧ or Siemens

For combinational R, L, and C see Fig. 6.22.

$$Z = R + JX_L$$

$$Z = R - JX_c$$

$$Z = R + J(X_L - X_c)$$

$$Y = G + J(B_c - B_L)$$

$$Z = \frac{1}{Y}$$

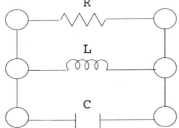

Figure 6.22a

Example

Determine the impedance across a and b of the series circuit shown in Fig. 6.22b.

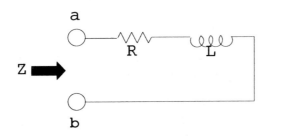

Figure 6.22b

Solution

$$Z = R + JX_L = |Z|\underline{/\theta°}$$

$$|Z| = \left(R^2 + X_L^2\right)^{1/2} \;;\; \theta = \mathrm{Tan}^{-1}\left[+\frac{X_L}{R}\right]$$

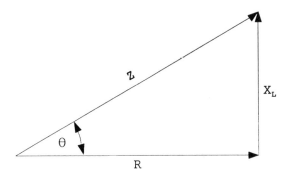

Figure 6.23a

Note: If C is substituted for L

$$Z = R - jX_c = |Z| \angle \theta°$$

$$|Z| = \left(R^2 + X_c^2\right)^{1/2} \; ; \; \theta = \tan^{-1}\left[-\frac{X_c}{R}\right]$$

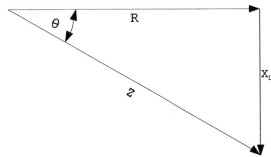

Figure 6.23b

Example

Determine the equivalent impedance across a and b of the following circuit.

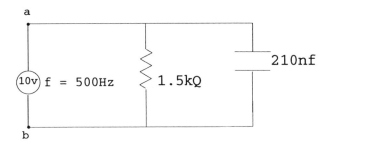

Figure 6.24a

Solution

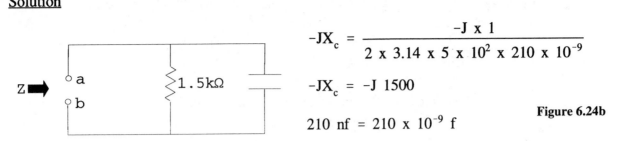

Figure 6.24b

$$-JX_c = \frac{-J \times 1}{2 \times 3.14 \times 5 \times 10^2 \times 210 \times 10^{-9}}$$

$$-JX_c = -J\,1500$$

$$210\text{ nf} = 210 \times 10^{-9}\text{ f}$$

Impedances in parallel are treated like resistors in parallel.

$$Z_{eq} = \frac{Z_1 Z_2}{Z_1 + Z_2} = \frac{(1500)(-J1500)}{1500 - J1500}$$

One method of solution is to simplify the denominator
$Z = 1500 - J1500$ by multiplying the numerator and denominator
by its Complex Conjugate, $Z^* = 1500 + J1500$

$$Z_{eq} = \frac{-J225 \times 10^4}{1500 - J1500} \times \frac{1500 + J1500}{1500 + J1500}$$

$$Z_{eq} = \frac{3375 \times 10^6 - J3375 \times 10^6}{1500^2 + 1500^2}$$

$$Z_{eq} = \frac{(3375 - J3375)\,10^6}{450 \times 10^4}$$

$$Z_{eq} = 750 - J750 = 1061\,\underline{|-45°}$$

An alternate method is to compute Y_{eq}.

$$Y_{eq} = Y_1 + Y_2$$

$$Y_{eq} = \frac{1}{1500} + J\frac{1}{1500}$$

$$Y_{eq} = 6.67 \times 10^{-4} + J6.67 \times 10^{-4} = 9.428 \times 10^{-4} \underline{|+45°}$$

$$Z_{eq} = \frac{1}{Y_{eq}} = \frac{1}{9.428 \times 10^{-4} \underline{|+45°}}$$

$$Z_{eq} = 1061 \underline{|-45°}$$

DC AND AC POWER

In the DC circuit the power dissipated in a resistance is

$$P_{average} = VI = I^2R = \frac{V^2}{R} \text{ in watts}$$

In an AC circuit the equivalent power dissipated in a resistive load is VI.

Where V and I are effective or RMS values

If the power source consists of a DC source and several series AC sources

$$I = \sqrt{I_{d.c}^2 + I_{1\,RMS}^2 + I_{2\,RMS}^2 + \text{---}}$$

$$V = \sqrt{V_{d.c}^2 + V_{1\,RMS}^2 + V_{2\,RMS}^2 + \text{---}}$$

In the AC circuit the power in watts dissipated in either a pure capacitor or pure inductor is zero because over any one AC cycle of a sine wave the voltage and current are 90° out of phase. Therefore, the average power is zero. Since the reactances store energy there is a reactive power associated with the L and C components. In R, L, and C circuits a power vector diagram is used to express the power relationships as follows (Fig. 6.25):

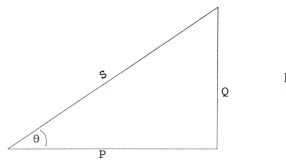

Figure 6.25

Real Power = P = VI Cos θ in watts
Reactive Power = Q = VI Sin θ = S Sin θ in VARS
Apparent Power = S = VI = $\sqrt{P^2 + Q^2}$ in volt amperes
V and I are the source or line factors.

Example

A small oven requires 10 amperes over a 15-minute period to properly heat its 120 volt DC element. How much heat is generated in BTU?

Solution

   P = VI = 120 x 10 = 1200 watts = 1.2 KW
   BTU = (1.2)(.9483)(15)(60) = 1024

Example

An impedance consists of a series R, L circuit each of 10 Ω in magnitude (Fig. 6.26). What is the resistor wattage consumed when the circuit is connected to a 120 volt, 400Hz line? What is the circuit apparent power?

Solution

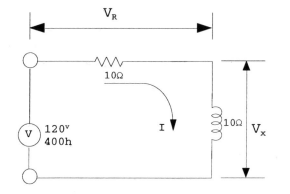

Figure 6.26

$Z = R + JX_L = 10 + J10 = 14 \angle 45°$

$I = V/Z = 120 \angle 0° / 14 \angle 45° = 8.57 \angle -45°$

$P = VI \cos \theta = (120)(8.57)\cos 45° = 727.2$ watts

$Q = VI \sin \theta = (120)(8.57)\sin 45° = 727.2$ vars

$S = (P^2 + Q^2)^{½} = 1028$ VA

The apparent power of the circuit is 1028 VA. The circuit P = 726.7 watts and circuit. Q = 726.7 Vars in accordance with the power vector triangle shown below in Fig. 6.27.

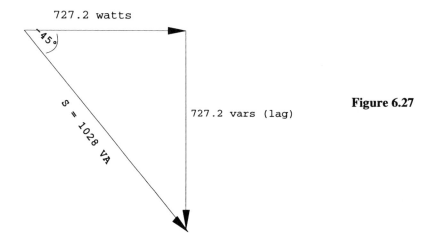

Figure 6.27

The power consumed by the 10 ohm resistor is given by 727.2 watts.

The circuit apparent power is 1028 VA

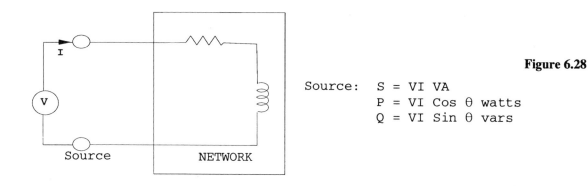

Figure 6.28

Source: S = VI VA
P = VI Cos θ watts
Q = VI Sin θ vars

COMPLEX CIRCUITS

Complex circuits or networks are formed by R, C and L components in various combinations of series, parallel and series parallel loops. The circuit may also contain one or more active current and/or voltage sources. In most instances a simplification of the circuit leads to an easier technique for solution. A few of the more frequently used methods will be demonstrated.

A general two loop complex circuit is shown in Fig. 6.29.

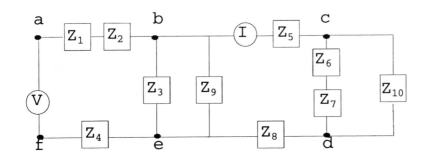

Figure 6.29

    a. Loop I = a b e f
    b. Loop II = b c d e
    c. V = voltage source
    d. I = current source
    e. Z_1 may be equal to R or $\pm$ JX or R $\pm$ JX etc

1. Circuit Impedance Simplifications

    a. Z series equivalent = $Z_1 + Z_2 + - - - - Z_n$

$$Z_{ab} = Z_1 + Z_2$$

$$Z_{cd} = Z_6 + Z_7$$

    b. Z parallel Equivalent = $\dfrac{1}{\dfrac{1}{Z_1} + \dfrac{1}{Z_1} - - - - \dfrac{1}{Z_n}}$

$$Z_{be} = \frac{Z_3 Z_9}{Z_3 + Z_9}$$

$$Z_{cd} = \frac{Z' Z_{10}}{Z' + Z_{10}} \text{ and } Z' = Z_6 + Z_7$$

c. Since $Y = \dfrac{1}{Z}$ it is often easer to use

$$Y \text{ parallel equivalent} = Y_1 + Y_2 + \cdots - Y_n$$

$$Y_{be} = \dfrac{1}{Z_3} + \dfrac{1}{Z_9} \text{ and } Z_{be} = \dfrac{1}{Y_{be}}$$

2. Series–Parallel Conversion (Fig. 6.30)

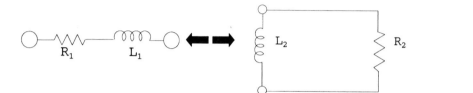

Figure 6.30

$Z_1 = R_1 + JX_{L_1}$ $\qquad Y_2 = G_2 - JB_{L_2}$

$R_2 = \dfrac{R_1^2 + X_{L_1}^2}{R_1}$ $\qquad X_{L_2} = \dfrac{R_1^2 + X_{L_1}^2}{X_{L_1}}$

3. Parallel–Series Conversion (Fig. 6.31)

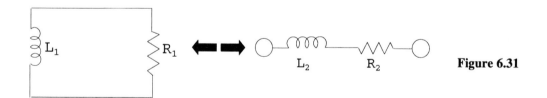

Figure 6.31

$Y_1 = G_1 - JB_{L_1}$ $\qquad Z_2 = R_2 + JX_{L_2}$

$R_2 = \dfrac{R_1 X_{L_1}^2}{R_1^2 + X_{L_1}^2}$ $\qquad X_{L_2} = \dfrac{R_1^2 X_{L_1}}{R_1^2 + X_{L_1}^2}$

4. Delta–WYE Transformation (Fig. 6.32)

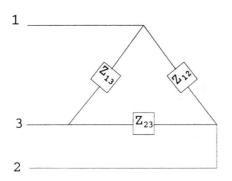
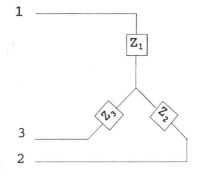

Figure 6.32

Δ to Y

$$Z_1 = \frac{Z_{12} \, Z_{13}}{Z_n}$$

$$Z_2 = \frac{Z_{12} \, Z_{23}}{Z_n} \qquad Z_n = Z_{12} + Z_{13} + Z_{23}$$

$$Z_3 = \frac{Z_{13} \, Z_{23}}{Z_n}$$

Y to Δ

$$Z_{12} = Z_0/Z_3$$
$$Z_{13} = Z_0/Z_2 \qquad Z_0 = Z_1 Z_2 + Z_1 Z_3 + Z_2 Z_3$$
$$Z_{23} = Z_0/Z_1$$

Special Case

If $\qquad Z_{12} = Z_{13} = Z_{23}$ or $Z_1 = Z_2 = Z_3$

Then $\qquad Z_y = \dfrac{Z\Delta}{3}$ and $Z\Delta = 3\, Z_y$

5. Voltage Source (Fig. 6.33)

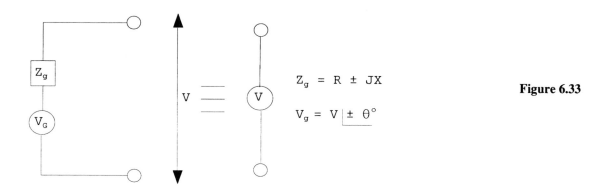

$Z_g = R \pm JX$

$V_g = V \underline{|\pm \theta°}$

Figure 6.33

6. Current Source (Fig. 6.34)

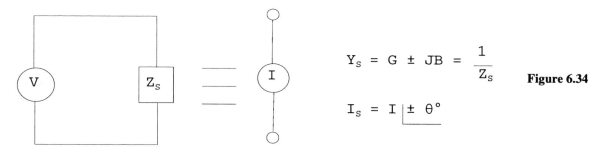

$Y_S = G \pm JB = \dfrac{1}{Z_S}$

$I_S = I \underline{|\pm \theta°}$

Figure 6.34

7. <u>Thevenin's Theorem</u>

Replace a complex active network at a selected pair of terminals by a voltage source with a series impedance. The technique is demonstrated by the following example.

<u>Example 16</u>

Replace indicated network by Thevenin circuit at A - B (Fig. 6.35a).

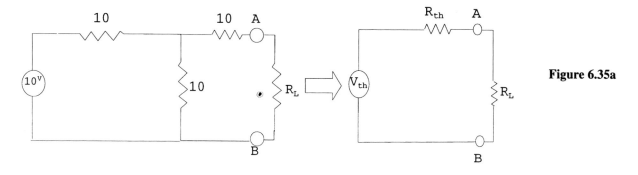

Figure 6.35a

Solution

1. Replace V = 10 V by V short circuit. Find Z at AB (Fig. 6.56b).

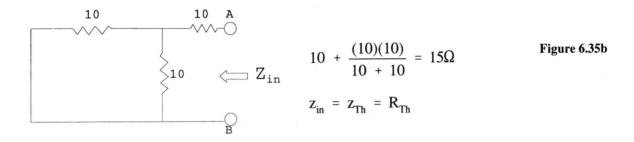

$$10 + \frac{(10)(10)}{10 + 10} = 15\Omega$$

$$Z_{in} = Z_{Th} = R_{Th}$$

Figure 6.35b

2. Restore V = 10V. Find E_{AB} without R_L (Fig. 6.35c).

$$V_{AB} = V_{OC} = V_{Th} = \frac{10}{10 + 10} \times 10 = 5V$$

(OC = open circuit)

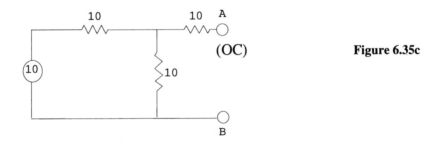

Figure 6.35c

Norton's Theorem

Replace a complex active network at a selected pair of terminals by a current source with a parallel impedance. The technique is demonstrated by the following example.

Example

Replace the indicated network by a Norton circuit (Fig. 6.36a).

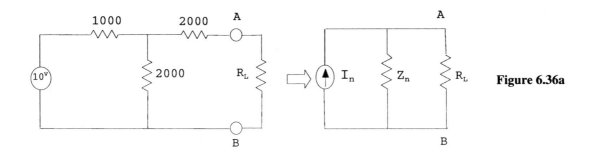

Figure 6.36a

1. $Z_n = z_{Th}$; Determined in the same manner as shown for the Thevenin example.

$$Z_n = 2000 + \frac{(2000)(1000)}{2000 + 1000} = 2670\,\Omega$$

2. $I_n = I_{\text{short circuit}}$

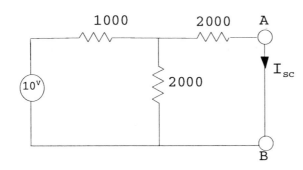

Figure 6.36b

$$I = \frac{10}{1000 + \frac{(2000)(2000)}{2000 + 2000}} = 5 \times 10^{-3}$$

$$I_{sc} = \frac{2000}{4000} \times 5 \times 10^{-3} = 2.5 \times 10^{-3}$$

$I_n = 2.5$ MA ; $Z_n = 2670\,\Omega$

Note: Thevenin and Norton circuits are duals.

$$V_{TH} = I_n Z_n \text{ and } Z_{Th} = Z_n$$

Power Transfer

In a power transmission line it is desired to get the maximum power delivered to the load. Ideally the internal impedance of the source is zero and all power is delivered to the load (efficiency is 100%).

For many other applications it is necessary to match the load impedance equal to the internal impedance of the source. Ideally the real part of the source impedance and the real part of the load impedance are made equal (efficiency = 50%) and the imaginary part of the source $\pm JX$ is equal to the complex conjugate imaginary part of the load $\mp JX$ as shown in Fig. 6.37.

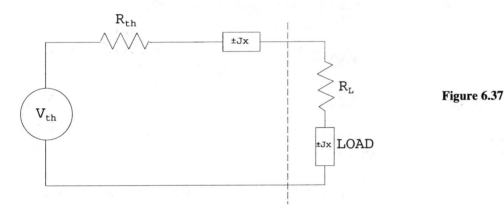

Figure 6.37

At Load: $P_{max} = \dfrac{V_{Th}^2}{4R_{Th}}$; $R_L = R_{Th}$ and $\mp JX_L = \pm JX$

Mesh Arrangement

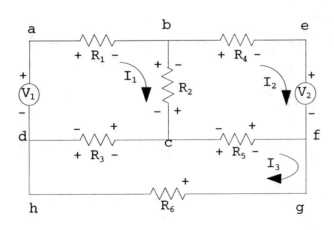

Figure 6.38

The network shown in Fig. 6.38 is a three loop network or mesh. Assume Z = R.

1. Assign a clockwise current to each loop.

    loop abcd, I_1

    loop befc, I_2

    loop d,f,g,h, I_3

2. If not shown, assign ± to each active (voltage) source or direction of current to each current source.

3. Using Kirchoff's Laws determine sum of voltage rises and drops in each loop.

Loop I, abcd: $V_1 = I_1 R_1 + I_1 R_2 + I_1 R_3 - I_2 R_2 - I_3 R_3$

In this loop the voltage drop across each resistor due to current flow is determined as follows.

$$V_{R2} = (I_1 - I_2) R_2$$

$$V_{R3} = (I_1 - I_3) R_3$$

Loop I, abcd: $V_1 = I R' - I_2 R_2 - I_3 R_3$

where $R' = R_1 + R_2 + R_3$

In the same manner

Loop 2, befc: $-V_2 = -I_1 R_2 + I_2 R_2'' - I_3 R_5$

where $R_2'' = R_2 + R_4 + R_5$

Loop 3, d,f,g,h: $0 = -I_1 R_3 - I_2 R_5 + I_3 R'''$

where $R''' = R_3 + R_5 + R_6$

(1) $V_1 = I_1 R' - I_2 R_2 - I_3 R_3$

(2) $-V_2 = -I_1 R_2 + I_2 R_2'' - I_3 R_5$

(3) $0 = -I_1 R_3 - I_2 R_5 + I_3 R'''$

See Appendix 6.3 for the calculation procedure for I_1, I_2, I_3.

Example

Determine the loop currents in the circuit shown in Fig. 6.39a.

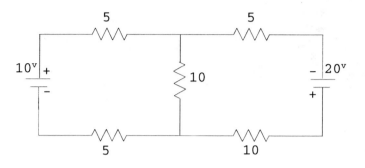

Figure 6.39A

Solution

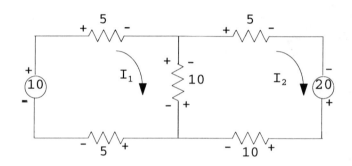

Figure 6.39B

(1) $10 = 20 I_1 - 10 I_2$

(2) $20 = -10 I_1 + 25 I_2$

(3) $I_1 = \dfrac{\begin{vmatrix} 10 & -10 \\ 20 & 25 \end{vmatrix}}{\Delta} = \dfrac{250 + 200}{\Delta} = \dfrac{450}{\Delta}$

(4) $I_2 = \dfrac{\begin{vmatrix} 20 & 10 \\ -10 & 20 \end{vmatrix}}{\Delta} = \dfrac{400 + 100}{\Delta} = \dfrac{500}{\Delta}$

(5) $\Delta = \begin{vmatrix} 20 & -10 \\ -10 & 25 \end{vmatrix} = 500 - 100 = 400$

(6) $I_1 = 450/400 = 1.125 A$

(7) $I_2 = 500/400 = 1.25 A$

Check

(1) $10 = 20 I_1 - 10 I2$

   $10 = 20 (1.125) - 10 (1.25)$

   $10 = 22.5 - 12.5 = 10$

(2) $20 = -10 I_1 + 25 I_2$

   $20 = -10 (1.125) + 25 (1.25)$

   $20 = -11.25 + 31.25 = 20$

Resonance

Series Resonance

In any electrical circuit consisting of a series arrangement of R, C and L such as shown in Fig. 6.40 the impedance Z is a function of frequency since $X_L = 2\pi f L$ and $X_c = \dfrac{1}{2\pi f C}$.

At some frequency, f_o, the magnitude of X_L becomes equal to the magnitude of X_c. This is the series resonant frequency, f_o.

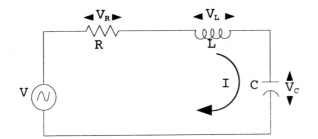

Figure 6.40

At f_o the following terms and cirucit relationships are valid.

a. $f_o = \dfrac{1}{2\pi\sqrt{LC}}$ in Hertz

where L is in Henry's and C is in Farads.

b. $Q = \text{Figure of Merit} = \dfrac{X_L}{R} = \dfrac{X_c}{R}$

c. $Z = R + J(X_L - X_c) = R$ at f_o

d. $I = \dfrac{E}{R}$ at f_o

e. $V_L = IX_L = \dfrac{V}{R} X_L = QV$ at f_o

f. f_1, f_2 are the frequencies above and below f_o at which the magnitude is .707 of the voltage at f_o and Series

Circuit Bandwidth, $BW = f_2 - f_1$ also $BW = \dfrac{f_o}{Q}$.

(g) at f_o, $P = I^2 R$.

These relationships and others are displayed below.

Figure 6.41

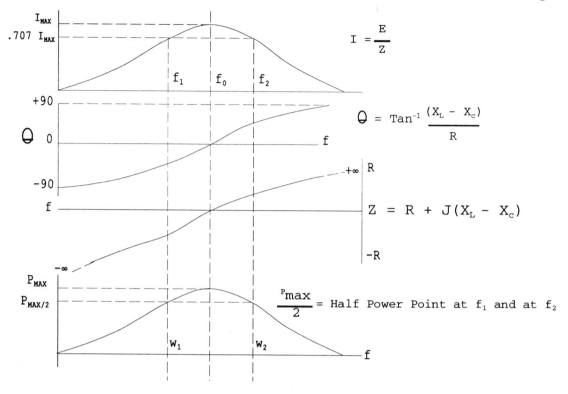

Example

A series resonant circuit consists of a coil L = 4mH, a resistor R = 5Ω and a capacitor C = 0.1u f. What is the resonant frequency of the circuit?

Solution

$$f_o = \frac{1}{2\pi\sqrt{LC}} = \frac{1}{2\pi\sqrt{4 \times 10^{-3} \times 0.1 \times 10^{-6}}}$$

$$f_o = 7.95 \text{ KHz}$$

It is of interest to note that if V were 100 volts then at f_o

$$I = \frac{100}{5} = 20A$$

Also $Q = \dfrac{X_L}{R} = \dfrac{2 \times 3.14 \times 7950 \times 4 \times 10^{-3}}{5}$

$$Q = 40 \text{ at } f_o$$

And

$$V_R = IR = 20 \times 5 = 100 \text{ volts}$$

$$V_L = V_L = QV = 40 \times 100 = 4000 \text{ volts}$$

At Series Resonance the voltage rise across the coil is equal and opposite to the voltage across the capacitor.

Parallel Resonance

In a circuit consisting of three parallel elements, R, L, and C (Fig. 6.42) the resonant frequency is approximately equal to

$$f_o = \frac{1}{2\pi\sqrt{LC}}$$

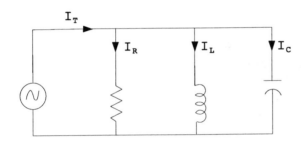

Figure 6.42

In the Tank Circuit, it is convenient to express Z in terms of Y.

$$Y = G + J(B_c - B_L)$$

and at resonance the following terms and circuit relationships are valid.

$$Z_o = \frac{L}{CR} = QX_L = Q^2R = \frac{X_L^2}{R} = \frac{X_L X_c}{R}$$

and $Z_o = \text{Max at } f_o$

$$I_T = \frac{V}{Z_o} \qquad I_L = I_c = QI_T$$

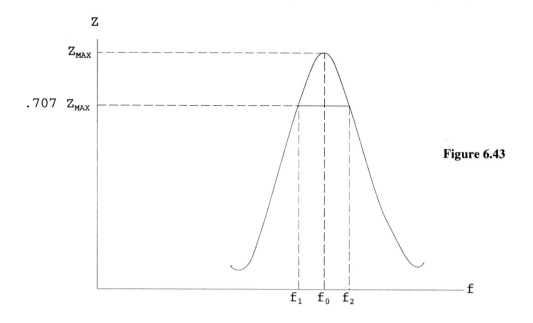

Figure 6.43

SIMPLE FILTERS AND ATTENUATORS

A filter is a network designed to discriminate between frequency regions. A pass band has substantially constant amplitude response. A reject band has low amplitude response.

High-Pass Filter (Fig. 6.44)

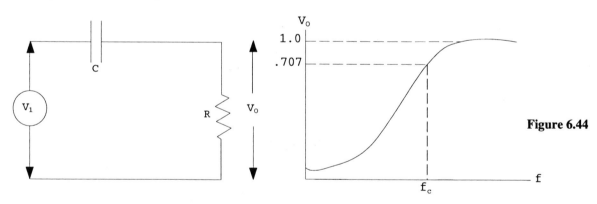

Figure 6.44

$$K = \frac{V_o}{V_i} = \frac{1}{\sqrt{1 + \left(\frac{X}{R}\right)^2}}$$

$$\text{for } |x| = R \;;\; \left|\frac{V_o}{V_i}\right| = .707$$

$$\text{at } |x| = R \;;\; f_c = \frac{1}{2\pi RC}$$

Low-Pass Filter (Fig. 6.45)

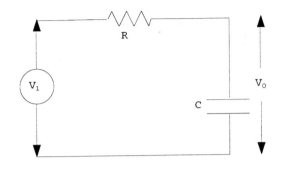
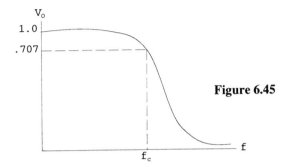

Figure 6.45

$$K = \frac{V_o}{V_i} = \frac{1}{\sqrt{1 + \left(\frac{R}{X}\right)^2}}$$

$$\text{for } |x| = R \;;\; \frac{V_o}{V_i} = .707$$

$$\text{at } |x| = R \;;\; f_c = \frac{1}{2\pi RC}$$

Band Pass Filter Characteristic (Fig. 6.46)

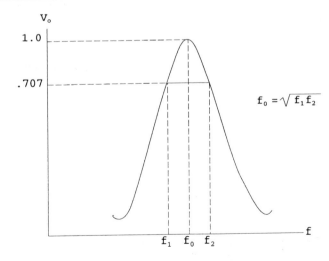

$f_0 = \sqrt{f_1 f_2}$

Figure 6.46

Band Reject Filter Characteristic (Fig. 6.47)

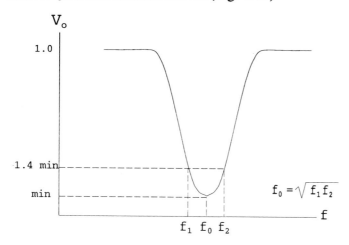

Figure 6.47

$$f_0 = \sqrt{f_1 f_2}$$

An attenuator is a network designed to isolate one network from another but preserve a match between networks (Impedance Match) (Fig. 6.48).

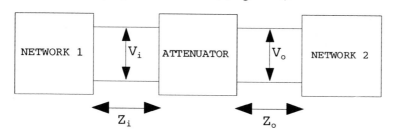

Figure 6.48

Simple L Pad (Fig. 6.49)

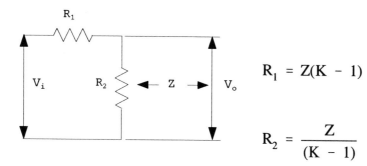

$$R_1 = Z(K - 1)$$

$$R_2 = \frac{Z}{(K - 1)}$$

Figure 6.49

$$\frac{V_o}{V_i} = \frac{R_2}{R_1 + R_2} = K$$

Amplitude loss $K = 20 \, \text{Log}_{10} \dfrac{V_o}{V_i} = 20 \, \text{Log}_{10} \dfrac{R_2}{R_1 + R_2}$

Example

Given an attenuator as shown in Figure 6.49 with K = 12 db. What is the ratio V_o to V_i?

Solution

$$\text{Attenuation } K = \frac{V_o}{V_i}$$

$$20 \log_{10} \frac{V_o}{V_i} = 12 \text{ db}$$

$$\frac{V_o}{V_i} = \text{anti log} \frac{12}{20} = 3.98$$

A decibel expressed as db, is defined as shown below:

$$db = 20 \log \frac{V_2}{V_1} = 10 \log \frac{P_2}{P_1} = 20 \log \frac{I_2}{I_1}$$

where P_1 = input circuit power or voltage or current
P_2 = output circuit power or voltage or current

Neper is the power ratio expressed in Naperian logs.

$$\text{Neper} = \frac{1}{2} \ln \frac{P_2}{P_1} = \ln \frac{V_2}{V_1} = \ln \frac{I_2}{I_1} = 8.686 \text{ db}$$

PRACTICE PROBLEMS WITH SOLUTIONS

1. A 600 volt supply provides a system with a load of 400 amperes. The system is comprised of three parallel elements. If two of the elements are each 6 ohms, determine the resistance of the third element.

   (A) 9Ω   (B) 6Ω   (C) 3Ω   (D) 1.5Ω   (E) 1.0Ω

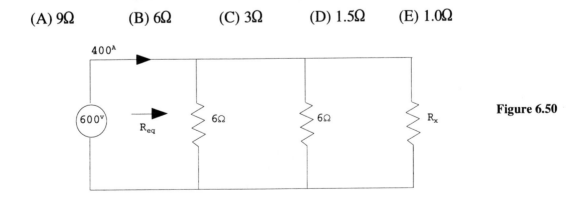

Figure 6.50

$R_{EQ} = 600/400 = 1.5$

$$\frac{1}{R_{EQ}} = \frac{1}{1.5} = \frac{1}{6} + \frac{1}{6} + \frac{1}{R_x}$$

$R_x = \frac{1}{.333} = 2.9 \approx 3\Omega$ The answer is (C).

2. Determine the equivalent admittance of a coil with a Figure 6.of merit equal to 1.0 and a resistance of 6 ohms.

   (a) $0.818 \angle -45° \mho$ (b) $0.118 \angle 45° \mho$ (c) $60 - J60 \mho$

   (d) $0.118 \angle -45° \mho$ (e) $.818 \angle +45° \mho$

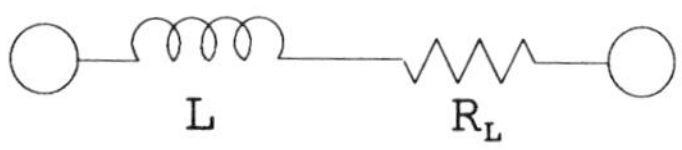

   [diagram: L — R_L]

   $Q = 1.0, R = 6\Omega$

   $Q = \frac{X_L}{R}$ and $X_L = QR = 1.0 \times 6 = 6\Omega$

   $Z = R + JX_L = 6 + J6 = 8.48 \angle 45° \Omega$

   $Y = \frac{1}{Z} = \frac{1}{8.48 \angle 45°} = 0.118 \angle -45° \mho$ The answer is (D)

3. An electrical motor draws 2 amperes of current from a 120 volt line. Determine the power factor if the motor dissipation is 200 watts.

   (A) 40°   (B) 1.2°   (C) 0.556   (D) 0.833   (E) 0.707

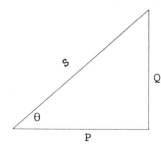

Figure 6.51

P = 200 watts

S = 120V x 2A = 240VA

$pf = \frac{P}{S} = \frac{200}{240} = 0.833$ The answer is (D).

4. Capacitors of 5u F, 10u F and 12u F are connected in series across a 300 volt source. What is the voltage across the 5 u F unit?

(A) 156   (B) 261.7   (C) 300   (D) 1620   (E) 1920

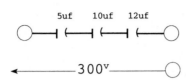

Q Total = C Total x V Total

$Q_T = Q_1 = Q_2 = Q_3$

$V_T = V_1 + V_2 + V_3$

$V_T = \dfrac{Q_1}{C_1} + \dfrac{Q_2}{C_2} + \dfrac{Q_3}{C_3} = \dfrac{Q_T}{C_T}$

$\dfrac{1}{C_T} = \left(\dfrac{1}{5} + \dfrac{1}{10} + \dfrac{1}{12}\right) \dfrac{1}{10^{-6}} = .383 \times 10^{+6}$

$C_T = \dfrac{1}{.383 \times 10^{+6}} = 2.6 \text{ uf}$

$Q_T = Q_1$ and $C_T V_T = C_1 V_1$

$$V_T = \left(\dfrac{C_T}{C_1}\right) V_T = \left(\dfrac{2.6 \text{ uf}}{5 \text{ uf}}\right) 300$$

$V_1 = 156$ volts   The answer is (A)

5. A power plant uses 10KW at a pf of 0.85. It is desired to correct the pf to 0.90. Determine the correction in power needed to improve the pf.

(A) 1400 vars   (B) 4800 vars   (C) 1400 watts

(D) 6200 vars   (E) 4800 watts

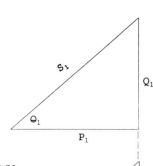
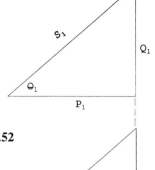

Figure 6.52

$P_1 = 10KW$

$Q_1 = P_1 \tan \theta_1 = 10KW \tan 31.8° = 6.2$ KVARS

$pf_1 = 0.85 \quad \theta = 31.8°$

$P_2 = P_1 = 10KW$

$pf_2 = 0.9 \; ; \; \theta_2 = 25.84°$

$Q_2 = P_1 \tan \theta_2 = 10KW \tan 25.84°$

$Q_2 = 4.8$ Kvars

Correcting Power $Q_1 - Q_2 = (6.2 - 4.8)$ Kvars or 1400 vars The answer is (A).

6. A rectanglar bus bar measures one half inch by one eighth inch in cross-section. The bus bar is to be replaced by a copper cable of equivalent cross-sectional area. Determine the resistance of 1000 ft of cable if the resistivity is 10.37 ohms per circular mil foot.

(a) 0.13Ω (b) 1.3Ω (c) 0.013Ω (d) 13Ω (e) 13 x 10^{-4}Ω

Bus cross-section area

$$1/8" \times 0.5"$$

Area bus = .0625 in.^{2} or 62,500 mil^2

Area circular mils = $\dfrac{\text{Area Mil}^2}{.7854}$

$A_{CM} = \dfrac{62500}{.7854} = 79580$ CM

$R = \dfrac{\rho L}{A} = \dfrac{(10.37)(1000)}{79580} = 0.13Ω$ The answer is (A).

7. In the circuit shown in Fig. 6.53, what is the resistance of the load for maximum power transfer?

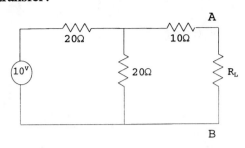

Figure 6.53

(A) 10Ω (B) 20Ω (C) 30Ω (D) 40Ω (E) 50Ω

Use Theveiun Theorem at A-B

$$Z_{Th} = 10 + \frac{20 \times 20}{20 + 20} = 20\Omega$$

$$E_{Th} = \frac{20}{40} \times 10 = 5V$$

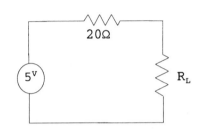

Figure 6.54

For maximum Power $R_L = 20\Omega$ The answer is (B).

8. A series circuit resonates at 5KHZ. The coil has a Q of 50 and resistance of 100. What is the circuit capacitance?

(A) 1600uu f   (B) 4600uu f   (C) 6380uu f   (D) 638u f

(E) 46 uf

$$f_o = \frac{1}{2\pi\sqrt{LC}} \quad ; \quad Q = \frac{X_L}{R}$$

$X_L = QR = 50 \times 100 = 5000$

$X_L = 2\pi fL = 5000 = (2)(3.14)(5000) L$

$L = 0.159H$

$$f_o^2 = \frac{1}{(6.28)^2 LC} \quad C = \frac{1}{(6.28)^2(5000)^2(0.159)} = 6380 \times 10^{-12} F$$

or 6380 uu f   The answer is (C).

9. A high pass filter has a 3 db roll off frequency of 1.5 KHZ. If the filter resistance is 1MΩ, what is the capacitance?

   (A) 106 u f   (B) 916 u f   (C) 106 uu f   (D) 916 uu f

   (E) none of these

   High Pass filter

   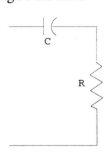
   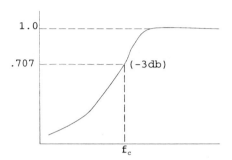

   Figure 6.55

   $f_c = \dfrac{1}{2\pi RC}$ $C = \dfrac{1}{2\pi Rf} = \dfrac{1}{(6.28)(1500)(10^6)} = 106\text{uuf}$ The answer is (C)

10. A Tee network consists of a set of resistors each of value 3 ohms. The equivalent P_I set of resistors would each be of value ---.

    (A) 3Ω   (B) 6Ω   (C) 9Ω   (D) 1Ω   (E) 12Ω

    Tee = Y ; P_I = Δ

    ZΔ = 3ZY

    RΔ = 3 R_y = 3 · 3 = 9Ω   The answer is (C).

11. What is the force on a charge of 2×10^{-10} Coulombs placed 3 cm away from a test charge?

    (A) 3×10^4 (B) 2.5×10^{-8} (C) 10×10^{-8}

    (D) 40×10^{-8} (E) $.8 \times 10^{-8}$

    $E = k(Q_1/d^2) = (9 \times 10^9)(2 \times 10^{-10})/(.03)^2 = 18 \times 10^{-1}/9 \times 10^{-4} = 2 \times 10^3$

    $F = EQ = (2 \times 10^3)(2 \times 10^{-10}) = 4 \times 10^{-7} = 40 \times 10^{-8}$ newtons   The answer is (D).

12. A steady current of 10 amperes is maintained for 15 minutes and deposits 2.9068 grams of zinc at the cathode. Determine the equivalent weight of the zinc.

    (A) 467.55   (B) 31.17   (C) 3096.18   (D) 37.37   (E) 46.76

    One Faraday deposits 1 gram-equivalent weight of a substance. 1 Faraday = 96,500 coulombs.

    10 amperes X (15 x 60) seconds = 9000 Coulombs deposits 2.9068 grams then 1 Faraday would deposit

    $$\frac{96,500}{9000} \times 2.9068 = 31.17g$$

    Equivalent weight of zinc, $Z_n = 31.17$ The answer is (B).

13. What is the equivalent inductance of the network shown in Fig. 6.56?

    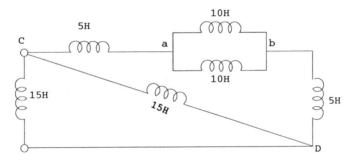

    Figure 6.56

    (A) 15H   (B) 2.5H   (C) 10H   (D) 5H   (E) 12.5H

    at a - b $L_{EQ} = \dfrac{10 \times 10}{10 + 10} = \dfrac{100}{20} = 5H$

    $L = \dfrac{L}{N} = \dfrac{15H}{3} = 5H$ Answer

    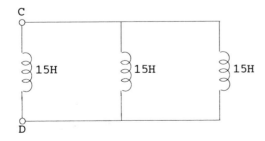

    Figure 6.57

Special Note: The energy stored in the equivalent inductance is
$$W = \tfrac{1}{2} L_{EQ} I^2 \text{ Joules}$$

14. A 120 volt power line has a resistance of 0.2 ohms. If the load is 100 ohms. What is the efficiency of transmission?

   (A) 0.98   (B) 0.985   (C) 0.995   (D) 0.998   (E) 0.988

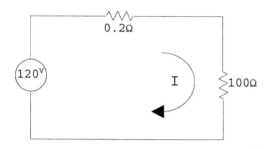

Figure 6.58

$$I = \frac{120}{100.2} = 1.198$$

$P_{IN} = 120 \times 1.198 = 143.71$ VA

$P_o = P_{IN}$ - line loss

$P_o = 143.71 - 1.198^2 \times .2 = 143.71 - .2870$

$P_o = 143.42$ VA

Efficiency $\eta = \dfrac{P_o}{P_i}$

$$\eta = \frac{143.42}{143.71} = .998 \text{ or } 99.8\%$$ The answer is (D)

15. A parallel tank circuit resonates at 1500 Hz. If the 3 db bandwidth is 100 H, what is the coil inductance if its resistance is 10 ohms?

   (A) 15.9 mH   (B) 150 mH   (C) 1.59 mH   (D) 0.159 mH   (e) 15.0 mH

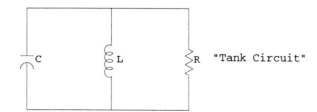

Figure 6.59

$$BW = \frac{F_o}{Q} \quad ; \quad Q = \frac{f_o}{BW} = \frac{1500}{100} = 15 \quad ; \quad Q = \frac{X_L}{R}$$

$X_L = QR = 15 \times 10 = 150$

$2\pi fL = 150 \quad ; \quad L = \dfrac{150}{6.28 \times 1500} = .0159H \quad$ The answer is (A)

APPENDIX 6.1

Selected Unit Relationships

Parameter	Conversion
Length	1 kilometer = 0.6214 miles 1 meter = 3.28 feet
Area	1 square mile = 640 acres 1 acre = 43,560 square feet 1 square meter = 10.76 square feet
Volume	1 Liter = 0.264 US gal = 1.057 quart = 0.03532 ft^3 = 1000 milliliters
Speed	1 kilometer/hour = 0.6214 miles/hour
Force	1 Newton = 0.2248 lb = 10^5 dynes (CGS)
Energy	1 Joule = 1 Newton-meter (N-m) = 10 ERGS (CGS) = 0.2389 1 calorie = 0.736 ft-lb = 0.0009481 BTU
Power	1 watt = 1 Joule/sec = 10^7 ergs/sec (CGS) = 0.2389 calories/sec = 1 horsepower/746 1 horsepower (hp) = 746 watts = 33,000 ft-lbs/min
Pressure	1 atmosphere (atm) = 29.92mm Mercury = 1.013 x 10^5 Nt/m^2 = 14.7 Lbs/in^{-2}

Physical Constants

Speed of light in free space = 3 x 10^8 m/sec.
Normal acceleration due to gravity = 9.807 m/sec.^{2}
Avogadro's number = 6.025 x 10^{23} particles/mole
Stefan-Boltzmann Constant = 5.6686 x 10^{-8} watt/m^2 per K
Electron - volt = 1.6021 x 10^{-19} Joule

APPENDIX 6.2

Review of Complex Numbers

By definition

$$j = \sqrt{-1}$$

Any number, z, can be represented by a real part x, and an imaginary part y and $z = x + jy$ can be plotted in a rectangular coordinate system as shown in Fig. 6.60.

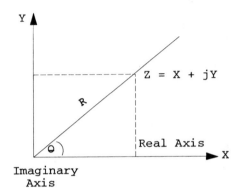

Figure 6.60

The number can also be plotted in a polar coordinate system r, θ where

$$r = \sqrt{x^2 + y^2} \; ; \; \theta = \text{Tan}^{-1} \frac{y}{x}$$

The number z then is $z = x + jy = r\underline{|\theta} = r\cos\theta + j\sin\theta$ note that $x = r\cos\theta$ & $y = r\sin\theta$

Still another form is the exponential form $z = r\,e^{j\theta}$ since $e^{j\theta} = \cos\theta + j\sin\theta$ (Euler's Theorem)
$e^{j\theta} = 1\underline{|\theta}$

Some examples:

$$e^{j\pi/2} = 1 \underline{|\pi/2} = j$$

$$e^{j\pi} = 1 \underline{|\pi} = -1$$

$$e^{j3\pi/2} = 1 \underline{|3\pi/2} = -j$$

$$e^{j2\pi} = 1 \underline{|2\pi} = 1 \quad \text{and}$$

$$e^{j(\theta \pm 2\pi)} = e^{j\theta}$$

Addition, Subtraction, and Multiplication of Complex Numbers

Let $z_1 = x_1 + jy_1 = r_1 \underline{|\theta_1} = r_1 e^{j\theta_1}$ and

$z_2 = x_2 + jy_2 = r_2 \underline{|\theta_2} = r_2 e^{j\theta_2}$

Then

$z_1 + z_2 = (x_1 + x_2) + j(y_1 + y_2)$

$z_1 - z_2 = (x_1 - x_2) + j(y_1 - y_2)$

$z_1 \times z_2 = r_1 \times r_2 \underline{|\theta_1 + \theta_2} = r_1 \times r_2\, e^{j(\theta_1 + \theta_2)}$

$\dfrac{z_1}{z_2} = \dfrac{r_1}{r_2} \underline{|\theta_1 - \theta_2} = \dfrac{r_1}{r_2}\, e^{j(\theta_1 - \theta_2)}$

Also

$z_1 \times z_2 = (x_1 + jy_1)(x_2 + jy_2)$

$z_1 \times z_2 = (x_1 x_2 - y_1 y_2) + j(x_1 y_2 + x_2 y_1)$

The Conjugate of complex numbers

If $z = x + jy = re^{j\theta}$ then its conjugate is $z^* = x - jy = re^{-j\theta}$ (by definition. See Fig. 6.61.)

(Read z* as "z conjugate")

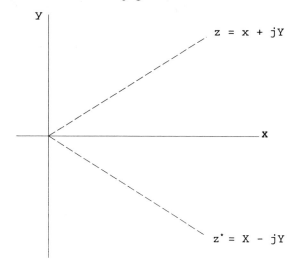

Figure 6.61

Re (z) = 1/2 (z + z*) (Read "Re (z)" as "real part of z")

Im (z) = (1/2 j) (z - z*) (Read "Im (z)" as imaginary part of z")

From Euler Theorem

$$e^{j\theta} = \cos\theta + j\sin\theta$$

$$e^{-j\theta} = \cos\theta - j\sin\theta$$

The following can be obtained:

$$\cos\theta = \frac{e^{j\theta} + e^{-j\theta}}{2} \quad ; \quad \sin\theta = \frac{e^{j\theta} - e^{-j\theta}}{2j}$$

Consider the function e^{jwt} (Fig. 6.62).

$$e^{jwt} = e^{j\theta}$$

where $\theta = wt$

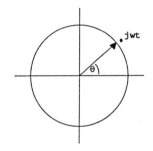

Figure 6.62

Because e^{jwt} can be viewed as a CCW rotating vector, then $e^{j(wt + \theta)}$ can also be viewed as rotating, but leading e^{jwt} by θ; correspondingly $e^{j(wt - \theta)}$, lags e^{jwt} by θ.

A useful property of e^{jwt} is that its derivatives also contain e^{jwt}.

Note:

$$\frac{d}{dt} e^{jwt} = jw\, e^{jwt}$$

$$\frac{d^2}{dt^2} e^{jwt} = (jw)^2\, e^{jwt}$$

$$\frac{d^3}{dt^3} e^{jwt} = (jw)^3\, e^{jwt}$$

APPENDIX 6.3

Algebra of Determinants

Case I. Two-Loop Mesh (Fig. 6.63),

Let Z = R

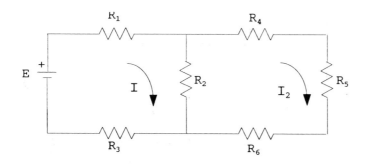

Figure 6.63

(1) $E = I_1 R' - I_2 R_2$; $R' = R_1 + R_2 + R_3$

(2) $0 = -I_1 R_2 + I_2 R''$; $R'' = R_4 + R_5 + R_6 + R_2$

(3) $I_1 = \dfrac{N_1}{\Delta}$

(4) $I_2 = \dfrac{N_2}{\Delta}$

(5) $N_1 = \begin{vmatrix} E & -R_2 \\ 0 & R'' \end{vmatrix}$

$N_1 = +ER'' - (-R_2)(0) = ER''$

(6) $N_2 = \begin{vmatrix} R' & E \\ -R_2 & 0 \end{vmatrix}$

$N_2 = (R')(0) - (E)(-R_2) = ER_2$

(7) $\Delta = \begin{vmatrix} R' & -R_2 \\ -R_2 & R'' \end{vmatrix}$

$$\Delta = R'R'' - (-R_2)(-R_2)$$

$$\Delta = R'R'' - R_2^2$$

Case II Three-Loop Mesh (Fig. 6.64)

Let $Z = R$

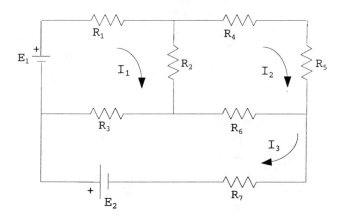

Figure 6.64

(1) $E_1 = I_1R' - I_2R_2 - I_2R_3$; $R' = R_1 + R_2 + R_3$

(2) $0 = -I_2R_2 + I_2R'' - I_3R_6$; $R'' = R_2 + R_4 + R_5 + R_6$

(3) $E_2 = -I_1R_3 - I_2R_6 + I_3R'''$; $R''' = R_3 + R_6 + R_7$

(4) $I_1 = \dfrac{N_1}{\Delta}$

(5) $I_2 = \dfrac{N_2}{\Delta}$

(6) $I_3 = \dfrac{N_3}{\Delta}$

(7) $N_1 = \begin{vmatrix} E_1 & -R_2 & -R_3 \\ 0 & R'' & -R_6 \\ E_2 & -R_6 & R''' \end{vmatrix}$

$$(8)\ N_2 = \begin{vmatrix} R' & E_1 & -R_3 \\ -R_2 & 0 & -R_6 \\ -R_3 & E_2 & R''' \end{vmatrix}$$

$$(9)\ N_3 = \begin{vmatrix} R' & -R_2 & E_1 \\ -R_2 & R'' & 0 \\ -R_3 & -R_6 & E_2 \end{vmatrix}$$

$$(10)\ \Delta = \begin{vmatrix} R' & -R_2 & -R_3 \\ -R_2 & R'' & -R_6 \\ -R_3 & -R_6 & R''' \end{vmatrix}$$

Evaluate N_1, N_2, N_3 in the manner as shown for Δ.

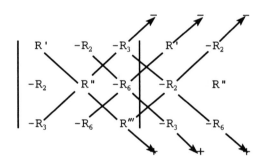

$\Delta = + (R'R''R''') + (-R_2)(-R_6)(-R_3) + (-R_3)(-R_2)(-R_6)$

$\quad - (-R_3)(R'')(-R_3) - (-R_6)(-R_6)(R') - (R''')(-R_2)(-R_2)$

ELECTRICITY II

THREE-PHASE CIRCUITS

Three phase circuits are characterized by three voltages (or currents) generated at 120° intervals as shown in Fig. 6.65.

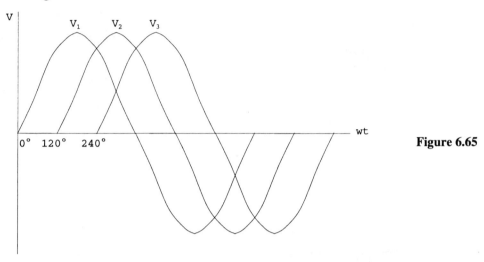

Figure 6.65

$V_1 = V \sin(wt + \theta°)$; $v_2 = V \sin(wt + 120°)$; $v_3 = V \sin(wt + 240°)$

The two basic generator configurations are the wye (Y) and the Delta (Δ). Usually the impedance load is connected like the generator. If not than a y - Δ or Δ - y conversion can be used if needed to convert impedence load.

Delta Circuit (Fig. 6.66)

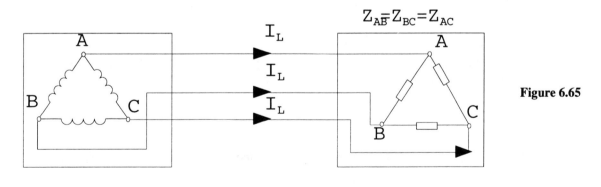

Figure 6.65

3ØΔ   3Ø Impedance Load

In the Δ system generator V coil = V line = $V_{AB} = V_{BC} = V_{CA}$

The voltages are vectorially separated by intervals of 120°, i.e. V_{CA} 120°, V_{BC} 0°, V_{AB} 240° as shown in the vector triangle of Fig. 6.67.

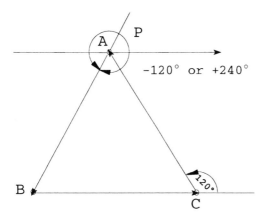

Figure 6.67

This arrangement is an ACB system, meaning that an observer at point P will see the sequence as a counterclockwise rotation.

In the Δ system the coil or phase currents are related to the line current as follows.

$$I_{CA} = \frac{I_L}{\sqrt{3}} \quad I_{BC} = \frac{I_L}{\sqrt{3}} \quad I_{AB} = \frac{I_L}{\sqrt{3}}$$

The power, in a balanced system, is

$$P_\Delta = \sqrt{3} \; V_L \; I_L \; \cos \theta \quad \text{watts}$$

Example

A three phase 120 volt system supplies a balanced three phase delta load comprised of 10 ohm resistors. Determine the line currents and total delivered power.

Solution

Using Fig. 6.66 and arbitrarily assuming an ACB system (does not matter for this problem) with the generator configured as Δ.

$$I_A = I_{AB} + I_{AC} = I_{AB} - I_{CA}$$

$I_B = I_{BA} + I_{BC} = I_{BC} - I_{AB}$

$I_C = I_{CB} + I_{CA} = -I_{BC} + I_{CA}$

$I_{AB} = \dfrac{V_{AB}}{Z} = \dfrac{120 \angle 240°}{10} \qquad I_{BC} = \dfrac{V_{BC}}{Z} = \dfrac{120 \angle 0°}{10} \qquad I_{CA} = \dfrac{V_{CA}}{Z} = \dfrac{120 \angle 120°}{10}$

Line Currents

$I_A = 12 \angle 240° - 12 \angle 120° = -6 - J10.4 + 6 - J10.4 = 20.8 \angle -90°$

$I_B = 12 \angle 0° - 12 \angle 240° = 12 - J0.0 + 6 + J10.4 = 20.8 \angle 30°$

$I_C = 12 \angle 0° + 12 \angle 120° = -12 - J0.0 - 6 + J10.4 = 20.8 \angle 150°$

Power Delivered

$P_T = P_A + P_B + P_C = \dfrac{V_{AB}^2}{10} + \dfrac{V_{BC}^2}{10} + \dfrac{V_{CA}^2}{10} = \dfrac{3}{10}(120)^2 = 4320 \text{ watts}$

Power Check

$P_L = \sqrt{3}\, V_L\, I_L \cos\theta = \sqrt{3}\,(120)\,(12 \times \sqrt{3})\,(1) = 4320 \text{ watts}$

The voltage - current vector diagram of Fig. 6.68 illustrates the various phase relationships.

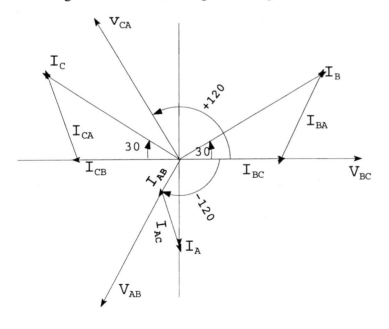

Figure 6.68

WYE Circuit

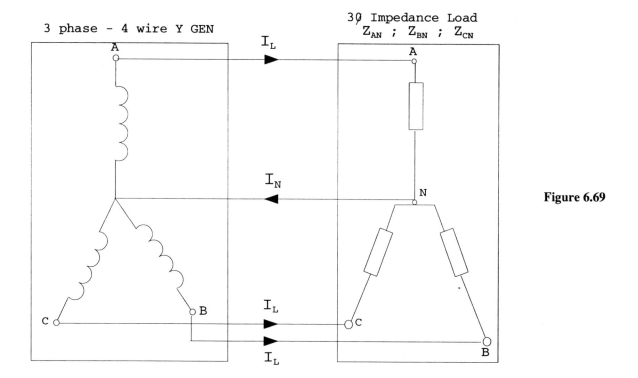

Figure 6.69

In the Y system generator, line current equals coil current (Fig. 6.69) and $I_A + I_B + I_C = I_N$ where N = neutral. The voltage phasors are separated by intervals of 120°; $V_{AB} \,\underline{|+120°}$, $V_{CB} \,\underline{|0°}$, $V_{AC} \,\underline{|-120°}$, as shown in the vector triangle of Fig. 6.70 for an ABC system.

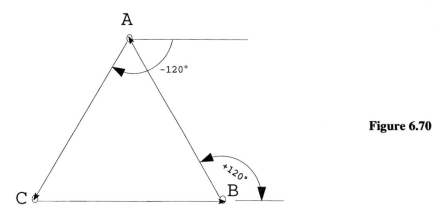

Figure 6.70

In the WYE system the coil or phase voltages are related to the line voltage as follows.

$$V_{coil} = \frac{V_L}{\sqrt{3}} \text{ in magnitude}$$

The voltage to neutral is frequently used as shown in Fig. 6.71 where

$$V_{coil} = \frac{V_L}{\sqrt{3}} \angle 90°, \quad V_{BN} = \frac{V_L}{\sqrt{3}} \angle -30°, \quad V_{CN} = \frac{V_L}{\sqrt{3}} \angle -150°$$

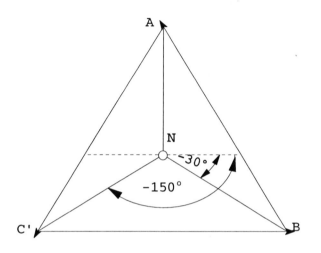

Figure 6.71

The power, in a balanced system, is

$$P_y = \sqrt{3}\, V_L\, I_L \cos\theta \text{ watts. Note it is the same as } P_\Delta.$$

Example

A three-phase four-wire 208 volt system provides power to a load comprised of 3 y connected resistors; 6, 8, and 10 ohms. Determine the line currents, the neutral current and the delivered power.

Solution: Assume a Y, ABC connected generator as shown in Fig. 6.69.

$$I_A = \frac{V_{AN}}{R_{AN}} = \frac{120}{6} \angle 90° = 20 \angle 90° \text{ Amperes}$$

$$I_B = \frac{V_{BN}}{R_{BN}} = \frac{120}{8} \underline{|-30°} = 15 \underline{|-30°} \text{ Amperes}$$

$$I_C = \frac{V_{CN}}{R_{CN}} = \frac{120}{10} \underline{|-150°} = 15 \underline{|-150°} \text{ Amperes}$$

Neutral Current

$$I_A = I_A + I_B + I_C$$

$$I_N = (0.0 + J20) + (13 - J7.5) + (10.4 - J6)$$

$$I_N = 2.6 + J6.5 = 7.0 \underline{|68.0°}$$

If the loads were balanced ($R_A = R_B = R_C$) $I_N = 0.0$

Delivered Power

$$P_A = I_A^2 R_A = 20^2 \times 6 = 2400$$

$$P_B = I_B^2 R_B = 15^2 \times 8 = 1800$$

$$P_C = I_C^2 R_C = 12^2 \times 10 = \underline{1440}$$

$$P_T = P_A + P_B + P_C = 5640 \text{ watts}$$

A system called the two wattmeter method can be used to measure total power. In this method a wattmeter is put into one line and power measured. Then the power is measured in a second line. In this example

$$P_{AB} = (208)(20) \cos \theta\,]_{90°}^{120°} \text{ and } \theta = 30°$$

$$P_{AB} = 3603 \text{ watts}$$

$$P_{BC} = (208)(12) \cos \theta\,]_{-180°}^{-150°} \text{ and } \theta = 30°$$

$$P_{BC} = 2161 \text{ watts}$$

$P_T = 3603 + 2161 = 5764$ compared with $P = 5640$ for total power for three phases the comparison yields a difference of 124 out of 5640 or about a 2% error. The two watt meter method can be expressed as

Tan $\theta = \sqrt{3}\,\dfrac{W_2 - W_1}{W_1 + W_2}$ if the load is balanced (Fig. 6.72)

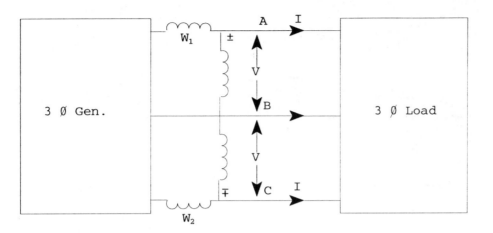

Figure 6.72

MAGNETICS

In the simple electric circuit, the resistance is calculated by Ohm's Law, as the ratio of V to I.

The analogy in a magnetic circuit is that the reluctance is the ratio of magnetic force, $\mathcal{F}$, to the flux flow, θ, as illustrated in Fig. 6.73.

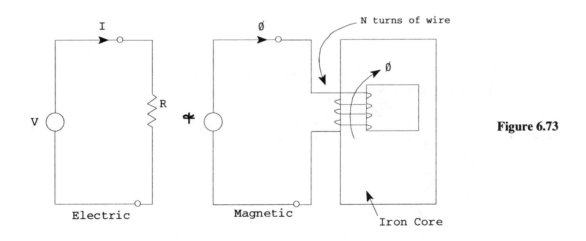

Figure 6.73

Many of the concepts of magnetics can be illustrated by comparison to the electric terminology as shown in Table 6.1.

TABLE 6.1

Electric/Magnetic Analogies

Term	Electric	Magnetic
Resistance	$R = \dfrac{\rho L}{A}$	$\mathcal{R} = \dfrac{L}{\mu A}$
Current density	$J = \dfrac{I}{A} = \dfrac{E}{\rho}$	$B = \dfrac{\phi}{A} = \mu H$
Field intensity	$\epsilon = \dfrac{E}{L} = \dfrac{E}{2\pi r}$	$H = \dfrac{\mathcal{F}}{L} = \dfrac{\mathcal{F}}{2\pi r}$
Ohm's Law for potential drop	$V_{1-2} = IR_{1-2}$	$\mathcal{F} = \phi \mathcal{R}_{1-2}$

Magnetic Units and Terminology are given in Appendix 6.4.

Example

An iron core has 300 turns of wire (Fig.6.74). The cross-sectional area is 0.0001 m² square and the magnetic path is 2 m. Determine the current necessary to develop a magnetic flux of 1.5×10^{-4} webers if the permeability is given as 0.005. What is the reluctance of the magnetic path?

Solution

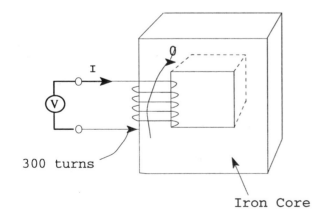

Figure 6.74

$$B = \frac{\emptyset}{A} = \frac{1.5 \times 10^{-4}}{0.0001} = 1.5 \; \frac{\text{webers}}{m^2} \quad \text{(MKS)}$$

$$H = \frac{B}{\mu} = \frac{1.5}{0.005} = 300 \; \frac{\text{ampere turns}}{m} \quad \text{(MKS)}$$

$$H = \frac{NI}{L} \; \text{and} \; I = \frac{HL}{N} = \frac{300 \times 2}{300} = 2 \; \text{amperes}$$

The magnetic circuit is equivalent to a simple series circuit.

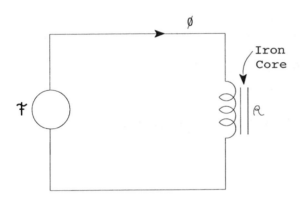

Figure 6.75

$$\Re\text{el} = \frac{L}{\mu A} = \frac{2}{(0.005)(0.0001)} = 4 \times 10^6 \; \text{rels.}$$

If the core were of two dissimilar materials the total reluctance is evaluated like two resistors in series.

$$\Re\text{el}_T = \Re\text{el}_{iron} + \Re\text{el}_{steel}$$

If the circuit had two parallel flux paths then $\frac{1}{\Re\text{el}_T} = \frac{1}{\Re\text{el}_{iron}} + \frac{1}{\Re\text{el}_{steel}}$ evaluated like two resistors in parallel.

TRANSFORMERS

The transformer is an electromagnetic device consisting of two or more windings wound on a core. It can be represented as a four terminal box as shown in Fig. 6.76.

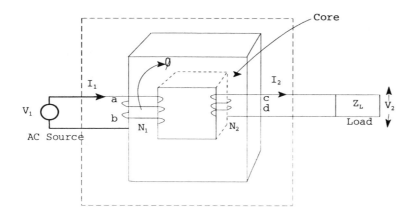

Figure 6.76

Input power, $P_i = V_1 I_1$ volt-amperes and output power, $P_o = V_2 I_2$ volt amperes. N_1 and N_2 are the number of turns of wire in each winding. Usually $a \equiv \dfrac{N_1}{N_2}$ and is called the turns ratio.

In power applications the core is iron and the windings are heavy copper. These components introduce heat losses so that transformer efficiency,

$$\eta = \frac{P_o}{P_i} = \frac{P_i - \Sigma \text{ losses}}{P_i} = \frac{P_o}{P_o + \Sigma \text{ losses}}$$

Transformers may be rated in VA or KVA. For example a single phase 100 KVA transformer, can deliver 100 KW if the power factor is unity. A balanced three phase 100 KVA transformer will also deliver 100 KW if the power factor is unity and the power delivered per phase is 100 KW/3.

The induced voltage in either the input or output side is given by

$$V = 4.44 \, f \, N \, \emptyset_{max} \times 10^{-8} \text{ effective volts}$$

where
- f = frequency in hertz
- n = number of turns
- $\emptyset$ = flux in lines

From V the following ratios can be made

$$\frac{V_1}{V_2} = \frac{I_2}{I_1} = \frac{N_1}{N_2} = a$$

Transformers are used for various purposes. Some are:

1. Power transmission and distribution
2. Impedance changing
3. Voltage and current sensing for relays and indicating and recording instruments
4. Circuit isolation
5. Lighting and control purposes

Some additional terms in use are:

> High voltage winding (or high tension winding)
> Low voltage winding (or low tension winding)
> Step-up transformer, step-down transformer
> Voltage ratio (the ratio of primary to secondary voltages)
> Auto transformer (single winding transformer)
> Tertiary transformer (three winding transformer)

Two basic types of construction of transformers are shown in Fig. 6.77.

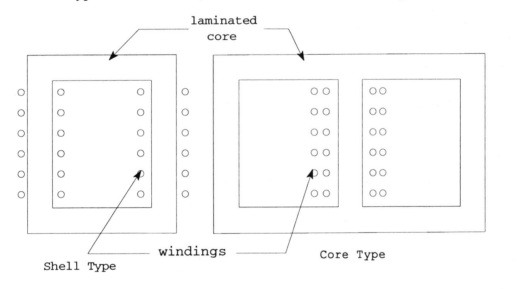

Figure 6.77

Transformer Models

Transformers have winding resistances and leakage reactance (caused by leakage flux). Three fluxes exist in a transformer, they are:

1. Mutual flux, produced by a portion of the primary current, which links N_1 and N_2.

2. Primary leakage flux, produced by the primary current and links N_1 only.

3. Secondary leakage flux, produced by the secondary current and links N_2 only.

Each of these fluxes produces a voltage and each voltage, caused by leakage fluxes, can be considered equivalent to an IX drop, i.e. for the primary:

$$V_1 = 4.44 f \emptyset_m \ N_1 \times 10^{-8} = I_1 X_1$$

where x_1 is called the primary leakage reactance. Similarly for the secondary, x_2 would be the secondary leakage reactance and represented as shown in Fig. 6.78.

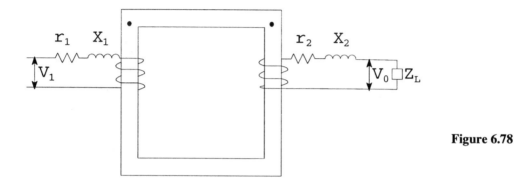

Figure 6.78

The (•) in Fig. 6.78 means that a voltage change is in the same direction.

Impedance Changing

Consider a transformer model connected as shown in Fig. 6.79.

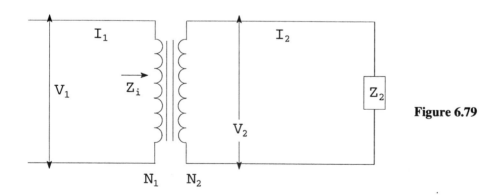

Figure 6.79

$$I_2 = \frac{V_2}{Z_2} \text{ and } I_2 = (I_1)(a)$$

Therefore the load Z_2 reflected from the output winding to the input is

$$Z_2 = \frac{V_2'}{I_2'} = \frac{V_{1/a}}{a\,I_1} = \frac{z1}{a^2}$$

$$Z_1 = a^2 z_2$$

The usual terminology is to state that the load z_L reflected into the primary is given by

$$z_i = a^2 z_L \text{ so that from the primary side the load is multiplied by } a^2 = \left(\frac{N_1}{N_2}\right)^2$$

Transformer Tests

The following tests are performed to determine the parameters of a transformer.

Short Circuit Test (Fig. 6.80)

The input voltage is increased until rated current flows through either winding.

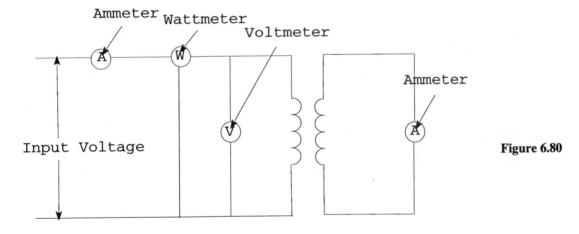

Figure 6.80

The following transformer parameters are obtained:

Copper losses ≈ the wattmeter reading

Equivalent impedance = Zeq. = $\dfrac{V}{I_1}$

Equivalent resistance = Req. = $\dfrac{W}{I_1^2}$

and

Equivalent reactance = Xeq. = $\sqrt{Zeq^2 - Req^2}$

These values are referred to the primary. To refer them to the secondary, multiply by the factor $\frac{1}{a^2}$

Note: Transformer losses consist of:

(a)   copper losses, in primary and secondary; and

(b)   core loss (or iron loss) due to eddy current and hysteresis

Open Circuit Test (Fig. 6.81)

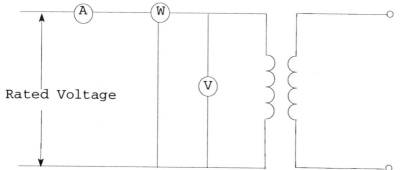

Figure 6.81

In this test the W reading   =   Core loss + I_{NL}^2 Req.

$\approx$ Core loss (I_{NL}^2 Req. usually neglected)

Per-Unit Notation

Per-Unit notations are definitions applied to transformers and other equipments for the purpose of normalizing quantities to a common base. The definitions follow:

Base current in amperes = $\dfrac{\text{base KVA}}{\text{base volt in KV}}$

Base impedance = $\dfrac{\text{base voltage in volts}}{\text{base current in amperes}}$
(in ohms)

$$= \frac{\text{(base voltage in KV)}^2 \times 10}{\text{base KVA}}$$

Base power in KW = base KVA

Per unit impedance of a circuit element $= \dfrac{\text{actual impedance in ohms}}{\text{base impedance in ohms}}$

Quantity = PU quantity x 100

Current, when stated, means line current. Voltage, when stated, means volts to neutral and KVA means KVA per phase.

Example

A step up transformer provides 480 volts at 60 hertz to a 50 ohm load. Determine the equivalent impedance reflected into the primary if the input voltage produces a flux of 1.5×10^{-4} webers through the primary coil of 3000 turns.

Solution

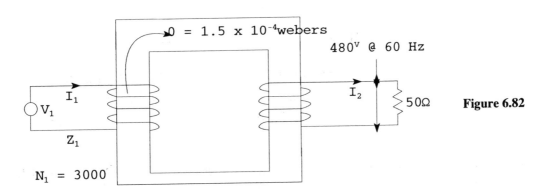

Figure 6.82

$V_1 = 4.44\ f_1\ N_1\ \phi = (4.44)(60)(3000)(1.5 \times 10^{-4}) = 120$ volts

Note that f is usually the same at primary and secondary windings.

$$a = \frac{V_1}{V_2} = \frac{120}{480} = \frac{N_1}{N_2}$$

$$Z_1 = a^2 Z_L = \left(\frac{120}{480}\right)^2 50 = 3.125\Omega$$

Autotransformer

If the windings are serially connected the transformer is called an autotransformer as shown in Fig. 6.83.

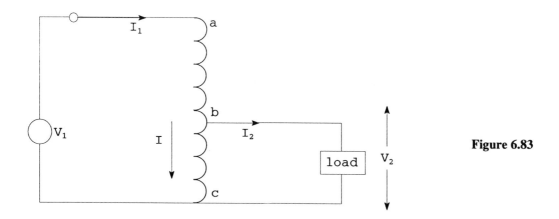

Figure 6.83

N_1 = winding a b c $\qquad$ $a = V_1/V_2 - I_2/I_1 + Nac/Nbc$
N_2 = winding b c $\qquad$ $I = I_1 - I_2$

Voltage, current and load relationships are similar to the transformer with separate windings

Example

A single phase 1:1 transformer is rated at 10KVA, 208 volts. What is the rating when operated as a step-up autotransformer?

Solution

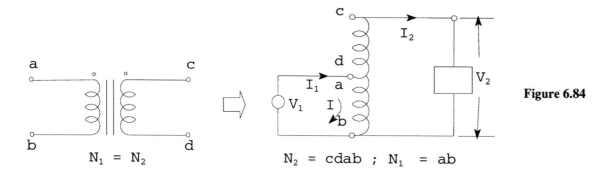

Figure 6.84

$N_1 = N_2$ $\qquad$ N_2 = cdab ; N_1 = ab

$$I_1 = \frac{10{,}000}{208} = 48A \quad I_2 = \frac{10{,}000}{416} = 24A$$

$I = I_1 - I_2 = 24A$

KVA Rating as Autotransformer is:

$V_1 I = 208 \times 24 = 4.992$ KVA (ab) or

$(V_1 - V_2) I_2 = 208 \times 24 = 4.992$ KVA (cd)

ROTATING MACHINES

All rotating electrical machines use the principle of electromagnetic induction, i.e., whenever there is a change in the magnetic flux interlinking a coil there is an induced electrical force. This force is expressed by Faraday's Law as

$$e = -\frac{d(N\emptyset)}{dt}$$ volts with the negative sign indicating that for a closed circuit the current produced by the induced emf will react to oppose the change of flux linkages (Lenz's Law).

In Figure 6.85, a single loop of wire is located in the field of a permanent magnet. If the field has a flux density B, maximum flux will link the loop when the plane of the loop is normal to the direction of the field. For the situation shown

$\emptyset_{max} = BA$ where $A = Ld$

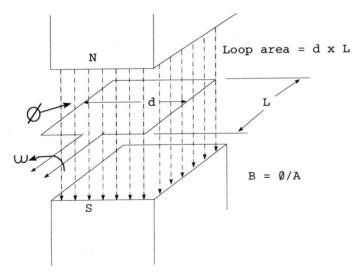

Figure 6.85

As the loop turns, three orthogonal effects occur: F, B, and I. In the generator, mechanical force is applied and electrical output is produced. In the motor electrical input is injected and mechanical output is produced. The effects are best illustrated by Flemming's Rules as shown in Fig. 6.86.

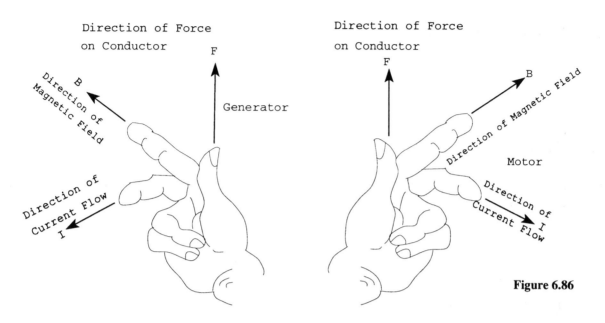

Figure 6.86

Basic Relationships

The general relationship for $\emptyset$ is $\emptyset = \emptyset_{max} \cos \omega \tau$ and since $e = -\dfrac{d\emptyset}{dt}$ then $e = BA\omega \sin \omega \tau$. Which results in $E_{MAX} = BA\omega$ at $\omega \tau = 90°$ using angular velocity, ωx perpendicular distance, r = linear velocity, $v \Rightarrow E_{max} = BL\, v$.

Assuming a fully efficient machine, mechanical power, $T\omega = VI$ the electrical power. By Substitution $T = \emptyset_{max} I \sin \omega t$.

Example

A rectangular coil of 10 turns with each turn measuring 8 in. by 4 in. is placed in a uniform magnetic field of 400 gauss. If the coil is rotated at an angular velocity equivalent to 60 Hz what is the induced voltage? What is the maximum flux encountered for one loop?

Solution

In the CGS system B = 400 gauss. In MKS this is equivalent to $B = 400 \times 10^{-4}$ webers/m². Assume Fig. 6.87 for one loop.

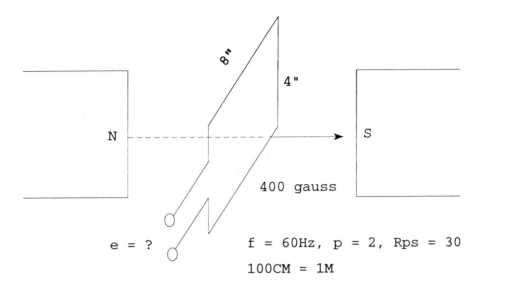

Figure 6.87

$f = 60\text{Hz}, \ p = 2, \ \text{Rps} = 30$

$100\text{CM} = 1\text{M}$

$e = B\,L\,\text{vel} = -\dfrac{d\varnothing}{dt} = AB\omega \sin \omega t$

$e = -\dfrac{d\varnothing}{dt} = \dfrac{(8 \times 2.54)(4 \times 2.54)}{100 \times 100}(400 \times 10^{-4})(2\pi 60)(\sin \omega t)$

$\approx .31 \sin \omega t$ volts for one loop

$e = 3.1 \sin \omega t$ volts for 10 loops

$\varnothing_{max} = A \times B = 8 \times 4 \times 400 = 12{,}800$ lines in English or Engineering Units.

DC GENERATORS

The basic DC generator has the features illustrated in Fig. 6.88.

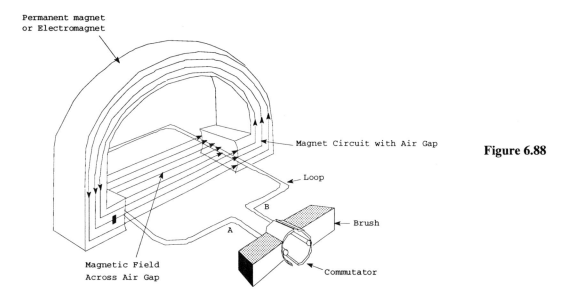

Figure 6.88

Mechanical power rotates the loop in the magnetic field. As the loop turns it cuts through the magnetic flux and is minimum when in the plane of the flux and maximum when perpendicular. The effect is to produce an AC voltage at a speed related to the angular velocity of the loop. The commutator, attached to the loop, commutes the voltage to the stationary brushes at A and B. As the loop transverses through 180° of the 360° turn, A is at the B side and vice versa. The net effect is to output a pulsating DC voltage. By increasing the loops and commutator segments, the pulsating DC amplitude and pulses become less and the output DC level has less ripple as shown in Fig. 6.89.

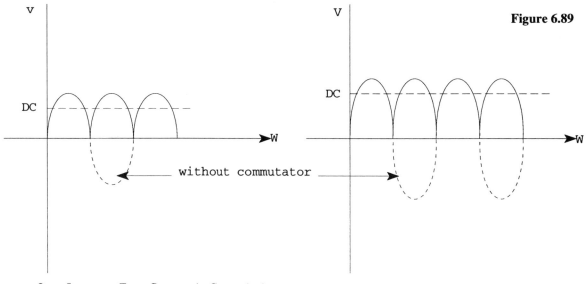

Figure 6.89

One Loop - Two Segment Commutator Two Loop - Four Segment Commutator

A commercial generator will have many distributed alternating N-S poles provided by an electromagnetic system, a multisegmented set of loops with commutator sections called the armature or rotor, and a fixed stator to provide the magnetic field path as illustrated in Fig. 6.90.

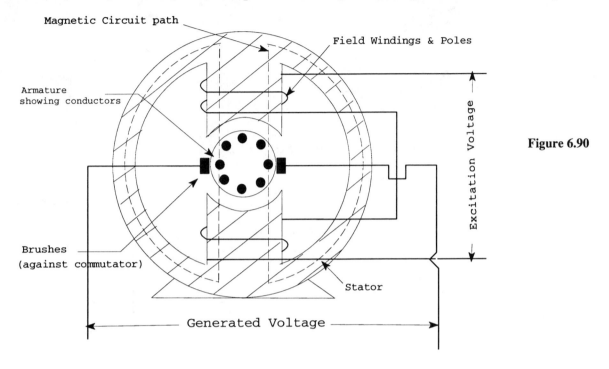

Figure 6.90

Generator Configurations

The generator is operated in any one of three basic configurations as shown in Fig. 6.91.

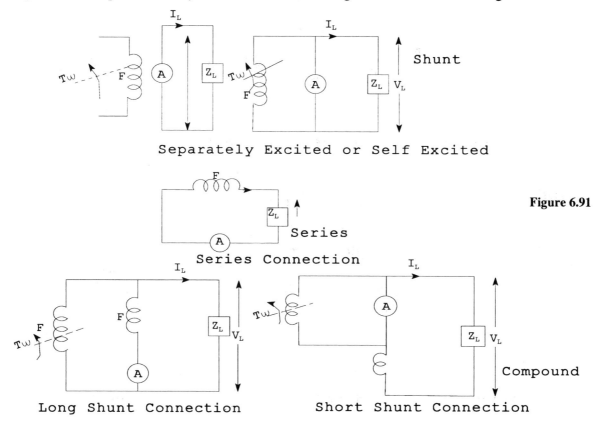

Figure 6.91

In compound machines there are sometimes additional interpoles which are field windings on the stator used to improve generator performance.

External rheostats, in series with the shunt field, are used to adjust field current.

If shunt and series field fluxes are additive compounding is "cumulative."

If shunt and series field fluxes are subtractive compounding is "differential."

The effective value of the output voltage of the generator is $V = K\emptyset_w$ and the torque required, $T = K\emptyset I_A$ to the conductors corresponds to conductor developed voltage $e = BL_w$. The term K is a machine constant $= \dfrac{Zp}{a} \times 10^{-8}$

where z = number of conductors

p = number of poles
a = number of parallel paths through the armature.

If a = 2, winding is Simplex Wave; if a = p winding is Simplex Lap.

w = speed in revolutions per second

Flux ∅ is produced by field current I_f, and plots of V vs I_f at various speeds are called saturation curves. Typical curves are shown in Fig. 6.92.

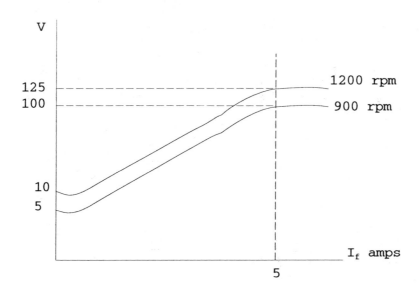

Figure 6.92

A plot of field voltage vs field current on the same plot, called the field resistance line, determines no load voltage for a self excited shunt generator.

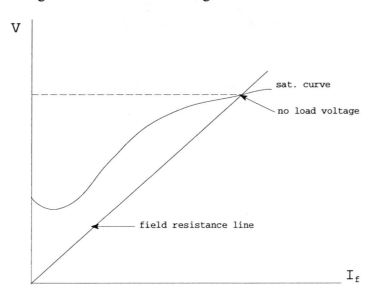

Figure 6.93

When load is added to a basic shunt generator, i.e., a generator without an external voltage regulator, terminal voltage decreases. This is due to armature circuit voltage drop and field degradation caused by armature reaction. This effect can be altered with compound generators since total field strength could increase with load because of the series field winding. Typical terminal voltage vs load curves are shown in Fig. 6.94.

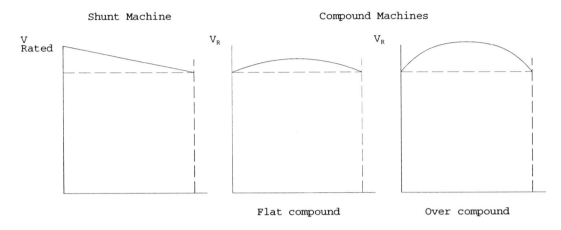

Figure 6.94

Generator Computations

Generally only the resistive values of the various windings are considered. A useful model for the compound generator is shown in Fig. 6.95. It should be noted that the model becomes a series generator if $R_{SH} = \infty$ and a shunt generator if $R_S = 0$. The armature acts to produce a voltage that opposes the condition that causes the rotation. This effect acts like a back emf and prevents the coil voltage from instantly dropping to zero when the coil is cutting no flux lines.

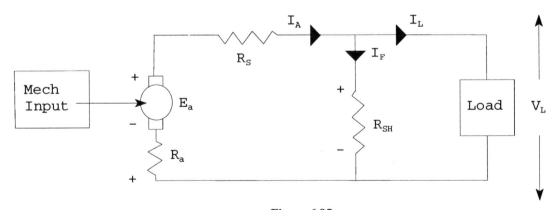

Figure 6.95

By Kirchoff's Rules, $I_A = I_F + I_L$
$V_L = E_A - I_A (R_S + R_a) = I_F R_{SH}$

Example 7

A 10 KW load is supplied with 200 volts by a DC shunt generator driven at 1000 rpm. The generator has rotational losses of 480 watts. Assuming an armature resistance of 0.15 ohms, a field current of 1.48 amperes, and a terminal voltage of 200 volts, find the induced armature voltage, the developed torque and the efficiency.

Solution
In the shunt generator, the series field is zero. An equivalent Shunt Generator model is used.

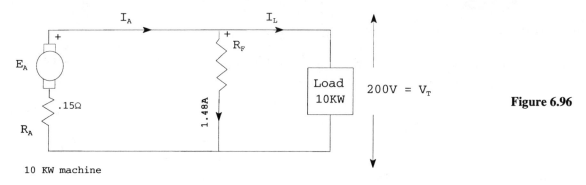

Figure 6.96

a. Induced armature voltage, E_A

$$I_T = E_A - I_A R_A$$

$$E_A = V_T + I_A R_A = V_T + (I_L + I_F) R_A$$

$$I_L = 10,000/200 = 50A$$

$$E_A = 200 + (50 + 1.48) .15 = 207.7 \text{ volts}$$

b. Developed torque, T

$$T = \frac{E_A I_A}{w} = \frac{(207.7)(51.48)}{2\pi \frac{1000}{60}} = \frac{10,692}{105} = 102 \text{ N - M}$$

c. Efficiency, η

$$\eta = \frac{P_o}{P_i} = 1 - \frac{\text{Sum of the losses}}{P_i}$$

Sum of the losses = (480 + 396 + 296) watts = 1172 watts

$$\eta = 1 - \frac{1172}{11,172} = 1 - .105 = .895$$

$$\eta = 89.5\%$$

The total torque required is T = 11174/105 = 106.4 N-M. The torque needed to overcome rotational losses is 106.4 - 102 = 4.4 N-m.

DC MOTORS

The motor is similar to the generator in many respects, but bear in mind that the motor is driven by electric power and converts this to mechanical power.

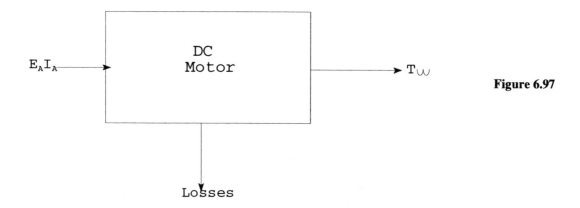

Figure 6.97

The rotation of the loop is the result of flux due to loop current stressing the magnetic field which produces the torque.

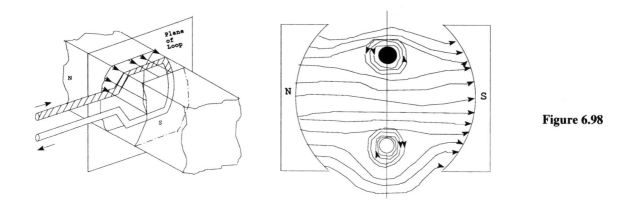

Figure 6.98

The turning force $F = \dfrac{BIL}{10}$ lbs and the torque T = F x distance in feet measured perpendicularly from direction of F to the center of rotation. The motor torque, T is proportional to $K \emptyset I_A$ for the shunt motor but $T \propto K \emptyset I_A^2$ for the series motor.

Motor Computations

Computations are similar to that for the generator except for current direction as shown for the compound motor model in Fig. 6.99.

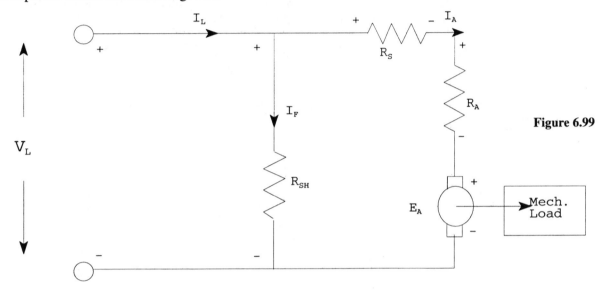

Figure 6.99

$I_L = I_A + I_P$

$V_L = E_A + I_A (R_S + R_A) = I_F R_{SH}$

Example

A 4.5 hp, 200 volt shunt motor runs at 1200 rpm and draws 20 amperes. The armature resistance is 0.18 ohms and the field resistance is 180 ohms. Find the developed counter emf, the electromagnetic torque, the efficiency and the shaft torque.

6-90 FUNDAMENTALS OF ENGINEERING EXAM REVIEW WORKBOOK

Solution

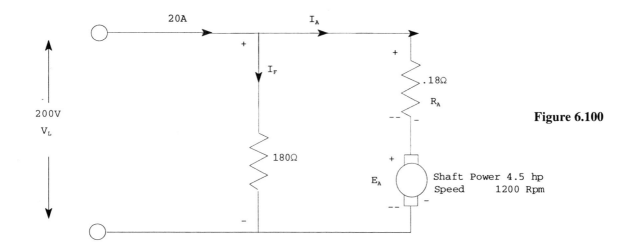

Figure 6.100

a. Developed counter emf, E_A

$$V_L = E_A + I_A R_A$$

$$E_A = V_L - I_A R_A = V_L - (I_L - I_F) R_A$$

$$I_F = \frac{200}{180} = 1.11 A$$

$$E_A = 200 - (20 - 1.11).18 = 196.6 \text{ volts}$$

b. Electromagnetic torque, T

$$T = \frac{E_A I_A}{w} = \frac{(196.6)(18.9)}{2\pi \frac{1200}{60}} = \frac{3714}{126} = 29.5 \text{ N-m}$$

c. Efficiency, η

$$\eta = \frac{P_o}{P_i} = \frac{P_i - \Sigma \text{ losses}}{P_i} = 1 - \frac{\Sigma \text{ losses}}{P_i}$$

$$P_i = V_L I_L = (200)(20) = 4000 \text{ watts}$$

Losses = armature heating, $I_A^2 R_A = (18.9^2)(.18) = 64$ watts

Field power, $V_L I_F = (200)(1.11) = 222$ watts

Rotational losses, $E_A I_A$ - shaft 4.5hp = 3714 - 3357 = 357 watts

Sum of losses = 643 watts

Other machine losses are iron (hysteresis and eddy current) friction and windage (bearings, brushes = 2 volts/pair), and stray losses (approx. 1% of output load). These are all neglected in this example. Thus, efficiency is:

$$\eta = \left(1 - \frac{643}{4000}\right) = .839 \text{ or } 83.9$$

d. Shaft Torque, T_{sh}

$$T_{sh} = \frac{P_o}{w} = \frac{(4.5)\,746}{126} = 26.6 \text{ N-m}$$

The torque needed to overcome rotational losses is 2.9 N-m.

Motor Speed

The speed of a DC motor is derived from $E_a = k\emptyset N$ as $N = \dfrac{V_L - I_A R_A}{k\emptyset}$ Rpm. When the load on a DC motor is removed the speed of a series motor rises very rapidly so that a dummy load is usually switched in. For the shunt and compound motors the effect is not as severe.

Example

At 230 volts, 5 hp shunt motor is driven at 1200 rpm. When idling (no load) the armature current is 2 amperes. By what percentage does the speed change from idle to full load? Assume an armature resistance of 0.15 ohms and rotational losses of 390 watts

Solution

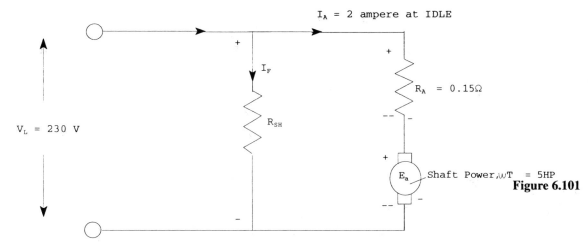

Figure 6.101

N = 1200 Rpm at full load; $P_{rot.}$ = 390 watts

At full load

T_w = 5 × 746 = 3730 watts

P_{rot} = 390 watts

T_{elect} = 4120 watts

T_{elect} = $E_A I_A$ = 4120

$$\frac{T_{elect}}{E_A} = I_A = \frac{4120}{230} = 17.9A$$

$$N_{FL} = \frac{230 - (18 \times .15)}{k\emptyset} = \frac{227.3}{k\emptyset}$$

$$N_{NL} = \frac{230 - (2 \times .15)}{k\emptyset} = \frac{229.7}{k\emptyset}$$

$$\frac{N_{FL}}{N_{NL}} = \frac{227.3}{229.7} = .989 \text{ OR } 98.9\%$$

Also $N_{NL} = \frac{1200}{.989} = 1213$ rpm and change is $\frac{13}{1200}$ or 1.1%

The solution is based on the fact that, unlike the series motor, the flux in the shunt motor is fairly constant.

A summary of general characteristics of a few types of DC motors is presented in Appendix 6.5.

AC GENERATORS

The AC machine or dynamo is a generator (alternator) when converting mechanical to electrical energy and is a motor when converting electrical to mechanical energy. In the simple case, replacing the commutator segments by slip rings will result in an AC machine as shown in Fig. 6.102 for a generator.

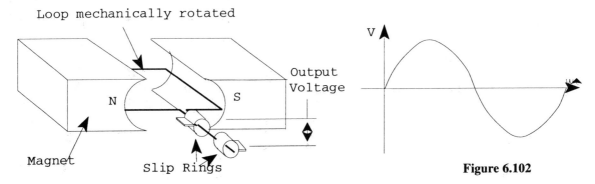

Figure 6.102

In the AC machine the frequency of operation is given by $f = \dfrac{pN}{2}$ in Hz where p = number of poles and N = speed in revolutions per second, sometimes called the synchronous speed.

A functional type block diagram of a typical large scale generator that outputs a three-phase voltage is shown in Fig. 6.103.

Figure 6.103

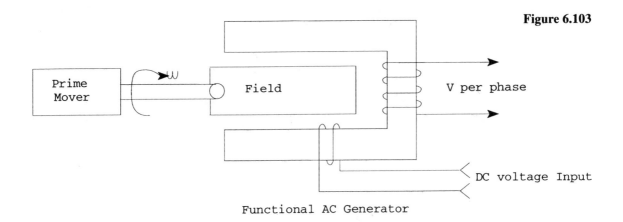

Functional AC Generator

The prime mover (steam) drives an alternator field (rotor) that electromagnetically transfers power to a fixed armature with a DC excited winding (stator). The induced voltage, on a per phase basis, is

$$V = 4.44 F N \emptyset \, 10^{-8} \text{ volts}$$

Since the system is assumed to be balanced, calculations can be made on a per phase basis using the equivalent circuit of Fig. 6.104.

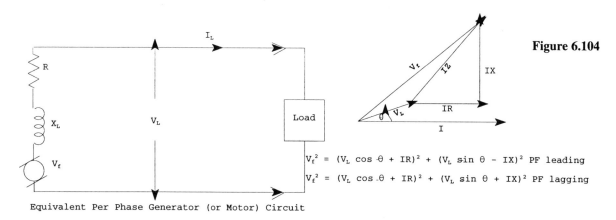

Figure 6.104

$V_f^2 = (V_L \cos \theta + IR)^2 + (V_L \sin \theta - IX)^2$ PF leading
$V_f^2 = (V_L \cos \theta + IR)^2 + (V_L \sin \theta + IX)^2$ PF lagging

Equivalent Per Phase Generator (or Motor) Circuit

In Fig. 6.104 R and X_L are the impedance components per generator phase and X_L is called the synchronous reactance because its flux change is the same as the changing magnetic field. V_f is the voltage generated and V_L is the terminal voltage available on a per phase basis.

Example

A three phase 2500 KVA, 2300 volt generator has a 400 ampere resistive load. The internal armature DC resistance is 0.05 ohms and the synchronous reactance is 2.5 ohms. What is the voltage regulation?

Solution

Compute all data on a per phase basis using the model of Fig. 6.104. Since the load is resistive, the power factor is unity. Assume a WYE connected generator output,

$$\text{Vphase} = \frac{2300}{\sqrt{3}} = 1329 \text{ volts}$$

Iphase = I Line = 400 amperes

$$V_f^2 = (V_L \cos \theta + IR)^2 + (V_L \sin \theta + IX)^2$$

cos θ = 1 and Sin θ = 0 Therefore

$V_f^2 = (V_L + IR)^2 + (IX)^2$

$V_f^2 = (1329 + 400 \times .05)^2 + (400 \times 2.5)^2$

$V_f = 1676$ volts, $V_L = 2899$ (no load)

Voltage Regulation $= \dfrac{V_{NL} - V_{FL}}{V_{FL}} = \dfrac{V_{oc} - V_{rated}}{V_{rated}}$

% Voltage Regulation $= \dfrac{2899 - 2300}{2300} \times 100 = 26\%$

In computing the IR drop, generally effective AC resistance $\approx \dfrac{R_{dc}}{2} \times k$ is used. However, the IR drop is negligible compared with terminal voltage so that the use of R_{DC} does not result in any large error. k is a value depending on phase wiring of the machine.

Efficiency

The efficiency of the generator can be written as

$$\eta = \dfrac{P_o}{P_i} = \dfrac{3 V_\phi I_\phi \cos \theta}{3 V_\phi I_\phi \cos \theta + P_{rot} + I_f^2 R_f + 3 I_a^2 R_a}$$

where V_ϕ, I_ϕ are the per phase voltages and currents

P_{ROT} are rotational losses in watts

R_f is the excitation field resistance

I_f is the excitation current

R_a is the armature resistance

I_a is the armature current

Other losses to be included, if given, are stray power loss, friction and windage, hysteresis, and eddy current. The maximum efficiency is given by the condition that

$P_{ROT} = I_f^2 R_f = 3 I_a^2 R_A$

Example

A 2500 KVA, three-phase 2300 volt alternator operates at rated power with a pf - .85. The field requires 60 amperes at 110 volts D.C. from the exciter. The following parameters are given: Armature resistance is 0.04 ohms, F & W loss is 15 KW, Iron losses are 28 KW, and stray losses are 2.5 KW. Determine the efficiency. Neglect rotational losses.

Solution

Efficiency $\eta = \dfrac{P_o}{P_i} = 1 - \dfrac{\Sigma \text{ losses}}{P_o + \Sigma \text{ losses}}$

$P_o = \sqrt{3} \text{ VI pf} = 2500 \text{ KVA} \times .85 = 2125 \text{ KW}$

$\text{KVA} = (\sqrt{3}) \, 2300 \, I_L = 2{,}500{,}000 \text{ VA}$

$I_L = \dfrac{2500000}{(\sqrt{3}) \; 2300} = 628 \text{A}$

Effective R_A (per phase) = .078 (See Note 1)

Σ losses are:

F and W	15KW
Iron	28KW
Field (110 x 60)	6.6KW
Armature 3 x (628)² x .078	92KW
Stray Power	2.5KW
Σ Losses	144KW

$\eta = 1 - \dfrac{144}{2125 + 144} = .937 \text{ or } 93.7\%$

Note 1: If the armature were delta wired then

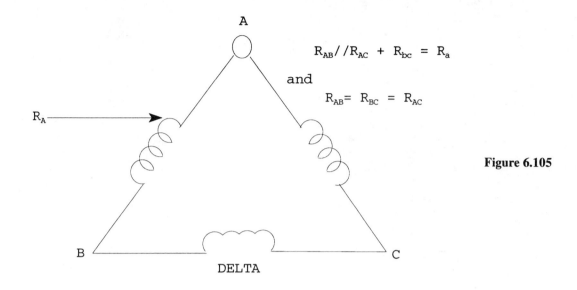

Figure 6.105

$$R_{AB}//R_{AC} + R_{bc} = R_a$$

and

$$R_{AB} = R_{BC} = R_{AC}$$

$$R_A = \frac{R_{AB}(2R_{AB})}{R_{AB} + 2R_{AB}} = \frac{2R_{AB}^2}{3R_{AB}}$$

$$R_{AB} = \frac{3R_A}{2} = \frac{3 \times .04}{2} = .06\Omega/\text{phase}$$

Usually the effective AC resistance rather than the DC resistance is used. A k factor of 1.3 is used. Therefore $R_{EFF/phase} = .06 \times 1.3 = .078$. If $R_{oc} = .04\Omega$ then the armature power loss would be 47KW and Σ losses = 100KW which results in $\eta = 95.5\%$; error is about 2%.

AC MOTORS

The AC motor, like the generator, utilizes a rotating magnetic field as the rotor that carries the permanent magnet armature with it. The speed at which the field rotates is the synchronous speed. To generate a starting torque the single-phase motor system is designed so that the input appears to be two phases as illustrated for the split-phase motor in Fig.6.106

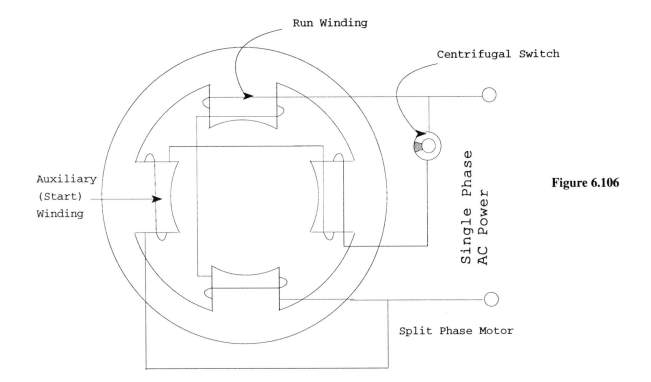

Figure 6.106

The inductance of the start winding is much less than the run winding which results in a phase difference. Their respective magnetic fields are also out of phase so that a revolving magnetic field is created. When the motor is up to speed the centrifugal switch opens the start winding.

Most single-phase motors are of fractional horsepower and use some scheme to split the phase to create the revolving magnetic field. Appendix 6.6 lists several motor types and their characteristics.

Three-phase induction motors such as the squirrel cage and wound rotor are widely used. Three phase synchronous motors are also used but need additional features to generate a starting torque.

Example

A split-phase motor runs at 1800 rpm from a 120 volt, 60 Hz source. How many poles does this machine have?

Solution

$$P = \frac{120 \times 60}{1800} = 4$$

Example

A small motor develops 18 ounce - inches of torque. What is the horsepower of the motor if the speed is 1800 rpm?

Solution

$$\text{In English Units } hp = \frac{2 \pi N}{33000} \times T$$

where
- hp = horsepower
- N = speed in rpm
- T = torque in lb-ft

Therefore

$$hp = \frac{2 \pi \cdot 1800}{33000} \times 18 \text{ ounce-in.}$$

$$\times \frac{1}{16} \frac{lb}{ounce} \times \frac{1}{12} \frac{ft}{in.} \approx \frac{1}{30} hp$$

THREE-PHASE INDUCTION MOTOR

The three-phase induction motor generates the revolving field by virtue of three out of phase input voltages. In operation the rotor speed never reaches the synchronous speed. This speed difference or slip is defined as

$$S \equiv N_{\text{sync speed}} - N_{\text{machine speed}} \text{ in rpm}$$

and on a per unit basis is

$$S = \frac{N_s - N}{N_s}$$

S has values between 0 and 1. At $S = 0$, $N_s = N$ and at $S = 1$, $N = 0$ (blocked rotor).

Example

A six-pole 60 Hz motor runs at 1100 rpm. What is the per unit slip.

Solution

It is assumed that the rotor speed is 1100. The speed of the revolving field is the synchronous speed.

$$N_s = \frac{120 \times 60}{6} = 1200$$

$$\%S = \frac{1200 - 1100}{1200} \times 100 = 8.3\%$$

In the induction motor, the conversion of the electrical power to mechanical energy takes place in the rotor where the power input to the rotor is inductively transferred across the air gap. The key to operation is to consider electrical relationships in the rotor.

Let SV_R = generated rotor voltage per phase

Z_R = rotor impedance per phase

Therefore $I_R = \dfrac{SV_R}{Z_R}$ = rotor current per phase

and $I_R = \dfrac{SV_R}{\sqrt{R_R^2 + (SX_R)^2}} = \dfrac{V_R}{\sqrt{\left(\dfrac{R_R}{S}\right)^2 + X_R^2}}$

with $\dfrac{R_R}{S} = R_R + R_R \left[\dfrac{1-S}{S}\right]$ which is a useful way of illustrating that the rotor resistance is best shown as two components. The second term represents the mechanical load on the rotor. Therefore on a per phase basis the rotor may be represented by the equivalent circuit shown in Fig. 6.107.

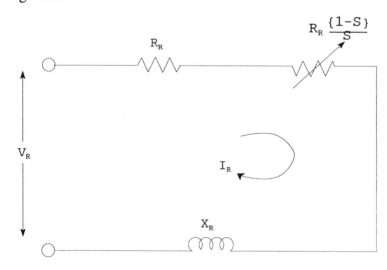

Figure 6.107

The stator and rotor can be combined as a single equivalent circuit as shown in Fig. 6.107,

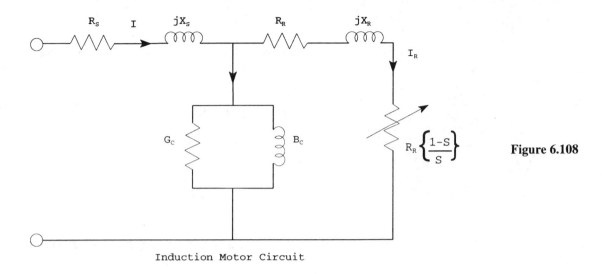

Figure 6.108

Induction Motor Circuit

Where $Y_0 = G_C - JB_L$ represents the admittance of the magnetic circuit: G_C represents iron loss and B_L represents reactive eddy current effects. In most cases G_C and B_L are negligible. The total power delivered to the rotor (RPI) on a per phase basis consists of the rotor copper loss (RCL) plus the rotor power developed (RPD): RPI = RCL + RPD

$$RPI = I_R^2 \left(\frac{R_R}{S} \right)$$

$$RCL = I_R^2 R_R = SRPI$$

$$RPD = I_R^2 R_R \left[\frac{1-S}{S} \right]$$

The torque developed is $T = 7.04 \dfrac{RPI}{N_S}$ lb=ft. At the instant of starting S = 1 so that the starting torque, $T = 7.04 \dfrac{RPI}{N_S}$ start lb-ft.

The stator is virtually independent of the rotor conditions and only depends on the supply frequency and voltage. The voltage induced in the rotor depends on the slip and the rotor induced voltage $V_R = SV_{R\,Blocked}$.

Also, the rotor frequency $fr = Sf_{supply}$. Note that f_r is the rotor frequency of the changing magnetic field, whereas machine speed is the actual rotor speed. Figure 6.108 represents a transformer equivalent circuit where the rotor becomes the secondary and (1-S)/S becomes the turns ratio, a, as indicated in Fig. 6.109.

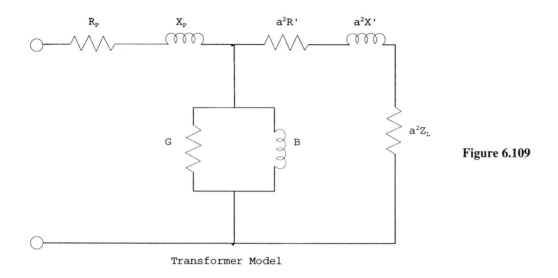

Figure 6.109

Example

A three-phase, WYE connected, 60 Hz, six pole, 220 volt induction motor has a speed of 1150 rpm. Find (a) rotor frequency and (b) Rotor voltage between terminals.

(a) $N_S = \dfrac{120\ f}{P} = \dfrac{120 \times 60}{6} = 1200$ RPM Synch Speed

Per Unit Slip $= \dfrac{N_S - N_R}{N_S} = \dfrac{1200 - 1150}{1200} = .04$

$f_R = S\ f_S = (.04)\ 60 = 2.4$ Hz.

(b) $V_R = S\ V_{BR}$

At V_{BR}, $S = 1$. Also at standstill the same flux at the stator must also be at the rotor. Therefore V stator = V rotor assuming stator and rotor each have the same turns (like $N_1 = N_2$).

$V_R' = V_S = 220 = V_{SR}$ at $S = 0$

Then $V_R = S\ V_{BR} = (.04)(220) = 8.8$, rotor generated voltage per phase at motor slip.

$V_{Rline} = \sqrt{3} \times 8.8 = 15.224$ volts

This example illustrates that the rotor voltage and frequency is very low relative to nameplate ratings.

Power Factor Correction

Synchronous motors have a leading power factor and are used to correct a plant power factor wherein the motor load results in a large power factor lag. The synchronous motor acts like a capacitor.

Example

A manufacturing facility requires 5000 KVA at a pf = .75 lag. A three-phase synchronous motor having a 1500 KVA rating is used in the plant to improve the pf and has an efficiency of 88% with a load of 800 hp and operated at the rated KVA. What is the overall power factor?

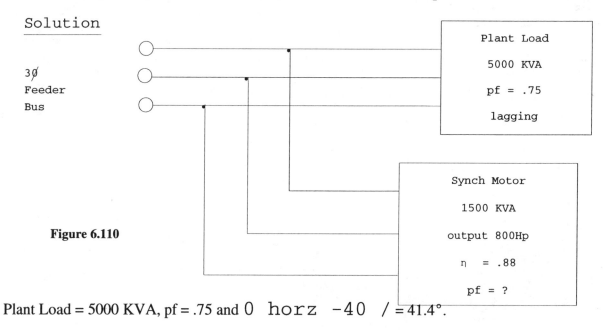

Figure 6.110

Plant Load = 5000 KVA, pf = .75 and $\cos^{-1} .75 = 41.4°$.

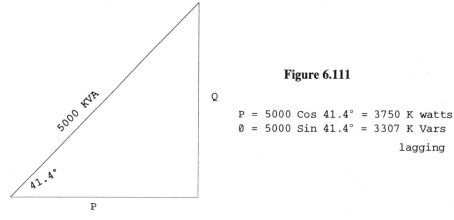

Figure 6.111

P = 5000 Cos 41.4° = 3750 K watts
Q = 5000 Sin 41.4° = 3307 K Vars
                           lagging

Synchronous Motor Input = $\dfrac{800 \times 746}{.88}$ = 678 KW

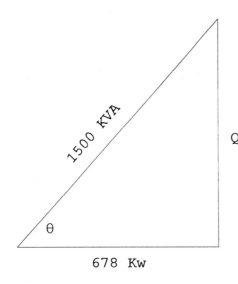

Figure 6.112

$Q_M = \sqrt{1500^2 - 678^2}$ = 1338 K Vars leading

$\emptyset = \text{Tan}^{-1} \dfrac{1338}{678}$ = 63°

The overall system is shown in Fig. 6.113.

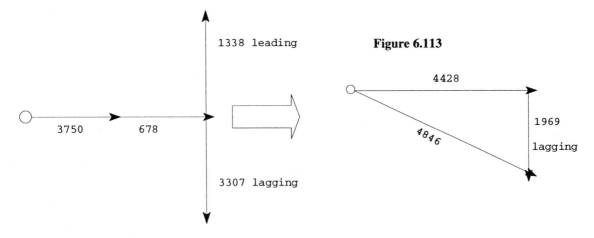

Figure 6.113

overall pf = $\cos^{-1}$ ($\text{Tan}^{-1} \dfrac{1967}{4428}$)

= $\cos^{-1}$ (23.97°) = 0.914

MEASUREMENTS

DC Measurements

The basic meter movement is the D'Arsonval meter and works on the principle of a simple machine; a pointer attached to a coil located in a magnetic field will rotate with a torque proportional to a DC current. The meter is usually specified in current ohms. To measure current larger than the specified meter value a shunt resistor is needed. To measure voltage larger than the meter IR a series resistor is needed.

Example

A 50 microampere, 20 Kohm meter is to be used to measure one quarter of a millampere. What shunt resistor is required?

Solution

An ammeter in series with a circuit element and the meter circuit is shown in Fig. 6.114.

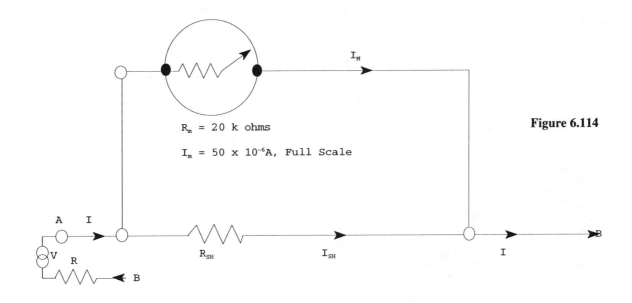

Figure 6.114

$I = \frac{1}{4}$ MA $= 250 \times 10^{-6}$ A. (Assume full scale)

$I = I_{sh} + I_m$ (Kirchoff's Current Law)

$I_m R_M = I_{sh} R_{sh}$

$I_{sh} = I - I_m = (250 - 50) \, 10^{-6} = 2000 \times 10^{-6}$

$$R_{sh} = \frac{I_m R_m}{I_{sh}} = \frac{50 \times 10^{-6}}{200 \times 10^{-6}} \times 20K = 5K\Omega$$

AC Measurements

The basic meter movement is the iron vane meter which works on the principle that two like magnets repel. Using two electromagnets, one fixed and one movable with a pointer attached, the rotor will have a torque proportional to AC current squared and can be calibrated to read RMS. A more complex type, the electrodynameter reads the RMS directly of a periodic AC waveform and is also the basic meter for the wattmeter. AC current and voltmeter resistor values are designed, for low frequencies in the same manner as a DC meter resistor is calculated.

Resistance

Very precise resistors can be passively determined using a bridge balancing technique as shown in Fig. 6.115 for a Wheatstone Bridge.

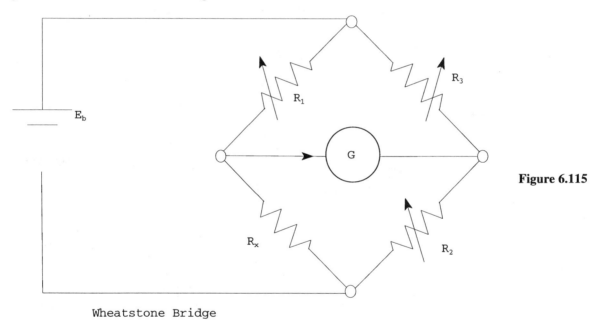

Figure 6.115

Wheatstone Bridge

When current through the sensitive DC meter (galvanometer) is zero.

$$R_1 R_2 = R_3 R_x$$

$$R_x = \frac{R_2}{R_3} \times R_1 = \text{Ratio Arm} \times \text{Balance Arm}$$

Example

A 1.5 DC battery is available. Show how a battery can be used to measure the value of an unknown resistor, R_x. What is the value of R_s for a midscale reading on the meter?

Solution

One method is to use a bridge circuit like a Wheatstone Bridge. This sort of circuit yields a resistor measurement accurate to better than .01% if very precise ratio arm and balance arms are used. If only 10% accuracy is required an ohm meter can be used.

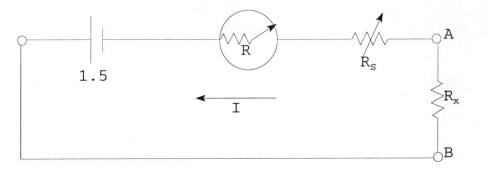

Figure 6.116

Simple Ohmeter

Assume D'Arsonval meter, R = 50, I = 1MA Short R_s. For full scale $R = \dfrac{1.5}{10^{-3}} = 1.5K$. Therefore $R_s = 1500 - 50 = 1450 \Omega$. With R_s in circuit and $I_m = I = \dfrac{10^{-3}}{2}$.

$$E = I(R_m + R + R_x)$$

$$1.5 = \dfrac{10^{-3}}{2}(1500) + \dfrac{10^{-3}}{2} R_x$$

$$R_x = \dfrac{1.5 - \dfrac{10^{-3}}{2} 1500}{\dfrac{10^{-3}}{2}}$$

$$Rx = \dfrac{1.5 - .75}{.0005} = 1500 \Omega$$

The meter scale can be calibrated directly in ohms by inserting different known values for R_x and noting the meter deflection.

PRACTICE PROBLEMS WITH SOLUTIONS

1. A three-phase, four wire 2500 KVA system supplies a 100 ampere load. Which answer best identifies the generator coil voltage?

   (A) 10KV   (B) 15 KV   (C) 20 KV   (D) 25 KV   (E) 30 KV

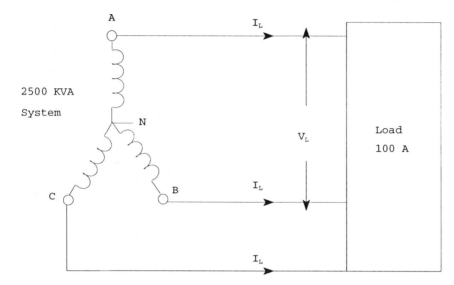

Figure 6.117

$SI = \sqrt{3}\ VI$ KVA (SI = system input)

$2500 = \sqrt{3}\ V_L\ 100$

$V_L = \dfrac{2500}{\sqrt{3}\ 100} = 14.45$ KV

$V_{AN} = V_{BN} = V_{CN} = V_C = \dfrac{V_L}{\sqrt{3}} = \dfrac{14.45 KV}{\sqrt{3}} = 8.35$ KV

Answer is (A) 10 KV, closest to 8.35 KV

2. The axis of a 0.5 meter length of conductor intersects 3×10^2 lines of flux, at an angle of 30°, of a magnet. The conductor has a current of 5 amperes. The cross-sectional area of the magnet is $0.005 M^2$. What is the magnetic field intensity at a point .25 CM from the coil axis center.

   (A) 0.06   (B) 3.75   (C) 0.15   (D) 7.5   (E) 0.375

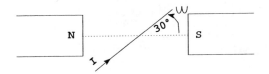

Figure 6.118

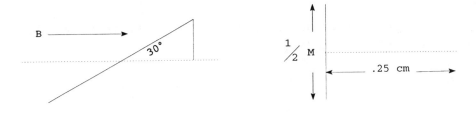

$T = \emptyset I \sin \omega t = \emptyset I \sin \theta$

$T = \dfrac{3 \times 10^2}{10^8} (5) \sin 30° = 7.5 \times 10^{-6}$ N-M

$T = F \times L$ and $F = \dfrac{T}{L} = \dfrac{7.5 \times 10^{-6}}{.5} = 15 \times 10^{-6}$ N-M

$H = \dfrac{F}{S} = \dfrac{15 \times 10^{-6}}{.0025} = .06 \dfrac{N}{M}$ The answer is (A)

3. A rectangular iron with a hollow center has μ=320. The iron path is 10 cm and has a cross-section of .001M x .05M. What is the developed magnetomotive force if the flux density is 40,000 Teslas?

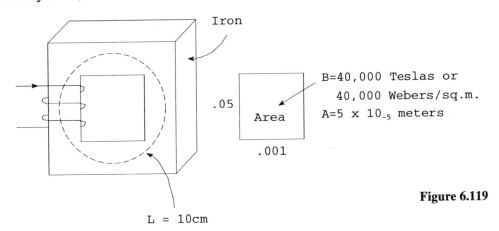

Figure 6.119

(A) 12.5 (B) 1250 (C) 25000 (D) 2500 (E) none of these

$\mathcal{F} = \Re el \times \emptyset$ $\Re el = \dfrac{L}{\mu A} = \dfrac{.1}{320(.001 \times .05)} = 6.25$

$\mathcal{F} = \Re el \times BA$

$\mathcal{F} = (6.25)(40000)(5 \times 10^{-5}) = 12.5$ The answer is (A).

4. A transformer is designed to transfer a load from 50 ohms to 5 ohms. What is the turns ratio of the transformer?

(A) 0.316 (B) 10 (C) 3.16 (D) 9.1 (E) 1.0

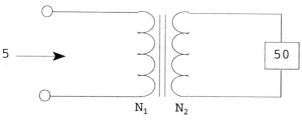

$5 = a^2 \, 50$ **Figure 6.120**

$a^2 = \dfrac{5}{50}$ $a = \dfrac{N_1}{N_2}$

$a = \sqrt{\dfrac{5}{50}} = .316$ Answer is (a)

5. A 120/180 autotransformer is used in a system and supplies 2KW to a unity power factor load. What current is required from the input line?

(A) 2.7A (B) 7.4A (C) 33A (D) 16.65A (E) 8.25A

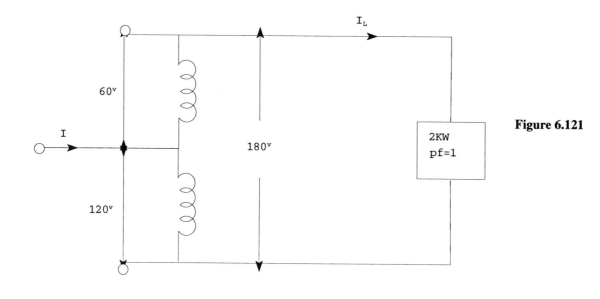

Figure 6.121

EI = P; 180 I = 2000 and I - 11.1A

$$a = \frac{N_1}{N_2} = \frac{E_1}{E_2} = \frac{120}{180} = \frac{2}{3}$$

$$\frac{I_2}{I_1} = \frac{I_L}{I} = a \text{ and } I = \frac{I_L}{a} = \frac{11.1}{.667} = 16.65A \quad \text{The answer is (D).}$$

6. A 1000 ohm per volt meter is available and it is required that the meter measure 250 volt DC with a midscale deflection. What resistor is required to ensure that the meter indicate correctly?

(A) 500K shunt (B) 250K series (C) 500K series
(D) 250K shunt (E) Can't be done

1000 ohm per volt, is the same as an IMA meter since

$$I = \frac{E}{R} = \frac{1}{R/E} = \frac{1}{1000} = IMA$$

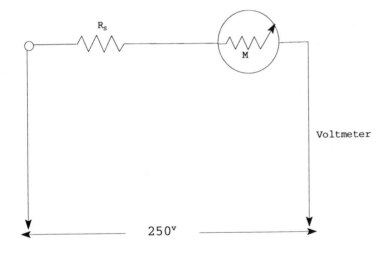

Figure 6.122

To measure 500V, the meter reads full scale or 1 MA. Since R_M is negligible compared with R_s

$$R_s = \frac{500}{1MA} = 500K$$

To measure 250V, the meter deflection will be $I = \dfrac{250}{500K}$ or ½ MA with 500K in series.

The answer is (C).

7. A 50 hp motor has an efficiency of 87.8% when delivering three quarters of its rated output. What are the total machine losses?

(A) 3.9 KW   (B) 7.8 KW   (C) 9.3 KW   (D) 8.2 KW   (E) 11.7 KW

$$\eta = \frac{P_o}{P_i} = \frac{(.75)(50)(746)}{P_i} = .878$$

$$P_i = \frac{27975}{.878} = 31862$$

Losses $= P_i - P_o = 31862 - 27975 = 3887 \approx 3.9$ KW

The answer is (A).

8. A series DC machine is rated at 50 amperes at 110 DC line voltage. The armature resistance is 0.045 ohms and the series (field) resistance is 0.125 ohms. What is the total per unit value of the winding resistance?

(A) 9%   (B) 0.45   (C) 0.077   (D) 3.6%   (E) 0.583

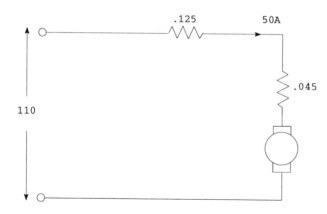

Figure 6.123

Per unit impedance $= \dfrac{\text{Actual Z}}{\text{Base Z}}$

$$\text{PUR} = \frac{\text{Actual R}}{\text{Base R}} = \frac{\frac{IR}{I}}{\frac{V}{I}} = \frac{50(.125+.045)}{110} = \frac{8.5}{110} = .077$$

The answer is (C).

9. A short circuit test of an autotransformer measures 2 KW and an open circuit test measures 0.5 KW. The no load copper loss is

   (A) 2 KW (B) 0.5 KW (C) 1.5 KW (D) 1.0 KW (E) None of these

   A short circuit test indicates copper loss. Open circuit test indicates iron loss. No load copper loss is negligible

   The answer is (E).

10. A shunt generator has 800 turns on each of its eight poles. The armature has 600 conductors arranged in two parallel paths. Air gap flux is 4×10^4 lines with an excitation of 2700 ampere turns per pole. What is the induced voltage if the speed is 1800 rpm?

    (A) 38V (B) 228V (C) 76V (D) 224V (E) 115V

    $$V = K \emptyset w$$

    $$K = \frac{Zp}{a} \times 10^{-8} = \frac{600 \times 8}{2} \times 10^{-8} = 2400 \times 10^{-8}$$

    $$V = (2400 \times 10^{-8})(4 \times 10^4)\frac{(1800)}{60} = 27.6 \text{ volts}$$

The answer is (A).

Generally $\emptyset$ is the flux per pole so that units of $\emptyset$ can be webers or lines in the equation. w is the speed in rps.

In the motor a similar equation is used and is expressed as torque.

$$T = .117 \frac{P}{P'} Z I_A \emptyset \times 10^{-8} \text{ in lb-ft}$$

11. Varying the field strength of a synchronous motor will vary.

    (A) speed (B) horsepower (C) power factor

(D) input voltage   (E) affect nothing

The synchronous motor is similar in connection to the AC generator in that it has both AC and DC connections. The AC is connected to a polyphase winding on the stator which produces a rotating magnetic field. The DC is connected to the rotor and produces a magnetic field in the rotor. Interaction between the two magnetic fields produces rotation of the rotor which will run at a constant speed equal to the speed of the rotating magnetic field, or synchronous speed. Synchronous speed is found from frequency = pole pairs x rps.

Increasing the rotor field strength causes the power factor of the AC connection to become more capacitive (lead). Decreasing the rotor field strength causes the power factor of the AC connection to become more inductive (lag). These effects are related to armature reaction. Synchronous motors are sometimes called synchronous condensers.

The answer is (C).

12. A 3∅ delta generator supplies a 3∅ 10 KVA motor load with a pf = .85. A 5 KVA resistive, 1∅ load with unity power factor is added. What is the resulting overall power factor?

(A) .87   (B) .93   (C) .81   (D) .85   (E) .99

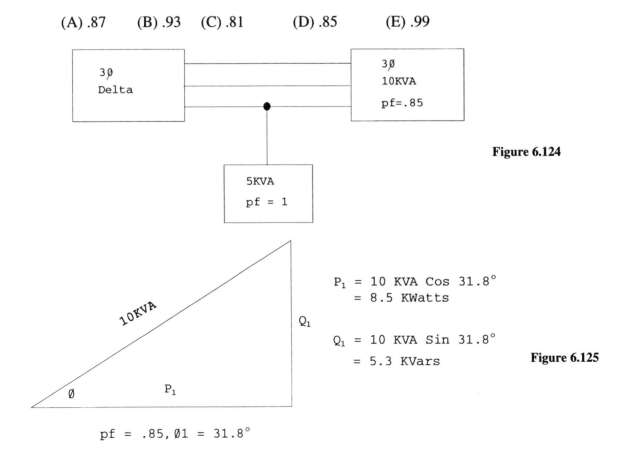

Figure 6.124

P_1 = 10 KVA Cos 31.8°
   = 8.5 KWatts

Q_1 = 10 KVA Sin 31.8°
   = 5.3 KVars

Figure 6.125

pf = .85, ∅1 = 31.8°

Add 5KVA load = 5KW Since pf = 1

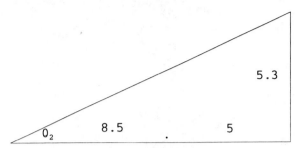

Figure 6.126

$O_2 = \text{Tan}^{-1} \dfrac{5.3}{13.5} = 21.43$

$pf_2 = \cos 21.43 = .93$ Answer is (b)

13. A series motor is operated such that the armature current is twice normal. The expected torque change is

   (A) 1   (B) 4   (C) 3   (D) 2   (E) none

   Torque $\propto I_A^2$; for series motor y I_A doubles T quadruples.

   The answer is (B).

14. A shunt generator has a counter emf of 300 volts. The armature resistance is 0.2 ohms. What is the line voltage when the armature delivers 150 amperes.

   (A) 300V   (B) 220V   (C) 270V   (D) 110V   (E) 115V

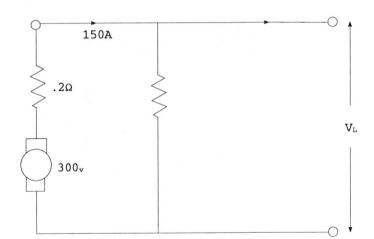

$V_L = 300 - (150 \times .2)$

$= 300 - 30 = 270$ volts

The answer is (C).

Figure 6.127

15. A motor takes 20 amperes at 220 volts. It is run at 1200 RPM for a Prony Brake Torque test. The net load measured is 48 lbs and the brake arm set at 4.8 in. What is the efficiency?

(A) 0.9   (B) 0.885   (C) 86%   (D) 74%   (E) 50%

Motor output is Tw = F x S x 2πN/33000 in hp

$$Tw = \frac{(48)\frac{4.8}{12}(2)(3.14)\,1200}{33000}$$

Tw = 4.38 HP = P_o

$$\eta = \frac{P_o}{P_i} = \frac{4.38}{20 \times 220/746} = \frac{4.38}{5.89} = .74 \text{ or } 74\%$$ The answer is (D).

APPENDIX 6.4
MAGNETIC UNITS AND TERMINOLOGY

Quantity	MKS	CGS	English
$\mathcal{F}$, MMF	AMP-turn	Gilbert	AMP-turn
ϕ, Flux	Weber	Maxwell	Line
B, Flux density	Tesla or Webers/M^2	Gauss or lines/cm^2	Lines/in.^{2}
H, Magnetic field intensity	Amp-turn per meter	Oersteds or Gilbert/cm	$\dfrac{\text{AMP-turn}}{\text{in.}}$
$\mathfrak{R}$, Reluctance	$\mathfrak{R}$els	$\mathfrak{R}$els x 10^{+7}	$\dfrac{\mathfrak{R}\text{els x } 10^{+7}}{3.19}$

Conversions

1 Oersted	= 79.6 AT/m
1 Gauss	= 6.45 Lines/m^2
1 Tesla	= 10^4 Gausses
1 Weber	= 10^8 Lines
1 AT	= 0.8 Gilbert
1 m^2	= 1550 in.^{2}
1 $\dfrac{\text{AT}}{\text{IN}}$	= 2 Oersteds

Terminology

$\mu = \dfrac{B}{H}$ Permeability

Hysteresis ≡ Curve of B vs H indicating residual magnetic effects

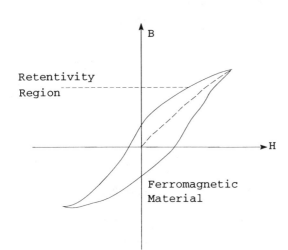

Figure 6.128

APPENDIX V
D.C. MOTOR CHARACTERISTICS

Type	Torque	Speed	Speed Control	Comment
Series	High start 500% Normal	Variable High at noload	Rheostat	• $T \propto I^2$ • Series machine used for hoists, cranes • Load directly on shaft
Shunt	Medium 100% Normal	Fairly constant with load	Not usually needed	• $T \propto I$ • Used for pumps, fans • Load can be connected by pulley
Compound	Series/shunt combination depending on ratio	Combined series and shunt features		

APPENDIX VI
A.C. MOTOR CHARACTERISTICS

Single Phase Type	Torque	EFF	HP	USE
Capacitor	High start 500% Normal	~ 60%	To 3/4	Compressor, pump
Split-phase	Moderate 250% Normal	~ 50%	To 1/3	Washing machine, Oil burner
Universal	Moderate 150% Normal	~ 50%	To 3/8	Runs on DC or AC Drill press Vacuum cleaner
Shaded pole	Low 50% Normal	~ 35%	To 1/20	Small fan, Toys

ELECTRICITY III

TIME DOMAIN ANALYSIS

Introduction

As the word transient (on transition) implies, a change from one state to another is taking place; in addition, this change cannot occur instantaneously, that is, in zero time. A finite time is associated with every transition.

Figure 6.129 shows the transient period diagrammatically.

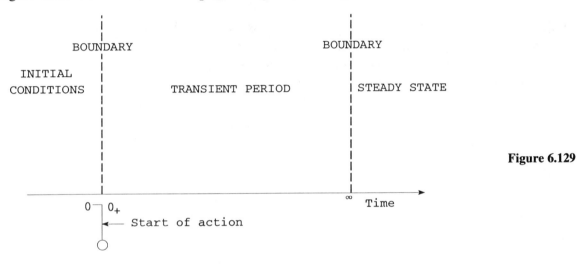

Figure 6.129

To denote the start of the transient period, a relative time of t=0 is assigned to this point, which corresponds to the time that any action is started. The action can be the opening or closing of a switch, or the reduction of loading on a motor, etc. Basically, it is the start of the period whereby a physical system must readjust itself to reflect the new conditions.

Immediately before the start of the transient period, the system in question is usually considered to be stabilized, i.e., no changes or variations occur. This time is referred to as $t=0^-$, and all constant quantities are termed the "initial conditions."

Mathematically the transient period is considered to be of infinite duration. Therefore to denote the end of the transient period, a point of $t=\infty$ is shown as the time the transient period ends. Beyond this $t=\infty$, all conditions are unchanging and "steady state" is said to have been reached.

Therefore, it is seen from Fig. 6.129 that the transient period exists from the boundary separating $t=0^-$ from $t=0^+$, to the boundary at $t=\infty$. These conditions play a significant role in solving boundary problems, and will be discussed later.

Physical Components

The three basic electrical components are the resistor, the inductor and the capacitor. The resistor is a device that has no means for energy storage, i.e., within a magnetic field or an electric field as with respectively, an inductor or a capacitor. Consequently, all electrical energy entering a resistor is converted into heat irreversibly; hence there are no transients associated with this type of element. This is not true with the inductor and capacitor. The stored energy in these latter devices prevents changes form occurring instantaneously, i.e., in zero time. Also, it is this stored energy that gives rise to transients.

Inductors

In an inductor a current is flowing from left to right. This current is the movement of electric charges (electrons) and by definition, if one moving electric charge exerts a force upon another moving electric charge, perhaps in another circuit or on the atoms of a material, then a magnetic field is said to exist. This magnetic force field is shown dotted in Fig. 6.130.

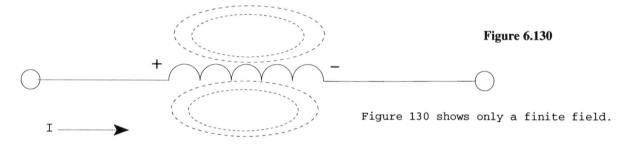

Figure 6.130

Figure 130 shows only a finite field.

It should be mentioned that the field extends from the center of the inductor, i.e., from zero radius, out to infinity, i.e., to infinite radius. Figure 6.130 shows only a finite field.

However, the energy stored in an inductor is contained within the magnetic field and is equal to:

$$W = \frac{1}{2} LI^2$$

where W = Energy (Joules)

L = Inductance (Henrys)

I = Current (Amperes)

or $\quad W = \frac{1}{2} L(I_2^2 - I_1^2)$ Joules

where in this case, W is the energy stored between two different current levels.

Another relationship applicable to an inductor deals with the voltage across it. This relationship (Fig. 6.131) is:

$$v_L = L \frac{di}{dt} \text{ in terms of differentials}$$

$$v_L = L \frac{\Delta i}{\Delta t} \text{ in terms of infintesimals}$$

where

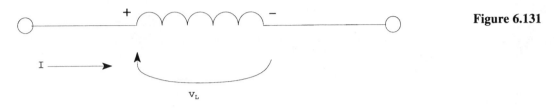

Figure 6.131

v_L = voltage across inductor (volts)

L = inductance (Henrys)

$\left.\begin{array}{r}\dfrac{di}{dt} \\ \dfrac{\Delta i}{\Delta t}\end{array}\right\}$ Rate of change of current $\left(\dfrac{\text{Amperes}}{\text{Second}}\right)$

What this equation states is that there is no voltage across an inductor if the current flowing through it is not changing with time. Accordingly, an inductor is a short circuit to a direct current (DC).

What is also implied from this equation is that the current flowing through an inductor cannot be changed in zero time, i.e., $\Delta t \to 0$. For in the expression

$$v_L = L \frac{\Delta i}{\Delta t}$$

v_L must be ∞ volts in order for Δi to reach a finite value in zero time. Therefore because no such forcing function exists, in practical cases, an inductor will resist abrupt changes in current. This is a characteristic of an inductor and is the result of the energy that is stored in its magnetic field.

To illustrate, suppose a steady current is flowing in the inductor. At some time, say t=0, the circuit is opened. The current flowing will tend to go to zero and, naturally, the magnetic field produced by this current can no longer be supported. Consequently, it will collapse. As it does, it induces a voltage into the inductor in such a direction, thereby resisting the current change. Effectively, the inductor becomes a source - it's effectively driving the current. This action is sometimes referred to as Lenz's Law.

Accordingly, if reference is made to Fig. 6.129, the boundary condition for an inductor is that the current at t=0⁻ (before the action) equals the current at t=0⁺ (after the action). Or stated differently:

$$i(0^-) = i(0^+)$$

Capacitors

In a capacitor (Fig. 6.132) the current is flowing downward.

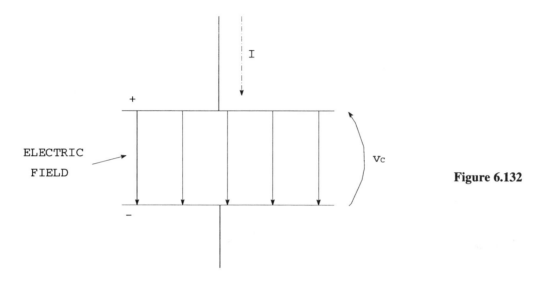

Figure 6.132

Actually, since the current flow is the movement of electrons, which is opposite to conventional current flow (by definition), as electrons leave the top plate, it becomes positively charged. Also, electrons accumulating on the bottom plate make that plate negatively charged. Consequently, the voltage across the capacitor is in direction shown, and it is the result of a current flow. Note that the current precedes the voltage.

When an equilibrium condition exists, i.e., when no more electrons are in motion, an electric field is said to exist between the plates. This is a condition whereby a stationary electric charge exerts a force upon another stationary electric charge.

As with the inductor, energy is stored in a capacitor within its electric field and the relationship si:

$$W = \frac{1}{2} CV^2$$

where W = Energy (Joules)
C = Capacitance (Farads)
V = Voltage (Volts)

or $\qquad W = \frac{1}{2} C(V_2^2 - V_1^2)$ Joules

where, in this case, W is the energy stored between two different voltage levels.

Another relationship applicable to a capacitor deals with the current flow through it. It is:

$$i_c = C \frac{dv_c}{dt} \text{ in terms of differentials}$$

or

$$i_c = C \frac{\Delta v_c}{\Delta t} \text{ in terms of infintesimals}$$

where

i_c = Current in capacitor (Amperes)

C = Capacitance (Farads)

$\frac{dv}{dt} \approx \frac{\Delta v}{\Delta t}$ = rate of change of voltage ($\frac{volts}{sec}$)

As a dual of the inductor, the above equation states that there is no current flowing through a capacitor if the voltage across it is not changing with time. Accordingly, a capacitor is an open circuit to direct current (DC). Also implied, the voltage across a capacitor cannot be changed in zero time for this would require an infinite current flow into it. Since each electron has mass (9.109×10^{-31} kg), regardless of how small, they cannot be moved any distance in zero time. Accordingly, if reference is made to Fig. 6.129, the boundary condition for a capacitor is that the voltage at t=0⁻ equals the voltage at t=0⁺. Or stated differently,

$$v(0^-) = v(0^+)$$

Total or Complete Response

The behavior of an RL or RC network is the result of two different sources of energy. One source is due to the internal energy storage, i.e., from a magnetic or electric field, and is called the natural response. The other response is called the forced response and it is produced by external energy sources such as batteries or generators. Naturally, the response due to an external energy source is dependent upon the form of the forcing function. For example, a sinusoidal voltage source will always produce sinusoidal currents in a device, etc. This response is determined Ohm's Law or other circuit theorems and the natural response is considered ended. Accordingly, this response, solved for at t=∞, is also called the steady-state response.

Accordingly, the method of solving a transient problem is to consider the natural and forced responses singularly, independent of each other, and then add them together to obtain the total or complete response.

The Series R-L Circuit (Initial Conditions Zero)

The circuit shown in Fig. 6.133 is to be energized by closing the switch at t=0. The excitation or forcing function is shown below the circuit diagram. It is a unit step multiplied by V.

At time, $t=0^+$, after the action, Kirchoff's voltage law is:

$$v_L + v_R = V$$

or $L \dfrac{di}{dt} + Ri = V$

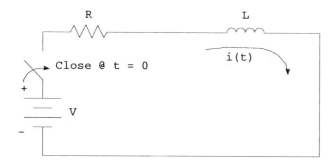

Figure 6.133

Setting the differential equation equal to zero and solving will result in the complementary solution in mathematical terms, or the natural or transient solution in electrical terms. Accordingly,

$$L \dfrac{di}{dt} + Ri = 0$$

Rearranging and using operator notation where $D = \dfrac{d}{dt}$,

$$(D + \dfrac{R}{L}) i = 0 \qquad \text{and} \qquad D = -\dfrac{R}{L}$$

The transient portion of the solution is:

$$i_t(t) = Ke^{Dt} \qquad \text{or} \qquad i_t(t) = Ke^{\dfrac{-Rt}{L}}$$

where K is to be determined from initial conditions, i.e., at t=0, and the subscript t (sometimes n is used) denotes that $i_t(t)$ is the "transient" or "natural" response.

The next step is to determine the particular solution in mathematical terms, or the steady-state solution in electrical terms.

For the circuit in Fig. 6.33, when all quantities are stablized, the final current, or $i_{ss}(t)$ is, from Ohm's Law:

$$i_{ss}(t) = \frac{V}{R} \text{ AMPS}$$

since the inductance has no effect when a constant current flows.

The sum of the two responses is the total or complete response and is:

$$i(t) = i_t(t) + i_{ss}(t) \quad \text{or} \quad i(t) = Ke^{\frac{-R_t}{L}} + \frac{V}{R}$$

where the value of K is to be determined.

Since the switch is open to zero time (t=0), the current in the circuit must be zero during that time. Hence, just prior to closing the switch, the current is zero. This is written as $i(0^-) = 0$.

Furthermore, if finite voltages are applied, the current through the inductor cannot change suddenly. Therefore, the current is zero at t=0 and shortly just afterward. This is written as $i(0^+) = 0$.

Therefore, $i(0^-) = i(0^+) = 0$

Incorporating this boundary condition at t=0 into

$$i(t) = Ke^{\frac{-R_t}{L}} + \frac{V}{R}$$

$0 = K[1] + \frac{V}{R}$, K is found equal to $-\frac{V}{R}$

And the final solution is:

$$i = \frac{V}{R}(1 - e^{\frac{-R_t}{L}}) \text{ amperes.}$$

This equation may also be used to determine the voltage across R and/or the voltage across L.

$$v_R = IR = \frac{V}{R}(1 - e^{-\frac{R_t}{L}})R$$

or

$$v_R = V(1 - e^{-\frac{R_t}{L}}) \text{ volts}$$

and

$$v_L = L\frac{di}{dt} = L\left(-\frac{V}{R} - e^{-\frac{R_t}{L}}\right)\left(-\frac{R}{L}\right)$$

or

$$v_L = V e^{-\frac{R_t}{L}} \text{ volts}$$

The validity of the three equations can be checked by utilizing the boundary conditions at t=0 and at t=∞.

Note that at t=0,

$$i(t) = 0 \text{ amperes}$$

$$v_R = 0 \text{ volts}$$

$$\text{and } v_L = V \text{ volts}$$

Also at t=∞,

$$i(t) = \frac{V}{R} \text{ amperes}$$

$$v_R = V \text{ volts}$$

$$\text{and } v_L = 0 \text{ volts}$$

which checks what is physically happening in the circuit.

A plot of all three equations is shown in Fig. 6.134.

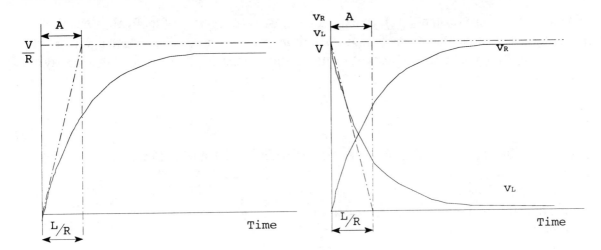

Figure 6.134

The initial slope is an important feature of the exponential function. Consider the derivative di/dt:

$$\frac{di}{dt} = \frac{V}{L} e^{\frac{-Rt}{L}}$$

which at t=0 is just $\frac{V}{L}$. This can be used to calculate the intercept A in Fig. 6.134. The slope is:

$$\frac{V}{R}/A = \frac{V}{L} \quad \text{or} \quad A = \frac{L}{R}$$

This is one interpretation of the quantity $\frac{L}{R}$ (its unit is seconds). It is the time it would take for the current to reach steady-state value if it were to continue to change at its original rate. L/R is called the "time constant" of the circuit (symbol τ).

Another interpretation of this quantity arises when t = L/R or one time constant.

Substituting in the equation for current i and inductor voltage

$$i = \frac{V}{R}(1 - e^{-1}) = 0.63 \frac{V}{R} \quad \text{where } e = 2.718$$

and

$$v_L = Ve^{-1} = 0.37V$$

Accordingly, at one time constant (τ), an exponential rise reaches 63% of its final value; correspondingly, an exponential decay falls to 37% of its peak or initial value. And, even though the final values are approached asymptotically, after 5 time constants, it is assumed that steady state is reached.

Example

The RL circuit in Fig. 6.135 is energized when the switch is closed at t=0. Find:

a) The current at t=½ msec
b) The time constant and the value of current at that time
c) Rate of current increase at $t=0^+$

Solution

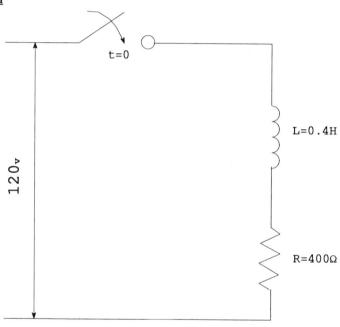

Figure 6.135

a) For initial conditions equal to zero

$$i = \frac{V}{R}(1 - e^{\frac{-R_t}{L}})$$

At t = 0.0005 sec

$$i = \frac{120}{400}(1 - e^{-(\frac{400}{0.4})(0.0005)})$$

$$= 0.3(1 - e^{-0.5})$$

$$= 0.3(1 - \frac{1}{e^{0.5}})$$

$$= 0.3(1 - 0.607)$$

$$i = 0.118 \text{ amperes} \quad \underline{\text{Answer}}$$

b) $\tau = \dfrac{L}{R} = \dfrac{0.4}{400} = 0.001 \text{ sec} \quad \underline{\text{Answer}}$

$i \; @ \; t = \tau$

$$i = 0.3(1 - e^{\frac{-R_t}{L}})$$

$$= 0.3 \, (0.63)$$

$$= 0.189 \text{ amperes} \quad \underline{\text{Answer}}$$

c) $\dfrac{di}{dt} = \dfrac{V}{L} e^{\frac{-R(t)}{L}} = \dfrac{V}{L}$ at $t = 0^+$

$$= \frac{120}{0.4} = 300 \text{ amperes/sec} \quad \underline{\text{Answer}}$$

The Series R-L Circuit (Initial Conditions Nonzero)

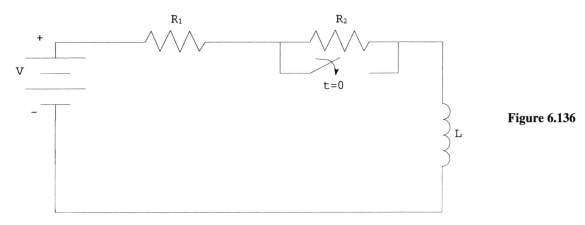

Figure 6.136

In the circuit shown in Fig. 6.136 the switch is closed at t = 0.

Kirchoff's Voltage Law: (t>0)

$$L\frac{di}{dt} + R_1 i = V$$

As before,

$$i_t(t) = Ke^{\frac{-R_1(t)}{L}} \text{ amperes}$$

$$i_{ss}(t) = \frac{V}{R_1} \text{ amperes}$$

$$i(t) = i_t(t) + i_{ss}(t)$$

$$i(t) = Ke^{\frac{-R_1(t)}{L}} + \frac{V}{R_1}$$

At $t = 0^-, 0$ and 0^+

$$i = \frac{V}{R_1 + R_2}$$

Therefore

$$\frac{V}{R_1 + R_2} = K + \frac{V}{R_1}$$

Or

$$K = V\left(\frac{1}{R_1 + R_2} - \frac{1}{R_1}\right)$$

$$K = V\left(\frac{R_1 - R_1 - R_2}{R_1(R_1 + R_2)}\right)$$

$$K = V\left(-\frac{R_2}{R_1(R_1 + R_2)}\right) = \frac{V}{R_1}\left(-\frac{R_2}{R_1 + R_2}\right)$$

$$i(t) = \frac{V}{R_1}\left(1 - \frac{R_2}{R_1 + R_2} e^{\frac{-R_1(t)}{L}}\right)$$

A plot of this response is shown in Fig. 6.137.

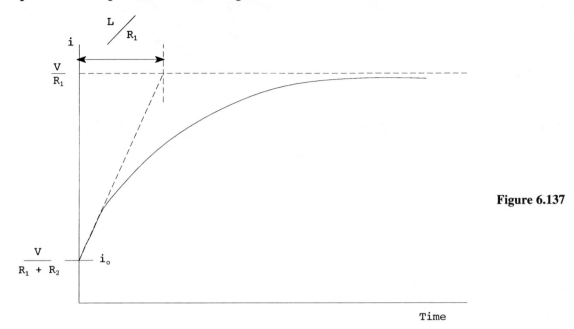

Figure 6.137

Example

The R-L circuit shown in Fig. 6.138 is shorted when the switch is closed at t=0. Find:

a) The current at t=0.01 sec.

b) The circuit time constant and the value of current at that time.

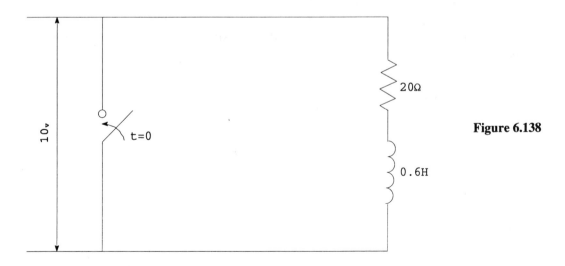

Figure 6.138

6-132 FUNDAMENTALS OF ENGINEERING EXAM REVIEW WORKBOOK

Solution

KVL:

$$0.6\frac{di}{dt} + 20\,i = 0$$

$$i = ke^{\frac{-20(t)}{0.6}} \qquad \text{NOTE: } i_{ss} = 0$$

At $t = 0$, $i = \dfrac{10}{20} = 0.5$ amperes

$\therefore 0.5 = K = I_o$

$i = 0.5e^{-33.33t}$ amperes

a) At t=0.01 sec

$i = 0.5e^{-(33.33)(0.01)}$
$= 0.5e^{-0.333}$
$= 0.358$ amperes <u>Answer</u>

b) $\tau = \dfrac{L}{R} = \dfrac{0.6}{20} = 0.003$ sec

$i = 0.5e^{-}$
$= (0.5)(0.37)$
$= 0.184$ amperes <u>Answer</u>

The Series R-C Circuit (Initial Conditions Zero)

The circuit shown in Fig. 6.139 is to be energized by closing the switch at t=0. The forcing function is a unit step multiplied by V.

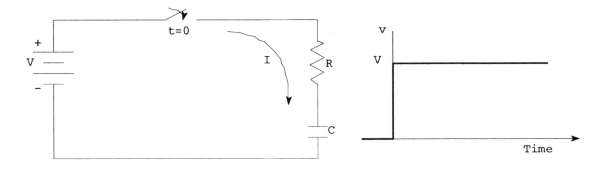

Figure 6.139

KVL:
$v_R + v_c = V$

$$iR + \frac{1}{C}\int_0^t i\,dt = V$$

$$(\frac{1}{CD} + R)\,i = V \quad \text{where} \quad \int = \frac{1}{D}$$

NOTE: $i_c = C\dfrac{dv}{dt}$

$$v_c = \frac{1}{C}\int_0^t i\,dt$$

The transient portion of the solution is:

$$i_t(t) = Ke^{Dt} \quad \text{where} \quad D = -\frac{1}{Rc} \quad \text{with} \quad V = 0$$

At steady state, the capacitor is fully charged to V volts, therefore $i_{ss} = 0$.

$$i(t) = Ke^{-\frac{1}{RC}t}$$

At t=0 $\quad i = \dfrac{V}{R}$ because $v_c(0^-) = v_c(0^+) = 0$.

Note the $v_c = 0$ is a short circuit (no voltage on the capacitor) and the current is maximum (V/R).

$$\frac{V}{R} = Ke^0 \quad \text{and} \quad K = \frac{V}{R}$$

$$i(t) = \frac{V}{R}e^{-\frac{1}{RC}t} \text{ amperes}$$

The capacitor voltage can be found from

$$v_c = \frac{1}{C}\int_0^t i\,dt$$

$$= \frac{1}{C}\int_0^t \frac{V}{R}e^{-\frac{1}{RC}t}\,dt$$

$$= \frac{V}{RC}\int_0^t e^{-\frac{1}{RC}t}\,dt$$

$$= -V \int_0^t t e^{-\frac{1}{RC}t}(-\frac{1}{RC})dt$$

$$= -V e^{-\frac{1}{RC}t}\Big]_0^t$$

$$v_c = V(1 - e^{-\frac{1}{RC}t}) \text{ volts}$$

Also $V_R = iR$; $V_R = V e^{-\frac{1}{RC}t}$ volts

A plot of all three equations is shown in Fig. 6.140.

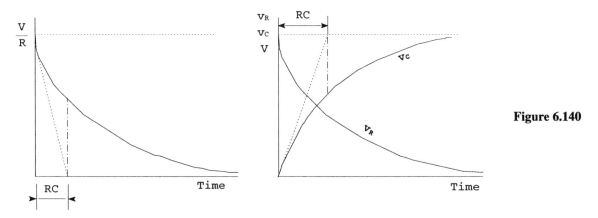

Figure 6.140

The same result can be obtained by converting the integral equation into a differential equation by differentiating

$$\frac{d}{dt}\left[iR + \frac{1}{C}\int idt\right] = V\frac{d}{dt}$$

$$R\frac{di}{dt} + \frac{1}{C}i = \frac{dV}{dt} = 0 \quad \text{For } V = \text{constant}$$

Or $\dfrac{di}{dt} + \dfrac{1}{RC} i = 0$

$$(D + \frac{1}{RC})i = 0$$

$$D = -\frac{1}{RC}$$

And $i_t(t) = Ke^{-\frac{1}{RC}t}$ as before

As an exercise, instead of solving for the current and then integrating to find v_c, find v_c directly.

Accordingly,

$$iR + \frac{1}{C}\int i\,dt = V$$

becomes

$$RC\frac{dv_c}{dt} + V_c = V$$

Solving this equation by first solving for $v_{c_t}(t)$, then $v_{c_{ss}}(t)$, etc. one obtains

$$RC\frac{dv_c}{dt} + V_c = 0$$

$$\frac{dv_c}{dt} + \frac{1}{RC}v_c = 0$$

$$(D + \frac{1}{RC})v_c = 0$$

$$D = -1/RC$$

And $v_{c_t}(t) = Ke^{-\frac{1}{RC}t}$

$v_{ss}(t) = V$ (capacitor is fully charged at t=∞)

$$v_c(t) = Ke^{-\frac{1}{RC}t} + V$$

At $t=0$ $v_c=0$

$0 = K + V$ and $K = -V$

$$v_c = V(1 - e^{-\frac{1}{RC}t})\quad \text{volts}$$

Also

$$i = C\frac{dv_c}{dt} = c[\frac{V}{RC} e^{-\frac{1}{RC}t}]$$

$$i = \frac{V}{R} e^{-\frac{1}{RC}t} \text{ amperes, as before}$$

The Series R-C Circuit (Initial Conditions Nonzero)

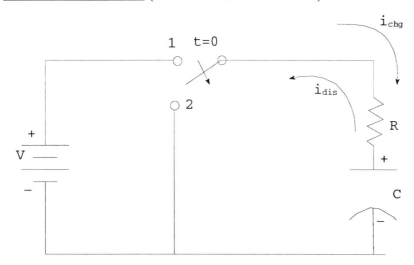

Figure 6.141

The switch shown in Fig. 6.141 has been in position 1 for a very long time. At t=0, the switch is moved to position 2. To solve for the circuit response, Kirchoff's Voltage Law is written and the differential equation solved.

NOTE:
$$Ri + \frac{1}{C} \int_0^t i\, dt = 0 \quad \text{At } t=0^+$$

Or $$(D + \frac{1}{RC})i = 0 \quad \text{And} \quad i = Ke^{-\frac{1}{RC}t}$$

To determine K, all that is known is that the capacitor is fully charged to V volts, the switch being in position 1 for a long time. Accordingly, at $t=0^-$, $v_c = V$ volts; therefore at $t=0^+$, $v_c=V$ because v_c cannot change. Consequently, relating this fact to current, the current at $t=0^-$ is zero since the capacitor is fully charged. However, at $t=0^+$, with the capacitor acting as a source, the current flowing is

$$i = -\frac{V}{R}. \quad \text{(Current discharging capacitor)}$$

This can also be determined by voltage balance at $t=0^+$,

$$V + iR = 0$$

And $$i = -\frac{V}{R} \text{ as before}$$

The solution for i is:

$$-\frac{V}{R} = K\{1\}$$

And

$$i = -\frac{V}{R} e^{-\frac{1}{RC}t}$$

It should be noted that $i_{ss}(t) = 0$, the capacitor being fully discharged at $t=\infty$.

The corresponding transient voltages are:

$$v_R = iR = -V e^{-\frac{1}{RC}t}$$

And

$$v_C = \frac{1}{C}\int_0^t i\,dt = V e^{-\frac{1}{RC}t}$$

All equations are plotted in Fig. 6.142.

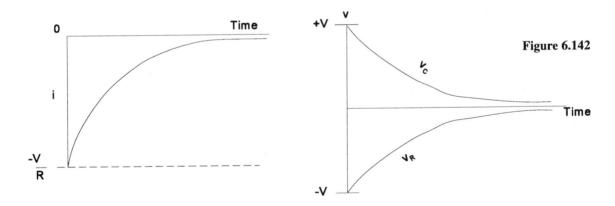

Figure 6.142

Example

In the circuit shown in Fig. 6.143, find the current i, at t=4.5 sec. after the switch is closed.

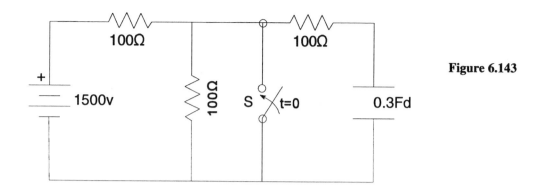

Figure 6.143

Solution

It must be assumed that the switch has been open for a long time. Accordingly, the capacitor is charged to 750 volts, since $i_c=0$. This can be verified via the Thevenin equivalent circuit which is:

$$v_{oc} = 750 \text{ volts}; \quad R_{thev} = 50 \text{ ohms}$$

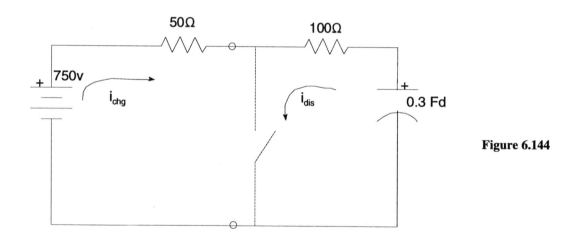

Figure 6.144

Note also that, from the Thevenin equivalent circuit, the charging time constant is

$$RC = (150)(0.3) = 45 \text{ sec}$$

which cannot easily be determined from Fig. 6.143.

When the switch is closed at t=0,

$$\frac{1}{C}\int_0^t i\,dt + Ri = 0$$

$$\left(\frac{1}{DC} + R\right)i = 0$$

$$\left(D + \frac{1}{RC}\right)i = 0$$

$$D = -\frac{1}{RC}$$

$$i = K\,e^{-\frac{1}{(100)(0.3)}t}$$

At t=0, $i = -\dfrac{750}{100} = -7.5$ amperes

$$i = -7.5\,e^{-\frac{1}{30}(4.5)}$$

(Note: The minus sign denotes the discharging current where the capacitor is the source and it is opposite to the original charging current.)

$$i = -7.5\,e^{-0.15}$$

$$= -\frac{7.5}{1.162} = -6.46 \text{ amperes.} \quad \underline{\text{Answer}}$$

RLC Series Circuit

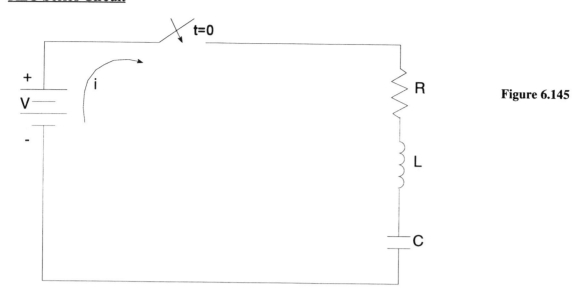

Figure 6.145

The application of Kirchoff's Voltage Law to Fig. 6.145 results in the following differential equation

$$Ri + L\frac{di}{dt} + \frac{1}{C}\int i\,dt = V$$

Differentiating results in a second order differential equation

$$L\frac{d^2i}{dt^2} + R\frac{di}{dt} + \frac{1}{C}i = 0$$

or in operator notation

$$\left(D^2 + \frac{R}{L}D + \frac{1}{LC}\right)i = 0$$

Solving for D

$$D_1 = \frac{-\frac{R}{L} + \sqrt{\left(\frac{R}{L}\right)^2 - \frac{4}{LC}}}{2} \quad \text{And} \quad D_2 = \frac{-\frac{R}{L} - \sqrt{\left(\frac{R}{L}\right)^2 - \frac{4}{LC}}}{2}$$

If $\alpha = -\dfrac{R}{2L}$ and $\beta = \sqrt{\left[\dfrac{R}{2L}\right]^2 - \dfrac{1}{LC}}$

then $\quad D_1 = \alpha + \beta \quad \text{and} \quad D_2 = \alpha - \beta$

In the typical case, C is initially uncharged and L carries no initial current. So, depending on the radicand of β, the current flowing in the RLC circuit can be over damped, critically damped, or under damped (oscillatory).

Case 1: $\left(\dfrac{R}{2L}\right)^2 > \dfrac{1}{LC}$. Roots D_1 and D_2 are real and distinct.

(over damped)

$$i_t(t) = K_1 e^{D_1 t} + K_2 e^{D_2 t}$$

$$i_t(t) = e^{-\alpha t}(K_1 e^{\beta t} + K_2 e^{-\beta t})$$

Case 2: $\left(\dfrac{R}{2L}\right)^2 = \dfrac{1}{LC}$. Roots D_1 and D_2 are real and equal.

(critically damped)

$$i_t(t) = e^{\alpha t}(A_1 + K_2 t)$$

Case 3: $\left(\dfrac{R}{2L}\right)^2 < \dfrac{1}{LC}$. Roots D_1 and D_2 are complex conjugates.

(under damped)

$$i_t(t) = e^{-\alpha t}(K_1 \cos \beta t + K_2 \sin \beta t)$$

or $\quad i_t(t) = e^{-\alpha t} \sin(\beta t + \theta)$

For any case, there are two constants to be determined from the initial conditions. In this section, $i_L = 0$ and $v_c = 0$ at $t = 0^+$. Then, the initial conditions are:

$$i = i_t + i_{ss} = 0 \quad \text{and} \quad L\dfrac{di}{dt} = V \quad \text{and} \quad \dfrac{di}{dt} = \dfrac{V}{L}$$

(From $V = iR + L\dfrac{di}{dt} + \displaystyle\int \dfrac{i\,dt}{c}$, where i=0). And, these two

equations provide two conditions for evaluating the two unknown constants.

Figure 6.146 depicts the current response for all three cases.

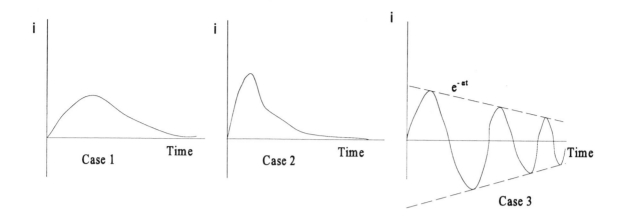

Figure 6.146

Review

In the previous sections, the solution of various type transient problems involved basically four steps. The steps are:

Step 1: The writing of Kirchoff's Voltage Law for the situation immediately after switch operation, i.e., at $t=0^+$.

Step 2: Obtaining the solution for the transient or natural response. This is obtained by setting the forcing or excitation function equal to zero and solving the resulting homogeneous equation. The solution is always of the form:

$$i = Ke^{-\frac{t}{\tau}} \text{ where } \tau \text{ is the time constant.}$$

Step 3: Determining the forced or steady-state response using conventional circuit theorems, i.e., Ohm's Law, etc.

Step 4: Adding the transient and steady-state responses together and solving for the undetermined coefficients using boundary conditions.

INTRODUCTION TO LAPLACE TRANSFORMS

In addition to the classical methods of differential equation solutions, another technique is available wherein transformations are utilized to solve Kirchoff's voltage laws in differential equation form. This latter technique is called the Laplace transform method of analysis and it accomplishes the following:

1. It transforms differential equations, in the time domain, into alegraic equations, in the frequency domain. These algebraic equations, now in terms of a new variable S, include the boundary or initial conditions.

2. These transformed equations, which are algebraic in nature, are solved for the new variable S algebraically, after which an inverse transformation is made. This effectively transfers the frequency domain solution back into the time domain where it is now the solution of the original differential equation, thus solving the transient problem.

The Laplace Transform

The Laplace Transform of f(t) denoted by $\mathcal{L}[f(t)]$ is defined by:

$$\mathcal{L}[f(t)] = F(s) = \int_0^\infty f(t)\, e^{-st}\, dt$$

Note that when the integration is performed, the function f(t) becomes a function of S—the Laplace variable.

To illustrate this transformation, consider a step function of voltage V. Find its Laplace Transform.

$$f(t) = V \text{ a constant voltage}$$

$$\mathcal{L}(V) = \int_0^\infty V e^{-st}\, dt = \left[-\frac{V}{S} e^{-st} \right]_0^\infty = \frac{V}{S}$$

Fortunately, it is not necessary to integrate f(t) whenever F(s) is needed. There exist tables of Laplace Transform pairs for all conceivable time functions. Several frequently used pairs are tabulated on p. 6-146. Use of this table will normally result in a great savings in time and effort.

To illustrate Laplace transform usage, consider the following two examples.

Example

Solve for i(t) using Laplace transforms.

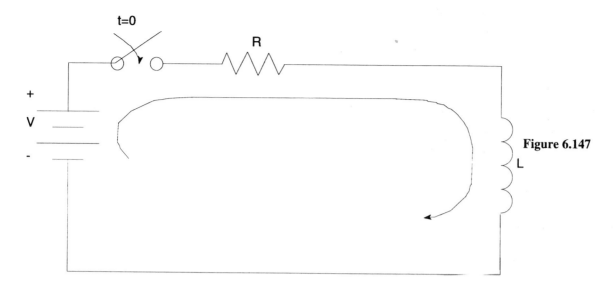

Figure 6.147

Solution

Kirchoff's voltage law $\mathcal{L} R\, i(t) + \mathcal{L} L\, \dfrac{di}{dt}(t) = \mathcal{L} V$.

After taking the Laplace using transforms 13, 11 and 1, the transformed equation becomes: [Note i(t) becomes I(s)]:

$$R\, I(S) + L[S\, I(s) - i(0^+)] = \dfrac{V}{S}$$

$\qquad\qquad\qquad\qquad\uparrow$
$\qquad\qquad\qquad\quad$ Initial current at t=0$^+$

which reduces to $(R + SL)I(s) = \dfrac{V}{S}$ since $i(0^+) = 0$

And $I(s) = \dfrac{V}{S}\, \dfrac{1}{SL + R}$ which is not in standard format.

Laplace Transforms

	$f(t)$	$F(s)$
1	1 or V $t \geq 0$	$\dfrac{1}{S}$ or $\dfrac{V}{S}$
2	Vt $t \geq 0$	$\dfrac{V}{S^2}$
3	e^{-at}	$\dfrac{1}{S+a}$
4	$t\,e^{-at}$	$\dfrac{1}{(S+a)^2}$
5	$\sin wt$	$\dfrac{W}{S^2+W^2}$
6	$\cos wt$	$\dfrac{S}{S^2+W^2}$
7	$\sin(wt+\theta)$	$\dfrac{S\sin\theta + W\cos\theta}{S^2+W^2}$
8	$\cos(wt+\theta)$	$\dfrac{S\cos\theta - W\sin\theta}{S^2+W^2}$
9	$e^{-at}\sin wt$	$\dfrac{W}{(S+a)^2+W^2}$
10	$e^{-at}\cos wt$	$\dfrac{S+a}{(S+a)^2+W^2}$
11	$\dfrac{df(t)}{dt}$	$SF(s) - f(0^+)$
12	$\int f(dt)\,dt$	$\dfrac{F(s)}{S} + \dfrac{f^{-1}(0^+)}{S}$
13	$f_1(t) + f_2(t)$	$F_1(s) + F_2(s)$

6-146 FUNDAMENTALS OF ENGINEERING EXAM REVIEW WORKBOOK

Further algebraic manipulation results in:

$$I(s) = \frac{V}{L} \left(\frac{1}{S}\right)\left(\frac{1}{S + R/L}\right).$$

This expression for I(s) must be converted into factors that can be looked up in the Table of Transforms. Accordingly, the method of partial fractions must be used.

Note

$$\frac{1}{S(S + R/L)} = \frac{A}{S} + \frac{B}{S + R/L} \quad \text{where A and B are to be determined}$$

One technique is to multiply both sides of the equation by S and S + R/L, respectively, and let S→0 and -R/L for each case.

(S)
$$\frac{1}{S(S + R/L)} = \frac{A(S)}{S} + \frac{B}{S + R/L}(S)$$

$$\frac{1}{S + R/L} = A + \frac{BS}{S + R/L}$$

As S→0

$$A = \frac{L}{R}$$

And

(S + R/L)
$$\frac{1}{S(S + R/L)} = \frac{A(S + R/L)}{S} + \frac{B}{S + R/L}(S + R/L)$$

$$\frac{1}{S} = \frac{A}{S}(S + R/L) + B$$

As S→ -$\frac{R}{L}$

$$B = -\frac{L}{R}$$

Therefore

$$\frac{1}{S(S + R/L)} = \frac{L}{R}\left(\frac{1}{S}\right) - \left(\frac{1}{S + R/L}\right)$$

which can now be looked up in the Table.

The inverse Laplace, i.e. $\mathcal{L}^{-1} F(s) = f(t)$, is taken using transforms 13, 3 and 1.

$$\mathcal{L}^{-1} \frac{1}{S(S + R/L)} = \mathcal{L}^{-1} \frac{L}{R} (\frac{1}{S}) - \mathcal{L}^{-1} \frac{L}{R} (\frac{1}{S + R/L})$$

$$= \frac{L}{R} [\mathcal{L}^{-1}(\frac{1}{S}) - \mathcal{L}^{-1}(\frac{1}{S + R/L})]$$

$$= \frac{L}{R} [1 - e^{-\frac{R_t}{L}}]$$

And

$$i(t) = \frac{V}{L} [\frac{L}{R} (1 - e^{-\frac{R_t}{L}})]$$

$$i(t) = \frac{V}{R} (1 - e^{-\frac{R_t}{L}}) \quad \underline{\text{Answer}}$$

Example

Solve for the current in the circuit of shown in Fig. 6.143 at t = 4.5 sec after switch closure using Laplace transforms.

Solution

Kirchoff's Voltage Law

$$\mathcal{L} \frac{1}{C} \int_0^t i \, dt + \mathcal{L} Ri = \mathcal{L} 0$$

Taking the Laplace using transforms 13 and 12, the transformed equation becomes:

$$\frac{1}{C} [\frac{I(s)}{S} + \frac{f^{-1}(0^+)}{S}] + R\, I(s) = 0$$

where

$f^{-1}(0^+)$ is the initial value of the $\int i \, dt \big|_{0+}$

i.e. $f^{-1}(0^+) = \int i \, dt \big|_{0+}$

This value can be determined from the expression:

$$i = \frac{dq}{dt} \text{ where q is the charge on the capacitor}$$

in coulombs.

and $q = \int i\, dt$

which is $q(0^+)$ where $\int i\, dt \Big|_{0+}$

Note also that from

$$q(0^+) = Cv_c(0^+)$$

that $v_c(0^+) = \frac{q(0^+)}{C}$ where $v_c(0^+)$ is the initial voltage on the capacitor.

Accordingly, the transformed equation, after inserting values, becomes

$$\frac{1}{0.3}\left[\frac{I(s)}{S}\right] + \frac{750}{S} + 100\, I(s) = 0$$

which reduces to:

$$\left(\frac{1}{0.3S} + 100\right) I(s) = -\frac{750}{S}$$

or

$$(30S + 1)\, I(s) = -225$$

And $\quad I(s) = -\dfrac{225}{1 + 30S} = -\dfrac{7.5}{S + 1/30}$

The $\mathcal{L}^{-1}[I(s)] = i(t) = -7.5 e^{-\frac{t}{30}}$ using transform 3.

And, at t = 4.5 sec

$i(t) = -7.5 e^{-\frac{4.5}{30}} = -7.5 e^{-0.15} = -6.46$ amperes.   **Answer**

SYSTEM ENGINEERING

Introduction

A system can be defined as a combination of diverse interacting elements that are integrated together to achieve an overall objective. The objective can be varied such as guiding a space vehicle or controlling a manufacturing process. To achieve this end, there are no constraints on the interacting elements. They can consist of mechanical devices, electrical devices, hydraulic devices, pneumatic devices, etc. In fact, a human being can also be considered an element in a system.

The systems engineer takes an overall view. His focus is on the system's external characteristics, predominately performance. Does it meet specifications? Is it reliable? Is it efficient and economical? The systems engineer does not concern himself with component design. To him a device or subsystem can be represented just by a simple block that is labeled to indicate the function it performs. A combination of such blocks can represent a complete system. This presentation will use blocks extensively because it removes excess detail from the picture and shows only the functional operation of the system.

Automatic Control Systems

In general there are two types of automatic control systems, open loop and a closed loop. An open loop system is one in which the output quantity has no effect upon the input action which is performed. A input signal or command is applied, perhaps amplified, and a power output is obtained. A block diagram illustrating this type system is shown in Fig. 6.148.

Figure 6.148

The command input actuates this system and specifies the desired output. However, the command input must first be translated into a signal that is understood by the system; and this accepted signal is termed the reference input. The reference selector is the device that converts the command input into the reference input and the reference input causes the dynamic unit to be activated thereby producing the controlled output.

An example of an open loop system is the simple toaster. The command input to the toaster is the desired darkness. The reference selector is a dial which is calibrated from light to dark. This produces the reference input which controls the time that the heating coils (the dynamic unit) are on. The degree of toasting of the bread is the controlled output. However, consistent with the definition of open loop, the toaster cannot by itself adjust itself to compensate for thick or thin slices of bread. A human being provides this information via the command input.

A closed loop (feedback) control system is one in which the output quantity influences the input to the dynamic unit. Consequently, it influences the control action. A block diagram illustrating this type system is shown in Fig. 6.149.

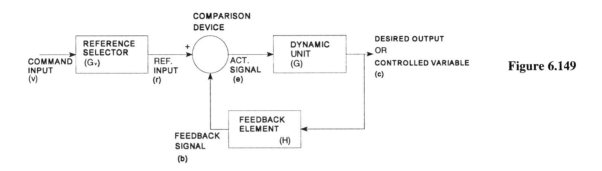

Figure 6.149

The feedback signal, which is a function of the output signal, is compared in the comparison device with the reference input. The difference between the reference input and the feedback signal produces an actuating signal. This is sometimes called an error signal. The actuating signal is the input to the dynamic unit and it produces the output. Accordingly, the controlled output is a function of both the reference input and the controlled output. This type of system is known as a feedback control system.

Closed loop systems are sometimes classified according to the manner in which the error signal is used to control the power output. The most common classifications are:

On-Off Controllers in which the power is turned on or off whenever the error signal reaches a predetermined value. Room temperature control via a thermostat or temperature control of a refrigerator are some examples of this type system. The power is continuously applied until the desired temperature is reached.

Step Controllers which are a type of on-off system whereby the power is applied in steps. The power can be applied at preset time intervals whenever an error signal is applied and it remains on for a definite time. In this manner, for example, the temperature of a furnace can be raised with minimum overshoot. Sometimes the duration of the power application is in variable pulses depending on the magnitude of the error.

Servomechanisms are a type of closed loop system where the output is measured and when compared to the reference, the error (or some function of it) controls not only the magnitude of the power applied but also its direction. An automatic pilot on an aircraft is an example of this type system.

Standard Block Diagram Symbols

The AIEE Subcommittee on Terminology and Nomenclature of the Feedback Control Systems Committee has proposed the following letter symbols as standard throughout the industry.

(v)   Is the command and is the input which is established by external means. It is independent of the feedback control system.

(r)   Is the reference input which is derived from the command and is the actual signal input to the system.

(c)   Is the controlled variable and is the quantity that is directly measured and controlled. It is the output of the system.

(b)   Is the primary feedback signal which is a function of the controlled variable and is subsequently compared with the reference input to obtain the actuating signal (e).

(e)   Is the actuating signal which is obtained from a comparison measuring device and is equal to the reference input (r) minus the primary feeback (b). Usually this signal is at a low energy level and it is usually applied to control elements that drive the control system. The driving signal usually is at a much higher energy level.

(G_v)   Is the reference input elements and it physically produces a signal proportional to the command.

(G)   Can be a combination of control elements and control systems. It comprised the dynamic unit.

(H)   Is the feedback elements that produces the primary feedback from the controlled variable. Usually this is a proportionally device and it may also modify the characteristics of the controlled vehicle.

Methods of System Analysis

There are basically two methods by which servomechanism systems can be analyzed. The first method involves writing the differential equation of motion for the system and solving for its roots. The performance may then be predicted by plotting the appropriate curves and inspecting them. The second method involves transfer functions. In brief, a transfer function describes the effect of the component upon a signal's amplitude and phase as the signal passes through the component—from its input to its output. This result is obtained by multiplying the signal by the transfer function of the component. All that is necessary for this method is that the characteristics of the component be expressed mathematically. Then, by algebraic manipulation, whereby transfer functions are combined, in series or parallel, depending on the system arrangement, the performance of the overall system is arrived at by the algebraic combination of the equations of the component parts. The net result is a savings in labor.

Differential Equations of Feedback Control Systems

The mathematical equations that describe system characteristics of a feedback control system are usually of the same form regardless of whether the component parts are electrical, mechanical, chemical, etc. The basic equation is:

$$A_n \frac{d^n \theta_0}{dt^n} + \ldots + A_2 \frac{d^2 \theta_0}{dt^2} + A_1 \frac{d\theta}{dt} + A_0 \theta_0 + A + \int \theta_0 \, dt = F(e)$$

Where θ_0 = output quantity

θ_i = input quantity

e = error $(\theta_i - \theta_0)$

and the "A_n's" are functions of system parameters.

Because the simplest type of feedback system may contain components with inertia, friction, etc., the above equation reduces to at least a second order differential equation. Therfore, if the differential equation of motion of a system is solved for its roots, its perfomance may be predicted. However, if some components are changed, the labor involved in repeated solutions is excessive and it is difficult to notice any trends. Therfore, the transfer function analysis method is preferred.

It should be mentioned that a majority of the problems in the NCEE publication dealing with automatic control systems give the transfer function and requests system performance and/or analysis from it. Accordingly, this section will concern itself with this latter method of system analysis.

Transfer Functions of Feedback Control Systems

The transfer function of a system may be obtained very simply from the Laplace Transform of the differential equation. However, the system must be assumed to be initially at rest. This makes all the initial conditions equal to zero.

The Laplace Transform

The Laplace Transform is a mathematical operation used to transform functions of time into functions of the complex variable S. This variable S is an arbitrary complex variable and is of the form σ and $j\omega$. By taking the Laplace of a differential equation, the subsequent solution is obtained by using simple algebra. The Laplace automatically takes into consideration initial or boundary conditions and the solution simultaneously contains both the transient and steady state components. Also, labor is reduced by using a table of transforms.

The Laplace is defined by:

$$F(s) \equiv \mathcal{L} f(t) = \int_0^\infty f(t) e^{-st} dt$$

where $\mathcal{L} \equiv e^{-st} dt$

Without going through the calculus,

$$\mathcal{L}\left\{\frac{d}{dt}\{f(t)\}\right\} = S F(s) - f(0^+)$$

Where $f(0^+)$ is the initial value of f(t), evaluated as t→0 from positive values.

Control system input signals can be of a variety of types. Some are step functions, ramp functions, sinusoidal functions, etc. The $\mathcal{L}$ of these functions with their corresponding F(S)'s are shown in Fig. 6.150.

After the transform of the differential equation is performed tht response function is easily found using simple algebra. However, to complete the solution, the inverse of the transform must be found. By definition

$$f(t) = \mathcal{L}^{-1}\{F(S)\}$$

where $\mathcal{L}^{-1}$ stands for the inverse Laplace. Most inverse transforms can be obtained by direct reference to transform tables. If a transform of a response cannot be found in a table, a procedure is used whereby F(S) is expressed as the sum of partial fractions with constant coefficients. The complete inverse transform is the sum of the inverse transforms of each fraction. As a review two cases are studied:

Case 1: Single Roots

$$F(S) = \frac{N(S)}{D(S)} - \frac{N(S)}{S(S-S_1)(S-S_2)} = \frac{A_0}{S} + \frac{A_1}{S-S_1} + \frac{A_2}{S-S_2}$$

where the A's have to be determined.

And

Case 2: Repeated Roots

$$F(S) = \frac{N(S)}{D(S)} - \frac{N(S)}{S(S-S_1)^n(S-S_2)} = \frac{A_0}{S} + \frac{A_1}{(S-S_1)^n} + \frac{A_2}{(S-S_2)^{n-1}} + \ldots + \frac{A_n}{(S-S_1)} + \frac{A_0}{(S-S_2)}$$

where all A's have to be determined.

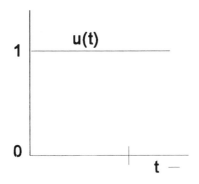

Unit Step Function

$$F(S) = \frac{1}{S}$$

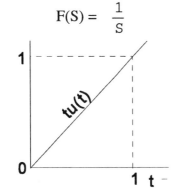

Unit Ramp Function

$$F(S) = \frac{1}{S^2}$$

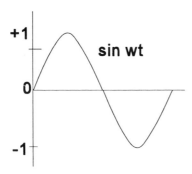

Sine Function

$$F(S) = \frac{\omega}{S^2 + \omega^2}$$

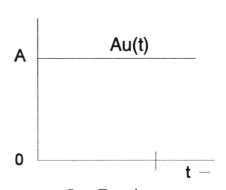

Step Function

$$F(S) = \frac{A}{S}$$

Ramp Function

Figure 6.150

$$F(S) = \frac{A}{S^2}$$

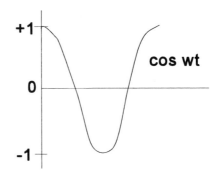

Cosine Function

$$F(S) = \frac{S}{S^2 + \omega^2}$$

If the $\mathcal{L}^{-1}$ is taken for Case 1 then

$$f(t) = A_0 + A_1 \epsilon^{S_1 t} + A_2 \epsilon^{S_2 t}$$

Correspondingly, for Case 2

$$f(t) = A_1 \frac{t^{n-1}}{2} \epsilon^{S_1 t} + A_2 t^{n-2} \epsilon^{S_1 t} + \ldots + A_n \epsilon^{S_1 t} + A_0 \epsilon^{S_2 t}$$

It should be noted the roots in the S domain becomes an exponential of the factor ϵ in the time domain. Consequently, the nature of the S domain roots definitely affects the time domain response. This will be further discussed in the system stability section.

Transfer Functions

A transfer function is defined as the complex ratio of the output function to the input function with zero initial conditions. To illustrate consider the example in Fig. 6.151; it is simple RC network. What is its transfer function?

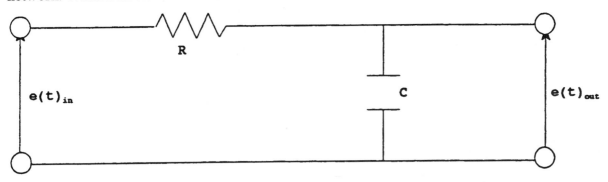

Figure 6.151

Writing Kirchoff's Voltage Law, the ration of output and input are solved for in one equation.

$$e(t)_{in} = e(t)_R + e(t)_C$$

Note:

$$e(t)_C = e(t)_{out}$$

Accordingly

$$e(t)_{in} = i(t) R + \frac{1}{C} \int i(t) \, dt$$

To eliminate the integral, the equation is differentiated.

$$\mathcal{L}\left[\frac{de(t)}{dt}\right] = \mathcal{L}\left[R\frac{di(t)}{dt} + \frac{i(t)}{C}\right]$$

and the Laplace is taken, yielding

$$SE(S)_{in} = RSI(S) + \frac{1}{C}I(S)$$

$$SE(S)_{in} = (SR + \frac{1}{C})I(S)$$

Where initial conditions are zero.

$$[e(0^+) \text{ and } i(0^+) \text{ are zero.}]$$

For the output.

$$e(t)_{out} = e(t)_C = \frac{1}{C}\int i\,dt$$

Again differentiating and taking the Laplace

$$\mathcal{L}\left[\frac{de(t)_{out}}{dt}\right] = \mathcal{L}\left[\frac{i(t)}{C}\right]$$

$$SE(S)_{out} = \frac{I(S)}{C}$$

$$I(S) = SCE(S)_{out}$$

Substituting I(S) into the above equation

$$SE(S)_{in} = (SR + \frac{1}{C}) SCE(S)_{out}$$

$$SE(S)_{in} = \left[\frac{SRC+1}{C}\right] SCE(S)_{out}$$

The following is obtained:

$$\frac{E(S)_{out}}{E(S)_{in}} = \frac{1}{1+SRC} \qquad \frac{E(S)_{out}}{E(S)_{in}} = \frac{\frac{1}{RC}}{S + \frac{1}{RC}} = \frac{1}{RC}\frac{1}{\left(S + \frac{1}{RC}\right)}$$

Accordingly, the transfer function can be represented as shown in Fig. 6.152 and it is independent of the form of the input.

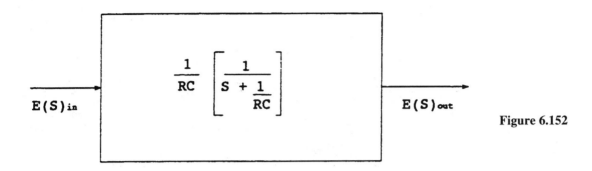

Figure 6.152

The same technique can be applied to amplifiers, motors, generators, etc. Consequently, a control system can simply be represented by blocks.

Overall Transfer Function

A block diagram of a control system with negative feedback is shown in Fig. 6.153.

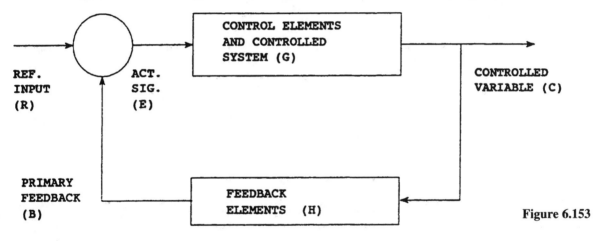

Figure 6.153

The equations describing this sytem can either be in terms of the complex variables (the Laplace operator), or the frequency parameter $j\omega$, e.g., $G(s)$ or $G(j\omega)$, which indicates that it is a transformed quantity.

Accordingly, the equations describing this system are:

$$C(s) = G(s)\ E(s)$$
$$B(s) = H(s)\ C(s)$$
and
$$E(s) = R(s) - B(s)$$

Further algebraic manipulation results in the overall system transfer function:

$$\frac{\text{Output}}{\text{Input}} = \frac{C(s)}{R(s)} = \frac{G(s)}{1+G(s)H(s)}$$

The denominator of the above expression when set equal to zero, i.e.,

$$1+G(s)H(s)=0$$

is referred to as the characteristic equation of the closed loop system, and, the system stability and response can be determined from the analysis of this equation.

Other ratios obtainable from the above equations are:

$$\frac{B(s)}{E(s)} = G(s)H(s)$$

which is called the open loop transfer function and

$$\frac{C(s)}{E(s)} = G(s)$$

which is called the forward transfer function.

The quantity G(s), in addition, to being made up of single elements or elements in cascade, may also contain minor feedback loops as well. Therefore, for analysis, these minor loops must be combined to get an equivalent forward transfer function.

Block Diagram Algebra

Three basic components of feedback systems together with their defining equations are shown in Figs. 6.154 to 6.156.

(a) Transfer function

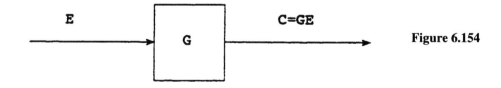

Figure 6.154

(b) Summing Point

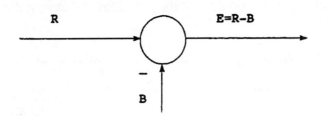

Figure 6.155

(c) Pickoff Point

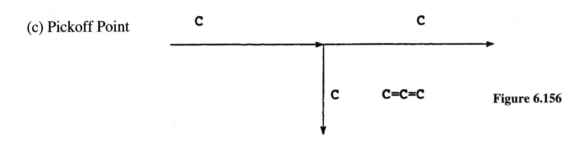

Figure 6.156

The following rules are applicable to the above blocks.

1. Any closed loop system can be replaced by an equivalent open loop system (Fig. 6.157).

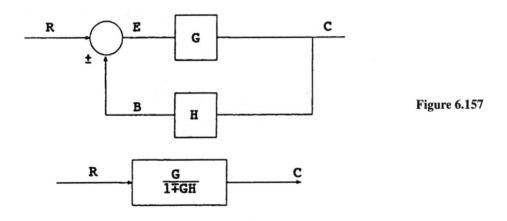

Figure 6.157

2. The gain of cascaded blocks is the product of the individual gains (Fig. 6.158).

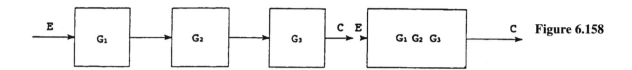

Figure 6.158

3. The order of summing does not affec the sum (Fig. 6.159).

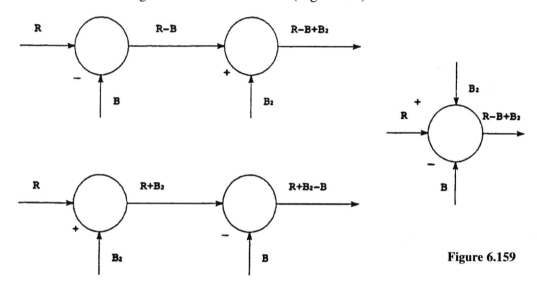

Figure 6.159

4. Shifting a summing point beyond a block of gain G requires the insertion of G in the variable added (Fig. 6.160).

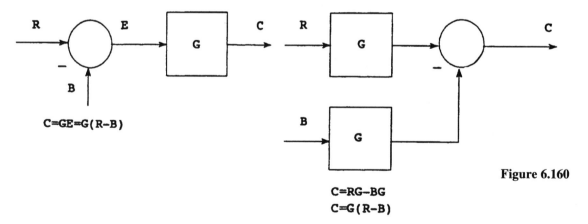

$C = GE = G(R-B)$

$C = RG - BG$
$C = G(R-B)$

Figure 6.160

5. Shifting a pickoff point beyond a block of gain G requires the insertion of 1/G in the variable picked off (Fig. 6.161).

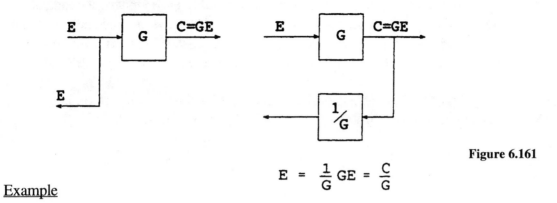

$$E = \frac{1}{G} GE = \frac{C}{G}$$

Figure 6.161

Example

The system shown in Fig. 6.162 has parallel loops. Reduce it to a single closed loop system.

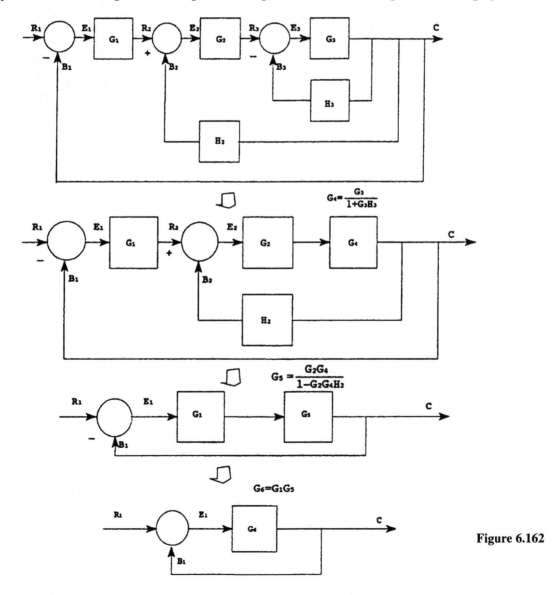

Figure 6.162

Stability

The feedback signal, in addition to providing a comparison to the reference input, if it contains amplification, it may give rise to system instability. In a stable feedback system, the response to an impulse disturbance will die away as time increases, whereas in an unstable system, a sudden disturbance can give rise to sustained oscillations or to an uncontrolled output.

From a mathematical point of view, the characteristic equation of a closed loop system, i.e., $1 + G(S)H(S)$ is what determines system stability. The overall gain with feedback is:

$$G_F(S) = \frac{G(S)}{1 + G(S)H(S)}$$

If the denominator is zero, the gain increases without limit and the system is unstable. There is an output with no input. Another way of stating this same criteria is if the open loop transfer function $G(S)H(S)$ is just equal to $-1 + j0$, the system is again unstable.

Therefore, to analyze a system for stability, if the transfer function is not given, the Laplace transform is taken of the describing differential equation. This results in an algebraic equation in (S) that can be solved for in terms of $C(s)/E(s)$, the forward transfer function. Generally, this transfer function has one of the following forms:

$$G(S) = \frac{K_0(1 T_1 S)\ldots}{(1+T_a S)(1+T_b S)}$$

$$G(S) = \frac{K_1(1 T_1 S)}{S(1+T_a S)(1+T_b S)\ldots\ldots}$$

$$G(S) = \frac{K_2(1 T_1 S)}{S^2(1+T_a S)(1+T_b S)}$$

Note: All the above expressions, for simplification, have unity feedback, i.e., $H(S)G(S) = G(S)$.

Note the powers of S alone in the denominator of the above expressions. In the first expression it is zero. In the second expression it is one. In the last expression it is two. Accordingly, feedback systems are classified by the exponent of S: a Type 0 system, a Type 1 system, a Type 2 system, a Type 3 system, etc.

To obtain the time domain response, the inverse Laplace Transfrom is utilized. The factors of the denominator (the roots of the characteristic equation) give rise to time functions in terms of ϵ^{st} where s are the roots of the denominator terms. These exponential factors are the transient terms in the time domain and for these terms to vanish with increasing values of time, it is necessary for the real part of s, the roots of the denominator, be negative in sign. This is the mathematical condition for stability. If this criterion is considered from the point of view of a plot of the roots in the complex plane, then for stability, all roots must lie on the left hand plane, i.e., all roots must lie to the left of the origin.

Location of Poles and Stability

The values of the roots, in the finite plane, that make the denominator equal to zero are called the "zeroes" of the denominator. At these values, which may be either real or complex, the overall function is equal to infinity. These values are called the "poles" of the complete function. Several examples will follow each of which will demonstrate a particular stability case.

Example 1: Consider the following transfer function (Fig. 6.1) representing a feedback system. Is it stable? Why?

$$G(S) = \frac{K}{(S+1)(S-6)(S+2)} \qquad H(S) = 1$$

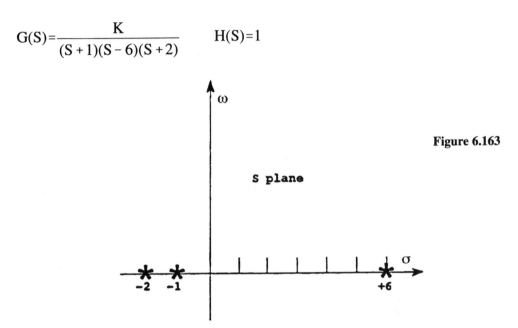

Figure 6.163

Solving for $1 + G(S)H(S) = 0$, it is seen that the system is not stable. There are three poles as plotted above on the S plane diagram. Note that the pole at +6 is in the right half plane. When the inverse Laplace is taken,

$$f(t) = A_0 e^{-t} + A_1 e^{-2t} + A_2 e^{+6t}$$

The poles at -1 and -2 gives rises to terms that approach zero as $t \to \infty$. However, the pole aat +6 gives rise to an increasing function $t \to \infty$. This last term causes the instability.

Example 2: Given the transfer function with repeated roots. Is it stable?

$$G(S) = \frac{K}{(S+2)^2 (S+3)}$$

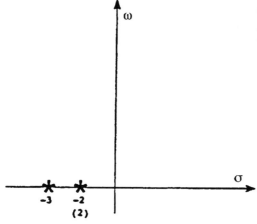

The poles are plotted as shown in Fig. 6.164. Note there is a double root at −2.

Figure 6.164

The $\mathcal{L}^{-1}$ is: $f(t) = A_0 t e^{-2} + A_1 e^{-2t} + A_3 e^{-3t}$

Note

$$\mathcal{L}^{-1}\left[\frac{1}{(S+a)^n}\right] = \frac{1}{(n-1)!} t^{n-1} e^{-\infty t}$$

The system is stable because all powers of e are negative and ther term goes to zero as $t \to \infty$.

Example 3: Given the transfer function $\qquad G(S) = \dfrac{K}{(aS^2+bS+C)}$

Is it stable?

The general case will be treated.

$$S_{1,2} = -\frac{b}{2a} \pm \frac{1}{2a}[b^2-4ac]^{\frac{1}{2}}$$

The poles of $as^2 + bs + c = 0$ are

Usually b is an energy dissipation or consumption element. Consequently it is called a "damping coefficient" and determines the rate of decay of the transient term. When $b = b'$ (critical) $= 2\sqrt{ac}$ the radical term is equal to zero and the two roots $S_{1,2}$ are equal. Accordingly another term

ζ (Zeta) called the "damping ratio" is defined as:

$$\zeta = \frac{\text{Actual Damping Coefficient}}{\text{Critical Damping Coefficient}} = \frac{b}{b'}$$

Therefore, depending on the value of ζ, i.e., $\zeta > 1$, $\zeta = 1$, or $\zeta < 1$, the transient response of the system can be known. Reference to Fig. 6.165 denotes the various responses.

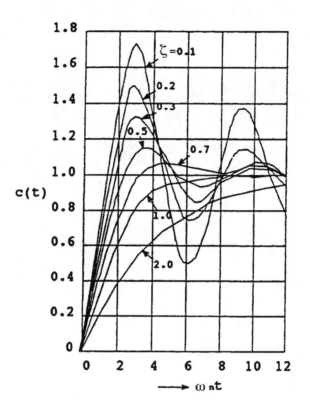

Figure 6.165

Standard Second-Order Transient Curve

Another term that is noted from the above figure is the "undamped natural frequency ω_n and is defined as the frequency of oscillation of the transient when the damping is zero. At this point the quadratic roots are:

$$S_{1,2} = \pm j\sqrt{c/a} = j\omega_n$$

and the response is a sine wave of constant amplitude that does not die out. Putting these terms ζ and ω_n into the original quadratic equation results in a new standard equation. This equation is:

$$aS^2 + bS + C = S^2 + \frac{b}{a}S + \frac{c}{a} = S^2 + 2\zeta\omega_N S + \omega_n^2$$

Lastly, ω_D is defined as the "damped natural frequency" of the transient and is equal to:

$$\omega_d = \omega_N \sqrt{1-\zeta^2}$$

For the two cases discussed above, namely the damped and undamped reponses the pole locations are as shown on the S plane diagrams.

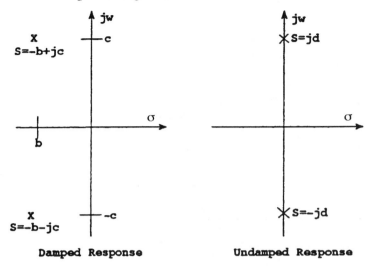

Figure 6.166

Damped Response   Undamped Response

Therefore, getting back to the problem of Example 3 where:

$$G(S) = \frac{K}{as^2 + bs + c}$$

Converting it into standard form

$$G(S) = \frac{K/a}{S^2 + 2\zeta\omega_N S + \omega_N^2}$$

The inverse Laplace is

$$f(t) = A_1 e^{(-\zeta\omega_n + j\omega_n\sqrt{1-\zeta^2})t} + A_2 e^{(-\zeta\omega_n + j\omega_n\sqrt{1-\zeta^2})t}$$

which can be simplified to:

$$f(t) = 2|A|\, e^{-\zeta\omega_n t}\, \sin(\omega_n\sqrt{1-\zeta^2}\, t + 0)$$

When the damping term is zero

$$G(S) = \frac{K/a}{S^2 + \omega_n^2} \quad \text{from which} \quad f(t) = A_1 e^{+j\omega_n t} + A_2 e^{-j\omega_n t}$$

which can be simplified to:

$$f(t) = 2|A| \sin(\omega_n + \emptyset)$$

Therefore, for the damped case, the system is stable, because $f(t) \to 0$ as $t \to \infty$. However, for the undamped case wherein the poles, are on the imaginary axis, a constant sinusoidal output is obtained independent of the forcing function. Obviously, this situation is unstable.

Routh's Criterion

A general form of transfer function is

$$G(S) = \frac{K_n(1 + a_1 S + a_2 S^2 + \ldots a_z S^Z)}{S^n(1 + b_1 S + b_2 S^2 + \ldots b_x S^X)}$$

To determine the system stability, the roots of the characteristic equation must be known. Sometimes this solution becomes exceedingly involved. However, whenever n = 0, 1, or 2 no particular difficulties may be encountered in obtaining the roots. But, for n = 3, 4, or 5 or larger, the labor involved can be considerable. Accordingly, Routh's Criterion can be applied to determine if the system is stable or not, without the necessity of evaluating the actual magnitude of the roots. However, it is still essential to determine whether there does exist any roots that are real and positive or that have positive real parts (if complex). This can be done quite directly by Routh's Criterion.

In the general form of the characteristic equation is

$$a_0 S^n + a_1 S^{n-1} + a_2 S^{n-2} + \ldots a_{n-1} S + a_n = 0$$

and an inspection reveals that one of the coefficients of the decreasing powers of S is missing, or if not all of the coefficients have the same algebraic sign, then it can immediately be determined that the system is unstable. Proof of this is beyond the scope of this section.

However, if the initial inspection reveals that the above two criteria are met, than an arrangement of terms in triangular array, as shown below, must be done. This is referred to as Routh's Criterion.

a_0 a_2 a_4 a_6 ...

a_1 a_3 a_5 a_7 ...

b_1 b_3 b_5 ... ...

c_1 c_3 ... ...

d_1 ... ...

where

$$b_1 = \frac{a_1 a_2 - a_0 a_3}{a_1}$$

$$b_3 = \frac{a_1 a_4 - a_0 a_5}{a_1}$$

$$b_5 = \frac{a_1 a_6 - a_0 a_7}{a_1} \quad \ldots \text{etc.}$$

$$c_1 = \frac{b_1 a_3 - a_1 b_3}{b_1}$$

$$c_3 = \frac{b_1 a_5 - a_1 b_5}{b_1} \quad \ldots \text{etc.}$$

$$d_1 = \frac{c_1 b_3 - b_1 c_3}{c_1} \quad \ldots \text{etc.}$$

In general each succeeding horizontal row will have fewer terms than the preceding row; thus the array is triangular. The procedure of forming additional horizontal rows must be carried out until no more rows can be formed by the procedure. Once the array is complete, it is necessary only to inspect the signs in front of the left hand column. If they all have the same sign, there are no positive roots and the system is stable. If a change in sign is noted, then positive roots exist and the system is unstable. If there are several changes in sign, the number of positive real roots corresponds to the number of changes in sign.

<u>Example</u> Is the system that has the following characteristic equation stable?

a_0 a_1 a_2 a_3 a_4

$1S^4 + 2S^3 + 6S^2 + 4S + 3 = 0$

The array is:

$$\begin{array}{c|cc}
 & a_0 \; a_2 \; a_4 \\
+ & 1 \; 6 \; 3 \\
 & a_1 \; a_3 \\
+ & 2 \; 4 \\
 & b_1 \; b_3 \\
+ & 4 \; 3 \\
 & c_1 \\
+ & 5/2 \\
 & d_1 \\
+ & 3
\end{array}$$

the system is stable since all terms in the first column are positive.

<u>Example</u> Is the system that has the following characteristic equation stable?

$$S^4 + 2S + S^2 + 4S + 2 = 0$$

The array is

$$\begin{array}{c|cc}
 & a_0 \; a_2 \; a_4 \\
+1 & 1 \; 1 \\
 & a_1 \; a_3 \\
+2 & 4 \\
 & b_1 \; b_3 \\
-1 & 2 \\
 & c_1 \\
+8 \\
 & d_1 \\
+2
\end{array}$$

The system is unstable since not all terms in the first column are positive. There are two positive roots.

Nyquist Stability Criterion

The Nuquist Stability Criterion is a graphical procedure for determining system stability. It is determined by plotting the frequency response of the open loop transfer function, i.e., $G(j\omega) H(j\omega)$. (Note that to get the frequency response replace all S's in the transfer function by $j\omega$.) The plot is done on polar coordinate paper.

The generalized form of the Nyquist Criterion to determine stability can be applied as follows:

Plot the G(S) H(S) function in the complex plan ($S = \sigma + j\omega$) with all values of ω from $-\infty$ to $+\infty$. The plot of $-\infty<\omega<0$ should be the conjugate of the plot from $+\infty>\omega>0$.

Next draw the vector from the $-1+j_o$ point to any point on the curve and observe the rotation of this vector as ω varies from $-\infty$ to $+\infty$. If the point $-1 + j_o$ is enclosed or encircled the system is unstable. If there is no encirclement, the system is stable.

Example

For the transfer function plots shown in Figs. 6.167 to 6.169, indicated whether the systems are stable or not.

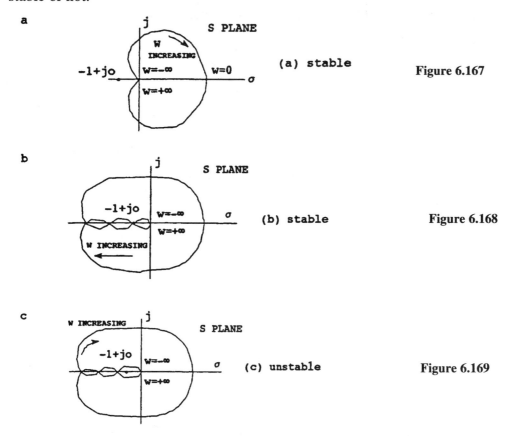

(a) stable — Figure 6.167

(b) stable — Figure 6.168

(c) unstable — Figure 6.169

Root Locus Method

A detailed in depth discussion of the Root Locus Method is beyond the scope of this section. However, suffice it to say that this method is a graphical procedure wherein the roots of the characteristic equation, as a function of system parameters, are plotted. From this plot, a designer can gain a clear indication of the effects of parameter adjustments or variations on a system with relatively small effort as compared to other methods. He can also obtain the system's transient and frequency responses. Accordingly, the Root Locus Method is a very powerful tool that a designer can use in the analysis and design of closed loop systems. However, for those students who desire more information of this method, it is recommended that they review this subject in the various Automatic Control Systems textbooks.

SYSTEMS ENGINEERING EXAMPLES

1. The transfer function of a system is:

$$F(S) = \frac{12(S + 4)}{S(S + 2)(S + 6)}$$

What is the corresponding time domain response f(t)? Use the method of partial fractions.

(a) $16 - 16e^{-6t}$
(b) $6e^{-2t}$
(c) $3e^{-2t} - e^{-6t}$
(d) $4 - 3e^{-2t} - e^{-6t}$
(e) $4 - 6e^{-6t}$

$$G(S) = \frac{12(S+4)}{S(S+2)(S+6)}$$

$$\frac{12(S+4)}{S(S+2)(S+6)} = \frac{A}{S} + \frac{B}{S+2} + \frac{C}{S+6}$$

Solving by equating coefficients:

$$\frac{12(S+4)}{S(S+2)(S+6)} = \frac{A(S+2)(S+6) + B(S)(S+6) + C(S)(S+2)}{S(S+2)(S+6)}$$

$$12(S+4) = AS^2 + 8AS + 12A + BS^2 + 6BS + CS^2 + 2CS$$

$$12S + 48 = (A + B + C)S^2 + (8A + 6B + 2C)S + 12A$$

$$A + B + C = 0$$

$$(8)(4) + 6B + 2C = 12$$

$$12A = 48$$

$$A = 4$$

$$4 + B + C = 0 \text{ or } B + C = -4$$

$$(8)(4) + 6B + 2C = 12$$

$$6B + 2C = -20, \quad -3B - C = 10$$

$$B + C = -4$$

$$\frac{-3B - C = +10}{-2B \quad = +6}, \quad B = -3$$

$$4 - 3 + C = 0, \quad C = -1$$

$$\frac{12(S+4)}{S(S+2)(S+6)} = \frac{4}{S} - \frac{3}{S+2} - \frac{1}{S+6}$$

$$f(t) = 4 - 3e^{-2t} - e^{-6t} \quad \text{Answer d)}$$

<u>Other Method</u>

$$\frac{12(S+4)}{S(S+2)(S+6)} = \frac{A}{S} - \frac{B}{S+2} - \frac{C}{S+6}$$

Multiply by S and let S→0

$$\frac{12(S+4)(S)}{S(S+2)(S+6)} = A + \frac{BS}{S+2} + \frac{CS}{S+6}$$

$$\frac{(12)(4)}{(2)(6)} = \frac{48}{12} = 4 = A$$

Multiply by (S+2) and let S→-2

$$\frac{12(S+4)(S+2)}{S(S+2)(S+6)} = \frac{A(S+2)}{S} + B + \frac{C(S+2)}{(S+6)}$$

$$\frac{12(-2+4)}{(-2)(-2+6)} = B = \frac{12(2)}{(-8)} = -3$$

Multiply by $S+6$ and let $S \to -6$

$$\frac{12(S+4)(S+6)}{S(S+2)(S+6)} = \frac{A(S+6)}{S} + \frac{B(S+6)}{S+2} + C$$

$$\frac{12(-2)}{(-6)(-4)} = \frac{-24}{+24} = -1 = C$$

2. Find the transfer function, $\dfrac{E(S)_{OUT}}{E(S)_{IN}}$ of the circuit shown in Fig. 6.170.

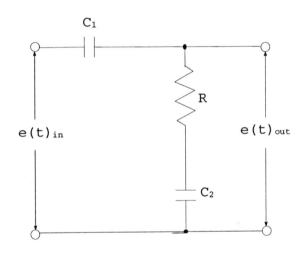

Figure 6.170

(a) $\quad S + \dfrac{1}{RC_2}$

(b) $\quad (S + \dfrac{1}{RC_1}) \Big/ S + \dfrac{1}{RC_2}$

(c) $\quad S + \dfrac{1}{\tau_1 \tau_2}$

(d) $\quad S + \dfrac{C_2 + C_1}{RC_1 C_2}$

(e) $\quad (S + \frac{1}{RC_2}) \Big/ (S + \frac{C_2 + C_1}{RC_1 C_2})$

Using Kirchoff's Voltage Law:

(1) $\quad e(t)_{IN} = \frac{1}{C_1} \int i(t) dt + R\, i(t) + \frac{1}{C_1} \int i(t)\, dt$

(2) $\quad e(t)_{OUT} = R\, i(t) + \frac{1}{C_1} \int i(t)\, dt$

Take derivative of (1) and (2)

(3) $\quad \dfrac{de(t)_{IN}}{dt} = \dfrac{i(t)}{C_1} + R\, \dfrac{di(t)}{dt} + \dfrac{i(t)}{C_2}$

(4) $\quad \dfrac{de(t)_{OUT}}{dt} = \dfrac{R\, d\, i(t)}{dt} + \dfrac{i(t)}{C_2}$

Take Laplace of (3) and (4) (Initial conditions are zero)

(5) $\quad SE(S)_{IN} = \dfrac{I(S)}{C_1} + RSI(S) + \dfrac{I(S)}{C_2}$

(6) $\quad SE(S)_{OUT} = RS\, I(S) + \dfrac{I(S)}{C_2}$

From (5)

$$SE(S)_{IN} = \left[\frac{1}{C_1} + \frac{1}{C_2} + RS \right] I(S)$$

$$SE(S)_{IN} = \left[\frac{C_2 + C_1}{C_1 C_2} + RS \right] I(S)$$

From (6)

$$SE(S)_{OUT} = \left[RS + \frac{1}{C_2} \right] I(S)$$

$$I(S) = \frac{SE(S)_{IN}}{(RS + \frac{C_2 + C_1}{C_1 C_2})} \text{ and } I(S) = \frac{SE(S)_{OUT}}{(RS + \frac{1}{C_2})}$$

$$\frac{SE(S)_{IN}}{(RS + \frac{C_2 + C_1}{C_1 C_2})} = \frac{SE(S)_{OUT}}{RS + \frac{1}{C_2}}$$

$$\frac{SE(S)_{OUT}}{SE(S)_{IN}} = \frac{(RS + \frac{1}{C_2})}{(R_S + \frac{C_2 + C_1}{C_1 C_2})}$$

$$\frac{E(S)_{OUT}}{E(S)_{IN}} = \frac{(S + \frac{1}{RC_2})}{(S + \frac{C_2 + C_1}{RC_1 C_2})} \text{ in standard form The answer is (E).}$$

3. A closed loop feedback system has the following characteristic equation

   $$S^3 + 7S^2 + 15S + 25 + K = 0$$

   For what range of K (positive and negative) is the system stable? Hint: Use Routh's Criterion.

   (a) $-25 < K < 80$
   (b) $K = 50$
   (c) $K > 40$
   (d) $-80 < K < 25$
   (e) $K = 0$

4. $S^3 + 7S^2 + 15S + 25 + K = 0$

   $a_0 = 1 \ a_1 = 7 \ a_2 = 15 \ a_3 = 25 + K$

   Routh's Array

   +1   +15

   +7   (25 + K)

   $\left[+15 - \frac{(25 + K)}{7} \right]$

+ (25 + K)

0

For the signs of the first row not to change

$$\frac{(25 + K)}{7} < 15 \text{ or } 25 + K < 105 \text{ and } K > -25$$
$$K < 80$$

$-25 < K < 80$

$$b_1 = \frac{a_1 a_2 - a_0 a_3}{a_1} = \frac{(7)(15) - (1)(25 + K)}{7} = 15 - \frac{(25 + K)}{7}$$

$$b_3 = \frac{a_1 a_4 + a_0 a_5}{a_1} = \frac{(7)(0) - (1)(0)}{7} = 0$$

$$c_1 = \frac{b_1 a_3 - a_1 b_3}{b_1} = \frac{\left[15 - \frac{(25 + K)}{7}\right](25 + K) - (7)(0)}{\left[15 - \frac{(25 + K)}{7}\right]} = 25 + K$$

c_3 and $d_1 = 0$ \hfill The answer is (A).

5. A system is configured as shown on the block diagram in Fig. 6.171. What is the overall loop transfer function? The overall closed loop transfer function?

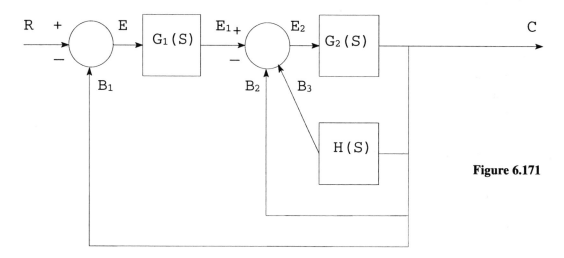

Figure 6.171

(a) $\dfrac{G(S)}{H(S)}$

(b) $G_1(S) G_2(S) / (1 + G_1(S) G_2(S) (1 + H(s)))$

(c) $G_1(S) G_2(S) / (1 + G_1(S) G_2(S) + G_2(S)(1 + H(s)))$

(d) $G(S) / (1 + G(S)(H(s)))$

(e) $G_1(S) / (1 + G_1(S) G_2(S) (1 + H(s)))$

The overall open loop transfer function is $G(S) H_2(S)$

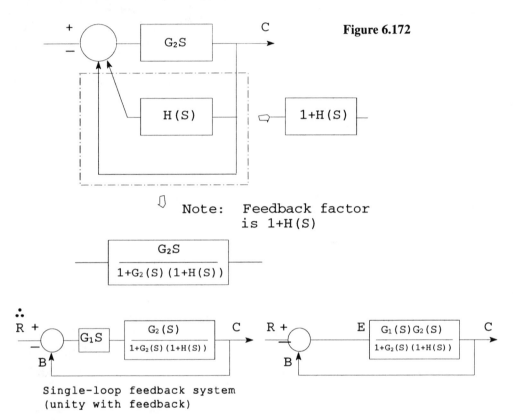

Figure 6.172

The open loop transfer function is $\dfrac{B(S)}{E(S)} = G(S)H(S)$

$G(S) H(S) \quad H(S) = 1$

The open loop transfer function is: $\dfrac{G_1(S)(G_2(S))}{1 + G_2(S)\ (1 + H(S))}$

The overall transfer function is: $\dfrac{C(S)}{R(S)} = \dfrac{G(S)}{1 + G(S)H_2(S)}$

$H_2(S) = 1$

The overall transfer function is:

$$\frac{\dfrac{G_1(S)\,G_2(S)}{1+G_2(S)(1+H(S))}}{1+\dfrac{G_1(S)\,G_2(S)}{1+G_2(S)(1+H(S))}} = \frac{\dfrac{G_1(S)\,G_2(S)}{1+G_2(S)(1+H(S))}}{\dfrac{1+G_2(S)(1+H(S))+G_1(S)\,G_2(S)}{1+G_2(S)(1+H(S))}}$$

$$= \frac{G_1(S)\,G_2(S)}{1+G_1(S)\,G_2(S)+G_2(S)(1+H(S))} \qquad \text{The answer is (C).}$$

ELECTRICITY IV

SEMICONDUCTORS

Introduction

In semiconductor theory, pure germanium and silicon are intrinsic semiconductors, i.e., they contain no impurities within their crystalline structure. These elements are in Column IV of the Periodic Table and each contains four electrons in the atom's outermost or valence shell. A Bohr model representation of these two elements is shown in Fig. 6.173.

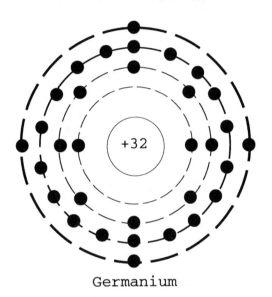

Figure 6.173

Germanium

When atoms link together or interconnect, they form molecules of matter, and this bonding takes place on the atomic level through the interaction of the valence electrons of each atom. When this occurs, each atom attempts to have a full complement of electrons in its outermost shell. Note that a full complement of electrons is eight electrons in the outermost ring or shell. If an atom has four or more valence electrons, it has a tendency to acquire additional electrons when the valence electrons of one particular atom join together with another atom to fill the particular atom's outer ring or shell. The force that holds these electrons together is electrostatic in nature and for semiconductors it is a covalent bond.

Covalent Bonding

Covalent bonding occurs when the outermost or valence electrons of neighboring atoms are shared between the atoms. This occurs when the atoms' electrons coordinate their motions so that an electrostatic binding force is produced. In this manner, say for a germanium atom in the center, when it shares one of its valence electrons with each of its four surrounding germanium atoms, each of which in turn shares one of its valence electrons with the center atom, then via this electron pairing, each atom appears to have an apparent eight valence electrons in its outer shell, or a full complement of electrons (Fig. 6.174).

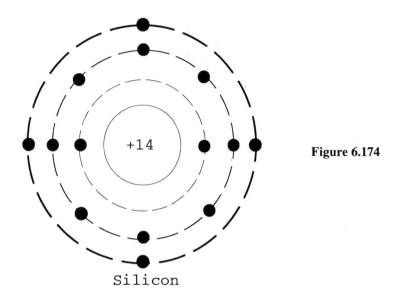

Figure 6.174

Silicon

A means of depicting this process schematically is shown in Fig. 175. The top part of the figure shows the covalent bonding of germanium atoms,

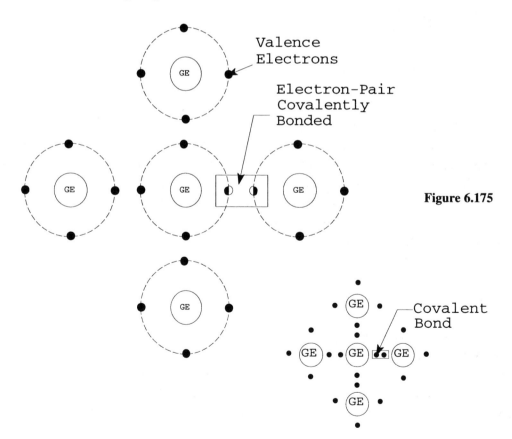

Figure 6.175

whereas the bottom part shows the symbolic picture.

Intrinsic Semiconductors

The symbol (GE) represents both the nucleus and the core electrons. Structures that are formed by atoms bonded together covalently are called crystal structures or crystal lattices. Usually, the repetitive crystal structure of germanium or silicon follows a tetrahedral pattern. (Crystal formation is discussed in Materials Science notes).

Because the outermost shell is apparently complete in a crystalline structure of germanium or silicon atoms covalently joined, the electronic arrangement is relatively stable. Accordingly, the valence electrons are held reasonably tightly to the parent atom, i.e., they cannot randomly wander throughout the crystal lattice. Consequently, a pure semiconductor cannot conduct an electric current, which is precisely the picture at a crystal temperature of absolute zero, i.e., -273°C or -460°F. Or, stated another way, an intrinsic semiconductor at a temperature of absolute zero is an insulator because it has no electrons available for current flow.

At higher temperatures, sufficient heat energy may be added to the crystal lattice whereby a covalent bond is broken. When this happens, a valence electron is effectively "excited" into the conduction band and it is available for current conduction under the influence of a potential difference. This electron is called a free electron and the vacant spot that the free electron leaves behind is called a hole. The combination of the freed electron and the hole it left behind is termed an electron hole pair.

When a hole is created by an electron breaking a covalent bond, the hole can easily be filled by another valence electron from a neighboring atom that breaks its own covalent bond. Naturally, this latter electron leaves behind its own hole, which may subsequently be filled by another electron from another atom. Thus it can be visualized that the hole is moving from atom to atom in the valence band and that it has a positive charge equal to that of an electron, i.e., 1.602×10^{-19} Coulombs.

Because the movement of holes also constitutes a current flow, the total current flow in a semiconductor is composed of both electron movement, in the conduction band, and hole movement, in the valence band. If a battery is placed across a semiconductor that contains free electrons and holes, the free electrons will move toward the positive battery terminal while the holes will move toward the negative battery terminal. So, it can be seen, that conduction in semiconductors is due to two separate and independent particles, carrying opposite charges and drifting in opposite directions when under the influence of an applied electric field.

The current density in a semiconductor with two charge carrying particles is:

$$J = (nu_n + Pu_p)\, e\varepsilon$$

where: J = current density (AMPS/m^2)

n = electronic concentration ($\frac{\text{number}}{M^3}$)

p = hole concentration ($\frac{\text{number}}{M^3}$)

u_n = electronic mobility ($\frac{M^2}{V\text{-Sec}}$)

u_p = hole mobility ($\frac{M^2}{V\text{-Sec}}$)

e = electronic charge (1.602×10^{-19} coulombs)

ε = electronic field strength (V/m)

Typical values for n, p, u_n and u_p are tabulated below, for a pure semiconductor at 300°K.

	Germanium	Silicon	
$n = p = n_i$	$2.4 \times 10^{+19}$	1.5×10^{16}	
u_n	0.39	0.135	n_i = Intrinsic Concentration
u_p	.19	0.048	

Extrinsic Semiconductors

From the preceding discussion it can be seen that pure germanium or silicon is of little use as a semiconductor device except for use perhaps as a heat-sensitive or light-sensitive resistive device. This is because free electrons and holes are created simultaneously in a pure semiconductor, with the number of free electrons just equal to the number of holes, thereby making the pure semiconductor a poor conductor.

However, to overcome the limitations of a pure semiconductor, most modern semiconductor devices contain semiconductor materials that are not pure, i.e. certain impurities have been added to pure germanium or silicon to give the material either a predominance of free electrons or holes. This process is called doping and the doped semiconductor is now referred to as an extrinsic semiconductor device.

N-Type Semiconductors

If a pentavalent element, i.e., one from the V Column of the Periodic table, such as antimony, phosphorus, or arsenic, is added, in controlled amounts, to an intrinsic semiconductor in its molten state, the following phenomenon results when the crystal lattice is reformed after cooling. The impurity atom, with its five electrons, has replaced either a silicon (or germanium) atom in the crystal lattice and four of its electrons form covalent bonds with the electrons of adjacent atoms. This bonding leaves one unbonded electron left over, and this fifth or extra electron is not tightly held to its parent atom. Consequently, to move this extra electron into the conduction band, considerably less energy is required. So, for each pentavalent impurity added, one loosely held electron is available for current flow. The figure in Fig. 6.176 depicts this schematically.

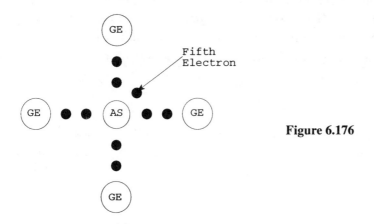

Figure 6.176

Note that for each impurity atom, one extra electron is donated to the semiconductor material. Accordingly, the electrons are classified as the majority carriers, and because they are negative, a doped semiconductor of this kind is called N-Type.

P-Type Semiconductors

If the same manufacturing process as for the N-Type material is done but using only a trivalent element, i.e., from the III Column of the Periodic Table, such as aluminum, boron, gallium, or indium when the impurity atom replaces a germanium (or silicon) atom in the crystal lattice, it can be seen that one covalent bond is not filled. This is shown schematically in Fig. 6.177.

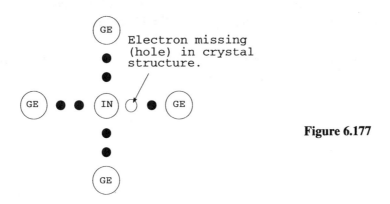

Figure 6.177

So for each trivalent impurity atom added, one hole (vacant electron) is available for current flow. Accordingly, the holes are classified as the majority carriers and, because they are considered positive, a doped semiconductor of this kind is called P-Type.

Minority Carriers

It should be mentioned that, in addition to the majority carriers, i.e., electrons in N-Type material and holes in P-Type material, minority carriers are also present. For example, holes in N-Type material and electrons in P-Type material are continuously being produced by the breaking of covalent bonds due to thermal energy at room temperature. These minority carriers are also available for current flow, sometimes to the detriment of the semiconductor.

The P–N Junction

The basic building block in the study of semiconductor devices is the P–N junction. Any type of semiconductor device must contain at least one of these junctions; the important characteristic of a P–N junction is its ability to allow current flow in only one direction, i.e., it is a unidirectional device.

A P–N junction is made up of separate P-Type and N-Type materials that are joined together on the atomic level. This joining together is not a mechanical type but joint, but is such that the inherent crystalline structure of the two materials is intact and continuous at all the joining points.

At the instant of P–N junction formation, the P side of the junction has a high concentration of holes, i.e., the majority carrier, and a low concentration of electrons, i.e., the minority carrier. Similarly, the N-side of the junction has a high concentration of electrons, i.e., the majority carrier, and a low concentration of holes, i.e., the minority carrier. Consequently, because of the concentration gradients, for both holes and electrons in the composite material, a current, called a diffusion current, will flow. This diffusion current is produced by the movement of charge, due to the concentration gradient, and tries to equalize the number of holes and electrons throughout the composite material and thereby eliminate the concentration gradient. Effectively, because of the diffusion current, electrons will flow from the N-region, across the junction into the P region while holes will flow from the P-region, across the junction, into the N-region.

While it might seem that at the end of this diffusion process, the electrons and holes would be evenly distributed throughout the composite material, such is not the case.

What happens is as follows.

Mobile electrons cross the junction and eliminate some holes in the P material by recombining with some of the many holes present. Similarly, holes diffuse across the junction and recombine in the N material with some of the many electrons present. The electrons that leave the N material (the majority carriers) leave their parent atoms with more positive protons than electrons, thus the remaining atoms, which are immobile, become positive ions. Similarly, the holes that leave the P material (the majority carriers) leave their parent atoms with more negative charges, thus the remaining atoms, which are immobile, become negative ions.

Another way of describing the above actions is to say that because of the diffusion of majority carriers across the junction, the uncovering of bound charges (ionic charges) occurs along the junction. However, these immobile charges exert a repelling force on the majority carrier diffusion that is taking place. Eventually, this repelling force becomes greater and greater as more charges diffuse across the junction, until the force is sufficient to stop further diffusion of the majority carriers. When this happens the N material is positively charged while the P material is negatively charged. This is shown in the Fig. 6.178.

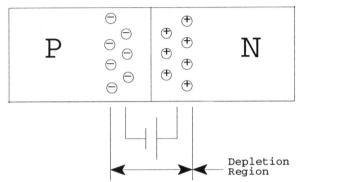

Figure 6.178

Note that on either side of the junction there is a lack of current carriers. Accordingly, this region is referred to as the depletion region. It is also called the space charge region. Note also that a battery is shown that symbolically represents the retarding force to the majority carrier motion. This battery is referred to as the barrier potential and can be likened to a hill (potential hill) that must be overcome by the majority carriers in order to move across the junction. Typical values for this carrier potential at room temperature are 0.3 volts and 0.7 volts respectively, for germanium and silicon. However, for minority carriers, i.e., electrons in the P material and holes in the N material, the retarding force battery is not retarding at all. The minority carriers flow easily across the junction. Consequently, a reverse current (usually in the order of microamps) flows dependent upon temperature.

SEMICONDUCTOR DEVICES

Semiconductor Diode

The simplest semiconductor device is the semiconductor diode. It consists of only one P–N junction and it is a two terminal device. Figure 6.179 depicts the configuration with the corresponding symbol. The arrowhead corresponds to the P type material.

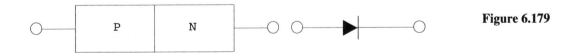

Figure 6.179

If an external battery is placed across the P–N junction, the semiconductor is said to be biased, and the battery voltage is referred to as the bias voltage; with no voltages applied, the semiconductor is considered unbiased.

If the external battery is applied as shown in Fig. 6.180, the semiconductor is considered forward biased.

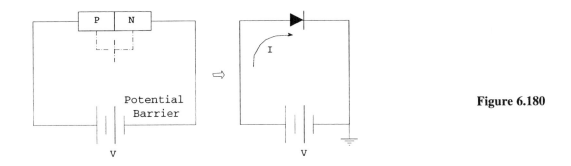

Figure 6.180

Note that because the positive terminal is connected to the P material, holes in the P material are forced toward the junction. Note also that the negative terminal is connected to the N material which forces the electrons toward the junction, neutralizes some of the donor and acceptor ions in the depletion or space charge region thereby reducing the potential barrier. This effectively allows some majority carriers to diffuse across the junction. If the battery voltage V is made sufficiently large, the potential barrier is reduced to zero and consequently a large forward current I will flow. Note that in the symbol I flows in the direction of the arrowhead.

If the external battery is applied as shown in Fig. 6.181 the semiconductor is considered reverse biased.

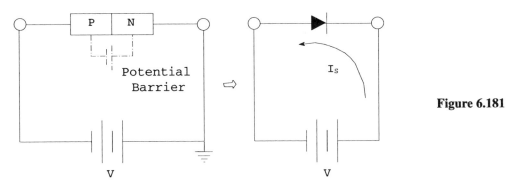

Figure 6.181

Note that the polarity of the battery is such as to increase the potential barrier, i.e., it pulls holes and electrons away from the junction. The only current that will flow is due to minority carriers and it is called the reverse saturation current I_s.

If the battery voltage is made variable whereby the voltage can be changed in increments, then a volt-ampere (VI) characteristic can be obtained for the semiconductor diode. This is shown in Fig. 6.182.

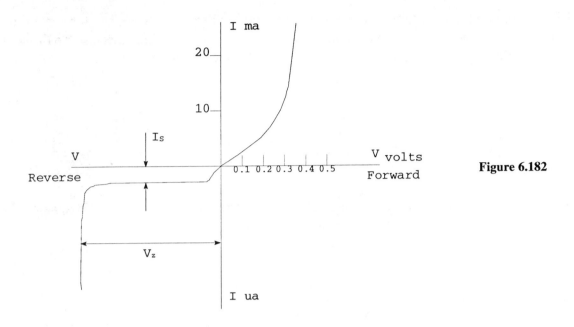

Figure 6.182

For quantitative analysis, the volt-ampere characteristic may be represented by the equation

$$I = I_s (e^{\frac{qV}{kT}} - 1)$$

where
- I = Diode current (Amps)
- I_s = Reverse saturation current (Amps)
- e = 2.718 Napier log base
- q = Electric charge (1.602×10^{-19} coulombs)
- V = Voltage across junction (volts; can be + or -)
- k = Boltzmann constant (1.381×10^{-23} Joules/°K)
- T = Absolute temperature (°K)

At room temperature, a convenient form of this equation is:

$$I = I_s (e^{40V} - 1)$$

It is important to note that for forward voltages, the curve lies in the first quadrant and the voltage across the semiconductor diode is approximately 0.3 volts for germanium or 0.7 volts for silicon. However, for reverse voltages, a current usually measured in microamperes, flows in the third quadrant. Consequently, the diode can be considered as a very large resistance R_R, in comparison to its forward resistance R_F. The ratio of R_R to R_F in inexpensive diodes is about 1000 to 1. In high quality diodes the ratio may exceed 100,000 to 1.

If the reverse voltage is still increased, at a reverse voltage of V_z (called the zener voltage), an abrupt change occurs signifying that a large reverse current flows. This is called the breakdown region and zener (or voltage regulating) diodes are specifically manufactured for operation herein.

This is because the large reverse current is nearly independent of voltage, i.e., the zener diode maintains a nearly constant voltage under conditions of widely varying currents. To differentiate between diode types, the zener diode has the symbol shown in Fig. 6.183.

Figure 6.183

Example 1

A semiconductor diode, at room temperature, has a current of 50 microamperes (50×10^{-6}) at a voltage of -4 volts. What is the current at a voltage of +0.1 volts?

Solution

At room temperature

$$I = I_s [e^{40V} - 1] \text{ or } I = I_s e^{40V} - I_s$$

At -4 volts

$$50 \times 10^{-6} = I_s e^{-160} - I_s$$

$$I_s \approx -50 \times 10^{-6} \text{ Amps}$$

At +0.1 volts

$$I = 50 \times 10^{-6} [e^{40 \times .1} - 1]$$

$$I = 50 \times 10^{-6} [e^4 - 1]$$

$$I = 50 \times 10^{-6} [53.6]$$

$$I = 2.68 \text{m amperes} \quad \text{Answer}$$

Semiconductor Diode Models

From observation, the volt-ampere (VI) characteristic of a semiconductor diode is nonlinear. Accordingly, for purposes of analysis, the diode is usually replaced by a linear model as shown in Fig. 6.184. If the diode is considered ideal, i.e. $R_F = 0$ and $R_R = \infty$, a simple switch can suffice.

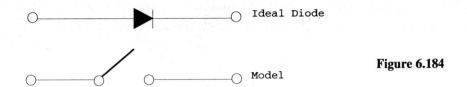

Figure 6.184

For forward biasing voltages, the switch is closed; for reverse biasing voltages, the switch is open.

However, for circuits wherein nonideal diodes are used, the usual case, other models are used to approximate the VI characteristics. Three such models are shown in Figs. 6.185 to 6.187 with their specific characteristics.

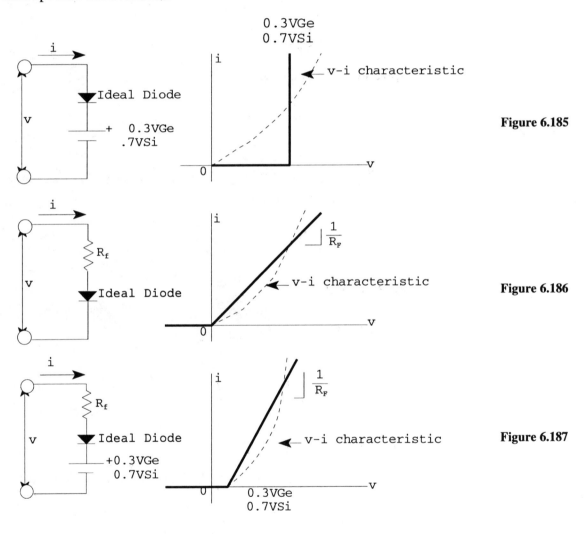

Figure 6.185

Figure 6.186

Figure 6.187

At the other end of the VI characteristic, i.e., at the breakdown or zener region, a corresponding model for a zener diode can also be developed. Such a model is shown in Fig. 6.188.

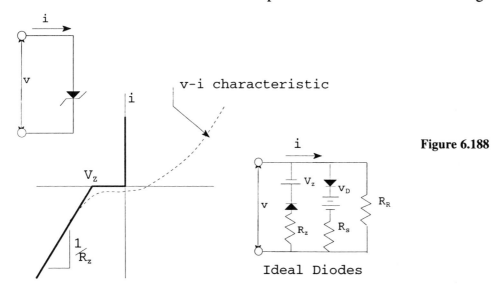

Figure 6.188

Example

A silicon diode is used in the half-wave rectifier circuit in Fig. 6.189. The voltage source supplies a voltage of $v_s = 4.0\sin 377t$. If $R_F = 10\Omega$, sketch the voltage v_{out}.

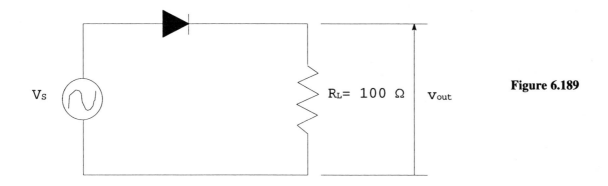

Figure 6.189

Solution

The diode model used is as shown in Fig. 6.189. Note that both "silicon" (0.7) and R_F are utilized in this model.

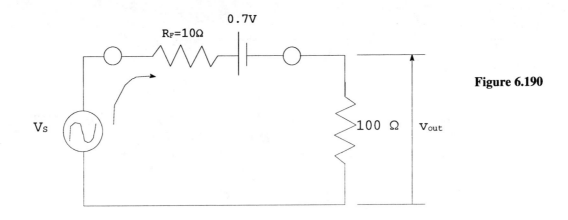

Figure 6.190

$$i = \frac{v_s - 0.7}{110}$$

$$v_{out} = 100\,i$$

$$= \frac{100}{110}[v_s - 0.7]$$

$$= 0.91[v_s - 0.7]; \text{ at } v_{s_{max}} = 4V, v_{out} = 3V$$

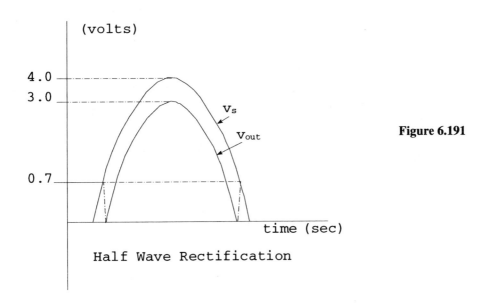

Figure 6.191

Half Wave Rectification

Example 3

A 9 volt zener diode is used in the circuit shown in Fig. 6.192. If $R = 400\Omega$ and $R_z = 25\Omega$, what is V_{out} for a $V_{in} = 15$ volts?

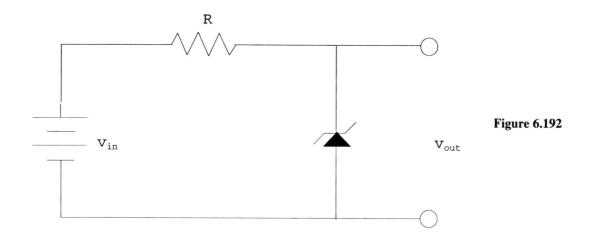

Figure 6.192

Solution

The zener diode is replaced by its model (Fig. 6.193).

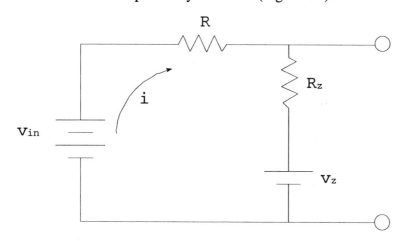

Figure 6.193

$$I = \frac{V_{in} - V_z}{R + R_z} = \frac{15 - 9}{425} = 0.014 \text{ Amps}$$

$$V_{out} = 9 + (0.014)(25)$$
$$= 9 + 0.35$$
$$= 9.35 \text{ volts} \quad \text{Ans.}$$

Dynamic Resistance (r_d)

The models utilized so far are sometimes referred to as large-signal models. However, if a small signal (i.e., in comparison to the total VI characteristic) is applied to the diode whereby the operation will be about a specific point, then the dynamic or AC resistance must be used in any analysis. This resistance is defined as the slope of the tangent line at the operating point.

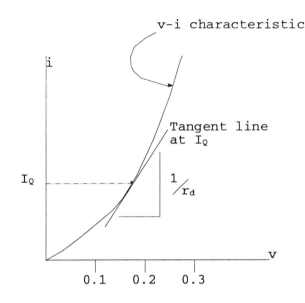

Figure 6.194

$$r_d \equiv \left. \frac{dv}{di} \right|_{Q_{point}}$$

since $i = I_s(e^{40V} - 1)$

$$\frac{di}{dv} = 40\, I_s e^{40V}$$

$$\frac{1}{r_d} = 40\,(i + I_s)$$

or $r_d \approx \dfrac{25m\ V_{out}}{I_Q}$

where I_Q is the current at any point i. If I_Q is in milliamperes, then

$$r_d \approx \frac{25}{I_Q} \text{ ohms.}$$

Note that r_d is a signal or small signal quantity and it can be determined from the DC current at the operating point.

FIELD EFFECT TRANSISTORS

Junction Field Effect Transistors (JFET)

The field effect transistor is a unipolar device, i.e., it uses only one type of current carrier - electrons or holes. It consists of a conducting channel of finite conductance that is established between two P–N junctions. The symbol and a physical arrangement are shown in Fig. 6.195 for a junction field effect transistor - JFET.

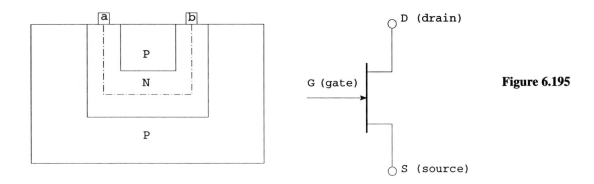

Figure 6.195

The channel is denoted by the dotted line from a to b. It is an N-Type channel. If the materials were interchanged, it would be a P-Type channel and the arrowhead on the symbol would be reversed.

Typical operation of the device is as follows:

The P-Type materials on either side of the N channel are thin and are very heavily doped (P^+), whereas the N channel is lightly doped, but is of large cross section.

If a battery is connected to points a and b with the positive terminal at b, then the current carriers (electrons) will flow from a to b in the N channel. Point a is called the source while point b is called the drain.

However, if another battery is connected to the P materials such that the P–N junctions are reverse biased, i.e., the battery's negative terminal is connected the P materials, then a depletion region will exist at the P–N junctions extending into the channel.

This region will be depleted of current carriers thereby making the channel's cross section somewhat smaller. Naturally, the higher the reverse potential, the smaller will be the channel. Accordingly, the potential on the P materials controls the current carriers flowing in the channel, and this potential is referred to as the gate voltage. It is important to note that small changes in this gate voltage, say in synchronism with a signal, will control the current flowing in the channel. It is also important to note that since the signal voltage varies the reverse voltage of the P–N junction, very little current flows in the gate circuit. This signifies a very large input resistance in the gate-source circuit that does not load down the signal source.

JFET Characteristics

To obtain the VI characteristics of a JFET, the following circuit is used. Usually, V_{GS}, the voltage from gate to source, is made the independent parameter whereas V_{DS}, the voltage from drain to source, and I_D, the drain current, are plotted for different values of V_{GS} (Fig. 6.196).

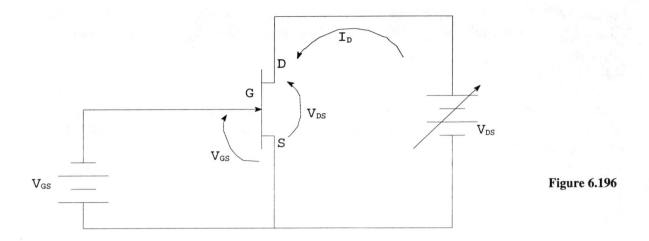

Figure 6.196

Typical characteristics obtained for a JFET are shown in Fig. 6.197.

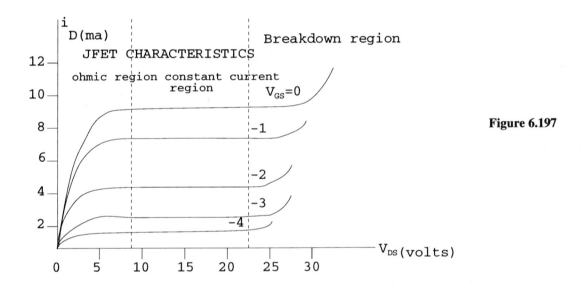

Figure 6.197

To understand why the curves have the form shown consider the curve where $v_{GS} = 0$, i.e. the gate is shorted to the source. Accordingly, at a low v_{DS}, the current i_D is limited by the resistance of the channel. This is considered the OHMIC or non-saturation region. However, as I_D increases, the voltage drop along the channel increases also and it is of such a polarity as to reverse bias the gate junction, i.e. the channel being positive with respect to the gate. Consequently, as the distance increases from the source, the conducting channel is being reduced in cross section. Eventually, a voltage v_{DS} is reached whereby the channel is considered "pinched off", i.e. further increases in v_{DS} does not increase the current, and the current approaches a constant value. This region is called the constant current region. And, if the voltage v_{DS} is still increased, a voltage is reached whereby the reversed biased junction breaks down. This is called the breakdown region.

If a gate voltage v_{GS} is now applied, additional reverse bias will be added. Consequently, pinch off occurs at a smaller value of v_{ds}. This can readily be seen from the JFET VI characteristics.

Transfer Characteristics

Because the output current i_D, is controlled by the input voltage v_{GS}, the transfer characteristic of a JFET is also an important quantity. It is obtained at a constant v_{DS}. A typical characteristic is shown in Fig. 6.198.

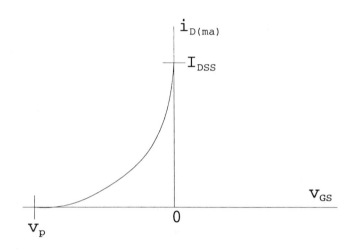

Figure 6.198

It is noted that the transfer characteristic is approximately parabolic. For the constant current region, the following relationship holds.

$$i_{DS} = I_{DSS} \left(1 - \frac{v_{GS}}{V_p}\right)^2$$

where i_{DS} = Drain current in the constant current region

I_{DSS} = The value of i_{DS} at $v_{GS} = 0$

V_p = The pinch-off voltage

Example 4

A JFET, when its gate terminal is connected to the source terminal a reading of $i_{DS} = 6$ milliamperes is obtained. What would i_{DS} be at $v_{GS} = -1$ volt. The pinch-off voltage is -5 volts.

Solution

With the gate connected to the source

$v_{GS} = 0$ ∴ $i_{DS} = I_{DSS}$ = 6m amperes

$$i_{DS} = 6\left(1 - \frac{-1}{-5}\right)^2$$

$$i_{DS} = 6\left(\frac{4}{5}\right)^2$$

$i_{DS} = 3.84$ m amperes   Answer

MOSFET (Metal Oxide Semiconductor Field Effect Transistor)

In the JFET, the high input resistance is due to the reverse bias between the gate and source. But, since the FET is still a semiconductor, the input resistance will depend on the minority carriers which in turn depends on temperature. Accordingly, another device, namely, the MOSFET is an improvement.

Basically, there are two types of MOSFETS: A depletion mode MOSFET, which behaves similar to a JFET, and an enhancement mode MOSFET. Either can be in an N channel or P channel configuration. The symbols for these devices, with an N Type channel, are shown in Fig. 6.199; note the arrow always points to N material.

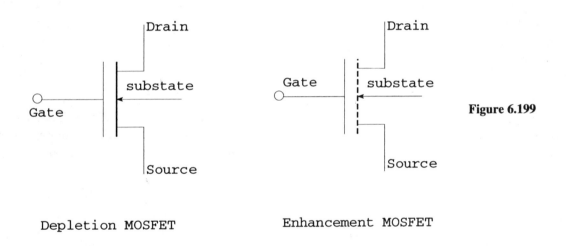

Depletion MOSFET   Enhancement MOSFET

Figure 6.199

The physical arrangements are very similar except for the channels between the source and drain. In one case, the depletion MOSFET, the channel is diffused between the two elements; in the other case, the enhancement MOSFET, the channel is induced between the source and drain. This is shown in Fig. 6.200.

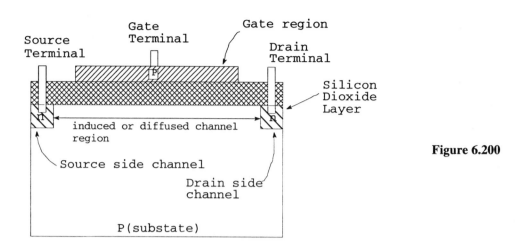

Figure 6.200

When the channel is diffused, i.e., with a lightly doped n type section, the operation is similar to the JFET in that at $v_{GS} = 0$, a large current will flow from the source to the drain. However, if a negative voltage is applied to the gate terminal, the drain current will be reduced. This results from the capacitive effect of the gate (conductor), the silicon dioxide layer (insulator), and the channel (conductor). The negative gate voltage induces positive charges into the channel, which in turn combine with negative charges in the channel, thus reducing the current flow. The reverse effect for a positive gate voltage occur also. A positive gate voltage induces negative charges into the channel which has the effect of providing more current carriers, thereby increasing the drain current. Accordingly, the input voltage can be both positive and negative whereby the drain current will be respectively enhanced or depleted.

The induced channel MOSFET operates similar to the diffused channel device with positive gate voltages. In fact, the induced N channel MOSFET operates only with positive voltages and it requires the exceeding of a threshold (V_T) to initiate the drain current. The VI characteristics shown in Fig. 6.201 depict the characteristics of these two devices.

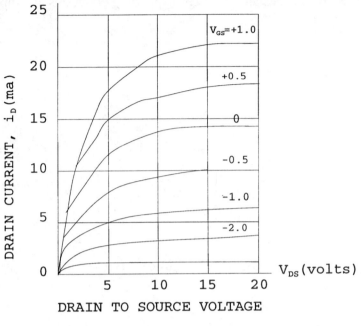

Figure 6.201

CHARACTERISTICS FOR A N-CHANNEL, DEPLETION TYPE MOSFET

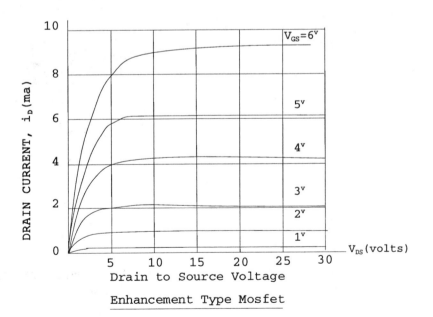

Figure 6.202

Enhancement Type Mosfet

The transfer characteristics of these two devices and their corresponding equations are shown below and in Fig. 6.203.

$$I_D = I_{DSS}(1 - \frac{V_{GS}}{V_P}) \qquad I_D = K(V_{GS} - V_T)^2$$

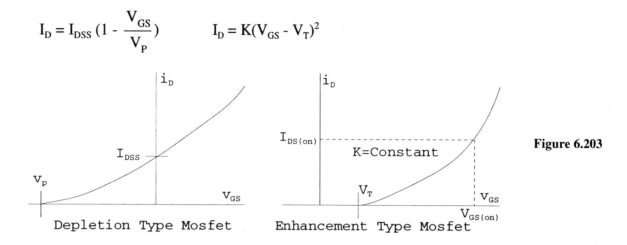

Figure 6.203

DC Conditions - FETS

The configuration of a common source amplifier is shown in Fig. 6.204.

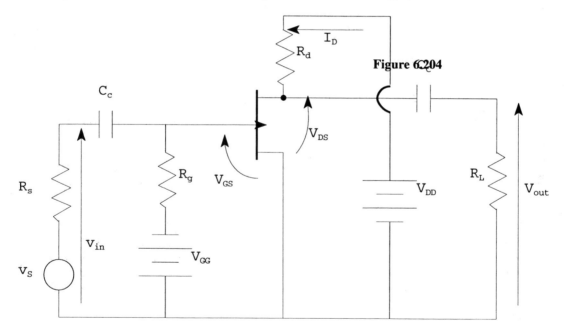

Figure 6.204

The analysis of the amplifier consists of two parts: DC and signal, the latter sometimes referred to as AC.

The DC part consists of setting up the bias voltages. V_{GG} reverse biases the gate circuit and V_{DD} sets up the quiescent (no signal) drain current I_G and V_{GS}. Because a capacitor blocks DC, R_S, R_L and V_s do not enter into this phase of the analysis. They will be discussed later during the signal analysis.

In the writing of Kirchoff's Voltage Laws, there are two loops: the gate source loop and the drain source loop.

Drain Source Loop

$$V_{DD} = I_D R_d + V_{DS}$$

Gate Source Loop

$$-V_{GG} = I_G R_g + V_{GS}$$

To set up a particular Q (no signal) point, i.e. I_D and V_{GS}, the above equations are utilized. Note that because of the reverse bias on the gate

$$I_G \approx 0, \text{ therefore}$$

$V_{GS} = -V_{GG}$ and the choice of battery for V_{GG} sets up V_{GS}.

Corresponding to V_{GS} and I_D is a specific V_{DS} (see characteristics).

Therefore, if a voltage source is given for V_{DD}, then R_d can be determined from

$$R_d = \frac{V_{DD} - V_{DS}}{I_D}$$

However, it is more economical to use only one battery. Accordingly, a most frequently used configuration is shown in Fig. 6.205.

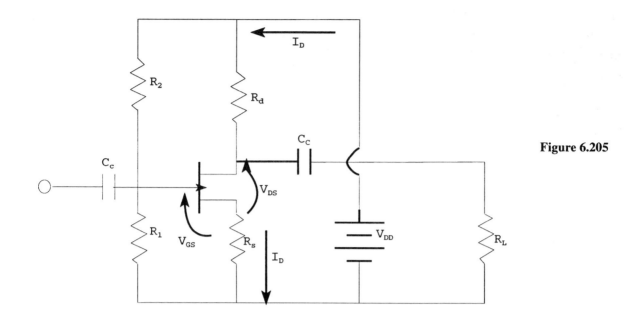

Figure 6.205

The $I_D R_S$ voltage drop places the source at a positive potential above ground. For a specific I_D, V_{GS}, the Thevenin equivalent of the gate circuit is used to determine all parameters. This circuit is shown in Fig. 6.206.

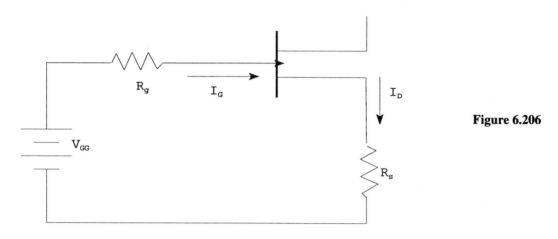

Figure 6.206

where $V_{GG} = \dfrac{R_1}{R_1 + R_2} V_{DD}$ and $R_G = \dfrac{R_1 R_2}{R_1 + R_2}$

For the gate source loop

$$V_{GG} - I_G R_G - V_{GS} - I_D R_S = 0$$

Since $I_G \approx 0.0$

$$I_D R_S = V_{GG} - V_{GS}$$

Then $R_S = \dfrac{V_{GG} - V_{GS}}{I_D}$; $R_2 = R_G \dfrac{V_{DD}}{V_{GG}}$; $R_1 = \dfrac{R_G R_2}{R_2 - R_G}$

Example 5

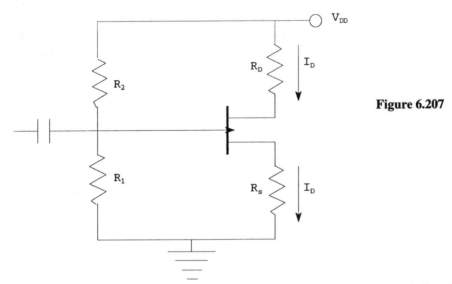

Figure 6.207

For the JFET, a Q point of $I_D = 5$ milliamperes, $V_{DS} = 6$ volts and $V_{GS} = -2.5$ volts. What are suitable values of R_1, R_2, R_D and R_S if $V_{GG} = 10$ volts, $V_{DD} = 25$ volts and $R_G = 6$ Meg Ω?

Solution

$$R_s = \dfrac{V_{GG} - V_{GS}}{I_D} = \dfrac{10 - (-2.5)}{5 \times 10^{-3}} = 2.5 \text{K}\Omega \quad \text{Answer}$$

$$R_2 = R_G \dfrac{V_{DD}}{V_{GG}} = (6 \times 10^{+6}) \dfrac{25}{10} = 15 \times 10^6 \Omega \quad \text{Answer}$$

$$R_1 = \dfrac{R_G R_2}{R_2 - R_G} = \dfrac{(6)(10^6)(15)(10^6)}{15 \times 10^6 - 6 \times 10^6} = 10 \times 10^6 \Omega \quad \text{Answer}$$

$$V_{DD} - I_D R_D - V_{DS} - I_D R_S = 0$$

For $V_{DS} = 6$

$$R_D = \frac{25 - 6 - (5)(10^{-3})(2.5)(10^3)}{5 \times 10^{-3}}$$

$$= \frac{6.5}{5 \times 10^{-3}} = 1.3 \text{K}\Omega \quad \text{Answer}$$

Signal Conditions - FETS

After the DC (or bias) conditions are set up, the analysis of the signal conditions can commence. At the established Q point, three (3) small signal parameters can be obtained. They are:

$$r_{ds} = \frac{v_{ds}}{i_{ds}} = \left.\frac{\partial v_{DS}}{\partial i_D}\right|_{V_{GS} = K} \quad \text{OHMS}$$

and is called the drain source resistance;

$$G_m = \frac{i_d}{v_{gs}} = \left.\frac{\partial v_d}{\partial v_{GS}}\right|_{V_{DS} = K} \quad \text{mhos}$$

and is called the transconductance;

and $$u = \frac{v_{ds}}{v_{gs}} = \left.\frac{\partial v_{DS}}{\partial v_{GS}}\right|_{i_D = K}$$

and is called the amplification factor. Note that

$$u = G_m r_{ds}.$$

Typical values of these parameters are given by the manufacturer and/or are listed in transistor manuals. If not given, they can be obtained by differentiating the transfer characteristics

$$i_D = I_{DSS}\left(1 - \frac{v_{GS}}{V_p}\right)^2 \quad \text{JFET and MOSFET (depletion)}$$

which yields

$$G_m = -\frac{2I_{DSS}}{V_p}\left(1 - \frac{V_{GS}}{V_p}\right)$$

$$G_m = G_m o \left(1 - \frac{V_{GS}}{V_p}\right)$$

$$G_m = G_m o \sqrt{\frac{I_D}{I_{DSS}}}$$

where $G_m o = -2 \dfrac{I_{DSS}}{V_p}$ the transconductance at $V_{GS} = 0$. The other above expressions are for a G_m at a DC bias voltage of V_{GS}.

It should be stressed that since these signal parameters are obtained by differentiation, they are non DC quantities. They only come into play when dealing with variations due to signals.

The equivalent circuit for signal conditions is shown in Fig. 6.208 for the JFET and the MOSFET (depletion).

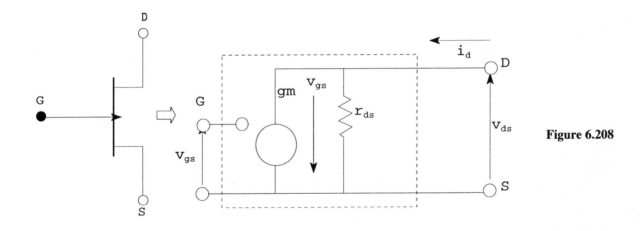

Figure 6.208

Note that all signal quantities have lower case subscripts, and that the FET is represented by the model inside the dotted box.

The application of the above small signal model to find the common source and common drain (source follower) amplifier voltage gain, Av, follows in Figure 6.209 and 6.210.

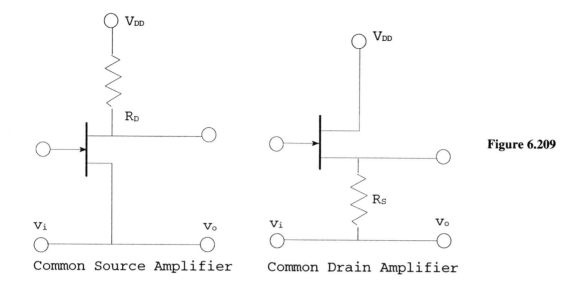

Figure 6.209

Common Source Amplifier   Common Drain Amplifier

Note that bias networks are not shown since the model applies only to signals.

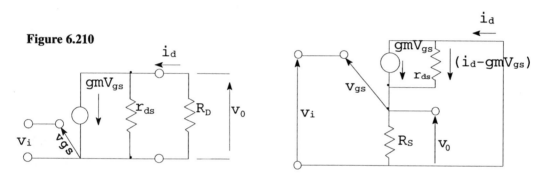

Figure 6.210

$Av = \dfrac{v_o}{v_i}$

Input circuit
$\quad v_i = v_{gs}$

Output circuit

$v_o = -Gm\, v_i\, \dfrac{r_{ds}\, R_D}{r_{ds} + R_D}$

$Av = -Gm\, \dfrac{r_{ds}\, R_D}{r_{ds} + R_D}$

For $r_{ds} \gg R_D$

$Av = -Gm\, R_D$

Input circuit

$$v_{gs} = v_i - i_d R_s$$

Output circuit

$$(i_d - G_m v_{gs}) r_{ds} + i_d R_s = 0$$

which reduces to

$$i_d = \frac{u\, v_i}{r_{ds} + (1 + u)R_S} \quad \text{using } u = G_m r_{ds}$$

$$v_o = i_d R_s$$

$$v_o = \frac{u\, v_i\, R_S}{r_{ds} + (1 + u)\, R_S}$$

$$\text{and } A_v = + \frac{u\, R_S}{r_{ds} + (1 + u)R_S}$$

Example 6

For the amplifier shown in Fig. 6.211, find the mid band voltage gain Av.

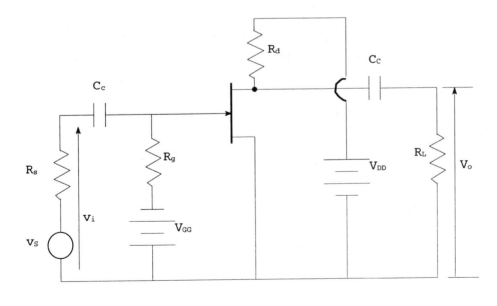

Figure 6.211

Solution

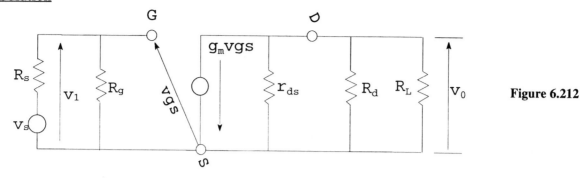

Figure 6.212

By definition, mid band means that the frequencies of interest are such the series capacitors are short circuits ($X_c = 0$) and shunt capacitors are open circuits ($X_c = \infty$). Accordingly, the small signal model is

$$v_{gs} = v_{in}$$

$$v_o = -G_m v_{in} \frac{1}{\frac{1}{r_{ds}} + \frac{1}{R_d} + \frac{1}{R_L}}$$

$$Av = \frac{v_o}{v_n} = -\frac{G_m \, r_{ds} \, R_D \, R_L}{R_d R_L + r_{ds} R_L + r_{ds} R_d} \quad \text{Ans.}$$

TRANSISTORS

Theory of Operation

The transistor, unlike the FET, is a bipolar device, meaning that both majority and minority carriers play roles in its operation. It has the designation BJT, which stands for "bipolar junction Transistor."

A transistor consists of two P–N junctions that are in very close proximity to each other. The configuration can be either p-n-p or n-p-n. The configuration of an n-p-n transistor together with its symbol is shown in Fig. 6.213.

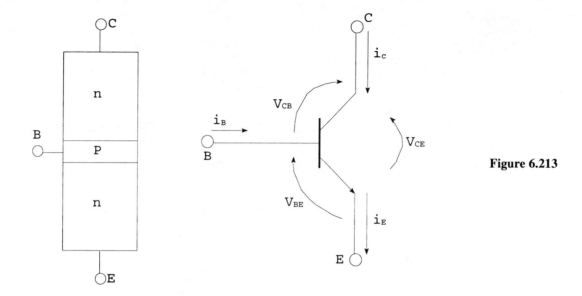

Figure 6.213

For a p-n-p symbol, the arrowhead is reversed.

The emitter E is usually heavily doped and together with the base B forms one P–N junction. The base region is lightly doped and it is made extremely thick somewhat akin to the thickness of this paper; and together with the collector C, which is made of the same material as the emitter, but not as heavily doped, forms the other P–N junction. For transistor action to occur, the emitter base junction is forward biased and the collector base junction is reverse biased.

In the symbol shown, all voltage polarities and current directions are standard, with the sometimes exception of the currents. In some textbooks the transistor is pictured as a current node with all currents flowing into it. In other textbooks, the emitter current is in the direction of the arrowhead, the actual direction of current flow and all other currents made consistent with this direction. In this text, actual current direction will be used, i.e., in the direction of the arrowhead. Accordingly, $i_E = i_B + i_C$.

To forward bias the base emitter junction, $V_{BS} > 0$. This will result in a large majority carrier current flow, similar to that of a semiconductor diode, and will consist mostly of electrons - the emitter being heavily doped. These electrons injected into the base or P material region are now the minority carriers therein. And, because the base is extremely thin and lightly doped, these "injected" electrons, from the emitter, travel across the base region to the collector junction, where the bias is such that the electrons are "collected," i.e., the reverse bias at the collector junction allows only minority carriers (in the P region) to flow into the collector.

To describe this action mathematically, we can say that the collector current i_c is some fraction of emitter current i_E.

Or
$$i_c = \alpha\, i_E$$

where α is a number less than 1. Typical values of α vary from 0.90 to 0.99. α is called the forward current transfer ratio, i.e., from collector to emitter.

Those electrons remaining in the base recombine with holes, thereby constituting the base current, i.e.,
$$i_B = (1 - \alpha)\, i_E$$

Note that the hole current is negligible.

Another parameter of interest in the transistor is the current ratio from the collector to the base. If i_E is replaced in the last expression by i_C/α, then
$$i_B = \frac{1 - \alpha}{\alpha} i_c \text{ or } i_c = \left(\frac{\alpha}{1 - \alpha}\right) i_B.$$

By definition $\beta \equiv \dfrac{\alpha}{1 - \alpha}$ From which $\beta + 1 = \dfrac{1}{1 - \alpha}$.

Accordingly,
$$i_c = \beta i_B \text{ and } i_E = (\beta + 1)\, i_B$$

which shows that a small i_B controls much larger currents, i.e., i_C and i_E, since $\beta \gg 1$.

The electron current αi_E is not the only current flowing across the collector junction. There is also a reverse current flowing which is due to thermally generated minority carriers. This latter current is called the cut off current, i_{co}, and is quite temperature sensitive. If I_{CBO} is the reverse saturation current, then
$$i_{co} = I_{CBO}\, (1 - e^{-Kv_{CB}})$$

$\underline{c}$ollector $\underline{b}$ase $\underline{e}$mitter $\underline{o}$pen

Figure 6.214

Similar to that of a diode.

For $Kv_{CB} \gg 1$ $i_{co} \approx I_{CBO}$

Accordingly,
$$i_c = \alpha\, i_E + I_{CBO}$$

If i_E is replaced by $i_C + i_B$, another relationship of i_C in terms of i_B is obtained.

This relationship is

$$i_C = \beta i_B + I_{CEO} \text{ where } I_{CEO} = (\beta + 1) I_{CBO}.$$

<u>c</u>ollector <u>e</u>mitter base <u>o</u>pen

Note that I_{CEO}, the collector cut off current, is much larger than the reverse saturation current I_{CEO}.

The Common Base Configuration

In the common base configuration, the base terminal is common to both the input and output circuits. This is shown in Fig. 6.215 for a pnp transistor.

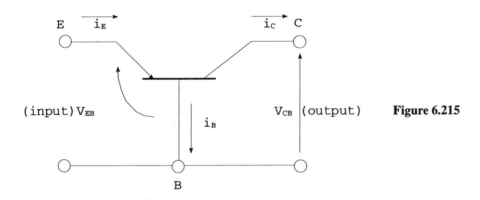

Figure 6.215

The input characteristics, for the emitter base junction forward biased, is similar to a semiconductor diode. Accordingly, $V_{EB} = 0.3$ volts for germanium or 0.7 volts for silicon.

The output characteristics, i.e. V_{CB}, V_S, I_C for different I_E is shown in Fig. 6.216.

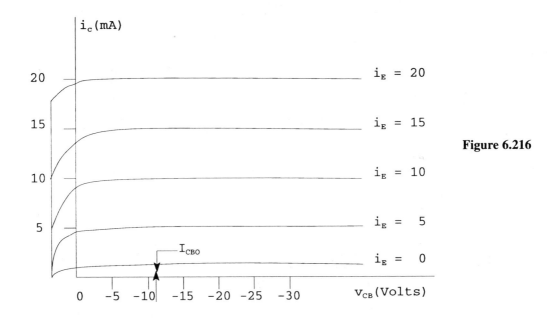

Figure 6.216

The curves lying in the first quadrant are of fairly constant current and the area is referred to as the active region. The equation $i_C = \propto i_E + I_{CBO}$ applies here.

That part of the curves lying in the second quadrant is referred to as the saturation region and it goes to zero. This is consistent for a semiconductor P–N junction that is forward biased.

<u>The Common Emitter Configuration</u>

In the common emitter configuration, the emitter terminal is common to both the input and output circuits. This is shown in Fig. 6.217 for a p-n-p transistor.

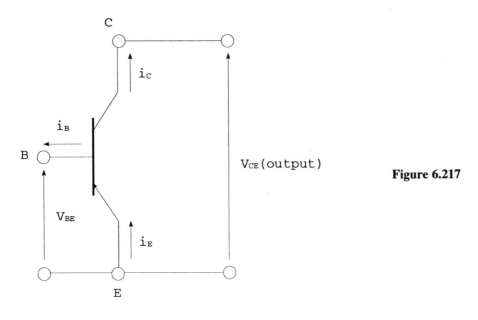

Figure 6.217

Because the base emitter junction is forward biased, $V_{BE} = 0.3$ volts for germanium or 0.7 volts for silicon.

The output characteristics, i.e. V_{CE} vs I_C for different I_B is shown in Fig. 6.218.

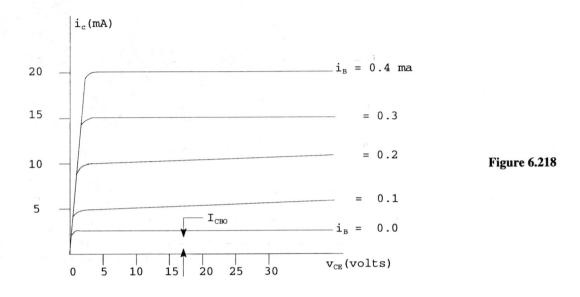

Figure 6.218

In the active region, the equation

$$i_C = \beta I_B + I_{CEO} \text{ applies.}$$

Also, note that from the base (i_B) to the collector (i_C), a very large current gain is obtainable.

<u>DC Conditions - BJT</u>

<u>Common Emitter Configuration</u>

The configuration of an n-p-n common emitter amplifier, using fixed bias, is shown in Fig. 6.219.

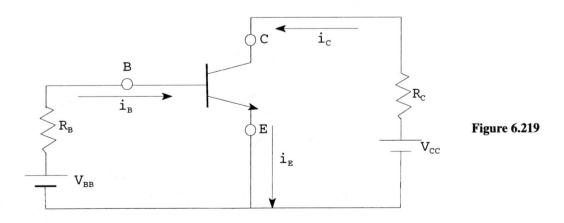

Figure 6.219

To establish the quiescent point (Q pt), a value of base current i_B must be selected, which is usually in the middle of the active region. After the selection is made, KVL for the input circuit is utilized.

$$V_{BB} - I_B R_B - V_{BE} = 0$$

where

$$V_{BE} = 0.3 \text{ volts for GE}$$

$$\text{or } 0.7 \text{ volts for Si}$$

Accordingly,

$$R_B = \frac{V_{BB} - V_{BE}}{I_B}$$

Note that V_{BB} is a separate supply (battery) and it represents the base bias voltage that helps set up I_B from which R_B can be calculated.

Example

A silicon transistor is used in the circuit shown in Fig. 6.220. If $\beta = 90$, find i_B and v_{CE}.

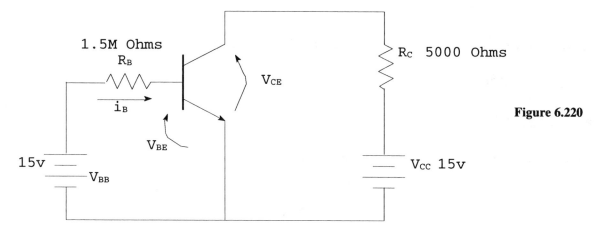

Figure 6.220

Solution

$$i_B = \frac{15 - 0.7}{1.5 \times 10^6} = 9.5 \text{ uAmps} \quad \text{Answer}$$

$$i_c = \beta i_B \text{ (Neglecting } I_{CEO}) = (90)(9.5)(10^{-6}) = 0.86 \text{ ma}$$

and $v_{CE} = 15 - (0.86)(5) = 10.7$ volts   Answer

Because two power supplies are needed for the fixed bias circuit, which is uneconomical, a more common circuit is used that employs only one power supply. The circuit is shown in Fig. 6.221.

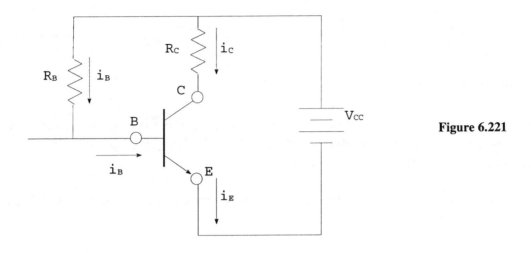

Figure 6.221

Using KVL the following expressions are obtained.

$$V_{CC} - I_B R_B - V_{BE} = 0$$

From which

$$R_B = \frac{V_{CC} - V_{BE}}{I_B}$$

Because transistors are temperature sensitive, stabilization circuits must be used, particularly for the common emitter configuration. Otherwise I_C, which contains the factor $(1+\beta) I_{CBO}$, will increase to such a level that the transistor will be destroyed.

One such stabilization circuit is shown in Fig. 6.222.

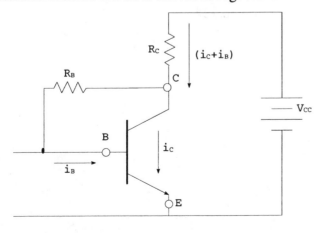

Figure 6.222

Writing KVL for the loop from V_{CC} through the base, the following is obtained.

$$V_{CC} - (I_C + I_B) R_C - I_B R_B - V_{BE} = 0$$

From which

$$I_B = \frac{V_{CC} - V_{BE} - I_C R_C}{R_C + R_B}$$

Stabilization occurs as follows:

If I_C increases, due to a temperature rise etc., I_B decreases; and this lower I_B causes the change in I_C to be somewhat less than what would have been without R_B being connected to the transistor's collector. hence thermal runaway is prevented.

Another, more frequently used circuit for stabilizing the operating point is the emitter self bias circuit shown in Fig. 6.223.

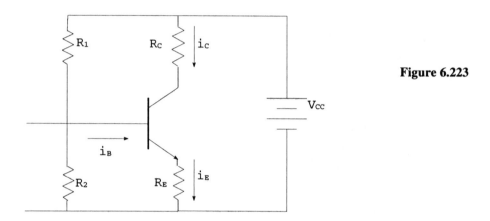

Figure 6.223

Quantitative analysis of the circuit is simplified if Thevenin's Theorem is applied to the resistors R_1 and R_2.

Accordingly, the circuit replaces to that shown in Fig. 6.224.

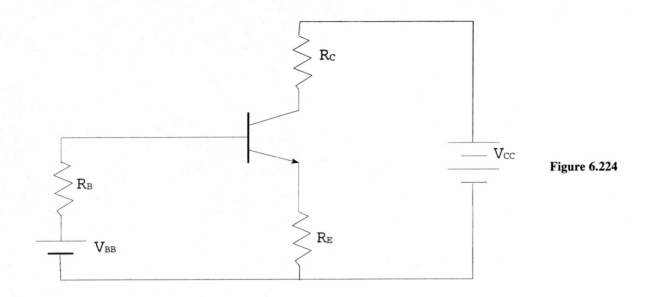

Figure 6.224

where $V_B = \left[\dfrac{R_2}{R_1 + R_2}\right] V_{CC}$ and $R_B = \left[\dfrac{R_1 R_2}{R_1 + R_2}\right]$

KVL for the input circuit is

$$V_B - I_B R_B - V_{BE} - (I_B + I_C) R_E = 0$$

From which

$$I_B = \dfrac{V_B - V_{BE} - I_C R_E}{R_B + R_E}$$

It can be seen that, as described previously, an increase in I_C results in a decrease of I_B etc.

Common Base Configuration

The configuration of an n-p-n common base amplifier is shown in Fig. 6.225.

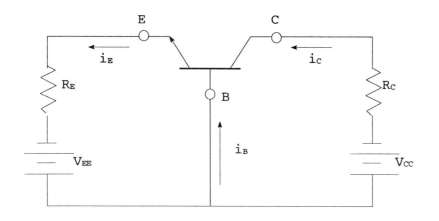

Figure 6.225

To set up the bias current, I_E, KVL is written for the input circuit.

$$V_{EE} - E_E R_E - V_{BE} = 0$$
$$I_E = \frac{V_{EE} - V_{BE}}{R_E}$$

If $V_{EE} \gg V_{BE}$, then
$$I_E \approx \frac{V_{EE}}{R_E}$$

Example

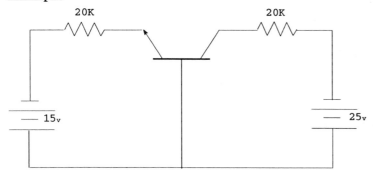

Figure 6226x

For the circuit shown, find v_{CB}. The transistor is germanium with an $\alpha = 0.97$.

Solution

$$v_{BE} = 0.3 \text{ volts}$$

$$I_E = \frac{15 - 0.3}{20} = \frac{14.7}{20} = 0.74 \text{ mA}$$

$$i_c = \alpha i_E + I_{CBO} \quad (\text{neglect } I_{CBO})$$

$$i_C = (0.97)(0.74)$$

$$i_C = 0.71 \text{ mA}$$

$$v_{CB} = 25 - (0.71)(20) = 10.8 \text{ volts} \quad \text{Ans.}$$

Common Collector Configuration

The configuration of an n-p-n common collector amplifier is shown in Fig. 6.227. This amplifier is sometimes called an emitter follower because the emitter voltage to ground follows the input signal voltage. A voltage gain of approximate unity is obtainable.

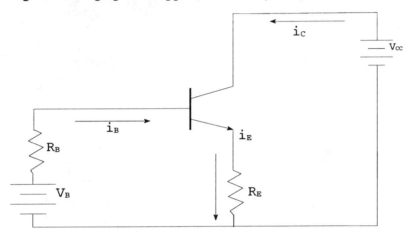

Figure 6.227

From KVL the following equations are obtained which will determine the operating point.

$$\text{Input } V_E = I_E R_E = (\beta + 1) I_B R_E$$

$$V_B = I_B R_B + (\beta + 1) I_B R_E + V_{BE}$$

$$I_B = \frac{V_B - V_{BE}}{R_B + (\beta + 1) R_E} \quad \text{where } V_{BE} = 0.3 \text{ for Ge}$$

$$0.7 \text{ for Si}$$

Signal Conditions - BJT

Because the VI characteristics of a BJT are nonlinear, small signals are assumed in the analysis of a transistor amplifier. In this manner, for small portions of the characteristics, the transistor is treated as a linear device. Accordingly, a signal or linear equivalent circuit can be utilized for the prediction of performance of a transistor amplifier wherein linear circuit analysis techniques are used.

Of the many models that can be used to represent a transistor, the most commonly used equivalent circuit is the one employing hybrid parameters. These parameters can be obtained from the VI characteristics, but they are usually tabulated in manufacturer's manuals.

The hybrid equivalent circuit (model) is shown in Fig. 6.228 to 6.230. Note that the same circuit is used for all transistor configurations; only the parameters are different.

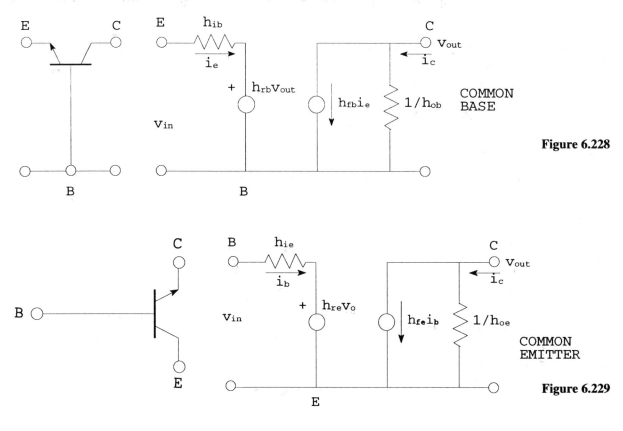

Figure 6.228

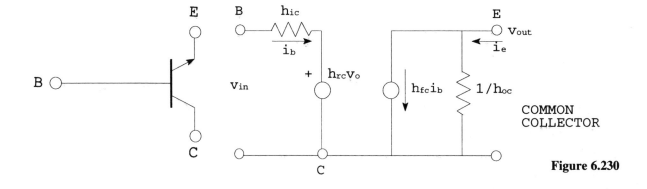

Figure 6.230

TRANSISTOR CHARACTERISTIC EQUATIONS

Symbol	Common Emitter	Common Collector	Common Base
h_{ie}	h_{ie}	h_{ic}	$\dfrac{h_{ib}}{1 + h_{fb}}$
h_{re}	h_{re}	$1 - h_{rc}$	$\dfrac{h_{ib} h_{ob}}{1 + h_{fb}} - h_{rb}$
h_{fe}	h_{fe}	$-1 - h_{fc}$	$\dfrac{-h_{fb}}{1 + h_{fb}}$
h_{oe}	h_{oe}	h_{oc}	$\dfrac{h_{ob}}{1 + h_{fb}}$
h_{ib}	$\dfrac{h_{ie}}{1 + h_{fe}}$	$\dfrac{-h_{ic}}{h_{fc}}$	h_{ib}
h_{rb}	$\dfrac{h_{ie} h_{oe}}{1 + h_{fe}} - h_{re}$	$h_{rc} - \dfrac{h_{ic} h_{oc}}{h_{fc}} - 1$	h_{rb}
h_{fb}	$\dfrac{-h_{fe}}{1 + h_{fe}}$	$\dfrac{-1 + h_{fc}}{h_{fc}}$	h_{fb}
h_{ob}	$\dfrac{h_{oe}}{1 + h_{fe}}$	$\dfrac{-h_{oc}}{h_{fc}}$	h_{ob}
h_{ic}	h_{ie}	h_{ic}	$\dfrac{h_{ib}}{1 + h_{fb}}$
h_{rc}	$1 - h_{re}$	h_{rc}	
h_{fc}	$-1 - h_{fe}$	h_{fc}	$\dfrac{-1}{1 + h_{fb}}$
h_{oc}	h_{oe}	h_{oc}	$\dfrac{h_{ob}}{1 + h_{fb}}$

The term hybrid arises because of the mixture of sources, i.e., a voltage source in the input and a current source in the output; hence the h in the nomenclature. In addition to the h are two-letter subscripts. The second subscript letter denotes the transistor configuration, i.e., b for common base, e for common emitter, and c for common collector. The first subscript letter denotes the parameter. A tabulation of these parameters are defined below.

h_i = Input resistance (OHMS)

h_f = Forward transfer current ratio (dimensionless)

h_r = Reverse transfer current ratio (dimensionless)

h_o - Output admittance

As an example, h_{oe} would be the output admittance for the common emitter amplifier, etc.

Understandably, a transistor type can be used in either of three amplifier configurations, i.e., common base, emitter, or collector. Therefore it is expected that an interrelationship of hybrid parameters exists. This inter-relationship is shown on p.

Example

A transistor type has the following hybrid parameters:

$h_{ie} = 700 \Omega$

$h_{oe} = 0.091 \mho$

$h_{fe} = 150$

$h_{re} = 0$

What are the corresponding common base parameters?

Solution

This involves the conversion from common emitter to common base.

$$h_{ib} = \frac{h_{ie}}{1 + h_{fe}} = \frac{700}{1 + 150} = 4.64 \Omega$$

$$h_{ob} = \frac{h_{oe}}{1 + h_{fe}} = \frac{0.091}{1 + 150} = 0.6 m\mho$$

$$h_{fb} = -\frac{h_{fe}}{1 + h_{fe}} = \frac{-150}{1 + 150} = -0.99$$

$$h_{rb} = \frac{h_{ie} h_{oe}}{1 + h_{fe}} - h_{re} = \frac{(700)(0.091)}{1 + 150} - 0 = 0.42$$

Important quantities associated with transistor amplifiers are:

Current Gain	A_i
Voltage Gain	A_v
Power Gain	A_p
Input Resistance	R_i
Output Resistance	R_o

The formulae for these quantities, in terms of hybrid parameters, are tabulated below for the various transistor amplifier configurations. No derivations will be given. Instead, examples of their usage will be presented.

Formulae for Transistor Quantities

Quantity	Common Base	Common Emitter	Common Collector
Current Gain (A_i)	$-\dfrac{h_{fb}}{1 + h_{ob} R_L}$	$\dfrac{h_{fe}}{1 + h_{oe} R_L}$	$\dfrac{h_{fc}}{1 + h_{oc} R_L}$
Voltage Gain (A_v)	$+\dfrac{h_{fb} R_L}{\Delta h_b R_L + h_{ib}}$	$-\dfrac{h_{fe} R_L}{\Delta h_e R_L + h_{ie}}$	$-\dfrac{h_{fc} R_L}{\Delta h_c R_L + h_{ic}}$
Power Gain (A_p) $A_p = A_i A_v$	$-\dfrac{(h_{fb})^2 R_L}{(1 + h_{ob} R_L)(h_{ib} + \Delta h_b R_L)}$	$-\dfrac{(h_{fe})^2 R_L}{(1 + h_{oe} R_L)(h_{ie} + \Delta h_e R_L)}$	$-\dfrac{(h_{fc})^2 R_L}{(1 + h_{oc} R_L)(h_{ic} + \Delta h_c R_L)}$
Input Resistance (R_i)	$\dfrac{R_L \Delta h_b + h_{ib}}{1 + h_{ob} R_L}$	$\dfrac{R_L \Delta h_e + h_{ie}}{1 + h_{oe} R_L}$	$\dfrac{R_L \Delta h_c + h_{ic}}{1 + h_{oc} R_L}$
Output Resistance (R_o)	$\dfrac{h_{ib} + R_g}{\Delta h_b + h_{ob} R_g}$	$\dfrac{h_{ie} + R_g}{\Delta h_e + h_{oe} R_g}$	$\dfrac{h_{ic} + R_g}{\Delta h_c + h_{oc} R_g}$
	$\Delta h_b = h_{ib} h_{ob} - h_{rb} h_{fb}$	$\Delta h_e = h_{ie} h_{oe} - h_{re} h_{fe}$	$\Delta h_c = h_{ic} h_{oc} - h_{rc} h_{fc}$

Example

A transistor type is to be used in a common base configuration and a common emitter configuration. The circuits are as shown. Calculate all quantities for these amplifiers.

Assume R_L is total load and V_o does not load R_L

Common Base

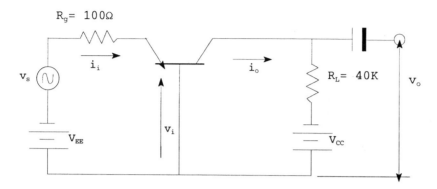

Figure 6.231

$h_{ib} = 60$

$h_{rb} = 4 \times 10^{-4}$

$h_{fb} = -0.98$

$h_{ob} = 10^{-6} \mho$

$\Delta h_b = 4.52 \times 10^{-4}$

Current Gain - A_i (Transistor Alone)

$$A_i = \frac{h_{fb}}{1 + h_{ob} R_L}$$

$$= -\frac{0.98}{1 + (10^{-6})(40000)}$$

$$= -0.942 \text{ (This is } \frac{i_o}{i_i}\text{)} \quad \text{Answer}$$

Voltage Gain - A_v (transistor alone)

$$A_v = -\frac{h_{fb} R_L}{\Delta h_b R_L + h_{ib}}$$

$$= \frac{(0.98)(40000)}{(4.52 \times 10^{-4})(40000) + 60}$$

$$= 500 \text{ (This is } v_o/v_i\text{)} \quad \text{Answer}$$

Power Gain = $A_p = (Av)(A_i) = (500)(-0.942) = -472$ (This is p_o/p_i) Ans.

Current Gain - A_{i_T} (complete circuit)

Same as for transistor alone ($\frac{i_o}{i_i}$)

$A_i = -0.942$ Ans.

Voltage Gain = a_{v_t} (complete circuit)

The signal equivalent circuit is:

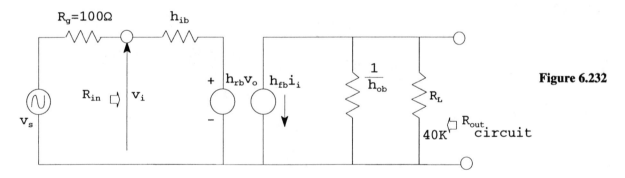

Figure 6.232

$$R_{in} = \frac{R_L \Delta h_b + h_{ib}}{1 + h_{ob} R_L}$$

$$= \frac{(40000)(4.52 \times 10^{-4}) + 60}{1 + (10^{-6})(40000)}$$

$$= 75\Omega \quad \text{Answer}$$

$$v_i = \frac{v_s}{(100+75)}(75)$$

$$= 0.43\, v_s$$

$$\frac{v_o}{v_s} = \frac{v_o}{v_i} \times \frac{v_i}{v_s} = (500)(0.43) = 215 \qquad \text{Answer}$$

Power Gain - A_{p_T} (complete circuit)

$$A_p = A_i A_v = (-0.942)(215) = -202 \qquad \text{Answer}$$

Output Resistance = R_o (transistor alone)

$$R_o = \frac{h_{ib} + R_g}{\Delta h_b + h_{ob} R_g}$$

$$= \frac{60 + 100}{4.52 \times 10^{-4} + (10^{-6})(100)}$$

Note: $R_s = R_g$

$$= 2.9 \times 10^5 \Omega \qquad \text{Answer}$$

Output Resistance = R_{ot} (complete circuit)

$$R_{ot} = \frac{R_o R_L}{R_o + R_L} = \frac{(2.9)(10^5)(40)(10^3)}{2.9 \times 10^5 + 40 \times 10^3} = 35000 \Omega \qquad \text{Answer}$$

Common Emitter

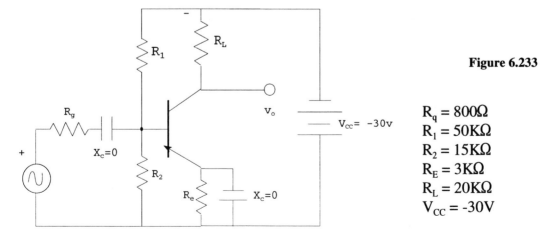

Figure 6.233

$R_q = 800\Omega$
$R_1 = 50K\Omega$
$R_2 = 15K\Omega$
$R_E = 3K\Omega$
$R_L = 20K\Omega$
$V_{CC} = -30V$

The signal equivalent circuit for the above amplifier is shown in Fig. 6.234.

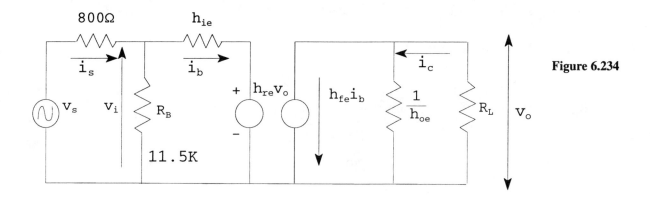

Figure 6.234

where $R_B = \dfrac{R_1 R_2}{R_1 + R_2} = \dfrac{(50)(15)}{65} = 11.5 \text{K}\Omega$

Because the transistor type is the same, the common base hybrid parameters can be converted into the common emitter parameters.

Accordingly, the conversion yields the following results.

$h_{ie} = 3000 \Omega$

$h_{re} = 26 \times 10^{-4}$

$h_{fe} = 49$

$h_{oe} = 50 \times 10^{-6} \mho$

And $\Delta h_e = 22.6 \times 10^{-3}$

Current Gain - A_i (transistor alone)

$$A_i = \dfrac{h_{fe}}{1 + h_{oe} R_L} = \dfrac{49}{1 + (50)(10^{-6})(20)(10^3)} = 24.5 \text{ Ans. } \left(\dfrac{i_c}{i_b}\right)$$

Voltage Gain - Av (transistor alone)

$$A_v = -\dfrac{h_{fe} R_L}{\Delta h_e R_L + h_{ie}} = -\dfrac{(49)(20000)}{(22.6)(10^{-3})(20)(10^3) + 3000}$$

$= -284$ Ans.

Power Gain = A_p (transistor alone)

$$A_p = A_v A_i = -(284)(24.5) = -6950 \quad \text{Answer}$$

Current Gain = A_{i_T} (complete circuit)

$$R_{in} = \frac{R_L \Delta h_e + h_{ie}}{1 + h_{oe} R_L} \quad \text{(transistor alone)}$$

$$= \frac{(20)(10^3)(22.6)(10^{-3}) + 3000}{1 + (50)(10^{-6})(20)(10^3)}$$

$$= 1726\Omega \quad \text{Answer}$$

Input Circuit

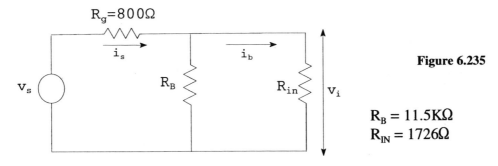

Figure 6.235

$R_B = 11.5 K\Omega$
$R_{IN} = 1726\Omega$

$$i_b = i_s \frac{11500}{11500 + 1726} \quad \text{current division}$$

$$i_b = 0.87 i_s \quad \text{or} \quad \frac{i_b}{i_s} = 0.87$$

$$A_{i_{Total}} = \frac{i_c}{i_s} = \frac{i_c}{i_b} \frac{i_b}{i_s} = (24.5)(0.87) = 21.3$$

Voltage Gain = a_{v_t} (complete circuit)

Input Circuit

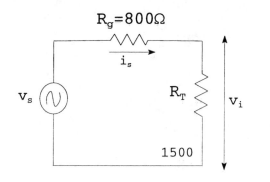

Figure 6.236

R_B is in parallel with R_{IN}

$$R_{Total} = \frac{(1726)(11500)}{1726 + 11500} \qquad R_{Total} = 1501$$

$$v_i = \frac{v_s}{(800 + 1501)}(1501)$$

$$\frac{v_i}{v_s} = 0.65$$

$$A_{v_{Total}} = \frac{v_o}{v_s} = \frac{v_o}{v_i}\frac{v_i}{v_s} = (-284)(0.65) = -184.6$$

Power Gain = A_{p_T} (complete circuit)

$$A_{p_T} = A_{i_T} a_{v_T} = (21.3)(-184.6) = -3932$$

Output Resistance = R_o (Transistor alone)

$$R_o = \frac{h_{ie} + R_g}{\Delta h_e + h_{oe}R_g} = \frac{3000 + 800}{22.6 \times 10^{-3} + (50 \times 10^{-6})(800)}$$

$$= 60703\Omega \quad \text{Ans.}$$

Output Resistance = R_{o_T} (complete circuit)

$$R_{o_T} = \frac{R_o R_L}{R_o + R_L} = \frac{(60703)(20000)}{(60703) + 20000} = 15044 \Omega \quad \text{Answer}$$

APPENDIX 6.7: OPERATIONAL AMPLIFIERS

An electronic amplifier is designed for voltage or current gain from unity to values of 100 or more. For example, an amplifier with a gain of 10 can be represented as shown in Fig. 6.237.

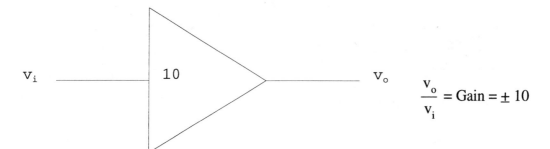

$$\frac{v_o}{v_i} = \text{Gain} = \pm\, 10$$

Figure 6.237

The gain is $\pm\, 10$ over the linear operating range of the amplifier. The $\pm$ sign depends on the phase shift characteristics of the amplifier configuration.

The operational amplifier is designed to have a very large gain in voltage or current and has parameters as shown in Fig. 6.238.

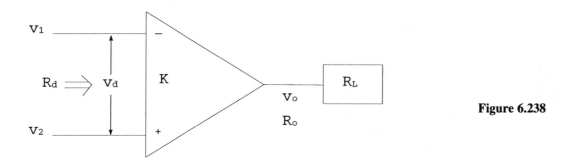

Figure 6.238

    a. $k = 10^6$ (very high)

    b. $v_d = v_1 - v_2$

    c. $R_d =$ input resistance and is very high implying that input current to amplifier approaches zero

    d. $R_o =$ output resistance and is very small implying that v_o is constant and independent of load R_L

Example

Determine the voltage gain, A_v for the following circuit.

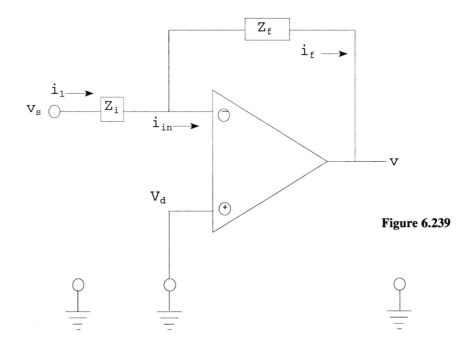

Figure 6.239

By Kirchoffs current law, $i_1 = i_{in} + i_f$, and

$$\frac{v_s - v_d}{Z_i} = \frac{v_d}{R_d} + \frac{v_o - v_d}{Z_f}$$

But:

$$i_{in} = \frac{v_d}{R_d} \to 0$$

$$K = \frac{v_o}{v_d} \to \infty$$

The voltage gain $A_v =$

$$\frac{V_o}{V_s} = -\frac{Z_f}{Z_s}$$

Where input is at the minus terminal and the plus terminal is called the circuit return. The minus terminal is called the inverting amplifier input, Z_f is the feedback impedance, and Z_i is the input impedance. If the input is at the plus terminal and the minus terminal is the circuit return, then the input is at the non-inverting amplifier terminal.

Three popular operational amplifier configurations are shown in Figs. 6.240 to 6.242.

Adder

Figure 6.240

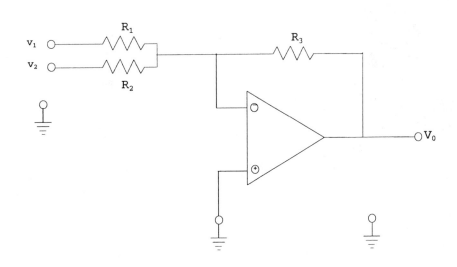

$$v_1 = -R_1 \left(\frac{v_1}{R_1} + \frac{v_2}{R_2} \right)$$

$$Z_1 = R_3$$

$$Z_1 = \frac{1}{R_1} + \frac{1}{R_2}$$

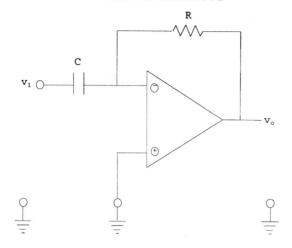

Figure 6.241

$$v_1 = -RC\frac{dv_1}{dt}$$

$$Z_1 = R$$

$$Z_i = C$$

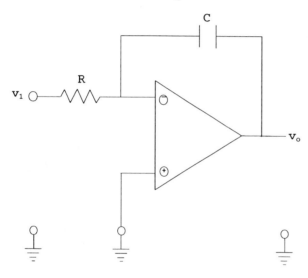

Figure 6.242

$$v_1 = -\frac{1}{RC}\int v_1\, dt$$

$$Z_1 = C$$

$$Z_i = R$$

7
Chemistry

TABLE OF CONTENTS

Selected References	7-ii
Periodic Law	7-1
Gas Laws	7-3
Chemical Reactions	7-6
Reaction Rate (Kinetics) and Equilibrium	7-8
Solutions	7-12
Nuclear Chemistry and Radioactivity	7-14
Thermochemistry	7-14
Organic Chemistry—Study of Carbon Compounds	7-15
Valence	7-19
Oxidation-Reduction (Redox) Reactions	7-22
Chemical Mathematics (Stoichiometry)	7-26
Illustrative Problems	7-27
Periodic Chart of the Elements	7-34

SELECTED REFERENCES

W.F. Ehret, "Smith's College Chemistry," D. Appleton-Century, New York

J.A. Babor, "General College Chemistry," Thomas Y. Crowell, New York

R.M. Whittaker, "Rudiments of Chemistry," The Ronald Press, New York

"Theory and Problems of College Chemistry," Schaum's Outline Series, McGraw-Hill, New York

P.R. Frey, "Chemistry Problems and How to Solve Them," Barnes & Noble, New York

CHEMISTRY 7-1

I. **PERIODIC LAW**

    A. The Periodic Chart is an orderly arrangement of elements according to atomic number. Properties of the elements are periodic functions of their atomic number.

        1. The Periods or Series is a horizontal row on the chart. All of the elements in any period have the same number of energy levels containing electrons. For example, all of the elements in period 4 have 4 energy levels containing electrons. The same for periods 5, 2, 1, 3, 6, or 7.

        2. The Group or Family is a vertical row on the periodic chart. All of the elements in any group have the same number of valence (outside shell) electrons. Since this determines chemical properties, all of the elements in any group have similar chemical properties. For example, which has chemical properties similar to those of Mg? (1) Na (2) Ca (3) Al (4) P. Answer is (2) Ca, since Ca is the only choice that is in the same group as Mg.

        3. Key (Fig. 7.1)

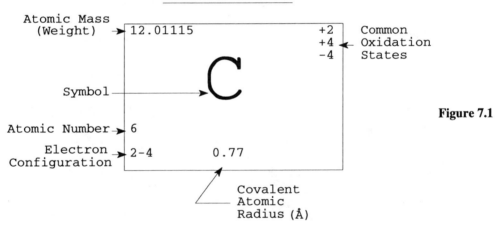

Figure 7.1

        (a) Atomic Number is the number of protons in the nucleus and this gives it its total positive charge.

        (b) Atomic Weight is the weighted average of the isotopes in a sample. For example, a sample of 10 atoms of Mg contains nuclides with the following atomic masses 4 at 24; 2 at 23; 3 at 25 and 1 at 26. The sum total of these is divided by 10 gives a relative average atomic weight of 24.3 for Mg. The atomic mass of a nuclide (single nucleus) is the sum of its proton and neutrons.

        (c) Electron configuration is the arrangement of electrons by level.

(d) The most common oxidation states are charges that the atoms of this element can have when they are found in compounds.

B. An <u>atom</u> is the smallest particle of an element that can exhibit the properties of that element. In the case of monatomic elements this is also a molecule.

C. A <u>molecule</u> is the smallest part of a substance, either an element or compound, containing all of the properties (both physical and chemical) of that substance.

D. <u>Atomic weight</u> is the mass of an atom from a chart. The atomic weight of an element is the relative weight of an atom of that element compared to the weight of an atom of carbon taken as 12.

E. <u>Molecular weight</u> is the sum of the atomic weights for all the atoms represented in the molecular formula.

<u>Example</u>

<u>Compound</u>	<u>Formula</u>	<u>Molecular Weight</u>
Hydrogen	H_2	2 x 1 = 2
Oxygen	O_2	2 x 16 = 32
Helium	He	1 x 4 = 4
Calcium hydroxide	$Ca(OH)_2$	Ca: 1 x 40 = 40 O : 2 x 16 = 32 H : 2 x 1 = 2 $\Sigma = 74$
Sulfuric acid	H_2SO_4	H : 2 x 1 = 2 S : 1 x 32 = 32 O : 4 x 16 = 64 $\Sigma = 98$

F. <u>Isotopes</u> are different weights of atoms of the same element due to different numbers of neutrons in the nucleus. The atoms of magnesium are all the same, but are isotopes because they contain different numbers of neutrons.

$$^{24}_{12}Mg, \quad ^{25}_{12}Mg, \quad ^{26}_{12}Mg$$

G. The <u>Rule of Dulong and Pettit</u> is an approximate empirical rule, not a law, which says that the product of the atomic weight of an elementary solid and the specific heat of that solid is approximately 6.4. Since the specific heat of a substance varies with temperature, application of this rule is restricted.

Example

Find the molecular weights of Al, C, Fe, Mg, Zn given their specific heat.

Element	Specific Heat-Cp cal/gm°C	Calculated Atomic Weight $6.4 \div Cp$	Actual Atomic Weight
Al	.214	29.8	26.97
C	.120	53.1	12.01
Fe	.107	59.5	55.85
Mg	.246	25.9	24.32
Zn	.094	68.0	65.38

II. GAS LAWS

A. Standard Temperaturess and Pressures

Standard temperature and pressure is defined as 0°C and 760 mm Hg or 32°F and 2116.2 psfa or 273°K and 1 atm.

1. Gram Molecular Volume

   The gram molecular volume is the volume occupied by the gram molecular weight of any gas at STP and is 22.4 liters.

2. Pound Molecular Volume

   The pound molecular volume is the volume occupied by the pound molecular weight of any gas at STP and is 359 cubic feet.

3. °C + 273 = °K

   °R = °F + 460

B. Gay–Lussac's Law

The pressure of a given mass of gas varies directly with the absolute temperature when the system is held at a constant volume.

$$\frac{P_1}{T_1} = \frac{P_2}{T_2}$$

C. <u>Avogadro's Law</u>

A volume of 22.4 liters of any ideal gas at STP contains the same number of molecules of the gas.

$$6.024 \times 10^{23} \frac{\text{Molecules}}{\text{Gm Molecular Volume}} \text{ or } \frac{\text{Molecules}}{22.4 \text{ l at STP}}$$

This is Avogadro's number.

D. <u>Ideal Gas Equation</u>

(1) PV - nRT

where: P = absolute pressure
V = gaseous volume
n = number of moles of gas
R = universal gas constant
T = absolute temperature

(2) $PV = WR_gT$

where: P = absolute pressure
V = gaseous volume
W = weight of gas
R_g = specific gas constant
T = absolute temperature

E. <u>Universal and Specific Gas Constant</u>

(1) PV = nRT (for 1 mole of a gas)

$$R = \frac{PV}{nT}$$

P = 1 atmosphere
T = 0°C = 273°K
V = 22.4 liters of gas
n = 1.0 mole
R = .082 l-atm/mole °K

(2) $R = \dfrac{PV}{nT}$

at STP: $T = 32°F + 460 = 492°R$
$P = 2116.2 \text{ \#f/ft}^2$
$V = 359 \text{ ft}^3$
$n = 1 \text{ lb. mole}$

$$R_g = 1544 \dfrac{\text{lbs}_f \text{ ft}}{\text{lbs}_{mole} \; °R} \times \dfrac{1}{MW} \dfrac{\text{lbs}_{mole}}{\text{lbs}_{mole}} = \dfrac{1544}{MW} \dfrac{\text{lbs}_f \text{ ft}}{\text{lbs}_{mole} \; °R}$$

Now to obtain the specific gas constant for a gas its molecular weight must be known. Therefore consider the following gases, Air, H_2, N_2, O_2.

$$R_g = 1545 \dfrac{\text{ft. lbs}}{\text{lb mole °R}}$$

Gas	Molecular Weight (MW)	$R_g \dfrac{\text{lbs}_f \text{ ft}}{\text{lbs}_{mole} \; °R}$
Air	29	53.3
H_2	2	772.5
N_2	28	55.3
O_2	32	48.5

F. <u>Graham's Law</u>

"The rate of diffusion of two gases (r_1 and R_2) are inversely proportional to the square roots of either their densities (ρ_1 or ρ_2) or molecular weights (MW_1 and MW_2)."

$$\dfrac{r_1}{r_2} = \sqrt{\dfrac{\rho_2}{\rho_1}} = \sqrt{\dfrac{MW_2}{MW_1}}$$

G. <u>Dalton's Law of Partial Pressures</u>

"The sum of the partial pressures of the gases of a mixture is equal to the total pressure of the mixture."

$$P_T = \sum_{i=1}^{n} P_i$$

Example

Given: A mixture of H_2, N_2 and O_2. The partial pressure of H_2 = .2 atm, N_2 = .4 atm, O_2 = .4 atm.

Find: The total pressure of the mixture

$$P_T = \sum_{i=1}^{3} P_i = P_1 + P_2 + P_3$$

$$P_T = P_{H_2} + P_{N_2} + P_{O_2} = .2 \text{ atm} + .4 \text{ atm} + .4 \text{ atm} = 1.0 \text{ atm}$$

$P_T = 1.0$ atm   Answer

III. **CHEMICAL REACTIONS**

    A. <u>Direct Combination, Synthesis, or Composition</u>

        There is only one product.

        Examples

        (1) $C + O_2 \rightarrow CO_2$

        (2) $2K + Cl_2 \rightarrow 2\,KCl$

        (3) $MgO + H_2O \rightarrow Mg(OH)_2$

    B. <u>Decomposition, or Analysis</u>

        There is only one reactant.

        Examples

        (1) $2HgO \rightarrow 2\,Hg + O_2$

        (2) $CaCO_3 \rightarrow CaO + CO_2$

C. **Single Displacement**

In this type of reaction a single element replaces a less active element, of the same type, from a compound.

1. **Metallic**

    (a) $Zn + 2HCl \rightarrow ZnCl_2 + H_2$

    (b) $3Mg + 2FeI_3 \rightarrow 3MgI_2 + 2Fe$

2. **Non-Metallic**

    (a) $3Cl_2 + 2AlBr_3 \rightarrow 2AlCl_3 + 3Br_2$

    (b) $F_2 + 2NaCl \rightarrow 2NaF + Cl_2$

D. **Double Displacement** (Exchange of Ions)

Double displacement reactions can be generalized as follows:

$$AB + CD \rightarrow AD + CB$$

This reaction is typical of the neutralization process between acids and bases.

Example

(1) $NaOH + HCl \rightarrow NaCl + HOH$

(2) $2\,NaOH + H_2SO_4 \rightarrow Na_2SO_4 + 2\,HOH$

(3) $Ba(OH)_2 + H_2SO_4 \rightarrow BaSO_4 + 2\,HOH$

E. **Reversible Reactions**

These are reactions that can go **BOTH** in the forward and reverse directions. They do NOT go to completion since none of the product are formed into molecules (H_2O), escape as a gas, or form insoluble solid (precipitate) and they can therefore be used as reactants and go in the reverse direction. ($\rightleftarrows$) means reversible.

IV. **REACTION RATE (KINETICS) AND EQUILIBRIUM**

A. **Equilibrium Condition**

When the rate of the forward reaction is equal to the rate of the reverse reaction, a condition exists known as dynamic equilibrium.

B. **Factors Affecting Reaction Rate**

(1) Concentration of reactants
(2) Temperature
(3) Pressure (only when there is at least one gaseous reactant)
(4) Activity of reactants
(5) Surface area
(6) Catalysts
(7) Concentration of products
(8) Reaction mechanism

C. **Law of Le Châtelier**

"When a system which is in equilibrium is subjected to a stress, there is a shift in the point of equilibrium so as to restore the original condition or to relieve the stress."

This principle is used to predict the effects of changes in temperature, pressure or concentration.

1. Application of Le Châtelier's Law

Note all stresses are applied from outside the system. The system then adjusts to relieve the stress.

$N_2(g) + 3H_2(g) \rightleftarrows 2NH_3(g) + 22.08$ K cal.

(a) <u>Pressure</u>. Rule: Increased pressure shifts to produce less gas; decreased pressure shifts to produce more gas. In this case there are 4 moles of gas on the left and 2 moles of gas on the right. An increase in pressure will make the reaction shift forward, decreasing the number of moles of N_2 and H_2 and increasing the number of moles of NH_3. A decrease in pressure will do the opposite.

(b) <u>Concentration</u>. Rule: An increase in concentration causes a shift to use up this material (away from it) and produce more product going in that direction. For example, increasing N_2 or H_2 shifts this reaction forward producing NH_3. It is important to note that when either N_2 or H_2 are increased, the other reactant will have to decrease in the process in order to use some of it to make more of the product. Decreasing either N_2 or H_2 will cause the reaction to shift in the reverse direction in order to replace what was removed, decreasing the NH_3.

(c) <u>Temperature</u>. Rule: When the temperature is increased the reaction shifts in the endothermic direction; when the temperature is decreased the reaction shifts in the exothermic direction. Changing the temperature on a system is the ONLY condition that changes the value of the equilibrium constant. In the reaction above 22.08 Kcal is given off going forward. Therefore the forward reaction is exothermic and the reverse is endothermic. Increasing the temperature on this system will make the reaction go in reverse; decreasing the temperature will make it go forward.

(d) <u>Catalysts</u> speed up forward and reverse reactions equally and therefore they cause no shifting.

D. <u>Equilibrium Constants</u>

1. Form $2A + B \rightleftharpoons 3C + 4D$

Consider the general equation for a reversible reaction (pure liquids, H_2O etc., and pure solids are NEVER shown in an equilibrium expression).

$$2A + B \underset{R_2}{\overset{R_1}{\rightleftharpoons}} 3C + 4D$$

Where R_1 is the rate of the forward reaction and R_2 is the rate of the reverse reaction.

At equilibrium $\quad R_1 = R_2$

and the equilibrium constant for the reaction at a specific temperature would be

$$K_{eq} = \frac{[C]^3 [D]^4}{[A]^2 [B]}$$

<u>Note</u>: [ ] represents the concentration of each constituent in moles/liter

2. Values

    (a) When K = 1 neither direction is favored, forward and reverse are equal

    (b) When K > 1 there are more products and the reaction is favored going forward. The greater the value of K the more forward the reaction goes.

    (c) When K < 1 there are more reactants than products and the reverse reaction is favored.

3. Types of Equilibrium Constants

    (a) Ionization constants show degree of ionization for weak acids and bases. The larger its value the stronger the acid or base is.

$$HA_c + H_2O \rightleftharpoons H_3O^+ + A_c^-$$

$$K_a = \frac{[H_3O]^+ [A_c]^-}{[HA_c]^-} = 1.8 \times 10^{-5}$$

$$CO_2 + 2H_2O \rightleftharpoons H_3O^+ + HCO_3^-$$

$$K_a = \frac{[H_3O]^+ [HCO_3]^-}{CO_{2\,(g)}} = 4.2 \times 10^{-7}$$

Carbonic acid is weaker than acetic acid.

    (b) The solubility product shows the solubility limit of slightly soluble salts. The larger its value the more soluble the salt.

$$AgCl_{(s)} \rightarrow Ag^+_{(aq)} + Cl^-_{(aq)}$$

$$K_{sp} = [Ag^+][Cl^-] = 1.7 \times 10^{-10}$$

s = solid   aq = ions in solution

Example:

$$PbCl_{2(s)} \rightleftharpoons Pb^{+2}_{(Aq)} + 2Cl^-_{(aq)}$$

Solid → ions in solution

Equilibrium constant can be expressed as:

$$K_{eq} = \frac{[Pb^{+2}][Cl^-]^2}{[PbCl_2]}$$

Since the concentration of an undissolved solid such as $PbCl_2$ is unity, the above expression becomes

$$K_{s.p.} = [Pb_{+2}][Cl^{-1}]^2$$

where $K_{s.p.}$ is called the solubility product

Precipitation of $PbCl_2$ will occur when

$$[Pb^{+2}][Cl^{-1}]^2 > K_{s.p.} \text{ for } PbCl_2$$

(c) **pH and pOH values of water solutions**

$$H_2O \rightleftharpoons H^+ + OH^-$$

$$K_W = [H^+][OH^-]$$

where K_W is known as the ionization constant for water

From experiments at room temperature (20°C)

10^7 liters of H_2O contain 1 mole of H^+, 1.008 g and 10^7 liters of H_2O contain 1 mole of OH^-, 17.008 g

Therefore

$$[H^+] = [OH^-] = \frac{1 \text{ mole}}{10^7 \text{ liters}} = 10^{-7} \frac{\text{mole}}{\text{liter}}$$

The ion product for water is defined as

$$K_w = [H^+][OH^-] = 10^{-7} \times 10^{-7} = 10^{-14} \text{ @ } 20°C$$

Now the acidity/alkalinity of a solution can be easily expressed as pH or pOH (Fig. 7.2).

$$pH = -\log[H^+]$$

$$pOH = -\log[OH^-]$$

$$pH + pOH = 14$$

[H⁺] mole/liter	pH	[OH⁻] mole/liter	pOH
10^0	0	10^{-14}	14
10^{-1}	1	10^{-13}	13
10^{-2}	2	10^{-12}	12
10^{-3}	3	10^{-11}	11
10^{-7}	7	10^{-10}	10
10^{-12}	12	10^{-7}	7
10^{-13}	13	10^{-1}	1
10^{-14}	14	10^{-0}	0

↑ Increasing Acidity   ↓ Increasing Basicity

Figure 7.2

V. **SOLUTIONS**

    A. Percentage

    Parts of solute (by weight) per 100 parts of total solution (by weight).

    Example:

    A 20% salt solution contains 20 g of salt dissolved in 80 g of water.
    Solution concentration in ppm = ml/kg = ml/L

    B. Molarity: $M = \dfrac{\text{moles of solute}}{\text{liters of solution}}$

    A one (1) molar solution contains one (1) gram molecular weight of solute to which enough solvent has been added to make a total of one (1) liter of solution.

    C. Molality: $m = \dfrac{\text{moles of solute}}{\text{Kg of solvent}}$

    A one (1) molal solution contains one (1) gram molecular weight of solute dissolved in 1000 grams of solvent.

    D. Normality: $(N) = \dfrac{\text{\# of equivalents}}{\text{liters of solution}}$

    A one (1) normal solution contains one (1) gram equivalent weight of solute to which enough solvent has been added to make a total of one (1) liter solution.

(1) Gram-equivalent weight

The gram molecular weight of a substance divided by the number of electrons the substance will gain or lose to fill its outer shell when combining with other atoms. (See p.XXX for further explanation of valence).

$$\text{eq. wt} = \frac{\text{mol wt.}}{\text{total pos valence}}$$

(2) Number of equivalents is equal to $\frac{\text{grams of solute}}{\text{equivalent weight}}$

E. Raoult's Law. The amount that the freezing point is lowered or the boiling point is raised for a one molal solution of a nonelectrolyte is called the freezing point depression (FPD) or boiling point elevation (BPE) constant for a solvent.

F. Freezing Point Depression and Boiling Point Elevation of Water Solutions

The following table illustrates the effect of dissolving 1 gram mole of solute in 1000 g of H_2O to produce a 1 molal (1 m) solution

Substance	Particles	Freezing Point Depression °C	Boiling Point Elevation °C
Water	-	-	-
1m Sugar	1	-1.86	+.52
1m Glycerine	1	-1.86	+.52
1m Na Cl	2	-3.72	+1.04
1m $CaCl_2$	3	-5.58	+1.56

VI. NUCLEAR CHEMISTRY and RADIOACTIVITY

A. Alpha decay is the spontaneous emission of an alpha particle (helium nucleus or 2 protons and 2 protons and 2 neutrons - $^{4}_{2}He$) from a nucleus. This decreases the atomic number by 2 and the mass number by 4.

$$^{226}_{88}Ra \rightarrow {}^{4}_{2}He + {}^{222}_{86}Rn$$

B. Beta decay is the spontaneous emission of a beta particle (a neutron changes into a proton and a beta particle and an electron $^{0}_{1}e$) from the nucleus. This causes the atomic number to increase by 1 and the mass number to remain the same

$$^{234}_{90}Th \rightarrow {}^{0}_{-1}e + {}^{234}_{91}Pa$$

C. Half-life and Radioactive Decay

If M_0 is the original mass of a substance and M is the mass remaining after time t, then:

$$m = m_0 e^{-\delta t}$$

Where δ is the decay constant and e is the base of natural logarithms.

The half life is the time for one half the substance to decay or $M = M_0/2$

By rearranging and taking logarithms, we arrive at

$$t = \frac{\ln 2}{\delta} = \frac{0.693}{\delta}$$

Where t is the time for half the substance remaining to decay.

VII. THERMOCHEMISTRY

The end product of most forms of energy is heat. One calorie of heat energy is equal to 4.185×10^7 ergs. One calorie is also the amount of heat required to raise the temperature of 1 gram of water 1°C.

An <u>endothermic process</u> is one in which heat is absorbed. An <u>exothermic</u> Process is one that releases heat energy. The <u>heat of fusion</u> is defined as the amount of heat energy required to change a given amount of solid to a liquid without giving an increase in temperature. For example, it requires 79.7 calories to melt one gram of ice. The solid turns to liquid at the same temperature.

The <u>heat of vaporization</u> is the amount of heat required to change a given amount of a liquid to a vapor, again with no increase in temperature. For water, 539.6 calories are required to vaporize one gram of water, that is, to turn it into steam.

<u>Bond energy</u> is a measure of the energy lost by an atom when it forms a bond with another atom. The same amount of energy is required to break the bond. In $H_2 \rightarrow 2H$ there is a change in total energy, ΔH, equal to 104.2 Kcal. It will require 104,200 calories of energy to break the bonds in one mole of hydrogen, H_2, to form 2 moles of atomic hydrogen, H.

VIII. <u>ORGANIC CHEMISTRY—STUDY OF CARBON COMPOUNDS</u>

A. Hydrocarbons are compounds of carbon and hydrogen. They are non-polar, covalent, and insoluble in water, with low melting and boiling points. They are nonconductors. The general formula is R-H, where R stands for the hydrocarbon portion of the molecule with an H added.

Aliphatic compounds are carbon compounds in which the carbon atoms are arranged in straight or branched chains.

Aromatic compounds are carbon compounds in which the carbon atoms are arranged in one or more rings.

1. A homologous series is a group or family of hydrocarbons with the same general formula, similar structures, and similar properties.

    a. Methane series: Alkanes
    General formula: C_nH_{2n+2}

    Structure: $-c-c-$ Saturated

Methane	CH_4
Ethane	C_2H_6
Propane	C_3H_8
Butane	C_4H_{10}
Pentane	C_5H_{12}
Hexane	C_6H_{14}
Heptane	C_7H_{16}
Octane	C_8H_{18}

b. Ethylene series: Alkenes - olefin series
General formula: C_nH_{2n}

Structure: $-\overset{|}{c}=\overset{|}{c}-$ Unsaturated

Ethylene	C_2H_4
Propene	C_3H_6
Butene	C_4H_8
Pentene	C_5H_{10}
Hexene	C_6H_{12}
Heptene	C_7H_{14}
Detene	C_8H_{16}

c. Acetylene series: Alkynes
General formula: C_nH_{2n-2}

Structure: $-c\equiv c-\overset{|}{c}-$ Unsaturated

Acetylene	C_2H_2
Propyne	C_3H_4
Butyne	C_4H_6
Pentyne	C_5H_8
Hexyne	C_6H_{10}
Heptyne	C_7H_{12}
Octyne	C_8H_{14}

d. Benzene series (Fig. 7.3) - C_nH_{2n-6} Unsaturated Bonds

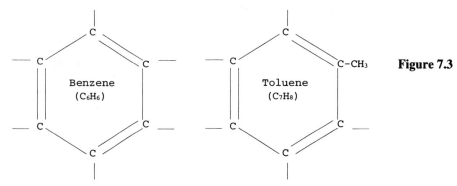

Figure 7.3

B. Substituted hydrocarbons are compounds where one or more of the H atoms has been substituted for by another atom, which gives it functional groups and determines its properties.

1. <u>Alcohols</u> R-OH

   Methyl Alcohol CH_3-OH

   Ethyl Alcohol C_2H_5-OH

   Propyl Alcohol C_3H_7-OH

2. <u>Aldehydes</u> $R-\underset{\underset{O}{\|}}{C}-H$

   Formaldehyde (HCHO) • $H-\underset{\underset{O}{\|}}{C}-H$

3. <u>Ketones</u> $R-\underset{\underset{O}{\|}}{C}-R$

   Acetone (CH_3COCH_3) • $CH_3-\underset{\underset{O}{\|}}{C}-CH_3$

4. <u>Ethers</u> R-O-R' (organic oxides)

   Ethyl Ether C_2H_5-O-C_2H_5

5. <u>Acids</u> $R-\underset{\underset{O}{\|}}{C}-OH$

   Acetic Acid (CH_3COOH) • $CH_3-\underset{\underset{O}{\|}}{C}-OH$

6. <u>Esters</u> $R-\underset{\underset{O}{\|}}{C}-OR'$

Esters are organic salts that are formed by the reaction between an alcohol and an acid. This reaction is called esterification.

$$R\text{-}OH + R'\text{-}\underset{\underset{O}{\|}}{C}\text{-}OH \rightarrow R'\text{-}\underset{\underset{O}{\|}}{C}\text{-}OR + H_2O$$

Esterification and neutralization are somewhat similar

Esterification: Alcohol + Acid $\rightleftarrows$ Ester + H_2O

Neutralization: Base + Acid $\rightarrow$ Salt + H_2O

Neutralization goes to completion while esterification is reversible.

7. <u>Soaps</u>

A soap is a metallic salt of a fatty acid. It is made by boiling vegetable oil or animal fat with a strong base such as sodium or potassium hydroxide. This reaction is called saponification and always results in the formation of a soap and glycerine.

Fat + Base $\rightarrow$ Soap + Glycerine

8. <u>Carbohydrates</u>

These are compounds of carbon, hydrogen, and oxygen in which the ratio of hydrogen to oxygen is 2:1. The table below gives the most common carbohydrates.

Compound	Formula	Commercial Name
Glucose	$C_6H_{12}O_6$	Dextrose or Grape Sugar
Sucrose	$C_{12}H_{22}O_{11}$	Saccharin
Starch	$(C_6H_{10}O_5)_x$	
Cellulose	$(C_6h_{10}O_5)_y$	

IX. **VALENCE**

The number of electrons an atom may gain or lose in completing its outer shell when it combines with another atom.

Note: An atom with a plus (+) will give up electrons to complete its outer shell. It becomes a positive ion (metals). An atom with a minus (-) valence will accept electrons to complete its outer shell. It becomes a negative ion (Nonmetals).

A. Types of Valence

(1) Electrovalence:

Compounds formed from either a gain or loss of electrons are called electrovalent, polar, or ionic compounds.

(2) Covalence:

Compounds formed in a way such that atoms share one or more pairs of electrons are called covalent.

B. Tables of Valences for Metallic Ions

Metallic Elements and Radicals - Cations			
+1	+2	+3	+4
Lithium Li^+	Barium Ba^{+2}	Bismuth Bi^{+3}	Carbon C^{+4}
Sodium Na^+	Cadmium Cd^{+2}	*Arsenic (Arsenous) As^{+3}	Silicon Si^{+4}
Potassium K^+	Calcium Ca^{+2}	*Antimony (Antimonous) Sb^{+3}	Tin (Stannic) Sn^{+4}
Copper (cuprous) Cu^+	Copper (Cupric) Cu^{+2}	Aluminum Al^{+3}	
Mercury (Mercurous) Hg^+	Mercury (Mercuric) Hg^{+2}	Chromium Cr^{+3}	
Hydrogen H^+	Lead Pb^{+2}	Iron (Ferric) Fe^{+3}	
Silver Ag^+	Iron (Ferrous) Fe^{+2}		
	Magnesium Mg^{+2}		
	Manganese Mn^{+2}		
	Zinc Zn^{+2}		
	Tin (Stannous) Sn^{+2}		

*Arsenic and Antimony also have +5 valence numbers.

Nonmetallic Elements and Radicals - Anions			
-1	-2	-3	-4
Borate BO_2^{-1}	Chromate CrO_4^{-2}	Phosphate PO_4^{-3}	Carbon C^{-4}
Nitrate NO_3^{-1}	Carbonate CO_3^{-2}	Arsenate AsO_4^{-3}	Silicon Si^{-4}
Acetate CH_3COO^{-1}	Sulfate SO_4^{-2}	Arsenite AsO_3^{-2}	
Hydroxide OH^-	Sulfite SO_3^{-2}	Nitrogen N^{-3}	
Chlorate ClO_3^-	Throsulfate $S_2O_3^{-2}$		
Fluorine F^{-1}	Sulfur S^{-2}		
Chlorine Cl^{-1}			
Bromine Br^{-1}			
Iodine I^{-1}			
BiCarbonate HCO_3^-			
Hypochlorite ClO^{-1}			
Thiocyanate SCN^{-1}			
Chlorite ClO_2^{-1}			
Perchlorate ClO_4^{-1}			
Nitrite NO_2^{-1}			

X. OXIDATION - REDUCTION (REDOX) REACTIONS

Oxidation: Loss of electrons in electrolysis always occurs at the anode (+) (AN OX)

Reduction: Gain of electrons in electrolysis always occurs at the cathode (-) (RED CAT)

During a REDOX reaction the oxidizing agent is reduced and the reducing agent is oxidized.

A. Balancing Redox Reactions

This is done by writing the half-reactions after assigning oxidation numbers to elements that change, equalizing the loss and gain of electrons and totaling up the equation.

Example:

(1) $Zn + 2\,AgNO_3 \rightleftarrows 2Ag^0 + Zn(NO_3)_2$

$\dfrac{2}{1}$ | $Ag^+ + e \rightarrow 2Ag^0$ Reduction Oxidation

     $Zn - 2e \rightarrow Zn^{+2}$ Oxidation

⇓

Multiply

$2Ag^+ + 2e \rightarrow 2Ag^0$ Reduction

$Zn - 2e \rightarrow Zn^{+2}$ Oxidation

⇓

Add

$2\,Ag^+ + Zn \rightarrow 2Ag^0 + Zn^{+2}$

For the above example the following can be stated

Oxidizing agent • Ag^+

Reducing agent • Zn^0

(2) $Hg_2Cl_2 + H_2S \rightarrow 2HgS + Hg^0 + 2HCl$

$Hg_2^{+2} + 2e \rightarrow 2 Hg^0$ REDUCTION

$Hg_2^{+2} - 2e \rightarrow 2 Hg^{+2}$ OXIDATION

⇓
Add

$2 Hg_2^{+2} \rightarrow 2 Hg^0 + 2 Hg^{+2}$

⇓
÷ by 2

$Hg_2^{+2} \rightarrow Hg^0 + Hg^{+2}$

Oxidizing Agent • Hg_2^{+2}

Reducing Agent • Hg_2^{+2}

Here is the unique case where an individual substance is both the oxidizing and reducing agent.

(3) Balance

$MnO_4^- + SO_3^{-2} + H^+ \rightarrow Mn^{+2} + SO_4^{-2} + H_2O^0$

$2 [MnO_4^- + 8H^+ + 5e \rightarrow Mn^{+2} + 4H_2O^0]$

REDUCTION

$5 [SO_3^{-2} + H_2O - 2e \rightarrow SO_4^{-2} + 2H^+]$

OXIDATION

⇓
Multiply

$2MnO_4^- + 16H^+ + 10e \rightarrow 2Mn^{+2} + 8H_2O$

$5SO_3^{-2} + 5H_2O - 10e \rightarrow 5SO_4^{-2} + 10H^+$

⇓
Add

$2MnO_4^- + 6H^+ + 5SO_3^{-2} \rightarrow 2Mn^{+2} + 3H_2O + 5SO_4^{-2}$

Final Balance

$2MnO_4^- + 5SO_3^{-2} + 6H^+ \rightarrow 2Mn^{+2} + 3H_2O + 5SO_4^{-2}$

Typical Reaction

$2KMnO_4^- + 5Na_2SO_3 + 6HCl \rightarrow 2MnCl_2 + 2KCl + 5Na_2SO_4 + 2H_2O$

B. Electromotive Force (Standard Oxidation Potentials)

This table lists a series of half-reactions in which atoms, molecules, or ions tend to lose electrons. This tendency is measured against a standard H_2 half-cell. As the oxidation potential becomes more positive, the relative ease to lose electrons increases. In brief the higher in the table the more active is the metal (more powerful reducing agent). The lower in the table, the more active is the nonmetal (more powerful oxidizing agent).

Half-Cell Reaction		E^0 (Volts)
Li	$Li^+ + e$	3.05
K	$K^+ + e$	2.93
Cs	$Cs^+ + e$	2.92
Na	$Na^+ + e$	2.71
Zn	$Zn^+ + e$	0.76
Fe	$Fe^{+2} + e$	0.44
H_2	$2H^+ + 2e$	0.00
Cu^+	$Cu^{+2} + e$	-0.15
Cu	$Cu^{+2} + 2e$	-0.34
$2I^-$	$I_2 + 2e$	-0.53
Fe^{+2}	$Fe^{+3} + e$	-0.77
Ag	$Ag^+ + e$	-0.80
$2Br^-$	$Br_2(L) + 2e$	-1.07
$2Cl^-$	$Cl_2 + 2e$	-1.36
$2F^-$	$F_2 + 2e$	-2.87

Reducing strength Increases (↑ left side)

Oxidation Strength Increases (↓ right side)

Examples

(a) $F_2 + 2KCl \rightarrow 2KF + Cl_2$

$F_2^0 + 2e \rightarrow 2F^-$ reduction $F_2 \rightarrow$ oxidizing agent

$2Cl^- - 2e \rightarrow Cl_2^0$ oxidation $Cl^- \rightarrow$ reducing agent

(b) $Na + ZnSO_4 \rightarrow Zn + Na_2SO_3$

2 (Na - e → Na⁺) oxidation Na → reducing agent

1 (Zn^{+2} + 2e → Zn⁰) reduction Zn^{+2}→ oxidizing agent

⇓

Multiply

2 Na - 2 e → 2 Na⁺

Zn^{+2} + 2e → Zn⁰

⇓

Add

2Na + Zn^{+2} → 2Na⁺ + Zn⁰

C. Solutions of Oxidants and Reductants Expressed in Terms of Normality

One gram-equivalent weight of an oxidant or reductant is equal to the formula weight divided by the change in oxidation number of the atoms expressed by the formula. Therefore, the gram equivalent weight of an oxidant or reductant is valid only when referred to a particular reaction.

D. Electrochemistry

1. Electrolysis and electroplating. In this case, the electricity produces the chemical change.

   a. Electrolysis is decomposing a compound either melted or in solution, to produce free elements. In the electrolysis of water solutions, if the cation is an active metal, it will NOT be deposited during electrolysis. Hydrogen will be liberated at the cathode. If the cation is an inactive metal, the metal will be deposited during electrolysis.

   One Faraday is the amount of electrical energy required to liberate one gram equivalent weight of an element from solution. One Faraday is equivalent to one mole of electrons.

   One Coulomb is 6.24×10^{18} electrons or elementary charges and is the amount of electricity required to deposit 0.001118 grams of silver from a solution containing Ag ions. One Faraday is equal to 96,490 coulombs.

One ampere is the rate of flow of one coulomb per second.

Electrolysis of NaF

anode $2F^- \rightarrow F_2 + 2e^-$

cathode $Na^+ + 2e^- \rightarrow 2Na$

$2NaF \rightarrow 2Na + F_2$

    b. Electroplating is the deposition of the metal that makes up the anode onto the surface of the cathode. The magnitude of current used and the length of time of its flow determines the amount of metal that can be deposited.

XI. CHEMICAL MATHEMATICS (STOICHIOMETRY)

    A. Moles

        1. $\dfrac{\text{grams}}{\text{mw}}$ = moles, moles x mw = grams

        2. $\dfrac{\text{liters}_{(stp)}}{22.4}$ = moles, 22.41 x moles = liters

        3. molecules - moles x (6.02×10^{23})

           6.02×10^{23} x moles = molecules

        4. $M = \dfrac{\text{moles solute}}{\text{liter of solution}}$

           liter of solution x molarity = moles

    B. Moles using equations: After an equation has been balanced, the following three steps can be used to solve any problem dealing with an equation.

        1. Change given quantity to moles

        2. Solve a proportion to find moles of unknown.

        3. Change moles of unknown to desired units.

CHEMISTRY 7-27

C. Empirical Formula Problems

The empirical formula of a compound is the simplest formula.

1. Divide % of the element by atomic weight to find gram atoms of that element in the formula.

2. The ratio of these gram atoms expressed in simplest whole number form is the empirical formula.

3. The weight of this formula is the empirical formula weight. If it is equal to the molecular weight, then this formula is also a molecular formula. If not, the molecular weight is a multiple of it and therefore the molecular formula is also the same multiple of the empirical formula. (NOTE: The molecular weight can be found from gas density:

$$\frac{g}{liter} = \frac{MW}{22.4}$$

or molarity information.)

ILLUSTRATIVE PROBLEMS

1. How many molecules of C_2H_6 are there in 250g of C_2H_6? C_2H_6 weighs $(12 \times 2) + (1 \times 6) = 30g$. Therefore,

$$250g \times \frac{1 \text{ mole}}{30g} \times 6.024 \times 10^{23} \frac{\text{Molecules}}{\text{mole}} = 50.2 \times 10^{23} \text{ Molecules}$$

See XI, A, 3 (This is Section XI, Item A, line 3 above)

2. In a mixture of 3 gases, X, Y and Z, there are 10 moles of X, 20 moles of Y, and 30 moles of Z. The pressure of the mixture is 1000 mm Hg. What is the partial pressure in mm Hg of gas Y?

$$\frac{20}{10 + 20 + 30} \times 1000 = \frac{20}{60} \times 1{,}000 = 333 \text{ mm Hg.}$$

See II, G

3. Gas X has a density of 1.96 moles/liter at 5°C. What is the pressure of that gas?

We can use the Characteristic equation or Perfect Gas Law. Here W = R.

$$P = \frac{n}{v} RT = 1.96 \frac{mole}{1} \times .0821 \frac{1 \text{ atm}}{mole \, °K} \times 278 \, °K$$

P - 44.7 atm

Remember that 22.4 liters contains one mole or 6.03×10^{23} molecules at STP.

$$R = \frac{PV}{nT} = \frac{22.4 \, 1 \times 1 \text{ atm}}{Mol \times 273°K} = .0821 \frac{1 \text{ atm}}{Mol \, °K}$$

When this value of R is used, P must be in atmospheres, V in liters, T in °K, and n in moles.

Other values of R in different units can be shown to be:

R = 0.730 ft³ atm/lb - mole °R or R = 1544 ft lb/lb mole °R

Rg = 1544/MW

See II, E

4. How many moles of carbon dioxide are formed when 5 moles of C_2H_6 are burned in excess air?

$C_2H_6 + 2.5O_2 \rightarrow 2CO_2 + 3H_2O$

5 moles of $C_2H_6 \rightarrow$ 10 moles of CO_2

See XI, B

5. Gases are allowed to diffuse through the same porous partition at the same temperature and pressure: (a) 3.5 L of CO_2; (b) 5 L of CO; (c) 4 L of HC1. Which will take the longest time and which will take the shortest time?

This is an application of Graham's Law. Let CO_2 be 1, CO be 2 and HC1 = 3.

$$r_2 = r_1 \sqrt{\frac{M_1}{M_2}} = r_1 \sqrt{\frac{44}{28}} = 1.25 \, r_1$$

$$r_3 = r_1 \sqrt{\frac{44}{36.5}} = 1.21 \, r_1$$

Since Time = Quantity/Rate,

$T_1 = 3.5/r_1$

$T_2 = 5.0/1.25\ r_1 = 4/r_1$

$T_3 = 4.0/1.1\ r_1 = 3.64/r_1$

Therefore, $T_1 < T_3 < T_2$ or T_2 is the longest time. T_1 is the shortest time.

See II, F

6. Atomic weight problem. It has been found that the relative abundances of the various isotopes of silicon in nature are 92.21% S_i^{28}, 4.70% S_i^{29}, and 3.09% S_i^{30}. The nuclidic masses are 27.977, 28.976, and 29.974 respectively. What is the atomic weight of silicon?

W = .9221 (27.977) + .047 (28.976) + .0309 (29.974) = 28.086

See I, F

7. A solution that contains one mole of a substance in 1000 g of solvent is said to be a _____ solution.

(A) Molar
(B) Molal
(C) Normal
(D) Supersaturated
(E) Saturation

By definition, the answer is (B), Molal.

See V, C

8. The MW of orthophosphoric acid (H_3PO_4) is 98. How many grams would be needed to make a one normal solution?

PO_4 is -3, a normal solution is 1 equiv/liter. Therefore, H_3PO_4 has an equivalent of 3. Therefore for one normal, we require 98/3 grams.

See V, D

9. $2KMnO_4 + 16HCl \rightarrow 2KCl + 1MnCl_2 + 8H_2O + 5Cl_2$

How many grams of $KMnO_4$ are required to make a 1.5 N solution? MW of $KMnO_4$ is 158.

Valence of   K is + 1
             O is - 2
             Mn is + 7 in KMnO₄ and is normally 2 +.

Therefore, we have a change of + 5. Therefore, for 1.5N solution, we need $\frac{158}{5}$ x 1.5 = 47.9 grams

See V, D

10. Which is <u>not</u> a ternary compound?

(A) H_2SO_4
(B) $KMnO_4$
(C) $CaCO_3$
(D) $C_{39}H2_8N_{12}$
(E) Water

All have 3 elements except water.

11. A piece of solder weighing 3 g was dissolved in dilute nitric acid and then treated with dilute H_2SO_4, sulfuric acid. This precipitated the lead as $PbSO_4$ which, after being washed and dried, was weighed at 2.93 g. The solution was then neutralized to precipitate stannic acid which was decomposed by heating. This yielded 1.27 g of SnO_2. Find the analysis of the solder in terms of percentage of tin and lead. The atomic weights are as follows:

Pb = 207.2, Sn = 118.7, O = 16.00, S = 32.07 and H = 1.00

$PbSO_4$ weighs (207.2 + 32.07 + 4 (16)) = 303.27

Now $\frac{207.2}{303.27}$ x 2.93 = 2.00 or 2.00 g of Pb in 2.93 g $PbSO_4$

SnO_2 weighs (118.7 + 2(16)) = 150.7

$\frac{118.7}{150.7}$ x 1.27 = 1.00 or 1 gram of Sn in 1.27 g of SnO_2.

Therefore, the original 3 g of solder contained 2 g of Pb and one g of Sn or 66.7% Pb and 33.3% Sn.

See XI, A

12. The half-life of a certain substance is 20 years. What percentage of the substance will have decayed after 60 years?

$$\frac{M_o/2}{M_o} = e^{-20k} = \frac{1}{2}$$

After 60 years, the percent decayed will be:

$$\frac{M_o - M_f}{M_o}$$

but $\frac{M_f}{M_o} = e^{-60k} = (e^{-20k})^3 = (½)^3 = 1/8$

$$\frac{M_o - M_o/8}{M_o} \times 100 = 87.5\%$$

See VI, C

13. A current of 50 milliamperes was allowed to pass through a solution of $AgNO_3$ for one hour. How much silver was deposited?

1 hour = 3600 sec
3600 sec x .050 amperes = 180 Coulombs
0.001128 g/ Coulomb x 180 Coulombs = .2 Ag deposited

One gram-atom of an element contains N atoms where N is Avogadro's number, 6.024×10^{23}. In electrolysis, N electrons pass through the solution for each Faraday used. Therefore, one Faraday contains 6.024×10^{23} electrons.

See X, D, 2b

14. How many Coulombs are on the N^{-2} ion?

6.024×10^{23} electrons = 96,490 coulombs

1 electron = $96,490/6.024 \times 10^{23}$ coulombs

Therefore, 2 electrons = $2 \times 1.6 \times 10^{-19} = 3.2 \times 10^{-19}$ Coulombs.

When a source of direct current is connected to electrodes in an aqueous solution of electrolyte, current will flow. The electrical energy supplied is used to make chemical changes. The + ions, called cations, are attracted to the anode where they give up electrons, an oxidation. The water molecules may undergo either oxidation or reduction.

$$2 H_2O \rightarrow 4 H^+ + O_2\uparrow + 4 e^-$$

$$2 H_2O \rightarrow 2 e^- + H_2\uparrow + 2 OH^-$$

During the electrolysis of a water solution, hydrogen is released at the cathode and oxygen is released at the anode.

See X, D, 2b

15. Show the electrolysis of a water solution of NaCl.

$$2 Na^+ + 2 Cl^- \rightarrow Cl_2\uparrow + 2 Na^+ + 2e^-$$

$$2 H_2O + 2 e^- \rightarrow H_2\uparrow + 2 OH^-$$

$$2 Na^+ + 2 Cl^- + 2H_2O \rightarrow Cl_2\uparrow + H_2^+\uparrow 2 Na^+ + 2OH^-$$

See X, D, 2

16. Show the electrolysis of a water solution of H_2SO_4.

$$4H_2O \rightarrow 4OH^- + 4H^+$$

$$4OH^- \rightarrow 2H_2O + O_2\uparrow + 4e^-$$

$$4H^+ + 4e^- \rightarrow 2H_2\uparrow$$

$$2H_2O \rightarrow 2H_2\uparrow + O_2\uparrow$$

See X, D, 2

17. How many pounds of NaOH would result from electrolysis of one ton of NaCl in water solution?

$$2 Na^+ + 2 Cl^- + 2 H_2O \rightarrow Cl_2 + H_2 + 2 Na^+ + 2OH^-$$

2 lb moles = 116.8 lb   2 lb moles = 10 lb

2000 lbs NaCl divided by 58.4 lbs NaCl per pound mole NaCl is equal to 100/58.4 lb mole NaCl.

2 moles of NaCl yields 2 moles of NaOH. Also, there are 80/2 or 40 lbs of NaOH per lb mole.

Therefore: $\dfrac{2 \times 1000}{58.4}$ lb mole NaOH $\times \dfrac{40 \text{ lb NaOH}}{\text{lb mole NaOH}} = 1368$ NaOH

See XI, B

18. Electrolysis of brine (salt water)

   anode $\quad\quad 2Cl^- \rightarrow Cl_2 + 2e^-$

   cathode $\;\; H_2O + 2Na^+ + 2e \rightarrow 2Na^+ + 2OH^- + H_2$

   (The active sodium metal produced here IMMEDIATELY reacts with the water to produce hydrogen gas.) The total reaction is

   $2Na^+ + 2Cl^- + 2H_2O \rightarrow 2Na^+ + Cl_2 + H_2 + 2OH^-$

   See X, D

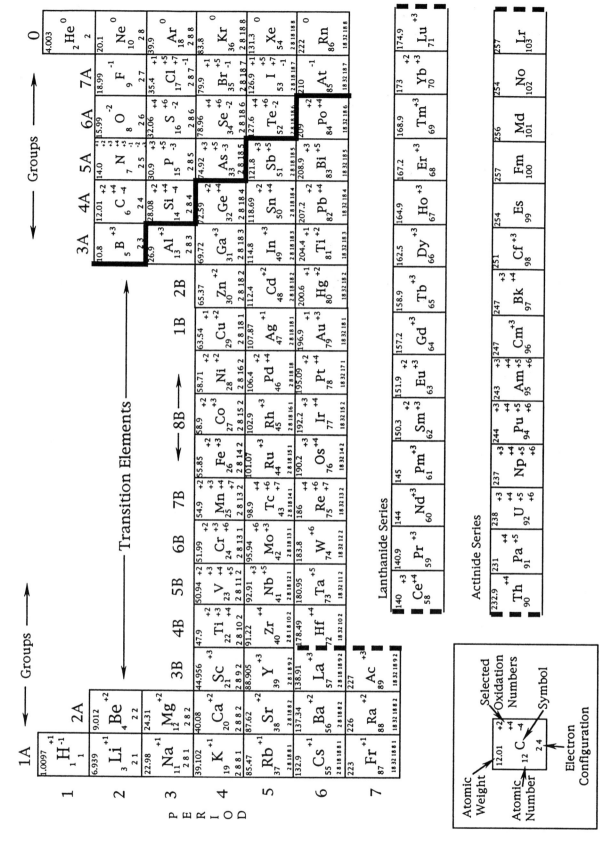

8

Thermodynamics

THERMODYNAMICS

TABLE OF CONTENTS

Selected References	8-ii
Definition of Symbols	8-iii
Thermodynamic Processes	8-1
General Energy Equation	8-3
Thermodynamic Cycles	8-10
Carnot Cycle	8-10
Ericcson Cycle	8-11
Stirling Cycle 12	8-12
Otto Cycle	8-12
Diesel Cycle	8-13
Illustrative Examples	8-18
Gas Mixtures	8-28
Volumetric Analysis	8-29
Gravimetric Analysis	8-29
Dalton's Law of Partial Pressures	8-30
Combustion	8-31
Heating Value	8-34
Liquids and Vapor	8-35
Throttling Calorimeter	8-38
Rankine Cycle	8-38
Efficiencies	8-39
Solved Problems	8-41
Psychrometrics	8-47
Power Plants	8-50
Reheat Cycle	8-54
Refrigeration	8-57
Heat Transfer	8-61
Radiation	8-66
Solved Problems	8-67
Steam Turbines and Nozzles	8-73

SELECTED REFERENCES

J.B. Jones, "Engineering Thermodynamics," John Wiley & Sons, New York

D. Mooney, "Mechanical Engineering Thermodynamics," Prentice-Hall, Englewood Cliffs, NJ

S.L. Soo, "Thermodynamics of Engineering," Prentice-Hall, Englewood Cliffs, NJ

S.E. Winston, "Thermodynamics," American Technical Society, New York

V.M. Faires, "Applied Thermodynamics," Macmillan, New York

Mackey, Barnard, & Ellenwood, "Engineering Thermodynamics," John Wiley & Sons, New York

G.M. Dusinberre, "Gas Turbine Power," International Textbook Co., Scranton, PA

E. Obert, "Internal Combustion Engines," International Textbook Co., Scranton, PA

Keenan & Keyes, Hill and Moore, "Steam Tables," John Wiley & Sons, New York

Keenan & Kayes, "Gas Tables," John Wiley & Sons, New York

R.H. Perry, "Perry's Chemical Engineers' Handbook, 7th Edition," McGraw-Hill, New York

THERMODYNAMICS 8-iii

Definition of Symbols

The following are commonly used acceptable symbols and will be used in the following discussion.

Symbol	Name	Common Units
C_n	specific heat for a polytropic process	Btu/lb °F
C_p	specific heat at constant pressure	Btu/lb °F
C_v	specific heat at constant volume	Btu/lb °F
h	specific enthalpy	Btu/lb
H	enthalpy (heat content)	Btu
J	mechanical equivalent of heat	778 ft lb/Btu
k	ratio of specific heats C_p/C_v	none
P	pressure (absolute)	psia
Q	heat energy transferred	Btu
R	ideal gas constant	Btu/lb °R
t	temperature	°F
T	temperature	°R
u	specific internal energy	Btu/lb
U	internal energy	Btu
v	specific volume	ft^3/lb
V	volume	ft^3
V	velocity	ft/sec
w	weight of substance	lb
W	work	foot-pounds
e	efficiency	none

Other symbols are defined as they are used in the text.

THERMODYNAMICS 8-1

THERMODYNAMIC PROCESSES

Boyle's Law

Boyle's Law states that if the temperature of a given quantity of gas is held constant, the volume of the gas will vary inversely with the absolute pressure (Fig. 8.1).

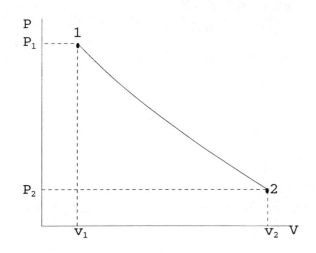

$$\frac{P_1}{P_2} = \frac{V_2}{V_1} \text{ or } P_1 V_1 = P_2 V_2$$

Figure 8.1

When $T = C$, $PV = C$

Charle's Law

Charles' Law tells us that if the pressure on a given quantity of gas is held constant, the volume will vary directly as the absolute temperature. If the volume of the gas is held constant, the pressure varies directly with the absolute temperature (Fig. 8.2).

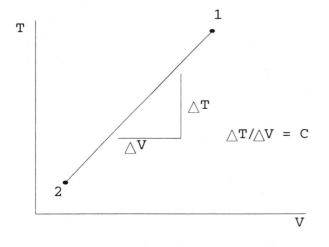

$$\left[\frac{T_1 - T_2}{V_1 - V_2}\right] = c$$

$\Delta T / \Delta V = C$

$V_1/V_2 = T_1/T_2$

Figure 8.2a

or $T_1/V_1 = T_2/V_2$

When $P = C$, $T/V = C$.

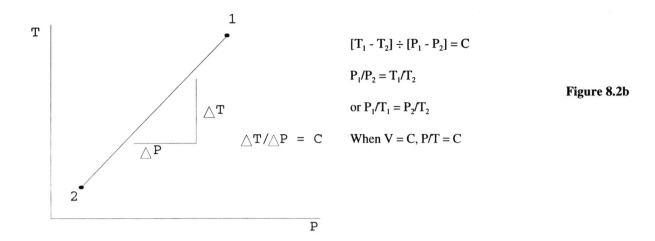

Figure 8.2b

$[T_1 - T_2] \div [P_1 - P_2] = C$

$P_1/P_2 = T_1/T_2$

or $P_1/T_1 = P_2/T_2$

$\triangle T/\triangle P = C$ When $V = C$, $P/T = C$

Equations of State

The Characteristic Equation, or Equation of State of a perfect gas, sometimes referred to as the Perfect Gas Law, is obtained by using a constant pressure process and a constant volume process (Fig. 8.3).

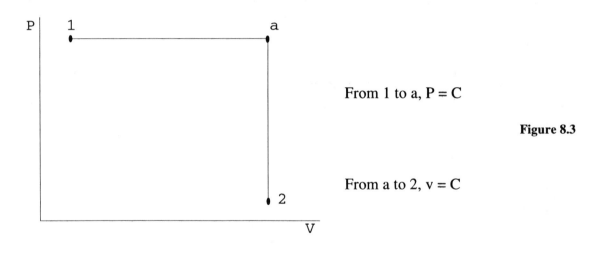

From 1 to a, $P = C$

From a to 2, $v = C$

Figure 8.3

For the process from 1 to a, $v_1/v_a = T_1/T_a$ or $T_a = T_1 (v_a/v_1)$
but $v_a = v_2$ therefore, $T_a = T_1 (v_2/v_1)$. Note: all volumes are <u>specific</u> volumes (e.g., ft³/lb).

For the process from a to 2, $P_a/P_2 = T_a/T_2$ or $T_a = T_2(P_a/P_2)$
but $P_a = P_1$, therefore, $T_a = T_2 (P_1/P_2)$

We can now equate the expressions for T_a as obtained in each process.

$T_1 (v_2/v_1) = T_2 (P_1/P_2)$ which yields $P_1v_1/T_1 = P_2v_2/T_2$ or $Pv/T = C$

This particular constant is R, the <u>specific</u> ideal gas constant.

Pv = RT now multiply each side by w (w = weight of gas)
    Pwv = wRT but wv = V
    PV = wRT which is the characteristic equation

For air, R has the value of 53.3. For other gases, the value of R can be calculated from R = 1544/molecular weight of the gas, where 1544 is the <u>universal</u> gas constant.

General Energy Equation

The General Energy Equation for an ideal, frictionless steady flow system states briefly that what energy goes into a system must be what comes out of the system, or Input = Output.

$$PE_1 + KE_1 + U_1 + FW_1 + Q_a = PE_2 + KE_2 + U_2 + W + FW_2$$

where PE = potential energy for elevation, z, above a datum.
    KE = kinetic energy or $V^2/2g$
    FW = flow work or pV term
    U = internal energy

Subscript 1 denotes entrance conditions and subscript 2 exit.

    Q_a = energy added to the system
    W = work output from the system

For w pounds of substance we can write:

$$\frac{wZ_1}{J} + \frac{wV_1^2}{2gJ} + U_1 + \frac{P_1V_1}{J} + Q_a = \frac{wZ_2}{J} + \frac{wV_2^2}{2gJ} + U_2 + \frac{P_2V_2}{J} + W$$

where J is the mechanical equivalent of heat = 778 ft lbs/BTU Joule's Constant.

For no flow conditions, KE and FW are both 0.

For no change in datum, $PE_1 = PE_2$.

Therefore, $U_1 + Q_a = U_2 + W$ or expressed differently, $Q_a = \Delta U + W$.

Now consider the piston and cylinder arrangement shown in Fig. 8.4.

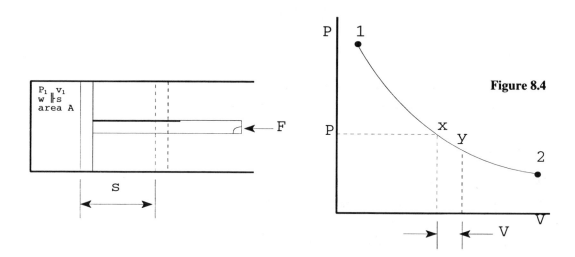

Figure 8.4

A quantity of working substance, w lb, is in the cylinder at a pressure P_1 and occupies a volume V_1. The substance then expands to a volume V_2 when the piston moves to the dotted line position while working against some resistance. Therefore work is done by the system (working substance) on the surroundings.

W = F x s = P x A x s

On the P-V diagram take a section of the curve where the pressure can be assumed constant. This is the section xy.

F = P x A and the piston moves a distance dL

dW = PA dL But A dL = dV so dW = P dV

$$W = \int_1^2 P\,(dV)$$ which is the area under the curve 1 - 2.

Note: Work depends on the process!

Specific Heat

The specific heat, C, is the amount of heat required to change the temperature of one pound of a substance one degree with no phase change (i.e., solid to liquid; liquid to gas). That is, if dQ amount of heat was transferred to w lb of a substance, the temperature will change by dT.

thus

$$C = \frac{dQ}{WdT}$$ $dQ = w\,C\,dT$ and If C is constant over the temperature range T_1 to T_2,

$$Q = wC \int_1^2 dT = wC\,(T_2 - T_1)$$

Internal Energy

Repeating the general energy equation for a nonflow process with a constant datum, $Q = \Delta U + W$. U, the internal energy of the substance, is a point function. The work, W, depends on the process. Therefore Q must depend on the path between state points and the path between state points will determine the value of C. We can have an infinite number of specific heats and paths between state points. However, two specific heats are so important as to have a definite name, C_p and C_v the specific heats at constant pressure and constant volume respectively. The ratio of the values of these specific heats is known as k where $k = C_p/C_v$. The commonly accepted values of these constants for air are $k = 1.4$, $C_p = 0.24$, and $C_v = 0.1715$.

The change in internal energy, $\Delta U = U_2 - U_1$, is equal to Q - W. For a constant volume process $W = 0$ and $Q = \Delta U$.

$$\Delta U = Q = w \int_1^2 C\,(dT) = w\,C_v\,(T_2 - T_1)$$

Since U is a point function that does not depend on the path (process), this equation holds for other processes. ΔU depends only on ΔT if there is no phase change. We may therefore select any convenient process as we have done, evaluate the change between two temperatures, and be assured that this change is the same no matter what the process.

Now let us look at the relationships between C_p and C_v. Let a gas undergo a non-flow process at constant pressure.

$$P = C \text{ and } PV = wRT; \quad W = \int P\,dV = P \int_1^2 dV = P\,(V_2 - V_1) \text{ ft. lbs}$$

But $P V_2 = wR\,T_2$ and $P V_1 = wR\,T_1$; $V_2 = w R\,T_2/P$ and $V_1 = w R\,T_1/P$

Substituting these values yields the following:

$$W = P\,(wRT_2/P - wRT_1/P) \text{ or } W = wR\,(T_2 - T_1)$$

$$\text{or } W = wR/J\,(T_2 - T_1) \text{ BTU}$$

Now, $Q = wC_p\Delta T$ for a constant pressure process

$$\Delta U = wC_v\Delta T$$

$$W = w\Delta T\ R/J$$

Repeating, $Q = \Delta U + W$. Substituting and rearranging:

$$w\Delta T\ C_p = w\Delta T\ C_v + w\Delta T\ R/J$$

$$C_p = C_v + R/J$$

Rewriting this as $C_p - C_v = R/J$ and noting that $C_p = kC_v$, then, $kC_v - C_v = \dfrac{R}{J}$

or $C_v = \dfrac{R}{J(k-1)}$

The value of k varies with temperature. When in doubt, check your reference text or handbook for the correct value of k. Generally, the exam problems give the value of k or give sufficient information to calculate the value of k.

Enthalpy

Enthalpy is defined as $H = U + PV/J$ for all processes, all fluids, and all flow conditions. Since we cannot measure U, a point function, we are unable to measure the absolute value of H. We can measure the change in enthalpy, ΔH.

$$\Delta H = H_2 - H_1 = (U_2 + P_2V_2/J) - (U_1 + P_1V_1/J)$$

$$\Delta H = \Delta U + \dfrac{P_2V_2 - P_1V_1}{J} \text{ but } \Delta U = w\ C_v\Delta T \text{ and } PV = wRT$$

$$\Delta H = w\ C_v\Delta T + w\Delta T\ R/J$$

$$\Delta H = w\Delta T\ (C_v + R/J) \text{ but } (C_v + R/J) = C_p$$

$$\Delta H = w\ C_p\Delta T$$

In the general form applicable to a gas only,

$$\Delta H = w\int_1^2 C_p\ dT$$

Entropy

Entropy changes often accompany heat transfer but can also occur without a transfer of heat. For a <u>reversible</u> transfer of heat, we define entropy as $dS = dQ/T$.

$$\Delta S = \int dQ/T \text{ and } \Delta S = S_2 - S_1$$

A process on the temperature entropy plane is plotted in Fig. 8.5.

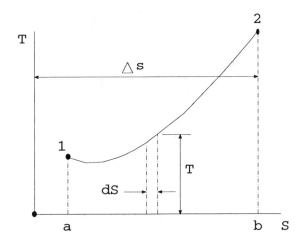

$dS = dQ/T$

$dQ = T\, dS$

$Q = \int T\, dS$

Q = area under the curve, area 1 - 2 - b - a

Figure 8.5

$Q = w\, C \int dT = w\, C\, (T_2 - T_1)$

$\Delta S = \int dQ/T = wC \int dT/T = w\, C \ln(T_2/T_1)$

Polytropic Processes

The polytropic process is defined by the expression:

$P V^n = C$ or $P_1 V_1^n = P_2 V_2^n$ or $P_1/P_2 = (V_2/V_1)^n$

and since $P_1 V_1/T_1 = P_2 V_2/T_2$ for any process,

$$\left[\frac{V_2}{V_1}\right]^n = \frac{T_1 V_2}{T_2 V_1} \quad \text{and} \quad \frac{T_1}{T_2} = \left[\frac{V_2}{V_1}\right]^{n-1}$$

In like manner we may arrive at $T_2/T_1 = [P_2/P_1]^{\frac{n-1}{n}}$

Repeating, $PV^n = C$ or $P = C/V^n$

$$W = \int P\, dV = C \int dV/V^n = \frac{CV_2^{-n+1} - CV_1^{-n+1}}{(-n + 1)}$$

and since $P_1 V_1^n = C = P_2 V_2^n$ $W = (P_2 V_2 - P_1 V_1) / [1 - n]$

Internal energy is still defined as $\Delta U = w\, C_v \Delta T$

Since the polytropic process is not specifically a constant pressure or a constant volume process, we denote C_n to be a specific heat to be known as the specific heat for a polytropic process.

$Q = w\, C_n \Delta T$ $\Delta U = w\, C_v \Delta T$ $W = (P_2 V_2 - P_1 V_1) / [1 - n]$

For a non-flow process we have $Q = \Delta U + W$ and by substitution of values we arrive at the following:

$$Q = w\, C_v (T_2 - T_1) + \frac{P_2 V_2 - P_1 V_1}{(1 - n)\, J}$$

$$Q = w\, C_v (T_2 - T_1) + \frac{wR}{J} \frac{(T_2 - T_1)}{1 - n}$$

$$Q = \frac{w\, C_v (T_2 - T_1)[(1 - n)J] + wR(T_2 - T_1)}{(1 - n)\, J}$$

$$Q = \frac{w(C_v - n C_v + R/J)(T_2 - T_1)}{(1 - n)}$$

This must, of course, be equal to the expression for Q using C_n

$$\frac{w(C_v - nC_v + R/J)(T_2 - T_1)}{1 - n} = wC_n (T_2 - T_1)$$

Therefore, $C_n = (C_v - nC_v) + R/J) / (1 - n)$

Since $C_p = C_v + R/J$ and $k = C_p/C_v$ we can show that: $C_n = \left[\dfrac{k-n}{1-n}\right] C_v$

The enthalpy, $\Delta H = w\, C_p (T_2 - T_1)$ as before.

The expression for ΔS becomes: $\Delta S = w\, C_n \ln(T_2/T_1)$

THERMODYNAMICS

P V T RELATIONSHIPS

Figure 8.6

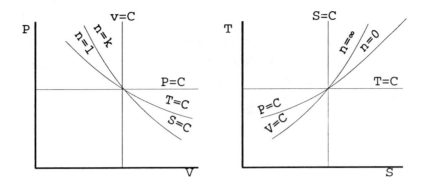

	ISOCHORIC V=C	ISOBARIC P=C	ISOTHERMAL T=C	ISENTROPIC S=C	POLYTROPIC PV^n=C
P,V,T	$\dfrac{T_2}{T_1} = \dfrac{P_2}{P_1}$	$\dfrac{T_2}{T_1} = \dfrac{V_2}{V_1}$	$P_1 V_1 = P_2 V_2$	$P_1 V_1^k = P_2 V_2^k$ $\dfrac{T_2}{T_1} = \left(\dfrac{V_1}{V_2}\right)^{k-1}$ $= \left(\dfrac{P_2}{P_1}\right)^{\frac{k-1}{k}}$	$P_1 V_1^n = P_2 V_2^n$ $\dfrac{T_2}{T_1} = \left(\dfrac{V_1}{V_2}\right)^{n-1}$ $= \left(\dfrac{P_2}{P_1}\right)^{\frac{n-1}{n}}$
$\int PdV = W$	0	$P(V_2 - V_1)$	$P_1 V_1 \ln \dfrac{V_2}{V_1}$	$\dfrac{P_2 V_2 - P_1 V_1}{1-k}$	$\dfrac{P_2 V_2 - P_1 V_1}{1-n}$
$U_2 - U_1$	$wCv(T_2 - T_1)$	$wCv(T_2 - T_1)$	0	$wCv(T_2 - T_1)$	$wCv(T_2 - T_1)$
Q	$wCv(T_2 - T_1)$	$wCp(T_2 - T_1)$	$\dfrac{P_1 V_1}{J} \ln \dfrac{V_2}{V_1}$ $Q_A = T \Delta S$	0	$wCn(T_2 - T_1)$
$H_2 - H_1$	$wCp(T_2 - T_1)$	$wCp(T_2 - T_1)$	0	$wCp(T_2 - T_1)$	$wCp(T_2 - T_1)$
$S_2 - S_1$	$wCv \ln T_2/T_1$	$wCp \ln T_2/T_1$	$\dfrac{wR}{J} \ln V_2/V_1$	0	$wCn \ln T_2/T_1$
n	∞	0	1	k	$-\infty \quad +\infty$
C	Cv	Cp	∞	0	$Cn = Cv \dfrac{k-n}{1-n}$

$k = Cp/Cv$ $\qquad Cv = \dfrac{R}{(k-1)J} \qquad Cp = \dfrac{kR}{(k-1)J}$

$h = u + \dfrac{PV}{J} \qquad R = \dfrac{1544}{M.W.} \qquad R = J(Cp - Cv)$

$R = \dfrac{ft.\,lb.}{lb\,°R} \qquad Cp = Cv + \dfrac{R}{J} \qquad PV = WRT$

THERMODYNAMIC CYCLES

Carnot Cycle

The Carnot cycle is composed of two reversible isothermal and two reversible adiabatic (isentropic) processes as shown in the P-V and T-S diagrams (Fig. 8.7).

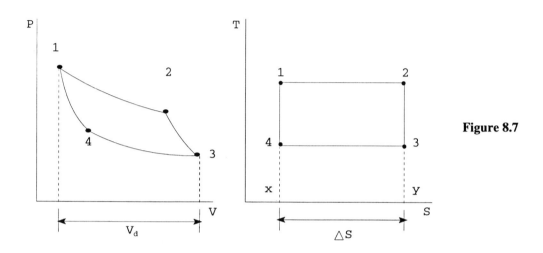

Figure 8.7

Q_a from 1 - 2, expansion from 2 - 3 with no Q_a or Q_r, Q_r from 3 - 4 and compression with no Q_a or Q_r from 4 - 1.

$$Q_a = \text{area } 1\text{-}2\text{-}y\text{-}x = T_1 (S_2 - S_1) = T_1 \Delta S$$

$$Q_r = \text{area } 3\text{-}4\text{-}x\text{-}y = T_3 \Delta S$$

$$W_{NET} = Q_a - Q_r = (T_1 - T_3) \Delta S$$

$\eta = W_{NET}/Q_a$ where η is the efficiency

$$\frac{W_{NET}}{Q_a} = \frac{(T_1 - T_3)(S_2 - S_1)}{T_1 (S_2 - S_1)} = \frac{T_1 - T_3}{T_1}$$

W is also the enclosed area 1-2-3-4-1 on the P-V diagram.

$$W = P_1 V_1 \ln(V_2/V_1) + (P_3 V_3 - P_2 V_2)/(1-k) + P_3 V_3 \ln(V_4/V_3)$$
$$+ (P_1 V_1 - P_4 V_4)/(1-k)$$

By substituting equivalent expressions we may obtain:

$$W = (T_1 - T_3)(wR \ln V_2/V_1) = (T_1 - T_3)\Delta S$$

The displacement volume from this process is V_d, $W = P_m V_d$

$$P_m = W/V_d = \frac{(T_1 - T_3) wR \ln V_2/V_1}{V_3 - V_1}$$

P_m (sometimes called mep) is the mean effective pressure.

r_k = compression ration, initial volume ÷ final volume

r_e = expansion ratio, final volume ÷ initial volume

Ericcson Cycle

The Ericcson cycle has two isobaric and two isothermal processes (Fig. 8.8). There is a regenerative effect during the constant pressure portion of the cycle. There is an interchange of heat within the cycle.

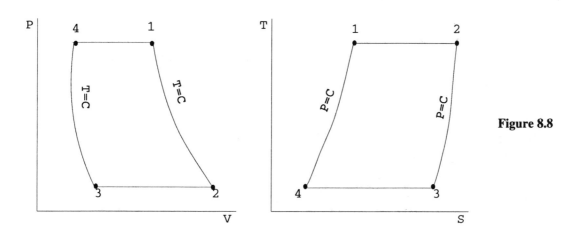

Figure 8.8

Heat is taken from storage during the cycle from 4 - 1.

Heat is stored during the cycle from 2 - 3.

$$Q_a = (P_1 V_1/J)(\ln V_2/V_1) = (wRT_1/J)(\ln V_2/V_1)$$

$$Q_r = (wRT_3/J)(\ln V_4/V_3) = (-wRT_3/J)(\ln V_3/V_4)$$

$$W = \Sigma Q = wR/J (T_1 - T_3) \ln V_2/V_1$$

$$e = W/Q_a = (T_1 - T_3)/T_1$$

$$P_m = W/V_d = \frac{wR/J\,(T_1 - T_3)\,\ln V_2/V_1}{V_2 - V_4}$$

Stirling Cycle

The Stirling cycle is similar to the Ericcson cycle except that it has two isothermal and two isochoric processes (Fig. 8.9). The regenerative action occurs during the constant volume process.

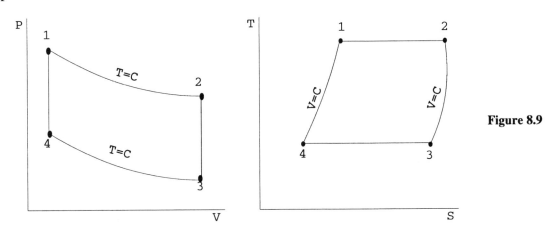

Figure 8.9

These cycles are used for a basis of comparison. By comparing thermal efficiencies we may determine relative merits.

The two internal combustion (IC) engine cycles are the two-stroke cycle and the four-stroke cycle. In the two-stroke cycle there are two strokes of the piston necessary for a cycle. The sequence is compression, ignition, power, scavenging. The four-stroke cycle requires four strokes of the piston to complete the cycle. We have suction, compression, ignition, power, and exhaust. (Now we can add the Wankel Engine.)

Otto Cycle

The Otto cycle is the prototype of most IC engines in use today. The cycle has two isentropic and two constant volume processes (Fig. 8.10).

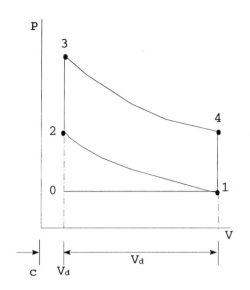
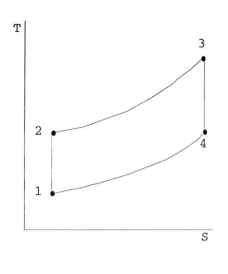

Figure 8.10

$Q_a = w\, C_v\, (T_3 - T_2); \quad Q_r = w\, C_v\, (T_4 - T_1)$

$W = \Sigma Q = wC_v\, (T_3 - T_2) - wC_v\, (T_4 - T_1)$

e, a new symbol for efficiency, $= W/Q_a$; $e = 1 - \dfrac{(T_4 - T_1)}{(T_3 - T_2)}$

but $T_4 = T_3\, (V_3/V_4)^{k-1} = T_3\, (V_2/V_1)^{k-1}$

$T_1 = T_2\, (V_2/V_1)^{k-1}$

substituting values yields $e = 1 - (V_2/V_1)^{k-1}$

But the compression ratio is defined as $r_k = V_1/V_2$

$e = 1 - 1/r_k^{k-1}$

The compression ratio r_k is varied by changing the clearance volume. The clearance volume is expressed as a percentage, c, of the displaced volume $r_k = V_1/V_2 = (cV_d + V_d)/cV_d = (1 + c)/c$.

Diesel Cycle

The Diesel cycle has two isentropic processes, a constant pressure process and a constant volume process (Fig. 8.11). There is an intake, an isentropic compression, a constant pressure burning, a power stroke, and an exhaust stroke.

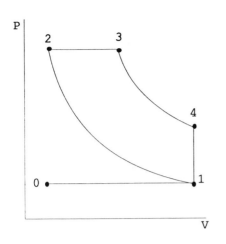 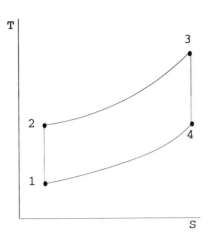

Figure 8.11

$$Q_a = wC_p (T_3 - T_2); \quad Q_r = w C_v (T_4 - T_1)$$

$$W = \text{sum of } Q_a - Q_r \text{ and } e = W/Q_a \text{ and } e = 1 - \frac{(T_4 - T_1)}{k(T_3 - T_2)}$$

During the compression from 1 - 2, $T_2 = T_1(V_1/V_2)^{k-1} = T_1 r_k^{k-1}$

During the constant pressure portion of the cycle (2 - 3) $T_3/T_2 = V_3/V_2 =$ where $V_3/V_2 = r_c$, the cutoff ratio. The fuel is injected into the cylinder from point 2 to point 3. At point 3 on the P - V diagram the fuel is cutoff.

$$T_3 = T_2 (V_3/V_2) = T_1 r_k^{k-1} r_c$$

$$T_4 = T_3 (V_3/V_4)^{k-1} = T_1 r_c^k$$

Therefore we can substitute values and obtain:

$$e = 1 - \frac{1}{r_k^{k-1}} \left[\frac{r_c^k - 1}{k(r_c - 1)} \right]$$

Note that this differs from the Otto cycle efficiency only by the bracketed term. This term will always be greater than unity since r_c is always greater than 1. Therefore, for a given r_k, the Otto cycle is more efficient.

In a modern diesel engine it is necessary to inject fuel before the piston reaches the top dead center (TDC). We then have a cycle with burning at constant volume as well as at constant pressure (Fig. 8.12). This is the dual combustion cycle.

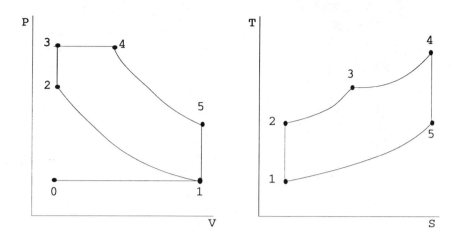

Figure 8.12

$$Q_a = w\,C_v\,(T_3 - T_2) + W\,C_p\,(T_4 - T_3)$$

$$Q_r = w\,C_v\,(T_5 - T_1)$$

$$w = Q_a - Q_r \qquad e = W/Q_a = 1 - \frac{(T_5 - T_1)}{(T_3 - T_2) + k(T_4 - T_3)}$$

Let r_p = a pressure ratio, P_3/P_2

And $r_k = V_1/V_2$ \qquad And $r_c = V_3/V_4$

Then $e = 1 - \dfrac{1}{r_k^{k-1}}\left[\dfrac{r_p r_c^k - 1}{r_p - 1 + r_p k(r_c - 1)}\right]$

Gas Turbine Cycle

The cycle for the gas turbine is the Brayton or Joule cycle composed of two constant pressure and two constant entropy processes (Fig. 8.13).

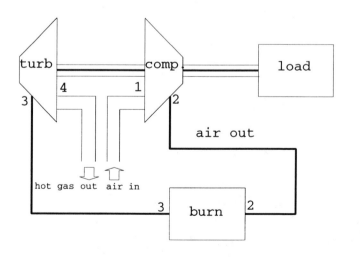

Figure 8.13a

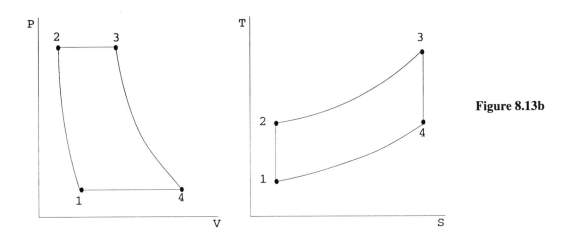

Figure 8.13b

We have isentropic compression from 1 to 2, combustion from 2 to 3, expansion from 3 to 4, and cooling from 4 to 1.

$$W = Q_a - Q_r = w\, C_p\, (T_3 - T_2) - w\, C_p\, (T_4 - T_1)$$

$$e\ W/Q_a = 1 - \frac{T_4 - T_1}{T_3 - T_2}$$

$$\eta_{th} = 1 - \left(\frac{1}{r_p}\right)^{\frac{k-1}{k}} \quad \text{where} \quad r_p = \frac{P_2}{P_1}$$

Until now, we have been dealing with ideal cycles. We do not have the internal losses. We will use the subscript c for compressor and t for turbine. An actual cycle is shown in Fig. 8.14.

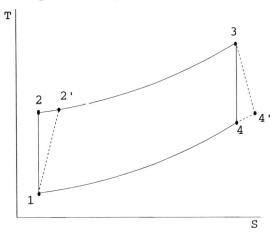

Figure 8.14

1 - 2 is ideal   1 - 2' is actual
3 - 4 is ideal   3 - 4' is actual

$W_c' = (h_1 - h_2') = w\, C_p\, (T_1 - T_2') = -w\, C_p\, (T_2' - T_1)$

$W_t' = (h_3 - h_4') = w\, C_p\, (T_3 - T_4')$

Q_a is from 2' to 3, $Q_a = w\, C_p\, (T_3 - T_2')$

$$e' = W'/Q'_a = \frac{w\, C_p\, (T_3 - T_4' - T_2' + T_1)}{w\, C_p\, (T_3 - T_2')}$$

η_t = turbine efficiency = $(h_3 - h_4')/(h_3 - h_4)$

η_c = compressor efficiency = $(h_2 - h_1)/(h_2' - h_1)$

where η is the adiabatic compression efficiency.

COMPRESSION OF AIR

a) <u>Polytropic compression</u> (Fig. 8.15)

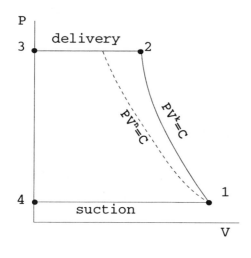

Figure 8.15

$$W = \frac{P_2 V_2 - P_1 V_1}{1 - n} + P_2 (V_3 - V_2) + P_1 (V_1 - V_4)$$

$$W = \frac{P_2 V_2 - P_1 V_1}{1 - n} - P_2 V_2 + P_1 V_1 \text{ since } V_3 \text{ and } V_4 = 0$$

$$W = \frac{n (P_2 V_2 - P_1 V_1)}{1 - n} = \frac{n\, w\, R\, (T_2 - T_1)}{1 - n} \text{ ft. lb.}$$

$$W = \frac{n \, w \, RT_1}{1-n}\left[\frac{T_2}{T_1} - 1\right] = \frac{n \, P_1 V_1}{1-n}\left[\left(\frac{P_2}{P_1}\right)^{\frac{n-1}{n}} - 1\right] \text{ ft. lb.}$$

b)   Isentropic compression is the same except that n = k.

c)   Isothermal compression

$$W = P_1V_1 \ln V_2/V_1 - P_2V_2 + P_1V_1 \text{ but } P_1V_1 = P_2V_2$$

$$W = P_1V_1 \ln V_2/V_1 = wRT_1 \ln V_2/V_1 = wRT_1 \ln P_1/P_2$$

Volumetric efficiency is shown in Fig. 8.16.

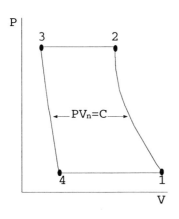

Figure 8.16

$$V_3 = c\,V_d;\ V_1' = V_1 - V_4;\ V_d = V_1 - V_3;\ V_1 = V_d + cV_d$$

η_v = volumetric efficiency = $V_1'/V_d = (V_1 - V_4)/V_d$

$V_4 = V_3\,(P_3/P_4)^{1/n} = cV_d\,(P_2/P_1)^{1/n}$

$\eta_v = (V_1 - V_4)/V_d$

$\eta_v = 1 + c - c\,(P_2/P_1)^{1/n}$ which is the conventional volumetric efficiency.

ILLUSTRATIVE EXAMPLES

Example

Given a constant temperature process of a perfect gas, the initial pressure is atmospheric and the initial volume is 4 ft^3. At the end of the process the pressure is 150 psig. There are 0.25 lb of gas used in the process. Find the final volume.

$$P_1V_1 = P_2V_2 \text{ or } V_2 = V_1\left[\frac{P_1}{P_2}\right]$$

$$V_2 = 4.0\left(\frac{14.7}{150 + 14.7}\right) = .357 \text{ ft}^3$$

Final volume is .357 ft³.   Answer

Note: P_2 was given as 150 psig. Always convert to psi<u>a</u>.

Example: 50 ft³ of a perfect gas at 540°F is compressed in an isobaric process to 25 ft³. What is the final temperature of the gas in °F?

$$\frac{P_1V_1}{T_1} = \frac{P_2V_2}{T_2} \quad \text{But } P_1 = P_2$$

$$T_2 = T_1\frac{V_2}{V_1} = (540 + 460)\frac{(25)}{50} = 500°R$$

$$T_2 = 500 - 460 = 40°F$$

Final Temperature is 40°F   Answer

Note: The units for the answer were specified as °F.

Example

A temperature of 700°F is required to fire a diesel fuel engine. Find the minimum compression ratio at which the engine will operate considering this to be the controlling factor. Assume the temperature at the beginning of the cycle to be 140°F and take the compression curve exponent to be 1.3 (Fig. 8.17).

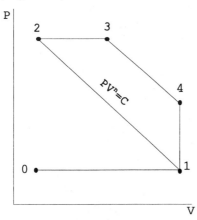

Figure 8.17

$r_k = V_1/V_2$

$T_1 = 140 + 460 = 600°R \qquad T_2 = 700 + 460 = 1160°F$

$T_2/T_1 = \left(\dfrac{V_1}{V_2}\right)^{n-1} = r_k^{n-1} \quad \text{or} \quad \dfrac{1160}{600} = r_k^{1.3-1} = 1.933 = r_k^{0.3}$

$0.3 \ln r_k = \ln 1.933$

$\ln r_k = \dfrac{0.660}{(0.3)} = 2.20$

$r_k = 9.02$ <u>Answer</u>

<u>Example</u>: The inlet temperature to a gas turbine is 1500°F. It has an expansion pressure ration of 6 and the gases leave the turbine at 955°F (Fig. 8.18). Calculate the adiabatic turbine efficiency.

$C_p = 0.26$

$C_v = 0.20$

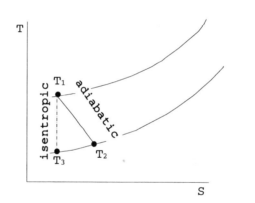

Figure 8.18

$T_1 = 1500°F = 1960°R;\ T_2 = 955°F = 1415°R$

$P_1/P_2 = 6;\ k - C_p/C_v = .26/.20 = 1.3$

By adiabatic efficiency is meant "how near is the cycle to an ideal isentropic process?

$\eta = \dfrac{h_1 - h_2}{h_1 - h_3} = \dfrac{wC_p(T_1 - T_2)}{wC_p(T_1 - T_3)} = \dfrac{T_1 - T_2}{T_1 - T_3}$

But $T_1/T_3 = (P_1/P_2)^{\frac{k-1}{k}}$ or $T_3 = \dfrac{T_1}{(P_1/P_2)^{\frac{k-1}{k}}} = \dfrac{1960}{(6)^{\frac{1.3-1}{1.3}}}$

$T_3 = 1295°R$

$$\eta = \frac{1960 - 1415}{1960 - 1295} = 0.82$$

Adiabatic efficiency = 82%   <u>Answer</u>

<u>Example</u>: Ten pounds of an ideal gas are compressed polytropically from a pressure of 20 psia and 400 F to 85 psia (Fig. 8.19). The gas constant R = 51. C_p = 0.275, n = 1.3. Find the initial volume, final temperature, work done, internal energy and enthalpy change, the transferred heat and the change in entropy.

$P_1 = 20$ psia

$P_2 = 85$ psia

$T_1 = 40°F = 500°R$

$w = 10$ lb

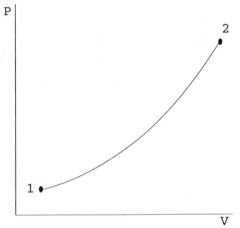

Figure 8.19

The characteristic equation is valid for all processes.

$$V_1 = \frac{w\,RT_1}{P_1} = \frac{10 \times 51 \times 500}{20 \times 144} = \underline{88.5 \text{ ft.}^3}$$

For the polytropic process: $T_2 = T_1 (P_2/P_1)^{\frac{n-1}{n}} = 500 \left(\frac{85}{20}\right)^{\frac{1.3-1}{1.3}}$

$$= \underline{698°R}$$

$$W = \frac{w\,R(T_2 - T_1)}{1 - n} = \frac{10 \times 51\,(698 - 500)}{1 - 1.3} = -336600 \text{ ft lb}$$

or $W = \dfrac{-336600}{778} = \underline{-432.6 \text{ Btu}}$ {minus indicates work done on the gas}

$\Delta H = w\,C_p\,(T_2 - T_1) = 10 \times .275\,(698 - 500) = \underline{544.5 \text{ Btu}}$

$C_v = C_p - R/J = 0.275 - 51/778 = 0.21$

$\Delta U = w\, C_v\, (T_2 - T_1) = 10 \times .21\, (698 - 500) = \underline{415.8\ \text{Btu}}$

$C_n = C_v \left[\dfrac{k-n}{1-n}\right]$ where $k = C_p/C_v$

$k = .275/.210 = 1.31$

$C_n = .21 \left[\dfrac{1.31 - 1.30}{1 - 1.30}\right] = -0.007$

$Q = w\, C_n\, (T_2 - T_1) = 10 \times -.007\, (698 - 500) = \underline{-13.86\ \text{Btu}}$

The minus sign indicates heat given off by the gas.

$\Delta S = w\, C_n\, \ln T_2/T_1 = 10 \times -.007\, \ln \dfrac{698}{500} = \underline{-.0234\ \text{units}}$

1. The pressure of the air in an automobile tire increased from 30 to 34 psi during a trip. If the initial temperature of the tire was 60°F, what was the final temperature?

   Approach

   The pressure is most likely in psig, the units used in measuring tire pressure. Use the characteristic equation:

   $P_1 = 44.7$ psia and $P_2 = 48.7$ psia

   $\dfrac{P_1 V_1}{T_1} = \dfrac{P_2 V_2}{T_2}$ since $w_1 = w_2$

   Also $V_1 \approx V_2$, ∴ assume $V_1 = V_2$

   $\dfrac{P_1}{T_1} = \dfrac{P_2}{T_2}$ or $T_2 = T_1 \left[\dfrac{P_2}{P_1}\right]$

   $T_2 = (460 + 60)(48.7/44.7) = 107°F$

   Final Temperature = 107°F   Answer

2. A volume of gas having an initial entropy of 2800 Btu/°R is heated at a constant temperature of 1000°F until the entropy is 4300 Btu/°R (Fig. 8.20). How much heat is added and how much work is done during the process?

Approach

Sketch the process on the T - S plane. Use the general energy equation to establish $Q = \Delta U + W$. Solve for work in mechanical units and for the heat added in thermal units. Draw the T - S diagram. This will make the solution almost self-explanatory.

$Q_A = T\Delta S$ $Q_A = 1460(4300 - 2800)$
$Q_A = 1460(1500)$ $Q_A = 2.19 \times 10^6$ Btu

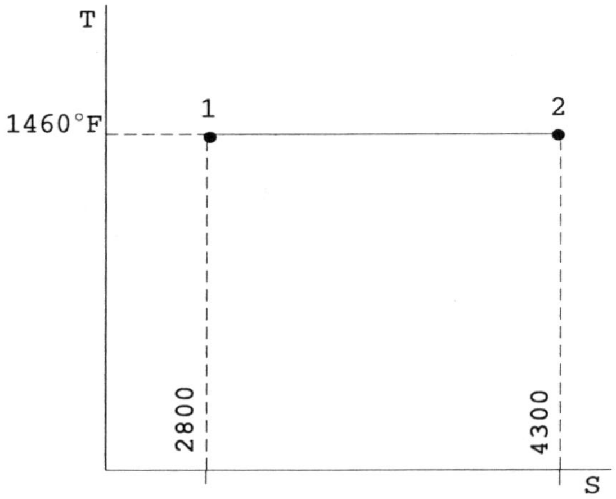

Figure 8.20

$Q = \Delta U + W$

but for T - C, $\Delta U = 0$

$Q = W$ or $W = 2.19 \times 10^6 \times 778$ ft-lbs/Btu $= 1.70 \times 10^9$ ft - lb

$Q_A = 2.19 \times 10^6$ Btu and $W = 1.70 \times 10^9$ ft lb   **Answer**

3. The constant volume specific heat of a certain gas is given by the formula: $C_v = 0.5 + 0.0002T$ (Fig. 8.21). If this gas is heated at constant volume form 60°F to 400°F and then expanded to its initial state in an engine in which the heat loss is 10% of the work done, what is the work done by the engine? What is the work efficiency of the process?

Approach

This is a problem in evaluating an integral once you recognize the improper wording of the problem.

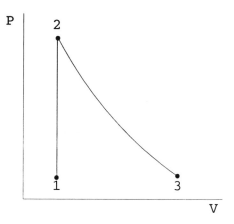

Figure 8.21

$T_1 = 60°F = 520°R$

$T_2 = 400°F = 860°R$

Expansion cannot happen by a system returning to its original temperature and volume, only to its original pressure.

$$Q_{1-2} = w \int C_v dT = 1 \times \int_{520}^{860} (.5 + .0002\,t)\, dt = \left. .5T + \frac{.0002T^2}{2} \right|_{520}^{860}$$

$Q_{1-2} = 170 + 46.92 = 217$ Btu/lb.

Assume all Q converted to W except for 10% loss.

$W + .1W = 217$ or $W = 217/1.1 = 197$ Btu/lb $\eta = W/Q_A = \dfrac{197}{217} = 91\%$

Work done = 197 Btu/lb, efficiency = 91%   <u>Answer</u>

4. Calculate the temperature and pressure at the end of compression in a diesel engine with a compression ratio of 14. Assume the pressure at the beginning of the compression stroke to be 13 psia, and the exponent of the curve to be 1.35 (Fig.8.22).

<u>Approach</u>

Sketch the cycle and use the basic gas laws to calculate the temperature and pressure.

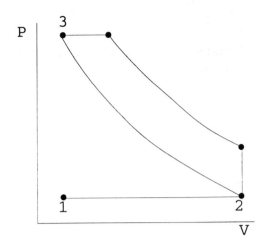

$P_1 = P_2 = 13$ psia

$n = 1.35$

Comp. ratio - 14

Figure 8.22

$$P_3 = P_2 \left[\frac{V_2}{V_3}\right]^k = 13(14)^{1.35} = 455 \text{ psia}$$

$T_3 = T_2 \left[\dfrac{V_2}{V_3}\right]^{k-1}$ Since T_2 is not given, it is not unreasonable to assume $T_2 = 150°F$.

This is a good approximation for the temperature under the hood of an engine which would be the intake temperature. Otherwise, express your answer in terms of T_2.

$T_3 = (460 + 150)(14)^{1.35 - 1.00} = 610(14)^{.35}$

$T_3 = 2.52\, T_2$

$T_3 = 1540°R = 1080°F$

Temperature and pressure at end of comp are 1080°F (or 2.52 x intake air temperature and 455 psia   **Answer**

5.   A theoretical heat engine using one pound of air as the working substance operates on a cycle consisting of four process, namely an isentropic compression, a combustion process at constant pressure, an isentropic expansion and a cooling process at constant volume. The temperatures at the beginning and end of compression are 80°F and 1046°F respectively. At the end of the combustion process the temperature is 1463°F. Assuming air to act as a perfect gas, what is the net work produced for this theoretical cycle?

Approach

As you sketch the cycle described (Fig. 8.23) you will recognize the Diesel cycle. The basic gas laws are used to calculate the temperature just prior to the exhaust valve opening. Then the work can be calculated.

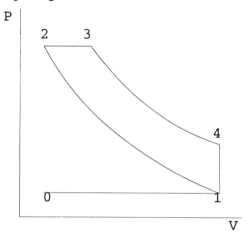

$T_1 = 80°F$

$T_2 = 1046°F$

$T_3 = 1463°F$

Figure 8.23

$W_{net} = W_{2-3} + W_{3-4} - W_{1-2}$

$W_{net} = R(T_3 - T_2) + J C_v (T_3 - T_4) - J C_v (T_2 - T_1)$

To find T_4 we need to make a few calculations.

$V_3 = V_2 (T_3/T_2);\ V_4 = V_3 (T_3/T_4)^{\frac{1}{k-1}}\ ;\ V_1 = V_2 (T_2/T_1)^{\frac{1}{k-1}}$

Since $V_4 = V_1$; $V_3 (T_3/T_4)^{\frac{1}{k-1}} = V_2 (T_2/T_1)^{\frac{1}{k-1}}$

$T_4 = T_1 (T_3/T_2)(T_3/T_2)^{k-1} = T_1 (T_3/T_2)^k$

$T_4 = 540 \left[\dfrac{1463 + 460}{1046 + 460}\right]^k = 762°R$ where $k = 1.4$

$W_{net} = R (T_3 - T_2) + 778 \times C_v(T_3 - T_2) - 778 \times C_v (T_2 - T_1)$

$W_{net} = 53.3(1923 - 1506) + 778 \times .173(1923 - 762) - 778 \times .173 (1506 - 540)$

$W_{net} = 48{,}472$ ft lb/lb of air   <u>Answer</u>

6. A water jacketed air compressor compresses 200 cfm of air with a density of 0.074 lb/ft³. Measurements at inlet and outlet show that the enthalpy of the air has increased 15 Btu/lb. The cooling water receives 380 Btu per minute from the air being compressed. Determine the horsepower required for the compression, neglecting kinetic and potential energy changes of the air.

Approach

Follow through the detailed explanation of this problem.

From the general energy equation $W = Q + \Delta H$; $W = Q + w(h_1 - h_2)$

But $w = 200 \dfrac{ft^3}{min} \times .074 \dfrac{lb}{ft^3} = 14.8$ lb/min. and $h_1 - h_2 = 15$ Btu/lb

Since the water receives Q, the gas is giving it up so Q is negative.

$W = -380 - (14.8 \times 15) = -602$ Btu/min = work done on the gas

$HP = 602 \dfrac{Btu}{min} \times 778 \dfrac{ft\ lb}{Btu} \times \dfrac{1\ HP}{33000\ ft\ lb/min}$

$HP = 14.2$ **Answer**

7. A diesel engine has a compression ratio of 16 and a cutoff at 1/15th the stroke. Find the efficiency and temperature of the exhaust using the cold air standard, starting at 14 psia and 100°F (Fig. 8.24).

Approach:

This is an analysis problem which requires thinking out the rationals.

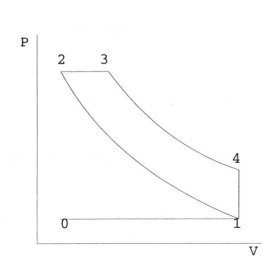

$\eta = 1 - \dfrac{1}{r_k^{k-1}} \left[\dfrac{r_c^k - 1}{k(r_c - 1)} \right]$

$r_k = \dfrac{V_1}{V_2} = 16 \quad P_1 = 14$ psia

$r_c = \dfrac{V_3}{V_2}$

$T_1 = 100°F$

Cold air standard: k = 1.4

Since $V_1 = 16 V_2$, $V_1 - V_2 = (16 - 1) V_2 = 15 V_2$

$V_3 - V_2 = 1/15$ the entire stroke

$V_3 - V_2 = 1/15 (V_1 - V_2) = 1/15 (16V_2 - V_2) = V_2$

or $V_3 = V_2$ then $r_c = V_3/V_2 = 2V_2/V_2 = 2$

$$\eta = 1 - \frac{1}{16^{1.4-1}} \left[\frac{2^{1.4} - 1}{1.4 (2-1)} \right] = .614$$

Efficiency = 61.4%   <u>Answer</u>

$T_4 = T_1 (P_4/P_1)$

$$P_4 = P_3 \left[\frac{V_3}{V_4}\right]^k = P_3 \left[\frac{2V_2}{V_1}\right]^k = P_3 \left[\frac{2}{16}\right]^k = P_2 \left[\frac{1}{8}\right]^k$$

but $P_2 = P_1 \left[\dfrac{V_1}{V_2}\right]^k = P_1 (16)^k$

$P_4 = P_1 (16)^k \left[\dfrac{1}{8}\right]^k = 36.9$ psia

$T_4 = (100 + 460) \dfrac{36.9}{14} = 1016°F$ <u>Answer</u>

GAS MIXTURES

Avogadro's Law states that all perfect gases when at a given pressure and temperature have the same number of molecules in a given volume.

$$\frac{\rho_x}{\rho_y} = \frac{M_x}{M_y}$$ where ρ is the density and M is the molecular weight of the gases.

The subscripts x and y denote two different perfect gases. z will be used to denote a further number of gases.

Since density is inversely proportional to specific volume, $M_x V_x = M_y V_y$ where M_v is called a mole volume.

$M_v = 359$ ft³ at 1 atmosphere and 32°F. M_v = the pound molecular volume, which is the volume occupied by 1 pound of any ideal gas at standard temperature and pressure (STP) (1 atmosphere, 32°F).

From the characteristic equation:

PV = W R T (Thermodynamic terms)

W = M V = Mv

P (Mv) = M R T (Chemistry terms)

$M R = P \dfrac{(Mv)}{T}$ where M R = 1544 = universal gas constant

Thus $R = \dfrac{1544}{M}$, the specific gas constant.

VOLUMETRIC ANALYSIS

Volumetric analysis is the % by volume of the total volume at a given temperature and pressure. This is also called the bulk analysis, hence the use of the symbol B.

$B_x + B_y + B_z = 100\%$ where $B = \dfrac{\text{Volume of component}}{\text{Volume of mixture}}$

GRAVIMETRIC ANALYSIS

Gravimetric analysis is the % by weight of the total weight of the mixture

$G_x + G_y + G_z = 100\%$ where $G = \dfrac{\text{Wt. of component}}{\text{Wt. of mixture}}$

To convert from volumetric to gravimetric analysis:

$$G_x = \dfrac{W_x}{W_x + W_y + W_z} = \dfrac{\rho_x B_x}{\rho_x B_x + \rho_y B_y + \rho_z B_z}$$

and since ρ is proportional to M: $G_x = \dfrac{M_x B_x}{M_x B_x + M_y B_y + M_z B_z}$

To convert from Gravimetric to volumetric analysis:

$$B_x = \frac{V_x}{V_x + V_y + V_z} = \frac{G_x/\rho_x}{G_x/\rho_x + G_y/\rho_y + G_z/\rho_z}$$

and since ρ is proportional to M:

$$B_x = \frac{G_x/M_x}{G_x/M_x + G_y/M_y + G_z/M_z}$$

Dalton's Law of Partial Pressures

All the components of a mixture occupy the same volume, the volume of the mixture, V_m, and are at the same temperature T_m the temperature of the mixture. The pressure of the mixture is the sum of the partial pressures of each of the components.

$P_m = P_x + P_y + P_z$ Since PV = WRT

$$P_x = \frac{W_x R_x T_m}{V_m} \text{ and } \frac{P_x}{P_m} = \frac{W_x R_x}{W_x R_x + W_y R_y + W_z R_z}$$

But $R = \dfrac{1544}{M}$; $\dfrac{P_x}{P_m} = \dfrac{G_x/M_x}{G_x/M_x + G_y/M_y + G_z/M_z} = B_x$

Therefore, $\dfrac{P_x}{P_m} = B_x$, $\dfrac{P_y}{P_m} = B_y$, $\dfrac{P_z}{P_m} = B_z$. Also from the characteristic equation,

$$P_m = P_x + P_y + P_z = (W_x R_x + W_y R_y + W_z R_z)\frac{T_m}{V_m}; \quad P_m = \frac{W_m R_m T_m}{V_m}$$

Combining these two equations and solving for R_m

$$R_m = \frac{W_x R_x + W_y R_y + W_z R_z}{W_m}$$

$$R_m = G_x R_x + G_y R_y + G_z R_z \; ; \; R_m = 1544\left[\frac{G_x}{M_x} + \frac{G_y}{M_y} + \frac{G_z}{M_z}\right]$$

$$R_m = \frac{1544}{B_xM_x + B_yM_y + B_zM_z} \; ; \; M_m = B_xM_x + B_yM_y + B_zM_z$$

Remember that $Q = W_m \int C_m dT$

$$W_m C_m dT = W_x C_x dT + W_y C_y dT + W_z C_z dT \; ; \; C_m = \frac{W_x C_x}{W_m} + \frac{W_y C_y}{W_m} + \frac{W_z C_z}{W_m}$$

$$C_m = G_x C_x + G_y C_y + G_z C_z$$

COMBUSTION

Three elements of a hydrocarbon fuel have heating values C, H, and S. The real reactions of the combustion are complex. We will take the combustion process as the simple chemical combination of oxygen and the combustibles. We assume the H_2O in the product to always be water vapor.

Combustion of Hydrogen

$$2 H_2 + O_2 \rightarrow 2 H_2O$$

Weight Balance 2 (2) + 32 → 2 (18) where the values in parentheses are the molecular weights or 4 lbs of H_2 + 32 lbs of O_2 → 36 lbs of H_2O.

For simplification we consider air to be composed of only nitrogen and oxygen, 21% O_2 and 79% N_2 by volume. 79/21 = 3.76 which means that for each unit of O_2 we carry along 3.76 units of N_2 in the air used for combustion.

$$2 H_2 + O_2 + 3.76 N_2 \rightarrow 2 H_2O + 3.76 N_2$$

Weight balance 2(2) + (32) + 3.78 (28) → 2(18) + 3.76(28)

Per pound of fuel 1 + 8 + 26.32 → 9 + 26.32

which reads: 1 lb of H_2 + 8 lbs of O_2 + 26.32 lbs of N_2 yield 9 lbs of water vapor + 26.32 lbs of N_2.

Since we are considering air as composed of N_2 and O_2, 8 lbs of O_2 + 26.32 lbs of N_2 = 34.32 lbs of air. Therefore, combustion of 1 lb of hydrogen requires 34.32 lbs of air to yield 9 lbs of water vapor plus 26.32 lbs of nitrogen.

In practice we use some excess air to ensure that all the fuel meets with the oxygen necessary for combustion. If we add 20% excess air to the equation for the combustion of H_2 we will use 1.2

(34.32) or 43.6 lbs of air per lb of H_2. Since, <u>by weight</u>, we have 23.1% O_2 and 76.9% N_2, of the 43.6 lbs of air used, 23.1% or 10.07 lbs is O_2 and 76.9% or 33.52 lbs is N_2.

$$2 H_2 + 1.2 O_2 + 3.99 N_2 \to 2H_2O + .2O_2 + 3.99 N_2 \text{ for 20\% excess air.}$$

Combustion of Carbon

$$C + O_2 \to CO_2$$

Weight balance $12 + 32 \to 44$

Per pound of fuel $1 + 2.67 \to 3.67$

Therefore $C + O_2 + 3.76 N_2 \to CO_2 + 3.76 N_2$

By weight $12 + 32 + 105.3 \to 44 + 105.3$

Per pound of fuel $1 + 2.67 + 8.78 \to 3.67 + 8.78$

1 lb of carbon + (2.67 + 8.78 = 11.45) lb. of air are required for complete combustion.

Combustion of Sulfur

$$S + O_2 + 3.76 N_2 \to SO_2 + 3.76 N_2$$

Weight balance: $32 + 32 + 105.3 \to 64 + 105.3$

Per pound of S $1 + 1 + 3.3 \to 2 + 3.3$

Relative Volume: $0 + 1 + 3.76 \to 1 + 3.76$

Note that sulfur as a solid has an insignificant volume. The same applies to solid carbon.

Combustion of Carbon Monoxide

In the combustion of carbon, there is sometimes some CO, carbon monoxide, formed. The CO combines with the O_2 in the air to form CO_2.

$$2CO + O_2 + 3.76 N_2 \to 2CO_2 + 3.76 N_2$$

Wt. Bal. $2(28) + (32) + 3.76(28) \to 2(44) + 3.76(28)$

Pr # of CO $1 + .571 + 1.88 \to 1.571 + 1.88$

Rel. Vol. $2 + 1 + 3.76 \to 2 + 3.76$

Combustion of a hydrocarbon (octane, C_8H_{18}). Note the usual formula is C_nH_{2n+2}.

$$C_8H_{18} + 12.5O_2 + 47.0 N_2 \rightarrow 9H_2O + 8CO_2 + 47.0 N_2$$

Weight Balance: $114 + 400 + 1316 \rightarrow 162 + 352 + 1316$

Per pound of Fuel $1 + 3.51 + 11.5 \rightarrow 1.42 + 3.09 + 11.5$

Relative volume: $1.1 + 12.5 + 47.0 \rightarrow 9 + 8 + 47.0$

For composition by weight of the products of combustion

$\rightarrow 1.42 + 3.09 + 11.5$ or 16.01 lbs

$G_{H2O} = 1.42/16.01 = 8.87\%$

$G_{CO2} = 3.09/16.01 = 19.30\%$

$G_{N2} = 11.5/16.01 = 71.83\%$

For composition by volume of the products of combustion (H_2O stays vapor)

$\rightarrow 9 + 8 + 47.0$ or 64.0 units of volume

$B_{H2O} = 9/64.0 = 14.06\%$

$B_{CO2} = 8/64.0 = 12.50\%$

$B_{N2} = 47.0/64.0 = 73.44\%$

Remember $P_x = B_x P_m$

If products of combustion in the above example are at 14.7 psia, atmospheric pressure: $P_{H2O} = .1406 (14.7) = 2.07$ psia

This is the vapor pressure of the water vapor.

From the combustion equation we find that $3.51 + 11.5 = 15.01$ lbs of air per lb of C_8H_{18} are required. To find the volume of air required we use the characteristic equation. Assume air at 14.7 psia and 70°F.

$$V = \frac{WRT}{P} = \frac{15.01 \times 53.3 (70 + 460)}{14.7 \times 144}$$

V = 200.3 ft³ of air per pound of C_8H_{18}

Let us take the volume of the product of combustion at 14.7 psia and 2000°R.

(Vol of H_2O) + (Vol of CO_2) + (Vol of N_2) = Vol of products

From the characteristic equation, $V = \dfrac{WRT}{P}$

for H_2O $V = \dfrac{1.42 \times 1544 \times 2000}{18 \times 14.7 \times 144} = 115.1 \text{ ft}^3$

for CO_2 $V = \dfrac{3.09 \times 1544 \times 2000}{44 \times 14.7 \times 144} = 102.4 \text{ ft}^3$

for N_2 $V = \dfrac{11.5 \times 1544 \times 2000}{28 \times 14.7 \times 144} = 599 \text{ ft.}$

The volume of the products of combustion at 14.7 psia and 2000°R are:

115.1 + 102.4 + 599 = 816.5 ft³/lb of C_8H_{18}

HEATING VALUE

The heating value of a fuel is sometimes quoted. The heating value of a fuel is the amount of heat given up by the products when they are cooled to the original temperature. Water vapor always appears in the products of combustion whenever the fuel contains hydrogen.

If the water vapor formed by the combustion is condensed, the maximum amount of heat is given up by the product of combustion and the heating value is called the higher heating value, HHV or qh. The lower heating value LHV or qL is equal to HHV less the latent heat of the condensed water vapor.

When excess air is used in combustion; % excess air = actual air supplied-theoretical air req. x 100/Theoretical air required.

Example:

If a fuel was burned with 15 lb. of air, 1 lb. of fuel and the theoretical air required was calculated as 7.5 lbs air/lbs fuel, what would be the % excess air?

% excess = $\dfrac{15.0 - 7.5}{7.5}$ x 100 = 100% excess air.

LIQUIDS AND VAPOR

When a gas is cooled or compressed it will condense. At this point, the molecules are closer together, have large forces of attraction and take up a significant part of the total space. A perfect gas has no forces of attraction and occupies no space.

Vapors have molecules that have mass and they do exert forces. If a vapor is heated to a temperature far above that at which condensation occurs or if the surrounding pressure is reduced to low values, the vapor is considered to be "superheated" and may be treated as a perfect gas.

For each pressure there is a precise temperature marking the boiling point of a liquid. This temperature is known as the saturation temperature.

A mixture of a vapor and its liquid is known as wet mixture. The liquid and vapor phases coexist.

The quality X of a mixture is the fraction by weight of the mixture that is vapor. Quality is generally denoted by the symbol X and is often quoted as a percent (100X). The % moisture of a mixture is the % by weight of the mixture which is liquid. The % moisture is symbolized by y. Note that y = 1 - x.

If x = 0.83, then in 1 lb of mixture, .83 lb is vapor and .17 lb is liquid.

A saturated vapor is of 1.0 (100%) quality at the saturation temperature only. If heat is added to a saturated vapor its temperature and volume will increase and the vapor will become superheated. The degree of superheat is the temperature difference between the temperature of the superheated vapor and the saturation temperature for the same pressure.

Therefore, at a given temperature and pressure a mixture of liquid water and water vapor can exist. Since a change in entropy usually accompanies an addition of heat, we can plot a temperature-entropy curve for water and for steam (Fig. 8.25).

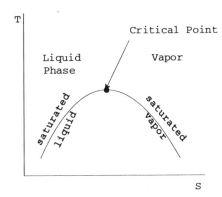

Figure 8.25

This T-S plot shows two separate and distinct curves, the saturated liquid curve and the saturated vapor curve. The two curves intersect at a particular point, the critical point. If a vapor is at a temperature greater than the temperature at the critical point, it is impossible to liquify vapor no matter what the pressure.

The region to the left of the saturated liquid line represents the liquid phase while the region to the right of the saturated vapor curve represents the vapor phase. The distance between the two curves can be subdivided to show the moisture fraction or the quality of the mixture between the two extremes.

Since every temperature has its associated pressure, the lines of constant pressure will be horizontal lines connecting the two saturated curves (Fig. 8.26).

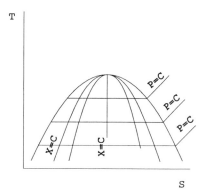

Figure 8.26

Above the saturated vapor line the vapor is superheated and acts like a gas and the constant pressure lines diverge from the lines of constant temperature (i.e., temperature and pressure become independent variables).

The steam tables have been prepared to show values of specific volume (v), specific internal energy (u), specific enthalpy (h) and specific entropy (s) for various conditions of temperature and pressure.

The subscripts f and g denote conditions at the saturated liquid line (fluid) and at the saturated vapor line (gas) respectively. The subscript fg is used to indicate between the liquid and vapor states for example,

$$v_f + v_{fg} = v_g$$

$$h_g + h_{fg} = h_g$$

It can be seen that h_{fg} represents what is known as the latent heat of evaporation, the amount of energy required to be added to transform a liquid into a gas, or extracted to transform a gas into a liquid.

For a mixture of quality x,

$$v = v_f + x\, v_{fg}$$

$$s = s_f + x\, s_{fg}$$

$$h = h_f + x\, h_{fg}$$

Table 1 of the steam tables gives indicated values as a function of saturation temperature while Table 2 uses saturation pressures. Table 3 concerns itself with the superheat region while Table 4 refers to the compressed liquid phase. These tables are readily available elsewhere.

Example:

Find the enthalpy of wet steam having 80% quality which is at 100 psia.

$h = h_f + x\, h_{fg}$

From Table 2, at 100 psia, find $h_f = 298.4^B/\#$ and $h_{fg} = 888.8^B/\#$

$h = 298.4 + .80(888.8)$
$h = 1009.4$ Btu/lb

Example

Water at 200°F has an entropy of .500 Btu/lb °R. Find the enthalpy of the water at this temperature.

$s = s_f + x s_{fg}$ only for the wet region. From Table 1 we find $s_g = 1.7762$ which is greater than 0.5000, therefore we have a wet mixture.

The values of s_f and s_{fg} are 0.2938 and 1.4824 respectively.

$$0.5000 = 0.2938 + X(1.4824)$$

$$X = \frac{.5000 - .2938}{1.4824} = \frac{.2062}{1.4824} = .1391$$

$h = h_f + .1391\, h_{fg}$

$h = 167.99 + .1391\,(977.9)$

$h = 167.99 + 136.03$

$h = 304.0^B/\text{lb}$

THROTTLING CALORIMETER

A <u>throttling calorimeter</u> is a device used to measure the quality of a wet steam mixture. Its performance is based on a throttling process in which the final enthalpy is the same as the initial enthalpy. The steam enters the calorimeter at an enthalpy of h_1 and leaves the calorimeter at an enthalpy of h_2, where $h_1 = h_2$. The value of h_2 can be found when T_2 and P_2 are known. This gives us the value of h_1.

$$\text{then } x = \frac{h_1 - h_{f1}}{h_{fg1}}$$

For enthalpy values in the compressed liquid range, Table 1 can be read at the appropriate temperature. This is discussed further in a later section when we do not make such assumptions. We can interpolate between the values in the superheat table, Table 3. In Tables 1 and 2 there is seldom need to interpolate.

A large chart or enthalpy-entropy chart known as the Mollier Chart has most of the information in the steam tables. The Mollier Chart can often be used to reduce the time necessary for interpolation and can be used with excellent results. It is advisable, however, that when the Mollier Chart is used, make a sketch of the chart and indicate how the necessary values were obtained.

RANKINE CYCLE

Steam generators and turbines operate on a Rankine Cycle. We use the steam tables to solve steam turbine problems.

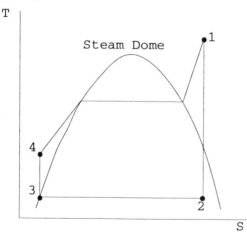

Figure 8.27

The T-S diagram shown in Fig. 8.27 is of a basic Rankine cycle. State point 1 indicates the conditions of the steam entering a steam turbine, state point 2 for the steam leaving the turbine and entering the condenser, state point 3 for the liquid leaving the condenser and state point 4 for the feedwater entering the boiler.

The work of the Rankine cycle is that done by the prime mover.

$$W_{net} = (h_1 - h_2) - (h_4 - h_3); \quad Q_a = (h_1 - h_4) \quad ; \quad e = \frac{W_{net}}{Q_a}$$

If pump work is considered negligible the equations reduce to

$$W_{net} = (h_1 - h_2) \qquad Q_a = (h_1 - h_3) \quad ; \quad e = \frac{W_{net}}{Q_a}$$

EFFICIENCIES

The indicated thermal efficiency e_i = indicated work/heat added

The <u>brake thermal efficiency</u> e_b = brake work/heat added

The <u>combined thermal efficiency</u> (when two machines are used and based on the combined output) e_{ct} = combined work/heat added or output/heat added or output/input

The <u>engine efficiency</u> e_e = ideal cycle effect / Rankine effect

Let e = the <u>ideal cycle efficiency</u>

Then:

<u>Brake engine efficiency</u> = e_{be} = $e_{b/e}$

<u>Combined engine efficiency</u> = e_{cte} = $e_{ct/e}$

<u>Mechanical efficiency</u> = e_m = e_b/e_i

<u>Volumetric efficiency</u> = e_v = $\dfrac{\text{Weight of mixture added to the cylinder}}{\text{Weight contained in cylinder disp. volume at atmosphere conditions}}$

Example

Given wet steam at 50 psia and 40% quality. Find T h, s. Using <u>Thermodynamic Properties of Steam</u> by Keenan & Keyes from Table 2, at 50 psia, read T = 281.01°F.

read h_f = 250.09

read h_{fg} = 924.0

$h = h_f + x\, h_{fg} = 250.09 + 0.40\,(924.0) = 619.69$ Btu/lb

From Table 2, at 50 psia, read $s_f = .4110$ read $s_{fg} = 1.2474$

$s = s_f + x\, s_{fg} = 0.4110 + 0.40\,(1.2474) = 0.9100$ units

Example:

Given wet steam at 300°F with h = 470 Btu/lb. Find p, s, v and x. From Table 1 at 300°F read p = 67.013 psia read $h_f = 269.59$ read $h_{fg} = 910.1$

Calculate $x = h - h_f/h_{fg}$ $x = \dfrac{470 - 269.59}{910.1} = 0.22$

From Table 1 at 300°F read: $S_f = 0.4369$ $v_f = 0.01745$

Read $s_{fg} = 1.1980$ $v_{fg} = 6.449$

Compute $xs_{fg} = 0.262$ $sv_{fg} = 1.418$

Compute $s = s_f + x\, s_{fg} = 0.4369 + 0.262 = 0.6989$

Compute $v = f_f + x\, v_{fg} = 0.01745 + 1.418 = 1.43545$

Example

A steam engine develops 50 indicated hp with dry saturated steam supplied at 150 psia and exhausted at 17 psia. The steam consumption is 1250 pounds per hour. Calculate or use a chart to obtain the following efficiencies: Carnot, Rankine, actual thermal, engine.

The steam rate is = 1250/50 = 25 lb/hp - hr.

For Carnot efficiency:

From Table 2 at 150 psia read T = 348.42°F = 808.42°R

From Table 2 at 17 psia read T = 219.44°F = 679.44°R

Efficiency $= \dfrac{T_1 - T_2}{T_1} = \dfrac{808.42 - 679.44}{808.42} = .160 = 16\%$

For Rankine efficiency:

From Mollier chart read h_1 at 150 psia saturated is found as 1194 Btu/lb. Read down the line of constant entropy to 17 psia to find h_2 = 1034 Btu/lb.

From Table 2 at 17 psia find h_{f2} = 187.56 Btu/lb

$$\text{efficiency} = (h_1 - h_2)/(h_1 - h_{f2}) = \frac{1194 - 1034}{1194 - 188} = 15.9\%$$

For actual thermal efficiency:

$$\text{Input} = \frac{1250 \text{ lb/hr} \times (1194 - 188) \text{ Btu/lb}}{2545 \text{ Btu/HP-hr}} = 494 \text{ hp}$$

Efficiency = output/input = 50/494 = 10.12%

For engine efficiency:

Efficiency = actual/Rankine efficiencies

Efficiency = 10.12/15.9 = 64%

SOLVED PROBLEMS

1. A steam throttling calorimeter receives steam at 120 psia and discharges it at 18 psia. If its thermometer reads 240°F, what is the percent quality of the entering steam?

   Approach:

   Check the steam table to find if the steam is saturated or superheated. Since throttling is ideally a constant enthalpy process we can use basic relationship to solve for quality.

   We are given P_1 = 120 psia, P_2 = 18 psia and T_2 = 240°F

   From our steam tables or Mollier diagram, Table 3 for superheated vapor, we find that at 18 psia and 240°F the steam is superheated. We find, from the table, that h_2 is 1163.0 Btu/lb.

   Since the throttling process results in h_2 = h, then h_1 = h_2 = 1163.0 Btu/lb.

But $h_1 = h_f + X h_{fg}$ or $X = \dfrac{h_1 - h_f}{h_{fg}}$

From Table 2 of the steam tables we find that at $P_1 = 120$ psia, $h_f = 312.44$ Btu/lb and $h_{fg} = 877.9$ Btu/lb.

Therefore $X = \dfrac{1163.0 - 312.44}{877.9} = .97$

Quality is 97%   <u>Answer</u>

2. A double acting steam engine has a bore of 20 in., a stroke of 24", and a piston rod of 2.5 in. diameter. The engine runs at 180 rpm. The area on the indicator card for the head end is 1.5 in² and for the crank end 1.6 in². The length of the indicator diagram is 3.0 in. The scale of the indicator spring is 60 psi/in. Calculate the indicated horsepower developed.

<u>Approach</u>:

Using the reciprocating engine horsepower equation, $\dfrac{\text{PLAN}}{550}$ or $\dfrac{\text{PLAN}}{33000}$, solve for horsepower after taking the data from the indicator card. Remember, the indicator card is a plot of pressure vs. stroke and the area of the diagram is a measure of work (Fig. 8.28).

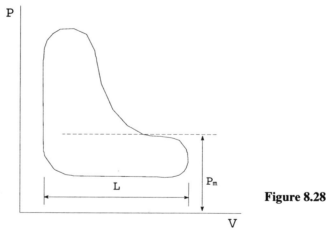

Figure 8.28

<u>Head end</u>

$P_{mi} = \dfrac{1.5 \text{ in}^2}{3 \text{ in}} = .5 \text{ in}$ $P_m = .5 \times 60 \text{ psi/in} = 30 \text{ psi}$

$HP_h = PLAN/33000$

$$HP_h = \frac{30 \times \frac{24}{12} \times \frac{\pi}{4} (20)^2 \times 180}{33000}$$

$HP_h = 103 HP$

Crank end

$$Pmi = \frac{1.6 \text{ in}^2}{3 \text{ in}} = .533" \qquad Pm = .533 \times 60 = 32 \text{ psi}$$

$HP_c = PLAN/330000$

$$HP_c = \frac{32 \times \frac{24}{12} \times \frac{\pi}{4} (20^2 - 2.5^2) \times 180}{33000}$$

$HP_c = 108 HP$

Indicated $HP = HP_h + HP_c = 103 + 108 = \underline{211 \text{ hp}}$

3. A steam turbo generator consumes 20 lb. of steam per kW-hr delivered. The steam enters at 250 psia and 550°F and discharges to the condenser at 2 psia. Calculate the following: Rankine efficiency, combined thermal efficiency, engine efficiency, and steam consumption of the Rankine engine.

Approach

This is strictly a problem in defining various efficiencies.

Actual steam rate is 20 lb/kW-hr.

$P_1 = 250$ psia $T_1 = 550°F$ $P_2 = 2$ psia

From Table 3 of the Steam Tables @ 250 psia and 550°F, $h_1 = 1291.4$ B/lb, $s_1 = 1.6242$

From Steam Tables or Mollier Chart find $h_2 = 943$ B/lb

From Table 2 of the Steam Tables @ 2 psia $h_{f2} = 94$ B/lb

Rankine efficiency $= (h_1-h_2)/(h_1-h_{f2}) = \dfrac{(1291.4 - 943)}{(1291.4 - 94)} = \underline{29.1\%}$

CTE = 3413/20 ($h_1 - h_{f2}$) = 3413/23948 = <u>14.3%</u>

Engine efficiency = CTE/R.E. = .143/.291 = <u>50.0%</u>

W_R = 3413/($h_1 - h_2$) = 3413/348.4 = <u>9.80 lb/kW-hr</u>

4. A steam engine develops 50 indicated HP with dry saturated steam supplied at 150 psia and exhausted at 17 psia. The steam consumption is 1250 lb/hr. Calculate or use a chart or tables to determine the following efficiencies: Carnot, Rankine, actual thermal, engine.

Approach

Note that "dry" saturated steam means that we are on the saturated vapor line where quality X = 1.0 (100% vapor).

Use the Mollier Chart to find the required enthalpies. Calculating the efficiencies is simply a matter of knowing what each efficiency means.

From Table 2 of the steam tables at 150 psia, T_1 = 358.4°F, h_1 = 1194 Btu/lb

From Table 2 of the Steam Tables @ 17 psia, T_2 = 219.4°F, h_{f2} = 187.6 Btu/lb

Carnot efficiency = ($T_1 - T_2$)/T_1 = (358.4 - 219.4)/(358.4 + 460) = <u>17%</u>

From Mollier Chart, h_2 = 1034 B/lb

Rankine efficiency = ($h_1 - h_2$)/($h_1 - h_{f2}$) = (1194 - 1034)/(1194 - 187.6) = <u>15.9%</u>

Thermal efficiency = $\dfrac{50 \quad HP}{w(h_1 - h_{f2})/2545}$ = $\dfrac{50 \times 2545}{1250(1194 - 187.6)}$ = <u>10.12%</u>

Engine efficiency = .1015/.159 = <u>64%</u>

5. A steam turbine carrying a full load of 50,000 kW uses 569,000 lbs of steam per hour (Fig. 8.29). The engine efficiency is 75%. The exhaust steam is at 1 psia and has an enthalpy of 950 Btu/lb. What are the temperature and pressure of the steam at the throttle?

Approach

For turbine problems, a Mollier Chart solution is generally the easiest and quickest method of solving the problem. However, watch the problem working, to see if a chart solution is acceptable. If the problem states "calculate," then the chart solution can only be used for checking your calculations.

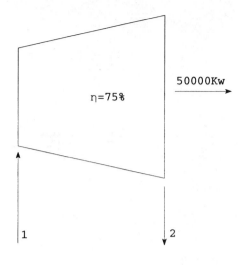

Figure 8.29

$h_1 - h_2'$ = Btu/lb required to supply 50,000 kw

$$h_1 - h_2' = \frac{50{,}000 \text{ kw} \times 3413 \text{ Btu/kw-hr}}{569{,}000 \text{ lbs/hr}} = 300 \text{ Btu/lb}$$

$h_1 = +950 + 300 = 1250$ Btu/lb

$$\frac{h_1 - h_2'}{h_1 - h_2} = 0.75 = \frac{1250 - 950}{1250 - h_2} \text{ or } h_2 = 850 \text{ Btu/lb}$$

From Mollier chart at $h_2 = 850$ Btu/lb and $P_2 = 1$ psia, follow constant entropy line to $h_1 = 1250$ Btu/lb to find

$P_1 = 430$ psia; $T_1 = 515°F$

6. Heat engine working on a reversible cycle receives 1000 Btu/min at 2540°F and discharges waste heat at 1040°F. If its mechanical efficiency is 60% and there are no heat losses, what is the net power output of the engine?

<u>Approach</u>

The key to this problem is to draw the T - S diagram (Fig. 8.30) and find the thermal efficiency of the cycle before working with the mechanical efficiency.

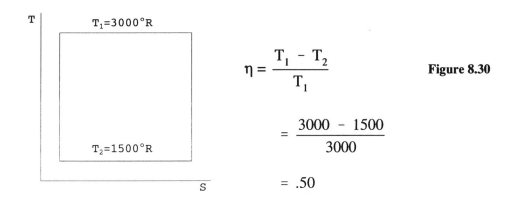

Figure 8.30

$$\eta = \frac{T_1 - T_2}{T_1}$$

$$= \frac{3000 - 1500}{3000}$$

$$= .50$$

$$\text{Power out} = \frac{.50 \times 1000 \text{ Btu/min} \times .60 \times 60 \text{ min/hr}}{2545 \text{ Btu/hp-hr}} = 7.07 \text{ hp}$$

7. A steam turbine receives 3600 lbs of steam per hour at 110 ft/sec velocity and 1525 Btu/lb enthalpy. The steam leaves at 810 ft/sec and 1300 Btu/lb. What is the hp output?

Approach

Set up the General Energy equation and solve for the work.

$\overline{V}_1 = 110$ fps $h_1 = 1525$ Btu/lb $w = 3600$ lb/HR $= 1$ lb/sec.

$\overline{V}_2 = 810$ fps $h_2 = 1300$ Btu/lb

From general energy equation: $\dfrac{w \overline{V}_1^2}{2g J} + \dfrac{w(U_1 + P_1 V_1)}{J} = \dfrac{w \overline{V}_2^2}{2g J} + \dfrac{w(U_2 + P_2 V_2)}{J} + W$

$$W = \frac{w}{2g}(\overline{V}_1^2 - \overline{V}_2^2) + w J (h_1 - h_2)$$

$$W = \frac{1}{2 \times 32.2}(110^2 - 810^2) + 1 \times 778 (1525 - 1300)$$

$W = 165,000$ ft lb/sec $= 300$ HP

8. How many pounds of steam at 212°F are required to heat the water in a swimming pool 60 ft. long by 20 ft wide by 5 ft deep from 40°F to 75°F? Assume all the condensed steam to remain in the pool and there is no loss of heat.

THERMODYNAMICS 8-47

Approach

Use the basic definition of specific heat. Remember, the steam gives up the latent heat of vaporization when it condenses.

Wt of H_2O in pool = 62.4 lbs/ft³ x 60 x 20 x 5 = 374,400 lbs

Q to heat water from 40 to 75° = wcΔT = 374,400 x 1 x 35 = 13.1 x 10⁶ Btu

hfg at 212°F = 970 Btu/lb

$$\text{Wt. of steam} = \frac{13.1 \times 10^6 \text{ Btu}}{970 + \frac{1 \times 11 (212 - 75)}{wc \Delta T}} = 13,505 \text{ lbs}$$

9. Assuming no losses, calculate how much warmer the water at the base of Niagara Falls should be than at the top after a fall of 165 ft.

Approach

Convert potential energy into thermal energy.

PE at top is converted to Q at bottom of falls.

PE = 165 ft-lbs/lb = QJ or Q = 165/778 = .212 Btu/lb

Q = w c Δ T = 1 x 1 x Δ T or T = .212°F

PSYCHOMETRICS

The vapor in a mixture is either superheated or saturated. Let P_v = the pressure of the vapor and P_m = the pressure of the mixture.

Relative humidity, $\emptyset$, is the ratio of the partial pressure of the water vapor in a mixture to the saturation pressure at the same temperature. P_w = partial pressure of the water vapor and P_g = saturation pressure at the same temperature.

$$\emptyset = \frac{P_w}{P_g}$$

Now, if the vapor can be treated as a perfect gas, $P_w = R_w T_w \rho_w$ (for water vapor) and $P_g = R_g T_g \rho_g$; $\rho_w/\rho_g = \dfrac{v_g}{v_w}$

The humidity ratio, γ, = $\dfrac{\text{mass of water}}{\text{mass of dry air}} = \dfrac{V_a}{V_w}$

$$\gamma = \dfrac{V_a}{V_w} = .622 \dfrac{P_w}{P_a} \text{ where subscript a is for dry air}$$

Therefore, $\emptyset = \dfrac{\gamma P_a}{.622 P_g}$

The dew point is the saturation temperature corresponding to P_w.

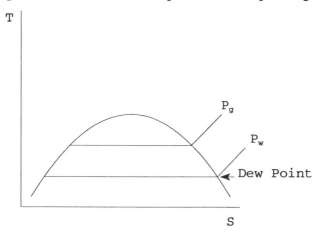

Figure 8.31

Example

A mixture of air and water vapor at 72°F and 14.7 psia has a relative humidity of 50%. Find the specific humidity and the dew point.

$$\gamma = .622 \dfrac{P_w}{P_a} \text{ where } P_w = \emptyset P_g$$

P_g is P_{sat} for 72°F from Table 1 of the steam table.

$P_w = .50 (.3886) = .1943$ psi

$P_a = P_m - P_w = 14.7000 - .1943$ psi $= 14.506$ psi

$\gamma = \dfrac{.622 \times .1943}{14.51} = .00833 \dfrac{\text{lbs wv}}{\text{lbs da}}$

The dew point is the saturation temperature corresponding to $P_w = 52.3°F$. Note the psychometric chart. This is used for a graphical solution of most psychometric or air conditioning problems. To illustrate, given a dry bulb temperature of 70°F and a wet bulb temperature of 60°F (Fig. 8.32). Find ∅, γ, h, and dew point temperature.

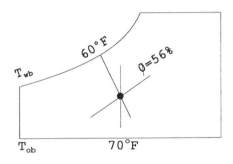

Figure 8.32

The intersection of 70°F db and 60°F vb is at = <u>56%</u>.

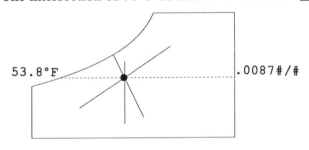

Figure 8.33

Follow horizontally from the intersection point to read <u>.0087 lb</u> moisture/lb dry air on the right hand scale and <u>53.8°F</u> dew point on the left hand scale.

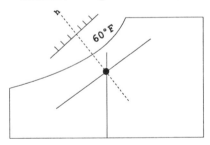

Figure 8.34

Read up from 60° w.b. to the enthalpy scale to read h = 26.5 Btu/lb at 60°F.

Example

Air at 1 atmosphere, 75°F and relative humidity of 70% is to be cooled, have some water removed, and heated to 70°F where it is to have 60% relative humidity (RH). To what temperature must the air be cooled? How much heat is removed in the cooling process and how much heat is added in the heating process?

From the psychometric chart at 75° db and 70% relative humidity:

Read $h_1 = 32.5$ Btu/lb $\gamma_1 = .0132$ lb water vapor/#dry air

From the chart at 70° db and $\emptyset = 60\%$

Read $h_2 = 27.2$ Btu/lb $\gamma_2 = .0095$ lb water vapor/dry air

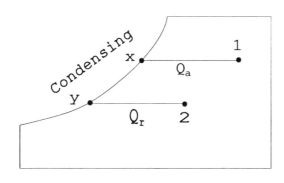

Figure 8.35

The mixture must be cooled to the dew point corresponding to the 70°F, 60% RH air. This is found as 56°F.

h_y is found to be 23.8 Btu/lb from the chart

h_f is found from the steam tables to be 24.06 Btu/lb at 56°F.

The water removed from the air is $\gamma_1 - \gamma_2$ or $.0132 - .0095 = .0037$ lb water vapor/lb dry air.

Therefore the heat removed in cooling is $Q_R = h_y - h_1 + (\gamma_1 - \gamma_2) h_f$

$Q_R = 23.8 - 32.5 + .0037 (24.06) = -8.61$ B/#

The head added in heating is $Q_A = h_2 - h_y = 27.2 - 23.8 = 3.4$ Btu/lb

POWER PLANTS

In steam power plants, a boiler converts the fuel or chemical energy into heat which provides high temperature steam to use in a prime mover such as steam engine or turbine.

The steam generator, commonly called the boiler, consists of the furnace or firebox where the chemical energy of the fuel is converted into heat energy. The boiler proper is composed of several banks of water filled tubes that absorb the energy to convert the water to saturated steam. The saturated steam can be passed through other banks of tubes where it receives the heat energy to bring it to its superheated state.

Auxiliary equipment to the steam generator may include an economizer. The economizer uses the exhaust flue gases to raise the temperature of the feedwater to the steam boiler. An air heater also uses the flue gases to reheat the air used for combustion. The reheater also uses flue gases to reheat steam bled off the turbine to add energy before returning the steam to the turbine.

If h_4 represents the enthalpy of one pound of entering feedwater and h_1 represents the enthalpy of one pound of steam leaving the boiler, then the heat added to the system is $Q_A = w(h_1 - h_4)$ where w is the pounds of steam generated per hour.

The steam passes through the prime mover and work is extracted from the cycle. This work, shaft work, is expressed as the difference in enthalpy between the entering steam and exhaust steam.

$$W_s = h_1 - h_2$$

The condenser is used to recover the exhaust steam and change it back into a liquid for feedwater. The heat rejected, Q_R, in condensing the steam is shown as $Q_R = h_3 - h_2$.

Lastly, a pump must be used to raise the pressure of the condensed steam to the pressure of the steam in the boiler so the feedwater can be added to the boiler to make steam. The pump work, $W_P = h_4 - h_3$.

This whole system operates on the Rankine cycle which is shown on the P - V and T - S diagram as well as in Fig. 8.36.

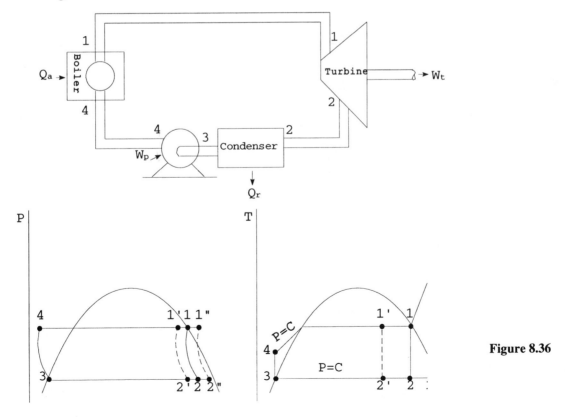

Figure 8.36

The steam entering the turbine may be at state points 1', 1, or 1" but the condenser output must be saturated liquid for the ideal cycle.

$$W_T = h_1 - h_2$$

$$h_2 + Q_R = h_3 \; ; \; Q_R = h_2 - h_3 \text{ (heat is rejected)}$$

$$W_P = h_4 - h_3 \; ; \; Q_A = h_1 - h_4$$

$$\eta = \frac{W_{net}}{Q_A} = \frac{W_T - W_P}{Q_A} = \frac{h_1 - h_2 + h_3 - h_4}{h_1 - h_4}$$

Assuming $h_4 = h_3$ (pump work is very small)

$$\eta = \frac{h_1 - h_2}{h_1 - h_4}$$

Steam rate $W_R = \dfrac{2545}{W_T - W_P}$ lb/hp-hr

$$W_R = \frac{3413}{W_T - W_P} \text{ lb/kw-hr}$$

Example

Given a Rankine cycle, steam is supplied at 200 psia and 750°F (Fig. 8.37). The condenser pressure is 1 psia. Find the cycle efficiency and steam rate.

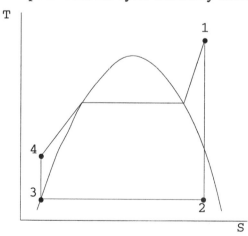

Figure 8.37

From Table 3 of the steam tables at 200 psia and 750°F.

$h_1 = 1399.2$ Btu/lb $s_1 = 1.7488$ Btu/lb°R

A Rankine cycle has a constant entropy process ideally, through the turbine. Therfore $s_2 = s_1$.

$s_2 = s_f + X s_{fg}$ Read s_f and s_{fg} from Table 2 at 1 psia.

$1.7448 = .1326 + X (1.8456)$

Therefore X = .874

$h_2 = h_f + .874 h_{fg}$ Read h_f and h_{fg} from Table 2 at 1 psia.

$h_2 = 69.70 + .874 (1036.3)$

$h_2 = 975$ Btu/lb.

Since state point 3 is on the saturation line, $h_3 = h_f$ at 1 psia or $h_3 = 69.70$ Btu/lb

TdS = dh - V dP. Ideally, the process from state points 3 to 4 is isentropic, dS = 0 and dh = V dP or Δh = VΔP where we assume that the water is incompressible.

So v is constant.

$$h_4 = h_3 + \left[v_3 \frac{(P_4 - P_3)}{J}\right]$$ where point 3 is saturated water

$$h_4 - h_3 = .0164 \times \frac{144}{778} (200 - 1) = .604 \text{ Btu/lb}$$

$h_4 = h_3 + .604 = 69.7 + .6 = 70.3$ Btu/lb

We now know the enthalpy at all four state points so we can compute the work and heat transferred.

$$\eta = \frac{W_T - W_P}{Q_A} = \frac{h_1 - h_2 + h_3 - h_4}{h_1 - h_4}$$

$$\eta = \frac{1399.2 - 975 + 69.7 - 70.3}{1399.2 - 70.3} = \frac{423.6}{1328.9} = .319$$

$$W_R = \frac{2545}{W_T - W_P} = \frac{2545}{623.6} = 6.01 \text{ \#/Steam}/_{\text{H-P hr.}}$$

We do have losses in a system which is not reversible adiabatic.

$$\text{let } \eta_T = \text{turbine efficiency} = \frac{W_T}{h_1 - h_{2s}} = \frac{h_1 - h_2}{h_1 - h_{2s}}$$

where h_1 is the actual enthalpy of the entering steam and h_{2s} is the enthalpy at a point having the actual exhaust pressure but lying on an isentropic path through state point 1.

$$\text{The actual steam rate } W_R = \frac{2545}{W_T}$$

$$\text{The actual steam rate } W_I = \frac{2545}{h_1 - h_{2s}}$$

$$\text{Therefore } \eta_T = \frac{2545}{W_R (h_1 - h_{2s})}$$

$$\text{Let } \eta_P = \text{Pump efficiency} \qquad \eta_P = \frac{h_{4s} - h_3}{W_P}$$

where h_3 is the enthalpy of the liquid entering the pump and h_{4s} is the enthalpy of the liquid at a state point having the actual pump discharge pressure and lying on an isentropic path through state point 3.

REHEAT CYCLE

From the Second Law of Thermodynamics, we know that efficiency of a cycle can be increased by supplying heat at higher temperature. In the reheat cycle we carry out the expansion of the steam in two or more steps (Fig. 8.38).

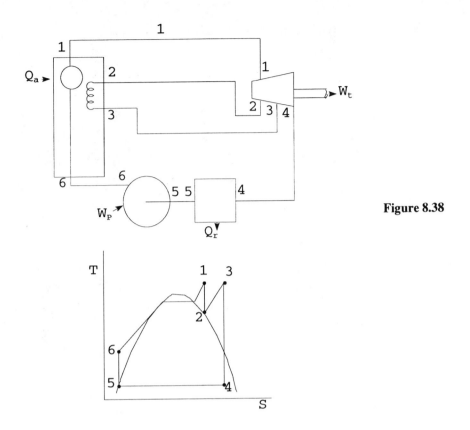

Figure 8.38

Let us show the effects of efficiency in the extraction of work by illustration of a multistage turbine (Fig. 8.39).

Let η_s = the efficiency of each stage

w_s = the work of each stage

$P_o, P_1, P_2 \text{ --- } P_n$ is the pressure at each stage.

The subscript s will denote an insentropic change.

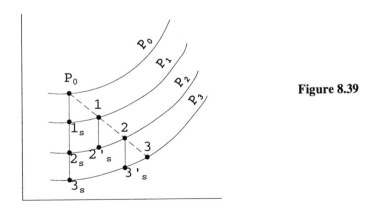

Figure 8.39

$$\eta_s = \frac{h_o - h_1}{h_o - h_{1s}} \text{ or } h_1 = h_o - \eta_s(h - h_{1s})$$

$$\eta_s = \frac{h_1 - h_2}{h_1 - h'_{2s}} \text{ or } h_2 = h_1 - \eta_s(h_1 - h_{2's})$$

$$W_T = \eta_s(h_o - h_{1s} + h_1 - h_{2's} + h_2 - h_{3's})$$

$$W_T = \eta_s(h_o - h_{3s})$$

Example

A 3 stage turbine is to operate between an initial state of 200 psia saturated steam and an exhaust pressure of 1 psia (Fig. 8.40). The intermediate stage pressures were chosen by evenly dividing the isentropic enthalpy drop. The stage efficiency is 70%. Neglect the velocity of approach. Find the turbine efficiency and reheat factor.

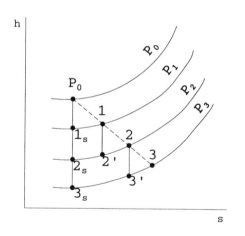

Figure 8.40

$h_o = 1198.4$ Btu/lb (hg @ 200 psia) $s_o = 1.5453$ Btu/lb °(sg at 200 psia)

$s_o = s_{3s} = s_f + X\, s_{fg} = 1.5453 = .1326 + X(1.8456)$

$X = .765 = 76.5\%$

$h_{3s} = h_f + X\, h_{fg} = 69.7 + .765(1036.3) = 862.5$ Btu/lb

$h_o - h_{3s} = 1198.4 - 862.5 = 335.9$ Btu/lb

Δh per stage $= 335.9/3 = 112$ Btu/lb

$h_{1s} = 1198.4 - 112 = 1086.4$ Btu/lb; $P_1 = 47$ psia

$h_{2s} = 1198.4 - 224 = 974.4$ Btu/lb; $P_2 = 8.2$ psia

$h_1 = h_o - .70 (h_o - h_{1s})$ $h_1 = 1198.4 - .70 (1198.4 - 1086.4) = 1020$ Btu/lb

$s_1 = 1.457$

$h_2' = 918$ Btu/lb $h_2 = h_1 - .70 (h_1 - h_2')$

$h_2 = 1020 - .70 (1020 - 918) = 949$ Btu/lb

$s_2 = 1.508$

$h_3' = 840$ Btu/lb

$h_3 = h_2 - .70 (h_2 - h_3')$

$h_3 = 949 - .70 (949-840) = 873$ Btu/lb

$$\eta_T = \frac{W_T}{h_o - h_{3s}} = \frac{h_o - h_s}{h - h_{3s}} = \frac{1198.4 - 873}{1198.4 - 862.5} = .969$$

Reheat factor = $\dfrac{.969}{.70} = 1.38$

REFRIGERATION

If heat is discharged to a high temperature region we have a heat pump. If heat is absorbed from a low temperature region, we have a refrigerator (Fig. 8.41).

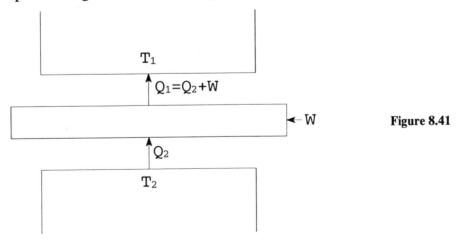

Figure 8.41

$$W = \left[\frac{T_1 - T_2}{T_1}\right] Q_1 = \left[\frac{T_1 - T_2}{T_2}\right] Q_2$$

Heat Pump COP = $\dfrac{Q_1}{W_{NET}}$ Refrigerator COP = $\dfrac{Q_2}{W_{NET}}$

For a <u>Carnot</u> cycle:

Heat Pump $COP_{max.} = \dfrac{T_1}{T_1 - T_2}$ Refrigerator $COP_{max.} = \dfrac{T_2}{T_1 - T_2}$

And: $COP_{Heat\ Pump} = COP_{refrig} + 1.0$

where COP is the Coefficient of Performance

COP = output/input = refrigeration/work required

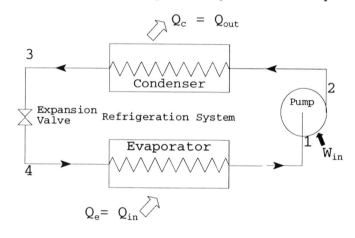

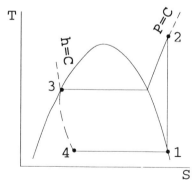

Figure 8.42

3-4 is NOT h = c but has the same end point as if it were. There is no heat transfer from 3 to 4.

One ton of refrigeration is the equivalent to 200 Btu/min. It is the rate of cooling obtained by melting one ton (2000 lbs) of ice in 24 hours.

$Q_E = h_1 - h_4$ Btu/lb; $Q_C = h_2 - h_3$ Btu/lb

$$\text{Refrigeration} = w \frac{(h_1 - h_4)}{200} \text{ tons}$$

$$\text{Refrigeration} = \frac{V_1}{v_1} \times \frac{(h_1 - h_4)}{200} = \frac{N \times PD \times \eta_v}{v_1} \left[\frac{(h_1 - h_4)}{200}\right]$$

where N is compressor speed in rpm and PD is the piston displacement in ft.^{3}

$$COP = \frac{200 \times \text{tons of refrig.}}{42.42 \text{ hp}}$$

$$\frac{HP}{\text{ton of refrig.}} = \frac{4.71}{COP}$$

Example

What HP is required to drive a refrigeration system with a Coefficient of Performance = 5 to remove the heat equivalent of 100 tons of ice per day?

$$COP = \frac{\text{Refrigeration}}{\text{work}} \quad \text{or} \quad W = \frac{\text{Refrigeration}}{COP}$$

$$\text{Refrigeration} = 100 \frac{\text{tons}}{\text{day}} \times 2000 \frac{\text{lb}}{\text{ton}} \times \frac{1 \text{ day}}{24 \text{ hrs.}} \times \frac{1 \text{ hr}}{60 \text{ min}} \times 144 \text{ Btu/lb.}$$

where 144 Btu/lb is the latent heat of fusion for ice.

Refrigeration = 20,000 Btu/min.

$$W = \frac{\text{Refrig.}}{COP} \quad \frac{20,000 \frac{BTU}{min}}{(42.42 \text{ Btu/hp} - \text{min}) \times 5}$$

$$W = \frac{472}{5} = 94.4 \text{ HP}$$

Example

An ammonia refrigeration plant is to operate between a saturated liquid at 120 psia at the condenser output and a saturated vapor at the evaporator output of 15 psia (Fig. 8.43). Thirty tons of refrigeration are required. On the basis of an ideal cycle compute a) the COP, (b) the work of compression, (c) the refrigeration, (d) flow rate of refrigerant required in lb/min, and (e) the hp/ton of refrigeration.

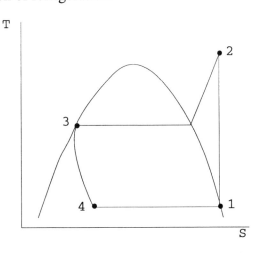

Figure 8.43

From <u>Properties of Saturated Ammonia,</u> pp 4-44, 4-45 and 4-46, of Marks Handbook, 6th Edition.

$P_1 = 15$ psia $s_1 = 1.3940$ $h_1 = 602.3$ Btu/lb

$P_2 = 120$ psia $s_2 = s_1$ $h_2 = 733.5$ Btu/lb

$P_3 = 120$ psia $h_3 = 116.0$ Btu/lb

$P_4 = 15$ psia $h_4 = h_3$

at 15 psia $= s_g = 1.3940$ and $h_1 = 602.3$ from Table 32

at 120 psia $= s_2 = 1.3940$, $h_2 = 733.5$ from Table 33.

at 120 psia, sat, $h_3 = h_f = 116.0$ Btu/lb from Table 32.

$h_4 = h_3$, end point of an apparent h = c process.

a) $COP = \dfrac{\text{Refrigeration}}{\text{Work}} = \dfrac{h_1 - h_4}{h_2 - h_1} = \dfrac{602.3 - 116.0}{733.5 - 602.3}$

$$COP = \frac{486.3}{131.2} = \underline{3.7}$$

b) Work of compression = $h_2 - h_1$ = (733.5 - 602.3) = <u>131.2 Btu/lb</u>

c) Refrigeration = $h_1 - h_4$ = (602.3 - 116.0) = <u>486.3 Btu/lb</u>

d) To find the flow rate.

$$\text{Rate of refrigeration} = \frac{200 \text{ Btu/min.}}{\text{ton}} \times 30 \text{ tons} = 6000 \text{ Btu/min}$$

$$\frac{6000 \text{ Btu/min}}{(h_1 - h_4)} = \frac{6000 \text{ Btu/min}}{486.3 \text{ Btu/lb}} = 12.3 \text{ lb/min}$$

$$\therefore \frac{hp}{ton} = \frac{4.71}{COP} = \frac{4.71}{3.7} = \underline{1.273 \text{ hp/ton}}$$

HEAT TRANSFER

Conduction

Of the three methods of heat transfer—conduction, convection, and radiation—conduction is by far the most common.

The basic heat transfer equation for conduction is:

$$Q = \frac{kA(T_1 - T_2)}{X} \quad \frac{Btu}{hr - ft^2 - °F}$$

where Q is the amount of heat transferred in Btu/hr, A is the area in ft^2 through which the heat flows, $(T_1 - T_2)$ is the temperature differential in °F causing the heat flow, X is the distance in feet for the length of the path of heat flow, measured in feet, and k is the thermal conductivity of the material through which the heat is flowing. The units for k are Btu-ft/hr-ft^2 °F. k is a physical property and is different for each material.

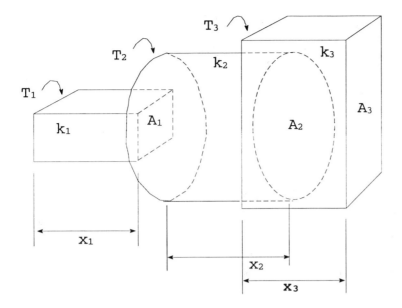

Figure 8.44

Given a composite as shown in Fig. 8.44.

$T_1 > T_4$. Dimensions for A, X and values for k are shown.

$$Q = \frac{T_1 - T_4}{\dfrac{X_1}{k_1 A_1} + \dfrac{X_2}{k_2 A_2} + \dfrac{X_3}{k_3 A_3}} \text{ Btu/hr.}$$

Let $\dfrac{X}{k} = \dfrac{1}{U}$ where U = overall coefficient heat transfer in

Btu/hr. - ft² - °F.

Then, $Q = UA (T_1 - T_4)$ Btu/hr.

In heat transfer between a fluid and a solid (Fig. 8.45) there is always a thin film that clings to the surface of the solid and acts as an additional resistance to heat flow. The film coefficient, h, is expressed in Btu/hr-ft²-°F and is independent of the film thickness.

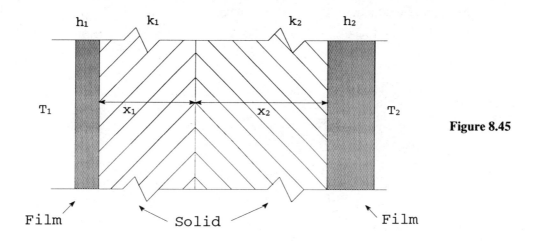

Figure 8.45

$$Q = \frac{A(T_1 - T_2)}{\frac{1}{h_1} + \frac{X_1}{k_1} + \frac{X_2}{k_2} + \frac{1}{h_3}} \text{ Btu/hr}$$

Where we have resistance due to a film, the value of $\frac{1}{U}$ includes the value for $1/h$.

or $\frac{1}{U} = \Sigma \frac{1}{h} + \Sigma \frac{X}{k} \quad \frac{\text{ft.}^2 - \text{hr. °F}}{\text{Btu}}$

We now write $Q = UA\Delta T_m$ for the general case.

ΔT_m is also called LMTD, or Log Mean Temperature Difference. When we do not have a linear variation in temperature we use ΔT_m. Examples are shown in Fig. 8.46.

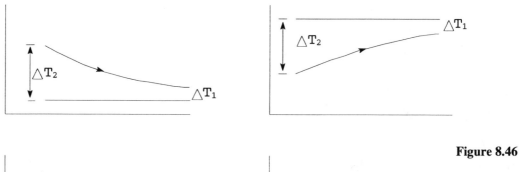

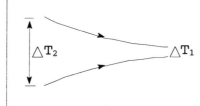
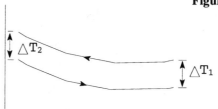

Figure 8.46

All plots are temperature vs. distance as in a heat exchanger.

$$T_m = \frac{\Delta T_2 - \Delta T_1}{\ln\left(\frac{\Delta T_2}{\Delta T_1}\right)} = \frac{\Delta T_2 - \Delta T_1}{2.3 \log_{10}\left(\frac{\Delta T_2}{\Delta T_1}\right)}$$

or $T_m = \dfrac{\text{larger } \Delta T - \text{smaller } \Delta T}{\ln \dfrac{(\text{larger } \Delta T)}{(\text{smaller } \Delta T)}}$

We may also use log mean areas.

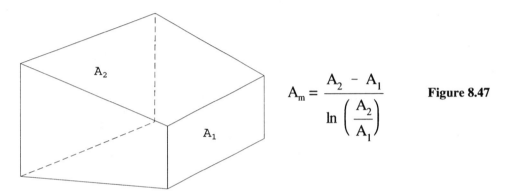

$$A_m = \frac{A_2 - A_1}{\ln\left(\frac{A_2}{A_1}\right)}$$

Figure 8.47

In a pipe, for example, the area across which the heat flows varies with the distance from the centerline of the pipe. If the overall resistance to heat flow is

$$\Sigma R \quad \text{and} \quad \frac{1}{U} = \Sigma R$$

Then $R = \dfrac{\ln D_o/D_i}{2 \pi Z k}$ where D_o and D_i are the outside and inside

diameters respectively and Z is the length of the pipe. k is the thermal conductivity of the material.

The most common case for heat flow through the walls of a pipe is when the pipe is lagged, or insulated.

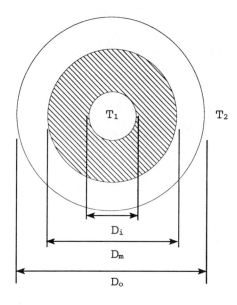

Figure 8.48

The subscript i is inside. The subscript o is outside.
Subscript m is the common interface between pipe and insulation.

$T_1 > T_2$

Subscript l is lagging.
Z is length of pipe and lagging.

$$Q = \frac{T_1 - T_2}{\dfrac{1}{A_i h_i} + \dfrac{\ln D_m/D_i}{2 \pi Z k} + \dfrac{\ln (S_o/S_m)}{2 Z \pi k_l} + \dfrac{1}{A_o h_o}}$$

Convection

The Nusselt equation, used for turbulent flow inside a pipe is

$N = 0.023\, R^{0.8}\, P^{0.4}$ where N = Nusselt number, R = Reynolds number,

and P = Prandtl number, all at the bulk temperature. Rewritten

$$\frac{h_i D_i}{k} = 0.023 \left(\frac{d_i V \rho}{\mu}\right)^{0.8} \left(\frac{C_p \mu}{k}\right)^{0.4}$$

Radiation

The Stefan - Boltzmann equation is used for radiation. The amount of radiation from a black body is

$$Q_R = 0.173 \, A \left[\frac{T}{100}\right]^4$$

where T is absolute temperature °R

$A = ft^2$

$Q_R = \dfrac{BTU}{hr-ft^2}$

Generally, radiation moves from a hot body to a cold body for heat exchange.

$$R = 0.173 \, Fe \, Fa \, A \left[\left(\frac{T_1}{100}\right)^4 - \left(\frac{T_2}{100}\right)^4\right]$$

where Fa is an angle factor and Fe is an emissivity factor. Values for these are given in handbooks.

A special case is when one body, 1, is small and inside a large enclosure 2,

$$\text{Then } Q_R = 0.173 \, \epsilon_1 \, A1 \left[\left(\frac{T_1}{100}\right)^4 - \left(\frac{T_2}{100}\right)^4\right]$$

where ϵ is the emissivity of the body.

Example

A counterflow economizer with 22,500 ft² of surface is used with a boiler that supplies steam at 1200 psia and 900°F (Fig. 8.49). Feedwater enters the economizer at 150,000 lb/hr and at a temperature of 350°F and is heated within 30 degrees of boiler water temperature. The flue gas temperature drops from 720°F to 400°F. Find the mean temperature difference and the overall heat transfer coefficient.

From Table 2 of steam tables by Keenan & Keyes

T_{sat} at 1200 psia = 567.22°F

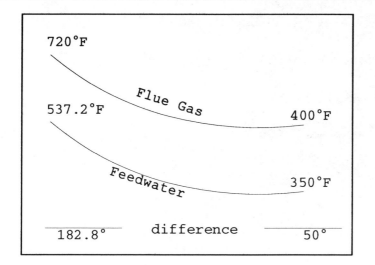

Figure 8.49

$$\Delta T_m = \frac{182.8 - 50}{\ln \frac{182.8}{50}} = \frac{132.8}{\ln 3.66} = \frac{132.8}{1.3} = 102°F$$

$$Q = U\,A\Delta T_m \text{ or } U = \frac{Q}{A\,\Delta T_m}$$

Since the incoming water is heated to within 30 degress of boiler water temperature, the ΔT is 537.2 - 350 = 187°F.

$$Q = wc\Delta T = 150,000\,\frac{lb}{hr} \times 1 \times 187°F = 28,050,000 \text{ Btu/hr}$$

$$\therefore U = \frac{28,050,000 \text{ Btu}}{22,500 \times 102 \text{ hr. ft.}^2 \text{ °F}} = 12.22\,\frac{\text{Btu}}{\text{hr - ft}^2\text{ °F}}$$

SOLVED PROBLEMS

1. A partly filled barrel contains 300 lb of water and 100 lb of ice at 32°F. How many lb of steam at 212°F must be run into the barrel to bring the contents to 80°F?

   Approach: Set up an energy balance. The heat energy given up by the steam must equal the energy required to melt the ice and warm the water.

   W_s = weight of steam ; h_{fg} = latent heat = 970 Btu/lb (h_{fg} at 212°F)

   h_f = heat of fusion = 144 B/lb

$W_s h_{fg} + W_s(212 - 80) = W_i (144) + W_i (80 - 32) + W_w (80 - 32)$

$970 W_s + 132 W_s = 100 (144) + 100 (80 - 32) + 300 (80 - 32)$

$970 W_s + 132 W_s = 14400 + 4800 + 14400$

$W_s = 30.5$ lb of steam   Answer

2. Butane, $2C_4H_{10}$, is burned with no excess air. Write the combustion equation and calculate the gravimetric analysis of the products of combustion.

   Approach: Write the combustion equation. From this set up a weight balance. Using the weights of the products of combustion, calculate the gravimetric analysis.

   $2 C_4H_{10} + 13 O_2 + 3.78 (13) N_2 \rightarrow 8CO_2 + 10 H_2O + 3.76 (13) N_2$

   by weight $2 (58) + 13 (32) + 48.88 (28) \rightarrow 8 (44) + 10 (18) + 48.88 (28)$

   $116 + 416 + 1375.6 \rightarrow 352 + 180 + 1368.6$

   $G\ CO_2 = \dfrac{352 \times 100}{352 + 180 + 1368.6} = \underline{18.5\%}$

   $G\ H_2O = \dfrac{180 \times 100}{1900.6} = \underline{9.47\%}$

   $G\ N_2 = \dfrac{1368.6 \times 100}{1900.6} = \underline{72\%}$

3. Calculate the area of a nozzle at the throat to permit passage of 300 lb of steam per hr when steam is supplied at 250 psia and 500°F. Coefficient of Discharge for the nozzle is 0.9. Nozzle outlet is at 2 psia.

   Approach: Use the Superheated Steam equation for weight flow.

   $v_1 = 2.1515$ ft³/lb from steam tables

   $Pc = .546\ P_1 = .546(250) = 136.5$ psia

   $w = .316\ A_t \sqrt{P_1/V_1}$ (superheated steam eq.)

   Actual steam flow = $\dfrac{300.0\ \text{lb/hr}}{3600\ \text{sec/hr}} = .0833$ lb/sec

   Ideal steam flow = $.0833/.90 = .0926$ lb/sec

$$A_t = \frac{w}{.316\sqrt{P_1/V_1}} = \frac{.0926}{.316\sqrt{250/2.1515}} = .0271 \text{ in}^2 = .0001882 \text{ ft}^2 \qquad \underline{\text{Answer}}$$

4. The inlet temperature to a gas turbine is 1500°F. The turbine has an expansion pressure ratio of 6. The gasses leave the turbine at 955°F. $C_p = .26$, $C_v = .20$. Calculate the adiabatic turbine efficiency.

Approach

Sketch the cycle on an h-s diagram. Using the basic relationship of $H = w C_p \Delta T$ and $T_2 = T_1 \left(\frac{P_2}{P_1}\right)^{\frac{k-1}{k}}$, calculate the enthalpies necessary to find the efficiency or ratio of output/input.

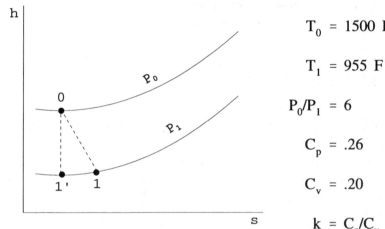

$T_0 = 1500$ F **Figure 8.50**

$T_1 = 955$ F

$P_0/P_1 = 6$

$C_p = .26$

$C_v = .20$

$k = C_p/C_v = 1.3$

$$\eta = \frac{\text{Actual}}{\text{Ideal}} = \frac{h_o - h_1}{h_o - h_1'} \text{ where } h_o - h_1 = w \, cp \, (T_0 - T_1)$$

$h_o - h_1 = 1 \times .26 \, (1500 - 955) = 141.6 \text{ B/lb}$

$h_o - h_1' = w \, C_p \, (T_0 - T_1')$ where $T_1' = T_0 \, (P_1/P_0)^{\frac{k-1}{k}} = 1295°R = 835°F$

$h_o - h_1' = 1 \times .26 \, (1500 - 835) = 173.0 \text{ B/lb}$

$$\eta = \frac{141.6}{173.0} = 82\%$$

5. What characteristics are necessary in a refrigerant to produce a high actual coefficient of performance? Assuming an actual coefficient of 3, calculate the indicated HP to produce one ton of refrigerating effect. With a temperature of 15°F in the cooler and condensing water available at 50°F, calculate the ideal coefficient of performance.

Approach

This problem is one strictly of knowing where to look in a book on thermodynamics or a handbook.

1) High latent heat of vaporization at low pressures
2) Low heat content at low pressures
3) Low specific volume at low volumes

$$COP = 4.71/\text{hp/ton or hp/ton} = \frac{4.71}{3} = \underline{1.57 \text{ hp/ton}}$$

$$COP = \frac{T_2}{T_1 - T_2} = \frac{460 + 15}{50 - 15} = \underline{13.57}$$

6. A hot water heater consists of a 20 ft length of copper pipe of ½ in. average diameter and 1/16 in. thickness. The outer surface is maintained at 212°F. What is the capacity of the coil in gallons of water per minute if water is fed into the coil at 40°F and is expected to emerge heated to 150°F. The conductivity of copper may be taken as 2100 Btu/ft³/deg/hr/in. of thickness.

Approach

Here we must recognize that we do need to account for the film coefficient. We must also use the LMTD in finding the heat transferred to the water. We then equate the heat transferred through the pipe and film to the heat gained in heating the water to 150°F.

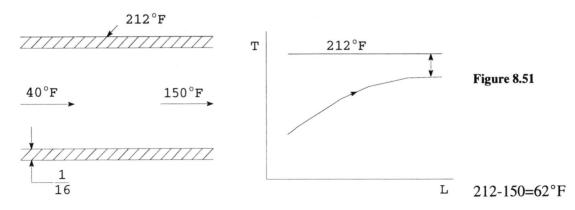

Figure 8.51

212-150=62°F

Find Q = UA Δ Tm and equate to Q = w cΔT

A = π d l = π(½/12) (20) = 2.62 ft²

From McAdams "Heat Transfer" h = 1000 Btu/ft² - °F - hr.

$$U = \frac{1}{\frac{1}{h}\text{ (water)} + \frac{x}{k}\text{ (copper)}} = \frac{1}{\frac{1}{1000} + \frac{1/16}{2100}} = 971 \text{ Btu/ft}^2\text{-F-hr}$$

$$\Delta Tm = \frac{172 - 62}{\ln 172/62} = 108°$$

Q = 971 x 2.62 x 108 = w x (150 - 40)

$$w = 2490 \text{ lb/hr} = \frac{2498 \text{ lb/hr}}{500 \text{lb/hr/gpm}} = \underline{5.0 \text{ gpm}}$$

7. A steam turbine exhausts at 1 psia into a condenser. 20,000 lb of steam is delivered per hour at 200 psia and 430°F. The turbine delivers 1000 HP. Neglecting heat loss, calculate the enthalpy of the steam entering the condenser. Is the steam wet, dry - saturated, or superheated?

Approach

This is simply a problem in <u>carefully</u> calculating quality. Remember, W = ΔH.

h_1 = 1228.9 Btu/lb (h at 200 psi and 430°F)

(hp) (Btu hp-hr)

$$W = \frac{1000 \times 2545}{20,000 \text{ lb/hr}} = 127.25 \text{ Btu/lb}$$

$h_2 = h_1 - W$ = 1228.9 - 127.25 = 1101.6 Btu/lb

$$X = \frac{1101.6 - 69.76}{1036.3} = 99.6\% \text{ where } (h_{fg} = 69 \text{ and hfg} = 1036.3)$$

Therefore, the steam is wet. <u>Answer</u>

8. An ideal Rankine cycle plant produces steam at 500 psia and 800°F. The steam drives an ideal turbine and is exhausted to a condenser at 2 in. of mercury absolute. Neglect the pump work and compute the efficiency of the cycle.

Approach

Sketch the process on the T-S plane. Find the enthalpy at each of the state points. Efficiency can then be calculated from the ratio of output/input.

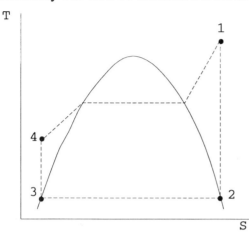

$P_1 = 500$ psia

$P_2 = 2$ in. Hg

$T_1 = 800°F$

Figure 8.52

$h_1 = 1412.1$ Btu/lb from steam tables

$h_2 = 964$ Btu/lb from Mollier Chart

$h_3 = h_4 = 69.1$ Btu/lb from Table 2 of Steam Tables

$$\eta = \frac{\text{output}}{\text{input}} = \frac{h_1 - h_2}{h_1 - h_4} = \frac{488}{1343} = \underline{36.3\%}$$

9. Atmospheric pressure air inlet to a cooling unit is at 80°F and has a specific humidity of 0.009 pounds of water vapor per pound of dry air. Calculate the relative humidity, vapor density, air density and dew point.

Approach

Here we must realize that we have air composed of dry air plus water vapor. The water vapor is also steam. With this in mind we can use our basic gas relationships and Steam Table data to calculate the required data.

$$\frac{w_1 R_1}{w_1 R_1 + w_2 P_2} = \frac{P_1}{P_m} \quad \text{where} \quad \begin{array}{l} \text{1 is water vapor and} \\ \text{2 is dry air} \end{array}$$

Since $\gamma = .009$ lb/lb, $w_1 = .009$ lb, $w_2 = 1.000$ lb

$R_1 = 1544/M_1 = 1544/18 = 85.8$

$R_2 = 53.3$

$$P_1 = \frac{.009 \ (85.8)}{.009 \ (86.2) + 1.00 \ (53.3)} \times 14.7 = .21 \text{ psi}$$

From Table 1 at 80°F $P = .5069$ psia

$$\emptyset = \frac{P_1}{P} = \frac{.21}{.5069} = 41.4\%$$

$$\rho \text{ vapor} = \frac{Pv}{RT} = \frac{.21 \times 144}{85.8 \times 540} = \underline{.00065 \text{ lb/ft}^3}$$

$$\rho \text{ air} = \frac{Pa}{RT} = \frac{(14.7 - .22)(144)}{53.3 \times 540} = .0725 \text{ lb/ft}^3$$

$T_{dp} = T_{sat}$ for .21 psia = $\underline{55.8°F}$

Steam Turbines and Nozzles

We will begin with two statements with regard to flow through a nozzle.

(1) If the exit pressure for a nozzle is less than some value know as the critical pressure, P_c, the flow through the nozzle will be less than the maximum attainable value.

(2) The flow through the nozzle is maximum for exit pressures equal to or less than P_o and P_c. P_o is the stagnation pressure.

For adiabatic flow we can write $KE_1 + h_1 = KE_2 + h_2$

where $KE = \dfrac{\overline{V}^2}{2g \ J}$ where $\overline{V}$ is velocity

Substituting values and solving for $\overline{V}_2$, the exit velocity, yields:

$$\overline{V}_2 = \sqrt{2gJ(h_1 - h_2) + \overline{V}_1^2} \text{ ft/sec}$$

Considering weight flow, the same weight of fluid enters the nozzle as leaves the nozzle.

$$w = \frac{A\overline{V}}{v} \text{ where v is specific volume.}$$

$$w = \frac{A_1 \overline{V}_1}{v_1} = \frac{A_2 \overline{V}_2}{v_2} \text{ or } \overline{V}_1 = \frac{A_2}{A_1} \frac{v_1}{v_2} \overline{V}_2$$

We can now write

$$\overline{V}_2 = \sqrt{2gJ(h_1 - h_2) + \left[\frac{A_2}{A_1} \frac{v_1}{v_2} \overline{V}_2\right]^2} \text{ ft/sec}$$

or $\overline{V}_2 = \sqrt{\dfrac{2gJ(h_1 - h_2)}{1 - \left(\dfrac{A_2 v_1}{A_1 v_2}\right)^2}}$ ft/sec

In most cases $\overline{V}_1$ is $<< \overline{V}_2$ and can be neglected. Then the equation can be further reduced.

$$\overline{V}_2 = \sqrt{2gJ(h_1 - h_2)} = 223.7\sqrt{\Delta h} \text{ ft/sec}$$

$$\overline{V}_2 = 223.7\sqrt{C_p \Delta T} \text{ ft/sec}$$

Flow through a nozzle can generally be considered as reversible, adiabatic, or isentropic. Then

$$\overline{V}_2 = 223.7\sqrt{C_p T_1 \left[1 - \frac{T_2}{T_1}\right]} \text{ ft/sec}$$

$$\overline{V}_2 = 223.7\sqrt{C_p T_1 \left[1 - \left(\frac{P_2}{P_1}\right)^{\frac{k-1}{k}}\right]} \text{ ft/sec}$$

Recalling our P, V, T relationships

$$v_2 = v_1 \left(\frac{P_1}{P_2}\right)^{1/k} = \left(\frac{P_1}{P_2}\right)^{1/k} \times \frac{R T_1}{P_1}$$

also, $w = \dfrac{A \overline{V}}{v}$, we can write

$$w = \frac{223.7 \, A_2}{\left(\dfrac{P_1}{P_2}\right)^{1/k} \dfrac{RT_1}{P_1}} \sqrt{C_p T_1 \left[1 - \left(\frac{P_2}{P_1}\right)^{\frac{k-1}{k}}\right]} \; \frac{lb}{sec}$$

which reduces to

$$w = \frac{223.7 \, A_2 \, P_1}{R} \sqrt{\frac{C_p}{T_1}\left[\left(\frac{P_2}{P_1}\right)^{2/k} - \left(\frac{P_2}{P_1}\right)^{\frac{k+1}{k}}\right]}$$

w will be maximum when the quantity in [ ] under the radical is a maximum.

Differentiating this expression and setting this to zero (an exercise in Mathematics) leads us to

$$P_2 = P_1 \left(\frac{2}{k+1}\right)^{\frac{k}{k-1}}$$

Remember, there is a specific point where $P_2 = P_c$

When $p_2 < p_c$, A_2 becomes the throat area. Taking k = 1.4 we obtain:

$$w = \frac{3.88 \, A_t \, p_1}{(RT_1)^{1/2}} \; lb/sec$$

For air, $p_c = .53 \, p_1$

$$w = \frac{0.532 \, A_t \, P_1}{(T_1)^{1/2}} \; lb/sec$$

For superheated steam

$$p_c = .546 \, p_1 \quad w = .316 \, A_t \sqrt{p_1/v_1} \text{ lb/sec}$$

For saturated steam

$$p_c = .577 \, p_1 \quad w = .304 \, A_t \sqrt{p_1/v_1} \text{ lb/sec}$$

For $p_2 > p_c$, $\overline{V} = 233.7 \sqrt{\Delta h}$

The coefficient of discharge for the nozzle will be C_d = actual weight flow/ideal weight flow.

The power in the jet issuing from a nozzle is

$$HP = \frac{w \, \overline{V}_a^2}{2g} \quad \frac{\text{ft lb}}{\text{sec}} \times \frac{1 \text{ HP}}{500 \text{ ft lb/sec}}$$

where $\overline{V}_a = c_d \, \overline{V}_{ideal}$

Example

Steam flows through the nozzle of a steam turbine at a rate of 3100 lb/hr. The steam is at 260 psia and 520°F. The exhaust pressure is 2 psia. C_d may be taken as .89. What is the throat area of the nozzle?

At 260 psia and 520°F the steam is superheated. From Table 3 of Keenan & Keyes v = 2.118 cu ft/lb.

For superheated steam $p_c = .546 \, p_1$ $p_c = .546 \times 260 = 141.96$ psia.

$w = .316 \, A_t \sqrt{p_1/v_1}$

$$w_{act} = \frac{3100}{3600} = .861 \text{ lb/sec} \quad w_{ideal} = \frac{.861}{.89} = .968 \text{ lb/sec}$$

$$A_t = \frac{.968}{(.316)/(260/2.118)^{½}} = \underline{.276 \text{ sq. in.}^3}$$

9
Material Science

TABLE OF CONTENTS

MATERIAL SCIENCE AND STRUCTURE OF MATTER 9-1

    Crystal structures 9-1

    Space lattices 9-2

    Phase diagrams 9-5

    Heat treating 9-6

    Corrosion 9-13

    Materials testing 9-13

    Impact tests 9-15

    Stress–strain relationships 9-18

    Physical properties 9-18

    Nonmetallic materials 9-23

PERIPHERAL SCIENCES 9-26

    Rectilinear motion 9-26

    Harmonic motion 9-31

    Wave motion 9-39

    Sound waves 9-47

    Intensity of sound 9-47

    Light and optics 9-50

    Spherical mirrors 9-51

    Refraction 9-54

    Prisms 9-56

    Lenses 9-57

    Keppler's Laws 9-64

MATERIAL SCIENCE AND STRUCTURE OF MATTER

All metallic structures are crystalline. Therefore, the characteristics of these materials are determined either by the nature of the crystals of which they are composed or by the manner in which these crystals are aggregated into larger masses. A metal composed of small grains will generally be harder and stronger than a metal composed of large grains. The general characteristics of a metal composed of only one type of crystal will differ from those of a metal composed of two or more different types of crystals.

Of the forty plus metals used commercially, iron, lead, copper, and tin have been used for thousands of years in many forms. Magnesium, zinc, nickel, and aluminum came into commercial use in the latter part of the 19th century. These eight metals are known as the Engineering Metals. Modern metallurgy deals with many new, specialized, sophisticated alloys.

These metals and their alloys are classified as ferrous if their properties are basically those of the metal iron an nonferrous if their properties are basically those of any metals other than iron.

Antimony, beryllium, cadmium, chromium, cobalt, columbium, manganese, mercury, molybdenum, titanium, tungsten, vanadium, and zirconium are known as the auxiliary metals. These metals are important in the metallurgical and chemical industries. Any of these metals, the lack of which would interfere with the economy, would be called a strategic metal as well.

The noble or precious metals include gold, silver, and platinum as well as iridium, osmium, palladium, rhodium, and ruthenium. Certain of these such as gold, silver, and platinum have monetary value in the international market.

The characteristics of metals are important in engineering. Some metals can be hardened and strengthened by deformation. Some have an ability to conduct electricity better than others. The properties of other metals can be improved by heat treatment. Most of all, metals can be formed into usable shapes by casting, forging, machining, etching, etc.

These properties are all related to the atomic structure of the metals—how the atoms form the crystals and crystal aggregates previously mentioned.

The crystal structures can be represented by fourteen point-lattices. Most of the metallic bond crystals are found in three of these forms. They are the face centered cubic, body centered cubic, and close packed hexagonal. These and the other types are shown in Fig. 9.1.

Note: Atom positions are represented by points.

Figure 9.1 Simple Metal Space Lattices

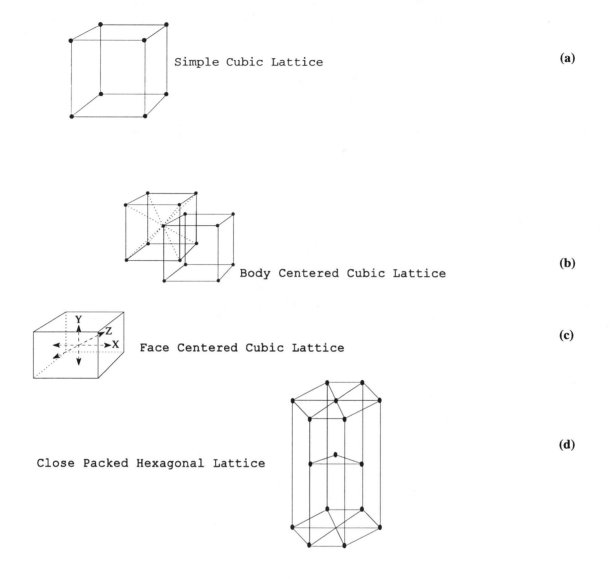

These space lattices can be determined by x-ray diffraction methods.

If two simple cubic lattices are interlocked so that a corner of one falls exactly in the center of another as shown in Fig. 9.1b, we have a body centered lattice. At room temperature we find many strong common metals in this lattice. Some of these metals are chromium, iron, molybdenum, tantalum, tungsten, and vanadium.

If simple cubic lattices are interlocked so that the corners of one lattice are located at the centers of the faces of another cubic such as shown in Fig. 9.1c, we have a face centered lattice. Ductile metals at room temperature crystalize in this lattice. Some of these metals are aluminum, copper, gold, iridium, lead, nickel, palladium, platinum, rhodium, and silver.

The close packed hexagonal arrangement is shown in Fig. 9.1d. At room temperature we will find beryllium, cadmium, magnesium, titanium, and zinc.

Because of the regular spacing of the atoms in a space lattice, it is possible to pass equidistant parallel planes through a lattice family. On any plane the atoms will be regularly, more or less densely, distributed. These planes are called the crystallographic planes. They are usually referred to by indices (Miller indices) secured by taking the reciprocals of their intercepts upon certain axes of symmetry in the lattice. Examples of close packed planes for the three simple metal space lattices are shown in Fig. 9.2.

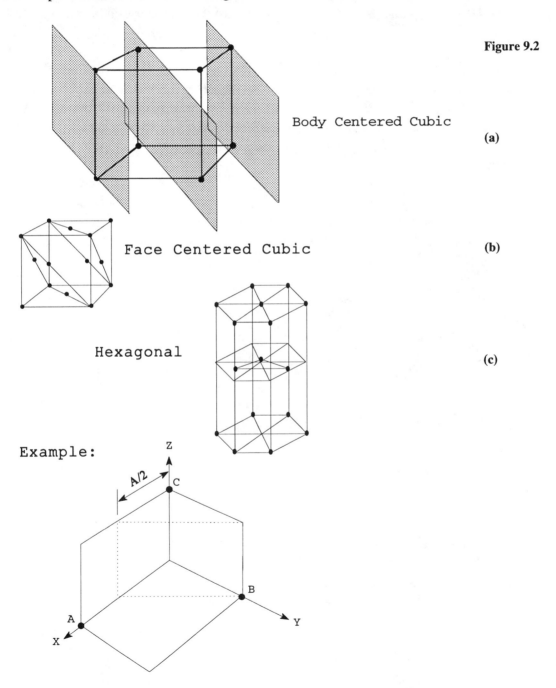

Figure 9.2

Body Centered Cubic (a)

Face Centered Cubic (b)

Hexagonal (c)

Example:

Unit cell dimensions are a, b, and c
The intercepts are a/2, b, ∞
The reciprocals are 2, 1, 0
The Miller indices of this plane are written (210).

Under the action of a sufficiently large stress, blocks of the crystalline materials slide over adjacent blocks along planes already described. This phenomenon is known as slip. Deformation by slip is apparently independent to stress normal to the slip plane; it proceeds only in certain directions; it proceeds only along certain planes; it is self stopping; it results in a definite work hardening. This work hardening also affects the inactive planes; the whole crystal is hardened and strengthened.

Under atmospheric pressure, a pure metal may exist in either the gaseous, liquid, or solid state, depending on the temperature. In changing from one state to another there will be a discontinuous change in energy called the latent heat of transformation. If the transformation is between the solid and liquid states the latent heat of transformation is specifically called the latent heat of fusion. The term latent heat of vaporization is used when the transformation is between the liquid and vapor states.

Some knowledge of these energies and of the change in the heat content of the metal when being heated or cooled may be found from a time history of the temperature as the metal passes from one state to another. [See Fig. 9.3. In the first part of the curve (a) the liquid solution cools uniformly. At (b) there is a change in slope where both solid and liquid coexist as the solution cools further. There is another change in slope for (c) where the solid cools uniformly. Different proportions of an alloy would give similar curves differing only in the temperature where the slopes change.]

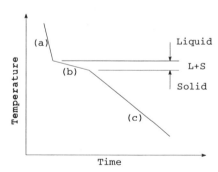

Figure 9.3

A binary constitutional diagram for a two-metal alloy is shown in Fig. 9.4. This diagram is made by plotting the freezing points of the two metals (the end point) and the two transformation points (freezing and melting points) for each of the alloys under consideration.

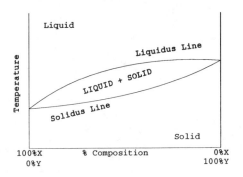

Figure 9.4

Above the liquidus line, the upper line, only one liquid solution exists. Below the lower or solidus line, only one solid solution exists.

Two statements or rules will now be presented. These refer to Fig. 9.5. The composition rule states that the makeup or composition of the solid and liquid in equilibrium in any two-phase field at any temperature can be obtained by drawing a constant temperature line at that temperature and determining its intercepts with the boundaries of this field (solidus and liquidus lines).

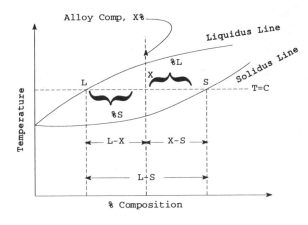

Figure 9.5

The lever rule is used to determine the relative amounts of the alloy that are liquid and solid at a given temperature.

$$\% \text{ liquid} = \frac{\text{distance X-S}}{\text{distance L-S}} \times 100\%$$

$$\% \text{ solid} = \frac{\text{distance L-X}}{\text{distance L-S}} \times 100\%$$

The elements of most alloys are not completely miscible in all phases and states. Some elements may be completely soluble in the liquid state but only partially soluble in the solid state. As a matter of fact, several solid or semisolid phases may occur. These would depend on the composition of the alloy. A simplified drawing of this is shown in Fig. 9.6.

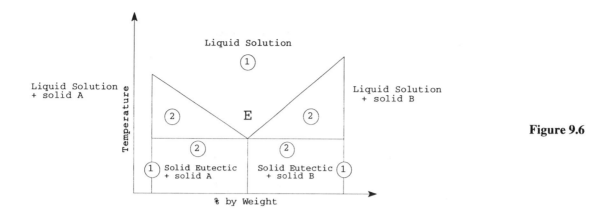

Figure 9.6

At point E, called the eutectic point, the constituents of the alloy are perfectly miscible. In this diagram the eutectic is considered as a separate constituent and treated as a phase strictly as a matter of convenience. The 1 and 2 indicate single and two-phase fields. Note that proceeding across the diagram at a constant temperature the sequence of the number of phases is 1:2:1. The colons indicate a line on the diagram and the numbers indicate the number of stable phases in each of the fields of the diagram. This is known as the 1:2:1 rule.

Heat treatment is used to remove or to balance internal stresses in a metal or alloy as well as to produce certain characteristics which cannot be obtained conveniently by other means. Chemical, mechanical, or physical characteristics may be affected by the heat treatment process. Energy is supplied to the metal and a reaction tends to occur. The higher the temperature, the greater the reaction, the faster the rate at which it will occur. This is true only if the temperature at which the reaction would occur under equilibrium conditions is exceeded. Other reactions or changes may occur on cooling such as recrystallization.

There are many different types of heat treating. The following is a description of some of the more common processes.

HEAT TREATING

Annealing

Heating to a temperature just above the critical point then cooling slowly through the critical range. This is used to relieve internal stresses and to soften the metal.

Case Hardening

Carburizing a low-carbon steel at an elevated temperature then quenching. This is used to produce an item having a soft ductile interior and a very hard surface.

Carburizing

Heating for a time of up to 24 hours at about 1650°F while in contact with carbonaceous material then rapid cooling. Sometimes called cementation.

Cyaniding

Heating to 1700°F in a cyanide-rich atmosphere or in a molten cyanide bath. This produces a superficial hard wear-resisting surface.

Flame Hardening

Local surface heating of steel to above the critical temperature then quenching to produce a hardened layer. Hardness depth can be controlled from 0.05 inches to 0.25 inches with minimum distortion.

Induction Hardening

Using an induction furnace or high-frequency current to heat the metal surface before quenching. This process is generally reserved for heating parts having a uniform cross-section.

Nitriding

Introduction of nitrogen into the outer surface of steel parts to give an extremely hard wear resisting surface. Often by heating for several days in an ammonia atmosphere followed by slow cooling. Since there is no quenching, the parts should be hardened and tempered before nitriding to provide the shallow, hard wear surface.

Normalizing

Heating the ferrous part to about 100°F above the critical temperature followed by air cooling in still air at normal temperatures. This process is similar to annealing which is generally preferred and is usually used to adjust mechanical properties.

Tempering

A heat treating process in which a metal is reheated after hardening to a point below the critical temperature. This process is also known as drawing, aging, or toughening depending on the type of alloy being treated and the temperature range of treatment. Length of stay at temperature and rate of cooling are not critical. The result of this process is a softer, tougher metal.

Steel

Steel is an iron carbon solid solution containing less than 2% carbon. It is known simply as carbon steel. Other alloying agents such as chromium, nickel, and molybdenum in small amounts change the designation to "low-alloy steel." These low-alloy steels can be heat treated to much higher strength levels than the low-carbon steels. These allows have toughness and resistance to fatigue.

When the low levels of alloy metals are increased an intermediate alloy steel results. Similar results can be obtained when the alloy agents are tungsten or vanadium.

Stainless Steels

These have the highest alloying contents. These may be chromium, nickel, molybdenum, tungsten, vanadium, titanium, copper, or aluminum. Many stainless steels can be age hardened by heat treatment. They become rust resistant as a result of a chromium oxide (10-12%) layer which forms on the surface and blocks contact with the surroundings.

Pig Iron

Iron tapped from the blast furnace and cast into cakes on a large bed of sand. The shape of these cakes and the runners attached to them resembled, in a fashion, a litter of piglets suckling on a sow. The cakes were therefore called pigs and the castings were called pig iron. Today the liquid iron is cast into "pigs" of a standard size and weight but the name remains.

Partially purified pig irons are called cast irons because their properties are such that they cannot be worked satisfactorily and must be cast into their final shape. The carbon content of cast irons varies from 2% to near 7%. The carbon in cast iron is found either as a flake graphite or a hard, brittle compound known as cementite, Fe_3C, or iron carbide, or as combined carbon. The form in which the carbon is found determines the properties and the type of the cast iron.

Gray Cast Iron

This has most of the carbon in the form of graphite and less than 8/10% of the carbon is in a combined form. When gray cast iron is fractured, the break tends to occur through the graphite flakes and the exposed surface appears dull gray in color.

If practically all the carbon is in the combined form as iron carbide, the fracture will be through this material and the exposed surface will have a silvery white luster, giving rise to the term white cast iron.

Ductile and Malleable Iron

Lies between the gray cast iron and steel. Ductile iron properties are developed mainly in the melting process. Malleable cast iron is made by annealing white cast iron.

Allotropic Changes

These are the reversible changes which occur in an iron carbon mixture at the critical points. The composition remains constant but material properties such as atomic structure, electrical resistance, and magnetism change.

Critical Point

Is the temperature at which allotropic changes take place. There are three critical points for pure iron.

       2802°F - liquid iron to delta iron
       2552°F - delta iron to gamma iron
       1670°F - gamma iron to alpha iron

Alpha Iron

Is soft and strongly magnetic. It has a body centered cubic (BCC) structure. It exists in the temperature range of -460° to +1670°F.

Beta Iron

Is a nonmagnetic form of alpha iron which exists between 1420° and 1670°F.

Gamma Iron

Is stable between 1670° and 2550°F. It is nonmagnetic and has a face centered cubic (FCC) structure.

Delta Iron

Exists above 2550°F and is of the body centered cubic structure.

Ferrite

Is a magnetic, solid solution of carbon in alpha iron. Ferrite has the structure of a body centered cubic lattice.

Cementite

Is the hardest and most brittle form of iron. It is the chemical compound Fe_3C.

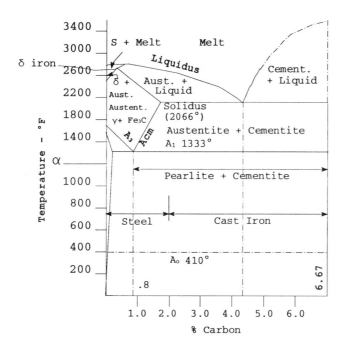

Figure 9.7

Pearlite

Gains its name from its lustrous appearance under the microscope. Pearlite exists in most cases as a lamellar constituent formed of alternate layers of ferrite and cementite.

Austentite

Is a solid solution of gamma iron and iron carbide. It is nonmagnetic.

Martinsite

Is an essential constituent of any hardened steel. It is obtained by rapid quenching carbon steels and is formed by the rapid decomposition of austenite. It is a mixture of very small crystals of alpha iron with carbon in the form of cementite. It is strongly magnetic.

Corrosion

Corrosion of metals is the result of chemical reactions between the metal and its environment. In some cases, such as that of stainless steel, the corrosion may form a surface layer which inhibits any further corrosion. In other cases, such as rust (iron oxide), the part may corrode until it is completely useless in form or load carrying ability. Corrosion is affected by six factors:

1. Acidity
2. Oxidizing characteristics
3. Electrolysis
4. Film formation

5. Agitation or rate of movement
6. Temperature

In engineering applications we find the majority of corrosion takes place either when the metal is exposed to the atmosphere or when it is exposed to an aqueous media. These factors listed above apply to these conditions. This does not mean, however, that they cannot be applied to conditions of chemical corrosion which are taken as special cases.

Acids form hydrogen ions in water solution. The hydrogen ions tend to be replaced by metal ions, freeing hydrogen gas.

Oxidation causes an increase in positive valence. Therefore, any metal whose valence is zero would cause a change to a compound having a positive valence. This acts in the same direction as the acid reaction. The two, acid reaction and oxidation, frequently act together.

Acidity and oxidation set up the tendency for corrosion. The rate of corrosion is generally determined by electrolysis, film formation, agitation, and temperature.

In electrolysis, the positive electrode called the anode will take up or absorb electrons. The metal in it, minus its electrons, will go into solution as a positive ion. At the negative electrode, called the cathode, hydrogen ions take on the electrons and become hydrogen gas. The result on the metal of this electrolysis is a pitting or roughening of the metal surface. Galvanic action, a type of electrolysis, is discussed in detail at a later point.

Film formation has already been mentioned with regard to stainless steel and again as rust. A thick film such as rust can deny free access to the corroding medium such as water. In atmospheric corrosion rust can retain moisture to accelerate the corrosion process. Cathodic films are extremely thin, adherent films of oxides that give protection to the base metal. It must be noted that some of these films resist the passage of ions giving rise to special electrical considerations. An example of this is a silicon rectifier diode.

Agitation will break down layers of concentration between the metal and the corroding medium. This causes the reaction to proceed more rapidly than it would have if the layers of concentration were not disturbed.

Temperature is one measure of heat energy. Adding energy to a system will increase its temperature. The higher the temperature the faster the rate of corrosion. At elevated temperature the corrosion can change from aqueous to chemical action. How this comes about and the effects have to be considered for each case.

In addition to the six factors affecting corrosion, we can categorize corrosion as galvanic corrosion, stress corrosion, and what has been called fretting corrosion.

Galvanic Action

Galvanic action occurs when two different metals having different tendencies to go into solution are in contact with a conducting solution and are connected electrically. The metal having the higher electro-chemical potential will act as the anode. This metal will corrode. The metal having the lower electro-chemical potential will be the cathode. The Galvanic Series is a classification of metals in terms of their anodic-cathodic relationship. Table 9.1 is a brief galvanic series table.

Table 9.1
Anodic or least noble metals (corroded end)
Magnesium
Zinc
Galvanized Steel or WI
Aluminum
Cadmium
Steel
Wrought iron
Cast iron
Lead
Tin
Copper
Nickel
Monel
Stainless steel
Gold
(protected end) Cathodic or most noble metals

To avoid problems with galvanic action certain precautions should be observed if dissimilar metals must be used. Use metals that are close neighbors in the galvanic series, for example, copper and monel rather than aluminum and monel. Use protective coatings or use inert spacers between dissimilar metals. Be careful of the use of plating such as zinc, cadmium, chromium, and nickel on steel for protective purposes. Porosities in the protective coating can cause subsurface corrosion problems that escape early detection. These are but a few of the precautions to be taken. Another obvious one is to avoid corrosive atmospheres. Sometimes it's good practice to provide what is known as a sacrificial metal. Those of you with power boats know the advantages of zinc plates on the keel or zinc fittings on the propeller shaft. The zinc corrodes away leaving the propeller and shaft free of corrosion. Of course, the zinc must be replaced periodically.

Stress Corrosion

Stress corrosion cracking causes failures due to the combined effects of corrosion and static stresses or corrosion and dynamic stresses. Generally the static stress condition caused by an external load causes the failure. Internal stresses caused by heat treatment, cold working, or welding are sometimes the culprit. Stress relief annealing is used to remove these stresses.

Stress corrosion failures are generally of an intercrystalline nature. Intergranular cracks propagate through grain boundaries until failure occurs.

Fretting Corrosion

Fretting corrosion takes place where two or more highly loaded members have a common surface where there is differential movement. This type of corrosion is a combination of chemical corrosion and wear. It is eliminated by proper lubrication and/or reducing or eliminating the differential movement.

Materials Testing

Radiography is a process by which a specimen is penetrated by x-rays or gamma rays and exposed to a photographic film or plate. Where the specimen is deficient, more radiation will pass through, thus producing a darker image on the film, and where the specimen is more dense penetration is less. It is the variation of the amount of image on the photosensitive material that produces useful information about the specimen.

Sound testing is generally acoustic or ultrasonic. The ultrasonic test passes high-frequency sound through a specimen. The sound waves are reflected by the material boundaries and discontinuities within the material. The reflected sound waves are detected by the ultrasonic tester. From the time-distance relationship and the velocity of sound in the material the material thickness depth of defect can be measured.

Acoustic emissions testing is based on the fact that when a material is stressed an active flaw will emit energy in the form of sound waves. A specimen is tested by instrumenting the specimen with sound detectors and stressing the material. Using triangulation the location of an active flaw can be pinpointed.

Magnetic particle testing is a nondestructive method which depends on the fact that a fault in a ferromagnetic material will cause a change in the magnetic reluctance. It is quite reliable and receives considerable use. A direct current is passed around the sample, causing it to become magnetically polarized. The surface of the piece is then covered with fine ferromagnetic particles which will then spread to outline the flaw.

Liquid penetrant testing is used to locate cracks and pores open to the surface. A dye in a penetrating fluid is applied to the test piece. The fluid will be drawn into cracks and pores by capillary action. The piece is examined under ultraviolet light to see the presence or absence of dye and thus flaws in the material.

Spectrographic testing requires a small sample of the test specimen. It is used for determining the position of atoms and the spacing in the space lattices of metal crystals. This method of testing is most often used for metal identification.

Hardness testing is used to measure the ability of a metal surface to resist deformation. This can be correlated with the strength of the metal as well as other properties. It is also used in inspection to verify heat treatment. Common hardness tests are the Brinell, Rockwell, Vickers, Herbert, scleroscope, cloudburst, and monotrone. All are described in detail in standard reference works. In the Rockwell, Brinell, and Vickers hardness testers a diamond cone, a hardened steel ball, or a diamond pyramid are pressed into the test specimen under a definite load and for a specified time. The Brinell hardness number

$$Bhn = \frac{\text{Load in kilograms}}{\text{Area of spherical impression in mm}^2}$$

$$Bhn = \frac{P}{\frac{\Pi D}{2}(D - \sqrt{D^2 - d^2})}$$

where
        P = load in kilograms
        D = diameter of hardened steel ball in mm
        d = diameter of impression in mm

The Vickers hardness number

$$Vhn = 1.854 \frac{W}{D^2}$$

where

    W = load in kilograms

    D = diagonal of diamond pyramid

The Rockwell hardness test is similar to the Brinell test but a special machine is used. Readout is directly on a scale (8 scales available A through H).

Destructive tests are used to measure impact strength and tensile strength. In these tests it is generally impractical to test an engineering model or a finished part so small test specimens are used.

The impact test assumes that resistance to shock depends on the ability to rapidly equalize concentrated stresses. Two common methods of impact testing already mentioned are the Izod and the Charpy methods of testing. In these tests, a pendulum of fixed weight is released from a predetermined height. This determines the energy the pendulum will have when it passes its lowest point. The test specimen is mounted at this point. Energy is absorbed in breaking the test specimen. Any energy remaining in the pendulum system is used in carrying the pendulum past this point. This energy is measurable.

The main differences in the tests lie in the test specimen. The Charpy test uses a simple horizontal beam supported by the ends as is shown in Fig. 9.8. The notch in the specimen is used to concentrate the stress. The Izod test uses a cantilever beam as shown in Fig. 9.9. Here, too, a notch is used to concentrate the stress.

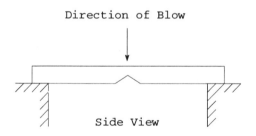

Figure 9.8

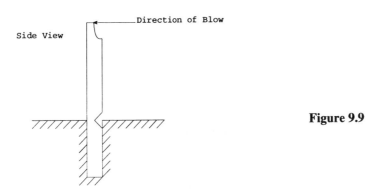

Figure 9.9

Since the purpose of this test is to determine the susceptibility of a material to brittle fracture, the test samples are notched. Brittle fracture occurs under tensile stress when the material has a preexisting flaw or stress concentration. The purpose of the notch is to simulate the flaw.

Tensile impact testing is accomplished by attaching one end of a tensile specimen to the pendulum. The other end is stopped as the pendulum goes through its maximum energy position. The test specimen setup is shown in Fig. 9.10.

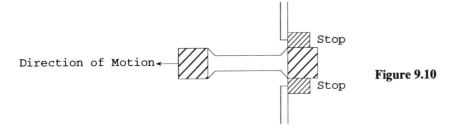

Figure 9.10

Torsion impact testing may be simply accomplished as in Fig. 9.11.

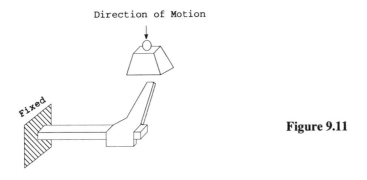

Figure 9.11

The ordinary tensile test is used to secure information on elastic and plastic properties of a material (See Mechanics of Materials notes). A specimen when stressed in tension will elongate. A plot of applied stress (load divided by cross-section area) versus elongation is shown in Fig. 9.12. An elastic material will obey Hooke's Law and the strain will be proportional to the stress. At some point it will be found that if the load is released, the specimen has taken a permanent set. This point is known as the <u>elastic limit</u>. This point can be determined only by successively loading and unloading a specimen, increasing the load slightly each time, until a load is reached where a measurable, permanent set results. This is shown also in Fig. 9.12. If the load is taken up to produce a stress of 82 ksi and then removed, the stress will return to zero along the dashed line PQ. The result is a 2% offset.

At some point close to but below the elastic limit there is a deviation from Hooke's Law but no apparent set. This point is called the <u>proportional limit</u>. The proportional limit is the point of a maximum stress for which the unit strain increases at the same rate as the unit stress.

In the range where Hooke's Law is valid, the slope of the stress strain curve is called the <u>modulus of elasticity</u>.

$$E = \frac{\text{Unit stress}}{\text{Unit strain}} = \frac{P/A}{\delta/L} = \frac{PL}{A\delta}$$

Where P = the load in pounds
       L = the original length in inches
       A = the cross section area in square inches
       δ = the change in L due to P expressed in inches

<u>E</u> is expressed in pounds per square inch. Calculation of E is shown in Fig. 9.12.

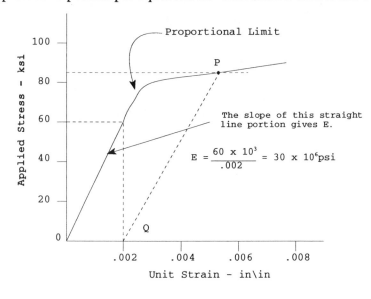

Figure 9.12

Typical Stress-Strain Curve for Steel

The yield strength is the stress at which the permanent set is .2% (.002 in/in). On Fig. 9.12 this is shown as the horizontal line from point P, 84,000 psi.

The ultimate tensile strength is the maximum stress point on the stress strain curve. From this point on an increasing load will result in decreasing stress and increasing strain until failure occurs. The ultimate tensile strength point is shown as point U in Fig. 9.13. The failing load is indicated at point X on the same figure.

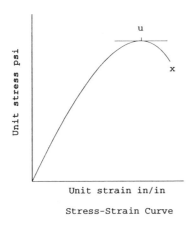

Figure 9.13

Stress-Strain Curve

PHYSICAL PROPERTIES

Thermal Conductivity

Thermal conductivity expresses how heat will flow through metals. It is expressed at Btu/in^2/in/sec/°F or Calories/cm^2/cm/sec/°C. It may be noted here that:

$$1 \text{ cal/cm}^2\text{/cm/sec/°C} = 0.0054 \text{ Btu/in}^2\text{/in/sec/°F}$$

Relative and absolute thermal conductivities are shown in Table 9.2. Copper is taken as the base for the relative calculations. The units for the absolute thermal conductivities are cal/cm^2/cm/sec/°C.

Table 9.2		
Metal	Relative Thermal Conductivity	Thermal Conductivity
Silver (Ag)	106	0.97
Copper (Cu)	100	0.92
Aluminum (Al)	52	0.48
Magnesium (Mg)	40	0.37
Zinc (Zn)	30	0.27
Iron (Fe)	18	0.16
Tin (Sn)	17	0.16
Nickel (Ni)	15	0.14
Lead (Pb)	9	0.08

Electrical Conductivity

Electrical conductivity is related to the type of structure of the metal. Generally, we do not talk of conductance (units of mhos) but of resistance expressed on ohms.

$R = \rho L/A$ ohms

R is the resistance to electrical flow
ρ is the resistivity in ohm-cm
L is the length of the electrical conductor in cm
A is the cross sectional area in cm^2

As in the case of thermal conductivity we will find the relative electrical conductivity to be in almost the same order for the metals as for thermal conductivity. Relative electrical conductivity and electrical resistivity are shown in Table 9.3.

Table 9.3			
Metal	Relative Electrical Conductivity Per Unit Cross Section Area	Per Unit Weight	Electrical Resistivity
Silver	108	92	1.468
Copper	100	100	1.6
Aluminum	61	201	2.6
Magnesium	37	180	4.4
Zinc	28	35	5.8
Nickel	23	23	6.3
Iron	18	20	8.8
Tin	12	14	13.0
Lead	8	6	20.4

Factors Affecting Thermal and Electrical Conductivity

Electricity and heat seem to be conducted through metals by the free or valence electrons that exist in the material structure. The lattice itself could aid in conducting heat. This would explain why the thermal and electrical conductivities differ. A perfect or exact atomic arrangement for a metal would lower its resistance to the flow of electricity or heat. Impurities, effects of strain, and increased temperature all tend to increase the resistance to heat and/or electrical flow. This is what gives alloys of metals some of their unusual characteristics.

Thermal Expansion

Thermal expansion takes place in two ways. The length (or a dimension) of a metal price changes with temperature. This is known as <u>linear thermal expansion</u>. The volume also changes with temperature. This is known as <u>volumetric thermal expansion</u>. Linear thermal expansion can be expressed as:

$$L_t = L_o (1 + \alpha [T - T_o])$$

where L_t is the length at temperature T.
L_o is the length at Temperature T_o.
α is the linear coefficient of thermal expansion.

Some values for α are given in Table 9.4.

Table 9.4	
Metal	(in/in/°F x 10^6)
Iron	6.5
Nickel	7.4
Copper	9.2
Silver	10.9
Tin	13.0
Aluminum	13.3
Magnesium	14.0
Zinc	15.2
Lead	16.3

For volumetric expansion,

$$V_t = V_o (1 + \alpha_v [T - T_o])$$

where
V_t is the volume at temperature T.
V_o is the volume at temperature T_o.
α_v is generally taken as 3 x α

Specific Gravity

Specific Gravity is the ratio of the mass of a body to the mass of an equal volume of water at 60°F. (Physicists may use water at 4°C as their reference. Most engineers use the 60°F reference). The customary units are lb/ft^3 or bm/cc. Some specific gravities are shown in Table 9.5.

Table 9.5	
Metal	Sp. Gr.
Lead	11.3
Silver	10.5
Copper	8.9
Nickel	8.9
Iron	7.9
Tin	7.3
Zinc	7.2
Aluminum	2.7
Magnesium	1.7

Denseness

Denseness is compactness, lack of porosity. It is <u>not</u> weight per unit volume which is density. To illustrate the difference gray cast iron, which may weigh 490 lb/ft^3 may be porous enough to leak under heavy hydraulic pressure. Aluminum, however, having a density of 170 lb/ft^3, is dense and compact enough to resist leaking.

Porosity

Porosity is the quality of containing pores and voids. It is the opposite of denseness.

Fusibility

Fusibility is the ease with which a metal may be melted. In general, softer metals are easily fusible and harder metals require higher temperatures to melt. Table 9.6 lists some values.

Volatility

Volatility is the ease with which some substances may be vaporized. It is clear from Table 9.6 that the order of fusibility and volatility of these metals is not the same. However, in general, most metals of low melting point are volatilized more easily than those of a high melting point.

Table 9.6				
Metal	Melting Point		Boiling Point	
	°C	°F	°C	°F
Tin	232	439	2260	4100
Lead	327	621	1620	2948
Zinc	419	787	905	1661
Magnesium	651	1204	1097	2007
Aluminum	660	1220	1800	3272
Silver	960	1761	2210	4010
Copper	1083	1981	2350	4259
Nickel	1455	2651	2900	5252
Iron	1535	2795	3000	5432

NONMETALLIC MATERIALS

In engineering and construction, the most common non-metallic materials are concrete and wood and synthetic organic materials. Polymers comprise the bulk of synthetic materials in engineering applications.

Concrete

Concrete has been called plastic stone because of its resemblance to stone in its cured state and its ability to be cast and formed in its first mixed or fluid state. It is made by mixing a cement paste with stones or aggregate. The simplest components are cement, sand, aggregate, and water.

Concrete is good in compression and poor in tension. Its performance in tension is improved by casting the concrete around steel rods or mesh. The steel takes the tension load rather than the concrete.

Even greater strengthening is obtained if the concrete is put permanently in a state of compression. This is done by keeping the steel reinforcement in tension while allowing the cast concrete to set. When the concrete has set, the external tensile load on the reinforcement steel is removed, producing compressive stresses in the concrete. This is known as <u>prestressed concrete</u>.

The ratio of the mix of cement/sand and aggregate/water is very important. Too little water makes the mix stiff and hard to cast around the forms and reinforcement. Air will be trapped in the concrete making it weaker. Too much water will cause the cement to settle away from the sand and aggregate, giving a poor bond and weak concrete.

The cement is made by grinding lime and clay to a powder, burning the powder, and regrinding the resulting substance. The sand is mined from pits and rivers. It can also be made by grinding stone. The sand used in concrete must be free of organic matter and must be as clean as possible. The aggregate, which generally ranges in size from 3/4 in. to 2½ in. can be sieved from the mined sand or can be crushed stone. The water used for mixing concrete must be clean and clear. It should be fit for drinking.

The usual proportions for concrete are 1:2:4 by volume. That is 1 ft^3 of cement to 2 ft^3 of sand to 4 ft^3 of aggregate. To this we add 5½ gallons of clean water (0.1337 ft^3/gal is a useful conversion factor to remember).

As previously mentioned the amount of water is important. Too much water will cause a too rich mixture on the bottom and too lean on the top. When the water evaporates it will leave voids, weakening the concrete. Water content is checked by a slump test. Fresh concrete is placed on a cone shaped mold as shown in Fig. 9.14. The mold is removed and the concrete is allowed to slump or sag. The difference in heights of the piles of concrete is a measure of slump which is also a measure of the water content of the concrete.

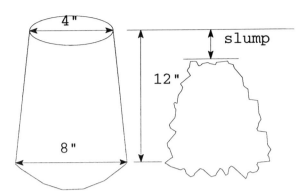

Figure 9.14

Find the weight and volume of each material required to yield 3 yds³ of 1:2:4 mix concrete. Cement weighs 94 lb/ft³ sack and has a sp. gr. = 3.1. Sand has a sp. gr. of 2.7 and weighs 105 pcf. Aggregate weighs 100 pcf and also has a sp. gr. of 2.7. The water will be 5½ gallons per sack of cement.

First, we'll calculate absolute volumes and weights for a one bag mix.

Cement

$$\frac{1 \text{ bag} \times 94 \text{ lbs/bag}}{3.1 \times 62.4 \text{ pcf}} = 0.486 \text{ ft}^3$$

Weight is 1 bag x 94 lbs/bag = 94 lbs

Sand

$$\frac{2 \text{ cu ft}^3 \times 105 \text{ pcf}}{2.7 \times 62.4 \text{ pcf}} = 1.246 \text{ ft}^3$$

Weight is 2 cu.ft. x 105 pcf = 210 lbs

Aggregate

$$\frac{4 \text{ cu ft}^3 \times 100 \text{ pcf}}{2.7 \times 62.4 \text{ pcf}} = 2.374 \text{ ft}^3$$

Weight is 4 cu. ft. x 100 pcf - 400 lbs

Water

5.5 gal x .1337 ft³ cu. ft./gal = .735 ft³

Weight is .735 x 62.4 = 45.88 lb

Now, sum up the weights and volumes.

Volume is 4.841 ft³ and weight is 749.88 lbs.

Now, if a one bag mix gives 4.841 ft³, we need 3 x 27/4.841 = 16.732 times as much to yield 3 yds³ or 81 ft³. The number 16.732 is known as the yield factor.

Therefore, a 3 yd³ mix contains

Cement	.486 x 16.732 =	8.132 ft³
Sand	1.246 x 16.732 =	20.848 ft³
Aggregate	2.374 x 16.372 =	39.722 ft³
Water	.735 x 16.732 =	12.298 ft³
		81.000 ft³
Cement	94.0 x 16.372 =	1572.81
Sand	210.0 x 16.732 =	3513.72
Aggregate	400.0 x 16.832 =	6692.80
Water	45.88 x 16.732 =	767.66
		12546.99 lb

This concrete weighs $\dfrac{12546.99}{81.0}$ = 154.9 pcf

Standard concrete is assumed to weigh 150 pcf.

<u>Wood</u>

The most common use of wood as an engineering material is as a structural material. Wood also often provides the cellulose required to make some polymers such as cellophane and rayon among others.

The mechanical properties of wood are greatly affected by grain orientation and by moisture content. A dry or seasoned wood might have twice the strength of the same size and wood type of green wood.

Tables of mechanical properties of woods are available in most handbooks. The following are <u>average</u> values. Flexural strengths ≈ 12,500 psi. Tensile strength (perpendicular to the grain) ≈ 550 psi, modules of Elasticity ≈ 1.66 x 10⁶ psi.

Synthetic Organic Material

The most important engineering synthetic materials are the polymers. Polymers are large molecules made up of long chains of repeating units, the basic unit being the mer; hence the term polymer.

The degree of the polymer relates to the number of units in the chain. It is also a function of the molecular weight. Therefore, the degree of polymerization is the molecular weight of the polymer divided by the molecular weight of the mer.

Some of the more common polymers and their repeating units are given in Table 9.7.

Table 9.7	
Polymer	Repeater
Polyvinylchloride	CH_2CHCl
Polyvinylacetate	$CH_2CH(OCOCH_3)$
Polystyrene	$CH_2CH(C_6H_5)$
Polymethl methacrylate	$CH_2C(CH3)(COOCH_3)$

The Peripheral Sciences section which follows the Material Science and Structure of Matter section contains Physics which <u>may</u> be found on the exam. Dynamics from the physicists point of view is given as well as Sound, Light, and Optics. The student will do well to make himself familiar with this section as such problems can appear as part of another problem in one of the well-defined disciplines.

PERIPHERAL SCIENCES

Rectilinear Motion

<u>Motion along a straight line</u>. The position of the body is specified by giving the distance from an origin on the straight line to the body (Fig. 9.15).

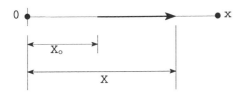

Figure 9.15

In Fig. 9.15, OX is the straight line along which the motion is occurring; X_o is the position of the body at some reference time t_o, usually 0, and X is the position of the body at time t. The following general expressions giving the relations between space, time velocity, and acceleration for rectilinear motion are useful.

Given X - f(t) $\quad V = \dfrac{dX}{dt} = \dfrac{df}{dt}$ and

$$a = \dfrac{dV}{dt} = \dfrac{d^2X}{dt^2} = \dfrac{d^2f}{dt^2}$$

Given V = f(t) $\quad X = x_o + \displaystyle\int_{t_o}^{t} V\, dt$ and $a = \dfrac{dV}{dt}$

Given V - f(x) $\quad t = t_o + \displaystyle\int_{x_o}^{x} \dfrac{dX}{V}$ and $a = V \dfrac{dV}{dX}$

Given a - f(t) $\quad V = V_o + \displaystyle\int_{t_o}^{t} a\, dt$ and $X = X_o + \displaystyle\int_{t_o}^{t} V\, dt$

Given a - f(X) $\quad V = \pm \sqrt{V_o^2 + 2\displaystyle\int_{x_o}^{x} a\, d_x}$

and $t = t_o + \displaystyle\int_{x_o}^{x} \dfrac{dX}{V}$

Given a - f(V) $\quad X = x_o + \displaystyle\int_{V_o}^{V} \dfrac{V}{a}\, dV$ and $t = t_o + \displaystyle\int_{V_o}^{V} \dfrac{dV}{a}$

The average velocity is given by $<V> = \dfrac{1}{t - t_o} \displaystyle\int_{t_o}^{t} V\, dt$

In these equations X_o and V_o are the position and velocity of the body at the reference or initial time t_o and X, V and a are the position, velocity and acceleration at the final time t.

Uniform Rectilinear Motion

Rectilinear motion with the velocity constant; hence the acceleration is zero. The average velocity and the instantaneous velocity are the same. For this motion we have

$$X = X_o + Vt, \quad V = <V> = \text{constant} \quad \text{and} \quad a = 0$$

Rectilinear Motion with Constant Acceleration

With the acceleration constant the velocity increases (or decreases in the case of constant deceleration) linearly with time. Freely falling bodies exhibit nearly this type of motion. The effect of air resistance and the decrease in acceleration with altitude cause departures from an exact constant acceleration situation.

Taking X_o and t_o to be zero the following equations give the relationship between space, time, velocity, and acceleration for rectilinear motion with constant acceleration:

$$a = \text{constant} \quad\quad X = <V>t$$

$$V = V_o + at \quad\quad X = \left(\frac{V_o + V}{2}\right) t$$

$$V = \pm\sqrt{V_o^2 + 2aX} \quad\quad X = V_o t + \tfrac{1}{2} at^2$$

$$<V> = \frac{V_o + V}{2}$$

For freely falling objects the acceleration is given approximately by

$$g = 32.2 \, \frac{\text{ft}}{\text{sec}^2} = 9.807 \, \frac{\text{meters}}{\text{sec}^2}$$

Some care is required with the sign affixed to the acceleration term. Use + if the body is actually accelerating, i.e., increasing in velocity with time and use − if the body is decelerating or slowing up. For problems involving the acceleration due to gravity note that the acceleration is always towards the earth, then if the positive direction of your coordinate system is down, affix a + sign to g; if the positive direction of your coordinate system is up, affix a − sign to g.

Example

Consider the following problem:

An object is projected vertically upward from the edge of a building with an initial velocity of 96 ft/sec. The object just misses the building on the way down and falls to the street 240 feet below the edge (Fig. 9.16).

Find: (A) The maximum height reached
       (B) The velocity of the object upon impact with the street
       (C) The elapsed time from projection to impact

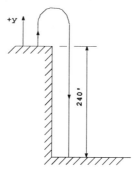

Figure 9.16

The origin has been selected at the point where the object was projected, with the up direction as positive. Therefore the initial velocity is positive.

(a) The maximum height reached can be calculated using

$$V = \pm \sqrt{V_o^2 + 2\,ay}$$

Noting that the velocity V is zero at the point of maximum height we have

$$V_o^2 - 2(32.2)y = 0$$

or

$$y = \frac{96^2}{2(32.2)} = 143.1 \text{ ft above building}$$

and

383.1 ft above the ground

(b) The velocity of the object upon impact with the street can be calculated using the same formula. Note that the distance y is the total distance traveled by the object.

Velocity upon impact

$$V = \pm \sqrt{0 + 2(-32.2)(383.1)}$$

$$V = -186 \text{ f/s}$$

The negative sign has been selected to show that the velocity is down.

(c) The elapsed time from projection to impact is the sum of the time it takes the object to reach the top of the trajectory plus the time it takes the object to fall from the top of the trajectory to the ground.

$$t_{total} = t_{up} + t_{down}$$

To find t_{up}

$$S = V_o t - \tfrac{1}{2} at^2$$

$$143.1 = 96t - \tfrac{1}{2}(32.2)t^2$$

or $16.1t^2 - 96t + 143.1 = 0$

$$t_{up} = \frac{-(-96) \pm \sqrt{(-96)^2 - (4)(16.1)(143.1)}}{2(16.1)} = 1.79 \text{ sec}$$

To find t_{down}

$$S = V_o t + \tfrac{1}{2} at^2$$

$$383.1 = (0)t + \tfrac{1}{2}(32.2)t^2$$

$$t_{down} = \sqrt{\frac{383.1 \times 2}{32.2}} = 4.88 \text{ sec}$$

$t_{total} = 1.79 + 4.88 = 6.67$ sec

Harmonic Motion

Characteristics:

Equations of motion involve sines or cosines, hence the name harmonic motion.

Motion is about the equilibrium position of the body.

Restoring force is proportional to displacement X from the equilibrium position, e.g., $F = -kX$

To obtain a graphical representation of simple harmonic motion we may consider point traveling with constant angular velocity ω around a circle (Fig. 9.17). The motion of its projection upon any diameter is a simple harmonic one.

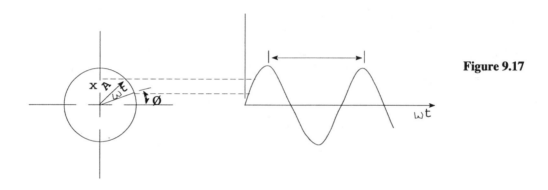

Figure 9.17

The motion is described by the following simple equation

$$X = A \sin(\omega t + \phi)$$

where

- X = displacement at any instant from the equilibrium position or center of the path
- A = amplitude of the motion, i.e., the maximum displacement
- ω = angular velocity of the motion
- t = time
- T = period, time for 1 cycle or 1 vibration

$$\omega T = 2\pi \quad \text{or} \quad T = \frac{2\pi}{\omega}$$

f = frequency, number of complete cycles or vibrations per unit time

$$f = \frac{1}{T} = \frac{\omega}{2\pi}$$

V = velocity of point at time t, $V = \dot{X} = A\omega \cos(\omega t + \emptyset)$

= $\omega\sqrt{A^2 - X^2}$

    Note: X = 0 (i.e., when $\omega t = 0, \pi, 2\pi, ---$). Specifically, the maximum velocity occurs when the particle is in its equilibrium position

a = acceleration of the point at time t, $a = \dot{V} = \ddot{X}$

= $-A\omega^2 \sin(\omega t + \emptyset) = -\omega^2 X$

    Note: $a_{max} = \pm A\omega^2$. The maximum acceleration occurs when $X - \equiv A$ which is when V = 0. Specifically, the maximum acceleration occurs when he particle is at the extremes of its motion where the restoring force is a maximum.

$\emptyset$ = constant associated with initial conditions of motion - usually expressed in radians.

Some common examples of harmonic motion are given in Table 9.8. Note the similarity in the equations for the translational, rotational, and torsional motion. Simple substitutions such as replacing X with θ, m with I, k with K, etc., transform the equations from being applicable for one system to those applicable for another system. Care must be exercised to use the proper units when working with the equations of Table 9.8. Table 9.9 has been prepared as an aid in this regard.

Consistent units for the individual terms appearing in Table I are given in Table 9.9 for the four most popular systems of units.

Example

An 8 lb weight is placed upon the lower end of a coil spring suspended from a ceiling. The weight comes to rest in its equilibrium position, thereby stretching the spring 6 in. The weight is then pulled down 3 in. below its equilibrium position and released at time t = 0. Neglect the resistance of the medium and assume that no external forces are present. Determine the amplitude, period, and frequency of the resulting motion. Also find the maximum values of the force, velocity and acceleration experienced by the weight.

First find the spring constant F = -Kx

or $\quad k = \dfrac{|F|}{|x|} = \dfrac{8 \text{ lbs}}{6/12 \text{ ft}} = 16 \text{ lb}_f/\text{ft}$

The amplitude equals the maximum displacement of the weight from its equilibrium position. Since there are no external forces and since there is no initial velocity, i.e., the weight is merely let go of at time t = 0, the amplitude equals the initial position of the weight. Therefore we have

$$A = 3/12 = 1/4 \text{ ft}$$

Using Tables 9.8 and 9.9 it is a straightforward matter to find the remaining quantities.

Period

$$T = 2\pi \sqrt{\dfrac{m}{kgc}} = 2\pi \sqrt{\dfrac{1 \text{ lbm}}{16 \dfrac{\text{lb}_f}{\text{ft}} \times 32 \dfrac{\text{lbm} - \text{ft}}{\text{lb}_f - \text{sec}^2}}}$$

$$T = \dfrac{\pi}{4} \text{ sec/cycle}$$

Frequency

$$f = 1/T = \dfrac{4}{\pi} \dfrac{\text{cycles}}{\text{sec}}$$

Max Force

$$F_{max} = \pm KA = \pm 16 \dfrac{\text{lb}_f}{\text{ft}} \times \dfrac{1}{4} \text{ ft} = \pm 4 \text{ lb}_f$$

The maximum force occurs when the spring is at its maximum extension and compression is in the direction of the equilibrium position.

Max Velocity

$$V_{max} = A\omega$$

First note $\omega = 2\pi f = 2\pi \times \dfrac{4}{\pi} = 8$ radians/sec

Then $V_{max} = \dfrac{1}{4}$ ft $\times$ 8 radians/sec = 2 ft/sec

The maximum velocity occurs as the weight passes through the equilibrium position of the system. Bear in mind that the unit of radians is dimensionless and can be added or deleted at will from the units of a term. It is generally added when you want to convey the notion of an angular velocity as in the case of ω.

Max Acceleration

$$a_{max} = \pm A\omega^2 = \pm \dfrac{1}{4} \text{ ft} \times (8 \text{ radians/sec})^2$$

$$a_{max} = \pm 16 \text{ ft/sec}^2$$

The maximum acceleration occurs simultaneously with the maximum force.

Table 9.8 Common Examples of Simple Harmonic Motion

System	Assumptions	Remarks	Period T	Frequency f	Angular Frequency ω
(spring-mass)	★ Linear spring ★ No damping	The results are valid indep of gravitation field. Results apply to system in horizontal position.	$T = 2\pi\sqrt{\dfrac{m}{kgc}}$ $T = 1/f$ $T = \dfrac{2\pi}{\omega}$	$f = \dfrac{1}{2\pi}\sqrt{\dfrac{kgc}{m}}$ $f = 1/T$ $f = \omega/2\pi$	$\omega = \sqrt{\dfrac{kgc}{m}}$ $\omega = 2\pi f$ $\omega = \dfrac{2\pi}{T}$
(pendulum)	★ Point mass ★ No damping ★ Small angles	Results are independent of amount of mass. Results depend on local value of g.	$T = 2\pi\sqrt{\dfrac{L}{g}}$	$f = \dfrac{1}{2\pi}\sqrt{\dfrac{g}{L}}$	$\omega = \sqrt{\dfrac{g}{L}}$

System	Assumptions	Remarks	Period T	Frequency f	Angular Frequency ω
	* No damping * Small angles	The simple pendulum with equal period has length $L_e = \dfrac{I}{mL}$	$T = 2\pi\sqrt{\dfrac{I}{mgL}}$	$f = \dfrac{1}{2\pi}\sqrt{\dfrac{K}{mgL}}$	$\omega = \sqrt{\dfrac{K}{mgL}}$
	* Linear spring $\tau = -K\theta$ * No damping	The results are valid independent of gravitation field or orientation of system.	$T = 2\pi\sqrt{\dfrac{I}{mgL}}$	$f = \dfrac{1}{2\pi}\sqrt{\dfrac{K}{I}}$	$\omega = \sqrt{\dfrac{K}{I}}$
	* Point mass * No damping	<u>NOT HARMONIC MOTION</u>	$T = 2\pi\dfrac{\sqrt{L\cos\Psi}}{g}$	$f = \dfrac{1}{2\pi}\sqrt{\dfrac{g}{L\cos\Psi}}$	$\omega = \sqrt{\dfrac{g}{L\cos\Psi}}$

General Notes

The constants A (amplitude) and ∅ (phase angle) are established from the initial conditions of the problems, i.e., initial displacement and velocity.

V_{max} occurs when X or θ = 0, which is when a = 0.

a_{max} occurs when X or θ = ± A, which is when V = 0.

Displacement	Velocity	Acceleration	Restoring Force
$X = A\sin(\omega t + \emptyset)$ $X_{max} = A$	$V = A\omega\cos(\omega t + \emptyset)$ $V_{max} = A\omega$	$a = A\omega^2\sin(\omega t + \emptyset)$ $a - \omega^2 x$ $a_{max} = \pm A\omega^2$	$F = -kx$ $F = \dfrac{ka}{\omega^2}$ $F_{max} = \pm kA$
$\theta = A\sin(\omega t + \emptyset)$ $\theta_{max} = A$	$\theta = A\omega\cos(\omega t + \emptyset)$	$a = A\omega^2\sin(\omega t + \emptyset)$ $a = \omega^2\theta$ $a_{max} = \pm \omega^2 A$	$F = -\dfrac{mg}{g_c}\sin\theta$

TABLE 9.9 Units

Quantity	System			
	Engineering	Absolute Engineering	CGS	MKS
a Acceleration	ft/sec2	ft/sec^2	cm/sec^2	m/sec^2
E Modulus of Elasticity or Young's Modulus	lb$_f$/in.^{2}	lb$_f$/in.^{2}	dynes/cm^2	Newtons/m^2
f Frequency	$\frac{\text{cycles}}{\text{sec}}$	$\frac{\text{cycles}}{\text{sec}}$	$\frac{\text{cycles}}{\text{sec}}$	$\frac{\text{cycles}}{\text{sec}}$
g Acceleration due to gravity at 45° Lat and sea level	$32.174 \frac{\text{ft}}{\text{sec}^2}$	$32.174 \frac{\text{ft}}{\text{sec}^2}$	$980.665 \frac{\text{cm}}{\text{sec}^2}$	$9.80665 \frac{\text{m}}{\text{sec}^2}$
gc Universal constant which emerges in Newton's Law as a result of the arbitrary choices for standards of measure	$32.174 \frac{\text{ft-lbm}}{\text{lb}_f\text{-sec}^2}$	1	1	1
I Moment of Inertia	lbm-ft2	slug-ft^2	g-cm^2	kg-m^2
k Spring constant	lb$_f$/ft	lb$_f$/ft	dynes/cm	Newtons/m
K Torsional spring constant	$\frac{\text{lb}_f\text{-ft}}{\text{radian}}$	$\frac{\text{lb}_f\text{-ft}}{\text{radian}}$	$\frac{\text{dyne-cm}}{\text{radian}}$	$\frac{\text{Newton-m}}{\text{radian}}$
m Mass	lbm	slug	g	kg
T Period	$\frac{\text{sec}}{\text{cycle}}$	$\frac{\text{sec}}{\text{cycle}}$	$\frac{\text{sec}}{\text{cycle}}$	$\frac{\text{sec}}{\text{cycle}}$

Quantity	System			
	Engineering	Absolute Engineering	CGS	MKS
V Velocity	ft/sec	ft/sec	cm/sec	m/sec
$\ddot{\alpha}$ ($\alpha = 0$) Angular acceleration	radians/sec²	radians/sec²	radians/sec²	radians/sec²
ρ Mass density	lbm/ft³	slug/ft³	g/cm³	kg/m³
θ Angular position	radians	radians	radians	radians
$\dot{\theta}$ Angular velocity	radians/sec	radians/sec	radians/sec	radians/sec
Torque	lb$_f$-ft	lb$_f$-ft	dyne-cm	Newton-m
μ (= M/L) Linear density of string	lbm/ft	slugs/ft	g/cm	kg/m

Example

Pendulums can be used to measure g-fields as the following example illustrates. Consider two simple pendulums of equal length. One is on Earth and the other is on planet X. The pendulum on planet X has a period that is twice that of the one on Earth. How does the g-field on planet X compare with the one on Earth.

From Table 9.8

$$T = 2\pi \sqrt{\frac{L}{g}}$$

Now since $T_x = 2 T_e$

we have $2\pi \sqrt{\dfrac{L}{g_x}} = 4\pi \sqrt{\dfrac{L}{g_e}} g_e$

or $g_x = \dfrac{1}{4}$

Therefore the acceleration due to gravity on planet X would be approximately 8 ft/sec².

Example

The moment of inertia of a body of any complex shape may be found by suspending the body as a physical pendulum and measuring its period of vibration. The location of the center of gravity can be found by balancing.

Consider the 2-lb object in Fig. 9.18 pivoted about a horizontal axis. The object is found to make 30 complete cycles in 2 minutes. Find the moment of inertia of the object.

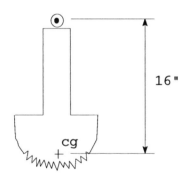

Figure 9.18

$$T = \dfrac{2 \text{ min} \times 60 \dfrac{\text{sec}}{\text{min}}}{30 \text{ cycles}} = 4 \dfrac{\text{sec}}{\text{cycle}}$$

From Table 9.8 we have

$$T = 2\pi \sqrt{\dfrac{I}{mg L}} \quad \text{or} \quad I = \dfrac{T^2}{4\pi^2} mg L$$

Using the Engineering system of units from Table 9.9 we have:

$$I = \frac{\left(4 \frac{\text{sec}}{\text{cycle}}\right)^2}{4\pi^2} \times 2 \text{ lbm} \times 32 \frac{\text{ft}}{\text{sec}^2} \times \frac{16 \text{ in.}}{12 \text{ in./ft}}$$

$$I = \frac{1024}{3\pi^2} \text{ lb}_m \text{ - ft}^2$$

In absolute Engineering units the answer would be

$$I = \frac{32}{3\pi^2}$$

Note that the moment of inertia so calculated is about the axis of rotation. The moment of inertia I_o about the center of gravity can be found from the formula

$$I_o = I - m X_o^2 \text{ where } X_o \text{ is distance}$$

of center of gravity from the axis of rotation. See your Mechanics of Materials notes.

Example

The torsion pendulum in a clock is displaced 60 degrees from its equilibrium position. The angular acceleration is 30 radians/sec². Is the clock keeping accurate time?

From Table 9.8
$$f = \frac{1}{2\pi}\sqrt{\frac{|\alpha|}{|\theta|}} = \frac{1}{2\pi}\sqrt{\frac{30 \frac{\text{radians}}{\text{sec}^2}}{60 \text{ degrees} \times \frac{\pi \text{ radians}}{180 \text{ deg}}}} \text{ slug-ft}^2$$

$$f = .85 \frac{\text{cycles}}{\text{sec}}$$

Therefore the clock is running slow. It should make one cycle per second to keep correct time.

Wave Motion

Wave motion in a medium is a method of transferring energy through a medium by means of a distortion of the medium which travels away from the place where the distortion is produced. The medium itself moves only a little bit. For example, a pebble dropped into quiet water disturbs it. The water near the pebble does not move far, but a disturbance travels away from that spot.

This disturbance is the wave. As a result some of the energy lost by the pebble is carried through the water by a wave and a cork floating at some distance away can be lifted by the water; thus the cork gets some energy. We can set up a succession of waves in the water by pushing a finger rhythmically through the surface of the water, much like a vibrating tuning fork produces waves in air. We distinguish two types of waves in media, transverse and longitudinal.

Transverse Waves

A transverse wave in a medium is a wave in which the vibrations of the medium are at right angles to the direction in which the wave is traveling. A transverse wave is readily set up along a string. Consider the string shown in Fig. 9.19.

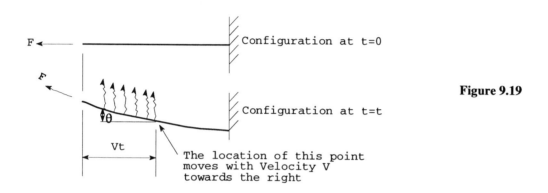

Figure 9.19

The string is under tension F, has mass M, length L, and a mass per unit length or linear density of $\mu = M/L$. Imagine the left end of the string moving with constant velocity u in a direction perpendicular to the string itself. This transverse displacement is propagated along the string with speed V so that as time goes on more and more of the string becomes bent at the angle θ and achieves the perpendicular velocity u.

The velocity of propagation of a transverse pulse in a string is given by:

$$V = \sqrt{\frac{FL\, g_c}{m}} = \frac{\sqrt{F\, g_c}}{\mu}$$

Table II gives consistent units for the four major systems.

Example

Calculate the velocity of a transverse pulse in a string under a tension of 10 lb if the string weighs .003 lb/ft.

In Engineering Units we have

$$V = \sqrt{\frac{20 \text{ lb}_f \times 32.17 \frac{\text{lb}_m - \text{ft}}{\text{lb}_f - \text{sec}^2}}{.003 \text{ lb}_m/\text{ft}}} = 462 \text{ ft/sec}$$

Continuing with the idea of a transverse wave consider that one end of a stretched string is forced to vibrate periodically in a transverse direction with simple harmonic motion with amplitude A and period T. Assuming the string to be long enough so that any effects at the far end need not be considered, a continuous train of transverse sinusoidal waves then advances along the string (Fig. 9.20).

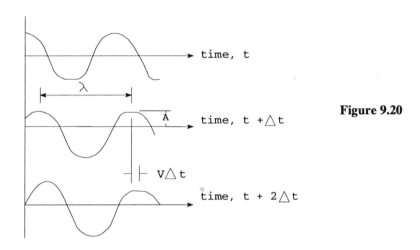

Figure 9.20

Motion of a point on the string is sinusoidal and up and down. However, the motion of the wave front is transverse to this. The wave front moves with V along the string. Since in one time period T, the waveform advances one wave length, and we have

$VT = \lambda$ or $V = f\lambda$

where V = speed of the wave

f = frequency of the wave

λ = wavelength

Vibration of a String Fixed at Both Ends (Standing Waves) (Fig. 9.21)

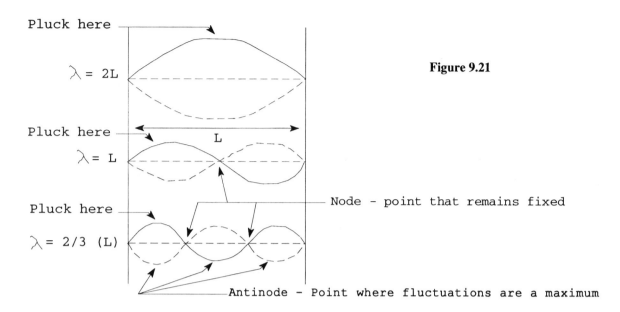

Figure 9.21

Since the nodes are one half a wavelength apart, for a string of length L standing waves may be set up in the string of wavelength.

$$\lambda = 2L, \frac{2L}{2}, \frac{2L}{3}, \frac{2L}{4}, \ldots \frac{2L}{n}$$

To find what frequencies can be set up recall $V = f\lambda$

or $f = \dfrac{V}{\lambda} = \dfrac{V}{2L} = \dfrac{V}{2L}, \dfrac{2V}{2L}, \dfrac{3V}{2L}, \dfrac{4V}{2L}, \ldots n\dfrac{V}{2L}$

but $V = \sqrt{\dfrac{FL\, g_c}{m}}$

Therefore

$f_1 = \dfrac{1}{2L}\sqrt{\dfrac{FL\, g_c}{m}}$ (fundamental frequency or first harmonic)

$f_2 = 2 f_1$ (first overtone or second harmonic)

$f_3 = 3 f_1$ (second overtone or third harmonic)

etc.

Note that by increasing the tension, decreasing linear density of the string (L/m) or increasing the length you cause in increase in pitch of the sound.

An interesting difference between the string and the string mass combination is that the string has many possible frequencies of vibration while the spring mass has only one.

Longitudinal Waves

The second basic type of wave is the longitudinal wave. In this case the wave is such that the particles of the medium vibrate back and forth along the path that the wave travels. Sound waves are longitudinal waves in the frequency range 20 to 20,000 cps.

To visualize how longitudinal waves are set up consider a plunger moving back and forth in a tube containing air or a tuning fork beating the air or even a violin string moving in air (Fig. 9.22).

Figure 9.22

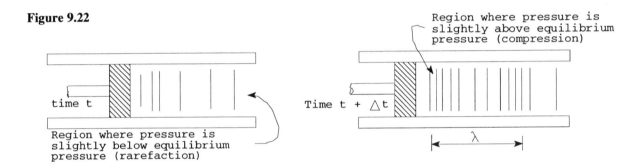

The wavelength of longitudinal wave is the distance between two successive compressions or rarefaction.

As with transverse waves we have

$V = f\lambda$

The velocity of the wave V is often called the velocity of sound when the frequency f is in the 20 to 20,000 cps range.

In general the velocity of a longitudinal wave in a medium is given by

$V = \sqrt{B_{ad}/\rho}$

where B_{ad} is the adiabatic bulk modulus of the medium

$$B_{ad} = \left.\frac{\text{change in pressure}}{\text{change in volume/unit volume}}\right|_{\text{adiabatic}}$$

For solids the above equation becomes

$$V = \sqrt{144\ E\ g_c/\rho\ [(1-\mu)/(1+\mu)(1-2\mu)]}$$

or approximately

$$V = \sqrt{144\ E\ g_c/\rho}$$

where

E is the elastic (Young's) modulus, psi

ρ is the mass density - lbm/ft^3

μ is Poissons ratio

$$g_c = 32.17\ \frac{lb_m - ft}{lb_f - sec^2}$$

V = velocity in ft/sec

For gases the above equation becomes

$$V = \sqrt{k\ g_c\ RT}$$

where

k = ratio specific heats C_p/C_v (1.4 for air)

R = gas constant ($53.3\ \dfrac{lb_f - ft}{lb_m = °R}$ for air)

T = absolute temperature of gas - degrees Rankine

V = velocity in ft/sec

For air this equation becomes $V = 49\ \sqrt{T}$

For liquids

Empirical equations are commonly used in place of theoretical ones since the velocity varies in a complex way with temperature, pressure and other factors.

Typical values for solids, gases, and liquids follow

Material	Velocity of Sound, ft/sec
Aluminum	16,700
Brass	11,500
Copper	11,700
Iron and soft steel	16,400
Lead	4,000
Brick	12,000
Cork	1,600
Wood	10,000 - 15,000
Water	4,800
Sea water	5,100
Air	1,100
Hydrogen	4,200
Water vapor	1,300

Example

Find the speed of sound in granite. From a handbook such as Mark's <u>Handbook For Mechanical Engineers</u> we find

$$E = 7.3 \times 10^6 \text{ psi}$$

Specific Gravity = 2.67

Poisson's ratio is not given so we use the approximate formula

$$V = \sqrt{144 \times E \times g_c / \rho}$$

The density of granite is found from

$$\rho = 2.67 \times 62.4 = 166.5 \ \frac{lb_m}{ft^3}$$

Then

$$V = \sqrt{144\frac{\text{in.}^2}{\text{ft}^2} \times 7.3 \times 10^6 \frac{\text{lb}_f}{\text{in.}^2} \times 32.2 \frac{\text{lb}_m\text{-ft}}{\text{lb}_f\text{-sec}^2} / 166.5 \frac{\text{lb}_m}{\text{ft}^3}}$$

$V = 14{,}300$ ft/sec

Example

Find the wavelength a 10,000 Hz sound wave traveling in helium gas at 70°K at 6 atmospheres pressure.

Although the temperature of the gas is quite cold (-334°F) the speed of sound can still be calculated from

$$V = \sqrt{k\, g_c\, RT}$$

From a handbook we find $k = C_p/C_v = 1.67$

(Note nomatomic gases usually have $k = 1.67$ while diatomic gases usually have $k = 1.4$)

The gas constant for helium is obtained by dividing the universal gas constant R_o by the molecular weight of helium, hence

$$R = \frac{R_o}{MW} = \frac{1545.3 \frac{\text{lb}_f - \text{ft}}{\text{lb}_{mole} - °R}}{4 \frac{\text{lbm}}{\text{lbm - mole}}} = 386.3 \text{ lb}_f\text{-ft/lbm-}°R$$

The temperature of the helium in degrees Rankine is obtained by multiplying by 1.8

$$T = 1.8\, \frac{°R}{°K} \times 70°K = 126.0°R$$

Then the velocity is found from

$$V = \sqrt{1.67 \times 32.17 \frac{\text{lbm - ft}}{\text{lb}_f - \text{sec}^2} \times 386.3 \frac{\text{lb}_f - \text{ft}}{\text{lbm} - °R} \times 126°R}$$

$V = 1610$ ft/sec

Since $f\lambda = V$

$$\lambda = \frac{1610 \text{ ft/sec}}{10000 \text{ cycles/sec}} = .161 \frac{\text{ft}}{\text{cycle}}$$

Nomenclature

C	Velocity of sound wave
V_L	Velocity of listener
V_s	Velocity of source
f	Frequency of sound at source
f'	Apparent frequency as heard by listener

Assumptions

1. Speed of source and listener are less than speed of sound.
2. Medium through which sound is traveling is stationary.

Doppler Effect

Consider a sound source at location A emitting waves of frequency f that move toward a listener at B. If neither the source nor the listener are moving, the frequency of the sound heard by the listener will be the same as that emitted by the source. However, if either or both the source and the listener are moving the frequency heard will be higher or lower in pitch than the emitted sound. This is known as the Doppler effect. The Doppler effect applies to electromagnetic waves as well as sound waves. The light from a star that is approaching the Earth is of somewhat higher frequency or shorter wavelength than it would be if the two were at relative rest. Table 9.10 will prove valuable for problems dealing with the Doppler effect.

Intensity of Sound

The human ear responds to an extremely wide range of sound intensities. It has therefore become customary to express the level of sound intensity or sound power by the logarithm of the ratio comparing the intensity or power to that of an arbitrary sound. The bel was originally used as the unit to express this level; the decibel is now used. The intensity level α of a sound wave is defined by the equation

$\alpha = 10 \log I/I_o$ db

TABLE 9.10 Doppler Effect

Sketch	Velocity of Source	Velocity of Listener	Apparent Freq. Mathematically	Apparent Frequency Qualitatively
0 0 Source Listener	$V_{s_{stationary}} = 0$	$V_{L_{stationary}} = 0$	$f = f$	No change in frequency
V_s 0→ 0 Source Listener	$V_{s_{towards\ listener}}$	$V_{L_{stationary}} = 0$	$f = \dfrac{c}{c - V_s} f$	Higher Pitch
V_s ←0 0 Source Listener	$V_{s_{away\ from\ listener}} = 0$	$V_{L_{stationary}} = 0$	$f = \dfrac{c}{c - V_s} f$	Lower Pitch
V_L 0 ←0 Source Listener	$V_{s_{stationary}} = 0$	$V_{L_{towards\ source}}$	$f = \dfrac{C - V_L}{C} f$	Higher Pitch
0 0→ Source Listener	$V_{s_{stationary}} = 0$	$V_{L_{away\ from\ source}}$	$f = \dfrac{C - V_L}{C} f$	Lower Pitch
V_s V_L 0→ ←0 Source Listener	$V_{s_{towards\ listener}}$	$V_{L_{away\ from\ source}}$	$f = \dfrac{C - V_L}{C - V_s} f$	If $V_L > V_s$ Lower Pitch If $V_L < V_s$ Higher Pitch
V_s V_L 0→ 0→ Source Listener	$V_{s_{towards\ listener}}$	$V_{L_{towards\ source}}$	$f = \dfrac{C + V_L}{C - V_s} f$	Higher Pitch
V_s V_L ←0 0→ Source Listener	$V_{s_{away\ from\ listener}}$	$V_{L_{towards\ source}}$	$f = \dfrac{C + V_L}{C + V_s} f$	If $V_L > V_s$ Higher Pitch If $V_L < V_s$ Lower Pitch
V_s V_L ←0 ←0 Source Listener	$V_{s_{away\ from\ listener}}$	$V_{L_{away\ from\ source}}$	$f = \dfrac{C - V_L}{C + V_s} f$	Lower Pitch

where I_o is an arbitrary reference intensity which is taken as 10^{-16} watt/cm^2, corresponding roughly to the faintest sound which can be heard. Typical sound levels are given in Table 9.11.

TABLE 9.11 Sound Levels of Common Sounds		
Description of Source	Qualitative Description	Decibels
Thunder, artillery, proximity of jet engine exhaust	Threshold of Feeling	120
Riveter, elevated train	Deafening	110
Noisy factory	Very loud	90
Average street noise, radio	Loud	70
Average office, conversation	Quiet	30
Average auditorium, quiet conversation	Quiet	30
Rustle of leaves, whisper	Faint	20
Soundproof room	Very faint	10

The intensity of a sound wave is the amount of wave energy transmitted per unit time per unit area perpendicular to the direction of sound propagation. If I is the intensity, f the frequency, and A the amplitude then the intensity if given by

$$I = 2 \pi^2 f^2 A^2 V \rho$$

where V is the speed of sound and ρ is the density of the medium through which the sound is being propagated.

Example

What is the intensity of a sound wave in air at standard conditions if its frequency is 1 kilohertz and its amplitude is 0.1 cm?

Using the above formula we have

$$I = 2 \pi^2 \times \left(10^3 \frac{\text{cycles}}{\text{sec}}\right)^2 \times \left(\frac{.1 \text{ cm}}{\text{cycle}} \times \frac{1 \text{ m}}{100 \text{ cm}}\right)^2 \times \left(331 \frac{\text{m}}{\text{sec}}\right) \times \left(1.293 \frac{\text{Kg}}{\text{m}^3}\right)$$

$$I = 8380 \frac{\text{Kg}}{\text{sec}^3}$$

Recall that 1 watt = $1 \frac{\text{Kg} - \text{m}^2}{\text{sec}^3}$, therefore

$$I = 8380 \frac{\text{watts}}{\text{m}^2}$$

LIGHT AND OPTICS

Reflections - Mirrors

Law of reflection. When a light wave is reflected the angle of reflection equals the angle of incidence; furthermore the incident ray, the reflected ray and the normal lie in one plane.

Image in Plane Mirror (Fig. 9.23)

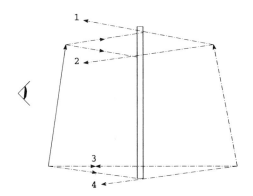

Figure 9.23

Image and extension are shown dotted.

Note that the reflected rays 1, 2, 3, and 4 appear to come from behind the mirror, thereby creating an image

Some notes on the image:

    The image is same size as the object.

    The image is erect (if the object is erect).

    The image is as far behind the mirror as the object is in front.

    The image is virtual, i.e., the image is formed by rays that do not actually pass through it; if a screen were placed at the position of the image, no image would be seen.

    The image is laterally reversed - this can be thought of in terms of two people shaking hands: their arms extend diagonally between them.

Image in Curved Spherical Mirrors (Fig. 9.24)

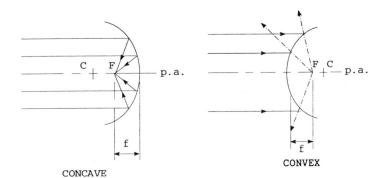

Figure 9.24

Spherical mirrors employ spherical surfaces as the reflecting surfaces.

A concave spherical mirror is one whose reflecting surface is the inside of a spherical shell.

A convex spherical mirror is one whose reflecting surface is the outside of a spherical shell.

The principal axis is the line through C and midpoint of mirror.

The principal focus (F) - incident rays parallel to the principal axis pass through this point after reflection by a concave mirror. (For a convex mirror the focus is a virtual one).

Focal length (f) is the distance form principal focus to mirror f = R/2 where R = radius of curvature for a spherical mirror.

To determine whether the image formed by a mirror will be inverted or erect, real or virtual, magnified or diminished. Table 9.12 will be of assistance. The two governing factors are seen to be the type of mirror and the location of the object with respect to the mirror.

Spherical Mirror Equations

The following equations can be used to locate images, determine whether they are real or virtual, and calculate the magnification of the mirror.

$$\frac{1}{S} - \frac{1}{S'} = -\frac{2}{R} = \frac{1}{f}$$

$$\text{magnification} = m = \frac{y'}{y} = \frac{S'}{S}$$

where

> f = focal length of mirror
> m = magnification of mirror (size of image/size of object)
> R = radius of curvature of mirror
> S = distance of object from mirror
> S' = distance of image from mirror
> y = height of object
> y' = height of image

Any consistent unit of length may be used, meters being the most common unit.

<u>Sign Conventions</u>

With reference to the figures in Table 9.12, the following sign conventions should be used with the spherical mirror equations.

Draw all diagrams with the incident light traveling from left to right.

S is plus on left side of surface, minus on right side.

S' is plus on right side of surface, minus on left side (a plus value for S' denotes a virtual image, a minus value denotes a real image)

Radii of curvature R are positive if C lies to the right side of surface.

y and y' are positive above the axis.

Image is erect if m is plus; inverted if m is minus.

<u>Example</u>

A small object lies 4 in. to the left of the vertex of a concave mirror with radius of curvature 12 in. (Fig. 9.25). Find the position and magnification of the image.

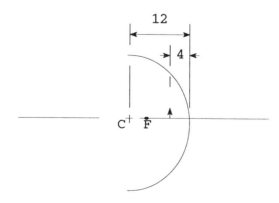

Figure 9.25

S = 4 in. R = -12

Using $\dfrac{1}{S} - \dfrac{1}{S'} = \dfrac{2}{R}$

we have $\dfrac{1}{4} - \dfrac{1}{S'} = -\dfrac{2}{12}$

or s' = $\dfrac{12}{5}$ = 2.4

Therefore M = $\dfrac{s'}{s}$ = $\dfrac{2.4}{4}$ = 0.6

TABLE 9.12 Images Formed by Spherical Mirrors

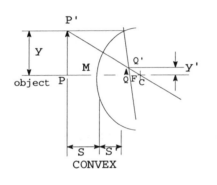
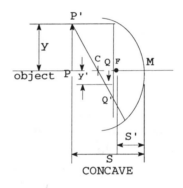

Figure 9.26

TYPE OF MIRROR	POSITION OF OBJECT (PP')	POSITION OF IMAGE (QQ')	CHARACTER OF IMAGE
Concave	At ∞	At F	Real (Point)
	Between ∞ and C	Between F and C	Real, Inverted, Diminished
	At C	At C	Real, Inverted, Same size
	Between C and F	Between C and ∞	Real, Inverted, Magnified
	At F	At ∞	___, ___, ___

TYPE OF MIRROR	POSITION OF OBJECT (PP')	POSITION OF IMAGE (QQ')	CHARACTER OF IMAGE
Concave	Between F and M	Between ∞ (Behind M) and M	Virtual, Erect, Magnified
	At M	At M	___, Erect, Same Size
Convex	At ∞	At F	Virtual (Point)
	Between ∞ and M	Between F and M	Virtual, Erect, Diminished
	At M	At M	___, Erect, Same Size

The image is therefore 12" to the right of the mirror (S' is +), is virtual (S' is +) and is 3 times the height of the object. The image is erect since m is positive.

Table 9.12 provides a qualitative check on these answers.

Refraction

Snell's Law is the phenomenon of refraction as the Law of Reflection is to the phenomenon of reflection.

Snell's Law

The ratio of the sine of the angle of incidence to the sine of the angle of refraction is equal to the ratio of the velocities of light in the two media.

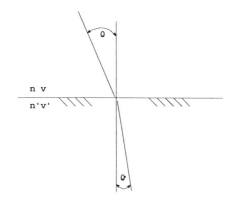

Figure 9.27

$$\frac{\sin \theta}{\sin \theta'} = \frac{V}{V'}$$

Index of Refraction

The ratio of the velocity of light in a vacuum to that in a substance gives the index of refraction for the substance.

$$n = \frac{C}{V}$$

where

n = index of refraction of a substance
C = velocity of light in vacuum
V = velocity of light in the substance

Since the velocity of light is a maximum in a vacuum, the index of refraction is always greater than 1. Typical values are between 1 & 2. The following is a sampling:

Material	n
Air	1.0003 (usually taken as 1)
Glass	1.46 to 1.96
Quartz	1.544
Ethyl Alcohol	1.361
Water	1.333
Diamond	2.42

Employing the index of refraction of two media Snell's Law can be rewritten in the more popular form

$n \sin \theta = n' \sin \theta'$ Snell's Law

Referring to Fig. 9.27, if the upper medium is air, then light will almost always bend towards the vertical when it enters the lower medium. This is true since n' will almost surely be greater than n for air.

Total Internal Reflection

Consider a source of light below the surface of water shown in Fig. 9.28.

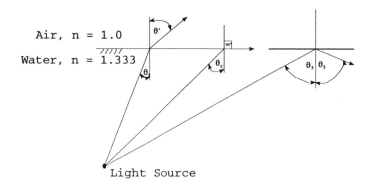

Figure 9.28

Three situations are shown. The first is where the angle of incidence is small and the light passes through to the air. The second situation is where the angle of incidence is such that the angle of refraction is equal to 90° and hence the light wave travels along the surface of the water. This angle of incidence is referred to as the <u>Critical Angle, θ_c</u>. Using Snell's Law the critical angle can be calculated. For the water/air problem, Snell's Law gives:

$n \sin \theta_c = n' \sin \theta$

$1.333 \sin \theta$

$1.333 \sin \theta c \quad 1 \cdot \sin (90°)$

or $\quad \theta_c = 48.6°$

The third situation in the above sketch shows θ greater than θ_c and hence the light is reflected internally, i.e., since θ' is greater than 90° the light does not emerge from the water.

Prisms

The phenomenon of total internal reflection forms the basis of prisms in optics. Figure 9.29 illustrates this.

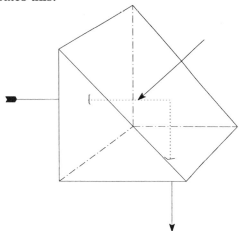

Angle of incidence = 45° which is greater than critical angle for glass with n = 1.5, θ_c = 42°)

Figure 9.29

Dispersion

Dispersion occurs when a light source is a mixture of waves of different wavelengths, e.g., white light, and passes through a material for which n depends on wavelength. For example n varies in the following way for quartz:

Color	λ	n
Violet	.430 microns	1.552
Blue	.490 microns	1.550
Yellow	.590 microns	1.545
Red	.660 microns	1.540

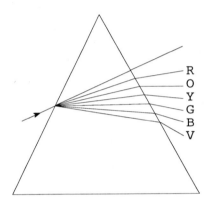

Therefore a beam of white light passing through a quartz prism would produce a dispersion pattern similar to that shown in Fig. 9.30.

Note how the wavelength with the largest n gets refracted the most.

Figure 9.30

The brilliance of a diamond is due in part to its large dispersion.

Lenses

An optical system bounded by two refracting surfaces having a common axis. Its purpose is to converge or diverge a beam of light transmitted through it.

Some common lenses are shown in Fig. 9.31.

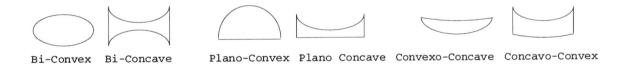

Bi-Convex   Bi-Concave   Plano-Convex   Plano Concave   Convexo-Concave   Concavo-Convex

Figure 9.31

Basically there are two types of lenses (Fig. 9.32):

<u>Convex or converging lens</u> - thicker in the middle than at the edges.

<u>Concave or diverging lens</u> - thinner in the middle than at the edges.

The line that passes through the center(s) of curvature of a lens is called the principal axis. The point of convergence of parallel rays after passing through a lens is called its focal point or principal focus. The focal length is usually denoted by the letter f.

Figure 9.32

CONVEX or Converging Lens   CONCAVE or Diverging Lens

To calculate the focal length of a lens we may use the Lensmaker's Equation:

$$\frac{1}{f} = (n - 1)\left(\frac{1}{R_1} - \frac{1}{R_2}\right)$$

where R_1 and R_2 are the radii of curvature of the left and right lens surfaces, respectively. They are signed numbers positive if the center of curvature is to the right of the lens and negative if to the left.

For example, a biconvex lens with radii of curvature 10 cm and 8 cm and an index of refraction of 1.5 (Fig. 9.33) has a focal length given by

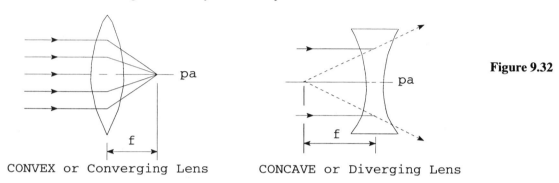

$$\frac{1}{f} = (n - 1)\left(\frac{1}{R_1} - \frac{1}{R_2}\right)$$

$R_1 = +8$ cm

$$\frac{1}{f} = (1.5 - 1)\left(\frac{1}{8} + \frac{1}{10}\right)$$

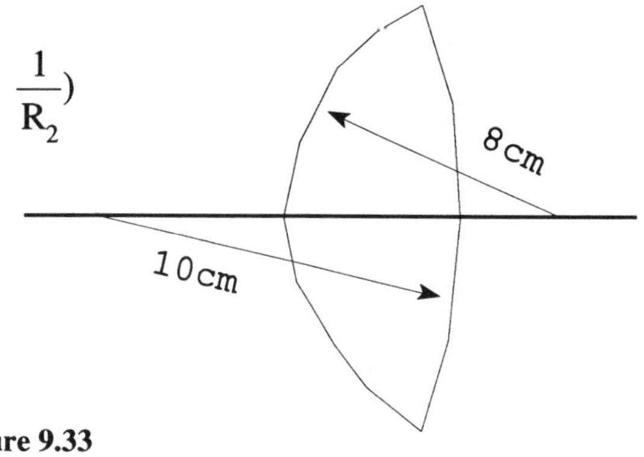

$R_2 = -10$ cm

Figure 9.33

$$f = \frac{80}{9} \text{ cm}$$

Consider now a biconcave lens with radii of curvature and index of refraction the same as the above biconvex lens. The focal length is now given by:

$R_1 = -10$ cm

$R_2 = +8$ cm

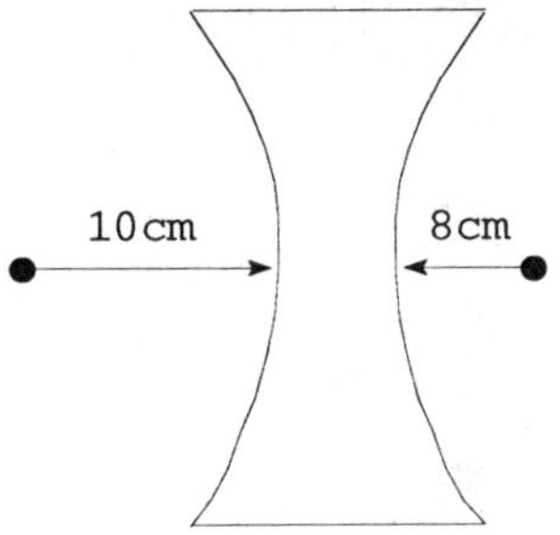

Figure 9.34

$$\frac{1}{f} = (n - 1)\left(\frac{1}{R_1} - \frac{1}{R_2}\right)$$

$$\frac{1}{f} = (1.5 - 1)\left(\frac{1}{-10} - \frac{1}{8}\right)$$

$$f = \frac{80}{9} \text{ cm}$$

Thus it is seen that the magnitude of the focal lengths are the same, however, the biconcave or diverging lens has a negative value.

Diverging lenses always have a negative focal length.

Two lenses separated by a distance d have an effective focal length f given by

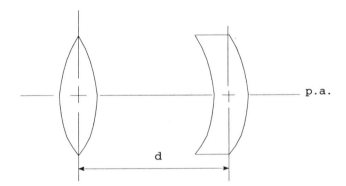

Figure 9.35

$$\frac{1}{f} = \frac{1}{f_1} + \frac{1}{f_2} - \frac{d}{f_1 f_2}$$

When the two lenses are in contact with each other the above reduces to

$$\frac{1}{f} = \frac{1}{f_1} + \frac{1}{f_2}$$

Lens Power

Lens power is defined as the reciprocal of the focal length with f measured in meters. The units for lens power are diopters which is equivalent to meters^{-1}.

Lens Power = $\frac{1}{f}$ (in meters) diopters

Example

Find the effective lens power of a two lens system where one lens is biconvex with a focal length of 25 cm and the other lens is biconcave with a focal length of 100 cm. Assume the two lenses are in contact with each other.

$$f_1 = 25 \text{ cm} = .25 \text{ meters}$$

$$f_2 = -100 \text{ cm} = -1 \text{ meter}$$

Then

$$\frac{1}{f} = \frac{1}{f_1} + \frac{1}{f_2}$$

$$\frac{1}{f} = \frac{1}{.25} - \frac{1}{1}$$

$$f = 3 \text{ meters}$$

Effective lens power $= \frac{1}{f} = \frac{1}{3}$ diopters

It should be noticed that adding the lens power of the individual lenses would be more direct in this case.

Images can be formed with lenses just as was the case with mirrors. Whether the image is real or virtual, inverted or erect, magnified or diminished in size depends on the type of lens and where the object is located with respect to the lens. Table 9.13 will be an aid in establishing the answers to such questions.

Spherical Lens Equations

The following equations can be used to locate images, determine whether they are real or virtual, and calculate the magnification of the lens.

$$\frac{1}{S} + \frac{1}{S'} = \frac{1}{f}$$

$$\text{magnification} = m = \frac{y'}{y} = -\frac{S'}{S}$$

Where

    f = focal length of lens
    m = magnification of lens (size of image/size of object)
    S = distance of object from lens
    S' = distance of image from mirror
    y = height of object
    y' = height of image

Any consistent unit of length may be used, meters being the most common unit.

Sign Convention

With reference to the figures in Table 9.13, the following sign convention should be used with the spherical lens equations.

    Draw all diagrams with the incident light traveling from left to right.

    S is plus on left side of lens, minus on right side

    S' is plus on right side of surface, minus on left side (a plus value for S' denotes a real image, a minus value for S' denotes a virtual image)

    y and y' are positive above the axis

    Image is erect if m is plus; inverted if m is minus.

Example

An object is located 25 mm in front of a convex lens of power 10.0 diopters. Describe the image.

Solution

To find the focal length f and hence the position of F we use

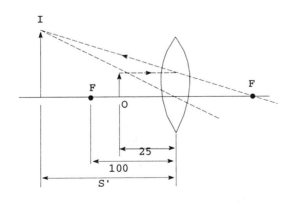

$$f = \frac{1}{\text{Lens power}} = \frac{1}{10} = .1 \text{ meters} = 100 \text{ mm}$$

Figure 9.36

Now $\frac{1}{S} + \frac{1}{S'} = \frac{1}{f}$ or $\frac{1}{25} + \frac{1}{S'} = \frac{1}{100}$

Therefore S' = -100/3 mm

Since S' is negative, the image is virtual and on the left side of the lens.

The magnification is given by $m = -\frac{S'}{S} = -\frac{100/3}{25} = +\frac{4}{3}$

Therefore the image is magnified and erect.

Table 9.13 can be used to obtain the qualitative answers of virtual erect, magnified and on the same side as the object.

Some Important Definitions and Terms Dealing with Light

- A luminous body is one that emits light of its own (e.g., firefly).

- An incandescent body is one that emits light because it has been heated (e.g., filament in light bulb).

- An illuminated body is one that is visible due to the light it reflects.

- Diffraction of light is the bending of light around edges. This is especially noticeable with small openings. A diffraction grating is made to have thousands of lines to the inch.

- Polarized light is light whose direction of vibration has been restricted in some way.

- Interference occurs when light from the same source travels two different paths to arrive a body. If the two waves arrive in phase they reinforce one another; if they arrive out of phase they annul each other.

From the modernized metric system or the International System of Units (SI) we have the following two definitions:

- The candela (cd) is defined as the luminous intensity of 1/700,000 of a square meter of a blackbody at the temperature of freezing platinum (2045°K).

- The lumen (lm) is the SI unit of light flux. A source having an intensity of 1 candela in all directions radiates a light flux of 4π lumens. A 100 watt light bulb emits about 1700 lumens.

TABLE VI IMAGES FORMED BY SPHERICAL LENSES

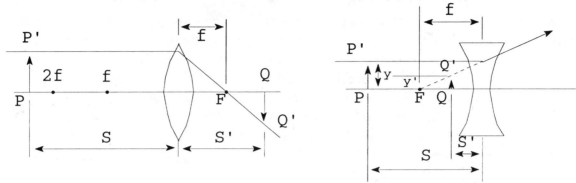

Figure 9.37

TYPE OF LENS	POSITION OF OBJECT	POSITION OF IMAGE	CHARACTER OF IMAGE
Convex	At ∞	At F	Real (Point)
	Between ∞ and 2f	Between f and 2f	Real, Inverted, and Diminished
	At 2f	At 2f	Real, Inverted, and Same Size
	Between 2f and f	Between 2f and ∞	Real, Inverted, and Magnified
	At f	At ∞	—, —, —
	Between f and lens	Between lens and -∞ (i.e., same side as object)	Virtual, Erect, and Magnified
Concave	At ∞	At F	Virtual (Point)
	Between ∞ and lens	Between F and lens	Virtual, Erect, and Diminished

Keppler's Laws

A topic that has most recently appeared involves Keppler's Laws. Keppler's Laws describe the motion of the planets in our solar systems. One law states that the sun is located at one focus of an ellipse, which describes the path of each planet in our solar system. The second law states that areas swept by two radius vectors in a given time interval are equal. Let us refer to Fig. 9.38. The orbit shown is only for one planet in the solar system. The sun is at one focus. In time period t_1 we will sweep the area A_1 and in time period t_2 we will sweep the area A_2. If $t1 = t_2$, then $A_1 = A_2$. We can also see from the sketch that the velocity of the body in orbit must be constantly changing. V_1 must be greater than V_2.

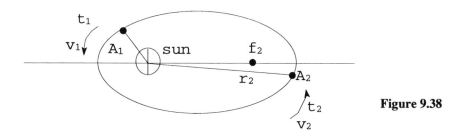

Figure 9.38

The third of Keppler's Laws states that for any two planets, the ratio of the squares of the orbital periods to the cubes of the mean distances of the planets from the sun are the same. If T is the orbital period and d is the mean distance from the planet to the sun then:

$$\frac{T_1^2}{T_2^2} = \frac{d_1^3}{d_2^3}$$

10

Engineering Economics

TABLE OF CONTENTS

Introduction 10-1

Time Value of Money 10-2

Cash Flow Diagrams 10-3

Financial Calculations 10-4

    Equivalency Formulas 10-5
    Bond Calculations 10-21

Depreciation Calculations 10-27

    Straight Line Depreciation Method (SLD) 10-28
    Sinking-Fund Depreciation Method (SFD) 10-28
    Declining Balance or Fixed Percentage Depreciation Method (DBD) 10-28
    Double Declining Balance Depreciation Method (DDBD) 10-29
    Sum of the Years Digits Depreciation (SYD) 10-30

Comparative Economic Analysis Methods 10-34

    Profit in Business 10-34
    Annual Cost (TAC) Method 10-35
        Basic Approach 10-37
        Approach Including Risk 10-44
    Present Worth Method 10-46
    Capitalized Cost Method 10-56
    Rate of Return Method 10-58
        Evaluation of an Investment 10-58
        Corporate Rate of Return Calculations 10-65
        Incremental Rate of Return (Comparing Alternatives) 10-71
    Benefit to Cost Ratio Method 10-76

Other Topics 10-81
    Replacement Studies 10-81
        Determination of Economic Replacement
            Interval for the Same Asset 10-81
        Determination of Economic Replacement for an Improved Asset .. 10-82
    Break Even Cost Analysis 10-84
    Minimum Cost Analysis/Economic Lot Size 10-87
    Homework Problems and Solutions 10-87

ENGINEERING ECONOMICS 10-1

INTRODUCTION

The purpose of these review notes is to provide a refresher in the major topics of Engineering Economics. Engineering economics deals with determining from a number of technically acceptable alternatives the one that provides the best economic result.

These notes cover the following topical areas:

> Time value of money
> Cash flow diagrams
> Financial calculations
> Depreciation methods
> Evaluation of alternatives
> Bond calculations
> Personal investment problems
> Corporate considerations - before/after tax profit

These major areas cover the main topics covered in both the Fundamentals of Engineering Examination and the Principles and Practices Examination. Engineering economics represents 8% of the morning section and 14% of the afternoon section of the Fundamentals of Engineering Examination. Economics represents 12.5% of the Principles and Practices Examination.

Interest Tables are included at the end of these notes to assist in problems requiring interest calculations. The following references are listed to supplement the material presented in these notes:

> E.L. Grant and W.G. Ireson, "Principles of Engineering Economy," The Ronald Press Company, New York
>
> G.A. Taylor, "Managerial and Engineering Economy," Van Nostrand, New York
>
> H.G. Thruesen and W.J. Fabrycky, "Engineering Economy," Prentice-Hall, Englewood Cliffs, New Jersey
>
> G.L. Bach, "Economics," Prentice-Hall, Englewood Cliffs, New Jersey
>
> E. P. DeGarmo and J.R. Canada, "Engineering Economy," Macmillan, New York

It is beneficial for the reader during preparation for either part of the Professional Engineers Examination to become familiar with only one economics reference book, which should to taken to the exam. Any reference material taken to the exam that the reader is not reasonably familiar with cannot be efficiently used during the exam.

TIME VALUE OF MONEY

Money has value because it is used to produce or purchase goods and services. As an example; if Mr. Jones was offered $100 now or in one year he would choose the $100 now because it is worth more than the $100 in a year. Why is it worth more? Mr. Jones could invest the $100 received now in a 6%/year bank account that would earn him $6 in one year. Therefore, $100 now is actually worth $106 in a year.

If Mr. Jones were offered $100 now or $106 in a year, for this example, the choices are now equivalent since in one year Mr. Jones would have $106 whether he takes the $100 now or waits and takes $106 in a year's time. The $6 interest payment Mr. Jones receives on his $100 deposit represents the fee the bank is willing to pay for the use of Mr. Jones' money for one year.

In a similar fashion, when money is borrowed, the borrower must return to the lender the amount borrowed, generally referred to as the principal and the lender's fee for the use of his money, generally referred to as interest. Typically then repayment of a loan has two components known as (1) return of the investment (principal) and (2) return on the investment (interest).

When performing an engineering economic analysis covering a time frame greater than one year, the time value of money must be considered. In order to perform time value or financial calculations, the following terms and symbols are defined:

> i—Interest Rate   An interest rate can be expressed in two ways. The first way is interest rate per interest period. If a problem states that interest is at 6% without any indication of the interest period, the period is <u>always taken as one year</u>. Interest stated on a per interest period basis, when the interest period is one year, is known as the Effective Interest Rate.
>
> When interest is stated as a percent per year compound over a specified period other than a year, for example 12%/year compounded monthly, the rate of interest is known as the nominal rate. In order to convert to an interest rate per interest period, the interest rate per year is divided by the number of interest periods in one year. For the example given, therefore, the interest rate per period would be 12%/12 or 1%/month.
>
> n—Number of Interest Periods   The number of interest periods is defined by the specific problem being solved. As an example of calculations based on monthly interest periods, a 5 year loan with a bank has 5 x 12 = 60 monthly payments or interest periods for calculations.
>
> P—Present Sum of Money   Present or current sum of money is a reference point for financial calculations and is therefore defined as appropriate for a specific problem. One or more cash flows can be expressed as an equivalent worth at a reference point in time defined as the present.

<u>F–Future Sum of Money</u>   A sum of money, n interest periods in the future, which is the equivalent to one or more cash flows plus the interest earned or paid during the n interest periods. The future point in time is relative to the defined present (reference) point in time.

<u>A–Uniform Equal Series (Annuity) of Payments or Receipts</u>   The payments or receipts must occur at the end of each interest period. Examples are loan payments or equal periodic deposits into a savings account.

<u>G–Uniform Gradient Series of Payments or Receipts</u>   The gradient is an arithmetic increase or decrease of payments or receipts occurring at the end of each interest period.

CASH FLOW DIAGRAMS

The use of cash flow diagrams is recommended as a simple way of visualizing or organizing a problem that includes various single payments (receipts) and/or series of payments (receipts) involved. The following conventions should be used when drawing cash flow diagrams, such as the diagram shown in Fig. 10.1:

1.   The horizontal line is a time scale with the progression of time proceeding from left to right. Each increment of time represents the duration of the interest period used in the problem. The scale can be represented as interest periods or specific points in time depending on the requirements of a problem. It should be recognized that the end of a period is also the beginning of the next period.

2.   In Fig. 10.1 the vertical arrows represent cash values at a specific point in time. Downward arrows are payments or cash outflows. Upward arrows represent receipts, deposits, or cash inflows

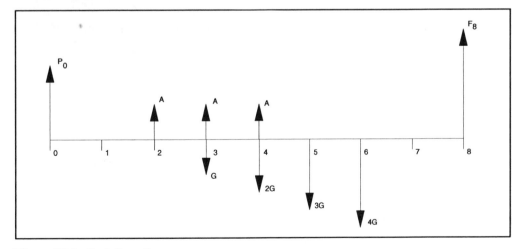

Figure 10.1   Cash flow diagram

FINANCIAL CALCULATIONS

Financial or interest calculations recognize the concept of time value of money and provide the means to determine equivalent values of cash flows at a specific point in time. The example of $100 now or a year from now demonstrated that $100 now is equivalent to $106 a year from now at 6% interest, therefore $100 now is not equivalent to $100 a year from now. When a problem's time frame exceeds one year, money can ONLY BE ADDED OR SUBTRACTED AT THE SAME POINT IN TIME. For example, how much money would you have if you received $100 today and another $100 a year from now, based on 6% interest? The answer $200 is of course incorrect because it would not account for the time value of money for the $100 received today. The correct answer is $206 ($100 today + $6 interest at the end of 1 year + the $100 received in 1 year).

Interest calculations can be based on simple or compound interest. Simple interest calculations assume interest is paid only on the present value of a deposit or principal. For example, if Mr Jones deposited $100 in a bank account at 6% simple interest, how much would he have in 5 years? The following formula is used to calculate the total amount of simple interest (I_T) paid:

$$I_T = P(i)(n)$$

For this example, P = $100, i = 6%/year, and n = 5 years

$$I_T = 100(0.06)(5) = \$30$$

The answer to the problem is therefore $130 ($100 + $30).

SIMPLE INTEREST CALCULATIONS ARE NOT GENERALLY USED IN ENGINEERING ECONOMIC CALCULATIONS.

Compound interest calculations recognize that once interest is earned, if left on deposit it becomes part of the principal for the next years calculations of interest. Therefore if Mr Jones were to deposit $100 at 6% interest compounded annually, the calculation would be as follows:

Year	Interest Calculation	Amount on Deposit
0	=	$100.00
1	100.00 + 100.00(.06) =	106.00
2	106.00 + 106.00(.06) =	112.36
3	112.36 + 112.36(.06) =	119.10
4	119.10 + 119.10(.06) =	126.25
5	126.25 + 126.25(.06) =	133.82

With compound interest $33.82 is received, while only $30.00 would be received with simple interest. The financial or interest formulas defined below are based on compound interest.

There are two types of financial calculations: calculations based on discrete compounding and calculations based on continuous compounding. Continuous compounding is not generally used, and is therefore not covered in these review notes. Discrete compounding means that interest is calculated and paid **at the end of each finite interest period**. Tables 10.1, 10.2, and 10.3 summarize the eight discrete compound interest factors, symbols, cash flows, and formulas used in interest calculations. These tables are based on functional notation for the defined factors. The formulas are divided by type of calculation: Table 10.1 for SINGLE PAYMENT FORMULAS, Table 10.2 for UNIFORM SERIES FORMULAS, Table 10.3 for GRADIENT SERIES FORMULAS.

The single payment formulas defined in Table 10.1 relate P—present value, F—future value, i - interest rate, and n - duration of calculation (number of interest periods) with the appropriate interest factor. Factors are defined functionally; for example, if a problem gives P and requires that F be calculated, the factor required would be the F/P factor or the "TO FIND/WHAT'S GIVEN" factor. These formulas can therefore only be used when a single amount of money P or F is to be moved to a different point in time.

TO FIND	GIVEN	FUNCTIONAL FORMULA	FACTOR NAME	FACTOR FORMULA	CASH FLOW DIAGRAM
F	P	$F = P(F/P, i, n)$	COMPOUND AMOUNT	$(1+i)^n$	
P	F	$P = F(P/F, i, n)$	PRESENT WORTH	$\dfrac{1}{(1+i)^n}$	

Table 10.1
Single Payment Interest Formulas

Tables 10.1, 10.2 and 10.3 provide functional formula notation for the interest factors based on the descriptive form (TO FIND/GIVEN, i%, n). The tables also give the factor name, for example the (F/P, i%, n) is the compound amount factor. This factor can be calculated using the formula $(1 + i)^n$ which is shown in Table 10.1. The purpose of interest tables is to avoid exponential factor calculations. Since each factor is a function of i and n, interest tables are organized by interest rates. Each interest rate table tabulates the various factors (refer to Tables 10.1, 10.2, and 10.3) for various values of n. The following is an example of the organization of an interest rate table:

$$i = 5.0\%$$

n	F/P	P/F	A/F	F/A	A/P	P/A
1	1.0500	0.9524	1.0000	1.0000	1.0500	0.9524
2	1.1025	0.9070	0.4878	2.0500	0.5378	1.8594
3	1.1576	0.8638	0.3172	3.1525	0.3672	2.7233
4	1.2155	0.8227	0.2320	4.3101	0.2820	3.5460
5	1.2763	0.7835	0.1810	5.5256	0.2310	4.3295
.						
.						
10	1.6289	0.6139	0.0795	12.5779	0.1295	7.7217
.						
.						
50	11.4674	0.0372	0.0048	209.3481	0.0548	18.2559

Suppose a deposit of $500 is made in a 5% interest savings account, and you want to know how much will be on deposit in 10 years. From Table 10.1, for single payment interest formulas, the appropriate equation would be either

$$F = P(F/P, i\%, n) \quad \text{OR} \quad F = P(1 + i)^n \text{ since the}$$

$$(F/P, i\%, n) = (1 + i)^n$$

USING THE FIRST FORMULA THE SOLUTION IS

$$F = 500(F/P, i = 5, n = 10) = 500(\text{VALUE FROM 5\% INTEREST TABLE}) = 500(1.6289) = 814.45$$

Uniform series formulas, shown in Table 10.2, relate A, a uniform equal series of payments, to either P, the present worth of the payments, or F, the future value of the payments, based on an i and n. These formulas should be used only when uniform equal, <u>end of year</u>, payments or deposits are involved.

The gradient series formulas (Table 10.3) are used to convert the following nonuniform series to an equivalent uniform series of payments (A) or a present value (P) as shown in Figs. 10.2 and 10.3 respectively.

The following example problems are presented to demonstrate the use of interest formulas and cash flow diagrams.

ENGINEERING ECONOMICS 10-7

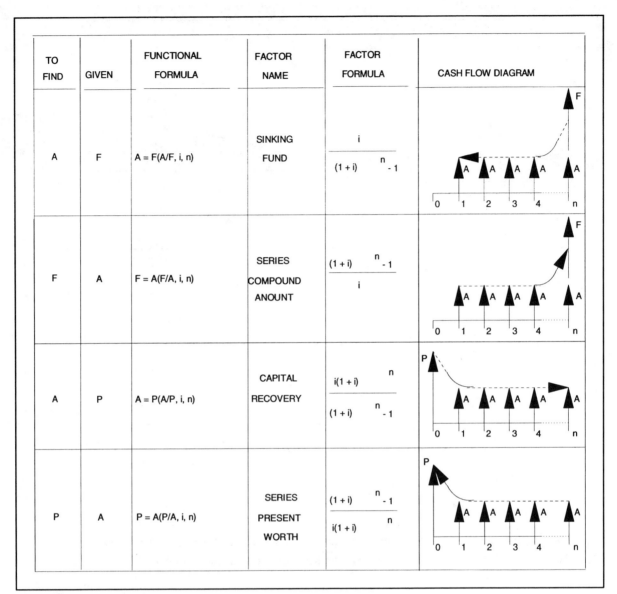

Table 10.2
Uniform Series Interest Formulas

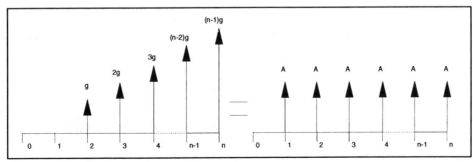

Figure 10.2 Gradient series to uniform series

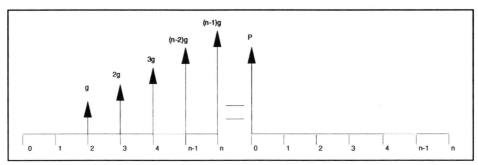

Figure 10.3 Gradient series to present worth

TO FIND	GIVEN	FUNCTIONAL FORMULA	FACTOR NAME	FACTOR FORMULA	CASH FLOW DIAGRAM
A	G	$A = G(A/G, i, n)$	GRADIENT TO UNIFORM SERIES	$\dfrac{1}{i} - \dfrac{n}{(1+i)^n}$	SEE FIGURE 2
P	G	$P = G(P/G, i, n)$	GRADIENT TO PRESENT WORTH	$\dfrac{1}{i}\left[\dfrac{(1+i)^n - 1}{i(1+i)^n} - \dfrac{n}{(1+i)^n}\right]$	SEE FIGURE 3

Table 10.3
Gradient Series Interest Formulas

Example 1 Interest is paid at 12%/year compounded annually, semi-annually, quarterly, and monthly. Determine the effective interest rate for each compounding period.

This problem is solved by looking up the F/P factor (compound amount factor) for the interest rate per period of compounding and the number of compounding periods in a year.

> 12% compounded annually is an effective interest rate since the compounding period is stated as one year.

> 12% compounded semi-annually, $i = 12\%/2 = 6\%$ (interest per period expressed as a decimal), $n = 2$ (2 interest periods per year).

> From the F/P interest table at $i = 6\%$, $n = 2$, F/P = 1.1236.

> The effective interest rate is determined by the relationship (F/P - 1); therefore $1.1236 - 1 = .1236$ or $\underline{12.36\%}$

12% compounded quarterly, i = 12% /4 = 3%/quarter, n = 4.
F/P at 3%, n = 4, F/P = 1.1255, effective interest rate is
1.1255 - 1 = .1255 or 12.55%

12% compounded monthly, i = 12% /12 = 1%/month, n = 12.
F/P at 1%, n = 12, F/P = 1.1268, effective interest rate is
1.1268 - 1 = .1268 or 12.68%

Example 2 A deposit $1050 is made in a bank account which pays 8% interest. How much will be on deposit at the end of 12 years?

Given: P = 1050, i = 8%/yr, n = 12 years   Find: F
F = P(F/P, 8%, 12 yrs) = 1050 (obtain from interest table)
  = 1050(2.518) = $2643.90

$2643.90   Answer

Example 3 Determine the amount on deposit after 12 years for $1050 with interest at 8% compounded quarterly.

Given: P 1050, i = 8%/yr compounded quarterly, i = 12 years   Find F

F = P(F/P, 8%/4 = 2%/period, n = 12 yrs X 4 periods/year = 48 periods)
  = 1050(2.587) = $2716.35

$2716.35   Answer

Example 4 How much is $14000 placed on deposit 20 years in the future worth today with interest at 12% ?

Given: F = 14000, i = 12%/yr, n = 20 yrs   Find P

P = F(P/F, 12%, 20 yr) = 14,000(0.1037) = 1,451.80

$1,451.80   Answer

Example 5 Fifteen years ago $1000 was deposited in a bank account and today it is worth $2368.74. The bank pays interest semi-annually. What interest rate compounded semi-annually was paid on this account?

Given: P = 1000, F = 2368.74, n = 15 years   Find: i

NOTE: P is taken as a point in time 15 years ago since this is a convenient reference point for this problem.

F = P(F/P, i = ?, n = 15 X 2 = 30)
2368.74 = 1000(F/P)   F/P = 2.3684

Using interest tables, the following data can be found:

i	F/P
2.0	1.8114
?	2.3684
3.0	2.4273

By interpolation:

$$\frac{X - 2.0}{3.0 - 2.0} = \frac{2.3684 - 1.8114}{2.4273 - 1.8114}$$

? = 2.0 + 1.0(0.90) = 2.90

<u>5.8%/yr compounded semiannually</u>   <u>Answer</u>

AN ALTERNATE SOLUTION WOULD USE THE P = F(P/F) FORMULA WITH THE SAME APPROACH.

<u>Example 6</u> How long will it take, to the nearest year, for money to triple at 9% interest?

Given: P = P, F = 3P, i = 9%/yr   Find: n
F = P(F/P, i = 9%/yr, n = ?)
3P = P(F/P) 3 = F/P

Using the 9% interest table the following can be found:

n	F/P
12	2.813
?	3.000
13	3.066

By interpolation:

$$\frac{n - 12}{13 - 12} = \frac{3.0 - 2.813}{3.066 - 2.813}$$

n = 12 + 0.739 = 12.739 yrs

<u>n = 13 yrs to the nearest year</u>   <u>Answer</u>

Example 7 It is planned to deposit $200 at the end of each month for 17 years at 12%/yr compounded monthly interest. How much will be available in 8 years?

NOTE: The problem states "end of period payments," if it had not been specific "end of year payments" should be assumed.

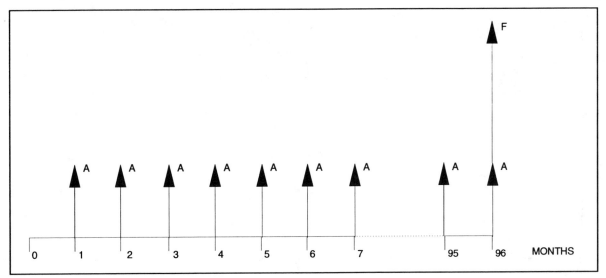

Figure 10.4 Cash flow diagram

Given: A = $200, i = 12%/yr compounded monthly, n = 8 yrs x 12 months/year = 96 interest periods   Find: F

F = A (F/A, i = 12%/12 = 1%, n = 96)
  = 200(159.927) = 31,985.40

$31,985.40   Answer

Example 8 If the deposits in example 7 are made at the beginning of each month, how much would be available in 8 years?

STOP AND REMEMBER! All financial calculations using interest tables are based on "end of year" payments, therefore an adjustment must be made to use the standard formulas.

FIRST DRAW A CASH FLOW (Fig. 10.5).

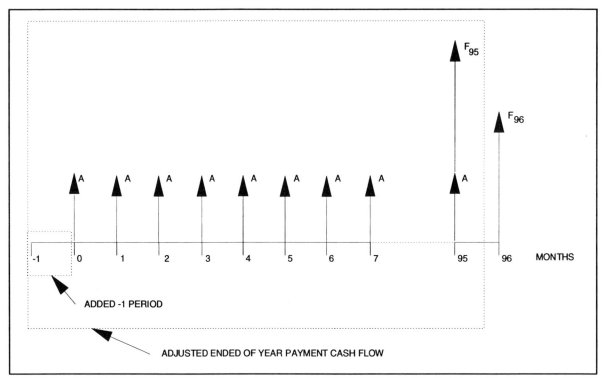

Figure 10.5 Cash flow diagram

Notice that although the cash flow now starts at the beginning of the first period rather than the end of the first period it still looks like an end of period cash flow by adding the -1 period and showing the last payment being made at the beginning of the 96th period rather than the end. Use of the F = A(F/A) formula will actually provide the value F_{95}, not F_{96} as was calculated in Example 7. The value required, however, is F_{96}, therefore F_{95} must be moved one period using the single payment formula F = P(F/P) were P = F_{95}. The solution is as follows:

F_{95} = A(F/A, i = 1%, n = 96) = 200(159.927) = 31,985.40
F_{96} = F_{95}(F/P, i = 1%, n = 1) = 31,985.40(1.01) = 32,302.22

$32,302.22   Answer

THE FOLLOWING SINGLE EQUATION CAN BE USED FOR BEGINNING OF YEAR PAYMENTS:

F = A(F/A, i, n)(1 + i)   NOTE: (1 + i) = F/P when n = 1

A second approach to "beginning of period" payments is based on the following cash flow diagram shown in Fig. 10.6:

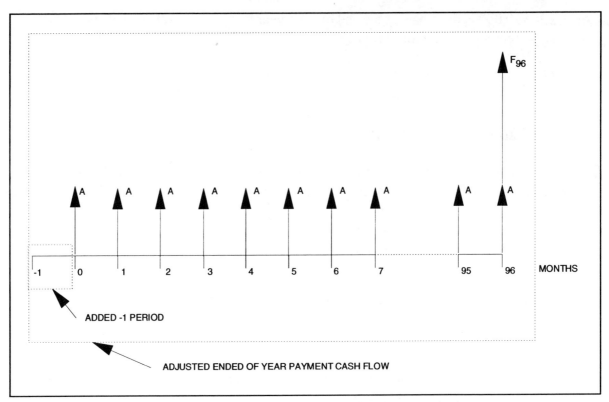

Figure 10.6

This cash flow adds the first payment at the beginning of the first year and also leaves the last payment at the beginning of the 96th period. With this cash flow F_{96} can be calculated directly, since it is an end of year cash flow starting in the -1 year and ending at the end of the 95th period! The last payment of this cash flow, however, would not be made if beginning of year payments are made, therefore it must be deducted from the F_{96} calculated with this cash flow. The deduction is made directly since the last payment of this cash flow receives no interest. The following equation is used with the cash flow:

$F = A(F/A, i, n + 1) - A$

Applied to this problem the equation is:

$F_{96} = 200(F/A, i = 1\%, n = 97) - 200 = 200(162.527) - 200$
$= 32,505.40 - 200 = \underline{32,305.40}$

$\underline{\$32,305.40}$ Answer

NOTE: Small difference in answers is the result of rounding off the F/A factor.

Example 9 $10,000 is to be paid at the end of each year for 7 years. What is this series of payments present worth if interest is at 8%?

Given: A = 10,000, i = 8%/yr, n = 7 years   Find: P
P = A(P/A, i = 8%, n = 7 yrs) = 10,000(5.206) = 52,060

$52,060   Answer

Example 10 A car costs $24,986. A down payment of 4,986 is made and the balance is borrowed from a bank at 9% interest for 5 years, with monthly payments required, how much are the payments?

Given: P = 24,986 - 4,986 = 20,000 (amount borrowed from bank),
i = 9%/12 = 0.75%, n = 12 x 5 = 60   Find: A

A = P(A/P, i = 0.75%, n = 60) = 20,000(0.0208) = 416.00

$416.00/MONTH   Answer

Example 11 Maintenance on a machine is forecast to be $500 the first year, and increase by $50 each year for 9 years. What present sum of money would be needed now to pay the maintenance for the 10-year period? Interest is 8% per year.

The cash flow diagram is shown in Fig. 10.7.

NOTE: This problems cash flow is broken into two cash flows based on the available formulas. The cash flow below the dotted line is an equal annual series with A = $500. The cash flow above the dotted line is a gradient with G = $50. The present sum of these cash flow are calculated and then added to determine the total present worth.

Given: A = 500, G = 50, i = 8%/yr, n = 10   Find: P

P = A(P/A, i = 8%, n = 10) + G(P/G, i = 8%, n = 10)

= 500(6.71) + 50(25.98) = 4,654

$4,654   Answer

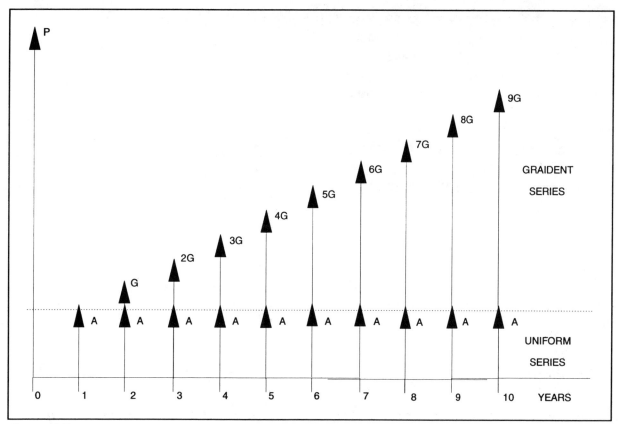

Figure 10.7

Example 12 Repeat Example 11 by calculating an equivalent equal annual series for the maintenance cost.

> NOTE: The 500 is already an equal annual series, therefore no calculation is required. The gradient series, however, needs to be converted to an equivalent equal annual series as shown in the cash flow diagram shown in Fig. 10.8.

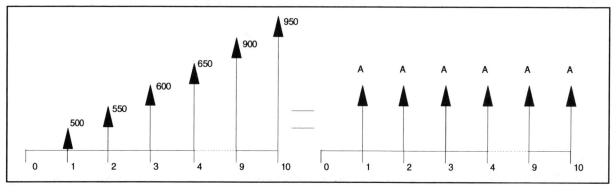

Figure 10.8

10-16 FUNDAMENTALS OF ENGINEERING EXAM REVIEW WORKBOOK

$A_{EQ} = A + G(A/G, i = 8\%, n = 10) = 500 + 50(3.87) = \underline{693.50}$

$\underline{\$693.50}$ Answer

As a check, if the present worth of 693.50 is calculated it should equal the answer to example 11, 4.654.00

$P = A(P/A, i = 8\%, n = 10) = 693.50(6.71) = \underline{4,653.39}$

ANSWERS CHECK WITHIN THE ROUNDOFF OF THE FACTORS

Example 13 In June, 1990 a firm purchased a used computer for $2,000. Repairs were needed in 1991 for $100, in 1993 for $165, and $200 in 1994. The computer was sold in 1996 for $650. The firm used the computer for payroll services and received $500 per year for each year it owned the computer. Calculate the future amount of money the firm had after sale of the computer. Interest is at 10%.

A CASH FLOW DIAGRAM FOR THIS PROBLEM IS A MUST!

NOTE: All cash flows must be moved to the 6th year and then the sum of all values expressed in the 6th year can be computed.

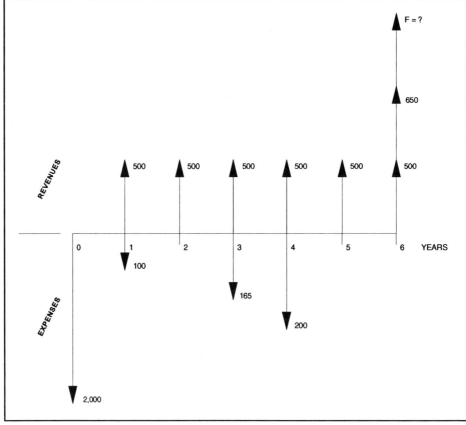

Figure 10.9

F = -2000(F/P, i =10%, n = 6) + [-100(F/P, i = 10%, n = 5)]
    +[-165(F/P, i = 10%, n = 3)] + [-200(F/P, i = 10%, n=2)]
    + 500(F/A, i = 10%, n = 6) + 650

F = -2000(1.772) - 100(1.611) - 165(1.331) - 200(1.210) + 500(7.716) + 650 = <u>341.29</u>

<u>$341.29</u>   <u>Answer</u>

<u>Example 14</u> An engineer desires to establish a fund for his newborn child's college education. The engineer estimates that the needs will be $15,000 for the first three years of college and $17,500 in the last year. The child will enter college on his 18th birthday, with payments for college due at the beginning of each year. The fund will receive lump sums of $5,000 on the day of the child's birth, $3,000 on the child's 10th birthday, and a fixed amount on the child's first through 17th birthdays, inclusive. If the fund earns 8% per annum, what should the annual deposit be?

<u>A CASH FLOW DIAGRAM FOR THIS PROBLEM IS A MUST !</u>

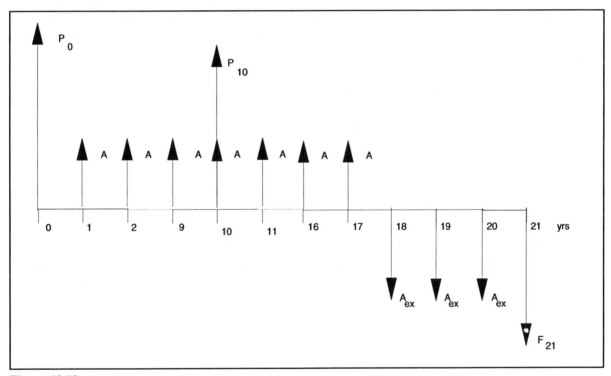

Figure 10.10

$P_0 = 5{,}000$, $P_{10} = 3{,}000$, $A = ?$, $A_{ex} = 15{,}000$, $F_{21} = 17{,}500$

The present worth of college expenses at year 17 must equal the future value of the college fund at year 17.

Future value of college fund = F_{17}

$F_{17} = P_0(F/P, i = 8\%, n = 17) + P_{10}(F/P, i = 8\%, n = 7)$
$\quad + A(F/A, i = 8\%, n = 17) = 5,000(3.700) + 3,000(1.714)$
$\quad + A(33.7502)$
$\quad = 23,642 + 33.7502(A)$

Present worth of college expenses = P_{17}
$P_{17} = A_{ex}(P/A, i = 8\%, n = 3) + F_{21}(P/F, i = 8\%, n = 4)$
$\quad = 15000(2.577) + 17500(0.735) = 51,517.50$

$F_{17} = P_{17}$ $23,642 + 33.7502(A) = 51,517.50$ $A = \underline{825.94}$

$\underline{\$825.94}$ Answer

Example 15 A person wants to buy a car that costs $16,500. The total value of the car can be financed at a special interest rate of 6% for four years or a dealer rebate of $1,500 and interest at 12% can be accepted. Payments are to be made quarterly. (A) Which payment approach should be take? (B) How much of the principal will be paid after 8 payments? (C) After 3 years the loan is to be retired, how much should be paid to the bank? (D) How much interest will be paid to the bank over the life of the loan? (E) Develop the cash flow schedule for the selected loan.

Part A:

    LOAN ALTERNATIVE A - SPECIAL FINANCE RATE

    $A = P(A/P, i = 1.5\%, n = 16) = 16,500(0.0708) = 1,168.20$

    LOAN ALTERNATIVE B - $1,500 REBATE

    $A = P(A/P, i = 3\%, n = 16) = (16,500 - 1,500)(0.0796) = 1,194.00$

    LOAN ALTERNATIVE A SHOULD BE SELECTED SINCE IT RESULTS IN
    LOWER MONTHLY PAYMENTS.   Answer

Part B:

    The present worth of the remaining series of payments are calculated and then subtracted from the principal

    $P = A(P/A, i = 1.5\%, n = 8) = 1,168.2(7.486) = 8,745.15$
    Principal paid after 8 payments = 16,500 - 8,745.15 = 7,754.85

    $\underline{\$7,754.85 \text{ of principal paid after 8 payments.}}$ Answer

Part C

    The present worth of the remaining payments equals the amount necessary to settle the loan.

$P = A(P/A, i = 1.5\%, n = 4) = 1,168.2(3.854) = 4,502.24$

$4,502.24 SHOULD BE PAID AT THE END OF THE THIRD YEAR TO SETTLE THE LOAN.   Answer

Part D

INTEREST PAID = AMOUNT PAID ON LOAN - PRINCIPAL
$= 1,168.20(16) - 16,500 = 2,191.20$

INTEREST PAID IS $2,191.20   Answer

Part E

PERIOD	INTEREST PAID	CUMULATIVE INTEREST PAID	PRINCIPAL PAID	CUMULATIVE PRINCIPAL PAID	LOAN BALANCE
0					16,500.00
1	247.50	247.50	920.70	920.70	15,579.30
2	233.69	481.19	934.50	1,855.21	14,644.79
3	219.67	700.86	948.53	2,803.74	13,696.26
4	205.44	906.30	962.76	3,766.50	12,733.50
5	191.00	1,097.30	977.20	4,743.70	11,756.30
6	176.34	1,273.64	991.86	5,735.56	10,764.44
7	161.47	1,435.11	1,006.73	6,742.20	9,757.71
8	146.37	1,581.48	1,021.83	7,764.12	8,735.88
9	131.04	1,712.52	1,037.16	8,801.20	7,698.72
10	115.48	1,828.00	1,052.72	9,854.00	6,646.00
11	99.69	1,927.69	1,068.51	10,922.50	5,577.49
12	83.66	2,011.35	1,084.54	12,007.05	4,492.95
13	67.39	2,078.74	1,100.81	13,107.86	3,392.14
14	50.88	2,129.62	1,117.32	14,225.18	2,274.82
15	34.12	2,163.74	1,134.08	15,359.26	1,140.74
16	17.11	2,180.85	1,151.09	16,510.35	-10.35

From this table the following comparison is made to the calculated values of Parts B, C, and D.

	CALCULATED	TABLE	DIFFERENCE
Part B	7,754.85	7,764.12	9.27
Part C	4,502.24	4,492.95	-9.29
Part D	2,191.20	2,180.85	-10.35

The difference is due to round off of interest table values. The final payment to the bank would actually be $1,157.85.

Series cash flow calculations based on the interest tables only apply to equal or gradient type series. The cash flow diagram in Fig. 10.11 cannot use the series formulas because it is an irregular series.

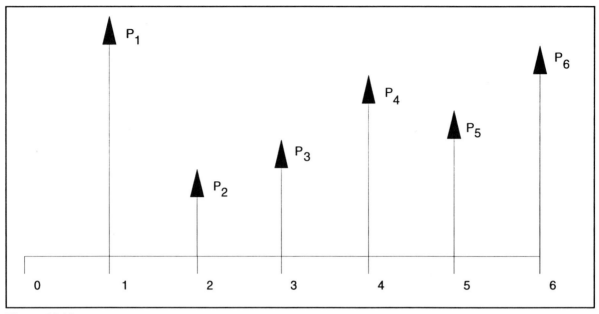

Figure 10.11

The levelized cost calculation will convert this irregular cash flow into an equal end of period cash flow. A_{wq} is the equal periodic cost which over n years is equivalent to the irregular cash flow.

A levelized cost calculation first calculates the present worth of the irregular cash flow and then calculates the equal annual payment using the A/p factor. The equation can be written as follows:

$$A_{eq} = \sum_{n=1}^{x} P_x(P/F, i, n = x) \ (A/P, i, n)$$

Example 16 The following maintenance costs are anticipated: 1991 - $200, 1992 - $1,000, 1993 - $750, 1994 - $1,600, and 1995 - $400. The company wishes to levelize these costs into equal annual payments, compute the levelized payments into a fund at 8% to cover these costs.

$$A_{eq} = \sum_{n=1}^{x} P_x(P/F, i, n = x) \ (A/P, i, n)$$
$$= [P_1(P/F, i, n = 1) + P_2(P/F, i, n = 2) + P_3(P/F, i, n = 3)$$
$$+ P_4(P/F, i, n = 4) \ldots + P_n(P/F, i, n = n)](A/P, i, n = n)$$

Given: $P_1 = 200$, $P_2 = 1{,}000$, $P_3 = 750$, $P_4 = 1{,}600$, $P_5 = 400$, $i = 8\%/YR$, $n = 5$

$$A_{eq} = [200(P/F, i = 8, n = 1) + 1000(P/F, i = 8, n = 2)$$
$$+ 750(P/F, i = 8, n = 3) + 1{,}600(P/F, i = 8, n = 4)$$
$$+ 400(P/F, i = 8, n = 5)](A/P, i = 8, n = 5)$$

$$= [200(0.926) + 1{,}000(0.857) + 750(0.794) + 1{,}600(0.735) + 400(0.681)](0.250)$$

= 3,086.10(0.250) = <u>771.53</u>

<u>EQUAL DEPOSITS OF $771.53</u>   Answer

The following table compares the required cash flow to the cash flow provided by the fund.

YEAR	REQUIRED CASH FLOW	INTEREST	DEPOSIT	PAYMENT	FUND BALANCE
1	200.00		771.53	200.00	571.53
2	1,000.00	45.72	771.53	1,000.00	388.78
3	750.00	31.10	771.53	750.00	441.41
4	1,600.00	35.31	771.53	1,600.00	-352.75
5	400.00	-28.14*	771.53	400.00	-8.36

* Interest paid to borrow $351.75 for one year to make up the short-fall.

Fund balance of -8.36 reflects the round-off in interest table factors used to determine fund deposits (A_{eq}).

Bond Calculations

Bonds are used by corporations to raise money. Investors need to evaluate the rate of return that can be realized from a bond versus other investment opportunities. A bond is a corporation's promise to pay the face value or principal at the end of the maturity period. The bond also pays the bond holder interest, <u>taken as semi-annually unless otherwise specified in a problem.</u> The following notation is used for bond calculations:

PW = Present worth or the price paid for the bond. The price paid for a bond is generally not the face value.

F = Face value or principal to be paid to the bond holder at the end of the maturity period.

S = The sum of money realized upon sale of the bond before the end of the maturity period or S = F when the bond is held to maturity.

N = the maturity period specified on the bond, usually expressed in years.

n = the number of interest periods associated with a bond. When interest is paid semiannually, n = 2N.

r = the interest rate, payable semiannually unless otherwise specified in a problem, on the bond face value payable to the bond holder.

i = the investment interest rate per period

The cash flow diagram in Fig. 10.12 depicts payments received from a bond:

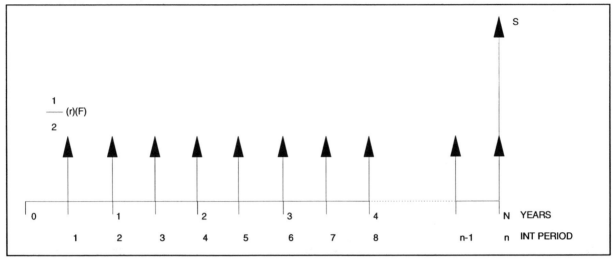

Figure 10.12

The present worth of a bond, or the value an investor would pay for a bond, is equal to the present worth of the maturity value plus the present worth of the series of equal interest payments. Using the appropriate financial formulas, the present worth of a bond is stated as follows:

$$PW = S(P/F, i, n) + [1/2(r)(F)](P/A, i, n)$$

Example 17 What should an investor pay for a $2,000 - 10 year bond paying 6%, if the investor requires an 8% rate of return on the investment?

Given S = F = $2,000, r = 6%/year or 3%/period based on semi-annual interest payments, i = 8%/year or   4%/period N = 10 years, n = 2(10) = 20 interest periods
Find   PW or price to pay for bond in order to receive   required rate of return

PW = S(P/F, i, n) + [1/2(r)(F)](P/A, i, n)
   = 2,000(P/F, i = 4%, n = 20) + [1/2(0.06)(2,000)](P/A,   i = 4%, n = 20)
   = 2,000(0.4564) + [1/2(0.06)(2,000)(13.5903)   = $1,728.22

$1,728.22   Answer

Example 18 What would the investor's rate of return be if the bond described in Example 17 is sold after 6 years for $1,850.00?

Given S = $1,850, F = $2,000, r = 6%/year or 3%/period, n = 2(6) = 12 interest periods
Find i the investors rate of return

Approach:

This problem can be solved by straight trial and error, however the following equation provides a reasonable estimate of the actual interest rate:

Approx. rate of return = $\dfrac{(S - PW)/N + rF}{(S + PW)/2}$

Solution:

Approx. rate of return = $\dfrac{(S - PW)/N + rF}{(S + PW)/2}$

$= \dfrac{(1{,}850 - 1{,}728.22)/6 + 0.06(2{,}000)}{(1{,}850 + 1{,}728.22)/2}$

= 0.0764 or 8%

Assuming 8% interest rate, apply the PW formula and determine if the PW equals the amount paid for the bond

PW = S(P/F, i, n) + [1/2(r)(F)](P/A, i, n)
   = 1,850(P/F, i = 4%, n = 12) + [1/2(0.06)(2,000)](P/A, i = 4%, n = 12)
   = 1,850(0.6246) + 60(9.3851) = 1,699.87

1,699.87 < 1,728.22, therefore the assumed interest rate is not 8% and a new assumption is necessary. RULE: TO RAISE THE PW, LOWER THE ASSUMED INTEREST RATE (RATE OF RETURN)

Assume 6% interest and recalculate the bond PW

PW = S(P/F, i, n) + [1/2(r)(F)](P/A, i, n)
   = 1,850(P/F, i = 3%, n = 12) + [1/2(0.06)(2,000)](P/A, i = 3%, n = 12)
   = 1,650(0.7014) + 60(9.9540) = 1,894.83

The actual rate of return is between 6 and 8% and is determined by interpolation as follows:

10-24 FUNDAMENTALS OF ENGINEERING EXAM REVIEW WORKBOOK

i	BOND PW
6	1,894.83
?	1,728.22
8	1,699.87

$$\frac{(? - 6) - (1,728.22 - 1,894.83)}{(8 - 6) - (1,699.87 - 1,894.83)}$$

$$? = 6 + \frac{(2)(-166.61)}{-194.96} = \underline{7.71\%}$$

<u>7.71% Actual rate of return</u>   Answer

Example 19 An investor is offered a bond for $7,000 with a face value of $10,000 at an interest rate of 5% with maturity in 15 years. What is the rate of return on the investment?

Approach

Calculate the approximate rate of return and the confirm the actual rate with the exact PW formula.

Approx. rate of return = $\frac{(S - PW)/N + rF}{(S + PW)/2}$ NOTE: S = F in this problem

$$= \frac{(10,000 - 7,000)/15 + 0.05(10,000)}{(10,000 + 7,000)/2}$$

$$= \underline{0.082 \text{ or } 8\%}$$

Assuming 8% or 4%/period
PW = S(P/F, i, n) + [1/2(r)(F)](P/A, i, n)
   = 10,000(P/F, i = 4%, n = 30) + [1/2(0.05)(10,000)](P/A, i = 4%, n = 30)
   = 10,000(0.3083) + 250(17.2920) = <u>7,406.00</u>

Since 7,406.00 > 7,000.00, must try a higher interest rate (rate of return to lower the calculated PW. Assume 10%.

PW = S(P/F, i, n) + [1/2(r)(F)](P/A, i, n)
   = 10,000(P/F, i = 5%, n = 30) +
     [1/2(0.05)(10,000)](P/A, i = 5%, n = 30)
   = 10,000(0.2314) + 250(15.3725) = <u>6,157.13</u>

The actual rate of return is between 8 and 10% and is determined by interpolation as follows:

i	BOND PW
8	7,406.00
?	7,000.00
10	6,157.13

$$\frac{(? - 8)}{(10 - 8)} = \frac{(7{,}000.00 - 7{,}406.00)}{(6{,}157.13 - 7{,}406.00)}$$

$$? = 8 + \frac{(2)(-406.00)}{-1{,}248.87} = 8.65\%$$

<u>8.65% Actual rate of return</u>   Answer

The last example indicates the investor will receive 8.65% for the investment. The cost of borrowing the money by the corporation is somewhat higher because of certain expenses associated with issuing bonds and paying the semiannual interest. When calculating the cost of money to the corporation, the following definitions apply:

PW - The amount of money realized by the corporation from the bond issue

AE - The annual expense for distribution of interest payments

S - Redemption value

F - Face value upon which interest is paid

r - Annual bond interest rate

i - Cost of money to the corporation

N - Maturity period of the bonds, years

The PW equation for the corporation is:

$$PW = S(P/F, i, n) + [(1/2)(r)(F) + AE/2](P/A, i, n)$$

This equation accounts for the corporation's expenses associated with making the periodic interest payments. The following example demonstrates the method used to determine the cost of money to a corporation from a bond issue.

<u>Example 20</u> A corporation is selling $1,000,000 of $10,000 face value bonds at an interest rate of 5% with maturity in 20 years. The investment banker handling the sale discounts the offering to $700,000 based on market conditions. Expenses for the sale of the bonds is 5% of the face value.

Annual expenses associated with the bond interest payments are $10,000. What is the cost of money to the corporation?

Approach:

This problem can be solved by straight trial and error, however the following equation provides a reasonable estimate of the actual interest rate:

Approx. rate of return = $\dfrac{(S - PW)/N + [rF + AE]}{(S + PW)/2}$

Solution:

The problem statement indicates that the bonds are sold for $700,000 by an investment banker whose (corporations expense) equals 5% of the face value of the bond issue or $50,000. The corporation therefore actual receives $650,000 for its promise to pay $1,000,000 at the end of 20 years plus semi-annual interest on the face value at 5%. PW = $650,000

Approx. rate of return = $\dfrac{(S - PW)/N + [rF + AE]}{(S + PW)/2}$

= $\dfrac{(1,000,000 - 650,000)/20 + [(0.05)(1,000,000) + 10,000]}{(1,000,000 + 650,000)/2}$

= 9.4%, therefore assume for calculation of PW 9.0%

PW = S(P/F, i, n) + [(1/2)(r)(F) + AE/2](P/A, i, n)
  = 1,000(P/F, i = 4.5, n = 40) + [(1/2)(0.05)(1,000) + 10/2](P/A, i = 4.5, n = 40)
  = 1,000(0.1719) + 30(18.4016) = 723.948

NOTE THIS EQUATION IS WRITTEN IN $1,000.

Since 723.948 > 650.000, the calculated PW must be lowered by raising the interest rate. Assume 10% and recalculate the PW.

PW = S(P/F, i, n) + [(1/2)(r)(F) + AE/2](P/A, i, n)
  = 1,000(P/F, i = 5, n = 40) + [(1/2)(0.05)(1,000) +10/2](P/A, i = 5, n = 40)
  = 1,000(0.1421) + 30(17.1591) = 656.873

Since 656.873 > 650.000, the calculated PW must be lowered by rasising the interest rate. Assume 11% and recalculate the PW.

$$PW = S(P/F, i, n) + [(1/2)(r)(F) + AE/2](P/A, i, n)$$
$$= 1{,}000(P/F, i = 5.5, n = 40) + [(1/2)(0.05)(1{,}000) + 10/2](P/A, i = 5.5, n = 40)$$
$$= 1{,}000(0.1175) + 30(16.0461) = 598.883$$

Now by interpolation:

i	PW
10	656.873
?	650.000
11	598.883

$$\frac{? - 10}{11 - 10} = \frac{650.000 - 656.873}{598.883 - 656.873}$$

$$? = 10 + \frac{1(-6.873)}{-57.99} = 10.12\%$$

<u>10.12% is the Corporate Cost of Money from the bond sale</u>   Answer

DEPRECIATION CALCULATIONS

Depreciation is the method used to recognize that an asset has a finite life and therefore its value decreases with the passage of time. As an asset is used, a company is permitted to depreciate (decrease) the asset's value for the purpose of determining profit and income tax. Depreciation calculations, on the Fundamentals of Engineering Examination, are considered independently to demonstration a working knowledge of the various accepted depreciation methods. This section defines the various depreciation methods and provides examples of questions likely to appear on the Fundamentals of Engineering Examination. Calculations, considering the tax implications of depreciation, which have appeared on the Principles and Practices Examination are covered in Corporate Considerations - Before/After Tax Profit.

There are five (5) methods of depreciation, which are summarized after the following notation definitions:

P is the first cost or purchase cost of an asset.

L is the salvage value of the asset at the end of its useful life.

n is the useful life of the asset. Useful life depends on the user of an asset. For example, a rental car company considers the useful life of a car 3 years with the resale value of the car considered its salvage value. A small company however may intend to keep its car for 10 years and then sell it for salvage (scrap) value.

D_n is the total amount to be depreciated over the useful life of the asset.

D_x is the depreciation charge in the Xth year of useful life. When the depreciation charge does not change with time the symbol D is used. Depreciation charges occur at the end of the year

D_{cum-x} is the cumulative depreciation charge through the Xth year of useful life.

B_x is the book value of the asset after X years. B_x equals the first cost (P) minus the cumulative depreciation through the Xth year (D_{cum-x}).

Straight Line Depreciation Method (SLD)

The SLD method assumes that the decrease in value of an asset is directly proportional to the age of the asset. The appropriate formulas are:

$D_n = P - L$

$D_x = D = (P - L)/n$ NOTE: D_x is constant over time and is therefore generally shown as D.

$D_{cum-x} = D_x(x)$

$B_x = P - D_{cum-x}$

Sinking-Fund Depreciation Method (SFD)

The SFD method assumes that a sinking fund is established in which depreciation charges accumulate. The total depreciation at any year X, is equal to the accumulated value of the sinking fund at the year X. The appropriate formulas are:

$D_n = P - L$

OR $D_x = D = (P - L)(A/F, i\%, n)$
$= (P - L)[i/(1 + i)^n - 1]$ NOTE: D_x is constant over time and is generally shown as D.

$D_{cum-x} = (P - L)(A/F, i\%, n)(F/A, i\%, x)$

OR $= (P - L)\left[\dfrac{(1 + i)^x - 1}{(1 + i)^n - 1}\right]$

$B_x = P - D_{cum-x}$

Declining Balance or Fixed Percentage Depreciation Method (DBD)

The previous two methods, SLD and SFD, resulted in constant or equal annual depreciation charges. The DBD method takes the cost of depreciation as a fixed percentage of the salvage or book value at the end of the previous year. The ratio of the depreciation in any year to the book value at the end of the previous year is constant for the entire useful life of the asset and is given the symbol f. The following formulas apply to this method:

$$f = 1 - \left[\frac{L}{P}\right]^{(1/n)} \qquad \text{f is the fixed percentage}$$

NOTE: Based on this formula an asset can never be depreciated to zero because a salvage value is required to calculate f. This however is not a significant problem since in practice a reasonable value of f is assumed and the calculation of f is therefore not required. If L = 0 in a problem, the value of f must be given in order to solve the problem. The use of this equation is based on the book value in the nth year equaling the salvage value.

$$D_x = f(B_{x-1})$$

$$D_{cum-x} = P[1 - (L/P)^{x/n}] \qquad \text{NOTE: THIS EQUATION SHOULD BE USED ONLY WHEN THE FIXED RATE f IS CALCULATED.}$$

$$B_x = P - D_{cum-x} \quad \text{or} \quad P(1 - f)^x$$

NOTE: THE FIRST EQUATION WHICH IS BASED ON D_{cum-x} SHOULD BE USED ONLY WHEN THE FIXED PERCENTAGE IS CALCULATED, OTHERWISE USE THE EQUATION $B_x = P(1 - f)^x$.

Notice that the DBD method can be used without knowing the useful life of an asset by using the D_x formula and $B_x = P(1 - f)_x$. Also when a fixed percentage is given and the useful life is known, depreciation charges continue until the salvage value is reached.

Double Declining Balance Depreciation Method (DDBD)

The DDBD method uses the following equation to determine the value of f - the fixed percentage:

$$f = 2/n$$

This rate is double the straight line depreciation rate that would occur for an asset having zero salvage value at the end of its useful life. Note that salvage value is not required to determine this rate. Additionally, the book value never stops decreasing, although the depreciation decreases in magnitude. Usually, depreciation continues until the remaining book value equals the salvage value. The equations that should be used for DDBD are:

$$D_x = f(B_{x-1})$$

$$B_x = P(1 - f)_x$$

Sum of the Years Digits Depreciation (SYD) Method

The SYD method requires that the digits corresponding to the number of each year of an assets useful life are listed in reverse order. The sum of these digits are determined and the depreciation factor for any year is the reverse digit for that year divided by the sum digits. The following example, for an asset with a useful life of 6 years, demonstrates the manual method of determining the depreciation factors.

YEAR	DIGITS IN REVERSE ORDER	SYD DEPRECIATION FACTOR
1	6	6/21
2	5	5/21
3	4	4/21
4	3	3/21
5	2	2/21
6	1	1/21
	21	

The depreciation for any year is the product of the SYD Depreciation Factor for the year and depreciable value D_n. The following equations aid in the calculations of SYD:

$$\text{SUM OF DIGITS} = n(n+1)/2$$

$$D_x = \frac{2(n+1-x)(P-L)}{n(n+1)}$$

where $\frac{2(n+1-x)}{n(n+1)}$ is the SYD depreciation factor.

$$D_{cum-x} = \frac{x(2n+1-x)(P-L)}{n(n+1)}$$

$$B_x = P - D_{cum-x}$$

<u>Example 21</u> A new asset is purchased for $1200, with an estimated salvage value of $200 and a useful life of 7 years. What will be the annual depreciation cost, the cumulative depreciation cost, and the book value of the asset at the end of the 4th year? Interest should be taken at 5%. Use the SLD, SFD, DBD, DDBD, and SYD methods. Also develop a depreciation schedule for each of the methods.

(a) SLD

$$D = (P - L)/n = (1200 - 200)/7 = \underline{\$142.86}$$

$$D_{cum-4} = D(x) = 142.86(4) = \underline{\$571.44}$$

$B_4 = P - D_{cum-4} = 1200 - 571.44 = \underline{\$628.56}$

The total depreciation schedule is:

BEGINNING OF YEAR	BEGINNING OF YEAR BOOK VALUE	END OF YEAR ANNUAL DEPRECIATION	CUMULATIVE DEPRECIATION	END OF YEAR BOOK VALUE
0				1,200.00
1	1,200.00	142.86	142.86	1,057.14
2	1,057.14	142.86	285.72	914.28
3	914.28	142.86	428.58	771.42
4	771.42	142.86	571.44	628.56
5	628.56	142.86	714.30	485.70
6	485.70	142.86	857.16	342.84
7	342.84	142.86	1,000.02	199.98*

* This value should be $200 but due to roundoff of the annual depreciation the calculation is not exact.

(b) SFD

$D = (P - L)(A/F, i = 5, n = 7) = (1,200 - 200)(0.123) = \underline{\$123.00}$

$D_{cum-4} = (P - L)(A/F, i = 5, n = 7)(F/A, i = 5, n = 4)$
$= (1,200 - 200)(0.123)(4.310) = \underline{\$530.13}$

$B_4 = P - D_{cum-4} = 1,200 - 530.13 = \underline{\$669.87}$

The total depreciation schedule is:

BEGINNING OF YEAR	BEGINNING OF YEAR BOOK VALUE	END OF YEAR ANNUAL DEPRECIATION	CUMULATIVE DEPRECIATION*	END OF YEAR BOOK VALUE**
0				1,200.00
1	1,200.00	123.00	123.00	1,077.00
2	1,077.00	123.00	252.15	947.85
3	947.85	123.00	387.82	812.18
4	812.18	123.00	530.13	669.87
5	669.87	123.00	679.70	520.30
6	520.30	123.00	836.65	363.35
7	363.35	123.00	1,001.47	198.53

* Cumulative depreciation must consider interest credited to the depreciation sinking fund, $D_{cum-x} = D(F/A, i, x)$ ** Book value equals P - Cumulative depreciation

(c) DBD

Since f is not given it should be determined

$$f = 1 - \left[\frac{L}{P}\right]^{(1/n)} = 1 - \left[\frac{200}{1200}\right]^{(1/7)} = \underline{0.226}$$

$$D_4 = f(B_{x-1}) = 0.226[P(1-f)^{x-1}] = 0.226[1200(1-0.226)]_3 = \underline{\$125.75}$$

$$D_{cum-4} = P[1-(L/P)_{x/n}] = 1200[1 - (200/1200)_{4/7}] = \underline{\$768.95}$$

$$B_4 = P - D_{cum-4} = 1200 - 768.95 = \underline{\$431.05}$$

The total depreciation schedule is:

BEGINNING OF YEAR	BEGINNING OF YEAR BOOK VALUE	END OF YEAR ANNUAL DEPRECIATION	CUMULATIVE DEPRECIATION	END OF YEAR BOOK VALUE
0				1,200.00
1	1,200.00	271.20	271.20	928.80
2	928.00	209.91	481.11	718.89
3	718.89	162.47	643.58	556.42
4	556.42	125.75	769.33	430.67
5	430.67	97.33	866.66	333.34
6	333.34	75.33	941.99	258.01
7	258.01	58.31	1,000.30	199.70

$$* D_x = f(B_{x-1})$$

(d) DDBD

for DDBD $f = 2/n = 2/7 = \underline{0.286}$

The IRS limits any annual depreciation charge, using the DDBD method to twice the charge that would be obtained using the SLD method. The first year's depreciation must be checked. If the limitation applies, then the depreciation formulas cannot be used and a depreciation schedule must be developed.

$$D_1 = f(B_{x-1}) = f[P(1-f)^{x-1}] = 0.286[1,200(1-0.286)^0] = 343.20$$

The SLD charge is 142.86, twice the rate is 285.72, therefore the limitation applies and the following schedule must be developed:

ENGINEERING ECONOMICS 10-33

BEGINNING OF YEAR	BEGINNING OF YEAR BOOK VALUE	END OF YEAR ANNUAL DEPRECIATION	CUMULATIVE DEPRECIATION	END OF YEAR BOOK VALUE
0				1,200.00
1	1,200.00	285.71*	285.71	914.29
2	914.29	261.49	547.20	652.80
3	652.80	186.70	733.90	466.10
4	466.10	133.30	867.20	332.80
5	332.80	95.18	962.38	237.62
6	237.62	37.62**	1,000.00	200.00
7	200.00	0.00	1,000.00	200.00

* Any year's depreciation charge is limited to twice the charge that would result from SLD, in this case $285.72

** Depreciation charge is limited to amount that results in cumulative depreciation of $1,000.00

(e) SYD

Sum of the Digits = $n(n + 1)/2 = 7(7 + 1)/2 = \underline{28}$

$$D_4 = \frac{2(n + 1 - x)(P - L)}{n(n + 1)} = \frac{2(7 + 1 - 4)(1200 - 200)}{7(7 + 1)}$$
$$= \$142.86$$

$$D_{cum-4} = \frac{x(2n + 1 - x)(P - L)}{n(n + 1)} = \frac{4[2(7) + 1 - 4](P - L)}{7(7 + 1)}$$
$$= \$785.71$$

$B_4 = P - D_{cum-4} = 1,200.00 - 785.71 = \underline{\$414.29}$

The total depreciation schedule is shown below:

BEGINNING OF YEAR	BEGINNING OF YEAR BOOK VALUE	FACTOR	END OF YEAR ANNUAL DEPRECIATION	CUMULATIVE DEPRECIATION	END OF YEAR BOOK VALUE
0					1,200.00
1	1,200.00	7/28	250.00	250.00	950.00
2	950.00	6/28	214.29	464.29	735.71
3	735.71	5/28	178.57	642.86	557.14
4	557.14	4/28	142.86	785.72	414.28
5	414.28	3/28	107.14	892.86	307.14
6	307.14	2/2	871.43	964.29	235.71
7	235.71	1/2	835.71	1,000.00	200.00

Example 22 A new $15,000 car will depreciate over 5 years based on SYD method of depreciation. The first year's depreciation is 3,400. (A) Find the end of year book value for each year. (B) If the car depreciates to the same value, over the same time period according to SFD at 8% interest, what would the depreciation charge be?

PART A

Sum of the Digits = $n(n + 1)/2 = 5(6)/2 = 15$

$D_1 = 5/15(P - L) = 3,400 = 5/15(P - L)$ by rearranging
$(P - L) = 10,200$

n	$D_{cum\text{-}x}$ $= \dfrac{x(2n + 1 - x)(P - L)}{n(n + 1)}$	$B_x = P - D_{cum\text{-}x}$ $= 15,000 - D_{cum\text{-}x}$
1	3,400	11,600
2	6,120	8,880
3	8,160	6,840
4	9,520	5,480
5	10,200	4,800

PART B

For SFD use

$D = (P - L)(A/F, i\%, n) = 10,200(A/F, i = 8, n = 5)$
$= 10,200(0.1705) = \underline{\$1,739.10}$

COMPARATIVE ECONOMIC ANALYSIS METHODS

Profit in Business

Investors of capital expect that their capital will be used profitably. In fact, a business must earn profits to assure that investors will continue to make capital available. The purpose of any economic study comparing alternatives is to maximize profit. Profit is defined as:

PROFIT = REVENUES - COSTS

Profit can be considered on a before tax (P_B) or an after tax (P_A) basis. Costs can be categorized as follows:

AE - annual expenses including operation and maintenance costs that occur each year. Examples are: utility costs, labor costs, supplies and administrative costs.

D - Depreciation, which is estimated using one of the depreciation methods discussed in the previous section. This cost reflects the utilization of an asset's useful life by the firm to obtain its revenue.

I - Interest paid for the use of borrowed capital. This is a financial cost associated only with borrowed capital. If only equity capital (owner's capital) is used by a business, then this cost does not exist. Profits are the pay back for the use of equity capital by the business.

T - Income taxes depends upon before tax profits.

R - Gross revenue received by a business for the sale of goods and services.

Profit, P_B and P_A, can therefore be expressed as follows:

$$P_B = R - (D + I + AE)$$

$$P_A = R - (D + I + AE) - T$$

When comparing engineering alternatives, it is assumed that gross revenues are not influenced by the means used to produce goods or services, and are therefore fixed. The maximization of profit is obtained when costs are minimized. Methods of comparing engineering alternatives therefore require no knowledge of gross revenues to select the best alternative from an economic viewpoint.

There are two (2) basic methods for comparing alternatives: the total annual cost or present worth methods. Both of these methods address before tax costs based on an appropriate depreciation method. Income tax considerations are addressed when considering the rate of return method of evaluating investments.

Total Annual Cost (TAC) Method

The total annual cost method is based on the following rearrangement of the before tax profit equation:

$$R = P_B + (D + I + AE)$$

remembering that P_B is the return on equity capital, an alternatives total annual cost is defined as:

$$TAC = (D + I) + AE$$

The quantity (D + I) is known as capital recovery and accounts for the depreciation of an asset and the interest paid on borrowed capital. There are two equations used to calculate TAC, the exact equation and the approximate equation.

The exact equation is based on sinking fund depreciation and is expressed as follows:

$$TAC = [(P - L)(A/P) + L(i)] + AE$$

where $[(P - L)(A/P) + L(i)]$ is defined as CR or capital recovery

The A/P factor is known as the capital recovery factor (CRF) because of the following relationship:

$$CRF = SFF + i$$
$$OR$$
$$(A/P) = (A/F) + i$$

Now:

$$CR = (P - L)(A/P) + L(i)$$

$$= P(A/P) - L(A/P) + L(i)$$

By substituting $(A/F) + i$ for (A/P)

$$= [P(A/F) + P(i)] - [L(A/F) + L(i)] + L(i)$$

$$= (P - L)(A/F) + P(i)$$

Therefore

$$CR = (P - L)(A/P) + L(i) \text{ or } (P - L)(A/F) + P(i)$$

The equation $CR = (P - L)(A/P) + L(i)$ is preferred because when $L = 0$ the equation for capital recovery simplifies to

$$CR = P(A/P)$$

The exact TAC method is based on the sinking fund depreciation method since the depreciation component of the capital recovery (CR) expression used is based on sinking fund depreciation. The exact total annual cost method should be used unless a problem specifically states use the approximate method or indicates that depreciation should be based on the straight line method.

The following approximate equation is used to calculate the TAC when straight line depreciation method is used:

$$TAC = \left[\frac{(P - L)}{n} + \frac{(P - L) i (N - 1)}{2n} + L(i)\right] + AE$$

The quantity in [ ] is the capital recovery of this equation.

ENGINEERING ECONOMICS 10-37

Notice that either the exact or approximate method requires the same information to solve for the TAC. The parameters are; P - first cost or investment cost, L - salvage value, AE - annual expenses, i - the interest rate, rate of return or corporate cost of money, and n - years of investments life.

To use the annual cost method to compare alternatives, all costs must be stated on an equal annual cost basis. The two depreciation methods used in the exact or approximate TAC methods, SFD or SLD respectively, yield equal annual depreciation costs. Annual Expenses (AE) must always be stated on an equal annual basis. Once all costs are on an equal annual basis the costs can be added to determine TAC.

The TAC method easily handles alternatives with different asset lives since capital recovery costs (D + I) are based on end of year payments and replacements are assumed to occur at the same cost. The following examples demonstrate the application of the TAC method.

Example 23 A city must convey sewage through or over a rock ridge to its treatment plant. Plan A calls for a 5,000 ft gravity sewer in rock tunnel, which costs $300/ft. Plan B requires the construction of a pumping station for $750,000, with a salvage value of $150,000, and a 5,000 ft forced main at a cost of $125/ft. The cost to maintain the gravity or forced main are the same, the pumping station has an annual operating and maintenance cost of $35,000.

Capital costs in each case will be financed by 30 yr, 5% bonds retired by equal annual payments combining interest and principal. Compute the annual cost for each plan, state which is more economical, and show the difference in annual cost.

Solution

The problem requires solution by the TAC method, and since it does not indicate the type of depreciation method (SFD or SLD) or the TAC method (exact or approximate) the exact method should be used, therefore:

$$TAC = [(P - L)(A/P) + L(i)] + AE$$

For PLAN A:

P = 5,000(300) = $1,500,000
L = 0, since no information is provided
AE = ?, no value is given, however, the cost of maintaining the gravity tunnel is equal to the cost of maintaining the forced main tunnel of Plan B, therefore the cost of maintaining the tunnels does not have to be considered in the comparative analysis.
n = 30 years, note the only information given relates to the length of the financing period which is therefore taken as the life of each asset in order to solve this problem. No other approach is reasonable since the economics problem is a machine graded problem.
i = 5% since this is the rate paid on the bonds used to finance either alternative.

since L = 0, TAC = P(A/P) + AE

$$= 1,500,000(A/P, i = 5, n = 30) + 0$$
$$= 1,500,000(0.0651) = \underline{\$97,650}$$

For PLAN B:

P = 750,000 for the pumping station and 5,000(125) for the forced main.
   = 750,000 + 5,000(125)
   = 1,375,000

NOTE: INVESTMENTS CAN BE ADDED BECAUSE THEY HAVE THE SAME USEFUL LIFE.

L = 150,000 for the pumping station
AE = 35,000 for the pumping station, which represents the additional cost associated with Plan B over Plan A since maintenance on the gravity or forced sewer mains are equal.
n = 30 yr, i = 5% the same as Plan A

$$\text{TAC} = [(P - L)(A/P) + L(i)] + AE$$

$$= [(1,375,000 - 150,000)(A/P, i = 5, n = 30) + 150,000(0.05)] + 35,000$$
$$= [(1,225,000)(0.0651) + 7,500] + 35,000$$
$$= \underline{\$122,248} \text{ (answer rounded to nearest dollar)}$$

Plan A is more economical than Plan B because Plan A's TAC is less than Plan B's. The difference in TAC is $24,598.

Example 24 In planning of a state police radio system, it is desired to maintain a specified minimum signal strength at all points in the state. Two plans for accomplishing this are proposed for comparison.

Plan 1 involves the establishment of six transmitting stations of low power. The investment at each of these in buildings, ground improvements, piping and tower, all assumed to have a life of 25 years is estimated as $35,000. The investment in transmitting equipment for each station, assumed because of the probability of obsolescence, to have a life of 8 years, is estimated as $25,000. The monthly disbursements for operation and maintenance of each station are estimated to be $1,000.

Plan 2 involves the establishment of only two transmitting stations of ultrahigh power. The investment in buildings, ground improvements, etc at these stations is $75,000. The salvage value of these stations, after considering the cost for removal is $15,000. The useful life of the stations are estimated at 25 years. The investment in state of the art transmitting equipment at each station is estimated at $200,000; and the life of this equipment is estimated to be 12 years. The monthly O & M cost is established as $1,400 per station.

Compare the annual cost of these plans based on straight line depreciation. The cost of money is 8% per annum.

Solution:

The approximate TAC method should be used because the problem states that straight line depreciation is to be used. The approximate TAC equation is

$$TAC = \left[\frac{(P-L)}{n} + \frac{(P-L)\,i\,(N-1)}{2n} + L(i)\right] + AE$$

Let P_B = investment for buildings
P_T = investment for transmitting equipment

Plan 1

$P_B = 35,000(6) = 210,000$; $L = 0$; $n = 25$ years
$P_T = 25,000(6) = 150,000$; $L = 0$; $n = 8$ years

NOTE: THE INVESTMENTS CANNOT BE ADDED SINCE THEY HAVE DIFFERENT USEFUL LIVES. THE CAPITAL RECOVERY FOR EACH INVESTMENT MUST THEREFORE BE CALCULATED SEPARATELY AND THEN ADDED AS ANNUAL COSTS.

AE = 6 STATIONS x 1,000 PER STATION PER MONTH x 12 MONTHS PER YEAR
= 72,000

$TAC = CR_B + CR_T + AE$

where $CR = \dfrac{(P-L)}{n} + \dfrac{(P-L)(i)(N-1)}{2(n)} + L(i)$

$$TAC = \frac{210,000}{25} + \frac{210,000(0.08)(26)}{2(25)} + \frac{150,000}{8} + \frac{150,000(0.08)(9)}{2(8)} + 72,000$$

= [17,136] + [25,500] + 72,000 = $114,636

Plan 2

$P_B = 75,000(2) = 150,000$; $L = 15,000(2) = 30,000$; $n = 25$ years
$P_T = 200,000(2) = 400,000$; $L = 0$; $n = 12$ year

AE = 2 STATIONS x 1,400 PER STATION PER MONTH x 12 MONTHS PER YEAR
= 33,600

$$TAC = CR_B + CR_T + AE$$

where $\quad CR = \dfrac{(P - L)}{n} + \dfrac{(P - L)(i)(N - 1)}{2(n)} + L(i)$

$$TAC = \dfrac{(150,000 - 30,000)}{25} + \dfrac{(150,000 - 30,000)(0.08)(26)}{2(25)} + 30,000(0.08)$$
$$+ \dfrac{400,000}{12} + \dfrac{400,000(0.08)(13)}{2(12)} + 33,600$$

$= [12,192] + [50,666] + 33,600 = \underline{\$96,458}$

Plan 2 is preferred to Plan 1 because its TAC is $18,178 less than Plan 1.

The next example problem includes assets with permanent or perpetual lives. Such an asset has no salvage value and an infinite life. Capital recovery is calculated as follows:

$\quad CR = P(i)$ when $n = \infty$
because
$\quad CR = (P - L)(A/P) + L(i)$
but $L = 0$, therefore
$\quad CR = P(A/P)$
when $n = \infty$ the A/P approaches i. As an example, looking at the 10% interest tables would yield the following:

$\underline{i = 10\%}$

n	A/P
10	0.1628
20	0.1175
40	0.1023
60	0.1003
80	0.1001
100	0.1000

<u>Example 25</u> Two alternative locations for a new aqueduct are being considered. If interest is 5%, determine which alternative location results in the lowest annual cost. Location 1 requires a rock tunnel 1,000 ft long which costs $100/ft to dig. The tunnel is considered permanent, with annual costs of $1,000/yr and an additional $5,000 every ten years for an inspection and repairs. A flume costing $45,000, with an estimated life of 20 years is also required. Maintenance of the flume will cost $2,000/yr.

Location 2 requires a pipe line costing $35,000 with a 50 year life and an annual maintenance cost of $1,200, a permanent earth canal costing $40,000, with annual costs of $1,000 and a concrete lining with a first cost of $20,000, a life of 25 years and maintenance costs of $300/yr.

Solution

This problem provides no information regarding salvage value, therefore L = 0 for all investments. Each alternative has multiple investments with different useful lives, therefore each investments capital recovery must be handled separately. Interest is given as 5%.

LOCATION 1

$$TAC = CR_T + CR_F + AE_T + AE_F$$

where the subscript T is for the tunnel and F is for the flume.

Recognizing $L = 0$, $n_T = ????$, and $n_F = 20$ years

$$TAC = [P_T(i)] + [P_F(A/P, i = 0.05, n = 20] + AE_T + AE_F$$

now

$P_T = 1,000(100) = 100,000$
$P_F = 45,000$
$AE_T = 1,000 + 5,000$ every 10 years

NOTE: THE 5,000 EVERY TEN YEARS MUST BE CONVERTED TO AN EQUAL ANNUAL COST AS FOLLOWS

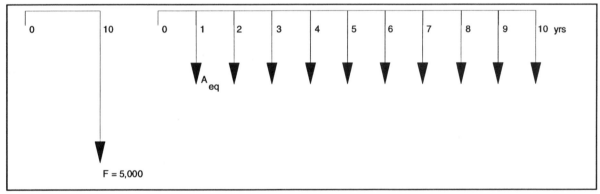

Figure 10.13

$A_{eq} = F(A/F, i = 0.05, n = 10)$
   $= 5,000(0.0795) = \underline{\$397.50}$

now

$AE_T = 1,000 + 397.50 = 1,397.50$
$AE_F = 2,000$

now

$$TAC = 100{,}000(0.05) + 45{,}000(0.0802) + 1{,}397.50 + 2{,}000$$
$$= \underline{\$12{,}006.50}$$

LOCATION 2

$$TAC = CR_P + CR_C + CR_L + AE_P + AE_C + AE_L$$

$P_P = 35{,}000, \ n = 50 \text{ yrs}, \ AE_P = 1{,}200$
$P_C = 40{,}000, \ n = \ , \ AE_C = 1{,}000$
$P_L = 20{,}000, \ n = 25 \text{ yrs} \ AE_L = \ 300$

$$TAC = [P_P(A/P, i = 0.05, n = 50)] + [P_C(i)] +$$
$$[P_L(A/P, i = 0.05, n = 25)] + AE_P + AE_C + AE_L$$

$$= 35{,}000(0.0548) + 40{,}000(0.05) + 20{,}000(0.0710) + 1{,}200 + 1{,}000 + 300$$
$$= 1{,}918 + 2{,}000 + 1{,}420 + 1{,}200 + 1{,}000 + 300$$
$$= \underline{\$7{,}838.00}$$

Location 2 is more economical and results in a $4,168.50 yearly savings.

Example 26 An underground gas main, placed in a corrosive environment was installed for $12,000/mile. The main is projected to develop leaks over time because of the corrosive environment. Leaks are not expected for the first 12 years. It is estimated that a mile of gas main will lose $75/mile worth of gas in the 13th year, and the loss will increase at $50/mile/year thereafter. The minimum acceptable rate of return is 10%. Determine if the main should be retired in 20 or 25 years.

Solution

NOTE: THIS PROBLEM CONTAINS A GRADIENT SERIES AND DELAYED ANNUAL COSTS BEGINNING IN THE 13th YEAR. THESE COSTS MUST BE CONVERTED TO EQUAL ANNUAL COSTS STARTING IN YEAR 1.

$P = 12{,}000$ for both alternative retirement dates, $L = 0$,
$i = 10\%$

20 YEAR ALTERNATIVE

$$AC = (P - L)(A/P) + L(i) + AE_{eq}$$

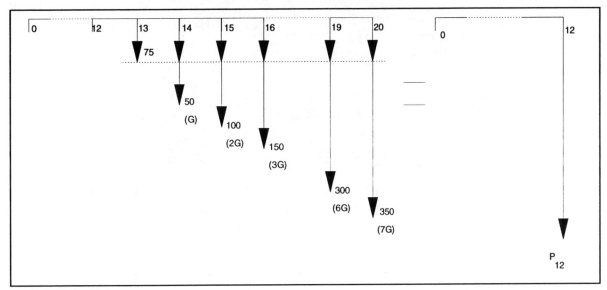

Figure 10.14

Convert delayed annual costs to equal annual costs as shown in Fig. 10.14.

$$P_{12} = A(P/A, i = 10, n = 8) + g(P/G), i = 10, n = 8)$$
$$= 75(5.3349) + 50(16.03)$$
$$= 400.12 + 800.24 = 1,200.36$$

in order to get equal annual payments starting in year 1 requires converting P_{12} to P_0 as follows:

$$P_0 = P_{12} (P/F, i = 10, n = 12)$$
$$= 1,200.36(0.3186) = 382.43$$

P_0 is now converted to AE_{eq} as follows:

$$AE_{eq} = P_0(A/P, i = 10, n = 20)$$
$$= 382.43(0.1175) = 44.94$$

now

$$TAC = 12,000(A/P, i = 10, n = 20) + 44.94$$
$$= 12,000(0.1175) + 44.94 = \underline{1,454.94}$$

25 YEAR ALTERNATIVE

since L = 0,
TAC = P(A/P) + AE_{eq}, using the same approach as for the 20 year alternative

$$P_{12} = A(P/A, i = 10, n = 13) + g(P/G, i = 10, n = 13)$$
$$= 75(7.1034) + 50(33.39)$$
$$= 532.76 + 1,669.30 = 2,202.06$$
$$P_0 = P_{12}(P/F, i = 10, n = 12)$$
$$= 2,206.06(0.3186) = 701.58$$
$$AE_{eq} = P_0(A/P, i = 10, n = 25)$$
$$= 701.58(0.1102) = 77.31$$

TAC = 12,000(A/P, i = 10, n = 25) + 77.31
= 12,000(0.1102) + 77.31 = 1,322.40 + 77.31
= <u>1,399.71</u>

The gas main should be retired in 25 years since this alternative results in a total annual cost $55.23 less than the 20 year alternative.

TOTAL ANNUAL COST INCLUDING RISK

Risk is expressed as a probability of loss or additional cost. For example, there is a 10% chance of the bridge collapsing or there is a 20% chance that a larger building will be required in 6 years. Evaluation of risk has a place in economics when determining the total annual cost of an alternative.

Risk is treated as an annual cost factor when the chance or probability of an event and the loss or additional cost associated with the occurrence of the event are known. The cost factor is calculated as the product of the probability of the event and the loss or additional cost associated with the occurrence of the event. The total annual cost equation with the risk component is expressed as follows:

TAC = (P - L)(A/P) + L(i) + AE + RISK

Example 27 The city of San Francisco is considering three alternatives for athletic field bleachers. A consulting engineer has prepared the following data:

Alternate 1 - Concrete bleachers with a first cost of $400,000, a life of 100 years, annual upkeep of 1,000/yr, and major repairs every 20 years costing $20,000. Should a mild earthquake occur, probability is 20% or one occurrence in five years, some structural damage would occur for which repair costs are estimated to be $75,000.

Alternate 2 - Steel bleachers with a first cost of $275,000, salvage value equals 10% of the first cost, a life of 50 years and annual up keep of $5,000/yr. Damage from a mild earthquake would be significant and repairs would cost $150,000.

Alternate 3 - Wooden bleachers costing $150,000, with a life of 25 years, and annual up keep of $8,000/year. A mild earthquake would require total replacement of the bleachers.

Determine the most cost effective alternative for the bleachers if the project will be financed with 8% bonds.

Solution

TAC = (P - L)(A/P) + L(i) + AE + RISK

Alternate 1

$$TAC = 400{,}000(A/P, i = 8\%, n = 100) + [1{,}000 + 20{,}000(A/F, i = 8\%, n = 20)]$$
$$\quad + 0.2(75{,}000)$$
$$= 400{,}000(0.0800) + [1{,}000 + 20{,}000(0.0219)] + 15{,}000$$
$$= 32{,}000 + 1{,}438 + 15{,}000$$
$$= \underline{\$48{,}438.00}$$

Alternate 2

$$TAC = [275{,}000 - 275{,}000(0.10)](A/P, i = 8\%, n = 50) + [275{,}000(0.10)](0.08) + 5{,}000$$
$$\quad + 0.2(150{,}000)$$
$$= 247{,}500(0.0817) + 2{,}200 + 5{,}000 + 30{,}000$$
$$= \underline{\$57{,}420.75}$$

Alternate 3

$$TAC = 150{,}000(A/P, i = 8\%, n = 25) + 8{,}000 + 0.2(150{,}000)$$
$$= 150{,}000(0.0937) + 8{,}000 + 30{,}000$$
$$= 14{,}055, + 8{,}000 + 30{,}000$$
$$= \underline{\$52{,}055.00}$$

The city should build the concrete bleachers since they result in the lowest Total Annual Cost.

Example 28 A small radio station is going to construct a new tower for their transmitting antenna. The wind design criteria has been reviewed by a consulting engineer, the pertinent data is summarized in the following table. If each tower has a useful life of 30 years, salvage value is offset by removal costs and interest is 8%. (A) Determine the most economical design criteria. (B) If insurance can be purchased for 5% of the potential repair/rebuild cost, which design criteria is best? (C) If Tower III is chosen, should the insurance be purchased?

TOWER	I	II	III
Wind Design Criteria - mph	100	125	150
Probable frequency of exceeding design wind	1 in 6 yrs	1 in 10 yrs	1 in 24 yrs
Cost of Tower -	$ 100,000	125,000	150,000
Annual Maintenance - $/yr	7,500	9,400	11,200
Estimated cost to repair/ rebuild due to wind damage	$ 75,000	95,000	100,000

Solution

Part A

L = 0, therefore,

TAC = P(A/P, i = 8, n = 30) + AE + RISK

TOWER I: TAC = 100,000(0.0888) + 7,500 + 1/6(75,000)
            = 28,883.00

TOWER II: TAC = 125,000(0.0888) + 9,400 + 1/10(95,000)
            = 30,003.75

TOWER III: TAC = 150,000(0.0888) + 11,100 + 1/24(100,000)
            = 28,591.16

The most economical design criteria without insurance, including the risk of potential tower damage is the 150 mph wind.

Part B

If insurance is purchased, the cost of insurance replaces the cost of the risk of damage due to wind.

TOWER I: TAC = 100,000(0.0888) + 7,500 + (0.05)(75,000)
             = 20,133.00

TOWER II: TAC = 125,000(0.0888) + 9,400 + (0.05)(95,000)
             = 25,253.75

TOWER III: TAC = 150,000(0.0888) + 11,100 + (0.05)(100,000)
             = 29,174.49

The most economical design criteria using the cost of insurance to eliminate the risk of damage due to wind is the 100 mph wind.

Part C

Tower III's design wind criteria is 150 mph. The potential cost of damage due to wind (risk) is 1/24(100,000) = 4,166.67. The cost of insurance to eliminate this risk is (0.05)(100,000) = 5,000. Therefore the insurance should not be purchased.

PRESENT WORTH METHOD

The present worth method considers all periodic or annual costs associated with an alternative and converts these recurring costs to their present value using the appropriate financial formulas. The present worth of the recurring costs are then added to the investment to get the total present worth of an alternative. The present worth of an alternative is sometimes referred to as the present worth of revenue requirements. The alternative with the lowest present worth of revenue requirements is the preferred alternative. Present worth analysis is best suited for comparison of alternatives with equal lives. Although the present worth method can be applied to the comparison of alternatives with different lives, the calculations are more complex.

The present worth method is better suited to deal with inflation or increase in future investment costs than the total annual cost method. The following examples will demonstrate the use of the present worth method. The general equation used to calculate the present worth of an alternative is:

$$PW = P + AE(P/A, i, n) - L(P/F, i, n)$$

Example 29 Two machines are to be compared for a 10 year service life using the present worth method. The cost of Machine A is $10,000. Annual disbursements for operation, maintenance, property taxes, and insurance are estimated as $2,000. The cost of Machine B is $7,000. Annual disbursements are estimated as $2,500. Both machines are estimated to have zero salvage value at the end of ten years. The minimum acceptable return is 8%.

Solution

$$PW = P + AE(P/A, i, n) - L(P/F, i, n)$$

For Machine A

$$P = 10,000, L = 0, n = 10 \text{ yrs}, i = 8, AE = 2,000$$
$$PW = 10,000 + 2,000(P/A, i = 8, n = 10)$$
$$= 10,000 + 2,000(6.7100) = 10,000 + 13,420$$
$$= \underline{\$23,420}$$

For Machine B

$$P = 7,000, L = 0, n = 10 \text{ yrs}, i = 8, AE = 2,500$$
$$PW = 7,000 + 2,500(P/A, i = 8, n = 10)$$
$$= 7,000 + 2,500(6.7100) = 7,000 + 16,775$$
$$= \underline{\$23,775}$$

Machine A should be purchased since its PW is $355 less than the PW of Machine B.

Example 30 Wood and steel structures are being considered for a foot bridge. The wooden structure requires a $10,000 first cost, annual painting and general maintenance is estimated to cost 1,800, a major overhaul will be required every 5 years and cost $3,500. The steel bridge will cost $25,000 with annual upkeep estimated to cost $1,000, and only one overhaul will be required in the 15th year and cost $6,000. The steel bridge will have a salvage value of $5,000 at the end of its life in 25 years. The wooden bridge will have no salvage value at the end of 25 years, but an expense of $4,000 will be required to remove the structure. Using present worth comparison determine the best alternative. Interest is 10%.

Solution

$$PW = P + AE(P/A, i, n) - L(P/F, i, n)$$

Wooden Foot Bridge

P = 10,000, L = 0, n = 25, i = 10, AE = 1,800 plus 3,500 every five years. NOTE: It should be recognized that the 3,500 is not spent in the 25th year, however the removal cost of 4,000 must be included.

The present worth of annual expenditures must account for all expenses, note how the expenses are accounted for in the following PW equation.

PW = 10,000 + 1,800(P/A, i = 10, n = 25) + 3,500[(P/F, i = 10, n = 5) + (P/F, i = 10, n = 10) + (P/F, i = 10, n = 15) + (P/F, i = 10, n = 20)] + 4,000(P/F, i = 10, n = 25)
= 10,000 + 1,800(9.0770) + 3,500[(0.6209) + (0.3855) + (0.2394) + (0.1486)] + 4,000(0.0923) = 10,000 + 16,338.60 + 4,880.40 + 369.20 = 31,588.20

Steel Foot Bridge

P = 25,000, L = 5,000, n = 25, i = 10, AE = 1,000 plus 6,000 in the 15th year.

PW = 25,000 + 1,000(P/A, i = 10, n = 25) + 6,000(P/F, i = 10, n = 15) - 5,000(P/F, i = 10, n = 25) = 25,000 + 1,000(9.0770) + 6,000(0.2394) - 5,000(0.0923) = 25,000 + 9,077 + 1,436.40 - 461.5 = 35,051.90

The wooden foot bridge should be built since it has the lower present worth.

Example 31 Solve Example 23 using the present worth method. Example 23 is restated as follows:

A city must convey sewage through or over a rock ridge to its treatment plant. Plan A calls for a 5,000 ft gravity sewer in rock tunnel, which costs $350/ft. Plan B requires the construction of a pumping station for $750,000, with a salvage value of $150,000, and a 5,000 ft forced main at a cost of $125/ft. The cost to maintain the gravity or forced main are the same, the pumping station has an annual operating and maintenance cost of $35,000.

Capital costs in each case will be financed by 30 yr, 5% bonds retired by equal annual payments combining interest and principal. Compute the present worth for each plan, state which is more economical, and show the difference in present worth.

Solution

$$PW = P + AE(P/A, i, n) - L(P/F, i, n)$$

For Plan A

$$P = 5,000(350) = 1,500,000, \ L = 0, \ AE = 0 \ n = 30$$
$$i = 5\%$$
$$PW = 1,500,000 + 0 + 0 = \underline{\$1,500,000}$$

For Plan B

$$P = 750,000 + 5,000(125) = 1,375,000, \ L = 150,000$$
$$AE = 35,000, \ n = 30, \ i = 5\%$$
$$PW = 1,375,000 + 35,000(P/A, i = 5, n = 30) - 150,000(P/F, i = 5, n = 30) = 1,375,000 + 35,000(15.3725) - 150,000(0.2314) = 1,375,000 + 538,037.50 - 34,710$$
$$= \underline{\$1,878,327.50}$$

Plan A is more economical than Plan B since Plan A's present worth is $378,327.50 less than Plan B's.

Present worth analysis is easiest to use when alternatives have the same economic lives. When alternatives have different economic lives the use of present worth analysis becomes more difficult than the previous examples. Use of present worth analysis when alternatives have different lives requires that the first point in time when replacement of both assets occurs be determined. As an example consider two assets, the first having a life of 3 years and the second having a 4 year life. The first point in time that a replacement of both assets would occur is in the 12th year as can be seen by the following cash flows. The replacement is not shown however because this is the point where the present worth analysis will end. Twelve years is the least common multiple for alternatives having 3 and 4 years of life. The number of replacements to be considered in the present worth analysis can be calculated as follows:

no. of replacements = (c/n - 1)

for c = 12 and n = 3, no. of replacements = (12/3 - 1) = 3

for c = 12 and n = 4, no. of replacements = (12/4 - 1) = 2

Asset A Cash Flow

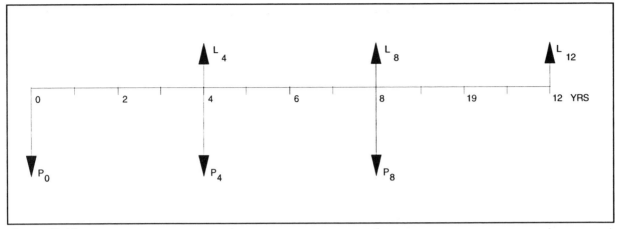

Figure 10.15

Asset B Cash Flow

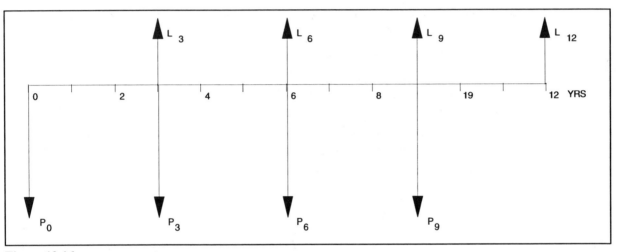

Figure 10.16

Note that these diagrams assume that each replacement costs the same as the first cost. If a problem indicates that the replacement costs are different than the first cost then the present worth method should be used. Problems dealing with escalating investment costs are therefore handled using present worth calculations. When the replacements are the same as the first cost the annual cost method is the preferred method unless the problem requires that present worth analysis be used.

Example 32 Two types of buildings are being considered for a new project. The following table provides the estimates prepared for each type of building. Determine which building is the economic choice when interest is at 9%.

DESCRIPTION	BUILDING A	BUILDING B
INVESTMENT COST -	$ 60,000	200,000
ESTIMATED LIFE - YEARS	15	25
NET SALVAGE VALUE -	$ 0	50,000
ANNUAL DISBURSEMENTS -	$ 11,000	5,000

Solution

The least common multiple is 75 years since this is the first point when both buildings would be replaced in the same year.

Building A, number of replacements = (75/15 - 1) = 4
Building B, number of replacements = (75/25 - 1) = 2

Building A cash flow is shown in Fig. 10.17.

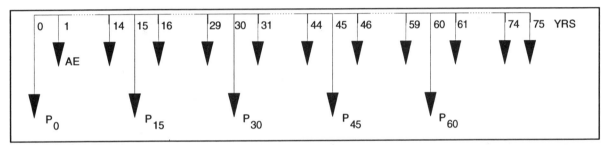

Figure 10.17

$$PW = P_0 + [P_{15}(P/F, i=9, n=15) + P_{30}(P/F, i=9, n=30) + P_{45}(P/F, i=9, n=45) + P_{60}(P/F, i=9, n=60)] + AE(P/A, i=9, n=75)$$

NOTE THAT THE PRESENT WORTH OF EACH REPLACEMENT IS CALCULATED IN THE [ ]

before calculating the PW also remember that all investments are equal therefore $P = P_0 = P_{15} = P_{30} = P_{45} = P_{60}$

$$PW = 60,000 + 60,000[(0.2745) + (0.0754) + (0.0207) + (0.0057)] + 11,000(11.0938)$$
$$= 60,000 + 22,578 + 122,031.80$$
$$= \underline{\$204,609.80}$$

Building B Cash Flow (Fig. 10.18)

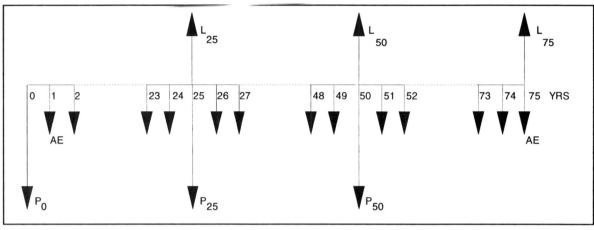

Figure 10.18

$$PW = P_0 + [(P_{25} - L_{25})(P/F, i=9, n=25) + (P_{50} - L_{50})(P/F, i=9, n=50)] + AE(P/A, i=9, n=75) - L_{75}(P/F, i=9, n=75)$$

NOTE THAT THE PRESENT WORTH OF EACH REPLACEMENT MINUS SALVAGE VALUE IS CALCULATED IN THE [ ]

Before calculating the PW also remember that all investments and salvage values are equal therefore $P = P_0 = P_{25} = P_{50}$; $L = L_{25} = L_{50} = L_{75}$

$$PW = 200,000 + [(200,000 - 50,000)(0.1160 + 0.0135)] + 5,000(11.0938) + 50,000(0.0016) = 200,000 + 19,425 + 55,469 - 80 = \underline{\$274,814.00}$$

Building A is better than Building B since A's PW is $70,204.20 less than Building B.

Below example 33 is example 30 repeated with the lives of the wood and steel bridge different.

Example 33 Wood and steel structures are being considered for a foot bridge. The wooden structure requires a $10,000 first cost, annual painting and general maintenance is estimated to cost 1,800, a major overhaul will be required every 5 years and cost $3,500. The steel bridge will cost $25,000 with annual upkeep estimated to cost $1,000, and only one overhaul will be required in the 15th year and cost $6,000. The steel bridge will have a salvage value of $5,000 at the end of its life in 25 years. The wooden bridge will have no salvage value at the end of 20 years, but an expense of 4,000 will be required to remove the structure. Using present worth comparison determine the best alternative, interest is 10%.

Solution

The least common multiple is 100 years since this is the first point when both bridges would be replaced in the same year.

Wooden Bridge, number of replacements = (100/20 - 1) = 4
Steel Bridge, number of replacements = (100/25 - 1) = 3

Wooden Bridge Cash Flow (Fig. 10.19)

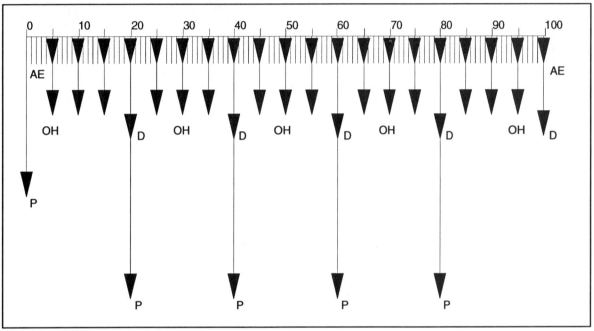

Figure 10.19

where P = 10,000 investment, AE = 1,800, OH = 3,500 occurring every 5th year except in the 20, 40, 60, 80 and 100th years when demolition of the structure is required, D = 4,000 which is the demolition cost

PW = 10 + [(10 + 4)((P/F, i=10, n=20) + (P/F, i=10, n=40) + (P/F, i=10, n=60) + (P/F, i=10, n=80))] + 1.8(P/A, i=10, n=100) + 3.5(A/F, i=10, n=5)(P/A, i=10, n=100) + 4(P/F, i=10, n=100)

= 10 + [(14)((0.1486) + (0.0221) + (0.0033) + (0.0005))] + 1.8(9.9993) + 3.5(0.1638)(9.9993) + 4(0.0001)

= 10 + 2.443 + 17.999 + 6.552 + 0 = $36.994 or $36,994   (NOTE EQUATION WAS WRITTEN IN $1,000)

Steel Bridge Cash Flow (Fig. 10.20)

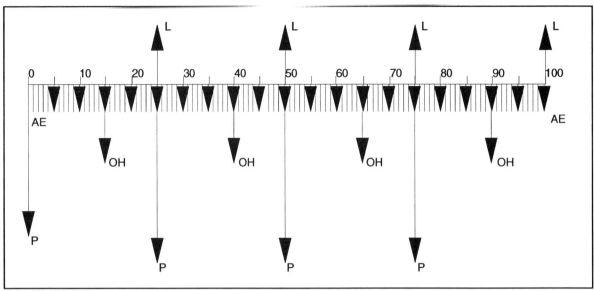

Figure 10.20

where P = 25,000 investment, L = 5,000, AE = 1,000, OH = 6,000 occurring in the 15, 40, 65, and 90th years

PW = 25 + [(25 - 5)((P/F, i=10, n=25) + (P/F, i=10, n=50) + (P/F, i=10, n=75))]
  +1(P/A, i=10, n=100) + 6[(P/F, i=10, n=15) + (P/F, i=10, n=40)
  + (P/F, i=10, n=65) + (P/F, i=10, n=90)] - 5(P/F, i=10, n=100)
= 25 + 20((0.0923) + (0.0085) + (0.0008)) + 1(9.9993) + 6[(0.2394) + (0.0221)
  + (0.0020) + (O.OOO2)] - 5(0.0001)
= 25 + 2.032 + 9.999 + 1.582 - 0.001 = $38.612 or 38,612

The wooden foot bridge should be built since it results in the lowest present worth.

The following problem demonstrates how the present worth method handles future or replacement investments which increase in cost. The increasing of future investment costs is sometimes referred to as inflation or escalation.

Example 34 A midwestern city needs to build more airplane hangers at the city airport. Forecasts show that 20 hangers will be needed but not for at least 9 years. Ten hangers are needed now and the remaining ten can be built in 9 years. All hangers will be of the same design and at today's prices each will cost $125,000. The cost to build hangers is expected to increase at the rate of 12%/year. Maintenance costs are estimated at $8,000 per year per hanger. Money to finance the hangers is to be borrowed at a 20% interest rate.

Should you build 10 hangers now and 10 in 9 years or build all 20 hangers now? All hangers will be required for 35 years from today and when retired their salvage value will be offset by removal costs.

Solution

Present worth analysis is the best method to use for this problem since increasing investment costs is involved in one of the alternatives to be considered.

Plan A - Build all 20 hangers now

REMEMBER THAT L = 0

PW = P + AE(P/A, i=20, n=35) = (125,000)(20) + 8,000(20)(4.9913) = 2,500,000 + 798,608 = $3,298,608

Plan B - Build 10 hangers now and 10 hangers in 9 years

THIS ALTERNATIVE IS MORE COMPLEX AND A CASH FLOW DIAGRAM IS ESSENTIAL (Fig. 10.21)

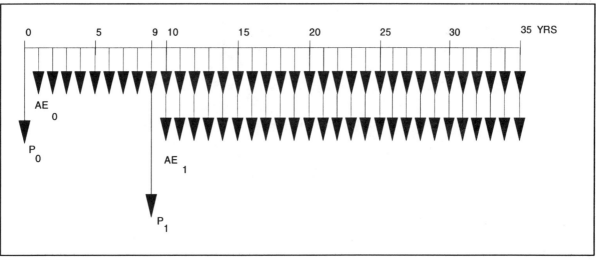

Figure 10.21

where: P_0 = investment for first 10 hangers
A_0 = maintenance for first 10 hangers
P_1 = investment for second 10 hangers
A_1 = maintenance for second 10 hanger

based on the cash flow diagram the following equation can be written:

PW = P_0 + AE_0(P/A, i=20, n=35) + P_1(P/F, i=20, n=9) + AE_1(P/A, =20, n=26)(P/F, i=20, n=9)

NOTE: 1. THAT THE P_1 AND THE PRESENT WORTH OF THE MAINTENANCE COST FOR THE SECOND 10 HANGERS ARE MOVED BACK TO TIME ZERO USING THE P/F FACTOR.
2. SINCE THE PROBLEM INDICATES THAT THE HANGERS TO BE BUILT 9 YEARS IN THE FUTURE WILL INCREASE IN COST AT A RATE OF 12%/YR P_1 IS CALCULATED AS FOLLOWS:

$$P_1 = P_0(F/P, i=12, n=9)$$

PW = P_0 + AE_0(P/A, i=20, n=35) + [P_0(F/P, i=12, n=9)](P/F, i=20, n=9) + AE_1(P/A, i=20, n=26)(P/F, i=20, n=9)
   = 125,000(10) + 8,000(10)(4.9915) + [125,000(2.7731)](10)(0.1938)
     + 8,000(10)(4.9563)(0.1938)
   = 1,250,000 + 399,320 + 671,783.48 + 76,842.48
   = $2,397,945.96

IT IS MORE ECONOMICAL TO BUILD TEN HANGERS NOW AND TEN HANGERS IN NINE YEARS.   Answer

CAPITALIZED COST METHOD

Capitalized cost comparisons are present worth comparisons when the investments have an infinite life. Recall the PW equation:

PW = P + AE(P/A, i, n) - L(P/F, i, n) but n = infinity   therefore

PW = P + AE/i, since L = 0 if n = infinity and (P/A) = 1/i, when n = infinity, THEREFORE

CAPITALIZED COST - CC = P + AE/i

When alternatives are compared which have finite lives, the Capitalized Annual Cost - CAC can be calculated by dividing the Annual Cost - AC by the interest rate i as follows:

$$CAC = \frac{P(i)}{i} + \frac{(P - L)(A/F, i, n)}{i} + \frac{AE}{i} = \frac{AC}{i}$$

Example 35 Two structures for an industrial building are under consideration. Using a minimum attractive rate of return of 5% which structure should be built based on a capitalized cost comparison. The following data is provided for each structure.

Given	STRUCTURE X	STRUCTURE Y
First Cost	$6,000	$20,000
Estimated Life	10 yrs	25 yrs
Estimated Salvage value at end of life	0	$5,000
Annual Expenses	$1,100/yr	$500/yr

$$CAC = \frac{P(i)}{i} + \frac{(P - L)(A/F, i, n)}{i} + \frac{AE}{i}$$

For Structure X

$$CAC = 6,000 + \frac{(6,000 - 0)(A/F, i=5, n=10)}{0.05} + \frac{1,100}{0.05}$$

$$= 6,000 + \frac{6,000(0.0795)}{0.05} + 22,000$$

$$= 6,000 + 9,540 + 22,000 = 37,540$$

For Structure Y

$$CAC = 20,000 + \frac{(20,000 - 5,000)(A/F, i=5, n=25)}{0.05} + \frac{500}{0.05}$$

$$= 20,000 + \frac{15,000(0.0201)}{0.05} + 10,000$$

$$= 20,000 + 6,030 + 10,000 = 36,030$$

Based on the CAC Structure Y should be built since its CAC is lower than Structure X's.

Example 36 A town is evaluating installing a water tank on a tower near the well, or as an alternative with a tank on a hill some distance away which will require a pumping station to store water. The cost to install the tank and tower is $120,000. Annual maintenance is estimated to be $2,000. The tank on a hill, including the pipeline, is $88,000, the required pumping station also requires an investment of $9,200 with a salvage value of $650 after a 15 year useful life. Annual maintenance costs for the tank and pipeline are estimated to be $1,650 and $900 for the pumping station. Determine the preferred plan on the basis of a capitalized cost analysis using a 10% rate of return.

Solution

$$CAC = \frac{P(i)}{i} + \frac{(P - L)(A/F, i, n)}{i} + \frac{AE}{i}$$

For the water tank on a tower:

Note that the problem does not indicate a useful life for the tank or the tower, the only logical approach is to assume they are permanent. The second term in the CAC equation is therefore zero.

$$CAC = 120{,}000 + 2{,}000/(0.10)$$
$$= 120{,}000 + 20{,}000 = 140{,}000$$

For the water tank, pipeline and pumping station:

Note since no useful life is given for the tank and pipeline they are assumed to be permanent. Also recognize that the investment for the pumping station must be handled separately from the tank and pipeline because it has a different useful life than the tank and pipeline.

$$CAC = \frac{P(i)}{i} + \frac{P(i)}{i} + \frac{(P-L)(A/F,\ i,\ n)}{i} + \frac{AE}{i}$$

$$= 88{,}000 + 9{,}200 + \frac{9{,}200 - 650)(A/F,\ i=10,\ n=15)}{0.10} + \frac{1{,}650 + 900}{0.10}$$

$$= 97{,}200 + \frac{8{,}550(0.0315)}{0.10} + 25{,}500$$

$$= 97{,}200 + 2693.25 + 25{,}500 = 125{,}393.25$$

The tank, pipeline, and pumping station is preferred because its CAC is lower than the tank and tower alternative.

RATE OF RETURN METHOD

EVALUATION OF AN INVESTMENT

The rate of return method is used to evaluate a potential investment by calculating the interest rate which equates the potential net income, revenue or benefits from an investment to the amount of the investment. It is shown graphically in Fig. 10.22.

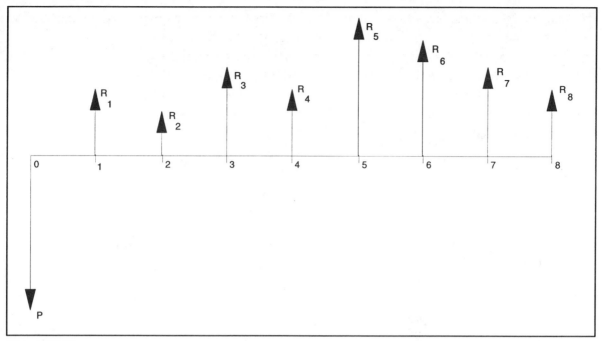

Figure 10.22

The equation that equates the present worth of the revenues -"R" to the investment P is:

$$P = R_1(P/F, i, n=1) + R_2(P/F, i, n=2) + R_3(P/F, i, n=3)$$
$$+ R_4(P/F, i, n=4) + R_5(P/F, i, n=5) + R_6(P/F, i, n=6)$$
$$+ R_7(P/F, i, n=7) + R_8(P/F, i, n=8)$$

Notice that a rate of return problem will provide the required investment cost, the projected or known net revenues, and the period over which the revenues expected to be realized. The interest rate that equates revenues to investment is determined using trial and error solution method. The following steps are used in solving rate of return problems:

1. Reduce data into investments, salvage value and net revenues by year.

2. Select the appropriate approximation formula and calculate an approximate rate of return.

3. Perform a detailed calculation to determine if the assumed rate of return is the exact value.

4. If required adjust rate of return assumptions and recalculate. NOTE THIS STEP AND THE FOLLOWING STEP 5 ARE GENERALLY REQUIRED.

5. Interpolate between appropriate assumed rates of return to determine exact answer.

10-60 FUNDAMENTALS OF ENGINEERING EXAM REVIEW WORKBOOK

Example 37 A machine has a first cost of $20,000. The annual maintenance costs associated with the machine will be $1,000. The yearly income generated by the machine is expected to be $8,000, 7,500, 5,000, 9,000 and 500 over the five year useful life of the machine. Part A: Determine the rate of return for this investment and if the investment should be made if the minimum acceptable rate of return is 10%. Verify the result by developing a yearly cash flow based on the calculated rate of return. Part B: Should the machine be operated for the fifth year, explain your answer.

Solution

PART A:

STEP 1 - In order to solve this problem first develop a cash flow table to determine the net revenues that will be realized if the machine is purchased as shown below:

YEAR	INCOME	DISBURSEMENTS	NET CASH FLOW
0		-20,000	-20,000
1	8,000	-1,000	7,000
2	7,500	-1,000	6,500
3	5,000	-1,000	4,000
4	9,000	-1,000	8,000
5	500	-1,000	(-)500

STEP 2 - Since this problem involves only an investment and net cash flows resulting from income less disbursements without any significant salvage value at the end of the problem, the following formula is used to determine the approximate rate of return:

$P = R_A(P/A, i=?, n=5)$,
where $R_A = [7,000 + 6,500 + 4,000 + 8,000 + (-500)]/5$
    $= 5,000$
$20,000 = 5,000(P/A, i=?, n=5)$
$(P/A, i=?, n=5) = 4.0$,   THE FOLLOWING DATA IS OBTAINED FROM INTEREST TABLES FOR n = 5

i	(P/A)
7.0	4.100
7.5	4.045
8.0	3.993

Based on this table 8.0% should be assumed for the detailed calculation as follows:

STEP 3

If the assumed rate of return is correct the following equation will be verified:

ASSUME i = 8.0%

$$P = R_1(P/F, i=8, n=1) + R_2(P/F, i=8, n=2) + R_3(P/F, i=8, n=3)$$
$$+ R_4(P/F, i=8, n=4) + R_5(P/F, i=8, n=5)$$

REARRANGING:

$$0 = -P + R_1(P/F, i=8, n=1) + R_2(P/F, i=8, n=2) + R_3(P/F, i=8, n=3)$$
$$+ R_4(P/F, i=8, n=4) + R_5(P/F, i=8, n=5)$$

$$0 = -20{,}000 + 7{,}000(0.9259) + 6{,}500(0.8573) + 4{,}000(0.7938) + 8{,}000(0.7350)$$
$$+ (-500)(0.6806)$$

0 = +768.65 INDICATES THAT THE PRESENT WORTH OF FUTURE REVENUES IS GREATER THAN THE INVESTMENT BY $768.65. THE INTEREST RATE ASSUMPTION MUST BE ADJUSTED TO REDUCE THE PW OF THE REVENUES. RULE: IN ORDER TO REDUCE THE PW OF REVENUES INCREASE THE INTEREST RATE.

STEP 4

ASSUME i = 10.0%

$$0 = -P + R_1(P/F, i=10, n=1) + R_2(P/F, i=10, n=2) + R_3(P/F, i=10, n=3)$$
$$+ R_4(P/F, i=10, n=4) + R_5(P/F, i=10, n=5)$$

$$0 = -20{,}000 + 7{,}000(0.9091) + 6{,}500(0.8265) + 4{,}000(0.7513) + 8{,}000(0.6830)$$
$$+ (-500)(0.6209)$$

0 = (-) 105.30 INDICATES THAT THE PRESENT WORTH OF FUTURE REVENUES IS LESS THAN THE INVESTMENT BY $105.30. SINCE THE FIRST CALCULATION RESULTED IN A PW GREATER THAN THE INVESTMENT AND THE SECOND CALCULATION IN A PW LESS THAN THE INVESTMENT, THE CORRECT RATE OF RETURN IS BETWEEN 8 AND 10%.

STEP 5, INTERPOLATION TO DETERMINE RATE OF RETURN

Set up the following table:

i	-P + PW
8.0	768.50
?	0.00
10.0	(-)105.30

$$\frac{?-8}{10-8} = \frac{0-768.50}{(-105.30)-768.50}$$

$? = 8 + 2(0.8795) = \underline{9.76\%}$

The investment in the machine would not be made if as stated in the problem the company requires a 10% minimum rate of return.

The following table of year by year calculations demonstrates that the calculated rate of return leads to the recovery of the invested capital.

END OF YEAR	YRS INT.@ 9.76% ON PREV YRS BAL	PREV YRS BAL PLUS YRS INT	END OF YR NET CASH REVENUES	END OF YR BALANCE@ 9.76% INT
0				20,000
1	20,000	(0.0976) 1,952=	20,000+	21,952- 7,000=
	-----------	----------		----------
	1,952	21,95	27,000	14,952
2	1,459	16,411	6,500	9,911
3	967	10,878	4,000	6,878
4	67	17,549	8,000	(-)450
5	(-)44	(-)494	(-)500	6

PART B

Although the machine has a useful life of 5 years, it should be noted that its operation in the 5th year results in a negative cash revenue. If the machine were not operated in the 5th year the total revenues from the investment would be higher and the rate of return would also be higher. It can therefore be concluded that the machine should not be operated in the 5th year. This can be verified by repeating the rate of return calculation for a 4 year period as follows:

YEAR	INCOME	DISBURSEMENTS	NET CASH FLOW
0		- 20,000	- 20,000
1	8,000	- 1,000	7,000
2	7,500	- 1,000	6,500
3	5,000	- 1,000	4,000
4	9,000	- 1,000	8,000

$P = R_A(P/A, i=?, n=4)$,
where $R_A = [7,000 + 6,500 + 4,000 + 8,000]/4$
$= 6,375$

$20,000 = 6,375(P/A, i=?, n=4)$

$(P/A, i=?, n=4) = 3.137,$ THE FOLLOWING DATA IS OBTAINED FROM INTEREST TABLES FOR n = 4

i	(P/A)
10	3.167
11	3.102

Based on this table either 10% or 11% could be

ASSUME i = 10%

$0 = -P + R_1(P/F, i=10, n=1) + R_2(P/F, i=10, n=2) + R_3(P/F, i=10, n=3) + R_4(P/F, i=10, n=4)$

$0 = -20,000 + 7,000(0.9090) + 6,500(0.8265) + 4,000(0.7513) + 8,000(0.6830)$

$0 = 204.45$ THE INTEREST RATE ASSUMPTION MUST BE ADJUSTED TO REDUCE THE PW OF THE REVENUES. RULE: IN ORDER TO REDUCE THE PW OF REVENUES INCREASE THE INTEREST RATE.

ASSUME i = 12.0%

$0 = -P + R_1(P/F, i=12, n=1) + R_2(P/F, i=12, n=2) + R_3(P/F, i=12, n=3) + R_4(P/F, i=12, n=4)$

$0 = -20,000 + 7,000(0.8929) + 6,500(0.7972) + 4,000(0.7118) + 8,000(0.6355)$

$0 = (-) 636.70$ INDICATES THAT THE PRESENT WORTH OF FUTURE REVENUES IS NOW LESS THAN THE INVESTMENT, THEREFORE THE CORRECT RATE OF RETURN IS BETWEEN 10% AND 12%.

Set up the following table:

i	-P + PW
10.0	204.45
?	0.00
12.0	(-)636.70

$? = 10 + 2(0.2431) = \underline{10.49\%}$

By not using the machine in the last year, the company's rate of return is increased to 10.49%, and the machine investment would be acceptable.

10-64 FUNDAMENTALS OF ENGINEERING EXAM REVIEW WORKBOOK

Example 38 An investor purchased a stock for $1,000. Dividends received through the current year are; 1st year $140, 2nd year $150, 3rd year $170, 4th year $50 and the current year $30. If the stock can be sold now for $700 what rate of return will the investor have received for this investment?

STEP 1 - CASH FLOW TABLE

YEAR	CASH FLOW
0	-1,000
1	140
2	150
3	170
4	50
5	30
5	700

STEP 2 - CALCULATION OF APPROXIMATE INTEREST RATE

This problem demonstrates the second method to calculate the approximate interest rate. Notice that this cash flow is weighted towards the 5th year because of the sale of the stock. When a cash flow is weighted towards the last year of the flow use the following equation for the approximate rate of return:

$P = F(P/F, i, n)$, where F = the sum of the total net revenues

$F = 140 + 150 + 170 + 50 + 30 + 700 = 1,240$

$1,000 = 1,240(P/F, i=?, n=5)$
$(P/F, i=?, n=5) = 0.8065$
From interest tables when i = 4.5, n = 5, (P/F) = 0.8025
therefore assume i = 4.5%

STEP 3 - EXACT CALCULATION BASED ON 4.5%

$0 = -1,000 + 140(P/F, i=4.5, n=1) + 150(P/F, i=4.5, n=2) + 170(P/F, i=4.5, n=3) + 50(P/F, i=4.5, n=4) + 730(P/F, i=4.5, n=5)$

$0 = -1,000 + 140(0.9569) + 150(0.9157) + 170(0.8763) + 50(0.8386) + 730(0.8025)$
= 48.05, SINCE PW OF REVENUES IS GREATER THAN INVESTMENT THE INTEREST RATE MUST BE INCREASED TO LOWER THE PW, THEREFORE ASSUME 6%

STEP 4 - EXACT CALCULATION BASED ON 6%

$0 = -1,000 + 140(P/F, i=6, n=1) + 150(P/F, i=6, n=2) + 170(P/F, i=6, n=3) +$

$50(P/F, i=6, n=4) + 730(P/F, i=6, n=5)$

$0 = -1,000 + 140(0.9434) + 150(0.8900) + 170(0.8396) + 50(0.7921) + 730(0.7473)$
$= -6.56$, SINCE PW OF REVENUES IS LESS THAN INVESTMENT
THE EXACT RATE OF RETURN IS BETWEEN 4.5 AND 6%

STEP 5 - EXACT RATE OF RETURN BY INTERPOLATION

i	-P + PW
4.5	48.05
?	0.00
6.0 (-)	6.56

$? = 4.5 + 1.5(0.8799) = \underline{5.82\%}$

CORPORATE RATE OF RETURN CALCULATIONS

Calculations regarding corporate rate of return can be considered on a before and after tax basis. Before tax cash flow items are:

1. Periodic gross receipts (R) representing income from the goods or services provided by the corporation.

2. Periodic disbursements representing annual expenses (operating and maintenance) associated with producing goods or providing services. If borrowed capital has been used for part or all of the investment, then the interest I paid to the lender should be included.

The before tax cash flow, sometimes referred to as the before tax earnings (E_B) is calculated as follows:

$E_B = R - (AE) - I$

Taxable income (TI) is defined as follows:

$TI = E_B - D$

After tax cash flow or after tax earnings (E_A) is defined as follows:

$E_A = E_B - T$

where T represents net taxes paid.

The PE exam has in various years covered three types of tax treatment for corporations. The treatments are (1) income taxes, (2) capital gains tax, and (3) investment tax credits.

The calculation of income tax (T) requires that the problem state the income tax rate (TR) which applies. Then the income tax, $T = TI(TR)$.

Capital gains tax results from the sale of a capital asset, usually the salvage value at the end of useful life when the asset is to be replaced. The tax treatment depends on the difference between the Book Value (BV) of an asset and the Selling Price (SP), money received from the sale, as summarized in the following table:

SITUATION	RESULT	TAX TREATMENT
SP>BV	CAPITAL GAIN	TAX PAYMENT
SP=BV	NO GAIN OR LOSS	NO TAX
SP<BV	CAPITAL LOSS	TAX CREDIT*

* Tax credit is a negative tax.

An investment tax credit, when available from the IRS, is a special tax credit applied to the value of a qualifying asset. The credit is taken in the year the investment is made based on the specific rules stated in the problem. The investment tax credit is used to reduce the tax liability of the corporation in the year of the investment. If the tax credit is greater than the tax liability the credit left after reducing the tax liability to zero is carried over to the next year.

Example 39 A corporation has purchased a new machine at a cost of $200,000 installed. It is expected that the machine will reduce the annual operation and maintenance cost by $40,000 per year for 10 years. The machine is expected to have a $35,000 salvage value after ten years of operation. Calculate the before tax and after tax expected rate of return associated with this investment based on straight line depreciation. The corporations income tax rate is 48%.

Solution

First set up a cash flow table as follows:

YEAR	E_B	D	TI	TAX	E_A
0	-200,000				-200,000
1-10	40,000	16,500	23,500	11,280	28,720
10	35,000				35,000

where:
$D = (200,000 - 35,000)/10 = 16,500$
$TI = 40,000 - 16,500 = 23,500$
$TAX = 23,500(0.48) = 11,280$
$E_A = E_B - TAX$

BEFORE TAX RATE OF RETURN CALCULATION:

$A_A = [40,000(10) + 35,000]/10 = 43,500$
$P = A_A(P/A, i=?, n=10)$
$200,000 = 43,500(P/A, i=?, n=10)$
$(P/A, i=?, n=10) = 4.598$
@ i = 17.5%, P/A = 4.575, THEREFORE ASSUME 17.5%

$0 = -200,000 + 40,000(P/A, i=17.5, n=10) + 35,000(P/F, i=17.5, n=10)$
$= -200,000 + 40,000(4.575) + 35,000(0.199)$
$= 10,035,$ which means that the PW of the revenues is greater than P, therefore assume i = 20%

$0 = -200,000 + 40,000(P/A, i=20, n=10) + 35,000(P/F, i=20, n=10)$
$= -200,000 + 40,000(4.192) + 35,000(0.162)$
$= -26,650,$ which means that the PW of the revenues is less than P and the correct i is therefore between 17.5 and 20%.

By interpolation:

$i = 17.5 + 2.5(0.274) = \underline{18.2\%}$

AFTER TAX RATE OF RETURN CALCULATION:

$A_A = [28,720(10) + 35,000]/10 = 32,220$
$P = A_A(P/A, i=?, n=10)$
$200,000 = 32,220(P/A, i=?, n=10)$
$(P/A, i=?, n=10) = 6.207$
@ i = 11%, P/A = 6.144, THEREFORE ASSUME 10%

$0 = -200,000 + 28,720(P/A, i=10, n=10) + 35,000(P/F, i=10, n=10)$
$= -200,000 + 28,720(6.144) + 35,000(0.386)$
$= -10,052,$ which means that the PW of the revenues is LESS than P, therefore assume i = 6%

$0 = -200,000 + 28,720(P/A, i=6, n=10) + 35,000(P/F, i=6, n=10)$
$= -200,000 + 28,720(7.360) + 35,000(0.558)$
$= 30,909,$ which means that the PW of the revenues is greater than P and the correct i is therefore between 6 and 10%.

By interpolation:

$i = 6 + 4(0.755) = \underline{9.02\%}$

10-68 FUNDAMENTALS OF ENGINEERING EXAM REVIEW WORKBOOK

<u>Example 40</u> Determine the after tax rate of return for Problem 39 based on fixed percentage depreciation over the 10 year life of the machine

<u>Solution</u>

First set up a depreciation table using fixed percentage depreciation:

Since f is not given it must be determined

$$f = 1 - \left[\frac{L}{P}\right]^{(1/N)} = 1 - \left[\frac{35{,}000}{200{,}000}\right]^{(1/10)} = \underline{0.16}$$

The total depreciation schedule is:

BEGINNING OF YEAR	BEGINNING OF YEAR BOOK VALUE	END OF YEAR ANNUAL DEPRECIATION*	CUMULATIVE DEPRECIATION	END OF YEAR BOOK VALUE
0				200,000
1	200,000	32,000	32,000	168,000
2	168,000	26,880	58,880	141,120
3	141,120	22,580	81,460	118,540
4	118,540	18,965	100,425	99,575
5	99,575	15,930	116,355	83,645
6	83,645	13,385	129,740	70,260
7	70,260	11,240	140,980	59,020
8	59,020	9,445	150,425	49,575
9	49,575	7,930	158,355	41,645
10	41,645	6,645	165,000	35,000

* $D_x = f(B_{x-1})$, values in table rounded off to nearest \$5.

Now the cash flow table is:

YEAR	E_B	D	TI	TAX	E_A
0	-200,000				-200,000
1	40,000	32,000	8,000	3,840	36,160
2	40,000	26,880	13,120	6,300	33,700
3	40,000	22,580	17,420	8,360	31,640
4	40,000	18,965	21,035	10,095	29,905
5	40,000	15,930	24,070	11,555	28,445
6	40,000	13,385	26,615	12,775	27,225
7	40,000	11,240	28,760	13,805	26,195
8	40,000	9,445	30,555	14,665	25,335
9	40,000	7,930	32,070	15,395	24,605
10	40,000	6,645	33,355	16,010	23,990
10	35,000				35,000

AFTER TAX RATE OF RETURN CALCULATION:

$A_A = 32,220$
$P = A_A(P/A, i=?, n=10)$
$200,000 = 32,220(P/A, i=?, n=10)$
$(P/A, i=?, n=10) = 6.207$
@ $i = 11\%$, $P/A = 6.144$, THEREFORE ASSUME 10%

$0 = -200,000 + 36,160(P/F, i=10, n=1) + 33,700(P/F, i=10, n=2) + 31,640(P/F, i=10, n=3) + 29,905(P/F, i=10, n=4) + 28,445(P/F, i=10, n=5) + 27,225(P/F, i=10, n=6) + 26,195(P/F, i=10, n=7) + 25,335(P/F, i=10, n=8) + 24,605(P/F, i=10, n=9) + 58,990(P/F, i=10, n=10)$

$0 = -200,000 + 36,160(0.909) + 33,700(0.826) + 31,640(0.751) + 29,905(0.683) + 28,445(0.621) + 27,225(0.564) + 26,195(0.513) + 25,335(0.467) + 24,605(0.424) + 58,990(0.386)$
$= -3,616$, which means that the PW of the revenues is LESS than P, therefore assume $i = 6$

$0 = -200,000 + 36,160(0.943) + 33,700(0.890) + 31,640(0.840) + 29,905(0.792) + 28,445(0.747) + 27,225(0.705) + 26,195(0.665) + 25,335(0.627) + 24,605(0.592) + 58,990(0.558)$
$= 35,584$, which means that the PW of the revenues is GREATER than P, therefore the correct i is between 6 and 10%.

By interpolation:

$i = 6 + 4(0.908) = \underline{9.63\%}$

Notice that the rate of return with the fixed percentage depreciation method, an accelerated method, is greater than for the straight line method.

Example 41 Repeat Example 39, based on tax regulations that permit the machine to be depreciated over 5 years using SLD. At the end of ten years the machine is sold for $70,000. Capital gains are taxed at 30%. The IRS has also enacted an investment tax credit of 7% based on the following table:

USEFUL LIFE	% OF ASSET QUALIFIED FOR CREDIT
GREATER THAN 10	100
7-10	67
4-7	33
LESS THAN 4	0

Solution

$$D = (200{,}000 - 35{,}000)/5 = 33{,}000$$
Investment Tax Credit $= 200{,}000(0.33)(0.07) = 4{,}620$
Capital gain in 10th year = selling price - book value $= 70{,}000 - 35{,}000 = 35{,}000$
Tax on capital gain $= 35{,}000(0.30) = 10{,}500$

YEAR	E_B	D	TI	TAX	E_A
0	-200,000				-200,000
1	40,000	33,000	7,000	0*	40,000
2	40,000	33,000	7,000	2,100**	37,900
3-5	40,000	33,000	7,000	3,360	36,640
6-10	40,000	0	40,000	19,200	20,800
10	70,000		35,000	10,500***	59,500

* Tax in first year on TI of 7,000 is 3,360, however the investment tax credit is 4,620, which reduces the tax to zero with a carry over of $4{,}620 - 3{,}360 = 1{,}260$.

** Tax in second year is reduced by the 1,260 carry over from the first year.

*** Since the selling price is greater than the book value a capital gains tax applys.

The after tax cash flow diagram is shown in Fig. 10.23.

AFTER TAX RATE OF RETURN CALCULATION:

$A_A = 35{,}132$
$P = A_A(P/A, i=?, n=10)$
$200{,}000 = 35{,}132(P/A, i=?, n=10)$
$(P/A, i=?, n=10) = 5.69$
@ $i = 12\%$, $P/A = 5.65$, THEREFORE ASSUME 12%

$0 = -200{,}000 + 40{,}000(P/F, i=12, n=1) + 37{,}900(P/F, i=12, n=2)$
$\quad + 36{,}640(P/A, i=12, n=3)(P/F, i=12, n=2) + 20{,}800(P/A, i=12, n=5)(P/F, i=12, n=5)$
$\quad + 24{,}500(P/F, i=12, n=10)$

$0 = -200{,}000 + 40{,}000(0.893) + 37{,}900(0.797)$
$\quad + 36{,}640(2.402)(0.797) + 20{,}800(3.605)(0.567) + 59{,}500(0.322)$
$\quad = -2{,}255$, therefore the PW of revenues is less than P, assume $i = 10\%$

$0 = -200{,}000 + 40{,}000(0.909) + 37{,}900(0.826)$
$\quad + 36{,}640(2.487)(0.826) + 20{,}800(3.791)(0.621) + 59{,}500(0.386)$
$\quad = 14{,}868$, the PW of revenues is greater than P, therefore the actual i is between 10 and 12%.

By interpolation:

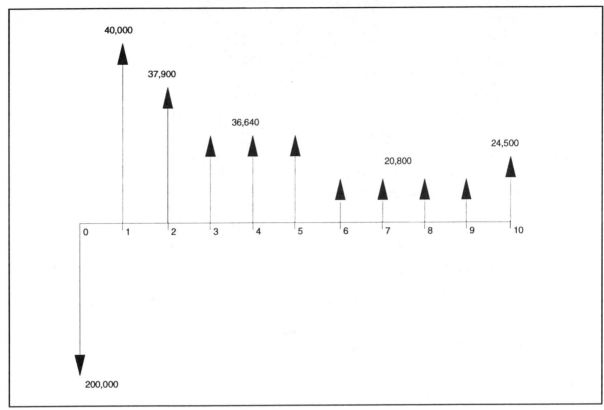

Figure 10.23

$$i = 10 + 2(0.868) = \underline{11.7\%}$$

INCREMENTAL RATE OF RETURN (COMPARING ALTERNATIVES)

Rate of return calculations can also be used to compare two alternatives by calculating the incremental rate of return between the two alternatives. This method considers that at a minimum the lower cost alternative is justified and the investment will be made and it must now be determined if the additional investment required for the higher investment alternative is justified by a sufficient return. This method uses the same calculations as presented in the section Evaluation of an Investment if the Alternatives have the same economic life once the appropriated cash flow table has been established to show the incremental investment and the associated benefits or positive cash flow resulting from the incremental investment. The flow table heading demonstrates how to set up the net cash flow table.

LOWER COST ALTERNATIVE	HIGHER COST ALTERNATIVE	DIFFERENCE HIGHER COST LESS LOWER COST ALTERNATIVE

Example 42 The following estimated data is provided for two investments, A and B, for which revenues as well as costs are known. If the minimum attractive rate of return is 15%, assuming the Alternative A has been justified, determine if Alternative B is justified.

	A	B
INVESTMENT	2,000	15,000
ANNUAL REVENUE	10,000	17,000
ANNUAL EXPENSES	8,200	12,700
ECONOMIC LIFE	10	10
NET SALVAGE	0	0

Solution

The net cash flow table is as follows:

YEAR	ALTERNATIVE A LOWER COST ALTERNATIVE	ALTERNATIVE B HIGHER COST ALTERNATIVE	DIFFERENCE HIGHER COST LESS LOWER COST ALTERNATIVE
0	-2,000	-15,000	-13,000
1-10	1,800	4,300	2,500

Estimate of the incremental rate of return:

$A_A = 2,500$
$P = A_A(P/A, i=?, n=10)$
$13,000 = 2,500(P/A, i=?, n=10)$
$(P/A, i=?, n=10) = 5.200$,
@ $i = 14\%$, $n = 10$, $P/A = 5.216$, therefore assume 14%

$0 = -13,000 + 2,500(P/A, i=14, n=10)$
$ = -13,000 + 2,500(5.216)$
$ = 40$, the PW of the net revenues is greater than the P, therefore assume $i = 16\%$

$0 = -13,000 + 2,500(P/A, i=16, n=10)$
$ = -13,000 + 2,500(4.833)$
$ = -917.50$, the correct i is therefore between 14 and 16%

By interpolation:

$i = 14 + 2(0.042) = \underline{14.1\%}$, SINCE THE MAR REQUIRED IS 15% ALTERNATIVE B WOULD NOT BE JUSTIFIED.

When the alternatives being compared have different economic lives, the incremental rate of return is calculated using the annual cost method. The annual cost equations are set up for each alternative and then set equal to each other. The interest rate for which equality is established is the incremental rate of return.

Example 43 The following estimated data is provided for two investments, A and B, for which revenues as well as costs are known. If the minimum attractive rate of return is 18%, assuming the Alternative A has been justified, determine if Alternative B is justified.

	A	B
INVESTMENT	15,000	25,000
ANNUAL EXPENSES	12,700	11,000
ECONOMIC LIFE	10	15
NET SALVAGE	0	2,500

Solution

Since the economic lives of the two alternatives are different the rate of return will be determined based on annual cost calculations.

An approximate rate of return is determined as the ratio of change in depreciation plus annual expenditures to the change in added investment. The equation is as follows:

$$\text{APPROX. RATE OF RETURN} = \frac{\Delta(D + AE)}{\Delta(P)}$$

The depreciation component D in this equation is SLD. The following table is set up to determine the APPROX RATE OF RETURN

	PLAN A	PLAN B
DEPRECIATION - D	1,500	1,500
AE	12,700	11,000
D + AE	14,200	12,500
P	15,000	25,000
Δ (D + AE)		1,700
Δ (P)		10,000
APPROX RATE OF RETURN		1,700/10,000 = 0.170

Assume 17% and perform an exact calculation using the AC equation:

$$AC = (P-L)CRF + Li + AE$$

PLAN A
$= (15,000)(A/P, i=17, n=10) + 0 + 12,700$
$= 15,000(0.215) + 12,700$
$= 15,925$

PLAN B
$= (25,000 - 2,500)(A/P, i=17, n=15) + 2,500(0.17) + 11,000$
$= 22,500(0.188) + 425 + 11,000$
$= 15,655$

AC OF PLAN A IS GREATER THAN AC OF PLAN B, THEREFORE INCREASE THE ASSUMED INTEREST RATE TO 20%

$$AC = (P-L)CRF + Li + AE$$

PLAN A
$$= (15,000)(A/P, i=20, n=10) + 0 + 12,700$$
$$= 15,000(0.239) + 12,700$$
$$= 16,285$$

PLAN B
$$= (25,000 - 2,500)(A/P, i=20, n=15) + 2,500(0.20) + 11,000$$
$$= 22,500(0.214) + 500 + 11,000$$
$$= 16,315$$

AC OF PLAN B IS NOW GREATER THAN AC OF PLAN A, THEREFORE THE RATE OF RETURN IS BETWEEN 17 AND 20%

By interpolation:

i	AC OF A	AC OF B	DIFFERENCE
17	15,925	15,655	270
?			0
20	16,285	16,315	- 30

$$? = 17 + 3(0.9) = \underline{19.7\%}$$

Example 44 Consider four mutually exclusive alternatives summarized as follows:

	A	B	C	D
INITIAL COST	400	100	200	500
UNIFORM ANNUAL BENEFIT	100.9	27.7	46.2	125.2

Each alternative has a five year useful life and no salvage value. If the minimum attractive rate of return (MAR) is 6%, which alternative should be selected. Use rate of return calculations to determine your answer.

Solution

When there are three or more mutually exclusive alternatives the procedure is generally the same as presented for two alternatives. The components of incremental analysis are:

1. Compute the rate of return for each alternative. Reject any alternative where the rate of return is less than the given MAR.
2. Rank the remaining alternatives in their order of increasing cost.
3. Examine the difference between the two lowest cost alternatives. Select the best of the two alternatives and reject the other.
4. Take the alternative selected in step 3 and compare it to the next higher alternative and proceed with another two alternative analysis.

5. Continue until all alternatives have been examined and the best of the multiple alternatives is identified. The reader should confirm the following table of rate of return results for each alternative evaluated individually.

	A	B	C	D
COMPUTED RATE OF RETURN	8.3%	12%	5%	8%

The problem requires a 6% MAR, therefore alternative C is eliminated from further consideration. The remaining alternatives are ranked as follows:

ALTERNATIVE	INITIAL COST	ANNUAL BENEFIT
B	100	27.7
A	400	100.9
D	500	125.2

The first two alternatives, B and A, are now compared:

	ALT B	ALT A	ALT A - ALT B
INITIAL COST	100	400	300
UNIFORM ANNUAL BENEFIT	27.7	100.9	73.2

Comparison of B to A
Approximate rate of return calculation:

$$P = A(P/A, i, n)$$
$$300 = 73.2(P/A, i=?, n=5)$$
$(P/A) = 4.098$, @ $i = 7\%$, n =5 P/A = 4.1, THEREFORE ASSUME $i = 7\%$

Detailed calculations
$$0 = -300 + 73.2(P/A, i=7, n=5)$$
$$= -300 + 73.2(4.1)$$
$= 0.12$, THE DIFFERENCE IS NEGLIABLE, THE EXACT RATE OF RETURN IS THEREFORE 7% AND ALTERNATIVE A IS THE PREFERRED ALTERNATIVE.

Now compare A to D
Approximate rate of return

	ALT A	ALT D	ALT D - ALT A
INITIAL COST	400	500	100
UNIFORM ANNUAL BENEFIT	100.9	125.2	24.3

$$P = A(P/A, i, n)$$
$$100 = 24.3(P/A, i=?, n=5)$$
$(P/A) = 4.12$, @ $i = 7\%$, n = 5, P/A = 4.10, THEREFORE ASSUME $i = 7\%$

Detailed calculations

$$0 = -100 + 24.3(P/A, i=7, n=5)$$
$$= -100 + 24.3(4.1)$$
$$= 0.37, \quad \text{THE DIFFERENCE IS NEGLIABLE, THE EXACT RATE OF RETURN IS THEREFORE 7\% AND ALTERNATIVE D IS THE PREFERRED ALTERNATIVE.}$$

CONCLUSION IS TO SELECT ALTERNATIVE D

NOTE: THE CORRECT ANSWER TO THIS PROBLEM HAS BEEN SHOWN TO BE ALT D, AND THIS IS TRUE EVEN THOUGH ALT A AND B HAVE RATE OF RETURN HIGHER THAN ALT D. THE HIGHER COST ALTERNATIVE SHOULD BE THOUGHT OF AS THE LOWER COST ALTERNATIVE PLUS THE DIFFERENCE BETWEEN THEM. LOOKED AT THIS WAY, THE HIGHER COST ALTERNATIVE IS EQUAL TO THE DESIRABLE LOWER COST ALTERNATIVE PLUS THE DESIRABLE DIFFERENCES BETWEEN THE ALTERNATIVES.

THE SIGNIFICANT CONCLUSION IS THAT COMPUTING THE RATE OF RETURN FOR EACH ALTERNATIVE DOES NOT PROVIDE THE BASIS FOR CHOOSING BETWEEN ALTERNATIVES. INCREMENTAL ANALYSIS IS REQUIRED WHEN COMPARING ALTERNATIVES.

BENEFIT TO COST RATIO METHOD

The benefit to cost (B/C) ratio method is the economic method most generally used in public works or governmental economic analysis. When considering public works or government projects the benefits accrue to the public. The costs are the costs associated with providing the project such as a new highway. The method is simply the ratio of benefits to the users divided by costs incurred by government, taking into account the time value of money. The following equation is used to define B/C:

$$B/C = \frac{\text{PW OF BENEFITS}}{\text{PW OF COSTS}} = \frac{\text{UNIFORM ANNUAL BENEFITS}}{\text{UNIFORM ANNUAL COST}}$$

This method is preferred by government agencies because it compares the benefits to the users (the public) to the cost required to provide the benefit. For example, consider the construction of a new elevated highway through an urban area. The costs of construction are incurred by the state and would represent the costs associated with the project. Potential benefits are (1) through traffic removed from local streets resulting in less delays for local citizens and fewer accidents; (2) vehicles do not have to wait at stop lights, reducing the operating cost of the vehicles using the elevated highway. Estimates of these benefits would be prepared for use in the B/C equation. There is also a disadvantage to the public in that fewer people will stop and use local restaurants. It is important to realize that this disadvantage is a negative benefit, not a cost. All benefits must be identified, both positive and negative. Benefits are therefore all advantages minus all disadvantages to the users. Costs are all disbursements minus savings paid or received by the government agency responsible for the project.

A specific problem must provide the criteria necessary to judge whether a B/C ratio is acceptable. If a problem does not provide specific criteria, it should be assumed that a B/C ratio greater than 1 reflects an acceptable project.

B/C analysis can also be used to compare two alternatives. The approach is parallel to that used in determining the incremental rate of return. The B/C ratio for the cash flow representing the difference (higher cost - lower cost alternative) should be calculated. Again, unless a specific criterion is provided, if the incremental B/C ratio is greater than 1, the higher cost alternative is acceptable. If the ratio is less than 1, the lower cost alternative should be selected.

Example 45 The Benefit to Cost Ratio method is used to evaluate a snow removal proposal by the Somewhere, New York Department of Public Works. A 10% cost of money is used in all economic analyses performed by this agency.

The proposal being considered is for the purchase of snow removal equipment costing $800,000. If purchased, the equipment will require $60,000 for annual operating and maintenance expenses. It has been estimated that the overall economic savings to the community resulting from efficient snow removal is $175,000 annually. The equipment is expected to last for 20 years at which time it will have no salvage value.

(A) Determine the benefit to cost ratio for this proposal and state whether it is acceptable.

(B) The County has offered to provide snow removal service to the City for a contract price of $235,000 per year as an alternative to the City purchasing its own snow removal equipment. The County would be able to cover a large snow removal area and the benefits to the City are estimated to be $284,000 per year. Determine, using the B/C ratio method, which alternative the City should choose.

Solution
Alternative A
Since the benefits to the community are stated on an annual basis the B/C ratio will be developed using annual costs.

Benefits = 175,000 as stated in the problem

Costs: AC = (P-L)(A/P, i=10, n=20) + Li + AE
= 800,000(0.1175) + 0 + 60,000
= 154,000

B/C = 175,000/154,000 = <u>1.14</u>   THE PURCHASE OF SNOW REMOVAL IS ACCEPTABLE

Alternative B

The incremental B/C ratio is calculated based on the following table:

	ALTERNATIVE 1	ALTERNATIVE 2	DIFFERENCE
COSTS	154,000	235,000	81,000
BENEFITS	175,000	284,000	109,000
B/C RATIO			1.34

The City should contract for snow removal from the County because the B/C ratio of 1.34 is greater than 1.

Example 46 Seven alternatives have been developed to solve a specific benefit. The first alternative is to take no corrective action and continue to accept an annual loss of $600,000. Each of the alternatives have been evaluated separately and the following table summarizes the benefits, costs and B/C ratio's calculated. Since all of the B/C ratio's exceed 1 it is not clear which alternative should be selected. Using incremental B/C ratio analysis determine which alternative should be selected.

	ALL COST IN $1,000		
ALTERNATIVE	ANNUAL BENEFITS	ANNUAL COSTS	BENEFITS TO COSTS RATIO
A	---	600	----
E	960	700	1.37
C	1,500	740	2.03
D	1,560	800	1.95
G	1,250	925	1.35
B	1,180	1,100	1.07
F	1,810	1,060	1.74

Solution

The problem has already ordered the alternatives by increasing cost, which is necessary for the incremental B/C analysis. Alternative A is the base alternative because it has the minimum investment/cost requirement. The next higher cost alternative - Alt E is compared to Alt A and the determination regarding Alt E is made. If E is justified then it becomes the base for the next comparison. If Alt E is not justified then Alt A remains the base for the next comparison. The following table is developed to determine the best alternative.

ALL COST IN $1,000

BASE/ALT	BASE BENEFIT	ALT BENEFIT	ALT-BASE BENEFIT	BASE COST	ALT COST	ALT-BASE COST	ALT-BASE B/C RATIO	ALT. ACCEPTED
A/E	0	960	960	600	700	100	9.6	YES
E/C	960	1,500	540	700	740	40	13.5	YES
C/D	1,500	1,560	60	740	800	60	1.0	YES
D/G	1,560	1,250	-310	800	925	125	NEGATIVE	NO
D/B	1,560	1,180	-380	800	1,100	300	NEGATIVE	NO
D/F	1,560	1,810	250	800	1,060	260	0.96	NO

The incremental B/C table indicates that the best choice is Alternative D since it is the highest incremental cost option with a B/C ratio equal to or greater than 1.0.

Example 47 Due to buildup in population, a rural highway along a historic scenic coastline has become inadequate. The highway cannot be upgraded without destroying historic buildings, cemeteries, etc and without destroying the scenic character of the coastline. Two alternative proposals for a new, parallel, inland highway are under consideration. Either proposal will provide positive benefits of $1,000,000 by preserving the character and historic nature of the coastline.

Highway A would be 9.7 miles long. The cost of obtaining the right of way and constructing the highway is $1,360,000. Expected traffic will cause the road to be resurfaced every 8 years at a cost of $572,000. Ordinary annual maintenance is estimated at $1,000 per mile.

Highway B would be 11.2 miles long in order to skirt around some hills that highway A goes over. This route also provides a more scenic panorama than highway A. Because this highway skirts the hills, even though its longer, will only cost $1,090,000. Resurfacing every 10 years will cost $700,000. Ordinary maintenance is estimated to be $800 per mile.

Interest on bonds to finance either project is 10% and the assumed life of either highway is 50 years. Estimated annual traffic is as follows:

Passenger cars	500,000
Light trucks/vans	65,000
Heavy trucks	30,000

The cost per mile to operate each of these vehicles is estimated as $0.09, $0.17, $0.60, respectively. The hills on highway A will increase the cost to operate heavy trucks by $0.05 per mile.

The 1.5 mile difference in length for highway A will save time. Assume a 45 mph average speed (30 mph for heavy trucks) over the length of the highway. The acceptable average cost for drivers wages and vehicle costs is $7.00/hr. The longer time spent on highway B will be offset by the time spent by drivers using the rest stop on highway A.

Using B/C ratio analysis which highway should be built?

Solution

The calculation of benefits and costs will be based on annual cost since all operating costs of the vehicles will be on an annual basis.

AC of Highway A

$$AC = (P-L)(A/P, i=10, n=50) + Li + AE$$
$$= 1,360,000(0.1009) + 0 + 572,000(A/F, i=10, n=8) + 1000(9.7)$$
$$= 137,224 + 572,000(0.0874) + 9,700$$
$$= 196,917$$

Benefits of Highway A
- POSITIVE BENEFITS
  - Environmental = 1,000,000
- NEGATIVE BENEFITS
  - Cost for cars - 500,000(0.09)(9.7) = -436,500
  - Cost for light trucks/vans - 65,000(0.17)(9.7) = -107,185
  - Cost for heavy trucks - 30,000(0.60)(9.7) = -174,600
  - Incremental cost for hills - 30,000(0.05)(9.7) = -14,550
- TOTAL BENEFITS TO USERS = 267,165

AC of Highway B

$$AC = (P-L)(A/P, i=10, n=50) + Li + AE$$
$$= 1,090,000(0.1009) + 0 + 700,000(A/F, i=10, n=10) + 800(11.2)$$
$$= 109,981 + 700,000(0.0628) + 8,960$$
$$= 162,901$$

Benefits of Highway B
- POSITIVE BENEFITS
  - Environmental = 1,000,000
- NEGATIVE BENEFITS
  - Cost for cars - 500,000(0.09)(11.2) = -504,000
  - Cost for light trucks/vans - 65,000(0.17)(11.2) = -123,760
  - Cost for heavy trucks - 30,000(0.60)(11.2) = -201,600
- TOTAL BENEFITS TO USERS = 170,640

HWY B BENEFITS	HWY A BENEFITS	A - B BENEFITS	HWY B COSTS	HWY A COSTS	A - B COSTS
170,640	267,165	96,525	162,901	196,917	34,016

B/C RATIO = 96,525/34,016 = <u>2.84</u>   Construction of Highway A is justified.

ENGINEERING ECONOMICS 10-81

OTHER TOPICS

Replacement Studies

Determination of Economic Replacement Interval for the Same Asset

This problem determines how long an asset should be kept by calculating the uniform annual cost to keep the asset from one to n years of life. The time to replace the asset is the end of the year resulting in the lowest uniform annual cost.

<u>Example 47</u> An automobile is purchased for $10,000 and has the following trade-in (salvage value) values and annual expenses for operation and maintenance. Determine when the automobile should be replaced. Interest is at 10%.

YEAR	TRADE-IN	AE
1	6,500	2,500
2	4,500	2,750
3	2,000	3,375
4	750	5,550
5	500	6,750

<u>Solution</u>

The uniform annual cost is determined based on ownership for 1, 2, 3, 4, and 5 years.

$$AC = (P-L)(A/P, i=10, n=x) + Li + AE_{EQ}$$

where x = the years of ownership and AE_{EQ} is the conversion of unequal annual expenses into uniform annual expenses.

$AC_{x=1}$ = (10,000 - 6,500)(A/P, i=10, n=1) + 6,500(0.10) + 2,500
  = 3,500(1.10) + 650 + 2,500
  = 7,000

$AC_{x=2}$ = (10,000 - 4,500)(A/P, i=10, n=2) + 4,500(0.10)
  + [2,500(P/F, i=10, n=1)+2,750(P/F, i=10, n=2)](A/P, i=10, n=2)
  = 5,500(0.5762) +450+[2,500(0.9091)+2,750(0.8265)](0.5762)
  = 3,169.10 + 450.00 + 2,619.19
  = 6,238.29

$AC_{x=3}$ = (10,000 - 2,000)(A/P, i=10, n=3) + 2,000(0.10) + [2,500(P/F, i=10, n=1)
  + 2,750(P/F, i=10, n=2) +3,375(P/F, i=10, n=3)](A/P, i=10, n=3)
  = 8,000(0.4021) +200 + [2,500(0.9091) + 2,750(0.8265) + 3,375(0.7513)](0.4021)
  = 3,216.80 + 200.00 + 2,847.38
  = 6,264.18

$AC_{x=4}$ = (10,000 - 750)(A/P, i=10, n=4) + 750(0.10) + [2,500(P/F, i=10, n=1) + 2,750
  (P/F, i=10, n=2) +3,375(P/F, i=10, n=3) + 5,550(P/Fi=10, n=4)](A/P, i=10, n=4)

$$= 9,250(0.3155) + 75 + [2,500(0.9091) + 2,750(0.8265) + 3,375(0.7513)$$
$$+ 5,550(0.6830)](0.3155)$$
$$= 2,918.38 + 75.00 + 3,430.09$$
$$= 6,423.47$$

$AC_{x=5} = (10,000 - 500)(A/P, i=10, n=5) + 500(0.10) + [2,500(P/F, i=10, n=1)$
$\quad + 2,750(P/F, i=10, n=2) + 3,375(P/F, i=10, n=3) + 5,550(P/Fi=10, n=4)$
$\quad + 6,750(P/Fi=10, n=5)](A/P, i=10, n=5)$
$\quad = 9,500(0.2638) + 50 + [2,500(0.9091) + 2,750(0.8265) + 3,375(0.7513)$
$\quad + 5,550(0.6830) + 6,750(0.6209)](0.2638)$
$\quad = 2,506.10 + 50.00 + 3,973.62$
$\quad = 6,529.72$

The results of these calculations are summarized as follows:

YEARS OF OWNERSHIP	UNIFORM ANNUAL COST	
1	7,000.00	
2	6,238.29	MINIMUM VALUE
3	6,264.18	
4	6,423.47	
5	6,529.72	

Based on this table the lowest uniform annual cost occurs based on two years of ownership, therefore the automobile should be replaced after two years.

Determination of Economic Replacement Interval for an Improved Asset

This type of problem evaluates if existing equipment should continue in operation or if it should be replaced with new equipment that will reduce the annual costs for providing goods or services.

The evaluation compares the savings resulting from not using the existing asset for the balance of its useful life to the costs of operating the new improved asset over its useful life based on uniform annual costs. If the savings from not continuing to use the existing equipment is greater than the cost to operate the new equipment then the replacement should be made. If the savings are less than the annual costs of the new equipment then the replacement should not be made.

The determination of the annual cost of the new equipment is based on the AC equation as follows:

$$AC = (P - L)(A/P, i, n) + Li + AE$$

NOTE ONLY DATA FOR THE NEW EQUIPMENT IS USED IN THIS EQUATION, SALVAGE VALUE FROM THE EXISTING EQUIPMENT IS CONSIDERED IN THE SAVINGS RESULTING FROM NOT USING THE EXISTING EQUIPMENT.

The savings from not using the existing equipment is calculated based on the following equation:

$$\text{SAVINGS} = (L_x - L)(A/F, i, n-x) + L_x(i) + AE_o$$

where: L_x = actual value (selling price) at the end of year x, which can be equal to book value
L = salvage value at the end of useful life
AE_o = existing equipment AE

Example 48 Determine if the existing machine should be replaced with a new more efficient machine based on the data in the following table. State what will be the net annual savings or profit if the machine is replaced. Interest is 5%.

	EXISTING MACHINE	NEW MACHINE
FIRST COST-$	50,000	60,000
USEFUL LIFE-YRS	15	20
PRESENT AGE-YRS	10	N/A
SELLING PRICE-$	15,000	N/A
SALVAGE VALUE-$	10,000	15,000
ANNUAL EXP.-$	15,000	10,000

Solutions

Savings if existing machine is replaced:

$$\text{SAVINGS} = (L_x - L)(A/F, i, n-x) + L_x(i) + AE_o$$
$(15,000 - 10,000)(A/F, i=5, 15-10) + 15,000(0.05) + 15,000$
$= 5,000(0.1810) + 750.00 + 15,000.00$
$= \underline{16,655.00}$

Annual cost for the new machine:

$AC = (P - L)(A/P, i, n) + Li + AE$
$= (60,000 - 15,000)(A/P, i=5, n=20) + 15,000(0.05) + 10,000$
$= 45,000(0.0802) + 750 + 10,000$
$= \underline{14,359.00}$

The machine should be replaced because the savings associated with not operating the existing machine is greater than the annual cost to operate the new machine. The profit or net annual savings due to the replacement is $16,655.00 - 14,359.00 = \underline{\$2,296.00}$

10-84 FUNDAMENTALS OF ENGINEERING EXAM REVIEW WORKBOOK

Break Even Cost Analysis

Breakeven analysis is used when comparing alternatives which are highly sensitive to a variable and the variable is difficult to estimate. By using breakeven analysis, the value of the variable can be calculated so the conclusion is a stand-off (i.e. either alternative is acceptable). The value of the variable is known as the breakeven point. Typical examples of variables subjected to breakeven analysis are:

1. Hours of Operation - Solve for the hours of use per year for which two alternatives are equivalent.

2. Number of Parts Produced - Solve for the production level that results in either manufacturing method being acceptable.

3. Equipment Life - Determine how many years of life are required from two alternative machines to make the choice a standoff.

The usual breakeven problem involving two alternatives can be most easily solved by equating the total annual costs or present worths of the two alternatives expressed as a function of that variable.

Example 49 A switchboard containing 150 pair cable can be made of either enameled wire or tinned wire. There will be 550 soldered connections. Cost to solder a connection is $0.02 for enameled wire and $0.015 for tinned wire. A 150 pair cable costs $0.65 per ft for enameled wire and $0.85 per ft for tinned wire. For what length of cable will the installation be the same.

Solution

Let L = the length of cable

Write an expression for the cost of enameled wire (C_E)
C_E = 550 connections($0.02/connection) + $0.65/ft (L ft)
     = 11.00 + 0.65(L)

Write an expression for the cost of tinned wire (C_T)

C_T = 550 connections($0.015/connection) + $0.85/ft (L feet)
     = 8.25 + 0.85(L)

Breakeven cost is determined by setting the cost of enameled and tinned wire equal and solving for L as follows:

$$11.00 + 0.65(L) = 8.25 + 0.85(L)$$

$$L = \frac{11.00 - 8.25}{0.85 - 0.65} = \frac{2.75}{0.20} = \underline{13.75 \text{ ft}}$$

Example 50 Two motors, each rated at 80 kW, are offered to a prospective buyer. Motor A will cost $900 and have a guaranteed full load efficiency of 90%. Motor B will cost $500 and have a guaranteed full load efficiency of 88%. Electric service costs $1.25 per month per kW of monthly demand and $0.03 per kWhr of use. For how many hours of full load operation per year would the motors just breakeven as to the economy. Interest is 10% and both motors are expected to last for 10 years.

Solution:

The total annual cost equation is written for both alternatives as a function of the hours of operation. The TAC equations are then set equal to each other and the hours of operation is determined.

$$TAC = (P - L)(A/P, i, n) + L(i) + AE$$

TAC for motor A

$$TAC = 900(A/P, i=10, n=10) + 0 + AE$$

NOTE AE in this problem has two components expressed as follows:
$$AE = [1.25(12)(80)]/0.90 + [H(80)/0.90](0.03)$$
where H = hours of operation per year

$$TAC = 900(A/P, i=10, n=10) + 0 + [1.25(12)(80)]/0.90 + [H(80)/0.90](0.03)$$
$$= 900(0.1628) + 1,330.30 + 2.67(H)$$
$$= 146.52 + 1,330.30 + 2.67(H)$$
$$= 1,479.82 + 2.67(H)$$

TAC for motor B

$$TAC = 500(A/P, i=10, n=10) + 0 + AE$$
$$AE = [1.25(12)(80)]/0.88 + [H(80)/0.88](0.03)$$
where H = hours of operation per year

$$TAC = 500(A/P, i=10, n=10) + 0 + [1.25(12)(80)]/0.88 + [H(80)/0.88](0.03)$$
$$= 500(0.1628) + 1,363.65 + 2.72(H)$$
$$= 81.40 + 1,363.65 + 2.72(H)$$
$$= 1,445.05 + 2.72(H)$$

For breakeven:

$$1{,}479.82 + 2.67(H) = 1{,}445.05 + 2.72(H)$$
$$(2.72 - 2.67)H = (1{,}479.82 - 1{,}445.05)$$
$$H = 34.77/0.05 = \underline{695.4 \text{ HOURS}}$$

Example 51 A new all-season motel is estimated to cost $1,000,000 to construct, exclusive of land and fixtures. The cost of land, which is not considered to depreciate in value, is $100,000. Fixtures will cost $200,000 and operations will require working capital of 30 days gross income at 100% capacity. The fixtures will have to be replaced every 6 years and the structure will be written off over 25 years.

The gross income will be $1,500 per day at 100% capacity. Operating expenses, exclusive of capital recovery and interest, will have a fixed component of $120,000/year and a variable cost, varying in direct proportion to capacity, with a cost of $80,000 per year for 100% capacity.

If interest is compounded annually at 8%, at what percent of capacity must the motel operate to breakeven?

Solution

In order to determine the per cent capacity for the motel to breakeven, an equations for income and expenses must be written and set equal to each other with the unknown being per cent capacity.

Organizing of the data by expense and income

EXPENSE DATA

Motel Cost $P_1 = 1{,}000{,}000$, $n_1 = 25$
Land Cost $P_2 = 100{,}000$, $n_2 =$
Fixture Cost $P_3 = 200{,}000$, $n_3 = 6$
Working Capital $= 1{,}500(30) = 45{,}000$, $n_4 =$
Annual Expenses $E = 120{,}000 + 80{,}000(X/100)$
where $X = \%$ of capacity

INCOME DATA

Annual income = Income at 100% Capacity x Operating % Capacity x 365 days/year
= $1{,}500(X/100)(365)$

To break even annual income = annual cost, therefore

$(1{,}000{,}000)(A/P, i=8, n=25) + 200{,}000(A/P, i=8, n=6) + (100{,}000 + 45{,}000)(0.05)$
$+ 120{,}000 + 80{,}000(X/100) = 1{,}500(X/100)(365)$

1,000,000(0.0937) + 200,000(0.2163) + 145,000(0.05) 120,000 + 800(X) = 5,475(X)

93,700 + 43,260 + 7,250 + 120,000 = (5,475 - 800)(X)
X = 264,210/4,675 = <u>56.5%</u>

Minimum Cost Analysis/Economic Lot Size

Economic studies are performed to minimize costs, where the total cost for an item is the sum of a fixed cost, a variable cost that increases directly with respect to a production function, and a second variable cost that is inversely proportional to the same production function. For this case the total cost C can be expressed as follows:

$$C = A(X) + B/(X) + k$$

where A, B, and k are constants and X is the variable production function. A number of real situations can be dealt with by relating them to the theoretical relationship expressed in the equation above.

The condition for minimum cost can readily be determined by differentiating the above equation with respect to the cost function x, equating to zero, and solving for the value x, thus:

$$\frac{dC}{dX} = A - \frac{B}{X^2} = 0$$

$$x = \sqrt{B/A}$$

Example 52 A proposed bridge is to be 980 ft long and have spans of equal length. The weight of the bridges steel super structure is expressed as W = 2,050 + 25(S), where W is weight in pounds per ft of bridge and S is the length of a bridge span in feet. The cost of steel is $0.10 per pound. The cost of center piers of the bridge are expressed as follows: C_c = 30,000 + 50(S) and for end piers C_e = 20,000 + 50(S) where C_c and C_e are the cost of one pier and S is the length of a bridge span. Determine how many spans are required for minimum cost.

Solution

The cost of the bridge consists of three components; the steel super structure, the center piers and two end piers.

The cost of these components are defined as follows:

C_s = weight of steel x price
    = [(2,050 + 25(S))980](0.10) = 200,900 + 2,450(S)

C_c = cost of one pier x number of center piers

NOTE THAT THE NUMBER OF CENTER PIERS EQUALS THE NUMBER OF SPANS LESS ONE THEREFORE NO. OF CENTER PIERS = (980/S - 1)

$= [30,000 + 50(S)](980/S - 1)$
$= 29,400,000/S + (-30,000) + 49,000 + (-50(S))$
$= 29,400,000/S + 19,000 - 50(S)$

C_e = cost of one pier x 2 end piers
$= [20,000 + 50(S)](2) = 40,000 + 100(S)$

$C_b = C_s + C_c + C_e$
$= [200,900 + 2,450(S)] + [29,400,000/S + 19,000 - 50(S)] + [40,000 + 100(S)]$
$= 2,500(S) + 29,400,000/S + 259,900$

NOTE THAT THIS EQUATION IS IN THE FORM OF THE MINIMUM COST EQUATION $C = A(x) + B/x + k$ THEREFORE THE MINIMUM COST EQUALS:

$S = \sqrt{B/A} = \sqrt{29,400,000/2,500} = \underline{108.5 \text{ ft}}$

Number of spans = 980/S = 980/108.5 = 9.03   **ANSWER IS THEREFORE 9 SPANS OF LENGTH OF 108.9 FT**

Example 53 The cost of an airport surveillance radar is denoted by the equation:

$C_r = 10,000(P_{av}) + 500(A_e) + 15,000$

Where: C_r = cost of the radar in dollars
P_{av} = average power output in kW
A_e = effective antenna aperture in square feet

If the required average power times aperture product is 500 (ie $P_{av} \times A_e = 500$), at what average power output in kW will the cost be at a minimum? What is the minimum cost of the radar in dollars?

In order to solve this problem the cost equation must be converted to the form:

$y = A(X) + B/X + k$, by using the relationship that
$P_{av} \times A_e = 500$ or $A_e = 500/P_{av}$.

now
$C_r = 10,000(P_{av}) + 500(A_e) + 15,000$
by substitution

$C_r = 10,000(P_{av}) + 500(500/P_{av}) + 15,000$

For minimum cost:

$$P_{av} = \sqrt{B/A} = \sqrt{250,000/10,000} = \underline{5kW}$$

The minimum cost is determined as follows:

$$\begin{aligned}C_r &= 10,000(P_{av}) + 500(500/P_{av}) + 15,000 \\ &= 10,000(5) + 250,000/5 + 15,000 \\ &= \underline{\$155,000}\end{aligned}$$

The same approach is used to determine the economic lot size for a manufacturing process type of problem.

Example 54 A manufacturer has to produce 50,000 parts a year and wishes to select the most economical lot size to be equally spaced during the year. The annual cost of storage and interest on the investment varies directly with the lot size and is $1,000 when the entire annual requirement is manufactured in one lot. The cost of setting up and dismantling the machine for each run is $30. What is the most economical lot size?

Solution

Let x = the lot size y = the cost to produce 50,000 parts

Annual storage and interest is determined by setting up the following direct ratio:

$$\frac{1,000}{50,000} = \frac{COST}{x}$$

$$COST = 1,000(x)/50,000 = 0.02(x)$$

The cost to set up and dismantle the machine is expressed as follows:

$$\begin{aligned}COST &= SETUP\ COST\ PER\ RUN \times NO.\ OF\ RUNS \\ &= 30(50,000/x)\ WHERE\ 50,000/x = NO\ OF\ RUNS\end{aligned}$$

COMBINING THE COST COMPONENTS GIVES:

$$y = 00.02(x) + 1,500,000/x$$

For minimum cost

$$x = \sqrt{B/A} = \sqrt{1,200,000/0.02} = \underline{8,660\ per\ lot}$$

NO OF LOTS = 50,000/8660 = 5.8, THEREFORE USE 6 LOTS WITH 8,333 PARTS PER LOT.

Example 55 A chemical is purchased for use in a manufacturing plant. The purchasing and accounting costs for making a purchase is $21 per purchase order regardless of the amount of chemical purchased. Throughout the year 3,000 gallons of the chemical are consumed at a fairly uniform rate. It is purchased and stored in 50 gallon drums. Its purchase price per gallon, including freight, is $3.30. Annual storage costs are estimated as 12% of the average inventory. To assure continuous operations, at least 200 gallons are maintained on hand at all times. What is the most economical lot size of drums to purchase.

Solution

Since the problem asks for the solution in drums of chemical, all quantities of chemical will be expressed in drums, therefore:

Annual amount used = 3,000/50 = 60 drums/year
Cost of a drum = 50(3.30) = $165.00

LET X = NO OF DRUMS/PURCHASED ORDER, THEREFORE
NO OF PURCHASE ORDERS = 60/X
Annual cost of purchase orders(PO's) = (no of PO's) x $21
= (60/X)(21) = 1,260/X

The problem states that the minimum inventory is 200 gallons or 4 drums. The maximum inventory is not directly defined, the reader must recognize that the maximum inventory is equal to the No of Drums Order/PO + 4 = X + 4. The average inventory is therefore (MAX INV - MIN INV)/2 + 4 = X/2 + 4.

Average inventory cost = (X/2 +4)(165)(0.12) = 9.9(X) + 79.20

Total Annual Cost = PO cost + Inventory cost
= 1,260/X + (9.9(X) + 79.20)

rearranging yields

= 9.9(X) + 1,260/X + 79.20

For min cost X = B/A = $\sqrt{1260/9.9}$ = 11.3 Drums,   use 12 Drums since 12 goes into 60 evenly.

11

Practice Problems

MATHEMATICS

1.
$$\frac{4\sqrt{3}}{3 + i\sqrt{3}}$$ will reduce to:

a. $\dfrac{4}{\sqrt{3} + 1/3}$ b. $3 - i\sqrt{3}$

c. $\sqrt{3} - i$ d. $2 + 2\sqrt{3}$

e. $\dfrac{\sqrt{3} - i\sqrt{3}}{4}$

2. The simultaneous equations:

$3x + y - z = 14$
$x + 3y - z = 16$
$x + y - 3z = -10$

Have the solution:

a. $x = -5, y = -6, z = -35$
b. $x = 5, y = 7, z = 8$
c. $x = -5, y = -6, z = -39$
d. $x = 6, y = 6, z = 10$
e. $x = 5, y = 6, z = 7$

3. What is the value of x if:

$$\log_4 (23)^{\frac{1}{x}} = 1$$

a. 3.5 b. 0.44
c. 0.17 d. 2.26
e. 5.75

4. What is the 5th term of the expansion:

$$\left[1 + \frac{1}{x}\right]^7$$

a. $\dfrac{25}{x^3}$ b. $\dfrac{35}{x^4}$

c. $\dfrac{15}{x^5}$ d. $\dfrac{5}{x^7}$

e. $\dfrac{n}{n!}$

5. Four men can build 3 benches in 2 days. How many days will it take one man to build 30 benches?

a. 12 days b. 20 days
c. 100 days d. 65 days
e. 80 days

6. Given: $3x - 2y = -3$
$-6x + y = -21$
Then:
a. $x = 3, y = 6$
b. $x = 5, y = 9$
c. $x = 5, y = 6$
a. $x = -3, y = 3$
a. $x = 9, y = -5$

7. A river flows at the rate of 3 mph downstream. It takes a person in a canoe the same time to row 22 miles downstream as it does or row 10 mi upstream. How fast can the person row in still water?

a. 11 mph b. 8 mph
c. 5 mph d. 16 mph
e. 4 mph

8. If -2 is a root of the equation:

$x^3 + 6x^2 + 11x + 6 = 0$

the other two roots are:

a. 4, 0 b. 2, -2
c. 1, 3 d. -1, -3
e. -3, -1

9. The expression:

$$a^{p\, \log_a x + \log_a y} =$$

a. $a^{xp\, y}$ b. $a^{xp} + y$
c. $x^p y$ d. xy^p
e. p^{xy}

10. The polar equation of the curve defined by the rectangular equation
$x^2 + y^2 = 4x$ is:

a. $r = r \cos q$ b. $r = 4x$
c. $r^2 = 4x$ d. $r = 2\sqrt{x}$
e. $r(\cos q + \tan q \sin q) = 4$

Problems 11 - 13

Given a reference line $y = 3x - 4$

11. What is the equation of the line parallel to the reference line passing through (3,4)?

    a. $y = 4x + 3$ b. $y - 3 = x - 4$
    c. $y = 3x - 5$ d. $y = -3x + 4$
    e. $3x = 5 + y$

12. What is the equation of the line perpendicular to the reference line which passes through the y axis at $y = 2$:

    a. $y = -(x/3) + 2$ b. $y = -3x + 2$
    c. $y = -2x + 3$ d. $y = -3x + 4$
    e. $y = x - (2/3)$

13. What angle does the line $y = -4x + 3$ make with the reference line?

    a. 48.64°   b. 38.64°
    c. 32.47°   d. -24.28°
    e. 63.64°

14. A function is defined by:

$$f(x) = \begin{vmatrix} -3x & 2 \\ x & -1 \end{vmatrix} + \begin{vmatrix} 2x & -4 \\ 5 & -3 \end{vmatrix}$$

When $f(x) = 0$, x is:

    a. 20   b. -20
    c. -4   d. 10
    e. 4

15. The center of a circle defined by:
$$x^2 + y^2 - 6x + 4y = 0$$

    a. (-6,4)   b. (-3,2)
    c. (3,-2)   d. (0,0)
    e. (3,4)

16. The length of a line with slope = 3/4, from the point (4,6) to the y axis is:
    a. 3   b. $\sqrt{53}$
    c. 5   d. 4
    e. 6

17. The shape of the curve defined by:

$$3x^2 - 12x - 2y^2 + 16y$$

    a. Circle   b. Parabola
    c. Ellipse   d. Hyperbola
    e. Cone

18. For $n = 7$, the value of the expression:
$$\frac{((n-1)!)^2 \, (0!) \, (n)}{(n-2)! \, (n!)}$$

    a. 7   b. 4320
    c. 6   d. 0
    e. 42

19. The value of A is:

$$\log_7 (101A + 40)^2 = 6$$

    a. 3   b. 9.5
    c. 5   d. 1164
    e. 1000

20. The rectangular coordinates for a point of polar coordinates $(4, \pi/6)$ is:

    a. $(2\sqrt{3}, 2)$ b. (0.64, 2)
    c. (4, 6)   d. $(2, 2\sqrt{3})$
    e. $(2\sqrt{3}, 2\sqrt{3})$

21. Evaluate the expression

$$\sin(1/2 \, \cos^{-1} 4/5)$$

    a. 4/5   b. 3/5
    c. $\sqrt{10}$ d. $1/\sqrt{10}$
    e. $\sqrt{10}/3$

22. For: A = |-6,-2,2,4,6,8|
            B = [x: $x^3 \le 125$]
            C = f

    (a $\cap$ B) $\cup$ C contains:

    a. |2,4| b. θ
    c. |-2,2,4| d. |-6,-2,2,4|
    e. |6,8|

23. One root of the equation:

$$f(x) = 2x^3 + 5x^2 - 13x - 30$$

lies between:

    a. -1 to 0   b. 0 to +2
    c. +2 to +4   d. +4 to +6
    e. +1 to +3

24. Given:

$$243 = 3^{(x^2 - 4)}$$

x is:

    a. 3   b. 3.5
    c. 4   d. 1
    e. 2

25. The center of the circle which passes through the points (9,6), (3,-2) and (10,-1) is:

    a. (6,3)   b. (6,2)
    c. (2,6)   d. (7,1)
    e. (4,4)

26. Find the average value of the amplitude of the half sin wave of $y = A \sin u$.

    a. A/π b. $A\pi$
    c. $2A/\pi$ d. $A/2$
    e. $A/2\pi$

27. A box contains 5 green balls and 4 red balls. The probability of selecting 2 green balls and 1 red ball with a random sample of 3 balls without replacement is:

    a. 0.119   b. 0.476
    c. 0.357   d. 0.238
    e. 0.333

28. The $\lim\limits_{\theta \to 0} \dfrac{\cos \theta - 1}{\sin \theta}$ is:

    a. 0   b. 1
    c. -1   d. ∞
    e. Undefined

29. For $3y^2 + 2yx^2 - x^3 + 6 = 0$, dy/dx is:

    a. $\dfrac{3x^2 - 4xy - 6y}{2x^2}$

    b. $\dfrac{3x^2}{6y + 2x^2}$

    c. $\dfrac{3x^2 - 2xy}{6y + 2x^2}$

    d. $\dfrac{3x^2 + 4xy}{6y + 2x^2}$

    e. $\dfrac{3x^2 - 4xy}{6y + 2x^2}$

Problems 30 - 32

The motion of a particle is described by

$$x = \cos \pi t$$
$$y = \sin \pi t$$

30. The coordinates of the position of the particle when t = 1 are:

    a. (0,-1)   b. (1,1)
    c. (0,0)   d. (-1, 0)
    e. (-1,-1)

31. The velocity in the x direction when t = 0.5 is:

    a. -1   b. 0
    c. $-\pi$ d. 1
    e. π

32. The acceleration in the x direction at t = 0 is:

    a. 0   b. $-\pi^2$
    c. 1   d. $-\pi$
    e. π

33. The slope of the curve defined by:

$$y = x^2 - 4x + 4, \text{ for } y = 0$$

is:

    a. -4   b. 4
    c. 2   d. 0
    e. π

34. The general solution of this differential equation is:

$$\frac{d^2y}{dx^2} - \frac{dy}{dx} - 12y = 0$$

a. $A e^{-4x} + B e^{3x}$

b. $A \cos 3x + B \sin 4x$

c. $e^{4x} + B e^{-3x}$

d. $A e^{4x} + B e^{-3x}$

e. $A \cos 4x + B \sin 3x$

35. The integral:

$$\int_0^3 \int_0^4 x \, dy \, dx$$

When $y = x^2$ is:

a. 24 b. 36
c. 32 d. 16
e. 8

Problems 36 - 37

Given the curve $x^2 + y = 10x$

36. The area bounded by the curve and the x axis from $x = 2$ to $x = 8$ is:

a. 30 sq units d. 126.67 sq units
b. 132 sq units e. 109.33 sq units
c. 166.67 sq units

37. What is the x coordinate of the centroid of the area under the curve between $x = 2$ and $x = 8$?

a. 8 b. 5.4
c. 5 d. 28
e. 132

38. If

$$z = \sqrt{2x^2 y^{-3}}, \quad \frac{\partial z}{\partial y} \text{ is?}$$

a. $-\frac{3\sqrt{2}}{2} x \, y^{-\frac{5}{2}}$ b. $-\frac{3\sqrt{2}}{2} x \, y^{-7}$

c. $-3x\sqrt{2} \, y^{-4}$ d. $-\frac{3\sqrt{2}}{2} x \, y^{-\frac{1}{2}}$

e. $-\frac{3\sqrt{2}}{2} x \, y^{-\frac{3}{2}}$

39. A cylinder with a closed top and bottom is to have a volume of 101 in³. If the cost of the material for the circular top and bottom costs three times as much as that for the curved surface, the diameter of the cylinder to minimize cost should be?

a. 4 in b. 5 in
c. 3.5 in d. 2.5 in
e. 3.8 in

40. if $z = 3x^2 \cos y$ and x and y are functions of t alone then dz/dt is?

a. $(6x \cos y)\frac{dy}{dt} - (3x^2 \sin y)\frac{dx}{dt}$

b. $(6x \cos y)\frac{dx}{dt} + (3x^2 \sin y)\frac{dy}{dt}$

c. $(6x \cos y) + (3x^2 \sin y)$

d. $(3x^2 \cos y)\frac{dx}{dt} + (3x^2 \cos y)\frac{dy}{dt}$

e. $(6x \cos y)\frac{dx}{dt} - (3x^2 \sin y)\frac{dy}{dt}$

Problems 41 - 43

For the reference curve $y = x^2$

41. The area bounded by the reference curve and the x axis from $x = 0$ to $x = 3$ is:

a. 9 b. 4.5
c. 11 d. 7
e. 9.5

42. The area moment of inertia of the area bounded by the reference curve and the x axis from x = 0 to x = 3 with respect to the x axis is?

    a. 381.7   b. 153.68
    c. 20.25   d. 48.6
    e. 104.143

43. The volume generated by rotating about the x axis the area bounded by the reference curve and the x axis from
x = 0 to x = 3 is:

    a. 381. 7   b. 152.68
    c. 63.62   d. 48.6
    e. 143.2

44.
$$\int_1^\infty xe^{-x^2}\,dx \text{ is}$$

    a. $-\dfrac{1}{2e}$ b. ∞

    c. $\dfrac{2}{e}$ d. $\dfrac{1}{2e}$

    e. $-\dfrac{1}{e}$

45. The standard deviation of 5,6,7,8,10,12 is:

    a. 3.2   b. 2.8
    c. 2.4   d. 3.0
    ~~e. 2.6~~

46.
$$L^{-1}\left\{\frac{a}{S(S+a)}\right\} \text{ is:}$$

    a. sin at   b. a sin at
    c. $1 - e^{at}$ d. cos $\sqrt{at}$
    e. e^{at}

47.
$$L^{-1}\left\{\frac{1}{S^2+a^2}\right\} \text{ is:}$$

    a. e^{-2at} b. a sin at
    c. $1/a$ (sin at)   d. $1/a$ (sinh at)
    e. ae^{at}

48. A stone dropped into a calm lake causes a series of circular ripples. The radius of the outer one increases at 3 ft/sec. How rapidly is the disturbed area changing at the end of 4 seconds.

    a. 24π b. 72π
    c. 48π d. 144π
    e. 64π

49. In one roll of a pair of dice, what is the probability of making a 7 or 11?

    a. 1/12   b. 2/12
    c. 2/10   d. 2/9
    e. 1/6

50. The general solution to the differential equation
$x^2y\,dx - dy = 0$ is:

    a. $y = Ce^{(x^3/3)}$

    b. $y = \dfrac{x^3}{3} y + C$

    c. $y^{-2} = \dfrac{-2x^3}{3} + C$

    d. $\ln y = \dfrac{x^3}{3}$

    e. $y = \dfrac{Ce^{(x^3/3)}}{3}$

STATICS

1. For the diagram below, what is the maximum tension in the rope?

    a. 200 lbs   b. 342 lbs
    c. 376 lbs   d. 394 lbs
    e. 412 lbs

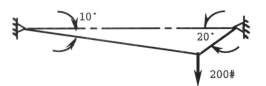

2. A rolled steel wide flange beam is subjected to non-concentric loads. Determine the torque about the centroids.

    a. 0.858 ft-k   b. 0.942 ft-k
    c. 0.991 ft-k   d. 1.252 ft-k
    e. 1.482 ft-k

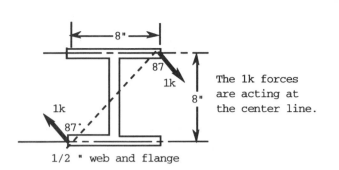

The 1k forces are acting at the center line.

1/2" web and flange

3. A 3200 lb car is raised at the rear by a jack. Assuming that the front wheels have no brakes, what is the total reaction where the jack meets the bumper. The CG is midway between the wheels.

a. 301 lbs
b. 1124 lbs
c. 1164 lbs
d. 2036 lbs
e. 2076 lbs

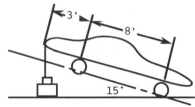

4. A triangular plate weighing 12 lbs rests on a circular table whose radius is 28 in. The plate has equal sides of 8 in. and is positioned as shown. Determine the distance between the center of the table and the CG of the plate when the plate will tend to fall off the table.

a. r/2
b. r/4
c. 2r
d. r
e. πr

Table Top r = 28"

5. Calculate the tension in the cable AB of the frame shown below:

a. 7.16k   b. 4.00k
c. 2.24k   d. 1.11k
e. 3.56k

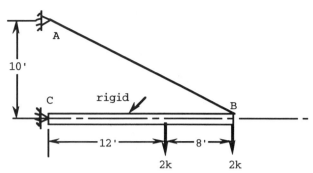

6. The supporting tower for a water tank has the loads and dimensions shown. The entire structure is pin connected. What is the load in member EF?

a. 26.82k   b. 42.69k
c. 52.64k   d. 64.24k
e. -4.743k

7. What is the vertical component of the reaction at A?

a. 200 lbs   b. 267 lbs
c. 400 lbs   d. 540 lbs
e. 450 lbs

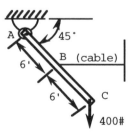

8. Using the figure for problem 7, what is the horizontal component of the reaction at A?

a. 200 lbs   b. 400 lbs
c. 600 lbs   d. 800 lbs
e. 1000 lbs

9. Determine the reaction at B for the beam loaded as shown.

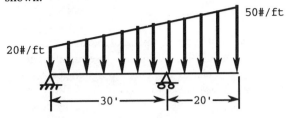

a. 500 lbs b. 1150 lbs
c. 1667 lbs d. 2450 lbs
e. 3300 lbs

10. The gate of a dam is 10 ft wide and held closed by cable AB. What is the tension in the cable? Assume the water has a density of 62.4 lbs/Ft^3.

a. 500 lbs b. 5325 lbs
c. 7425 lbs d. 15,245 lbs
e. 6724 lbs

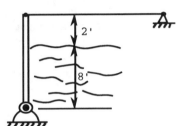

For questions 11 to 13 consider the truss loaded as shown.

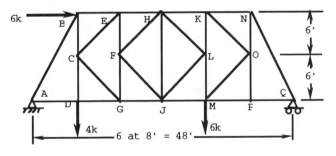

11. What is the reaction at Q?

a. 4.17k b. 6.17k
c. 8.23k d. 12.24k
e. 5.25k

12. Find the force in member HK.

a. 4.17k b. -6.17k
c. -8.23k d. 12.24k
e. 5.25k

13. Find the force in member KN.

a. -7.42k b. 10.67k
c. 13.38k d. 16.42k
e. -4.11k

For questions 14 - 18 use the frame shown which supports a load of 8k.

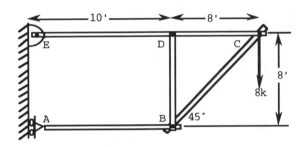

14. What is the vertical component of the reaction at E?

a. 2k b. 4k
c. 6k d. 8k
e. 10k

15. Determine the force acting on member EC at D.

a. 4k b. 8k
c. 10k d. 18k
e. 16k

16. What is the total reaction at C on member EC?

a. 12.84k b. 18.0k
c. 20.59k d. 24.62k
e. 22.46k

For questions 17 - 19 the following situation applies:

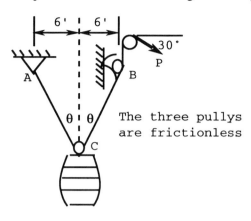

The three pullys are frictionless

17. If the water and barrel weigh 64 lbs and the angle θ is 30°, what is the tension in the rope at P?

    a. 64 lbs   b. 36.95 lbs
    c. 32 lbs   d. 55.46 lbs
    e. 73.9 lbs

18. If the barrel is raised so that loop C is at the same elevation as A and B, what is the tension in the rope at P?

    a. 0.00 lbs   b. 32 lbs
    c. 64 lbs   d. 128 lbs
    e. ∞ lbs

19. If the maximum force that the rope can sustain is 2000 lbs, how high can the rope be raised? Calculate the angle θ between the vertical and the rope.

    a. 88.08°   b. 89.94°
    c. 90.00°   d. 79.02°
    e. 89.08°

20. Determine the direction of the total frame reaction relative to the horizontal at B.

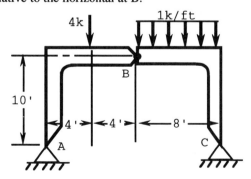

    a. 21.08°   b. 21.77°
    c. 22.62°   d. 19.02
    e. 24.15°

21. Given a thin circular plate of radius r, and mass M, determine the mass moment of inertia about an axis perpendicular to the plate, through the center of the plate.

    a. Mr^2 b. $0.75Mr^2$
    c. $0.5Mr^2$ d. $0.25Mr^2$
    e. $0.0833Mr^2$

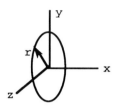

22. The angle of repose is the maximum angle to which a plane can be tilted before a block on the plane will begin to slide. If the coefficient of friction is 0.25 and a block weighs 10 lbs, what is the angle of repose?

    a. 14.04°   b. 14.48°
    c. 75.52°   d. 11.30°
    e. 13.62°

23. A hemp rope and pully are used to lift a 100 lb bucket of sand. For the conditions shown, what is the tension P required to lift the bucket? Let f = 0.2 and assume the pully is locked against rotation.

    a. 391 lbs   b. 231 lbs
    c. 208 lbs   d. 100 lbs
    e. 198 lbs

24. Derive a relationship for the brake and drum shown between the applied force F and the torque E. The drum rotates counterclockwise and f = 0.5

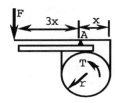

- a. T = 4.81 Fr
- b. T = 11.43 Fr
- c. T = 14.43 Fr
- d. T = 3F/r
- e. T = 12.43 Fr

25. Obtain the cross product A x B of the two vectors A = 2i +3j, and B = 5i + 2j + k.

- a. 1-i + 6j + k
- b. 7i + 5j + k
- c. 3i - 2j + 19k
- d. 3i -2j - 11k
- e. -3i + 2j + 11k

26. An 18-in. equilateral, 24-in. wide triangular truss supports a load of 128 lbs applied at D and in the plane of BDF. What is the force in member CD?

- a. zero
- b. 59.0 lbs
- c. 111.0 lbs
- d. 128 lbs
- e. 12.5 lbs

27. A Grooved block weighing 86 lbs rests on the inclined plane shown. What force must be applied to prevent its sliding down the incline.

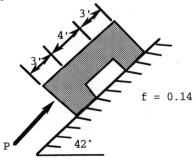

- a. zero
- b. 32.2 lbs
- c. 48.6 lbs
- d. 56.3 lbs
- e. 86 lbs

28. Determine the location of the y coordinate of the centroid of the extruded channel shown. The answer must be relative to the given datum.

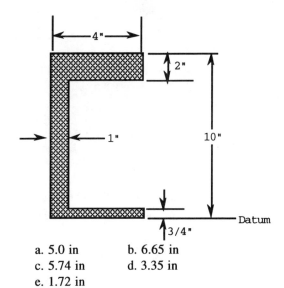

- a. 5.0 in
- b. 6.65 in
- c. 5.74 in
- d. 3.35 in
- e. 1.72 in

29. Compute the moment of inertia of the channel in problem 8 about the centroidal axis that is parallel to the datum line.

- a. 41.2 in^4
- b. 108.9 in^4
- c. 150.1 in^4
- d. 196.8 in^4
- e. 219.4 in^4

30. What is the minimum value of P that will cause sliding of the block.

- a. 400 lbs
- b. 78.3 lbs
- c. 69.5 lbs
- d. 60.2 lbs
- e. 71.6 lbs

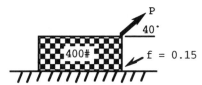

For questions 31 - 33

A 1 ft diameter drum weighing 20 lbs has a brake shoe device which is used to suspend a weight W. The drum axle is mounted in an oval shaped hole so that the full weight of the axle bears on a sliding block. The pully and axle are frictionless.

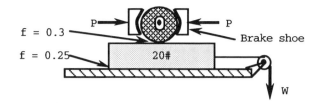

31. If the brake shoes can lock the drum, what is the maximum weight that can be suspended?

a. 6 lbs
b. 8 lbs
c. 12 lbs
d. 14 lbs
e. 16 lbs

32. What is the normal force P that must be applied to each brake shoe to stop the rotation of the drum when the maximum weight (from 11) is suspended?

a. 10 lbs
b. 8 lbs
c. 3 lbs
d. 1.5 lbs
e. 12 lbs

33. The capacity of the device could be expanded if a single brake shoe were mounted on top of the drum. (why?) If the normal force P is 10 lbs, what weight could be suspended?

a. 21.5 lbs
b. 15.5 lbs
c. 10.5 lbs
d. 8 lbs
e. 12 lbs

Questions 34 - 36

Equipment weighing 4k is packaged in the rectangular rigid container shown. The weight is uniformly distributed, and an 8k horizontal load applied. The supports are:

A. Ball roller, transmits vertical load only

G. Ball and socket, transmits 3 mutually perpendicular components of load.

F. Pinned, load along axis of cable

C. Pinned, load along axis of cable

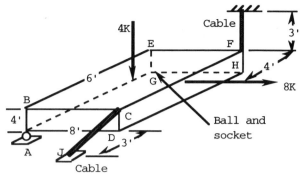

34. What is the component of the reaction at H which is parallel to the 8k load:

a. zero
b. 2k
c. 4k
d. 6k
e. 8k

35. Determine the magnitude of the reaction at H which is perpendicular to the EFGH plane.

a. zero
b. 2k
c. 4k
d. 6k
e 8k

36. Determine the force in the cable attached at F.

a. zero
b. 2.46k
c. 6.84k
d. 11.33k
e. 12.46k

Questions 37 - 39

A square threaded jackscrew is to be used to raise a load of 2 tons. The thread selected will have a mean diameter of 2 in and a lead of 0.5 in. The coefficient of static friction is 0.3 and the jack handle is 15 in long as measured from the screw center line.

37. What is the force required to raise the load?

a. 51.9 lbs
b. 103.8 lbs
c. 111.2 lbs
d. 123.1 lbs
e. 98.6 lbs

38. All other conditions being the same, what must the coefficient of friction be in order to raise the same load with a push on the jack handle of 75 lbs?

a. 0.15
b. 0.20
c. 0.24
d. 0.22
e. 0.18

39. Which one of the following statements is false?

a. The lead of the screw is the distance between similar points of adjacent threads.

b. The pitch is the distance between similar points of adjacent threads.

c. The friction force always acts in the direction of the motion of a body.

d. When the lead angle is less than the angle of friction, the screw is self locking.

e. The friction force between two rubbing surfaces is independent of the contact area.

40. A chain hoist used for installing auto engines is composed of two attached pulleys, 1 and 2. Each pulley has teeth to prevent slippage of the chain with respect to the pulley. If a 1 ton load can be lifted by a 100 lb pull, what is the ratio of r_2 to r_1. Neglect friction and other factors.

a. 0.99   b. .909
c. 1.16   d. 2.03
e. 1.05

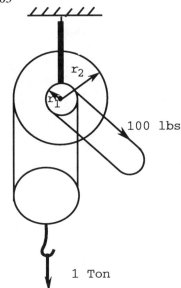

MECHANICS OF MATERIALS

Problem statement for questions 1-3

1. A retangular bar supports a 20k load attached as shown. The bar is steel and may be assumed to be weightless. What is the maximum tensile stress in the bar?

a. 1667 psi   b. 3333 psi
c. 16,667 psi   d. 41,667 psi
e. 2935 psi

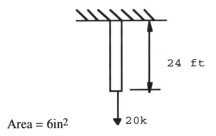

2. If the bar weight is to be considered, what then is the maximum stress? Take the density of the steel to be 500 pcf.

a. 1667 psi   b. 3417 psi
c. 6750 psi   d. 12167 psi
e. 4867psi

3. Accounting for the bar weight (500 pcf) and the applied load of 20k, what is the total elongation of the free end A? (solve to four decimal places.)

a. 0.0004 in.   b. 0.004 in.
c. 0.0032 in.   d. 0.0324 in.
e. 0.00366

4. A steel wire 8 m long is used to suspend a mass of 10 kg. If the wire is 1 mm in diameter and Young's modulus is 20×10^{10} N/M^2 what is the total elongation?

a. 5 cm   b. 2 mm
c. 2 cm   d. 5 mm
e. 6 mm

5. A standard wide flanged beam has been mounted between two rigid walls. Initially there are no forces on the beam. It is then heated uniformly to a temperature of 175°F. If the ambient, installation temperature was 70°F, what is the stress in the beam?

$a = 6.5 \times 10^{-6}$ in/in-°F and $E = 30 \times 10^6$ psi.

a. 20,475 psi   b. 34,125 psi
c. 13,650 psi   d. 23,450 psi
e. 20,475 psig

Problem statement for questions 6-8

Three columns X, Y and Z are used to support a load of 20 tons. Each column has the same length and diameter and deforms the same amount. The outer two columns are steel and the third is bronze.

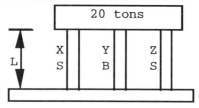

6. Which column has the maximum stress?
a. X   b. Y
c. Z   d. X and Z
e. all have the same stress

7. What would be the stress of each column if the bronze was heated to a temperature above that in the steel columns?

   a. The stress in the steel would decrease while the stress in the bronze would increase

   b. The stress in the steel would increase while the stress in the bronze would decrease

   c. The stress in all the columns would remain unchanged

   d. The stress in all the columns would increase

   e. The stress in all the columns would decrease

8. If the columns described above each have a cross-sectional area of 2 in² and are 2 ft long, what is the deflection of column y?

a. 3.1×10^{-3} in.   b. 5.8×10^{-3} in.
c. 3.2×10^{-3} in.   d. 7.8×10^{-3} in.
e. Cannot be determined

$\alpha_S = 6.5 \times 10^{-6} \frac{\text{in}}{\text{in-°F}}$, $E_S = 30 \times 10^6$ psi

$\alpha_B = 11.2 \times 10^{-6} \frac{\text{in}}{\text{in-°F}}$, $E_B = 15 \times 10^6$ psi

9. The pin connected device shown has a horizontal rigid bar supported by steel and aluminum bars. The cross-sectional area of the steel is 1.5 in² and that of the aluminum is 3/8 in². Use $E_a = 10.5 \times 10^6$ psi, $E_s = 30 \times 10^6$ psi. $a_a = 12.5 \times 10^{-6}$ in/in-°F and $a_s = 6.5 \times 10^{-6}$ in/in-°F.

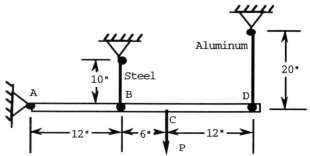

What is the stress in the steel bar if the force P = 500 lbs is acting downward?

a. 64 psi   b. 171 psi
c. 333 psi   d. 393 psi
e. 589 psi

Statement for problems 10 - 13

A compound bar fixed at its upper end, supports a 2k load axially. $E_s = 10 \times 10^6$, $E_a = 10^6$.

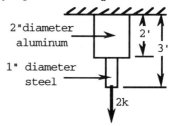

10. What is the ratio of the forces in each portion of the bar resulting from the 2k load?

　　a. 1/2　　　b. 1/1
　　c. 3/2　　　d. 2/1
　　e. 3/1

11. What is the ratio of stresses on each portion of the compound bar resulting from the 2k load?

　　a. 1/2　　　b. 1/1
　　c. 3/2　　　d. 2/1
　　e. 4/1

12. What is the total elongation of the bar?

　　a. 1.02×10^3 in　　　b. 1.53×10^3 in
　　c. 2.55×10^3 in　　　d. 5.61×10^3 in
　　e. 16.32×10^3 in

13. If Poisson's Ratio for aluminum is 1/3, what is the change in diameter of the aluminum portion resulting from the 2k load?

　　a. 6.375×10^{-5} in　　　b. 4.25×10^{-5} in
　　c. 2.125×10^{-5} in　　　d. 2.00×10^{-5} in
　　e. 1.775×10^{-5} in

14. A coupling connecting two chafts transmits a torque of 1000 in-lbs. Four bolts fasten the flanges together, each bolt has a cross-sectional area of 1/2 in and is on a 4" bolt circle. What is the stress in each bolt?

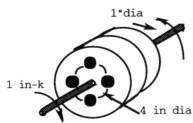

　　a. 250 psi　　　b. 500 psi
　　c. 1000 psi　　　d. 2000 psi
　　e. 2500 psi

15. A steel bolt and bronze tube are assembled as shown. The cross-sectional area of the bolt is 0.75 in.^{2} and the tube is 1.5 in.^{2}. The bolt has 32 threads per inch. Find the number of turns of the nut that will produce a stress of 40,000 psi in the bronze tube.

$$E_b = 12 \times 10^6, \ E_s = 30 \times 10^6$$

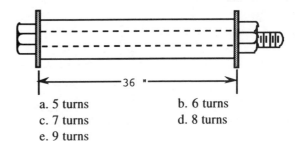

　　a. 5 turns　　　b. 6 turns
　　c. 7 turns　　　d. 8 turns
　　e. 9 turns

16 What is the maximum shear stress parallel to the axis of this shaft?

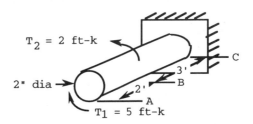

　　a. $10/\pi$ ksi　　　b. $60/\pi$ ksi
　　c. $120/\pi$ ksi　　　d. $240/\pi$ ksi
　　e. $80/\pi$ ksi

17. A bar is made of aluminum and steel rigidly attached to one another. The best description of the effect of uniform heating of the bimetallic bar is:

　　a. The bar will elongate

　　b. The bar will shrink

　　c. The bar will curve with the center of curvature toward the steel

　　d. The bar will curve with the center of curvature toward the aluminum

　　e. Not much

18. A stepped steel shaft has axial forces applied as shown. What is the maximum stress in the shaft?

　　a. 1 ksi　　　b. 2.5 ksi
　　c. $16/\pi$ psi　　　d. $42.67/\pi$ psi
　　e. $16/\pi$ ksi

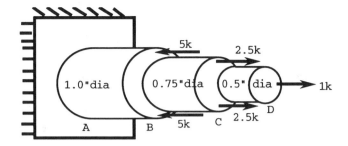

19. A shouldered circular compound shaft is made of bronze and steel. The shaft is rigidly attached to immovable walls and an unknown torque is applied where the shaft changes diameter. If the maximum allowable shear stress in the bronze is 8000 psi and in the steel 12,000 psi, what is the maximum value of the applied torque?

    a. 18,000 in-lbs   b. 53,700 in-lbs
    c. 71,600 in-lbs   d. 90,400 in-lbs
    e. 42,500 in-lbs

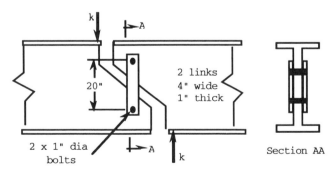

20. An expansion joint in the road bed of a bridge supports a load of K kips as shown. What is the limiting value of K governed by the 1-in. diameter bolts? The yield point of all steel is 36,000 psi.

    a. 20π k   b. 24π k
    c. 18π k   d. 40π k
    e. 36π k

21. In the structure shown, what is the bending moment directly under the load?

    a. 3 ft-k   b. 4 ft-k
    c. 5 ft-k   d. 6 ft-k
    e. 8 ft-k

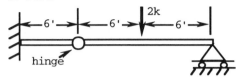

22. The maximum unit stress at any vertical section in a beam is obtained by dividing the moment at that section by:

    a. The cross-sectional area
    b. The radius of gyration
    c. The section modulus
    d. The distance to the point where the shear is zero

23. The stiffness of a rectangular beam is proportional to:

    a. The square of the depth
    b. The cube of the depth and directly to the length
    c. The cube of the depth and inversely as the length
    d. The square of the depth and inversely as the length
    e. The length and inversely as the square of the depth

Problems 24 and 25

A cantilever beam loaded as shown has shear and bending moment diagrams drawn by parts.

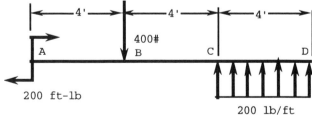

24. Select the correct shear diagram.

25. Select the correct bending moment diagram.

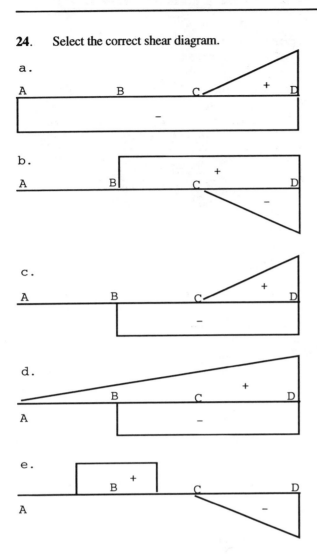
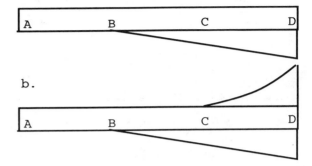
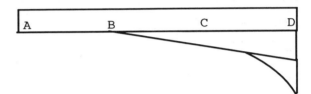
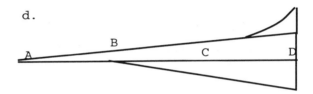
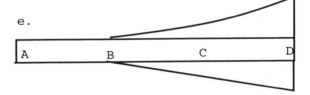

Problems 26 through 30

Given this hinged beam, loaded as shown, neglect the weight of the beam.

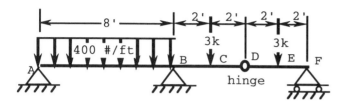

26. Is this beam statically determinate? If so, to what degree?

    a. It is statically determinate
    b. It is statically indeterminate to the 1st degree
    c. It is statically indeterminate to the 2nd degree
    d. It is statically indeterminate to the 3rd degree
    e. It will fall apart in a small wind

27. What is the hinge reaction at D?

    a. 1^k b. 1.5^k
    c. 2^k d. 2.5^k
    e. 3^k

28. Select the equation of the bending moment for any portion of the beam between point B and C.

a. $R_{Ay} x - 400 (x-4) = M_x$
b. $R_{Ay} x - 400 (x-8) + R_{BY} (x-8)$
c. $R_{Ay} x - 400 [8 (x-4)] = M_x$
d. $R_{Ay} x - 400 [8 (x-4)] - R_{BY} (x-4)$
e. $R_{Ay} x - 400[8 (x-4)] + R_{By} (x-8) = M_x$

29. What is the bending moment at D?

    a. zero
    b. 6000 ft-lbs
    c. 3000 ft-lbs
    d. 1500 ft-lbs
    e. ∞

30. In any beam where a splice has to be made, it is advisable to place the splice where the stress is a minimum. This condition generally occurs where the bending moment is zero. Where would you place the splice in section AD?

    a. .50 ft from left end
    b. .25 ft from left end
    c. .33 ft from left end
    d. .75 ft from left end
    e. .66 ft from left end

Problems 31 - 32

A beam is made up of the built-up section shown.

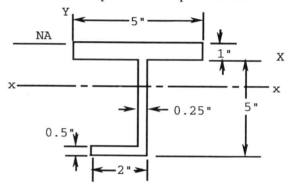

31. Locate the neutral axis from heavy end.

    a. 1.518 in.   b. 1.636 in.
    c. 2.125 in.   d. 2.878 in.
    e. 1.672 in.

32. Determine the moment of inertia I_{na} for the shape given.

    a. 42.443 in^4 b. 44.783 in^4
    c. 52.677 in^4 d. 66.129 in^4
    e. 28.617 in^4

33. Determine the maximum bending moment in this beam when it is loaded as shown.

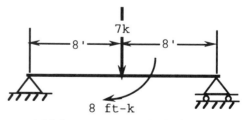

    a. 16 ft-k   b. 24 ft-k
    c. 32 ft-k   d. 40 ft-k
    e. 28 ft-k

34. A rectangular bar loaded as shown is made of steel and is assumed to be weightless. What is the maximum stress in the bar?

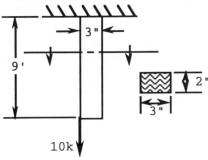

a. 1667 psi
b. 3333 psi
c. 5000 psi
d. 6667 psi
e. 6974 psi

35. Determine the maximum shear force in this simply supported beam.

a. 4k
b. 6.67k
c. 10.67k
d. 16.67k
e. 13.33k

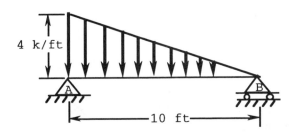

Questions 36-38

This bent beam has a rigid corner and supports a 6k load as shown.

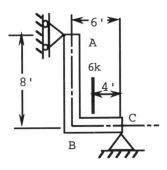

36. Determine the maximum bending moment in the beam.
a. 6 ft-k
b. 12 ft-k
c. 18 ft-k
d. 24 ft-k
e. 48 ft-k

37. If the cross-sectional area of the entire beam is rectangular, 6-in. deep by 2-in. wide, what is the maximum stress due to the load shown?

a. 16,250 psi
b. 24,000 psi
c. 24,250 psi
d. 2,020 psi
e. 24,500 psi

38. If the portion AB of the beam is fabricated from a material with a yield point stress of 36000 psi, and a factor of safety of 2.0 based on the yield point is used, what is the required section modulus?

a. 16 in^3
b. 12 in^3
c. 18 in^3
d. 8 in^3
e. 14 in^3

39. Determine the maximum stress in the beam resulting from the applied loads. The beam is steel and has a constant cross-section as shown.

a. 43,655 psi
b. 43,468 psi
c. 47,046 psi
d. 48,296 psi
e. 94,262 psi

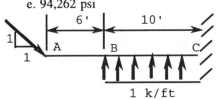

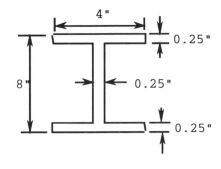

40. A hooked bracket, rigidly attached to a wall as shown has an inclined load, P, applied as shown. If the maximum stress permitted by the material is 20,000 psi, what is the maximum load P that can be applied? Base this load on the moment at the wall.

a. 25,280 lbs b. 26,700 lbs
c. 27,800 lbs d. 28,900 lbs
e. 31,768 lbs

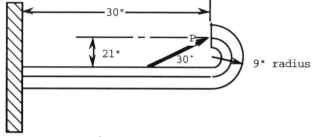

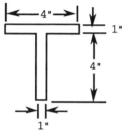

DYNAMICS

1. The total distance travelled by the particle whose velocity time graph is shown below in 0 to 8 sec is:

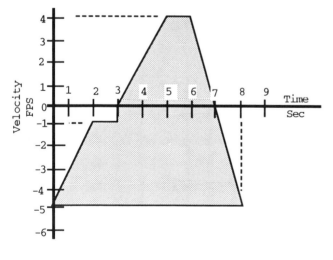

a. 4 ft b. 0 ft
c. 0.5 ft d. -2 ft
e. 2 ft

2. A 3200 lb auto is traveling along a level straight road at 55 mph. It has four wheel brakes that are equally distributed. The coefficient of friction between tire and road is 0.6 and the drivers reaction time (time from sighting to brake application) is 0.5 sec. How far will the car travel from the point at which an obstacle is sighted if deceleration is constant and uniform?

a. 208.76 ft b. 284.5 ft
c. 322 ft d. 240.5 ft
e. 168.42 ft

3. How many seconds will it take the auto in problem 2 to stop?

a. 0.5 sec b. 5.47 sec
c. 5.97 sec d. 4.68 sec
e. 5.18 sec

4. A balloon has been rising vertically from the ground at a constant speed for 12 sec. A stone is released from the balloon and reaches the ground after and additional 6 sec. Find the speed of the balloon at the instant of release of the stone.

a. 16.1 ft/sec b. 8 ft/sec
c. 26 ft/sec d. 48.3 ft/sec
e. 32.2 ft/sec

5. Find the height of the balloon when the stone reaches the ground.

a. 386.4 ft b. 193.2 ft
c. 579.6 ft d. 732 ft
e. 646 ft

6. A 200 lb weight rests on a plane (coefficient of friction 0.25) inclined 45° as shown. It is connected by a cable to a 100 lb weight. The cable goes over a frictionless, massless pulley. Upon release, what is the acceleration of the system?

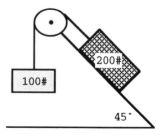

a. 28.5. ft/sec²   b. 16.1 ft/sec²
c. 8.05 ft/sec²   d. 6.49 ft/sec²
e. 0.649 ft/sec²

7. What is the velocity of the 100 lb weight 4 seconds after release?

a. 26 ft/sec   b. 2.6 ft/sec
c. 1.3 ft/sec   d. .13 ft/sec
e. 5.2 ft/sec

8. A carton is vertically dropped onto a conveyor belt moving at 10 ft/sec. For the carton/belt m = 0.3. How long will it take the carton to reach conveyor speed?

a. 1.04 sec   b. 1.25 sec
c. 2.08 sec   d. 10 sec
e. 0.5 sec

9. The cable is weightless and the pulley weightless and frictionless. The tension in the cable is:

a. 193.2 lbs   b. 42.93 lbs
c. 85.86 lbs   d. 171.76 lbs
e. 257.6 lbs

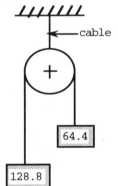

10. The coefficients of friction between blocks A and B and the 30° plane is 0.20. What is the tension in the weightless cable C?

a. 30 lbs   b. 25 lbs
c. 10 lbs   d. 64.4 lbs
e. 0 lbs

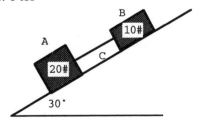

Problems 11 - 13

In the diagram below the block attains a velocity from rest to 88 fps in 4 sec. The truck is moving in the same direction as the block and doubles its velocity while moving through its own length of 27 ft. The truck is moving at 6 fps when the block starts to move.

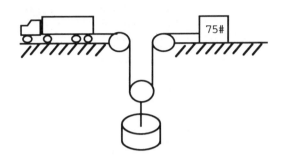

11. The acceleration of the block is:

a. 7.5 ft/sec²   b. 11 ft/sec²
c. 16.1 ft/sec²   d. 22 ft/sec²
e. 15 ft/sec²

12. The acceleration of the truck is:

a. 12 ips²   b. 22 ips²
c. 24 ips²   d. 18 ips²
e. 20 ips²

13. The acceleration of the bucket is:

a. 12 fps²↑   b. 12 fps²↓
c. 10 fps²↑   d. 10 fps²↓
e. 22 fps²↓

14. A 5-lb block is pulled up a smooth plane which makes an angle of 30° with the horizontal. If the constant pull is measured as 12 lbs along a line parallel to the plane, what is the acceleration of the block along the plane?

a. 110 ft/sec²   b. 55 ft/sec²
c. 61.17 ft/sec²   d. 32 ft/sec²
e. 28 ft/sec²

15. A 50-lb block is projected up a 23° incline. The initial velocity of the block is 20 fps. If the block moves 10 ft up the incline, what is the coefficient of friction between the block and the plane?

    a. 0.09   b. 0.90
    c. 0.25   d. 0.62
    e. 0.39

16. A space station revolves around the earth at an altitude of 25,000 mi. The radius may be taken as 4000 mi and that g is inversely proportional to the square of the distance from the center of the earth and is 32 ft/sec² at the surface. The acceleration due to gravity at the space station is:

    a. 32 ft/sec²   b. 0.609 ft/sec²
    c. 1682 ft/sec²   d. 3.05 ft/sec²
    e. 0.0609 ft/sec²

17. Find the speed of the space station in MPH.

    a. 25,000   b. 1800
    c. 6585   d. 4382
    e. 208,000

18. Assuming the spacecraft does not rotate about its own center, what is its center's angular velocity about the earth's center in rad/hr?

    a. 0.227 rad/hr   b. 15.95 rad/hr
    c. 4.54 rad/hr   d. 7.83 rad/hr
    e. 0.2634 rad/hr

19. A stunt driver (do not try this at home) rides around the hoop and across the moat as shown. For safety reasons she must exert a resultant force upward on the hoop when she is at the top equal to her weight, W. What is the entry velocity V_1. Given: $V_0^2 = V_i^2 = V_T^2 + 50g$ and $g = 32$ ft/sec²

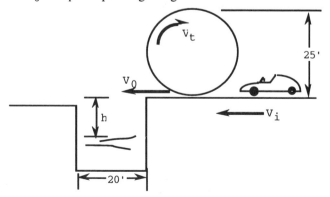

    a. 49 fps   b. 2400 fps
    c. 32 fps   d. 25 fps
    e. gh fps

20. How much lower must the far side of the moat be to make a safe landing?

    a. 9.67 ft   b. 8.4 in
    c. .47 ft   d. 5.83 ft
    e. 2.66 ft

21. A car is rounding a curve of 500 ft radius at a speed of 75 mph. If the coefficient of friction between tires and road is 0.55 what is the minimum angle of the bank of the curve?

    a. 8.11°   b. 28.82°
    c. 36.96°   d. 3.76°
    e. 65.78°

22. If the road was covered with ice ($\mu = 0$) what would the bank angle be?

    a. 12.37°   b. 28.82°
    c. 36.99°   d. 0
    e. 75.78°

23. An airplane diving at a horizontal speed of 200 mph and an altitude of 15,000 ft releases a bomb to hit a target on the ground directly in front of the flight path. At what horizontal distance before the target must the bomb be released to hit the target. Neglect air resistance and wind effects.

    a. 10,382 ft   b. 6104 ft
    c. 15,000 ft   d. 18,312 ft
    e. 8952 ft

Problems 24 - 26

A cannon with a projectile muzzle velocity of 1150 fps is fired at an angle of 37° from the horizontal.

24. How high is the projectile after 4 sec?

    a. 257.6 ft   b. 4600 ft
    c. 2760 ft   d. 2510 ft
    e. 1150 ft

25. What is the maximum height which it reaches?

    a. 15,044 ft   b. 2760 ft
    c. 10,523 ft   d. 7437 ft
    e. 4000 ft

26. What is the horizontal range at this angle?

    a. 19,200 ft   b. 9742 ft
    c. 39,364 ft   d. 98,482 ft
    e. 63,343 ft

27. A block is placed on a rough horizontal plate ($\mu = 0.65$) at a distance of 6 in from the vertical axis of rotation of the plate. What rpm will cause the block to just begin to slide?

    a. 6.47 rpm   b. 61.8 rpm
    c. 41.86 rpm   d. 30.4 rpm
    e. 123.6 rpm

28. If the plate starts from rest and accelerates at a constant rate of 5 rad/sec^2, how many seconds will be required to cause the block to begin to slip?

    a. 12.4   b. .29
    c. 2.6   d. 4.29
    e. 1.29

Problems 29 - 30

The rotating flywheel is as shown:

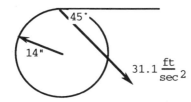

29. What is the angular acceleration of the flywheel in rad/sec^2?

    a. 1.57   b. 2.22
    c. 18.85   d. 26.66
    e. 36.28

30. What is the angular velocity of the flywheel?

    a. .334 rad/sec   b. 4.69 fps
    c. 1.16 fps   d. 4.34 rad/sec
    e. 13.92 rad/sec

Problems 31 - 33

A cylindrical shaped yo yo as shown is dropped freely. The cord maintains a constant radius.

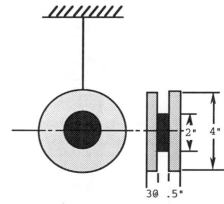

31. What is the mass moment of intertia of the yo yo direction of rotation in terms of its weight W^2.

    a. 0.4445 W lb-in-sec
    b. 0.5556 W lb-in-sec
    c. 0.00445 W lb-in-sec
    d. 0.1111 W lb-in-sec
    e. 0.05706 W lb-in-sec

32. What is its linear acceleration:

    a. 16.76 in/sec^2
    b. 386 in/sec^2
    c. 32.2 in/sec^2
    d. 210.5 in/sec^2
    e. 105.3 in/sec^2

33. The yo yo weighs 4 oz. What is the tension in the cord?

    a. .25 lb   b. .5 lb
    c. .162 lb   d. .024 lb
    e. .179 lb

Problems 34 - 36

A 5-lb weight connected to a spring moves upward at 5 fps from a height of 3 ft above the spring. It later comes down, impacts, and remains in contact with the spring. Neglect the usual neglects. The spring constant is 30 in/in.

34. What is the maximum deflection of the spring?

a. 4.37 in   b. .167 in
c. 2.09 in   d. .367in
e. 3.85 in

35. What is the final deflection of the spring?

a. 4.37 in   b. .167 in
c. 2.09 in   d. .367in
e. 3.85 in

36. What is the oscillation frequency of the mass as it is coming to rest. (Spring is massless)

a. 1.39 Hz   b. 12 Hz
c. 6.28 Hz   d. 2.21 Hz
e. 7.67 Hz

37. Two smooth 6oz discs sliding on a smooth surface collide as shown. (V_1 = 40 ft/sec, V_2 = 50 ft/sec, Coefficient of restitution = 0.6) What is the velocity of disc 1 after impact?

a. 12.64 ft/sec   b. 25.28 ft/sec
c. 31 ft/sec   d. 46.98 ft/sec
e. 21.7 ft/sec

38. Two weights are suspended from a frictionless 10-lb pulley with a radius of gyration of 0.8 ft. How long will it take to change the speed of the weights from 15 ft/sec to 25 ft/sec?

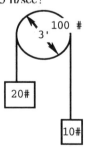

a. 1.82 sec   b. 2.17 sec
c. 3.60 sec   d. 4.34 sec
e. 0.98 sec

Problems 39 - 40

Part of a mechanism includes a weight sliding along a rod that rotates about one end as shown.

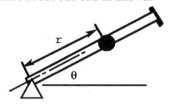

θ = 0.5 t^2 radians
r = 6.5 - 0.42 t^2 in
t = sec

39. When θ = 30°, r = ?

a. 5.05 in   b. 6.06 in
c. 7.07 in   d. 8.08 in
e. 9.09 in

40. The velocity of the block moving along the rod at r = 5 is ?

a. 1.89 in/sec   b. 3.16 in/sec
c. 0.95 in/sec   d. 1.58 in/sec
e. 4.22 in/sec

FLUID MECHANICS

Problems 1 - 6

Given the masonry dam as shown. Take a 1 ft crest for the solutions to the following problems.

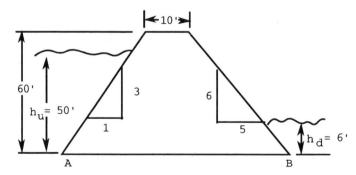

1. The upstream horizontal force is:

a. 3120 lbs   b. 122,320 lbs
c. 78,000 lbs   d. 1560 lbs
e. 93,600 lbs

2. The location of the center of pressure on the upstream side is:

   a. 20 ft above the base
   b. 33.33 ft below the water surface
   c. 40 ft below the top of the dam
   d. 25 ft below the water surface
   e. In the big middle

3. The overturning moment is:

   a. 1.3×10^6 ft lb
   b. 1.6×10^6 ft lb
   c. 3.8×10^6 ft lb
   d. 1.1×10^6 ft lb
   e. 2.3×10^6 ft lb

4. The restoring moment is:

   a. 18.45×10^6 ft lb
   b. 19.22×10^6 ft lb
   c. 15.32×10^5 ft lb
   d. 22.81×10^5 ft lb
   e. 20.30×10^6 ft lb

5. Where does the resultant force lie with respect to point B?

   a. 44 ft from B
   b. 55 ft from B
   c. 25 ft from A
   d. 36 ft from A
   e. 38 ft form B

6. What is the minimum required coefficient of friction between the dam and the bedrock if the dam is to be safe?

   a. 0.122   b. 0.150
   c. 0.178   d. 0.181
   e. 0.203

Problems 7 - 15

Given the pipeline as shown and that f = .02 for a 12-in. dia pipe.

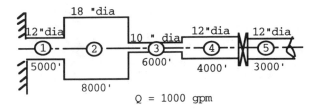

Q = 1000 gpm

7. The entrance loss is:

   a. 3.5 ft   b. .165 ft
   c. .228 ft   d. .062 ft
   e. 1.932 ft

8. The equivalent length of 12" dia pipe for this entrance loss is:

   a. 37.2 ft   b. 25 ft
   c. 11.4 ft   d. 8.25 ft
   e. 3.1 ft

9. The velocity in pipe 3 is:

   a. 4.07 ft/sec   b. 2.13 ft/sec
   c. 4.26 ft/sec   d. 5.52 ft/sec
   e. 0.98 ft/sec

10. The loss caused by the sudden enlargement at pipes 2 and 3 is:

   a. .044 ft   b. .123 ft
   c. .212 ft   d. .316 ft
   e. .429 ft

11. The equivalent length of 12 in dia pipe for the contraction between pipe 3 and 4 is:

   a. 11 ft   b. 22 ft
   c. 33 ft   d. 44 ft
   e. 55 ft

12. The head loss at the contraction of pipes 3 and 4 is:

   a. .054 ft   b. .014 ft
   c. .027 ft   d. .016 ft
   e. .032 ft

13. If f for the 10-in. pipe is .018 and f for the 18-in. pipe is .022 and pipe 5 discharges to the atmosphere, H is:

    a. 46.2 ft   b. 59.9 ft
    c. 69.7 ft   d. 76.6 ft
    e. 84.2 ft

14. Taking the minor losses as 2% of the friction losses, what horsepower could be developed in an 80% efficient turbine at the end of pipe 5 if H was measured as 250 ft.

    a. .721 hp   b. 1.05 hp
    c. 15.62 hp   d. 34.6 hp
    e. 63 hp

15. With conditions of head and flow in Problem 14, if the valve is closed, what is the hoop stress in pipe 4 if its thickness is 0.5 in.:

    a. 2599 psi   b. 1300 psi
    c. 3000 psi   d. 18,720 psi
    e. 187,200 psi

Problems 16 - 20

A cylindrical tank, 20 ft diamater and 20 ft tall is full of water. An orifice is located in the side of the tank. The head on the orifice is maintained at 15 ft at all times. One point in the trajectory of the stream issuing from this 3-in. diameter orifice is located 16 ft to the left and 4.5 ft down from the center of the orifice.

16. The theoretical velocity of the stream issuing from the orifice is:

    a. 30.24 fps   b. 43.27 fps
    c. 31.08 fps   d. 49.96 fps
    e. 386.4 fps

17. The actual velocity of the stream issuing from the orifice is:

    a. 30.24 fps   b. 43.27 fps
    c. 31.08 fps   d. 49.96 fps
    e. 386.4 fps

18. If the discharge from the orifice was measured as 1 cfs, the coefficient of discharge is.

    a. .598   b. .408
    c. .471   d. .655
    e. .725

19. The C_c is:

    a. .615   b. .419
    c. .484   d. .673
    e. .745

20. The head loss through the orifice is:

    a. .667 ft lb/lb
    b. .799 ft lb/lb
    c. .845 ft lb/lb
    d. 1.62 ft lb/lb
    e. 2.08 ft lb/lb

21. The hydraulic radius for 48 in diameter reinforced concrete pipe having flow 3 ft deep is:

    a. $2\pi + 3$ ft   b. $2/3\pi$ ft
    c. $2\pi + 3\pi$ ft   d. 6π ft
    e. $1 + [(3\sqrt{3})/8\pi]$

22. A 100% efficient pump can draw a suction of:

    a. 34 ft H_2O b. 29.92 in Hg
    c. 14.696 psi   d. 1 atm
    e. all of the above

23. A pressure of 3.34 psi is measured 11.2 ft below the surface of a liquid. The specific gravity of the liquid is?

    a. .298   b. .688
    c. 1.234   d. .798
    e. .573

24. In an open channel of small depth and great width, the hydraulic radius is most likely equal to:

    a. 1/3 depth   b. 1.2 depth
    c. 2/3 depth   c. 3/4 depth
    e. Full depth

25. A cylindrical buoy 3 ft diameter weighing 150 lbs floats vertically in sea water. The anchor is one yd³ of concrete. The mooring cable has no slack. What rise in tide will cause the anchor to lift from the bottom? Specific weights: Sea Water = 64 pcf; Concrete 150 pcf.

26. The actual velocity at the exit of an orifice is:

    a. $\sqrt{2gh}$ b. $C_d\sqrt{2gh}$
    c. $C_v\sqrt{2gh}$ d. $C_c\sqrt{2gh}$
    e. $C_v(2gh)$

27. A 10 Hp, 100 cfm pump is used to pump water to a storage tank. The efficiency of the pump is 62.4% and total systems losses are 3ft. With the pump level as datum, to what elevation can the tank be filled?

    a. 30 ft   b. 33 ft
    c. 36 ft   d. 39 ft
    e. 32 ft

Problems 28 - 30

A storm drain 12 ft diameter flows half full at a section where the slope is 1 ft/1000 ft. n for the pipe is 0.18. The drain empties into a canal with a 10 ft wide base and side slopes of 1 vertical to 5 horizontal. The slope of the canal is 50 ft/1000 ft. Depth of flow is 5 ft.

28. The hydraulic radius of the drain pipe is:

    a. 1.5 ft   b. 3 ft
    c. 6 ft   d. 9 ft
    e. 12 ft

29. The flow is:

    a. 5.5 cfs   b. 307 cfs
    c. 317 cfs   d. 55 cfs
    e. 395 cfs

30. The value of n for the earth canal is:

    a. .022   b. .383
    c. .0017   d. .417
    e. .012

31. A 2-in. diameter hose has a 1-in. nozzle. The pressure in the hose is 35 psi. How much force is required to hold the nozzle stationary when the flow is 1 cfs?

    a. 109 lbs   b. 266 lbs
    c. 157 lbs   d. 484 lbs
    e. 593 lbs

Problems 32 - 34

An open top rectangular tank is 5 ft wide, 12 ft long, and 5 ft deep. The tank is filled to a depth of 3 ft. It is then accelerated forward along its 12 ft axis at 10 ft/sec².

32. How deep is the water at the forward end of the tank?

    a. 0 ft   b. 1.14 ft
    c. 1.86 ft   d. 4.14 ft
    e. 4.86 ft

33. How deep is the water in the aft end of the tank?

    a. 0 ft   b. 1.14 ft
    c. 1.86 ft   d. 4.14 ft
    e. 4.86 ft

34. What is the force on the aft end of the tank?

    a. 5161 lbs   b. 3870 lbs
    c. 4297 lbs   d. 6248 lbs
    e. 3685 lbs

35. The pressure gage across a 2-in. diameter orifice in a water line indicates 20-in. Hg. The orifice coefficient is .72. The flow is?

    a. 6 cfs   b. 6 cfm
    c. 36 cfs   d. 36 cfm
    e. 9.76 csm

36. A suppressed wier:

    a. Has 2 end contractions
    b. Has 1 end contraction
    c. Has no end contractions
    d. Has 3 contractions
    e. Leaks around the ends

37. Given the venturi as shown.
        Pa - Pb is

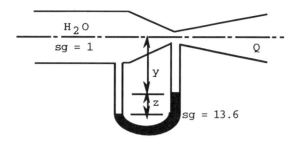

a. 13.6z b. 12.6z
c. 15.2 psi d. 18.6 psi
e. 14.7z

38. A 2-in. diameter nozzle on a 4-in. diameter hose passes a stream of water. Then the nozzle is held horizontally the coordinates of a point on the stream are x = 100 ft and y = 50 ft. The nozzle coordinates are 0,0. What is the flow in the 8-in.diameter main supplying the nozzle.

a. 1.24 cfs b. 2.48 cfs
c. 4.95 cfs d. 7.44 cfs
e. 19.84 cfs

39. A fluid having a kinematic viscosity of 1 ft²/sec in the English system has a viscosity of what in the metric system?

a. 687.9 poise b. 365.76 stokes
c. 11726 dynes d. 14,849 squ's
e. 15.298 dyne-cm/sec²

40. Water is flowing at 10 fps and is 6 ft deep in an open channel. The water passes to a second channel where the velocity doubles and the depth of flow halves. Assuming ideal conditions, what is the difference in elevation of the channel floors?

a. 2.84 ft b. 1.66 ft
c. 1.55 ft d. 4.45 ft
e. 6.21 ft

Problems 41 - 42

A rectangular wier with a crest of 10 ft is installed in a 10 ft wide channel. The lower edge of the wier is located 2 ft above the channel floor. The flow is 3,000,000 ft³/day.

41. What is the head on the wier?

a. 1 ft b. 2 ft
c. 3 ft d. 6-in.
e. 1.5 ft

42. What is the velocity of the approach?

a. 17.36 fps b. 33.7 fps
c. 1.14 fps d. 22.86 fps
e. .02 fps

Problems 43 - 44

An oil spill containment scheme for protecting beaches is attached to a sea wall as shown. The oil has a sg of 0.83, is 5 ft deep, and floats on top of the sea water.

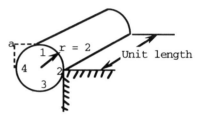

43. What is the weight of the cylinder per foot of length?

a. 685.2 lbs b. 758.7 lbs
c. 739.7 lbs d. 695.3 lbs
e. 712.9 lbs

44. What is the force against the sea wall?

a. 117.8 lbs b. 114.6 lbs
c. 110.3 lbs d. 106.2 lbs
e. 103.6 lbs

45. Q = 664 lbs/min: V = 40 ft/sec: The momentum force at a 90° bend is?

a. 1884 lbs b. 18.84 lbs
c. 13.33 lbs d. 11.31 lbs
e. 1131 lbs

ELECTRICITY/ELECTRONICS

Problems 1 and 2

A relay coil has a resistance of 100 Ω and operates from a 12 volt line. It is to be placed in a 120 volt system.

1. Determine the total circuit resistance required to run the 12 V system from the 120 V line.

a. 100Ω b. 1000Ω
c. 900Ω d. 925Ω
e. 850Ω

2. The power used from the 120 V line is?

a. 14.4 Watts   b. 144 Watts
c. 28.8 Watts   d. 1.44 Watts
e. 116 Watts

3. A series coil of 200 mH is connected to a parallel network made up of a 100 mH coil and a 1 μF capacitor. Determine the resonant frequency of the series-parallel circuit.

a. 380 Hz   b. 490 Hz
c. 616 Hz   d. 588 Hz
e. 640 Hz

4. Determine the current across coil L_1 at 1 Khz.
$R_1 = R_2 = R_3 = 100Ω$,
$C_2 = C_3 = 1.59\ \mu F$, $L_1 = 15.9$ mH

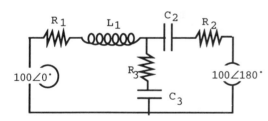

a. .52∠-26.6°   b. 1.48∠46.6°
c. .52∠-45°   d. .95∠-18.4°
e. .26∠45°

5. Determine R_{A-B} if each resistor is 15Ω.

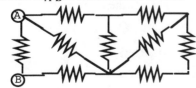

a. 7.5Ω   b. 15Ω
c. 24.6Ω d. 18.7Ω
e. 9.28Ω

6. A stockroom has several capacitors each rated at 10 μF, 150 V. It is desired to form a capacitor from these units which will be 10 μF at 300 V. What arrangement of capacitors is needed?

a. Two 10 μF in parallel, in series with two 10 μF in parallel

b. Two 10 μF in series

c. Two 10 μF in parallel

d. Two 10 μF in series, in parallel with two 10 μF in series

e. One 10 μF in series with two 10 μF in parallel

7. An R-C high pass filter is designed so that the 3 db frequency is 20 Khz. What is the ratio of filter output to input amplitude at 30 Khz?

a. .707   b. .92
c. 1.00   d. .85
e. .833

Problems 8 and 9

Given the following circuit:

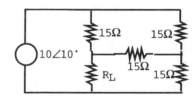

8. What value of R_L will provide maximum power transfer?

a. 5Ω   b. 6.67Ω
c. 9Ω   d. 2.5Ω
e. 18Ω

9. What is the maximum power delivered to R_L?

a. 2.8 Watts   b. 1.8 Watts
c. 0.14 Watts d. 1.0 Watt
e. 6.67 Watts

10. The resistance of a coil of wire at 20°C was found to be 48.656Ω. The temperature was raised to 50°C the resistance was 49.875Ω. What is the temperature coefficient of the wire?

  a. 0.000835   b. 0.000746
  c. 0.000498   d. 0.000816
  e. 0.000749

11. A coil has a resistance of 50Ω and an inductance of 10 mH. What time domain voltage must be applied to produce an RMS current of 20 milliamperes at a frequency of 400Hz?

  a. 1.41 sin wt
  b. 1.58 sin (2512t + 26.6°)
  c. 1.12 sin (2512t + 26.6°)
  d. 1.58 t
  e. 1.24 sin (2498t + 24.3°)

Problems 12 - 14

A 60 Hz, 480 V power line is terminated with a 25Ω load. The power line is equivalent to a series R-L circuit with R = 5Ω and L = 50 mH.

12. What is the current at the receiving end of the transmission line?

  a. 8.7 amperes   b. 96 amperes
  c. 13.56 amperes   d. 11.32 amperes
  e. 9.6 amperes

13. What is the voltage regulation of the line?

  a. 60%   b. 100%
  c. 56%   d. 84%
  e. 42%

14. What is the efficiency of transmission?

  a. 83%   b. 86%
  c. 100%   d. 79%
  e. 92%

15. Determine the impedence at terminals A-B when the frequency is 100 Khz?

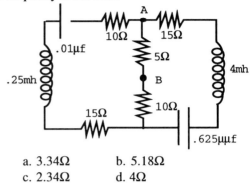

  a. 3.34Ω   b. 5.18Ω
  c. 2.34Ω   d. 4Ω
  e. 4.46Ω

Problems 16 - 18

16. It is required to supply a 14a load from a 24 V battery bank. Several 20 ampere-hr batteries rated at 3.5 amperes, 12 V, 0.2Ω internal resistance are available. How should they be configured?

  a. Two parallel plus one series
  b. Four parallel sets, each set two in series
  c. Two in series
  d. Four series sets, each set two in parallel
  e. Two in parallel

17. What is the load resistance?

  a. .4Ω   b. 1.6Ω
  c. 2.85Ω   d. 3.25Ω
  e. .2Ω

18. How long will the battery operate before a decrease in voltage is apparent?

  a. 1.6 hr   b. 1.4 hr
  c. 14 hr   d. 5.7 hr
  e. 3.5 hr

19. A composite conductor has a steel core and an aluminum shell. It's outside diameter is 200 mils. The resistivity of the steel and aluminum are 80 and 24Ω circular mill per ft respectively. The ratio of steel to aluminum by volume is 2:3. What is the resistance of 1000 ft of the conductor?

  a. 6Ω   b. 1Ω
  c. 1.2Ω d. .833Ω
  e. 11Ω

Problems 20 - 21

A 1Ω load is fed by two batteries connected in parallel. One battery is 6.3 V with an internal resistance of .004Ω and the other is 5.4 V with an internal resistance of .008Ω.

20. What is the current flowing in the 5.4 V battery?

    a. 6.0 amperes   b. 73 amperes
    c. 69 amperes   d. 3.66 amperes
    e. 3.23 amperes

21. What is the current flowing in the 1Ω load?

    a. 6.0 amperes   b. 73 amperes
    c. 69 amperes   d. 3.66 amperes
    e. 3.23 amperes

22. A 240 V, 20 KW single phase transformer has two secondary 120 V windings A and B. Winding A has a load which has a power factor of .6 lagging and B has a load with a power factor of .8 leading. What load impedance is seen from the 240 V winding if A and B are each 10 KW?

    a. 6.2   b. 10.16
    c. 12   d. 16
    e. 5.8

23. A three phase three wire A-B-C system supplies 440 V to an equal impedance delta load of 10Ω. Determine the magnitude of the line current I_B in amps.

    a. 14.6 amperes   b. 76 amperes
    c. 89 amperes   d. 26 amperes
    e. 44 amperes

24. What is the resistance in ohms looking into a 120/440 V single phase transformer connected to a load of $(30 + j30)Ω$?

    a. 4.44Ω   b. 2.22Ω
    c. 42.2Ω   d. 11.56Ω
    e. 3.15Ω

25. A 20,000 ohm/volt meter is designed to measure 150 V full scale. The meter is to be redesigned to measure 100 V full scale. The change should be:

    a. Soak in water for 1 hr
    b. Place 3 megohms in parallel with 3 megohms of meter
    c. Place 1.5 megohms in parallel with 1.5 megohms of meter
    d. Place 6 megohms in parallel with 3 megohms of meter
    e. Place 3 megohms in series with 3 megohms of meter

26. Mechanical power developed by an armature and mechanical power delivered by a motor are:

    a. Identical
    b. Different by the stray power loss
    c. Different by armature copper loss
    d. Related by the motor slip
    e. Dependent on starting torque

Problems 27 - 29

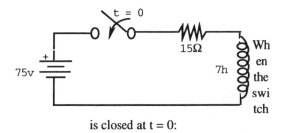

is closed at t = 0:

27. The voltage across the inductor is:

    a. 0 V   b. 25 V
    c. 50 V   d. 75 V
    e. 80 V

28. The charging current has a time constant of:

    a. 105 msec   b. .333 sec
    c. .467 sec   d. 105 sec
    e. .750 sec

29. The final energy in the inductor is:

  a. 175 Joules b. 75 Watts
  c. 87.5 Joules d. 87.5 Watts
  e. 75 Joules

Problems 30 - 32

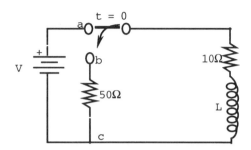

The switch has been at a for a very long time. At t = 0, it is moved to b, when a current of 8 amperes is flowing in the circuit. The voltage 2 sec later is 25 V across b-c.

30. The inductance of the coil is:

  a. 33 h b. 38 h
  c. 43 h d. 45 h
  e. 50 h

31. The maximum voltage across b-c will be?

  a. 80 V b. 480 V
  c. 400 V d. 444 V
  e. V_{volts} DC

32. The decaying time constant of the circuit is:

  a. 0.55 sec b. 0.76 sec
  c. 0.75 sec d. 0.72 sec
  e. 0.80 sec

Problems 33 - 35

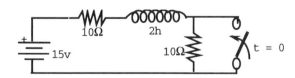

In the circuit shown, the switch has been open for a long time. At t = 0, it is closed.

33. The voltage drop across the inductor is:

  a. 15 V b. 7.5 V
  c. 1.5 V d. 0 V
  e. 10 V

34. The current through the inductor at t = 0 is:

  a. 0.75 amperes b. 0 amperes
  c. 1.5 amperes d. 15 amperes
  e. 1.0 ampere

35. The current at t = 0.1 after the switch is closed is:

  a. 1.04 ampere b. .75 ampere
  c. 1.25 ampere d. 1.5 ampere
  e. 1.75 ampere

Problems 36 - 38

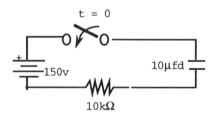

The switch in the circuit above is closed at t = 0 and opened at t = 0.03 sec.

36. The current flowing in the circuit at t = 0 is:

  a. 0 milliamperes b. 10 milliamperes
  c. 20 milliamperes d. 15 milliamperes
  e. 5 milliamperes

37. The voltage across the capacitor when the switch is opened is:

  a. 30 V b. 36 V
  c. 39 V d. 42 V
  e. 46 V

38. The energy stored in the capacitor during this period is:

  a. 4.5 Joules b. 5 x 10^{-3} Joules
  c. 7.6 x 10^{-3} Joules d. 10^{-3} Joules
  e. 6.6 x 10^{-3} Joules

Problems 39 - 41

The rise time of an RC circuit to a step input is defined as the difference in time between the 90% and 10% points of the charging curve.

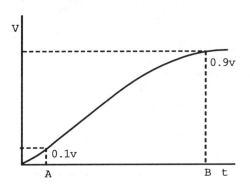

39. In terms of t (time constant) point B is:

   a. t = 3t   b. t = 2.5t
   c. t = 2.3t   d. t = t
   e. t = 2t

40. In terms of t (time constant) point A is:

   a. t = .1t   b. t = .15t
   c. t = .2t   d. t = .22t
   e. t = .25t

41. In terms of t (time constant) the rise time is:

   a. t = 2.9t   b. t = 2.4t
   c. t = 1.3t   d. t = t
   e. t = 2.2t

42. A 100 V step is applied to an RC circuit having R = 5Ω, C = 20 mfd, and t = 0. The expected rise time is:

   a. .22 msec   b. .3 msec
   c. .25 msec   d. .23 msec
   e. .15 msec

43. Determine the voltage transfer funcion ratio of the following circuit:

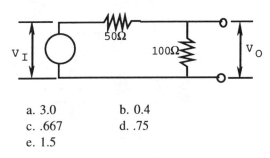

   a. 3.0   b. 0.4
   c. .667   d. .75
   e. 1.5

Problems 44 - 46

Two P-N semiconductor diodes are connected as shown below. For each diode at room temperature, V_Z = 6.75 V and I_S = 1.5 x 10^{-13} amperes.

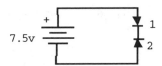

44. The state of each semiconductor diode is:

   a. Diodes 1 and 2 are forward biased
   b. Diodes 1 and 2 are reverse biased
   c. Diodes 1 is forward biased and diode 2 reverse biased.
   d. Diodes 1 is reverse biased and diode 2 forward biased.
   e. Both are reverse biased

45. The voltage across diode 1 is:

   a. 7.5 V   b. 0.75 V
   c. 3 V   d. 2.5 V
   e. 0.3 V

46. The voltage across diode 2 is:

   a. 7.5 V   b. 0.75 V
   c. 3 V   d. 2.5 V
   e. None of the above.

47. The current through the series circuit is:

   a. 1.0 amperes   b. 1.5 amperes
   c. 1.6 amperes   d. 1.7 amperes
   e. 2.0 amperes

Problems 48 - 51

A silicon transistor is used as shown:

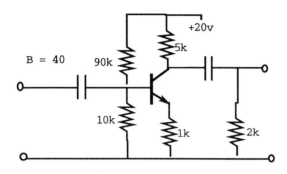

48. The quiescent value of base current I_B is:

   a. .050 amperes   b. .030 amperes
   c. .026 amperes   d. .025 amperes
   e. .020 amperes

49. The quiescent value of collector current I_C is:

   a. 2.0 milliamperes   b. 1.2 milliamperes
   c. 1.04 milliamperes   d. 1.5 milliamperes
   e. 1.5 milliamperes

50. The quiescent value of emmiter current I_E is:

   a. 2.05 milliamperes   b. 1.23 milliamperes
   c. 1.066 milliamperes   d. 1.025 milliamperes
   e. 1.055 milliamperes

51. The quiescent value V_{CE} is:

   a. 12 V   b. 13.7 V
   c. 14 V   d. 15 V
   e. 15.3 V

CHEMISTRY

1. The reaction below is:

$$3Mg + Cr_2(SO_4)_3 \rightarrow 3MgSO_4 + 2Cr$$

   a. Synthesis
   b. Analysis
   c. Single Replacement
   d. Double Replacement
   e. Decomposition

2. Calculate the pH of the resulting solution when 200 ml of 0.1M HCl and 300 ml of 0.05M Mg(OH)2 are reacted:

$$Mg(OH)_2 + 2HCl \rightarrow MgCl + H_2O$$

   a. 7   b. 1
   c. 13.3   d. 12.3
   e. 10.7

Problems 3 - 6

For a solution of $AlCl_3$ with a density of 1.1 g/L, and 10% by weight:

3. The molarity is:

   a. 8.3M   b. .83M
   c. 1M   d. 1.2M
   e. .75M

4. The molality of the solution is:

   a. 8.4m   b. .94m
   c. 1m   d. .75m
   e. .84m

5. The freezing point of the soluton will be:

   a. 6.25°C   b. -6.25°C
   c. 1.56°C   d. -1.56°C
   e. 0°C

6. The normality of the solution is:

   a. .83N   b. 84N
   c. 2.5N   d. 0.4N
   e. 2.5N

Problems 7 - 11

For the reaction:

$$2SO_{2(g)} + O_{2(g)} \leftrightarrow 2SO_{3(g)} + 44 \text{ Kcal}$$

What will be the effect of the stress applied to the reaction after the new equilibrium is attained?

1. Increase SO_2 2. Increase SO_3
3. Increase O_2 4. Decrease SO_2
5. Decrease SO_3 6. Decrease O_2

7. Add SO^2

   a. 1 only   b. 2 only
   c. 3 and 4   d. 4 and 5
   e. 2 and 6

8. Add a catalyst:

   a. 3 only   b. 1 and 3
   c. 4 and 5   d. 2 only
   e. Speed up the reaction

9. Increase the pressure:

   a. 2 only   b. 1 and 3
   c. 2,4, and 6   d. 7 only
   e. 4 and 5

10. Increase the temperature:

   a. 2 only   b. 1, 3 and 5
   c. 2,4, and 6   d. 7 only
   e. 4 and 5

11. Decrease the O_2:

   a. 1 only   b. 1 and 3
   c. 1 and 5   d. 5 only
   e. 4 and 5

12. Find the solubility of Pb^{+2} in moles/liter:

$$PbI_{2(s)} \rightarrow Pb^{+2} + 2I^{-1}$$
$$K_{sp} = [Pb^{+2}][I^{-1}]^2 = 1.4 \times 10^{-8}$$

   a. 4.66×10^{-6} b. 1.52×10^{-3}
   c. 1.18×10^{-4} d. 3.54×10^{-4}
   e. 0.59×10^{-4}

13. If 84g of $NaHCO_3$ are decomposed, what volume of CO_2 will be produced at 47°C and 570 torrs?

$$2NaHCO_3 \rightarrow Na_2CO_3 + H_2O + CO_2$$

   a. 0.51l   b. 11.21l
   c. 17.5l   d. 1.01l
   e. 22.4l

14. How many calories are needed to change 10g of ice at 0°C to steam at 100°C?

   a. 100   b. 720
   c. 1000   d. 7200
   e. 1800

15. Which is the general formula for a saturated hydrocarbon?

   a. C_nH_{2n} b. C_nH_{2n+1}
   c. C_nH_{2n+2} d. C_nH_{2n-2}
   e. C_nH_n

16. The correct name for Hg_2Cl_2 is:

   a. Mercuric chloride
   b. Mercurous chloride
   c. Mercury chloride
   d. Hydrogen chloride
   e. Sodium chloride

17. The correct formula for chromium III carbonate is:

   a. Cr_3CO b. $Cr_2(CO)_3$
   c. Cr_2CO_3 d. Cr_3CO_3
   e. $Cr_2(CO_3)_2$

18. When the equation:

$$?Cr_2O_7 + ?H_2O + ?S \rightarrow xSO_2 + yOH^- + zCr_2O_3$$

is correctly balanced, the coefficients x, y, z, will be:

   a. 2,2,3   b. 3,4,2
   c. 2.1.3   d. 2.3.4
   e. 3,3,3

19. How long must a plating cell run at a current of 1.0 ampere if it must deposit 16g of copper on a plate from a $CuSO_4$ solution?

   a. 96,500 sec   b. 24,125 sec
   c. 60 min   d. 13.4 hr
   e. 604 hr

20. Which of the following would be most effective in neutralizing the formic acid produced by a bee sting?

    a. Vinegar   b. Aspirin
    c. Lemon juice   d. Lye
    e. Baking soda

21. The element in period 4 with chemical properties most similar to carbon is:

    a. Silicon   b. Boron
    c. Nitrogen   d. Lead
    e. Germanium

22. How many molecules of O_2 are present in a mixture of gases in a one liter vessel if it is at 2 atm and 27°C and the partial pressure of the O_2 is 0.5 atm?

    a. 0.023×10^{23} b. 0.088×10^{23}
    c. 0.496×10^{23} d. 0.120×10^{22}
    e. 0.122×10^{23}

23. How many ml of 0.2N HCl are needed to react completely with 30 ml of a 0.1N solution of $Ba(OH)_2$?

    a. 15 ml   b. 10 ml
    c. 20 ml   d. 25 ml
    e. 30 ml

24. How many liters of SO_2 at 27° and 0.75 atm will be produced by reacting 10.4 g of $NaHSO_3$ according to the reaction:

$$NaHSO_3 + HNO_3 \rightarrow H_2O + SO_2$$

    a. 2.24l   b. 3.284l
    c. 0.01l   d. 22.4l
    e. 0.30l

25. Calculate the percentage ionization of a 0.1 M acetic acid solution.

    a. 0.134%   b. 1.34%
    c. 13.4%   d. 98.6%
    e. 86.6%

26. What is the pH of the ionized solution from problem 25?

    a. 2.87   b. -2.87
    c. -1.99   d. 1.99
    e. 2.00

Problems 27 - 29

A sample of nicotine was found to contain 72.73% C, 16.97% N, 10.3% H.

27. What is the empirical formula:

    a. $C_5NH_{8.5}$ b. $C_{15}N_3H_{22}$
    c. CNH d. $C_{10}N_2H_{17}$
    e. C_6NH_{10}

28. The freezing point of 100g of water containing 10g nicotine was -1.127°C. What is the molecular weight?

    a. 165   b. 162
    c. 168   d. 160
    e. 170

29. What is the molecular formula for nicotine?

    a. $C_5NH_{8.5}$ b. $C_{15}N_3H_{22}$
    c. CNH d. $C_{10}N_2H_{17}$
    e. C_6NH_{10}

THERMODYNAMICS

1. The value of specific heat for non gases is:

    a. A function of pressure
    b. A function of volume
    c. A function of pressure and volume
    d. A function of substance only
    e. Relatively independent of process

2. The Carnot efficiency for a system operating between the limits of 1000°R and 2000°F is:

    a. 22%   b. 27%
    c. 50%   d. 56%
    e. 59%

3. The most general thermodynamic process is?

   a. Isentropic
   b. Polytropic
   c. Isothermal
   d. Isochoric
   e. Isobaric

4. A reversible adiabatic process is called?

   a. Isentropic
   b. Polytropic
   c. Isobaric
   d. Isothermal
   e. Isochoric

5. Symbolically the first law of thermodynamics can be expressed as:

   a. $\Delta U = U_2 - U_1$ b. $\Delta Q = C_p \Delta T$
   c. $k = C_p/C_v$ d. $Q = \Delta U + W$
   e. $W = \int PV\, dw$

6. Which of the following is not true for an isentropic process?

   a. Energy gains or losses during the process can only result from work done on the process.
   b. The process is irreversible
   c. No heat is transferred
   d. P, V, and T, all vary
   e. $Q = \Delta U + W$

7. The temperature in a 60 ft³ tank of gas at 100°F. The molecular weight is 22 and the tank pressure is 25 psia. How many moles of an ideal gas are in the tank?

   a. 0.173   b. 0.204
   c. 0.245   d. 1.188
   e. 3.678

Problems 8 - 10

Two and one half pounds of air (R = 53.3) follow a polytropic process. At initial conditions the pressure is 85 psia and the temperature is 470°F. At the final conditions, volume is 16 ft³, and temperature is 140°F.

8. What is the final pressure?

   a. 30.1 psia   b. 34.7 psia
   c. 52.1 psia   d. 57.5 psia
   e. 62.2 psia

9. What is the value of n for the process?

   a. 1.41   b. 1.68
   c. 1.82   d. 1.96
   e. 2.15

10. The change in entropy from initial to final states is:

   a. -0.057 units   b. -1.09 units
   c. +0.057 units   d. +0.11 units
   e. -0.042 units

Problems 11 - 13

A four cylinder, four cycle IC engine has a bore of 3.3 in and a stroke of 4.1 in. At 1800 rpm, the engine is drawing 60 cfm of entering mixture.

11. The engine displacement is:

   a. 31.7 in³   b. 35.07 in³
   c. 126.7 in³   d. 140.3 in³
   e. 70.14 in³

12. The volume displaced per minute is:

   a. 36.53 cfm   b. 146.1 cfm
   c. 82.6 cfm   d. 46.2 cfm
   e. 73.06 cfm

13. The volumetric efficiency is:

   a. 44%   b. 58%
   c. 63%   d. 82%
   e. 90%

14. The enthalpy of 3.3 lb of a fluid with internal energy of 1000 Btu/lb, a pressure of 2 atm and a volume of 20 ft³ is:

   a. 1108.8 Btu   b. 3300 Btu
   c. 3408.8 Btu   d. 108.8 Btu
   e. 2208.8 Btu

15. A block of metal composed of 50% CU and 50% PV weighs 1Kg. The block is dropped into a tank of oil (sg = 0.82) from a height of 8 meters. The change in internal energy of the oil is:

    a. 34.52 cal   b. 18.74 cal
    c. 16.13 cal   d. 42.34 cal
    e. 20.56 cal

16. The heat of combustion of a fuel is 17,500 Btu/lb. It is used in an engine consuming the fuel at a rate of .35 lb/(hp- hr). The engine efficiency is:

    a. 17.5%   b. 25.5%
    c. 35.0%   d. 41.6%
    e. 56.2%

17. What volume in liters would 6.49g of CO_2 occupy at 730 mm Hg and 29°C?

    a. 3.807   b. 3.513
    c. 0.366   d. 0.387
    e. 168

18. Given: $C + H_2O \rightarrow CO + H_2$

How many lbs CO are produced per ton of fuel if the fuel contains 95% Carbon?

    a. 1900 lbs   b. 2850 lbs
    c. 4433 lbs   d. 323 lbs
    e. 2000 lbs

Problems 19 - 21

The hydrocarbon C_6H_{14} is burned with zero excess air.

19. The % by weight of CO_2 in the product is:

    a. 7.3%   b. 11.6%
    c. 18.9%   d. 44.6%
    e. 24.5%

20. If the water vapor in the product stream is condensed, what %CO_2 is in the product stream:

    a. 7.3%   b. 11.6%
    c. 18.9%   d. 44.6%
    e. 24.5%

21. If the water vapor in the product stream is condensed, what volume %CO_2 is in the product stream:

    a. 12.4%   b. 14.3%
    c. 22.4%   d. 30.6%
    e. 44.0%

Problems 22 - 24

500 lbs of a gas are held in a 3.5 ft diameter spherical tank made of thin wall, high strength stainless steel. The gas composition is 21% CH_4, 20% CO, 19% CO_2, 10% H_2S, 30% N_2. The tank pressure is 30 psia.

22. The average molecular weight of the gas is:

    a. 15.00   b. 22.44
    c. 29.12   d. 36.78
    e. 38.46

23. How many lb-moles of the gas are in the tank?

    a. 17.17   b. 29.12
    c. 22.40   d. 500
    e. 2912

24. What is the volume of the gas in the tank? (R = 1544/MW)

    a. 22.4 ft^3 b. 3252 ft^3
    c. 7.07 ft^3 d. 14.14 ft^3
    e. 3.1416 ft^3

25. 400 g of a metal at 212°F was dropped into an insulated tank containing 200cc of H_2O at 22°C. The temperature of the metal and water stabilized at 34°C. The specific heat of the metal is:

    a. 0.0037 cal/g   b. 0.0428 cal/g
    c. 0.0563 cal/g   d. 0.0773 cal/g
    e. 0.0909 cal/g

Problems 26 - 27

A fluid with SG = 1.7 and C = 2.84 enters a heat exchanger at 250°F and leaves at 90°F. Air enters at 40°F, and leaves at 75°F.

26. For HE parallel, the LMTD is:

a. 114°F b. 100°F
c. 74°F d. 145°F
e. 82.5°F

27. For 100% heat transfer, what flow rate of the liquid is required?

a. 0.24 lbs/hr b. 8.4 lbs/hr
c. .01849 lbs/hr d. .0845 lbs/hr
e. .0141 lbs/hr

28. If the actual COP is 5.3, the indicated horsepower for a compressor per ton of refrigeration is:

a. 4.71 hp/T b. 24.96 hp/T
c. 23.55 hp/T d. .889 hp/T
e. .94 hp/T

29. The latent heat of evaporation of Freon 12 at 5°C is 68.2 BTU/lb. How much refrigerant must be circulated to produce one ton of refrigeration?

a. 179.9 lb/hr b. 152.4 lb/hr
c. 126.3 lb/hr d. 3.62 lb/hr
e. 2.28 lb/hr

30. What is the ideal COP for a refrigerating system operating between the limits of 0°F and 500°R?

a. .92 b. 1.15
c. 11.5 d. 115
e. 15.1

31. The saturation temperature of a water air mixture at atmospheric pressure is 535°R. How many pounds of water are there in 4.5 lbs dry air.?

a. .0189 lb b. 0.835 lb
c. 132 Gr wv/lb da d. .019 lb wv/lb da
e. .848 lb

Problems 32 - 35

In cooling a liquid food, frozen water is added to the room temperature liquid while it is being stirred. Condensation appears on the outside of the container when the contents are at 52°F after being cooled from 74°F. Pressure is 14.696 psi.

32. What is the partial pressure of the dry air in the room?

a. 14.5 psia b. 14.6 psia
c. 14.7 psia d. .2 psia
e. .19182 psia

33. What is the ratio of the weight of air to the weight of water for the ambient air in the room?

a. 77 b. 84
c. 93 d. 102
e. 121.7

34. What is the relative humidity of the air in the room?

a. 42.3% b. 44.5%
c. 45.6% d. 46.1%
e. 47.1%

35. What is the specific humidity of the air?

a. .0130 lb wv/lb da b. .0119 lb wv/lb da
c. .0108 lb wv/lb da d. .0098 lb wv/lb da
e. .0085 lb wv/lb da

Problems 36-38

A thin flat plate has a surface temperature of 150°F when placed in sunlight. The ambient air is at 72°F. The plate is well insulated on its dark side. Emissivity of the plate is 0.99, and it receives 350 Btu/hr/ft^2.

36. What is the heat radiated from the plate to the surrounding air?

a. 1 Btu/hr/ft^2 b. 10 Btu/hr/ft^2
c. 100 Btu/hr/ft^2 d. 1000 Btu/hr/ft^2
e. 10,000 Btu/hr/ft^2

37. How much heat goes into the air by conduction?

a. 2.5 Btu/Ft^2 b. 25 Btu/Ft^2
c. 250 Btu/Ft^2 d. 2500 Btu/Ft^2
e. .25 Btu/Ft^2

38. What is the unit surface conductance?

a. 2.62 Btu/Ft^2/°F b. 3.21 Btu/Ft^2/°F
c. 5.14 Btu/Ft^2/°F d. 6.18 Btu/Ft^2/°R
e. 2.25 Btu/Ft^2/°F

39. A steam turbine receives 3500 lbs of steam per hour at 105 ft/sec and an enthalpy of 1530 Btu/lb. The exit velocity and enthalpy are 800 fps and 1250 Btu/lb, respectively. The Hp output of the turbine is?

    a. 123 Hp   b. 283 Hp
    c. 350 Hp   d. 450 Hp
    e. 549 Hp

Problems 40 - 42

The inlet temperature to a gas turbine is 1500°F. It has an expansion pressure ratio of 6.0. The gasses leave the turbine at 995°F. $C_v = .2$ and $C_p = .26$.

40. The ideal Δh is:

    a. 142 Btu/lb   b. 173 Btu/lb
    c. 185 Btu/lb   d. 192 Btu/lb
    e. 1.3 btu/lb

41. The actual Δh is:

    a. 142 Btu/lb   b. 154 Btu/lb
    c. 166 Btu/lb   d. 173 Btu/lb
    e. 185 btu/lb

42. The adiabatic efficiency of the turbine is:

    a. 38%   b. 59%
    c. 64%   d. 78%
    3. 82%

43. A steam jet (c = .9) expands adiabatically in a nozzle from 150 psia and 150° superheat into a condenser at 2-in. Hg abs. What is the velocity of the steam jet?

    a. 2812 fps   b. 3419 fps
    c. 3528 fps   d. 3767 fps
    e. 223.7 fps

44. Electric cable is used for melting snow on a driveway 8 ft wide and 45 ft long. What is the cost of melting 6 in. of snow (w = 10 lb/ft^3) at 32°F if the efficiency of operation is 50% and electricity costs $.05/kwh?

    a. $7.59   b. $6.42
    c. $5.83   d. 4.22
    e. $3.98

45. Saturated steam at (212°F) is bubbled into a liquid bath containing 2 cu yds of a mixture of SG = 2.0 and specific heat is 0.5. The liquid bath is initially at 35°F, and must be heated to 96.3°F. For an ideal process with no losses, how much steam is required?

    a. 63.4 lb   b. 1785.3 lb
    c. 190.2 lb   d. 375.6 lb
    e. 580.1 lb

46. A steam throttling calorimeter receives steam at 150 psia and discharges the steam at 15 psia. The calorimeter thermometer reads 250°F. The quality of the steam is:

    a. 97%   b. 93%
    c. 88%   d. 76%
    e. 63%

Problems 22 - 24

A steam turbogenerator uses 20lb of steam per kwh produced. The steam enters at 250°F and leaves at 2 psia.

$$h = 934 \text{ Btu/lb}, h_f = 94 \text{ Btu/lb}$$

47. The Rankine cycle efficiency is:

    a. 12%   b. 19%
    c. 27%   d. 31%
    e. 35%

48. Steam consumption is:

    a. 11 lbs/kwh   b. 10 lbs/kwh
    c. 9 lbs/kwh   d. 8 lbs/kwh
    e. 5 lbs/kwh

Problems 49 - 50

These concepts were not covered in the notes. Static pressure and temperature are measured in a high velocity air stream. At mach 0.68 the pressure is 42 psia and the temperature is 230°F.

49. The stagnation temperature is:

    a. 690 °R   b. 754 °R
    c. 284 °R   d. 251 °R
    e. 783 °R

50. The stagnation pressure is:

a. 42 psia   b. 57.23 psia
c. 54.7 psia d. 65.4 psia
e. 1.22 psia

MATERIAL SCIENCE

Problems 1 - 6

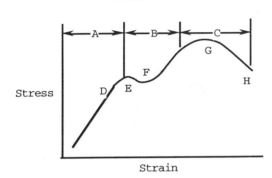

1. Range A on the stress-strain curve is:

a. Ultimate strength range
b. Plastic range
c. Elastic range
d. Ultimate strain range
e. Indeterminant range

2. The proportional limit is:

a. Point D   b. Point E
c. Point F   d. Point G
e. Point H

3. The yield point is:

a. Point D   b. Point E
c. Point F   d. Point G
e. Point H

4. The elastic limit is:

a. Point D   b. Point E
c. Point F   d. Point G
e. Point H

5. Failure occurs at or in:

a. B   b. F
c. C   d. G
e. E

Problems 7 - 10

Given the following equilibrium phase diagram for materials X and Y:

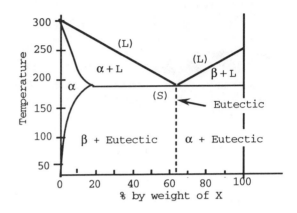

7. The eutectic composition is:

a. 18% X   b. 50% X
c. 80% X   d. 36% X
e. 62% X

8. The eutectic temperature is:

a. 50°   b. 100°
c. 180°   d. 250°
e. 325°

9. What is the lesser composition of the solid phase of X in equilibrium at the eutectic temperature?

a. 10%   b. 18%
c. 40%   d. 62%
e. 97%

10. For an alloy containing 70% by weight of X at 250°, what fraction exists as the α phase?

a. 10%   b. 18%
c. 40%   d. 40%
e. 0%

11. A solid is produced from two components. The equilibrium condition of this solid has α and β when:

a. α and β phases have the same crystal structure
b. P and Q have the same melt points
c. The internal energy of α is greater than β
d. The internal energy of α is less than β
e. None of the above

12. The eutectoid mixture of steel is:

a. Ferrite and cementite
b. Ferrite and austentite
c. Ferrous and cementite
d. Distentite and loostite
e. No tite at all

13. The molecular weight of vinyl chloride is 62.5. PVC has a degree of polymerization of 20,350. The molecular weight of PVC is:

a. 325.6
b. 3.07×10^{-3}
c. .999
d. 1.27×10^6
e. 10,250

14. What are the Miller Indices for the plane shown?

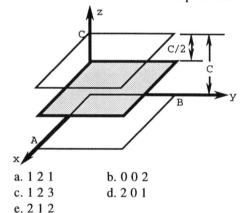

a. 1 2 1
b. 0 0 2
c. 1 2 3
d. 2 0 1
e. 2 1 2

15. What are the Miller indices for the plane:

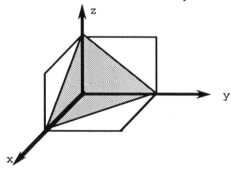

a. 0 0 0
b. 1 1 1
c. 2 2 2
d. 1 2 3
e. 3 2 1

16. A unit cell is:

a. A group of atoms in a cubic arrangement
b. An FCC
c. A unit cube containing the fewest number of atoms
d. The smallest group of atoms which when regularly repeated forms the crystal
e. A BCC

17. A test specimen is subjected to a stress and a crack appears. The stress is maintained at the level that caused the crack to appear.

a. The crack will continue to propagate in its original direction
b. The crack will stop growing
c. The crack will propagate in a direction normal to the original
d. The crack will continue to propagate in both normal and transverse directions.
e. The specimen will self destruct

18. A force of 8N is applied to a flat tensile specimen 10 cm long, 1 cm wide, and 0.2 cm thick. The stress is:

a. 40 N/M²
b. 8 N/M²
c. 8 N/cm²
d. 1.6 N/M²
e. 4 N/M²

19. Which of the following is not a Non-destructive test?

a. Rockwell hardness
b. Magnafluxing
c. Izod impact
d. Radiography
e. Acoustic testing

20. Molecules of a polymer are not held together by which of the following:

a. Hydrogen bonds
b. Primary bonds
c. Secondary bonds
d. Covalent bonds
e. Tertiary bonds

ENGINEERING ECONOMICS

1. A deposit of $1,500 is made in a savings account which pays 7.5% interest compounded annually. How much money will be available to the depositor at the end of 16 years?

   a. $4,438.35   d. $471.60
   e $5,129.10   e None of the above
   e $4,771.20

2. If a deposit of $1,000 is made today in a 8% per year compounded quarterly savings account it would grow to $2,208 at the end of 10 years. If the deposit is made two years from now how much will have to be deposited so the $2,208 terminal amount will still be realized?

   a. $1,171.56   d. $1,271.56
   b. $1,071.56   e $971.56
   c. $1,371.56

3. It was planned to leave $2,500 on deposit in a savings account for 15 years at 6.5% interest. It became necessary to withdraw $750 at the end of the 5th year. How much will be on deposit at the end of the 15 year period?

   a. $5,679.50   d. $6,021.71
   b. $5,379.50   e $5,021.71
   c. $4,679.50

4. Fifteen years ago a deposit of $875.00 was made in a commercial bank. Today the account has a balance of $2,069.38. The bank pays interest on a semiannual basis. Although the bank interest varied over the 15 year period based on economic conditions, determine the average interest rate compounded semi-annually the deposit earned.

   a. 6.27%   d. 5.84%
   b. 5.04%   e 3.77%
   c. 4.04%

5. How long will it take a deposit to triple at 18% interest compounded monthly?

   a. 68 Months   d. 70 Months
   b. 72 Months   e 76 Months
   c. 74 Months

6. A deposit of $650 is made at the end of each year in an account paying 10%. The deposits were made for the 5th and 6th years. What amount of money is on deposit at the end of 25 years?

   a. $57,557.23   d. $55,757.23
   b. $57,755.23   e $55,577.23
   c. $50,577.23

7. The U.S. sells Savings Bonds for $37.50, which are redeemable for $75.00 in 12 years. What is the annual interest rate paid by the U.S.?

   a. 5.94%   d. 7.04%
   b. 5.40%   e None of the above
   c. 6.40%

8. A child receives $50,000 as a gift which is deposited in a 6% bank account compounded semiannually. If 2,500 is withdrawn at the end of each half year, how long will the money last?

   a. 21.0 Years   d. 18.0 Years
   b. 15.5 Years   e 11.5 Years
   c. 25.0 Years

9. If $10,000 is borrowed with interest at 12% compounded quarterly for 5 years, what sum should be paid at the end of the 5 years to settle the debt and accumulated interest?

   a. $18,061   d. $18,601
   b. $16,801   e $17,816
   c. $16,108

10. A car loan for $12,500 is to be repaid in equal monthly installments over two years at a 12% interest rate How much of the principle will have been paid after 18 payments?

    a. $8,585.55   d. $9088.90
    b. $9,585.55   e None of the above
    c. $8,085.55

11. For the loan described in Problem 10, how much would have to be paid to the bank to settle the loan after 15 years?

    a. $4,053.23   d. $6,043.23
    b. $5,643.23   e $5,043.23
    c. $5,933.43

12. A debt of $300,000 due in 10 years is to be paid at maturity by means of a semiannual sinking fund. The borrower finds that only a 6% sinking fund is available at the present time, but feels that after 4 years 8% sinking fund semi-annual interest will be available Based on the borrowers assumption what semiannual equal payment into the sinking funds should be made?

    a. $13,781   d. $9,947
    b. $10,828   e Problem can't be solved
    c. $15,438

13. A new grinding machine cost $15,000 and has a useful life of 15 years. If the asset is depreciated at 18.6%, what is its book value after 12 years based on declining balance depreciation?

    a. Can't be solved   d. $1,069
    b. $1,469   e $1,269
    c. $984

For problems 14 thru 16 consider a company that has purchased a piece of capital equipment for $100,000. The useful life of the asset is estimated to be 15 years with a salvage value of $12,000 at the end of the estimated life

14. Using SLD what is the depreciation after 7 years?

    a. $6,666.67   d. $5,866.67
    b. $46,666.69   e None of the above
    c. $41,066.69

15. Using the sum of the years digits method, what is the depreciation after 7 years?

    a. $58,100   d. $60,100
    b. $63,600   e $61,600
    c. $65,100

16. What is the annual depreciation using the sum of the years digits depreciation method?

    a. $6,898.54   d. $7,345.98
    b. $7,209.03   e Can't be answered
    c. $7,951.57

17. A new machine costing $144,500 will have a salvage value of $15,500 at the end of 10 years. Using the declining balance method of depreciation what will the book value be at the end of 9 years?

    a. $23,459   d. $19,395
    b. $28,376   e $15,863
    c. $16,948

For problems 18 thru 20 consider buying a new home The selling price for the house is $375,000. You have enough money for a 15% down payment with the balance to be amortized at 12% compounded monthly for 15 years.

18. How much will the monthly payments be?

    a. $3,825   d. $3,892
    b. $3,486   e $3,582
    c. $3,977

19. How much interest is paid over the term of the loan?

    a. $396,570   d. $688,500
    b. $547,639   e None of the above
    c. $369,750

20. If you decided to increase your down payment by borrowing $75,000 at 8% interest compounded semiannually for 15 years, how much would be saved in total interest payments?

    a. $37,000   d. $7,000
    b. $27,000   e $32,000
    c. $17,000

21. The operating and maintenance cost for a mining machine are expected to be $11,000 in the first year, and increase by 800 a year during the 15 year life of the machine What equal year end series of payments would cover these expenses over the life of the machine, if interest is 10%/year?

    a. $11,000   d. $15,223
    b. $13,423   e $4,223
    c. $17,322

Your company has been given two take over offers. Corporations A's offer is for $1,500,000 in cash upon the agreement and 10 annual payments $150,000. Corporation B's offer is for $2,000,000 now, $750,000 in one year, $500,000 in two years, and $250,000 in five years. The cost value of money to your company is 10%.

22. What is the present worth value of Corporation A's offer?

   a. $1,500,000   d. $150,008
   b. $1,140,915   e $2,640,915
   c. $2,421,690

23. What is the present worth value of Corporation B's offer?

   a. $3,250,250   d. $2,000,000
   b. $1,250,250   e None of the above
   c. $4,250,000

24. Which offer would you recommend to the owner of your company?

   a. Corporation A   d. Neither Corporation
   b. Corporation B   e None of the above
   c. Either Corporation

Five years ago you bought a home for $325,000. You put down 15% of the purchase price of the house and obtained a 20 year mortgage for the balance The mortgage is at 9.5% and you make your payments annually. The closing costs on the house were $10,000 which you included in the mortgage amount. You have just made you seventh payment and have decided to purchase a new house

If you sell your house and pay off the existing mortgage you must pay the bank a penalty of 3.5% of the current mortgage balance In order to purchase the new house you need to put down $100,000.

25. How much are your current mortgage payments?

   a. $30,984   d. $34,289
   b. $32,489   e $23,489
   c. $38,249

26. How much must you pay the bank to settle your existing mortgage?

   a. $236,886   d. $228,595
   b. $422,357   e None of the above
   c. $245,178

27. How much must you sell the house for in order to make the down payment on the new house?

   a. $345,177   d. $475,177
   b. $395,177   e None of the above
   c. $425,177

28. How much did your house appreciate each year that you owned it?

   a. 6.2%   d. 4.2%
   b. 5.2%   e None of the above
   c. 7.2%

29. An investment property cost $62,000 to purchase, another $7,500 was required to improve the property. The property produced annual return after deducting all expenses of 11,500 per year for 10 years, after which the property is worthless. What is the annual rate of return from this investment?

   a. 10.40%   d. 11.40%
   b. 10.00%   e 9.40%
   c. 11.00%

Mr Engineer started working for the ABC Utility Company on his 25th birthday after graduating from college Each year 5% of his salary was withheld and placed in a retirement account. The company also deposited an amount equal to 25% Mr Engineer's contribution into the account which has earned 8% interest. Mr Engineer's starting salary was $10,000 per year, with annual increases of $1,250 each year. Mr. Engineer retired on his 65th birthday.

30. How much money did Mr. Engineer deposit in his retirement account during his employment, expressed in year of hire dollars?

   a. $5,962   d. $11,840
   b. $7,878   e $15,480
   c. $13,840

31. How much money did the company put into Mr Engineer's account?

a. $2,500.00
b. $3,460.00
c. $1,969.50
d. $312.50
e $1490.50

32. How much money, including interest will be in Mr Engineer's account when he retires?

a. $75,168
b. $250,583
c. $457,279
d. $300,667
e $375,834

33. Mr Engineer elects to receive his retirement money as a 20 year annuity with the first payment on his 65th birthday. How much will he receive each year?

a. $36,225
b. $39,124
c. $34,728
d. $38,297
e $35,460

A large manufacturing corporation pays $0.18 per mile to employees who use their personal automobiles for company business. Interest is at 9%. An engineer working for the company checked the records on his automobile and found the following data:

First cost - $12,500.00
Useful Life - 8 years
Trade in Value - $3,000
Cost of gas per mile travelled - $0.09
Average cost of maintenance - $22.15 per month
Insurance cost - $1,200/yr
Miles per year - 18,000

34. What is the total annual owning and operating cost of the car based on straight line depreciation?

a. $3,824.24
b. $4,758.44
c. $3,404.24
d. $4,273.30
e. $5,024.24

35. How many miles does the car above have to be driven, if 65% of the car's use is for the company, so the company allowance breaks even with owning the car?

a. 38,151
b. 18,551
c. 28,115
d. 43.251
e 22,454

A machine shop is considering a new piece of machinery. Three machines are under consideration. Use 14% interest and the following available data for your calculations:

	MACHINE X	MACHINE Y	MACHINE Z
FIRST COST	$15,000	$17,000	$20,500
ANNUAL EXPENSES	$625	$515	$400
FIVE YEAR OVERALL	$1,000	$1,070	$1,200
USEFUL LIFE - YEARS	$15	$20	$25
SALVAGE VALUE	$0	$0	$1,500

36. What is the annual cost of Machine X?

a. $3,067.00
b. $4,218.30
c. $2,593.30
d. $3,218.30
e $4,067.00

37. Which machine would you recommend for purchase?

a. MACHINE X
b. MACHINE Y
c. MACHINE Z
d. MACHINE X & Y
e MACHINE Y & Z

38. Considering unequal lives of the machines, if PW analysis was required, what period of time should the economic analysis be based on?

a. 15 yrs
b. 20 yrs
c. 25 yrs
d. 100 yrs
e 300 yrs

39. What is the total present worth of machine Z using a 100 year time period?

a. $21,400
b. $23,400
c. $25,400
d. $27,400
e $29,400

40. Using SFD, what is the book value of machine Y after 10 years?

a. $18,479
b. $12,070
c. $16,496
d. $7,835
e $13,384

12

Solutions to Practice Problems

MATHEMATICS

1.

$$\frac{4\sqrt{3}}{3 + i\sqrt{3}} \text{ will reduce to:}$$

a. $\dfrac{4}{\sqrt{3} + 1/3}$ b. $3 - i\sqrt{3}$

c. $\sqrt{3} - i$ d. $2 + 2\sqrt{3}$

e. $\dfrac{\sqrt{3} - i\sqrt{3}}{4}$

Multiply the numerator and the denominator by the complex conjugate of the denominator:

$$\frac{4\sqrt{3}}{3 + i\sqrt{3}} \times \frac{3 - i\sqrt{3}}{3 - i\sqrt{3}} = \frac{12\sqrt{3} - 12i}{3^2 + (\sqrt{3})^2}$$

$$\sqrt{3} - i \quad \underline{\text{Answer (c)}}$$

2. The simultaneous equations:

$$3x + y - z = 14$$
$$x + 3y - z = 16$$
$$x + y - 3z = -10$$

Have the solution:

a. $x = -5, y = -6, z = -35$
b. $x = 5, y = 7, z = 8$
c. $x = -5, y = -6, z = -39$
d. $x = 6, y = 6, z = 10$
e. $x = 5, y = 6, z = 7$

Using determinants:

$$x = \frac{\begin{vmatrix} 14 & 1 & -1 \\ 16 & 3 & -1 \\ -10 & 1 & 3 \end{vmatrix}}{\begin{vmatrix} 3 & 1 & -1 \\ 1 & 3 & -1 \\ 1 & 1 & -3 \end{vmatrix}}$$

$$= \frac{-126 + 10 - 16 - 30 + 14 + 48}{-27 - 1 - 1 + 3 + 3 + 3} = 5$$

$$y = \frac{\begin{vmatrix} 3 & 14 & -1 \\ 1 & 16 & 1 \\ 1 & -10 & -3 \end{vmatrix}}{-20} = 6$$

$$z = \frac{\begin{vmatrix} 3 & 1 & 14 \\ 1 & 3 & 16 \\ 1 & 1 & -10 \end{vmatrix}}{-20} = 7$$

<u>Answer (e)</u>

3. What is the value of x if:
$$\log_4 (23)^{\frac{1}{x}} = 1$$

a. 3.5 b. 0.44
c. 0.17 d. 2.26
e. 5.75

If $\log_4 (23)^{\frac{1}{x}} = 1$: $\dfrac{1}{x} \log_4 (23) = 1$

$$x = \log_4 (23) = \frac{\log_{10}(23)}{\log_{10} 4} = 2.26$$

<u>Answer (d)</u>

4. What is the 5th term of the expansion:
$$\left[1 + \left(\frac{1}{x}\right)\right]^7$$

a. $\dfrac{25}{x^3}$ b. $\dfrac{35}{x^4}$

c. $\dfrac{15}{x^5}$ d. $\dfrac{5}{x^7}$

e. $\dfrac{n}{n!}$

$$(a + b)^n = a^n + na^{n-1}b + \frac{n(n-1)a^{n-2}b^2}{2!}$$

$$+ \frac{n(n-1)(n-2)a^{n-3}b^3}{3!}$$

$$+ \frac{n(n-1)(n-2)(n-3)a^{n-4}b^5}{4!}$$

$$+ \frac{7(7-1)(7-2)(7-3)1^{7-4}\left(\frac{1}{x}\right)^5}{(4)(3)(2)(1)} = \frac{35}{x^4}$$

<u>Answer (b)</u>

5. Four men can build 3 benches in 2 days. How many days will it take one man to build 30 benches?

    a. 12 days   b. 20 days
    c. 100 days   d. 65 days
    e. 80 days

$$\frac{\text{days}}{\text{bench}} = \frac{4 \times 2}{3}$$

$$\frac{\text{days}}{30 \text{ benches}} = \frac{4 \times 2}{3} \times 30 = 80$$

<u>Answer (e)</u>

6. Given: $3x - 2y = -3$
             $-6x + y = -21$
Then:
    a. $x = 3, y = 6$
    b. $x = 5, y = 9$
    c. $x = 5, y = 6$
    d. $x = -3, y = 3$
    e. $x = 9, y = -5$

Multiply the 2d equation by 2 and add:

$$\begin{array}{l} 3x - 2y = -3 \\ \underline{-12x + 2y = -42} \\ -9x = -45 \quad x = 5 \end{array}$$

$$3(5) - 2y = -3 \quad y = 9$$

<u>Answer (b)</u>

7. A river flows at the rate of 3 mph downstream. It takes a person in a canoe the same time to row 22 miles downstream as it does or row 10 mi upstream. How fast can the person row in still water?

    a. 11 mph   b. 8 mph
    c. 5 mph   d. 16 mph
    e. 4 mph

If r = rate in still water:

$$\frac{r+3}{22} = \frac{r-3}{10}: r = 8 \text{ mph} \quad \underline{\text{Answer (b)}}$$

8. If -2 is a root of the equation:

$$x^3 + 6x^2 + 11x + 6 = 0$$

the other two roots are:

    a. 4,0   b. 2, -2
    c. 1,3   d. -1, -3
    e. -3,-3

Reduce the expression to a quadratic:

$$\begin{array}{r} x^2 + 4x + 3 \\ x+2 \overline{\smash{\big)}\, x^3 + 6x^2 + 11x + 6} \\ \underline{x^3 + 2x^2} \\ 4x^2 + 11x \\ \underline{4x^2 + 8x} \\ 3x + 6 \\ \underline{3x + 6} \\ 0 \end{array}$$

$$x = -4 \pm \frac{\sqrt{16 - (4)(1)(3)}}{2}$$

$$x = -1, -3$$

<u>Answer (d)</u>

9. The expression:

$$a^{p \log_a x + \log_a y} =$$

    a. $a^{x^p y}$ b. $a^{x^p} + y$
    c. $x^p y$ d. xy^p
    e. p^{xy}

$$\left(a^{\log_a x^p}\right)\left(a^{\log_a y}\right) = x^p y \quad \underline{\text{Answer (c)}}$$

10. The polar equation of the curve defined by the rectangular equation

$$x^2 + y^2 = 4x \text{ is:}$$

    a. $r = 4 \cos q$ b. $r = 4x$
    c. $r^2 = 4x$ d. $r = 2\sqrt{x}$
    e. $r(\cos q + \tan q \sin q) = 4$

Substitute $x = r \cos q$ and $y = r \sin q$:

$$r^2 \cos^2 q + r^2 \sin^2 q = 4 r \cos q$$
$$r^2 (\cos^2 q + \sin^2 q) = 4 r \cos q$$
$$r = 4 \cos q$$

<u>Answer (a)</u>

Problems 11 - 13

Given a reference line $y = 3x - 4$

11. What is the equation of the line parallel to the reference line passing through (3,4)?

    a. $y = 4x + 3$ b. $y - 3 = x - 4$
    c. $y = 3x - 5$ d. $y = -3x + 4$
    e. $3x = 5 + y$

The required line must have the same slope +3, as the reference line to be parallel. The y intercept is:

$$y = 3x + b$$
$$4 = 3(3) + b$$
$$b = -5$$

The required line is $y = 3x - 5$

                                    Answer (c)

12. What is the equation of the line perpendicular to the reference line which passes through the y axis at $y = 2$:

    a. $y = -(x/3) + 2$ b. $y = -3x + 2$
    c. $y = -2x + 3$ d. $y = -3x + 4$
    e. $y = x - (2/3)$

$$m_1 m_2 = -1$$
$$m_2 = -\frac{1}{m_1} = -\frac{1}{3}$$
$$y = -\frac{x}{3} + 2$$

                                    Answer (a)

13. What angle does the line $y = -4x + 3$ make with the reference line?

    a. 48.64°   b. 38.64°
    c. 32.47°   d. -24.28°
    e. 63.64°

$$\alpha = \tan^{-1} \frac{m_1 - m_2}{1 + m_1 m_2}$$

$$= \tan^{-1} \frac{3 - (-4)}{1 + (3)(-4)} = 32.47°$$

                                    Answer (c)

14. A function is defined by:

$$f(x) = \begin{vmatrix} -3x & 2 \\ x & -1 \end{vmatrix} + \begin{vmatrix} 2x & -4 \\ 5 & -3 \end{vmatrix}$$

When $f(x) = 0$, x is:

    a. 20   b. -20
    c. -4   d. 10   e. 4

$$f(x) = -3x(-1) - 2x + 2x(-3) - 5(-4) = 0$$

$$x = 4 \qquad\qquad \text{Answer (e)}$$

15. The center of a circle defined by:
$$x^2 + y^2 - 6x + 4y = 0$$

    a. (-6,4)   b. (-3,2)
    c. (3,-2)   d. (0,0)
    e. (3,4)

Group:
$$(x^2 - 6x) + (y^2 + 4y) = 3$$

Complete the squares:
$$(x^2 - 6x + 9) + (y^2 + 4y + 4) = 3 + 9 + 4$$

Factor:
$$(x - 3)^2 + (y + 2)^2 = 16$$

The center is (3,-2)   Answer (c)

16. The length of a line with slope = 3/4, from the point (4,6) to the y axis is:

    a. 3   b. $\sqrt{53}$
    c. 5   d. 4
    e. 6

The equation of the line is:

$$y = \frac{3}{4}x + b$$

The y intercept is:

$$6 = \frac{3}{4}4 + b: \; b = 3$$

The length of the line is:

$$L = \sqrt{(6-3)^2 + (4-0)^2} = 5 \quad \text{Answer (c)}$$

17. The shape of the curve defined by:

$$3x^2 - 12x - 2y^2 + 16y$$

   a. Circle   b. Parabola
   c. Ellipse   d. Hyperbola
   e. Cone

From the general equation:

$$Ax^2 + 2Bxy + Cy^2 + 2Dx + 2Ey + F = 0$$
$$B = 0, A = 3, C = -2$$
$$B^2 - AC = 24$$
$$30 > 0 \text{ Hyperbola} \qquad \text{Answer (d)}$$

18. For $n = 7$, the value of the expression:

$$\frac{((n-1)!)^2 (0!) (n)}{(n-2)! (n!)}$$

   a. 7   b. 4320
   c. 6   d. 0
   e. 42

$0!$ is 1 thus:
$$\frac{((n-1)!)^2 (0!) (n)}{(n-2)! (n!)} = \frac{((n-1)!)((n-1)!)(n)}{(n-2)! (n!)}$$

$$= \frac{((n-1)!)((n-1)!)}{((n-2)!)((n-1)!)}$$

$$= \frac{((n-1)!)}{((n-2)!)} = n - 1 = 6 \qquad \text{Answer (c)}$$

19. The value of A is:

$$\log_7 (101A + 40)^2 = 6$$

   a. 3   b. 9.5
   c. 5   d. 1164
   e. 1000

$$2 \log_7 (101A + 40) = 6$$
$$\log_7 (101A + 40) = 3$$
$$7^3 = 101A + 40: A = 3 \qquad \text{Answer (a)}$$

20. The rectangular coordinates for a point of polar coordinates $(4, \pi/6)$ is:

   a. $(2\sqrt{3}, 2)$ b. $(0.04, 2)$
   c. $(4, 6)$ d. $(2, 2\sqrt{3})$
   e. $(2\sqrt{3}, 2\sqrt{3})$

$$x = r \cos \theta = 4 \cos \frac{\pi}{6} = 2\sqrt{3}$$
$$y = r \sin \theta = 4 \sin \frac{\pi}{6} = 2$$
$$\text{Answer (a)}$$

21. Evaluate the expression

$$\sin (1/2 \cos^{-1} 4/5)$$

   a. 4/5   b. 3/5
   c. $\sqrt{10}$ d. $1/\sqrt{10}$
   e. $\sqrt{10}/3$

Let $x = \cos^{-1}(4/5)$

$$\sin \frac{x}{2} = \pm\sqrt{\frac{1 - \cos x}{2}} = \pm\sqrt{\frac{1 - \frac{4}{5}}{2}} = \frac{1}{\sqrt{10}}$$

$$\text{Answer (d)}$$

22. For: $A = |-6, -2, 2, 4, 6, 8|$
$B = [x : x^3 \le 125]$
$C = \theta$

   $(A \cap B) \cup C$ contains:
   a. $|2, 4|$ b. θ
   c. $|-2, 2, 4|$ d. $|-6, -2, 2, 4|$
   e. $|6, 8|$

The intersection of A and B consists of all elements belonging to both sets:

$$|-6, -2, 2, 4|$$

The union of $|-6, -2, 2, 4|$ with the null set consists of all elements belonging to both sets:

$$|-6, -2, 2, 4|$$
$$\text{Answer (d)}$$

23. One root of the equation:

$f(x) = 2x^3 + 5x^2 - 13x - 30$

lies between:

    a. -1 to 0   b. 0 to +2
    c. +2 to +4   d. +4 to +6
    e. +1 to +3

$f(-1) = -2 + 5 + 13 - 30 = -14$
$f(0) = -30$
$f(1) = 2 + 5 - 13 - 30 = -26$
$f(2) = 16 + 20 - 26 - 30 = -20$
$f(3) = 54 + 45 - 39 - 30 = 30$
$f(4) = 128 + 80 - 52 - 30 = 126$

The sign change is between 2 and 3

<u>Answer (c)</u>

24. Given:
$$243 = 3^{(x^2 - 4)}$$

x is:

    a. 3   b. 3.5
    c. 4   d. 1
    e. 2

Take the log of both sides:

$$\log 243 = \left(x^2 - 4\right) \log 3$$

$$x^2 - 4 = \frac{\log 243}{\log 3} = 5$$

$x = 3$ <u>Answer (a)</u>

25. The center of the circle which passes through the points (9,6), (3,-2) and (10,-1) is:

    a. (6,3)   b. (6,2)
    c. (2,6)   d. (7,1)
    e. (4,4)

Let the coordinates h,k be the center of the circle. The following equations must be satisfied:

$(9 - h)^2 + (6 - k)^2 = r^2$ (1)
$(3 - h)^2 + (-2 - k)^2 = r^2$ (2)
$(10 - h)^2 + (-1 - k)^2 = r^2$ (3)

Expand:

$117 - 18h + h^2 - 12k + k^2 = r^2$ (1)
$13 - 6h + h^2 + 4k + k^2 = r^2$ (2)
$101 - 20h + h^2 + 2k + k^2 = r^2$ (3)

Subtract (2) from (1)

$104 - 12h - 16k = 0$ (4)

Subtract (2) from (3)

$88 - 14h - 2k = 0: k = 44 - 7h$

Substitute this value into (4)

$104 - 12h - 16(44 - 7h) = 0$:

    $h = 6$

From (4)
$104 - 12(6) - 16k = 0$

    $k = 2$

The center is at (6,2)   <u>Answer (b)</u>

26. Find the average value of the amplitude of the half sin wave of $y = A \sin \theta$.

    a. A/π b. $A\pi$
    c. $2A/\pi$ d. $A/2$
    e. $A/2\pi$

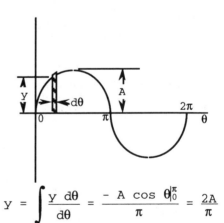

$$y = \int \frac{y \, d\theta}{d\theta} = \frac{-A \cos \theta \Big|_0^\pi}{\pi} = \frac{2A}{\pi}$$

<u>Answer (c)</u>

27. A box contains 5 green balls and 4 red balls. The probability of selecting 2 green balls and 1 red ball with a random sample of 3 balls without replacement is:

- a. 0.119
- b. 0.476
- c. 0.357
- d. 0.238
- e. 0.333

$$P = h/t$$

t = total combinations:
$$_9C_3 = \frac{9!}{(9-3)!\,3!} = 84$$

h = combinations of 2 green and 1 red:
$$(_5C_2)(_4C_1) =$$
$$\left(\frac{5!}{(5-2)!\,2!}\right)\left(\frac{4!}{(4-1)!\,1!}\right) = 40$$

$$p = 40/84 = .476 \quad \underline{\text{Answer (b)}}$$

28. The $\lim_{\theta \to 0} \dfrac{\cos\theta - 1}{\sin\theta}$ is:

- a. 0
- b. 1
- c. -1
- d. ∞
- e. Undefined

$$\lim_{\theta \to 0} \frac{\cos\theta - 1}{\sin\theta} = \lim_{\theta \to 0} \frac{-\sin\theta}{\cos\theta} = \frac{0}{1} = 0$$

$$\underline{\text{Answer (a)}}$$

29. For $3y^2 + 2yx^2 - x^3 + 6 = 0$, dy/dx is:

- a. $\dfrac{3x^2 - 4xy - 6y}{2x^2}$
- b. $\dfrac{3x^2}{6y + 2x^2}$
- c. $\dfrac{3x^2 - 2xy}{6y + 2x^2}$
- d. $\dfrac{3x^2 + 4xy}{6y}$
- e. $\dfrac{3x^2 + 4xy}{6y + 2x^2}$

Solution:

$$\frac{d3y^2}{dx} + \frac{d2yx^2}{dx} - \frac{dx^3}{dx} + \frac{d6}{dx} = 0$$

$$6y\frac{dy}{dx} + 2x^2\frac{dy}{dx} + 4yx - 3x^2 = 0$$

$$\frac{dy}{dx} = \frac{3x^2 + 4xy}{6y + 2x^2} \quad \underline{\text{Answer (e)}}$$

Problems 30 - 32

The motion of a particle is described by

$$x = \cos \pi t$$
$$y = \sin \pi t$$

30. The coordinates of the position of the particle when t = 1 are:

- a. (0,-1)
- b. (1,1)
- c. (0,0)
- d. (-1, 0)
- e. (-1,-1)

When t = 1, x = cos π = -1, y = sin π = 0.
$$\underline{\text{Answer (d)}}$$

31. The velocity in the x direction when t = 0.5 is:

- a. -1
- b. 0
- c. -π
- d. 1
- e. π

$$V = \frac{dx}{dt} = \frac{d(\cos \pi t)}{dt} = -\pi \sin \pi t$$

At $t = \frac{1}{2}$: $\dfrac{dx}{dt} = -\pi \sin \dfrac{\pi}{2} = -\pi$

$$\underline{\text{Answer (c)}}$$

32. The acceleration in the x direction at t = 0 is:

- a. 0
- b. $-\pi^2$
- c. 1
- d. -π
- e. π

$$a = \frac{d^2x}{dt^2} = -\pi^2 \cos t$$

At $t = 0$: $-\pi^2 \cos 0 = -\pi^2$

Answer (b)

33. The slope of the curve defined by:

$$y = x^2 - 4x + 4 \text{ for } y = 0$$

is:

a. -4   b. 4
c. 2   d. 0
e. π

$$\text{Slope} = \frac{dy}{dx} = 2x - 4$$

When $y = 0$, $x^2 - 4x + 4 = 0$
$(x - 2)(x - 2) = 0$: $x = 2$

Slope $= 2(2) - 4 = 0$ **Answer (d)**

34. The general solution of this differential equation is:

$$\frac{d^2y}{dx^2} - \frac{dy}{dx} - 12y = 0$$

a. $A e^{-4x} + B e^{3x}$

b. $A \cos 3x + B \sin 4x$

c. $e^{4x} + B e^{-3x}$

d. $A e^{4x} + B e^{-3x}$

e. $A \cos 4x + B \sin 3x$

This is a homogeneous second order linear differential equation with constant coefficients. The solution has the form $y = e^{ax}$. The auxiliary equation is:

$$a^2 - a - 12 = 0$$
$$(a - 4)(a + 3) = 0$$
$$a = (4, -3)$$

The general solution is:

$A e^{4x} + B e^{-3x}$ **Answer (d)**

35. The integral:

$$\int_0^3 \int_0^4 x \, dy \, dx$$

When $y = x^2$ is:

a. 24   b. 36
c. 32   d. 16
e. 8

$$\int_0^3 \int_0^4 x \, dy \, dx = \int_0^3 \int_0^4 y^{1/2} \, dy \, dx$$

$$\int_0^3 \left. \frac{y^{3/2}}{\frac{3}{2}} \right|_0^4 dx = \int_0^3 \frac{16}{3} dx = 16$$

Answer (d)

Problems 36 - 37

Given the curve $x^2 + y = 10x$

36. The area bounded by the curve and the x axis from $x = 2$ to $x = 8$ is:

a. 30 sq units
b. 132 sq units
c. 166.67 sq units
d. 126.67 sq units
e. 109.33 sq units

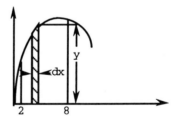

$$A = \int_2^8 (10x - x^2) \, dx = \left. \frac{10x^2}{2} - \frac{x^3}{3} \right|_2^8 = 132$$

Answer (b)

37. What is the x coordinate of the centroid of the area under the curve between x = 2 and x = 8?

 a. 8   b. 5.4
 c. 5   d. 28
 e. 132

$$\bar{x} = \frac{\int xy\, dx}{\int y\, dx} = \frac{\int_2^8 x(10x - x^2)\, dx}{A}$$

$$\bar{x} = \frac{\left.\frac{10x^3}{3} - \frac{x^4}{4}\right|_2^8}{132} = 5$$

Answer (c)

38. If

$$z = \sqrt{2x^2 y^{-3}}, \quad \frac{\partial z}{\partial y} \text{ is?}$$

 a. $-\frac{3\sqrt{2}}{2} x\, y^{-\frac{5}{2}}$ b. $-\frac{3\sqrt{2}}{2} x\, y^{-7}$

 c. $-3x\sqrt{2}\, y^{-4}$ d. $-\frac{3\sqrt{2}}{2} x\, y^{-\frac{1}{2}}$

 e. $-\frac{3\sqrt{2}}{2} x\, y^{-\frac{3}{2}}$

$$z = \sqrt{2x^2 y^{-3}} = (2x^2 y^{-3})^{1/2} = ((2x^2)^{1/2} y^{-3/2})$$

$$\frac{\partial z}{\partial y} = (x\sqrt{2} x^2)\left(-\frac{3}{2} y^{-5/2}\right)$$

$$-\frac{3\sqrt{2}}{2} x\, y^{-\frac{5}{2}}$$

Answer (a)

39. A cylinder with a closed top and bottom is to have a volume of 101 in³. If the cost of the material for the circular top and bottom costs three times as much as that for the curved surface, the diameter of the cylinder to minimize cost should be?

 a. 4 in   b. 5 in
 c. 3.5 in   d. 2.5 in
 e. 3.8 in

let D = diameter and h = height of the cylinder. h in terms of V and D is

$$h = \frac{4V}{\pi D^2}$$

The cost equation is:

$$C = (3)(2)\frac{\pi D^2}{4} + \pi D h$$

Substituting

$$= 6\frac{\pi D^2}{4} + \pi D \frac{4V}{\pi D^2} = \frac{3}{2}\pi D^2 + \frac{4V}{D}$$

$$\frac{dC}{dD} = 3\pi D - \frac{4V}{D^2} = 0 \; : \; 3\pi D^3 = 4V$$

$$D = \sqrt[3]{\frac{4V}{3\pi}} = \sqrt[3]{\frac{4(101)}{3\pi}} = 3.5$$

Answer (c)

40. If z = 3x² cos y and x and y are functions of t alone then dz/dt is?

 a. $(6x \cos y)\frac{dy}{dt} - (3x^2 \sin y)\frac{dx}{dt}$

 b. $(6x \cos y)\frac{dx}{dt} + (3x^2 \sin y)\frac{dy}{dt}$

 c. $(6x \cos y) + (3x^2 \sin y)$

 d. $(3x^2 \cos y)\frac{dx}{dt} + (3x^2 \cos y)\frac{dy}{dt}$

 e. $(6x \cos y)\frac{dx}{dt} - (3x^2 \sin y)\frac{dy}{dt}$

$$\frac{dz}{dt} = \frac{\partial z}{\partial x}\frac{dx}{dt} + \frac{\partial z}{\partial y}\frac{dy}{dt}$$

$$\frac{\partial z}{\partial x} = 6x \cos y \quad \frac{\partial z}{\partial y} = -3x^2 \sin y$$

$$\frac{dz}{dt} = (6x \cos y)\frac{dx}{dt} - (3x^2 \sin y)\frac{dy}{dt}$$

Answer (e)

Problems 41 - 43

For the reference curve $y = x^2$

41. The area bounded by the reference curve and the x axis from $x = 0$ to $x = 3$ is:
    a. 9   b. 4.5
    c. 11   d. 7
    e. 9.5

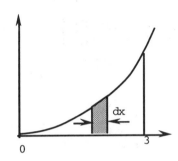

$$dA = y\, dx$$

$$A = \int_0^3 y\, dx = \int_0^3 x^2\, dx = \left.\frac{x^3}{3}\right|_0^3 = 9$$

<u>Answer (a)</u>

42. The area moment of inertia of the area bounded by the reference curve and the x axis from $x = 0$ to $x = 3$ with respect to the x axis is?

    a. 381.7   b. 153.68
    c. 20.25   d. 48.6
    e. 104.143

The choice of differential area is a key to this problem:

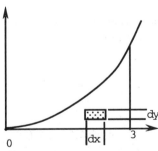

Using the differential area shown in the above figure, $A = dx\, dy$ becomes a 2d order integral.

$$I_x = \int_{y=0}^{y=9} \int_{x=\sqrt{y}}^{x=3} y^2\, dx\, dy$$

Integrate wrt x

$$I_x = \int_{y=0}^{y=9} y^2 (3 - \sqrt{y})\, dy =$$

$$\left.\frac{3y^3}{3} - \frac{y^{7/2}}{\frac{7}{2}}\right|_0^9 = 104\frac{1}{7}$$

<u>Answer (e)</u>

If the order of integration is reversed, Figure 2 must be used, and:

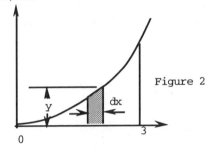

Figure 2

$$I_x = \int_{x=0}^{x=3} \int_{y=0}^{y=y} y^2\, dy\, dx$$

$$\int_{x=0}^{x=3} \left.\frac{y^3}{3}\right|_{y=0}^{y=y} dx = \int_{x=0}^{x=3} \left.\frac{y^3}{3}\right|_{y=0}^{y=y} y\, dx$$

$$\int_{x=0}^{x=3} \frac{x^6}{3}\, dx = \left.\frac{x^7}{3(7)}\right|_0^3 = 104\frac{1}{7}$$

<u>Answer (e)</u>

43. The volume generated by rotating about the x axis the area bounded by the reference curve and the x axis from $x = 0$ to $x = 3$ is:

    a. 381.7   b. 152.68
    c. 63.62   d. 48.6
    e. 143.2

$$V = \int_{x=0}^{x=3} \pi y^2\, dx = \int_{x=0}^{x=3} \pi x^4\, dx$$

$$\left.\frac{\pi x^5}{5}\right|_0^3 = 152.68$$

<u>Answer (b)</u>

44.

$$\int_1^\infty xe^{-x^2}\,dx \text{ is}$$

a. $-\dfrac{1}{2e}$ b. ∞
c. $\dfrac{2}{e}$ d. $\dfrac{1}{2e}$
e. $-\dfrac{1}{e}$

$$\int_1^\infty xe^{-x^2}\,dx = -\frac{1}{2}\int_1^\infty e^{-x^2}(-2x)\,dx$$

$$= -\frac{1}{2}e^{-x^2}\Big|_1^\infty = 0 - \left(-\frac{1}{2}e^{-1}\right) = \frac{1}{2e}$$

<u>Answer (b)</u>

45. The standard deviation of 5,6,7,8,10,12 is:

a. 3.2   b. 2.8
c. 2.4   d. 3.0
e. 2.6

The mean is:
$$\bar{X} = \frac{5+6+7+8+9+10+12}{6} = 8$$

$(x_i - \bar{X})$	$(x_i - \bar{X})^2$
-3	9
-2	4
-1	1
0	0
1	1
2	4
4	16

$$\sum(x_i - \bar{X})^2 = 34$$

$$S = \sqrt{\frac{\sum(x_i - \bar{X})^2}{n-1}} = \sqrt{\frac{35}{5}} = 2.6$$

<u>Answer (e)</u>

46.

$$L^{-1}\left\{\frac{a}{S(S+a)}\right\} \text{ is:}$$

a. sin at   b. a sin at
c. $1 - e^{at}$ d. $\cos\sqrt{a}t$
e. e^{at}

Let $\dfrac{a}{S(S+a)} = \dfrac{A}{S} + \dfrac{B}{S+a} = \dfrac{A(S+a) + BS}{S(S+a)}$

$a = A(S+a) + BS = (A+B)s + aA$

Thus $(A+B) = 0$ and $A = 1$

then $A = -B$ or $B = -1$

$$L^{-1}\left\{\frac{a}{S(S+a)}\right\} = L^{-1}\left\{\frac{1}{S}\right\} - L^{-1}\left\{\frac{1}{S+a}\right\}$$

$$= 1 - e^{-at}$$

<u>Answer (c)</u>

47.

$$L^{-1}\left\{\frac{1}{S^2 + a^2}\right\} \text{ is:}$$

a. e^{-2at} b. a sin at
c. 1/a (sin at)   d. 1/a (sinh at)
e. ae^{at}

$$L^{-1}\left\{\frac{1}{S^2+a^2}\right\} = L^{-1}\left\{\frac{1}{a}\cdot\frac{a}{S^2+a^2}\right\}$$

$$\frac{1}{a}L^{-1}\left\{\frac{a}{S^2+a^2}\right\} = \frac{1}{a}\sin at$$

<u>Answer (c)</u>

48. A stone dropped into a calm lake causes a series of circular ripples. The radius of the outer one increases at 3 ft/sec. How rapidly is the disturbed area changing at the end of 4 seconds.

a. 24π b. 72π
c. 48π d. 144π
e. 64π

The radius is increasing at a rate of 3 ft/sec, thus $dr/dt = 3$. At the end of 4 seconds:

$$r = \frac{dr}{dt}t = 3 \times 4 = 12$$

$$A = \pi r^2 \quad \frac{dr}{dt};\ \frac{dA}{dt}$$

$$= 2\pi r\frac{dr}{dt} = (2\pi)(12)(3) = 72\pi$$

<u>Answer (b)</u>

49. In one roll of a pair of dice, what is the probability of making a 7 or 11?

    a. 1/12   b. 2/12
    c. 2/10   d. 2/9
    e. 1/6

One roll of a pair of dice gives 36 possibilities. The combinations that generate either a 7 or an 11 are (1,6) (6,1) (2,5) (5,2) (4,3) (3,4) (6,5) and (5,6). Since there are 8 of 36 opportunities the probability is 8/36 = 2/9.   <u>Answer (d)</u>

50. The general solution to the differential equation $x^2 y\, dx - dy = 0$ is:

    a. $y = Ce^{(x^3/3)}$
    b. $y = \dfrac{x^3}{3} y + C$
    c. $y^{-2} = \dfrac{-2x^3}{3} + C$
    d. $\ln y = \dfrac{x^3}{3}$
    e. $y = \dfrac{Ce^{(x^3/3)}}{3}$

Rearrange

$$x^2\, dx = \frac{dy}{y}$$

Integrating

$$\frac{x^3}{3} + C' = \ln y$$

Thus

$$y = e^{(x^3/3 + C')} = e^{(x^3/3)} e^{C'} = Ce^{(x^3/3)}$$

<u>Answer (a)</u>

STATICS

1. For the diagram below, what is the maximum tension in the rope?

    a. 200 lbs   b. 342 lbs
    c. 376 lbs   d. 394 lbs
    e. 412 lbs

$\uparrow \Sigma F_Y = 0$

$T_A \sin 10° + T_B \sin 20° - 200 = 0$

$.174 T_A + .342 T_B - 200 = 0$ (1)

$\rightarrow \Sigma F_x = 0$

$-T_A \cos 10° + T_B \cos 20° = 0$

$-.985 T_A + .940 T_B = 0$ (2)

From (2),

$T_A = \dfrac{.940}{.985} T_B = .954 T_B$ (3)

Substituting (3) into (1);

$.174 (.954 T_B) + .342 T_B = 200$

$T_B = \dfrac{200}{.174(.954) + .342} = 393.66$

<u>Answer (d)</u>

2. A rolled steel wide flange beam is subjected to non-concentric loads. Determine the torque about the centroids.

    a. 0.858 ft-k   b. 0.942 ft-k
    c. 0.991 ft-k   d. 1.252 ft-k
    e. 1.482 ft-k

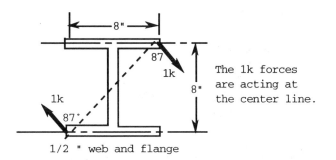

1/2 " web and flange

From symmetry of section, locate CG by inspection. Note that the dotted line shown is <u>NOT</u> perpendicular to the line of action of the 1k forces Use components of the 1k forces when summing moments.

$\circlearrowright \Sigma M = 0$

$= 2(1 \sin 48°) \times 4" + 2(1 \cos 48°) \times 4"$

$SM_c = 5.945\ 5.353 = 11.298$ in kips

$M_c = (11.298 \text{ in-k}) \times \dfrac{1 \text{ ft}}{12 \text{ in}}$

$= .9415$ ft-K

<u>Answer (b)</u>

3. A 3200-lb car is raised at the rear by a jack. Assuming that the front wheels have no brakes, what is the total reaction where the jack meets the bumper. The CG is midway between the wheels.

    a. 301 lbs   b. 1124 lbs
    c. 1164 lbs   d. 2036 lbs
    e. 2076 lbs

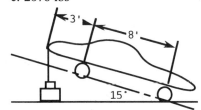

Take X axis along the plane of the auto.

$\circlearrowright \Sigma M = 0$

$R_{AY} \times 11 - 3200 (4 \cos 15) = 0$

$R_{Ay} = \dfrac{3200 (4 \cos 15)}{11} = 1123.99$ lbs

$\uparrow \Sigma F_y = 0$

$R_{Ay} - 3200 (\cos 15) + R_B (\cos 15) = 0$

$R_B = \dfrac{3200 (\cos 15) - 1123.99}{(\cos 15)} = 2036.4\#$

$\rightarrow \Sigma F_x\ 0$

$R_{Ax} + 3200 (\sin 15) - R_B (\cos 15) = 0$

$R_{Ax} = (2036.4 - 3200)(\sin 15)$
       $= -301.17$

$R_A = \sqrt{(R_{Ax})^2 + (R_{Ay})^2}$

$= \sqrt{(-301.17)^2 + (1123.99)^2}$

$= 1163.6$ lbs   <u>Answer (c)</u>

4. A triangular plate weighing 12 lbs rests on a circular table whose radius is 28-in. The plate has equal sides of 8-in. and is positioned as shown. Determine the distance between the center of the table and the CG of the plate when the plate will tend to fall off the table.

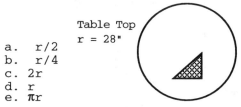

    a. r/2
    b. r/4
    c. 2r
    d. r
    e. πr

When the C.G. of the triangle passes over the edge of the table (28" from the center), the triangle will tend to fall off the table. The orientation of the triangle has no bearing on the problem.

<u>Answer (d)</u>

5. Calculate the tension in the cable AB of the frame shown below:

    a. 7.16k   b. 4.00k
    c. 2.24k   d. 1.11k
    e. 3.56k

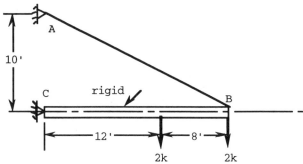

Draw an FBD of the system.

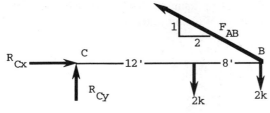

$$\sum M = 0$$

$$2(12) + 2(20) - F_{AB}\left(\frac{1}{\sqrt{5}}\right)(20) = 0$$

$$F_{AB} = \frac{2(12) + 2(20)}{(1/\sqrt{5})(20)} = 7.155^K \quad \underline{\text{Answer (a)}}$$

6. The supporting tower for a water tank has the loads and dimensions shown. The entire structure is pin connected. What is the load in member EF?
    a. 26.82k   b. 42.69k
    c. 52.64k   d. 64.24k
    e. -4.743k

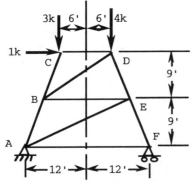

Look at the applied loads and the answers. Only Answer (e) is of the right order of magnitude. To calculate the numerical value of the load in EF:

$$\sum M = 0$$

$$1^K(18) + 3^K(6) + 4^K(18) = R_F(24)$$

$$R_F = \frac{18 + 18 + 72}{24} = 4.5^K$$

(No horizontal reaction due to the rollers)

At joint F, $\uparrow\Sigma\ F_y = 0$

$$EF = \frac{\sqrt{90}}{9}(-4.5) = -4.743^K$$

                                    <u>Answer (e)</u>

7. What is the vertical component of the reaction at A?
    a. 200 lbs   b. 267 lbs
    c. 400 lbs   d. 540 lbs
    e. 450 lbs

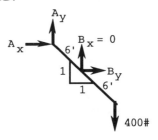

Sketch a FBD.

The cable at B can only take an axial load. Therefore $B_y = 0$

$\uparrow \Sigma\ F_y = 0;\ A_y = 400\#$ <u>Answer (c)</u>

8. Using the figure for problem 7, what is the horizontal component of the reaction at A?
    a. 200 lbs   b. 400 lbs
    c. 600 lbs   d. 800 lbs
    e. 1000 lbs

Refer to FBD for problem #7.

$$\sum M = 0$$

$$A_x(6 \cos 45) + A_y(6 \sin 45) + 400(6 \sin 45) = 0$$

$$A_x = \frac{400(6 \times .707) + 400(6 \times .707)}{(6 \times .707)}$$

$$= 800 \qquad \underline{\text{Answer (d)}}$$

9. Determine the reaction at B for the beam loaded as shown.

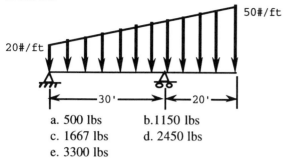

a. 500 lbs   b. 1150 lbs
c. 1667 lbs   d. 2450 lbs
e. 3300 lbs

Divide the beam loading into two loadings, one uniform loading and one uniformly increasing loading as shown.

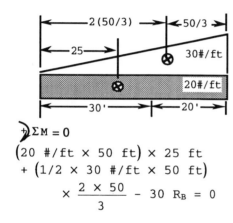

$\circlearrowleft \Sigma M = 0$

$(20 \text{ #/ft} \times 50 \text{ ft}) \times 25 \text{ ft}$
$+ (1/2 \times 30 \text{ #/ft} \times 50 \text{ ft})$
$\times \dfrac{2 \times 50}{3} - 30 R_B = 0$

$R_B = \dfrac{(20 \times 50 \times 25)}{30} +$
$\dfrac{\left(\dfrac{1}{2}\right) \times 30 \times 50 \times \dfrac{2}{3} \times 50}{30}$
$= 1666.67 \text{ lbs}$ Answer (c)

10. The gate of a dam is 10 ft wide and held closed by cable AB. What is the tension in the cable? Assume the water has a density of 62.4 lbs/Ft³.

a. 500 lbs   b. 5325 lbs
c. 7425 lbs   d. 15,245 lbs
e. 6724 lbs

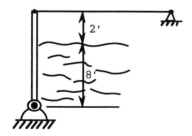

This problem is similar to #9 but requires some knowledge of Fluid Mechanics.

$p = w h = 62.4 \times 8 = 499.2 \text{ psf}$
$F = \text{average pressure} \times \text{area}$

$= \dfrac{0 + 499.2}{2} \times (8 \times 10) = 19968 \text{ lbs}$

From fluid statics, F is located h/3 above the base

$\circlearrowleft \Sigma M = 0$

$F \times \dfrac{h}{3} = F_{AB} \times 10$

$F_{AB} = \dfrac{19968 \times 8}{10 \times 3} = 5324.8 \text{ lbs}$

Answer (b)

For questions 11 to 13 consider the truss loaded as shown.

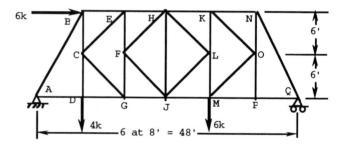

11. What is the reaction at Q?

a. 4.17k   b. 6.17k
c. 8.23k   d. 12.24k
e. 5.25k

$\circlearrowleft \Sigma M = 0$
$(4^K \times 8) + (6^K \times 32) +$
$(6^K \times 12) = R_Q \times 48$
from which $R_Q = 6.167^K$ Answer (b)

12. Find the force in member HK.

a. 4.17k   b. -6.17k
c. -8.23k   d. 12.24k
e. 5.25k

For interior members, the best approach is usually the Method of Sections.

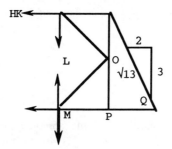

$\circlearrowleft \Sigma M = 0$

$(-HK \times 12) - (6.17 \times 16) = 0$; $HK = -8.23^k$

Answer (c)

13. Find the force in member KN.
    a. -7.42k   b. 10.67k
    c. 13.38k   d. 16.42k
    e. -4.11k

Use Method of Joints, starting at Q.

$\uparrow \Sigma F_y = 0$; $\dfrac{3}{\sqrt{13}} NQ + 6.17 = 0$

$\rightarrow \Sigma F_x = 0$; $-KN - 7.42 \times \dfrac{2}{\sqrt{13}} = 0$

$KN = -\dfrac{7.42 \times 2}{\sqrt{13}} = -4.11^K$

Answer (e)

For questions 14 - 18 use the frame shown which supports a load of 8k.

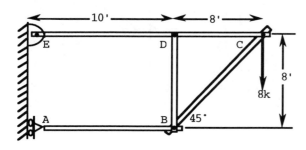

14. What is the vertical component of the reaction at E?
    a. 2k   b. 4k
    c. 6k   d. 8k
    e. 10k

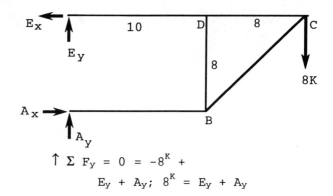

$\uparrow \Sigma F_y = 0 = -8^K + E_y + A_y$; $8^K = E_y + A_y$

However, member AB is not rigidly attached to the rest of the frame or the wall but can pivot at A & B (pin jointed). Therefore, there can be no vertical force at A:

$A_y = 0$ ∴ $8^k = E_y + 0$ or $E_y = 8^K$

Answer (d)

15. Determine the force acting on member EC at D.
    a. 4k   b. 8k
    c. 10k   d. 18k
    e. 16k

Using FBD from 14.

$\circlearrowleft \Sigma M = 0$

$8^K \times 18 - A_x \times 8 = 0$

or $A_x = \dfrac{8 \times 18}{8} = 18^K$

Therefore, member AB is in compression.

$\rightarrow \Sigma F_x = 0$; $18 - BC \cos 45 = 0$

$BC = \dfrac{18}{\cos 45} = 25.46^K$

$\uparrow \Sigma F_Y = 0$; $DB - BC \sin 45 = 0$

$DB = \dfrac{18}{\cos 45} \times \sin 45 = 18^K$

Answer (d)

16. What is the total reaction at C on member EC?

    a. 12.84k   b. 18.0k
    c. 20.59k   d. 24.62k
    e. 22.46k

$$BC_x = \frac{18}{\cos 45} \times \sin 45 = 18^K$$

$$BC_y = \frac{18}{\cos 45} \times \cos 45 = 18^K$$

$$\rightarrow \Sigma F_x = 18^K$$

$$\uparrow \Sigma F_y = 18^K - 8^K = 10^K$$

$$F = \sqrt{F_x^2 + F_y^2} = \sqrt{18^2 + 10^2} = 20.59^K$$
<u>Answer (c)</u>

For questions 17 - 19 the following situation applies:

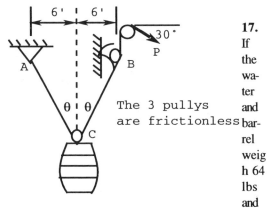

17. If the water and the barrel weigh 64 lbs and the angle θ is 30°, what is the tension in the rope at P ?

   a. 64 lbs   b. 36.95 lbs
   c. 32 lbs   d. 55.46 lbs
   e. 73.9 lbs

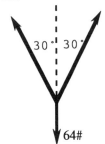

$\uparrow \Sigma F_y = 0;$
$P \cos 30 + P \cos 30 = 64$

$$P = \frac{64}{2 \cos 30} = 36.95\#$$ <u>Answer (b)</u>

18. If the barrel is raised so that loop C is at the same elevation as A and B, What is the tension in the rope at P
   a. 0.00 lbs   b. 32 lbs
   c. 64 lbs   d. 128 lbs
   e. ∞ lbs

From 17 above, $2P \cos\theta$ is the general expression for the load in the cable. When C is at the same level as A and B, $\theta = 90°$, $\cos 90° = 0$

$$P = \frac{64}{2 \cos \theta} = \frac{64}{0} = \infty$$ <u>Answer (e)</u>

19. If the maximum force that the rope can sustain is 2000 lbs, how high can the rope be raised. Calculate the angle θ between the vertical and the rope?
   a. 88.08°   b. 89.94°
   c. 90.00°   d. 79.02°
   e. 89.08°

Using the general expression from 18 above:

$$\cos \phi = \frac{64}{2P} = \frac{64}{2 \times 2000} = .016$$

$\theta = \cos^{-1} .016 = 89.08°$ <u>Answer (e)</u>

20. Determine the direction of the total frame reaction relative to the horizontal at B.

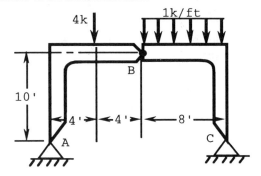

   a. 21.08°   b. 21.77°
   c. 22.62°   d. 19.02°
   e. 24.15°

Note that there is a pivot point at B. This is like a hinge. No moment can be transmitted through a hinge.

Consider member AB
$\circlearrowright \Sigma M = 0$
$-8B_Y + 10B_Y + (4^K \times 4) = 0$ (1)

Consider member BC

$\circlearrowright \Sigma M = 0$

$-8B_y - 10B_x - (1 \times 8 \times 4) = 0$ (2)

Adding (1) and (2) yields;

$-16B_y - 16 = 0$ or $B_y = -1^K$

Substituting back into (1) yields;

$-8(-1) + 10B_x + 16 = 0$ or

$B_x = -\dfrac{24}{10} = -2.4^K$

$B = \sqrt{B_x^2 + B_y^2} = \sqrt{(2.4) + 1} = 2.6^K$

$\phi = \tan^{-1} B_y/B_x = 1/2.4 = .4167$

$\theta = 22.62°$ <u>Answer (c)</u>

21. Given a thin circular plate of radius r, and mass M, determine the mass moment of inertia about an axis perpendicular to the plate, through the center of the plate.
    a. Mr^2 b. $0.75Mr^2$
    c. $0.5Mr^2$ c. $0.25Mr^2$
    e. $0.0833Mr^2$

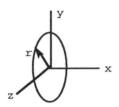

The fastest and easiest solution would be to look this up in your handbook.

    $I = 1/2\, M\, r^2$ <u>Answer (c)</u>

Analytically:

    ρ is density, t is thickness

$dM = \rho\, dV = \rho t\, dA$ where $dA = 2\pi r\, dr$
    $r^2 = y^2 + z^2$

$I_x = \int (y^2 + z^2)\, dm = \int r^2\, dM = \int r^2 \rho t\, 2\pi r\, dr$

    $I_x = 2\pi \rho t \int r^3\, dr = 2\pi \rho t\, \dfrac{r^4}{4}$

    $I_x = 1/2\, \pi \rho t\, r^4$ but $M = \pi r^2 \rho t$

    $I_x = \dfrac{1}{2} M r^2$ <u>Answer (c)</u>

22. The angle of repose is the maximum angle to which a plane can be tilted before a block on the plane will begin to slide. If the coefficient of friction is 0.25 and a block weighs 10 lbs, what is the angle of repose?
    a. 14.04°   b. 14.48°
    c. 75.52°   c. 11.30°
    e. 13.62°

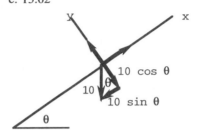

$\Sigma F_x = 0;\; -10 \sin\theta + F = 0;\; F = 10 \sin\theta$
But $F = mN = .25\,(10 \cos\theta)$

Equating values for F yields

    $10 \sin\theta = 2.5 \cos\theta$ or $\tan\theta = .25$

    and $\theta = 14.04°$ <u>Answer (a)</u>

23. A hemp rope and pully are used to lift a 100-lb bucket of sand. For the conditions shown, what is the tension P required to lift the bucket? Let f = 0.2 and the pully is locked against rotation.
    a. 391 lbs   b. 231 lbs
    c. 208 lbs   c. 100 lbs
    e. 198 lbs

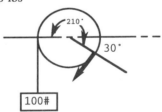

$\beta = 210° \times \dfrac{2\pi}{360°} = 3.665$ radians

$\dfrac{T_t}{T_s} = e^{fb}$ where $T_s = 100$, and $f = 0.2$

    $T_t = p = 100\, e^{0.2 \times 3.665} = 208.13$
    <u>Answer (c)</u>

24. Derive a relationship for the brake and drum shown between the applied force F, and the torque E. The drum rotates counter clockwise and f = 0.5

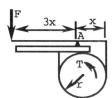

a. T = 4.81 Fr b. T = 11.43 Fr
c. T = 14.43 Fr d. T = 3F/r
e. T = 12.43 Fr

A free body diagram:

$\beta = 180° = \pi$ Radians

$T_t = T_s\ e^{f\beta} = T_s\ e^{0.5 \times \pi} = 4.81\ T_s$ (1)

$\Sigma M_o = 0 : (T_t - T_s)\ r = T$ (2)

$\Sigma M_A = 0 : (-3x)\ F + T_s\ (x) = 0 : T_s = 3F$ (3)

From 1
$T_t = 4.81\ T_s = 4.81\ (3F) = 14.43\ F$ (4)
Substituting 4 and 3 into 2

$(14.43F - 3F)r = T$ or $T = 11.43Fr$

Answer (b)

25. Obtain the cross product A x B of the two vectors A = 2i +3j, and B = 5i + 2j + k.
  a. 1-i + 6j + k   b. 7i + 5j + k
  c. 3i - 2j + 19k   d. 3i -2j - 11k
  e. -3i + 2j + 11k

$$A \times B = \begin{vmatrix} i & j & k \\ 2 & 3 & 0 \\ 5 & 2 & 1 \end{vmatrix} = 3i - 2j - 11k$$

Answer (d)

26. An 18" equilateral, 24" wide triangular truss supports a load of 128 lbs applied at D and in the plane of BDF. What is the force in member CD?

a. zero   b. 59.0 lbs
c. 111.0 lbs   d. 128 lbs
e. 12.5 lbs

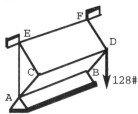

For equilibrium,
 $\Sigma F_x, \Sigma F_y, \Sigma F_z, \Sigma M = 0$
There are no forces in the plane of CD, so the force in member CD = 0.

Answer (a)

27. A grooved block weighing 86 lbs rests on the inclined plane shown. What force must be applied to prevent its sliding down the incline.

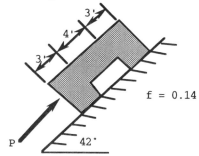

a. zero   b. 32.2 lbs
c. 48.6 lbs   d. 56.3 lbs
e. 86 lbs

A free body diagram:

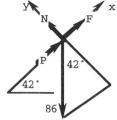

N = 86 cos 42 = 63.91 lbs
F = mN = .14 x 63.91 = 8.95 lbs
$S_x = 0$: F + P = 86 sin 42°
P = 86 sin 42° - F
P = 57.54 - 8.95 = 48.59 lbs   Answer (c)

28. Determine the location of the y coordinate of the centroid of the extruded channel shown. Answer must be relative to the given datum.

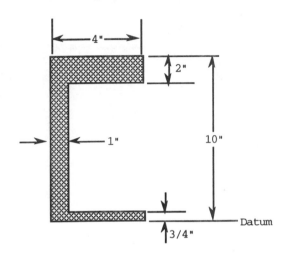

a. 5.0 in b. 6.65 in
c. 5.74 in d. 3.35 in
e. 1.72 in

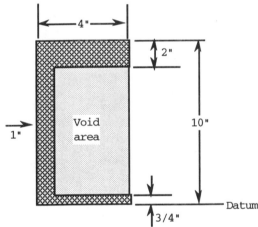

Total area = 4 x 10 = 40 Located 5" above, thus Ay = 200. Void area = 3 x -7.25 = -21.75 located 4.375 "above thus Ay = -95.16.

$$y = \frac{\Sigma A_y}{\Sigma A} = \frac{104.84}{18.25} = 5.74 \text{ in.}$$

Answer (c)

29. Compute the moment of inertia of the channel in problem 8 about the centroidal axis that is parallel to the datum line.

a. 41.2 in⁴ b. 108.9 in⁴
c. 150.1 in⁴ d. 196.8 in⁴
e. 219.4 in⁴

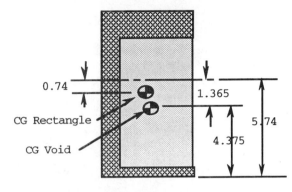

Part	Area	$I_o = \frac{bh^3}{12}$	d	d^2	Ad^2
Rectangle	40	333.3	0.74	0.548	21.92
Void	-21.75	-95.3	1.365	1.863	-40.52
Σ		238.0			-18.61

$$I = \Sigma I_o + \Sigma Ad^2 = 238.0 - 18.61$$
$$= 219.31$$

Answer (c)

30. What is the minimum value of P that will cause sliding of the block.

a. 400 lbs b. 78.3 lbs
c. 69.5 lbs d. 60.2 lbs
e. 71.6 lbs

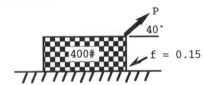

A free body diagram:

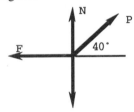

$\uparrow \Sigma F_y = 0;$ N + P sin 40 - 400 = 0
N = 400 - P sin 40 = 400 - .643P

F = μN = .15(400 - 643P) = 60 = .096P
→ $\Sigma F_x = 0;$ - F+P cos 40 = 0
- 60 + .096P + .766P = 0
P = $\frac{60}{.862}$ = 69.6 lbs

Answer (c)

For questions 31 - 33

A 1 ft diameter drum weighing 20 lbs has a brake shoe device which is used to suspend a weight W. The drum axle is mounted in an oval shaped hole so that the full weight of the axle bears on a sliding block. The pully and axle are frictionless.

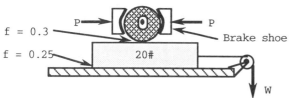

31. If the brake shoes can lock the drum, what is the maximum weight that can be suspended?
    a. 6 lbs   b. 8 lbs
    c. 12 lbs   d. 14 lbs
    e. 16 lbs

A free body diagram:

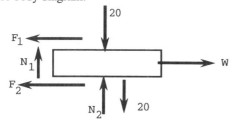

N_1 = 20 lbs: $F_1 = mN_1 = .3 \times 20 = 6$ lbs
N_2 = 40 lbs: $F_2 = m_2N_2 = .25 \times 40 = 10$ lbs
$SF_x = 0 = -F_1 - F_2 + W = 0$
$W = 6 + 10 = 16$ lbs   <u>Answer (e)</u>

32. What is the normal force P that must be applied to each brake shoe to stop the rotation of the drum when the maximum weight (from 11) is suspended?

    a. 10 lbs   b. 8 lbs
    c. 3 lbs   d. 1.5 lbs
    e. 12 lbs

A free body diagram:

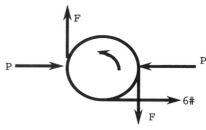

$\curvearrowright \Sigma M = 0$;

$(F \times 1) - (6 \times .5) = 0 \quad\quad F = 3$ lbs

$F = \mu_1 P \text{ or } P = \dfrac{F}{\mu_1} = \dfrac{3}{.3} = 10$ lbs

<u>Answer (a)</u>

33. The capacity of the device could be expanded if a single brake shoe were mounted on top of the drum. (why?) If the normal force P is 10 lbs, what weight could be suspended?

    a. 21.5 lbs   b. 15.5 lbs
    c. 10.5 lbs   d. 8 lbs
    e. 12 lbs

The capacity is increased because the normal force and the force P act in the same direction.

A free body diagram of the drum::

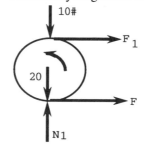

$F = mN = .3 \times 10 = 3$ lbs
$N_1 = 20 + 10 = 30$ lbs
$F_1 = m_1N_1 = .3 \times 30 = 9$ lbs

Although F_1 is greater than before due to the increased normal force, the drum will turn only if F_1 exceeds 3 lbs. SM = 0 must be satisfied using the limiting value of F_1 = 3 lbs.

A free body diagram of the block:

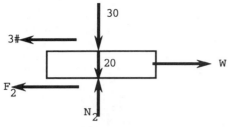

$N_2 = 30 + 20 = 50$

$F_2 = .25 \times 50 = 12.5$ lbs

$SF_x = 0 = -3 - 12.5 - W$

$W = 15.5$ lbs   Answer (b)

Questions 34 - 36

Equipment weighing 4k is packaged in the rectangular rigid container shown. The weight is uniformly distributed, and an 8k horizontal load applied. The supports are:

- A. Ball roller, transmits vertical load only
- G. Ball and socket, transmits 3 mutually perpendicular components of load.
- F. Pinned, load along axis of cable
- C. Pinned, load along axis of cable

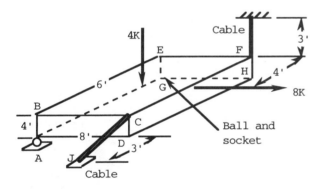

34. What is the component of the reaction at H which is parallel to the 8k load:
   a. zero   b. 2k
   c. 4k   d. 6k
   e. 8k

Call the direction of the 8k load the X axis. $SF_x = 0$, $8k - H_x = 0$, $H_x = 8k$

                   Answer (e)
(Cables can only take tension load)

35. Determine the magnitude of the reaction at H which is perpendicular to the EFGH plane.

   a. zero   b. 2k
   c. 4k   d. 6k
   e 8k

A free body diagram:

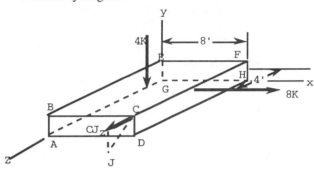

ΣM_G (about y axis) = 0

$(-8 \times 4) + (CJ_z \times 8) = 0: CJ_z = 4k$

$SF_z = 0 = CJ_z + H_z$ $H_z = -4k$

                   Answer (c)

36. Determine the force in the cable attached at F.

   a. zero   b. 2.46k
   c. 6.84k   d. 11.33k
   e. 12.46k

A free body diagram:

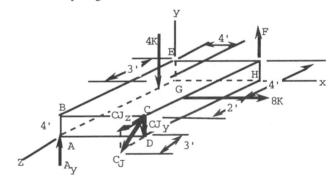

$CJ = \dfrac{5}{3} \times 4 = \dfrac{20}{3}$ $CJ_z = \dfrac{3}{5} CJ = 4k$

$CJ_y = \dfrac{4}{5} CJ = \dfrac{4}{5} \times \dfrac{20}{3} = \dfrac{16}{3}$

$\Sigma M_A = 0$ (about z axis):
$(-4 \times 4) - (8 \times 4) + (F_y \times 8) - (CJ_y \times 8) = 0$

$$F_y = \frac{16 + 32 + \left(\frac{16}{3} \times 8\right)}{8} = 11.33$$

Answer (d)

Problems 37 - 39

A square threaded jackscrew is to be used to raise a load of 2 tons. The thread selected will have a mean diameter of 2 in. and a lead of 0.5 in. The coefficient of static friction is 0.3 and the jack handle is 15 in. long as measured from the screw center line.

37. What is the force required to raise the load?
  a. 51.9 lbs   b. 103.8 lbs
  c. 111.2 lbs   d. 123.1 lbs
  e. 98.6 lbs

A free body diagram of the nut as a block being pushed up an inclined plane.

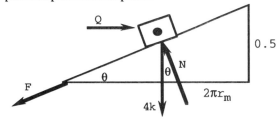

$$\text{Tan } \phi = \frac{.5}{2\pi \, r_m} = \frac{.5}{2\pi \times \frac{2}{2}} = .0796$$

For small angles, $\sin \theta = \tan \theta = \theta$ rad.

$P = mN = 0.3N$

$\uparrow \Sigma F_y = 0; \; N \cos\phi - 4000 - .3N \sin\phi = 0$

$N(.9968 - .0238) = 4000$

$N = \frac{4000}{.973} = 4111.0$ lbs

$\rightarrow \Sigma F_x = 0; \; Q - N \sin\phi - .3N \cos\phi = 0$

$Q = 4111.0(.0796 + .2991) = 1556.8$ lbs

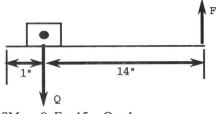

$SM_0 = 0; \; F \times 15 = Q \times 1$

$F = \frac{Q}{15} = \frac{1556.8}{15} = 103.8$ lbs

Answer (b)

38. All other conditions being the same, what must the coefficient of friction be in order to raise the same load with a push on the jack handle of 75 lbs?
  a. 0.15   b. 0.20
  c. 0.24   d. 0.22
  e. 0.18

A free body diagram:

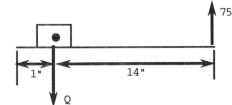

$\curvearrowleft \Sigma M = 0$

$Q = 75 \times 15 = 1125$ lbs

$\rightarrow \Sigma F_x = 0;$
$1125 - N(.0796) - \mu N(.9968) = 0$ (1)

$\uparrow \Sigma F_y = 0;$
$N(.9968) - 4000 - mN(.0796) = 0$ (2)

Multiply (2) by $\frac{\cos \phi}{\sin \phi}$ or $\cot \phi$

$-12.483N + 50,090 + mN(.9968) = 0$ (3)

Add 1 and 3 to remove the μN term:
$51,215.452 - 12.563N = 0$

$N = \frac{51215.452}{12.563} = 4076.82$ lbs

Use this value in 1 to solve for μ:

$$\mu = \frac{1125 - .0796N}{.9968N} = .1968 \approx 0.20$$

Answer (b)

39. Which one of the following statements is false?

    a. The lead of the screw is the distance between similar points of adjacent threads.

    b. The pitch is the distance between similar points of adjacent threads.

    c. The friction force always acts in the direction of the motion of a body.

    d. When the lead angle is less than the angle of friction, the screw is self locking.

    e. The friction force between two rubbing surfaces is independent the contact area.

Friction force always opposes the direction of the body.   <u>Answer (c)</u>

40. A chain hoist used for installing auto engines is composed of two attached pulleys, 1 and 2. Each pulley has teeth to prevent slippage of the chain with respect to the pulley. If a 1 ton load can be lifted by a 100 lbs pull, what is the ratio of r_2 to r_1. Neglect friction and other factors.

    a. 0.99   b. .909
    c. 1.16   d. 2.03
    e. 1.05

Draw an FBD of upper pulleys.

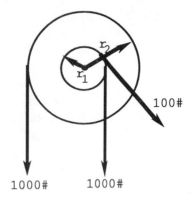

$\Sigma M_o = 0;$ $100 r_1 + 1000 r_1 = 1000 r_2$
$1100\ r_1 = 1000\ r_2$

The ratio asked for is

$$\frac{r_2}{r_1} = \frac{1000}{1100} = \frac{10}{11} = .909$$

<u>Answer (b)</u>

MECHANICS OF MATERIALS

Problem statement for questions 1-3

1. A rectangular bar supports a 20k load attachead as shown. The bar is steel and may be assumed to be weightless. What is the maximum tensile stress in the bar?

    a. 1667 psi   b. 3333 psi
    c. 16,667 psi   d. 41,667 psi
    e. 2935 psi

Area = 6in²   24 ft   20k

$$\sigma = \frac{P}{A} = \frac{20000}{6} = 3333\,\text{psi}$$

<u>Answer (b)</u>

2. If the bar weight is to be considered, what then is the maximum stress? Take the density of the steel to be 500 pcf.

    a. 1667 psi   b. 3417 psi
    c. 6750 psi   d. 12167 psi
    e. 4867psi

$$W = \frac{6 \times 24 \times 500}{144 \frac{\text{Sq in}}{\text{Sq Ft}}} = 500\#$$

$$\sigma = \frac{P}{A} = \frac{20000 + 500}{6} = 3417\,\text{psi}$$

<u>Answer (b)</u>

3. Accounting for the bar weight (500 pcf) and the applied load of 20k, what is the total elongation of the free end A? (solve to 4 decimal places.)

    a. 0.0004 in.   b. 0.004 in.
    c. 0.0032 in.   d. 0.0324 in.
    e. 0.00366

The elongation due to W = total weight of 500 lbs.

$$\delta_w = \frac{500\,\text{lbs.} \times 12\,\text{ft} \times 12\,\text{in/ft}}{2 \times 3\,\text{in}^2 \times 30 \times 10^6\,\text{lbs/in}^2} = .0004\,\text{in.}$$

$$\delta_{20k} = \frac{20,000\,\text{lbs.} \times 24\,\text{ft} \times 12\,\text{in/ft}}{6\,\text{in}^2 \times 30 \times 10^6\,\text{lbs/in}^2}$$

$$= 3.2 \times 10^{-2}\,\text{in}$$

<u>Answer (d)</u>

4. A steel wire 8 m long is used to suspend a mass of 10 kg. If the wire is 1 mm in diameter and Young's modulus is 20×10^{10} N/M² what is the total elongation?

    a. 5 cm   b. 2 mm
    c. 2 cm   d. 5 mm
    e. 6mm

$E_{al} = 20 \times 10^{-10}$ N/M²

$$\delta = \frac{PL}{AE} =$$

$$\frac{10.0\,\text{kg} \times 8\,\text{m} \times 9.81\,\frac{\text{m}}{\text{sec}^2} \times 1000\,\frac{\text{mm}}{\text{m}}}{\frac{\pi}{4}(1)^2\,\text{mm}^2 \times \frac{1\,\text{m}^2}{1 \times 10^6\,\text{mm}^2} \times 20 \times 10^{10}\,\frac{\text{N}}{\text{m}^2}}$$

$= 5\,\text{mm}$ <u>Answer (d)</u>

5. A standard wide flanged beam has been mounted between two rigid walls. Initially there are no forces on the beam. It is then heated uniformly to a temperature of 175°F. If the ambient, installation temperature was 70°F, what is the stress in the beam? $a = 6.5 \times 10^{-6}$ in/in-°F and $E = 30 \times 10^6$ psi.

    a. 20,475 psi   b. 34,125 psi
    c. 13,650 psi   d. 23,450 psi
    e. 20,475 psig

$$\delta = L\alpha\Delta T \text{ and } \delta = \frac{PL}{AE} = S\left(\frac{L}{E}\right) \therefore L\alpha\Delta T = S\left(\frac{L}{E}\right)$$

$$S = \alpha\Delta T\, E = 30 \times 10^6 \times 6.5 \times 10^{-6} \times 105°$$

<u>Answer (a)</u>

Problem statement for questions 6-8

Three columns x, y and z are used to support a load of 20 tons. Each column has the same length and diameter and deforms the same amount. The outer two columns are steel and the third is bronze.

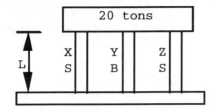

6. Which column has the maximum stress?
  a. x   b. y
  c. z   d. x and z
  e. all have the same stress

$$\delta = \frac{PL}{AE} \text{ find } P = \frac{\delta AE}{L}$$

since $s = P/A = \delta E/L$ and δ and L are the same for each column, then s_{max} is a funtion of E_{max}.

Answer (d)

7. What would be the stress of each column if the bronze was heated to a temperature above that in the steel columns?

  a. The stress in the steel would decrease while the stress in the bronze would increase
  b. The stress in the steel woul increase while the stress in the bronze would decrease
  c. The stress in all the columns would remain unchanged
  d. The stress in all the columns would increase
  e. The stress in all the columns would decrease

The bronze column would tend to elongate and as a result the stress in it would increase. Since

$$\text{Stress} = \frac{\text{Force}}{\text{Unit Area}}$$

and area remains constant, the load in the heated column increases. Since the total applied load, 2 tons, is constant, then the load and stress in the steel columns must be less.

Answer (a)

8. If the columns described above each have a cross sectional area of 2 in^2, are 2 ft long, what is the deflection of column y?

  a. 3.1 x 10^{-3} in.   b. 5.8 x 10^{-3} in.
  c. 3.2 x 10^{-3} in.   d. 7.8 x 10^{-3} in.
  e. Cannot be determined

$\alpha_S = 6.5 \times 10^{-6} \frac{\text{in}}{\text{in-}°F}$, $E_S = 30 \times 10^6 \text{psi}$

$\alpha_B = 11.2 \times 10^{-6} \frac{\text{in}}{\text{in-}°F}$, $E_B = 15 \times 10^6 \text{psi}$

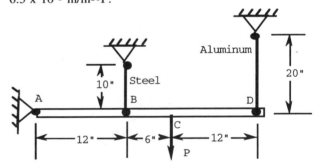

$$P_S = \frac{30 \times 10^6}{15 \times 10^6} P_B: \quad P_S = 2P_B$$

but

$$\delta_S = \delta_B \text{ (given)} \quad \frac{P_S L}{AE_S} = \frac{P_B L}{AE_B} \therefore P_S = \left(\frac{E_S}{E_B}\right) P_B$$

$$\Sigma F_Y = 0 = 2P_S + P_B = 20^K = 5P_b$$
$$P_b = 4k$$

$$\delta_B = \frac{(4000 \times 24 \text{in.})}{(2\text{in}^2 \times 15 \times 10^6 \text{psi})} =$$

3.2×10^{-3} in   Answer (c)

9. The pin connected device shown has a horizontal rigid bar supported by steel and aluminum bars. The cross-sectional area of the steel is 1.5 in^2 and that of the aluminum is 3/8 in^2. Use $E_a = 10.5 \times 10^6$ psi, $E_s = 30 \times 10^6$ psi. $\alpha_a = 12.5 \times 10^{-6}$ in/in-°F and $\alpha_S = 6.5 \times 10^{-6}$ in/in-°F.

What is the stress in the steel bar if the force P = 500 lbs. acting downward?

  a. 64 psi   b. 171 psi
  c. 333 psi   d. 393 psi
  e. 589 psi

A free body diagram:

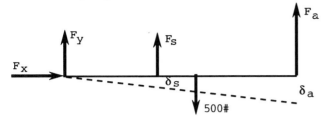

$$\Sigma F_y = 0 \;;\; F_y + F_s + F_a - 500 = 0$$

$$\Sigma M_A = 0 \;:\; F_s \times 12 + 500 \times 18 - F_a \times 30 = 0$$

$$\delta_s/12 = \delta_a/30 \text{ or } \delta_a = 2.5\,\delta_s$$

$$\frac{F_a L_a}{A_a E_a} = 2.5 \frac{F_s L_s}{A_s E_s}$$

$$\frac{F_a \times 20}{.375 \times 10.5 \times 10^6} = \frac{2.5\, F_s \times 10}{1.5 \times 30 \times 10^6}$$

$$F_a = \frac{2.5 \times 10 \times .375 \times 10.5}{1.5 \times 30 \times 20} F_s$$

$$0.109\, F_s = F_a$$

$\Sigma M_A = 0$;
$$-12 F_s + 9000 - 0.109 \times 30 F_s = 0$$
$$F_s = 589 \text{ lbs}$$

$$\sigma = \frac{F_s}{A_s} = \frac{589}{1.5} = 393 \text{ psi} \qquad \underline{\text{Answer (d)}}$$

Statement for problems 10 - 13

A compound bar fixed at its upper end, supports a 2k load axially. $E_s = 10 \times 10^6$, $E_a = 10^6$.

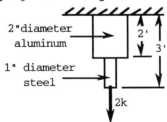

10. What is the ratio of the forces in each portion of the bar resulting from the 2k load?

    a. 1/2   b. 1/1
    c. 3/2   d. 2/1
    e. 3/1

Since the 2k load acts axially throughout the compound bar, each portion of the bar experiences the same load of 2k.   <u>Answer (b)</u>

11. What is the ratio of stresses on each portion of the compound bar resulting from the 2k load?

    a. 1/2   b. 1/1
    c. 3/2   d. 2/1
    e. 4/1

$$\frac{\sigma_s}{\sigma_a} = \frac{\frac{F_s}{A_s}}{\frac{F_a}{A_a}} = \frac{8}{2} = 4 \qquad \underline{\text{Answer (e)}}$$

12. What is the total elongation of the bar?
    a. 1.02×10^3 in   b. 1.53×10^3 in
    c. 2.55×10^3 in   d. 5.61×10^3 in
    e. 16.32×10^3 in

$$\delta_a + \delta_s = \delta_t$$

$$\left[\frac{FL}{AE}\right]_{\text{Steel}} + \left[\frac{FL}{AE}\right]_{\text{Aluminum}} = \delta_T$$

$$\delta_T = \left[\frac{2000 \times 12}{\frac{\pi}{4} \times 30 \times 10^6}\right]_{\text{Steel}} + \left[\frac{2000 \times 24}{\pi \times 10 \times 10^6}\right]_{\text{Aluminum}}$$

$$= 2.55 \times 10^{-3} \text{ in} \qquad \underline{\text{Answer (c)}}$$

13. If Poisson's Ratio for aluminum is 1/3, what is the change in diameter of the aluminum portion resulting from the 2k load?

    a. 6.375×10^{-5} in   b. 4.25×10^{-5} in
    c. 2.125×10^{-5} in   d. 2.00×10^{-5} in
    e. 1.775×10^{-5} in

From 12 the total deformation for the aluminum is 1.53×10^{-3} in.

$$\varepsilon_{\text{long}} = \frac{1.53 \times 10^{-3}}{24} = 6.375 \times 10^{-5}$$

$$\mu = \frac{\varepsilon_{\text{lat}}}{\varepsilon_{\text{long}}} \therefore \varepsilon_{\text{lat}} = \mu\, \varepsilon_{\text{long}}$$

$$\varepsilon_{\text{lat}} = (2)\left(\frac{1}{3}\right) 6.375 \times 10^{-5} = 4.25 \times 10^{-5}$$

The question calls for change in diameter, the 2 is required.   <u>Answer (b)</u>

14. A coupling connecting two chafts transmits a torque of 1000 in-lb. Four bolts fasten the flanges together, each bolt has a cross sectional area of 1/2 in. and is on a 4 in. bolt circle. What is the stress in each bolt?

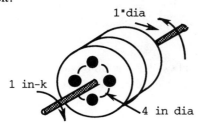

a. 250 psi
b. 500 psi
c. 1000 psi
d. 2000 psi
e. 2500 psi

$$P \times 4 \times 2 = 1000 \text{ in-lbs}$$
$$P = 125 \text{ lbs}$$

$$\sigma = \frac{P}{A} = \frac{125}{0.5} = 250 \text{ psi} \quad \underline{\text{Answer (a)}}$$

15. A steel bolt and bronze tube are assembled as shown. The cross sectional area of the bolt is 0.75 in.^{2} and the tube is 1.5 in.^{2}. The bolt has 32 threads per in. Find the number of turns of the nut that will produce a stress of 40,000 psi in the bronze tube.
$E_b = 12 \times 10^6$, $E_s = 30 \times 10^6$

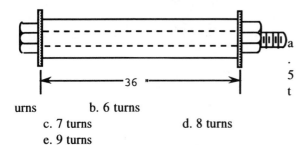

a. 5 turns
b. 6 turns
c. 7 turns
d. 8 turns
e. 9 turns

P = force in bolt = force in tube
$L_b = L_t$: n = number of turns

$$\delta_b + \delta_t = \left[\frac{PL}{AE}\right]_b + \left[\frac{PL}{AE}\right]_t = \frac{n}{32}$$

$$PL_B = \left[\frac{1}{AE}\right]_b + \left[\frac{1}{AE}\right]_t = \frac{n}{32}$$

$$P = \frac{n \times 10^6}{(32 \times 36)\left[\left(\frac{1}{.75 \times 30}\right) + \left(\frac{1}{1.5 \times 12}\right)\right]}$$

$$P = (0.87 \times 10^4)n$$

$$P = \sigma_T A_T = 40,000 \times 1.5 = 60,000$$

$$n = \frac{60,000}{.87 \times 10^{-4}} = 6.9 \text{ turns}$$
$$\underline{\text{Answer (c)}}$$

16. What is the maximum shear stress parallel to the axis of this shaft?

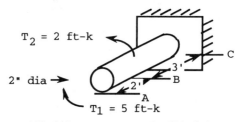

a. 10/π ksi
b. 60/π ksi
c. 120/π ksi
d. 240/π ksi
e. 80/π ksi

$$\tau = \frac{T\rho}{J} = \frac{(5000 \times 12 \times 1)}{\frac{\pi}{32} \times 16} = \frac{120,000}{\pi} \text{ psi}$$

Since axial shear = transverse shear,   <u>Answer (c)</u>

17. A bar is made of aluminum and steel rigidly attached to one another. The best description of the effect of uniform heating of the bimetallic bar is:

a. The bar will elongate
b. The bar will shrink
c. The bar will curve with the center of curvature toward the steel
d. The bar will curve with the center of curvature toward the aluminum
e. Not much

Because aluminum has a higher coefficient of expansion, it will expand more than steel for the same temperature change, resulting in the following shape.

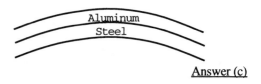

<u>Answer (c)</u>

18. A stepped steel shaft has axial forces applied as shown. What is the maximum stress in the shaft?

a. 1 ksi
b. 2.5 ksi
c. 16/π psi
d. 42.67/π psi
e. 16/π ksi

A Free Body Diagram:

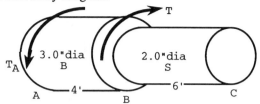

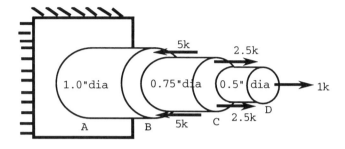

$*\tau_{AB} = \dfrac{P_{AB}}{A_{AB}} = \dfrac{1 + 2.5 + 2.5 - 5 - 5}{\pi \times \dfrac{1.0^2}{4}} = \dfrac{-16}{\pi}$

$*\tau_{BC} = \dfrac{P_{BC}}{A_{BC}} = \dfrac{1 + 2.5 + 2.5}{\pi \times \dfrac{0.75^2}{4}} = \dfrac{42.67}{\pi}$

$*\tau_{CD} = \dfrac{P_{CD}}{A_{CD}} = \dfrac{1}{\pi \times \dfrac{0.5^2}{4}} = \dfrac{16}{\pi}$

The angle of twist at section B with respect to A must equal the angle of twist at section B with respect to C, or the shaft will fail.

$\theta = \dfrac{TL}{GJ}$

$\theta_{BA} = \theta_{BC} = \left[\dfrac{TL}{GJ}\right]_{BA} = \left[\dfrac{TL}{GJ}\right]_{BC}$

$\left[\dfrac{T_A \times 48}{(6 \times 10^6)\left(\dfrac{\pi}{32}81\right)}\right] = \dfrac{T_C \times 72}{(12 \times 10^6)\left(\dfrac{\pi}{32}16\right)}$

$T_A = 3.8\, T_C$

$-T_A + T - T_C = 0$

If the bronze shaft governs:

$T_A = \dfrac{\tau J}{\rho} = \dfrac{8000\left(\dfrac{\pi}{32}\right)81}{1.5} = 42{,}412\text{ in-lb}$

$T_C = \dfrac{T_A}{3.8} = \dfrac{42{,}412\text{ in-lb}}{3.8} = 11{,}161\text{ in-lb}$

$T = T_A + T_C$

$T = 42{,}412 + 11{,}161 = 53{,}573\text{ in-lb}$

Answer (d)

19. A shouldered circular compound shaft is made of bronze and steel. The shaft is rigidly attached to immovable walls and an unknown torque is applied where the shaft changes diameter. If the maximum allowable shear stress in the bronze is 8000 psi and in the steel 12,000 psi, what is the maximum value of the applied torque?

a. 18,000 in-lbs
b. 53,700 in-lbs
c. 71,600 in-lbs
d. 90,400 in-lbs
e. 42,500 in-lbs

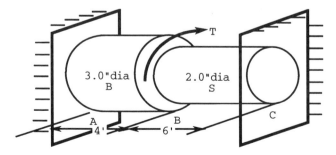

If the steel shaft governs:

$T_C = \dfrac{\tau J}{\rho} = \dfrac{12{,}000\left(\dfrac{\pi}{32}\right)16}{1.0} = 18{,}850\text{ in-lb}$

$T_A = 3.8\, T_A = 3.8 \times 18{,}850 = 71{,}630\text{ in-lb}$

$T = 18{,}850 + 71{,}630 = 90{,}480\text{ in-lb}$

The minimum T which yields at one of the maximum stresses is 53,753 in-lb.

Answer (b.)

20. An expansion joint in the road bed of a bridge supports a load of K kips as shown. What is the limiting value of K governed by the 1" diameter bolts? The yield point of all steel is 36,000 psi.

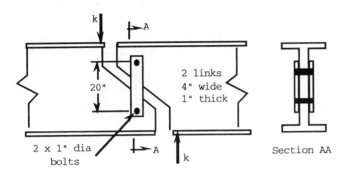

    a. 20π k   b. 24π k
    c. 18π k   d. 40π k
    e. 36π k

S = P/A, but this is a case of double shear.
$$K = 2\ AS = 2\left(\frac{\pi}{4}\right)36 = 18\pi\ k$$

<u>Answer (c)</u>

21. In the structure shown, what is the bending moment directly under the load?

    a. 3 ft-k   b. 4 ft-k
    c. 5 ft-k   d. 6 ft-k
    e. 8 ft-k

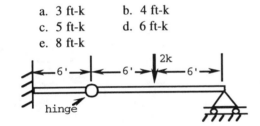

A free body diagram of the right hand portion of the beam:

$M = R_R \times 6 = 6$ ft-k   <u>Answer (d)</u>

22. The maximum unit stress at any vertical section in a beam is obtained by dividing the moment at that section by:

    a. The cross sectional area
    b. The radius of gyration
    c. The section modulus
    d. The distance to the point where the shear is zero

From the flexure formula, $\sigma = Mc/I$ where the quantity I/c is the Section Modulus.
$$\therefore \sigma_{max} = (M_{max})/(I/\sigma)$$ <u>Answer (c)</u>

23. The stiffness of a rectangular beam is proportional to:

    a. The square of the depth
    b. The cube of the depth and directly to the length
    c. The cube of the depth and inversely as the length
    d. The square of the depth and inversely as the length
    e. The length and inversely as the square of the depth

<u>Stiffness</u> of a beam is defined as a function of EI/L which may be expressed as:

$$\frac{E\frac{hh^3}{12}}{L}$$

for a rectangular beam. Clearly this is proportional to the cube of the depth and inversely proporational to the length.

<u>Answer (c)</u>.

Problems 24 and 25.

A cantilever beam loaded as shown has shear and bending moment diagrams drawn by parts.

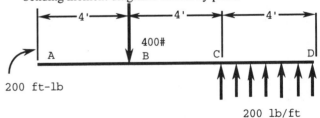

24. Select the correct shear diagram.

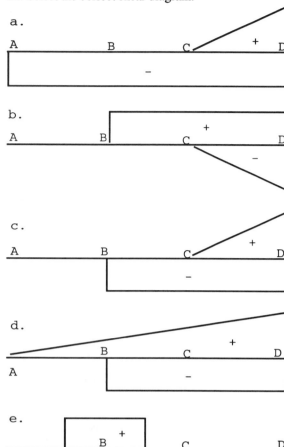

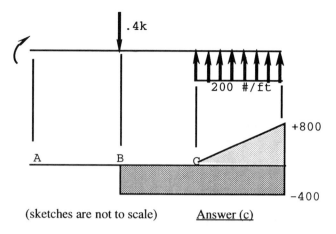

(sketches are not to scale)   Answer (c)

25. Select the correct bending moment diagram.

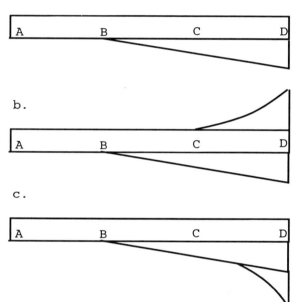

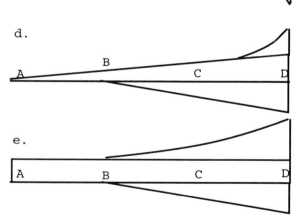

The problem statement says that the shear and bending moment diagrams were drawn by parts. That's a clue to do your solution using V & M diagrams 'by parts'. The concentrated BM does not produce any shear force. The 400 lb applied load produces negative shear while the uniformly distributed load produces positive shear.

From the shear diagram (answer b.) of problem #4, by taking the areas under the shear diagram we obtain:
Answer (b)

Problems 26 through 30

Given this hinged beam, loaded as shown, neglect the weight of the beam.

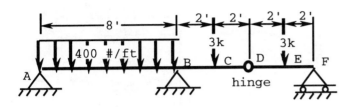

26. Is this beam statically determinate? If so, to what degree?

    a. It is statically determinate
    b. It is statically indeterminate to the 1st degree
    c. It is statically indeterminate to the 2nd degree
    d. It is statically indeterminate to the 3rd degree
    e. It will fall apart in a small wind

Sketching an FBD of each side of the hinge:

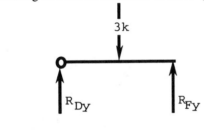

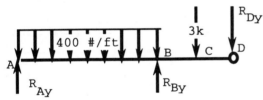

we see that each beam can be solved by the equations of equilibrium, hence the entire beam is statically determinate.

<u>Answer (a)</u>

27. What is the hinge reaction at D?

    a. 1^k b. 1.5^k
    c. 2^k d. 2.5^k
    e. 3^k

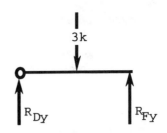

From the FBD of the right hand portion we see from symmetry that

$$R_{Dy} = 1.5^k \quad \underline{\text{Answer (b)}}$$

28. Select the equation of the bending moment for any portion of the beam between point B and C.

a. $R_{Ay} x - 400 (x-4) = M_x$
b. $R_{Ay} x - 400 (x-8) + R_{By} (x-8)$
c. $R_{Ay} x - 400 [8 (x-4)] = M_x$
d. $R_{Ay} x - 400 [8 (x-4)] - R_{By} (x-4)$
e. $R_{Ay} x - 400[8 (x-4)] + R_{By} (x-8) = M_x$

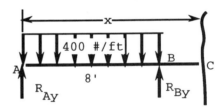

Let $8 < x < 10$

$$R_{AY} x = 400 \times 8 (x-4) - R_{BY} (x-8) = M_x$$
<u>Answer (e)</u>

29. What is the bending moment at D?
    a. zero
    b. 6000 ft-lbs
    c. 3000 ft-lbs
    d. 1500 ft-lbs
    e. ∞

The bending moment must be zero, because a bending moment cannot be transmitted through a hinge.
<u>Answer (a)</u>

30. In any beam where a splice has to be made, it is advisable to place the splice where the stress is a minimum. This condition generally occurs where the bending moment is zero. Where would you place the splice in section AD?

      a. .50 ft from left end
      b. .25 ft from left end
      c. .33 ft from left end
      d. .75 ft from left end
      e. .66 ft from left end

A free body diagram of AD:

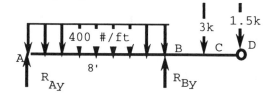

$\Sigma M_B = 0:$

$R_{Ay}(8) - 400(8)(4) + 3000(2) + 1500(4) = 0$

$R_{Ay} = 100$ lbs

Let x be the distance where M = 0

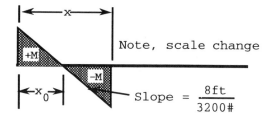

$\dfrac{8}{320} = \dfrac{x_0}{100}$; $x_0 = 0.25'$

Answer (b)

Problems 31 - 32

A beam is made up of the built-up section shown.

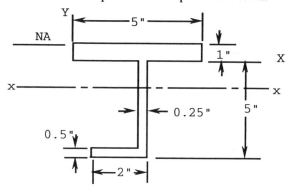

31. Locate the neutral axis from heavy end.

    a. 1.518 in.   b. 1.636 in.
    c. 2.125 in.   d. 2.878 in.
    e. 1.672 in.

Assign areas (A,B,C) of the shape:

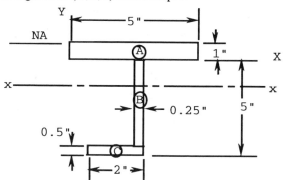

Part	A	Y	AY
A	5 x 1 = 5.00	0.50	2.50
B	4.5/4 = 1.125	3.25	3.66
C	1.00	5.75	5.75
Σ	7.125		11.91

$\overline{Y} = \dfrac{11.91}{7.125} = 1.672$ in

Answer (e)

32. Determine the moment of inertia I_{na} for the shape given.

    a.   42.443 in^4 b. 44.783 in^4
    c.   52.677 in^4 d. 66.129 in^4
    e.   28.617 in^4

Using the shape selections from problem 12:

Part	A	I₀	d	d²	Ad²
A	5.00	5/12 = .417	1.172	1.37	6.85
B	1.12	(.25×4.5)/12 = 1.898	1.578	2.49	2.80
C	1.00	2/96 = .021	4.078	16.63	16.63
Σ		2.336			26.28

$I_{na} = I_0 + ad^2 = 2.336 + 26.281 = 28.617 \text{ in}^4$

Answer (e)

33. Determine the maximum bending moment in this beam when it is loaded as shown.

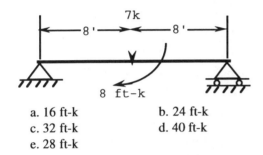

a. 16 ft-k   b. 24 ft-k
c. 32 ft-k   d. 40 ft-k
e. 28 ft-k

Using equations of equilibrium, solve for the reactions and draw the V and BM diagrams.

$\Sigma M_R = 0$; $R_L \times 16 - (7 \times 8) + 8 = 0$
or $R_L = (56-8)/16 = 3k$

$\Sigma F_y = 0$; $R_L - 7 + R_R = 0$ or $R_R = 4k$

The moment at the center is positive and is to be added to the area under the shear diagram to the left of center:

$3k \times 8' + 8\text{ft-k} = \underline{32 \text{ ft-k}}$ Answer (c)

34. A rectangular bar loaded as shown is made of steel and is assumed to be weightless. What is the maximum stress in the bar?

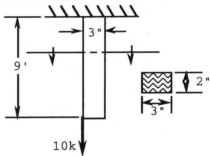

a. 1667 psi   b. 3333 psi
c. 5000 psi   d. 6667 psi
e. 6974 psi

$\sigma_{max} = \dfrac{Mc}{I} + \dfrac{P}{A}$,

bending stress + axial stress.

$M = 10,000 \times 1.5 = 15,000 \text{ in-lb}$
$c = 1.5 \text{ inches}$
$I = \dfrac{bh^3}{12} = \dfrac{2 \times 3^3}{12} = 4.5 \text{ in}^4$

$\therefore \sigma_{max} = \dfrac{15000 \times 1.5}{4.5} + \dfrac{10000}{6}$
$= 5000 + 1667 = \underline{6667 \text{ psi}}$

Answer (d)

35. Determine the maximum shear force in this simply supported beam.

a. 4^k b. 6.67^k
c. 10.67^k d. 16.67^k
e. 13.33^k

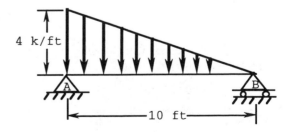

↻$\Sigma M_B = 0$

$(R_A \times 10) - 1/2 \times 4 \times 10 \times 2/3 \times 10$

from which

$(R_A \times 10) - \left(\frac{1}{2} \times 4 \times 10 \times \frac{2}{3} \times 10\right)$

$R_A = \frac{400}{30} = 13.33$ <u>Answer (e)</u>

Problems 36-38

This bent beam has a rigid corner and supports a 6k load as shown.

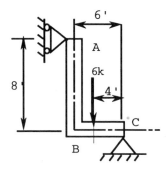

36. Determine the maximum bending moment in the beam.

a. 6 ft-k   b. 12 ft-k
c. 18 ft-k   d. 24 ft-k
e. 48 ft-k

↻$\Sigma M = 0$
$R_{AX}(8) - 6(4) = 0;$
or $R_{AX} = 3k$
↑$\Sigma F_y = 0;$ $R_{CY} - 6k = 0;$ or $R_{CY} = 6k$
$M_{max} = 3k \times 8 \text{ ft} = 24 \text{ k-ft}$ <u>Answer (d)</u>

37. If the cross-sectional area of the entire beam is rectangular, 6 in. deep by 2 in. wide, what is the maximum stress due to the load shown?

a. 16,250 psi   b. 24,000 psi
c. 24,250 psi   d. 2,020 psi
e. 24,500 psi

Using the value of the maximum bending moment from problem #16 we have in section BC:

$\sigma_{max} = \frac{Mc}{I} + \frac{P}{A}$

bending stress + axial stress

$\sigma_{max} = \frac{(24000)(12)(3)}{\frac{2}{12}(6)^3} + \frac{3000}{2 \times 6} = 24,250 \text{ psi}$

<u>Answer (c)</u>

38. If the portion AB of the beam is fabricated from a material with a yield point stress of 36000 psi, and a factor of safety of 2.0 based on the yield point is used, what is the required section modulus?

a. 16 in^3 b. 12 in^3
c. 18 in^3 d. 8 in^3
e. 14 in^3

$f_{max} = M/Z$ or $Z = M/f$
$M = 24000 \times 12$ and $f = 36000/2$

$f_{max} = \frac{24000 \times 12}{18000} = 16 \text{ in}^3$ <u>Answer (a)</u>

39. Determine the maximum stress in the beam resulting from the applied loads. The beam is steel and has a constant cross section as shown.

a. 43,655 psi   b. 43,468 psi
c. 47,046 psi   d. 48,296 psi
e. 94,262 psi

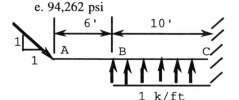

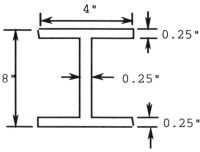

Calculate the cross sectional area and I_{na}.

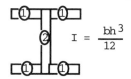

$I = \frac{bh^3}{12}$

	Area	I_0	d	d^2	Ad^2
Part 1 4 pcs	1.875 × .25 = .47 each	0.00244 each	3.875 each	15.02 each	7.06 each
Part 2	8 × .25 = 2	10.67	0	0	0
Σ	3.88	10.676			28.24

$I_{na} = \Sigma I_o + \Sigma Ad^2 = 38.91$

Construct the V and BM diagrams

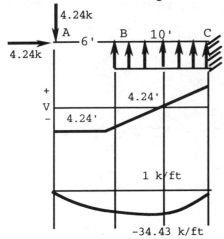

$M_{max} = 25.44 + \left(4.24 \times 4.24 \times \dfrac{1}{2}\right) = 34.43$

$\sigma_{max} = \dfrac{Mc}{I} + \dfrac{P}{A} =$

$\left(\dfrac{-34.33 \times 1000 \times 12 + 4}{38.91}\right) + \dfrac{(-4240)}{3.88}$

$= \underline{-43655 \text{ psi}} \qquad \underline{\text{Answer (a)}}$

40. A hooked bracket, rigidly attached to a wall as shown has an inclined load P, applied as shown. If the maximum stress permitted by the material is 20,000 psi, what is the maximum load P that can be applied? Base this load on the moment at the wall.

a. 25,280 lb
b. 26,700 lb
c. 27,800 lb
d. 28,900 lb
e. 31,768 lb

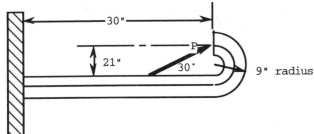

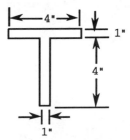

Resolve the force P into components. The vertical component will cause a bending stress with tension on the upper surface. The horizontal component will cause a tensile stress on the upper and lower surface. Locate the neutral axis:

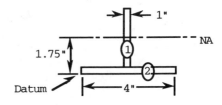

Part	A	y	Ay
1	4	3	12
2	4	0.5	2
	8		14

$\overline{Y} = \dfrac{\Sigma Ay}{\Sigma A} = \dfrac{14}{8} = 1.75''$

$SM_a = 0$

$-(P \sin 30°)30 + (P \cos 30°)(21 + 1.75) - M_a = 0$

$M_a = 4.7P$

Evaluate I_{na}

Part	A	Io	d	Ad²
1	4	$\dfrac{4^3}{12} = 5.33$	3 − 1.75 = 1.25	6.25
2	4	Neglect	1.75 − .5 = 1.25	6.25
		5.33		12.50

$I_{NA} = 5.33 + 12.5 = 17.83$

The error due to neglecting I_0 for part 2 is small.

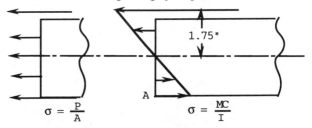

Determine the greater of the combinations.

On the top surface:

$$\sigma = \frac{MC}{I} + \frac{P}{A} = \frac{4.7P \times 1.75}{17.83} + \frac{.866P}{8} = .571P$$

On the bottom surface:

$$\sigma = \frac{MC}{I} - \frac{P}{A} = \frac{4.7P \times 3.25}{17.83} - \frac{.866P}{8} = .748P$$

The bottom surface governs:

$$20,000 = .748P$$
$$P = 26,700 \text{ lbs}$$

<u>Answer (b)</u>

DYNAMICS

1. The total distance travelled by the particle whose velocity time graph is shown below in 0 to 8 sec is:

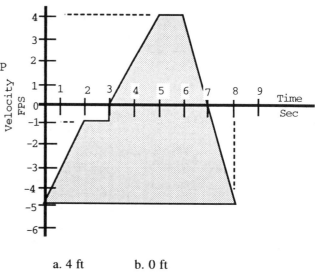

a. 4 ft   b. 0 ft
c. 0.5 ft   d. -2 ft
e. 2 ft

$$\text{Since } S = \int_0^8 v(t)\, dt$$

and the velocity is constant or changes at a constant rate, the distance travelled is the sum of the shaded areas.

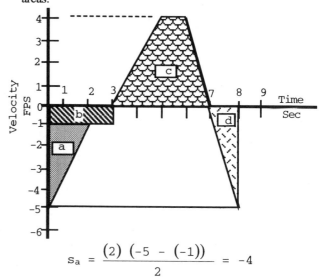

$$S_a = \frac{(2)(-5-(-1))}{2} = -4$$

$s_b = -1(3) = -3$

$s_c = \dfrac{(4)(1 + (7-3))}{2} = 10$

$s_d = \dfrac{(1)(-5)}{2} = -2.5$

Total distance traveled 0.5 ft,

<u>Answer (c)</u>

2. A 3200 lb auto is traveling along a level straight road at 55 mph. It has four wheel brakes that are equally distributed. The coefficient of friction between tire and road is 0.6 and the drivers reaction time (time from sighting to brake application) is 0.5 sec. How far will the car travel from the point at which an obstacle is sighted if deceleration is constant and uniform?

    a. 208.76 ft   b. 284.5 ft
    c. 322 ft   d. 240.5 ft
    e. 168.42 ft

Deceleration is dV/dt
Normal force/ wheel is 3220/4 = 805 lbs
$F = fN = 0.6 \times 805 = 483$
Total decelerating force is
$483 \times 4 = 1932$ lbs

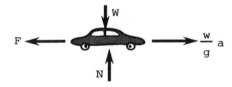

$\rightarrow \Sigma F_x = 0 \qquad F - \dfrac{w}{g} a = 0$

$a = \dfrac{Fg}{W} = \dfrac{1932 \times 32.2}{3220} = 19.32 \dfrac{ft}{sec^2}$

$V_F^2 = V_0^2 + 2aS$ thus $S = \dfrac{V_F^2 - V_0^2}{2a}$

$S_{Brake} = \dfrac{\left[\dfrac{(55)(5280)}{3600}\right]^2 - 0}{(2)(-19.32)} = 168.42$ ft

$S_{Reaction} = V_0 t = \left[\dfrac{(55)(5280)}{3600}\right] \dfrac{1}{2} = 40.34$ ft

Total distance traveled 208.76 ft

<u>Answer (a)</u>

3. How many seconds will it take the auto in problem 2 to stop?

    a. 0.5 sec   b. 5.47 sec
    c. 5.97 sec   d. 4.68 sec
    e. 5.18 sec

$V_F = V_0 + at$

$t_{Brake} = \dfrac{V_F - V_0}{a} = \dfrac{0 - 80.67}{-19.32} = 4.18$

Reaction time is 0.5, thus total time is 4.68 sec
<u>Answer (d)</u>

4. A balloon has been rising vertically from the ground at a constant speed for 12 sec. A stone is released from the balloon and reaches the ground after and additional 6 sec. Find the speed of the balloon at the instant of release of the stone.

    a. 16.1 ft/sec   b. 8 ft/sec
    c. 26 ft/sec   d. 48.3 ft/sec
    e. 32.2 ft/sec

Use up as the positive direction:
For the stone a = -g:

$-S = V_0 t + \dfrac{1}{2} gt^2$

$-S = V_0 6 + \dfrac{1}{2}(-32.2)(6)^2$

For the balloon:

$S = V_0 t = V_0(12)$

Equating:

$V_0(12) = -V_0 6 + \dfrac{1}{2}(32.2)(6)^2$

$V_0 = 32.2$ ft/sec   <u>Answer (e)</u>

5. Find the height of the balloon when the stone reaches the ground.

    a. 386.4 ft   b. 193.2 ft
    c. 579.6 ft   d. 732 ft
    e. 646 ft

$S = V_0 t = 32.2(12 + 6) = 579.6$

<u>Answer (c)</u>

6. A 200 lb weights rests on a plane (coefficient of friction 0.25) inclined 45° as shown. It is connected by a cable to a 100 lb weight. The cable goes over a frictionless, massless pulley. Upon release, what is the acceleration of the system?

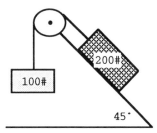

a. 28.5 ft/sec²   b. 16.1 ft/sec²
c. 8.05 ft/sec²   d. 6.49 ft/sec²
e. 0.649 ft/sec²

Free body diagrams:

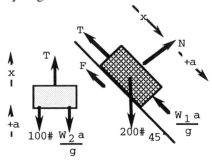

$$F = mN = (0.25)(.707)(W_1)$$

$\Sigma F_{x(200\ lbs)} = 0$

$$W_1 \sin 45° - T - F - \frac{W_1}{g} a = 0$$

$\Sigma F_{x(100\ lbs)} = 0$

$$T - \frac{W_2}{g} a - W_2 = 0$$

Adding:

$$-\frac{a}{g}(W_1 + W_2) - W_2 - F + .707\ W_1 = 0$$

$$-a = \frac{[(100) + (.25)(.707)(200) - (.707)(200)]\ 32.2}{200 + 100}$$

$$a = 0.649\ ft/sec^2 \quad \underline{Answer\ (e)}$$

7. What is the velocity of the 100 lbs weight 4 sec after release?

a. 26 ft/sec   b. 2.6 ft/sec
c. 1.3 ft/sec   d. .13 ft/sec
e. 5.2 ft/sec

$$V_F = V_0 + at = 0 + 0.649(4) = 2.6\ ft/sec$$

<u>Answer (b)</u>

8. A carton is vertically dropped onto a conveyor belt moving at 10 ft/sec. For the carton/belt m = 0.3. How long will it take the carton to reach conveyor speed?

a. 1.04 sec   b. 1.25 sec
c. 2.08 sec   d. 10 sec
e. 0.5 sec

A Free Body Diagram:

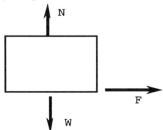

While the block is slipping N = W

$$F = mN = mW = .3W$$

$$.3\ W = \frac{W}{g} a$$
$$a = 9.66\ ft/sec^2$$

$$V_F = V_0 + at$$

$$10 = 0 + 9.66t$$

$$t = 1.04\ sec \quad \underline{Answer\ (a)}$$

9. The cable is weightless and the pulley weightless and frictionless. The tension in the cable is:

a. 193.2 lb   b. 42.93 lb
c. 85.86 lb   d. 171.76 lb
e. 257.6 lb

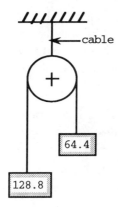

10. The coefficients of friction between blocks A and B and the 30° plane is 0.20. What is the tension in the weightless cable C?

a. 30 lb   b. 25 lb
c. 10 lb   d. 64.4 lb
e. 0 lb

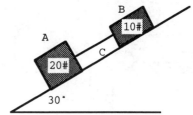

Free body diagrams:

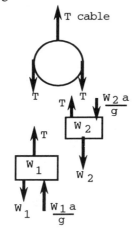

Free Body Diagrams

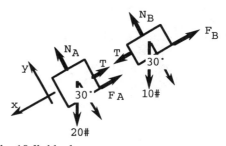

For the 10-lb block:

$$\Sigma F_y = 0: N_B = 10 \cos 30°$$
$$F_B = .2(10)(.866) = 1.732$$
$$\Sigma F_x = W_B \frac{a}{g}$$
$$T + 10 \sin 30° - F_B = 10 \frac{a}{g}$$
$$T + 5 - 1.732 = 10 \frac{a}{g} \quad (1)$$

For block 1: $W_1 - \frac{W_1}{g} a - T = 0$

For block 2: $T - \frac{W_2}{g} a - W_2 = 0$

Adding: $W_1 - W_2 - (W_1 + W_2) \frac{a}{g} = 0$

$$128.8 - 64.4 - (128.8 + 64.4) \frac{a}{32.2} = 0$$
$$a = 10.73 \text{ ft/sec}^2$$

From Block 1: $W_1 - \frac{W_1}{g} a - T = 0$

$$64.4 - \frac{64.4}{32.2} 10.73 = T = 85.88$$

$$T_{cable} = 2T = 171.76$$

Answer (d)

For the 20-lb block:

$$\Sigma F_y = 0: N_A = 20 \cos 30°$$
$$F_A = .2(20)(.866) = 3.464 \text{ lbs}$$
$$\Sigma F_x = W_A \frac{a}{g}$$
$$20 \sin 30° - T - 3.464 = 20 \frac{a}{g}$$
$$10 - T - 3.464 = 20 \frac{a}{g} \quad (2)$$

Adding (-2 x Eq 1) and (Eq 2) gives:

$$(-2 \times \text{Eq 1}) \quad -2T - 10 + 3.464 = -20 \frac{a}{g}$$
$$(\text{Eq 2}) \quad -T + 10 - 3.464 = 20 \frac{a}{g}$$

From which $-3T = 0$, $T = 0$ Answer (e)

Problems 11 - 13

In the diagram below the block attains a velocity from rest to 88 fps in 4 sec. The truck is moving in the same direction as the block and doubles its velocity while moving through its own length of 27 ft. The truck is moving at 6 fps when the block starts to move.

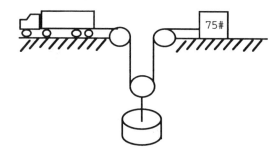

11. The acceleration of the block is:

    a. 7.5 ft/sec²   b. 11 ft/sec²
    c. 16.1 ft/sec²   d. 22 ft/sec²
    e. 15 ft/sec²

$$a_{block} = \frac{V_f - V_t}{t} = \frac{88 - 0}{4} = 22 \ \frac{ft}{sec^2}$$

Answer (d)

12. The acceleration of the truck is:

    a. 12 ips²   b. 22 ips²
    c. 24 ips²   d. 18 ips²
    e. 20 ips²

$$a_{Truck} = \frac{V_t^2 - V_0^2}{2S} = \frac{(12)^2 - (6)^2}{(2)(27)} = 2 \ \frac{ft}{sec^2}$$

Answer (c)

13. The acceleration of the bucket is:

    a. 12 fps²↑   b. 12 fps²↓
    c. 10 fps²↑   d. 10 fps²↓
    e. 22 fps²↓

$$a_{Bucket} = \frac{a_{Block} \downarrow + a_{Truck} \uparrow}{2}$$

$$\frac{22 - 2}{2} = 10 \ \frac{ft}{sec^2} \downarrow$$

Answer (d)

14. A 5-lb block is pulled up a smooth plane which makes an angle of 30° with the horizontal. If the constant pull is measured as 12 lbs along a line parallel to the plane, what is the acceleration of the block along the plane?

    a. 110 ft/sec²   b. 55 ft/sec²
    c. 61.17 ft/sec²   d. 32 ft/sec²
    e. 28 ft/sec²

$\Sigma F_x = a \frac{W}{g}$, let the x axis be the plane

$$a = \frac{(12 - 5 \sin 30°)(32.2)}{5} = 61.17 \ \frac{ft}{sec^2}$$

Answer (c)

15. A 50 lb block is projected up a 23° incline. The initial velocity of the block is 20 fps. If the block moves 10 ft up the incline, what is the coefficient of friction between the block and the plane?

    a. 0.09   b. 0.90
    c. 0.25   d. 0.62
    e. 0.39

Friction force = μ (50 cos 23°)
Initial energy = final energy + losses

$$\frac{V_0^2}{2g} = (W)(10 \sin 23°) + (\mu W \cos 23°)(10)$$

$$\mu = \frac{\frac{V_0^2}{2g} - (10 \sin 23°)}{10 \cos 23°}$$

$\mu = 0.25$ Answer (c)

16. A space station revolves around the earth at an altitude of 25,000 mi. The radius may be taken as 4000 mi and that g is inversely proportional to the square of the distance from the center of the earth and is 32 ft/sec² at the surface. The acceleration due to gravity at the space station is:

    a. 32 ft/sec²   b. 0.609 ft/sec²
    c. 1682 ft/sec²   d. 3.05 ft/sec²
    e. 0.0609 ft/sec²

r = 4000 mi

$$\frac{g_{station}}{g_{Earth}} = \frac{d_{Earth}^2}{d_{Station}^2}$$

$$\frac{g_{station}}{32.2} = \frac{4000^2}{(4000 + 25000)^2}$$

$g_{station} = 0.609$ ft/sec^2 Answer (b)

17. Find the speed of the space station in MPH.

   a. 25,000   b. 1800
   c. 6585   d. 4382
   e. 208,000

$$V^2 = (g_{station})(D)$$

$$V = \sqrt{(.609)\left(\frac{3600^2}{5280}\right)(29,000)} = 6585 \text{ mph}$$

Answer (c)

18. Assuming the spacecraft does not rotate about its own center, what is its center's angular velocity about the earth's center in rad/hr?

   a. 0.227 rad/hr b. 15.95 rad/hr
   c. 4.54 rad/hr   d. 7.83 rad/hr
   e. 0.2634 rad/hr

$$\omega = \frac{V}{d} = \frac{6585}{29,000} = 0.227 \frac{\text{rad}}{\text{hr}}$$

Answer (a)

19. A stunt driver (do not try this at home) rides around the hoop and across the moat as shown. For safety reasons she must exert a resultant force upward on the hoop when she is at the top equal to her weight, W. What is the entry velocity V_1. Given:
$V_0^2 = V_i^2 = V_T^2 + 50g$ and $g = 32$ ft/sec^2

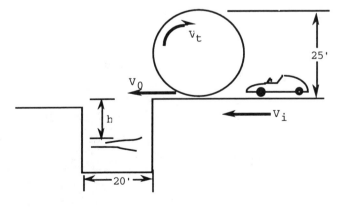

   a. 49 fps   b. 2400 fps
   c. 32 fps   d. 25 fps
   e. gh fps

A free body diagram at the top:

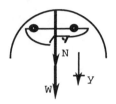

$$\Sigma F_y = \frac{W}{g} a_y \; ; \; N + W = \frac{W}{g} a_y; \; N = W$$

$$2W = \frac{W}{g} a_y; \; 2g = a_y = \frac{V_T^2}{R}$$

$$V_T^2 = 2gR = (64.2)(12.5) = 800$$

Given: $V_i^2 = V_T^2 + 50g;$

$$V_i = \sqrt{800 + (50)(32)} = 49 \frac{\text{ft}}{\text{sec}}$$

Answer (a)

20. How much lower must the far side of the moat be to make a safe landing?

   a. 9.67 ft   b. 8.4 in
   c. .47 ft   d. 5.83 ft
   e. 2.66 ft

$$S = V_0 t = 20 \text{ ft}; \; V_0 = V_1 = 49 \frac{\text{ft}}{\text{sec}}$$

$$t = \frac{20}{49} = 0.408 \text{ sec}$$

$$S_y = V_{0y} + \frac{1}{2}gt^2$$

$$0 + \frac{1}{2}(32.2)(.408)^2 = 2.66 \text{ ft}$$

Answer (e)

21. A car is rounding a curve of 500 ft radius at a speed of 75 mph. If the coefficient of friction between tires and road is 0.55 what is the minimum angle of the bank of the curve?

   a. 8.11°   b. 28.82°
   c. 36.96°   d. 3.76°
   e. 65.78°

Free body diagram and force polygon:

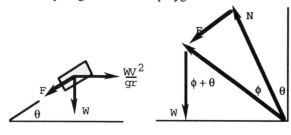

$$V = 75 \frac{5280}{3600} = 110 \frac{ft}{sec}$$

$$\tan \phi = 0.5, \phi = 28.82°$$

$$\tan \theta + \phi = \frac{V^2}{gr} = \frac{110^2}{(32.2)(500)} = .7516$$

$$\tan \theta + \phi = .7516: \theta + \phi = 36.93$$

$$\theta = 36.93 - 28.82 = 8.11° \quad \underline{Answer\ (a)}$$

22. If the road was covered with ice ($\mu = 0$) what would the bank angle be?

    a. 12.37°   b. 28.82°
    c. 36.99°   d. 0
    e. 75.78°

$$\tan \phi = 0.0, \phi = 0$$
$$\theta = 36.93 - 0 = 36.99° \quad \underline{Answer\ (c)}$$

23. An airplane diving at a horizontal speed of 200 mph and an altitude of 15,000 ft releases a bomb to hit a target on the ground directly in front of the flight path. At what horizontal distance before the target must the bomb be released to hit the target. Neglect air resistance and wind effects.

    a. 10,382 ft   b. 6104 ft
    c. 15,000 ft   d. 18,312 ft
    e. 8952 ft

$$V_h = 200 \frac{5200}{3600} = 293.33 \text{ fps}$$

The time to reach the ground:

$$S = \frac{1}{2} gt^2: t = \sqrt{\frac{2(15,000)}{32.2}} = 30.52$$

$$S_h = V_0 t = 293.33 (30.52) = 8952 \text{ ft}$$

$$\underline{Answer\ (e)}$$

Problems 24 - 26

A cannon with a projectile muzzle velocity of 1150 fps is fired at an angle of 37° from the horizontal.

24. How high is the projectile after 4 sec?

    a. 257.6 ft   b. 4600 ft
    c. 2760 ft   d. 2510 ft
    e. 1150 ft

The vertical distance traveled is the velocity of the projectile minus the effects of gravity.

$$S_y = V_{0y}t - \frac{1}{2} gt^2$$

$$1150 \sin 37°(4) - \frac{1}{2}(32.2) 4^2 = 2510 \text{ ft}$$

$$\underline{Answer\ (d)}$$

25. What is the maximum height which it reaches?

    a. 15,044 ft   b. 2760 ft
    c. 10,523 ft   d. 7437 ft
    e. 4000 ft

$$V_{fy}^2 = V_0^2 + 2ay_{max}$$

$$y_{max} = \frac{V_{fy}^2 - V_0^2}{2a} = \frac{0^2 - (1150 \sin 37°)^2}{2(-32.2)}$$

$$= 7437 \text{ ft} \quad \underline{Answer\ (d)}$$

26. What is the horizontal range at this angle?

    a. 19,200 ft   b. 9742 ft
    c. 39,364 ft   d. 98,482 ft
    e. 63,343 ft

Find the flight time of the projectile, then the distance traveled.

$$S_y = V_0 t - \frac{1}{2} gt^2 = 0$$

$$t = \frac{2 V_0}{g} = \frac{2(1150 \cos 37°)}{32.2} = 42.98$$

$$S_{Horizontal} = V_{ox} t = 1150 \sin 37°(42.98)$$

$$= 39,364 \text{ ft} \quad \underline{Answer\ (c)}$$

27. A block is placed on a rough horizontal plate ($\mu = 0.65$) at a distance of 6 in from the vertical axis of rotation of the plate. What rpm will cause the block to just begin to slide?

    a. 6.47 rpm   b. 61.8 rpm
    c. 41.86 rpm   d. 30.4 rpm
    e. 123.6 rpm

The force required to slip the block is:

$$F = \mu W$$

Angular velocity in terms of force:

$$\mu W = \frac{W}{g} a = \frac{W}{g} r\omega^2$$

$$\omega^2 = \sqrt{\mu \frac{g}{r}} = \sqrt{.65 \frac{32.2}{.5}} = 6.47 \frac{rad}{sec}$$

    rpm = 6.47 (60)/2p = 61.8   <u>Answer(b)</u>

28. If the plate starts from rest and accelerates at a constant rate of 5
rad/sec^2, how many seconds will be required to cause the block to begin to slip?

    a. 12.4   b. .29
    c. 2.6   d. 4.29
    e. 1.29

$$\omega_f = \omega_0 + at$$
$$6.47 = 0 + 5t$$
$$t = 1.29 \text{ sec} \quad \text{Answer (e)}$$

Problems 29 - 30

The rotating flywheel is as shown:

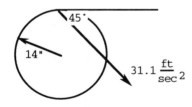

29. What is the angular acceleration of the flywheel in rad/sec^2?

    a. 1.57   b. 2.22
    c. 18.85   d. 26.66
    e. 36.28

Since the angle is 45°, $a_t = a_n = ra$

$$\alpha = \frac{a_t}{r} = \frac{a \cos 45°}{r}$$

$$a_t = \frac{(31.1)(.707)}{14/12} = 18.85 \frac{rad}{sec^2}$$

    <u>Answer (c)</u>

30. What is the angular velocity of the flywheel?

    a. .334 rad/sec   b. 4.69 fps
    c. 1.16 fps   d. 4.34 rad/sec
    e. 13.92 rad/ sec

$$\omega = \sqrt{\frac{a_t}{r}} = \sqrt{\frac{31.1 \cos 45°}{14/12}} = 4.34 \frac{rad}{sec}$$

    <u>Answer (d)</u>

Problems 31 - 33

A cylindrical shaped yo yo as shown is dropped freely. The cord maintains a constant radius.

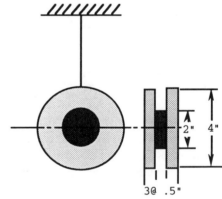

31. What is the mass moment of intertia of the yo yo direction of rotation in terms of its weight W^2.

    a. 0.4445 W lb-in-sec
    b. 0.5556 W lb-in-sec
    c. 0.00445 W lb-in-sec
    d. 0.1111 W lb-in-sec
    e. 0.05706 W lb-in-sec

$$W_{1 \text{ wheel}} = \frac{\pi (2)^2 (0.5)}{2 \pi (2)^2 (0.5) + \pi (1)^2 (0.5)} W = 0.445W$$

$$W_{axel} = \frac{\pi (1)^2 (0.5)}{2 \pi (2)^2 (0.5) + \pi (1)^2 (0.5)} W = 0.1111W$$

$$I_0 = \frac{1}{2}\left[\frac{2\,W_{wheel}}{g}\,2^2 + \frac{W_{axel}}{g}\,1^2\right]$$

$$I_0 = \frac{W}{2g}\left[2\,(.4455)\,2^2 + (.1111)\,1^2\right]$$

$$= .05706W \text{ lb-in-sec}^2$$

<u>Answer (e)</u>

32. What is its linear acceleration:

a. 16.76 in/sec^2
b. 386 in/sec^2
c. 32.2 in/sec^2
d. 210.5 in/sec^2
e. 105.3 in/sec^2

A Free Body Diagram

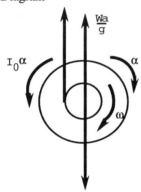

$(Tr) - (I_0)a = 0 = (Tr) - (.05706W)a$
$(Tr) = (.05706W)a$

Since the yo yo unwinds from the string without slipping, the instantaneous center of zero velocity is at IC and $a = ar$.

Using $\quad (T)(1) = (.05706W)\dfrac{a}{r}$

$$T - W + \frac{W}{g}\,a = 0$$

$$(.05706W)a - W + \frac{W}{g}\,a = 0$$

$$a = \frac{1}{(.05706) + \dfrac{1}{386}}$$

16.76 ft/sec^2

<u>Answer (a)</u>

33. The yo yo weighs 4 oz. What is the tension in the cord?

a. .25 lbs   b. .5 lbs
c. .162 lbs d. .024 lbs
e. .179 lbs

$$T = .00475\,Wa = .00475\,\frac{4}{16}$$

$$136.2 = .162 \text{ lbs}$$

<u>Answer (c)</u>

Problem 14 - 16

A 5 lb weight connected to a spring moves upward at 5 fps from a height of 3 ft above the spring. It later comes down, impacts, and remains in contact with the spring. Neglect the usual neglects. The spring constant is 30 in/in.

34. What is the maximum deflection of the spring?

a. 4.37 in   b. .167 in
c. 2.09 in   d. .367 in
e. 3.85 in

Since all the neglects have been neglected, the weight will rise a distance S such that energy is conserved.

$$WS = \frac{W}{2g}\,v_i^2$$

$$S = .388 \text{ ft}$$

PE of weight = PE of spring

$$[12\,(3.0 + .388) + \delta]5 = \frac{k}{2}\,\delta^2$$

$$\delta = .333 \pm \sqrt{\frac{-.333^2 - 4\,(-13.55)}{2}}$$

Using the positive root, $\delta = 3.85$
<u>Answer (e)</u>

35. What is the final deflection of the spring?

a. 4.37 in   b. .167 in
c. 2.09 in   d. .367in
e. 3.85 in

$$\Sigma F_y = 0 = k\delta - W$$
$$\delta = \frac{W}{k} = \frac{5}{30} = 0.167 \text{ in}$$
<u>Answer (b)</u>

36. What is the oscillation frequency of the mass as it is comming to rest. (Spring is massless)

a. 1.39 Hz   b. 12 Hz
c. 6.28 Hz   d. 2.21 Hz
e. 7.67 Hz

$$f = \frac{1}{2\pi} \sqrt{k \frac{g}{W}} = \frac{1}{2\pi} \sqrt{30 \frac{32.2}{5}} = 7.67 \text{ Hz}$$

<u>Answer (e)</u>

37. Two smooth 6oz discs sliding on a smooth surface collide as shown. ($V_1 = 40$ ft/sec, $V_2 = 50$ ft/sec, Coefficient of restitution = 0.6) What is the velocity of disc 1 after impact?

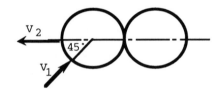

a. 12.64 ft/sec   b. 25.28 ft/sec
c. 31 ft/sec   d. 46.98 ft/sec
e. 21.7 ft/sec

The x axis is the common normal to both bodies. Since they are smooth, no perpendicular forces are acting.

$$V_{1x} = 40 \cos 45° = 28.3 \text{ ft/sec}$$
$$V_{1y} = 40 \sin 45° = 28.3 \text{ ft/sec}$$
$$M_1 V_{1x} + M_2 V_{2x} = M_1 U_{1x} + M_2 U_{2x}$$

Since $M_1 = M_2$

$$V_{1x} + V_{2x} = U_{1x} + U_{2x}$$
$$28.3 + (-50) + U_{1x} + U_{2x} \quad (1)$$

By Definition:
$$e = \frac{u_{2x} - u_{1x}}{V_{1x} - V_{2x}} = .6$$
$$-1 (u_{2x} - u_{1x}) = (.6)[-1(V_{1x} - V_{2x})]$$

Substituting:
$$(u_{1x} - u_{2x}) = (.6)[(-28.3) + (-50)] = -46.98 \quad (2)$$

Adding (1) and (2):
$$2 U_{1x} = -46.98 - 21.7$$
$$U_{1x} = 34.34$$

Because the bodies are smooth:
$$U_{1y} = V_{1y} = 28.3$$

$$U_1 = \sqrt{U_{1x}^2 + U_{1y}^2} = \sqrt{34.34^2 + 28.3^2} = 44.49$$
<u>Answer (c)</u>

38. Two weights are suspended from a frictionless 10-lb pulley with a radius of gyration of 0.8 ft. How long will it take to change the speed of the weights from 15 ft/sec to 25 ft/sec?

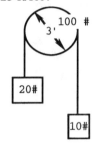

a. 1.82 sec   b. 2.17 sec
c. 3.60 sec   d. 4.34 sec
e. 0.98 sec

Use impulse momentum for the whole system;

$$\frac{\Delta_{\text{Angular Momentum}}}{r} = \Delta_{\text{Linear Momentum}}$$

$$Ft = \frac{W_{20}}{g} (V_2 - V_1) +$$

$$\frac{I_{100} (\omega_2 - \omega_1)}{r} + \frac{W_{10}}{g} (V_2 - V_1)$$

$$(\omega_2 - \omega_1) = \frac{(V_2 - V_1)}{2} \text{ and } I_{100} = \frac{W}{g} K^2$$

Thus
$$Ft = \frac{W_{20}}{g} (V_2 - V_1) +$$

$$\frac{\frac{W_{100}}{g} K^2 (V_2 - V_1)}{r^2} + \frac{W_{10}}{g} (V_2 - V_1)$$

Substituting and solving for t:

$$(20 - 10) t = \frac{20}{32.2} (25 - 15) +$$

$$\frac{\left(\frac{100}{32.2}\right)(.8^2)(25-15)}{1.5^2} + \frac{10}{32.2}(25-15)$$

t = 1.82 sec <u>Answer (a)</u>

Velocity is
$$v = \frac{dr}{dt} = \frac{d(6.5 - .42t^2)}{dt} = .84t$$
$$= .84(1.89)$$
$$= 1.58 \text{ in/sec} \quad \underline{\text{Answer (d)}}$$

Problems 39 - 40

Part of a mechanism includes a weight sliding along a rod which rotates about one end as shown.

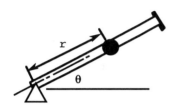

q = 0.5 t² radians
r = 6.5 - 0.42 t² in
t = sec

39. When q = 30°, r = ?

    a. 5.05 in   b. 6.06 in
    c. 7.07 in   d. 8.08 in
    e. 9.09 in

$$0.5t^2 = 30 \frac{2\pi}{360}$$
$$t^2 = 2(.524) = 1.048$$

r = 6.5 - .42 (1.048) = 6.06 in

<u>Answer (b)</u>

40. The velocity of the block moving along the rod at r = 5 is ?

    a. 1.89 in/sec   b. 3.16 in/sec
    c. 0.95 in/sec   d. 1.58 in/sec
    e. 4.22 in/sec

Distance traveled = 5 - r

5 - (6.5 - 0.42 t²)

Time required

$$t = \sqrt{\frac{5 - 6.5}{-.42}} = 1.89 \text{ sec}$$

FLUID MECHANICS

Problems 1 - 6

Given the masonry dam as shown. Take a 1 ft crest for the solutions to the following problems.

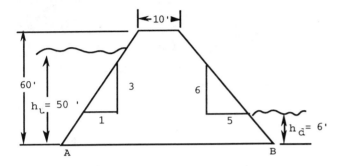

1. The upstream horizontal force is:

    a. 3120 lbs   b. 122,320 lbs
    c. 78,000 lbs   d. 1560 lbs
    e. 93,600 lbs

Pressure at the base = wh: $P_{av} = \dfrac{wh_u}{2}$

$$F = P_{av} A = \left(\dfrac{wh_u}{2}\right)(h_u \times 1)$$

$$\left(\dfrac{50}{2}\right)(62.4)(50) = 78,000\#$$ Answer (c)

2. The location of the center of pressure on the upstream side is:

    a. 20 ft above the base
    b. 33.33 ft below the water surface
    c. 40 ft below the top of the dam
    d. 25 ft below the water surface
    e. In the big middle

The hydrostatic center of pressure for the rectangular projection of the face of the dam is locates 2/3 of the depth of the water below the surface or 2/3 x 50 = 33.33 ft.   Answer (b)

3. The overturning moment is:

    a. 1.3 x 10⁶ ft lb
    b. 1.6 x 10⁶ ft lb
    c. 3.8 x 10⁶ ft lb
    d. 1.1 x 10⁶ ft lb
    e. 2.3 x 10⁶ ft lb

The dam will overturn at point B

$$M_u = 78,000 \times 1/3(50) = 1.3 \times 10^6$$
    Answer (a)

4. The restoring moment is:

    a. 18.45 x 10⁶ ft lb
    b. 19.22 x 10⁶ ft lb
    c. 15.32 x 10⁵ ft lb
    d. 22.81 x 10⁵ ft lb
    e. 20.39 x 10⁶ ft lb

a. Moment of water on upstream toe

$$\left[\left(\dfrac{1}{2}\right)\left(\dfrac{50}{3}\right)(50)(62.4)\right]$$
$$\left[50 + 10 + 20 - \left(\dfrac{1}{3}\right)\left(\dfrac{50}{3}\right)\right] = 1.94 \times 10^6 \text{ft }\#$$

b. Moment of masonry on the upstream toe:

$$\left[\dfrac{(20)(60)(150)}{2}\right]\left[50+10+\left(\dfrac{20}{3}\right)\right] = 6.0 \times 10^6 \text{ft }\#$$

c. Moment of masonry of the center portion:

$$[(10)(60)(150)][50 + 5] = 4.95 \times 10^6 \text{ft }\#$$

d. Moment of water on the downstream face:

$$\left[\dfrac{(6)(5)(62.4)}{2}\right]\left[\dfrac{2}{3}\,5\right] = .00312 \times 10^6 \text{ft }\#$$

e. Moment of masonry on the downstream heel:

$$\left[\dfrac{(50)(60)150}{2}\right]\left[\dfrac{2}{3}\,50\right] = 7.5 \times 10^6 \text{ft }\#$$

f. Hydrostatic forces on downstream heel

$$\left[\dfrac{(6)(5)62.4}{2}\right]\left[\dfrac{1}{3}\,6\right] = .002246 \times 10^6 \text{ft }\#$$

Sum a - f gives 20.395 ft lbs   Answer (e)

5. Where does the resultant force lie with respect to point B?

   a. 44.2 ft from B
   b. 55.6 ft from B
   c. 25.4 ft from A
   d. 36.9 ft from A
   e. 38.1 ft form B

F_v = Sum of all the downward weights from problem 4:

$$\left[\left(\frac{1}{2}\right)\left(\frac{50}{3}\right)(50)(62.4)\right] + \left[\frac{(20)(60)(150)}{2}\right] + \left[\frac{(50)(60)(150)}{2}\right]$$

$$+ \left[(10)(60)(150)\right] + \left[\frac{(6)(5)(62.4)}{2}\right] = 431,396\#$$

Moments at B (from problem 3 and 4)

$$M_B = (1.3 - 20.395)\,10^6 = -19.095 \times 10^6$$

$$\overline{X} = \frac{M_B}{F_v} = \frac{-19.005 \times 10^6}{.432 \times 10^6} = -44.2 \text{ ft}$$

<u>Answer (a)</u>

6. What is the minimum required coefficient of friction between the dam and the bedrock if the dam is to be safe?

   a. 0.122 b. 0.150
   c. 0.178 d. 0.181
   e. 0.203

$$\mu = \frac{\text{Net } F_h}{\text{Net } F_v} = \frac{\left(78,000 - \left[\frac{(6)(5)62.4}{2}\right]\right)}{431,936} = .178$$

<u>Answer (c)</u>

Problems 7 - 15

Given the pipeline as shown and that f = .02 for a 12-in. dia pipe.

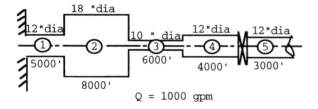

Q = 1000 gpm

7. The entrance loss is:

   a. 3.5 ft   b. .165 ft
   c. .228 ft d. .062 ft
   e. 1.932 ft

$$V = \frac{Q}{A}$$

$$= \frac{(1000)\left(\frac{\text{gal}}{\text{min}}\right)\left(\frac{\text{cu ft}}{7.5 \text{ gal}}\right)\left(\frac{60 \text{ sec}}{\text{min}}\right)}{\pi \, .5^2 \text{ ft}^2} = 2.837 \text{ fps}$$

$$h_e = \frac{1}{2}\frac{V^2}{2G} = \frac{1}{4}\frac{2.837^2}{32.2} = .062 \text{ FT}$$

<u>Answer (d)</u>

8. The equivalent length of 12-in. dia pipe for this entrance loss is:

   a. 37.2 ft b. 25 ft
   c. 11.4 ft d. 8.25 ft
   e. 3.1 ft

$$h_e = h_f: \frac{1}{2}\frac{V^2}{2g} = f\frac{L}{d}\frac{V^2}{2g}; \frac{1}{2} = f\frac{L}{d}$$

$$L = (d/2f) = [1/(2)(.02)] = 25\text{ft}$$

<u>Answer (b)</u>

9. The velocity in pipe 3 is:

   a. 4.07 ft/sec b. 2.13 ft/sec
   c. 4.26 ft/sec d. 5.52 ft/sec
   e. 0.98 ft/sec

$$V_3 = \frac{Q}{A_3}$$

$$= \frac{(1000)\left(\frac{\text{gal}}{\text{min}}\right)\left(\frac{\text{cu ft}}{7.5 \text{ gal}}\right)\left(\frac{\text{min}}{60 \text{ sec}}\right)}{\pi \left(\frac{5}{12}\right)^2 \text{ ft}^2} = 4.07 \text{ fps}$$

<u>Answer (a)</u>

10. The loss caused by the sudden contraction at pipes 2 and 3 is:

   a. .044 ft b. .123 ft
   c. .212 ft d. .316 ft
   e. .429 ft

$V_2 = \dfrac{Q}{A_2}$

$= \dfrac{(1000)\left(\dfrac{\text{gal}}{\text{min}}\right)\left(\dfrac{\text{cu ft}}{7.5 \text{ gal}}\right)\left(\dfrac{\text{min}}{60 \text{ sec}}\right)}{\pi \left(\dfrac{5}{12}\right)^2 \text{ ft}^2} = 4.07 \text{ fps}$

$H_x = \dfrac{(V_2 - V_3)^2}{2g} = \dfrac{(1.26 - 4.07)^2}{2(32.2)} = .123 \text{ ft}$

Answer (b)

11. The equivalent length of 12 in dia pipe for the enlargement between pipe 3 and 4 is:

a. 11 ft   b. 21 ft
c. 33 ft   d. 44 ft
e. 55 ft

As in problem 8, L = .83/[(2)(.02)] = 11 ft   Answer (b)

12. The head loss at the enlargement of pipes 3 and 4 is:

a. .054 ft   b. .014 ft
c. .027 ft   d. .016 ft
e. .032 ft

$h_c = K_c \dfrac{V^2}{2g} = .22 \dfrac{2.837^2}{2(32.2)} = .027 \text{ ft}$

Answer (c)

13. If f for the 10-in. pipe is .018 and f for the 18-in. pipe is .022 and pipe 5 discharges to the atmosphere, H is:

a. 46.2 ft   b. 59.9 ft
c. 69.3 ft   d. 76.6 ft
e. 84.2 ft

As pipe lengths are long, minor losses are neglected. From above:

$V_1 = V_4 = V_5 = 2.837 \text{ fps}$
$V_2 = 4.07 \text{ fps}$
$V_3 = 1.26 \text{ fps}$

$H = \Sigma \left[(f)\left(\dfrac{L}{d}\right)\left(\dfrac{V^2}{2G}\right)\right]$

$H = \left[(.02)\left(\dfrac{5000 + 4000 + 3000}{1}\right)\left(\dfrac{2.837^2}{2(32.2)}\right)\right] +$

$\left[(.018)\left(\dfrac{6000}{\dfrac{10}{12}}\right)\left(\dfrac{4.07^2}{2(32.2)}\right)\right] +$

$\left[(.022)\left(\dfrac{8000}{\dfrac{18}{12}}\right)\left(\dfrac{1.26^2}{2(32.2)}\right)\right] = 69.28 \text{ ft}$

Answer (c)

14. Taking the minor losses as 2% of the friction losses, what horsepower could be developed in an 80% efficient turbine at the end of pipe 5 if H was measured as 250 ft.

a. .721 hp   b. 1.05 hp
c. 15.62 hp   d. 34.6 hp
e. 63 hp

$Hp = \dfrac{Q \rho h \eta}{550} =$

$\dfrac{(2.22)(62.4)(250 - 76.6(1.02))(.8)}{550}$

$= 34.6 \text{ Hp}$ Answer (d)

15. With conditions of head and flow in Problem 14, if the valve is closed, what is the hoop stress in pipe 4 if its thickness is 0.5 in:

a. 2599 psi   b. 1300 psi
c. 3000 psi   d. 18,720 psi
e. 187,200 psi

$s = \dfrac{\rho h r}{t} = \dfrac{(62.4)(250)(6)\dfrac{1}{144}}{.5} = 1300 \text{ psi}$

Answer (b)

Problems 16 - 20

A cylindrical tank, 20 ft diamater and 20 ft tall is full of water. An orifice is located in the side of the tank. the head on the orifice is maintained at 15 ft at all times. One point in the trajectory of the stream issuing from this 3-in. diameter orifice is located 16 ft to the left and 4.5 ft down from the center of the orifice.

The diagram is included for the purposes of academic enlightenment:

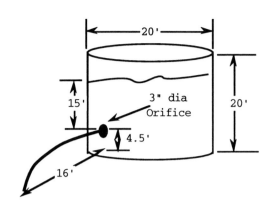

16. The theoretical velocity of the stream issuing from the orifice is:

    a. 30.24 fps   b. 43.27 fps
    c. 31.08 fps   d. 49.96 fps
    e. 386.4 fps

$$V_t = \sqrt{2gh} = \sqrt{(2)(32.2)(15)} = 31.08 \text{ fps}$$

<u>Answer (c)</u>

17. The actual velocity of the stream issuing from the orifice is:

    a. 30.24 fps   b. 43.27 fps
    c. 31.08 fps   d. 49.96 fps
    e. 386.4 fps

Since V_a must be less than V_t, <u>Answer (a)</u>

$$S = \frac{1}{2}gt^2: \quad 4.5 = \frac{1}{2}(32.2)t^2$$
$$t = .529 \text{ sec for the 4.5 ft drop}$$

For the horizontal flow $s = Vt$:
$V = s/t = 16/.529 = 30.24$ fps   <u>Answer (a)</u>

18. If the discharge from the orifice was measured as 1 cfs, the coefficient of discharge is.

    a. .598   b. .408
    c. .471   d. .655
    e. .725

$$C_d = \frac{Q_a}{Q_i} = \frac{Q_a}{A_o \sqrt{2gh}}$$

$$= \frac{1}{\pi \left(\frac{1.5}{12}\right)^2 \sqrt{(2)(32.2)(15)}} = .655$$

<u>Answer (d)</u>

19. The C_c is:

    a. .615   b. .419
    c. .484   d. .673
    e. .745

$$C_c = \frac{C_d}{C_v} = \frac{C_d}{\frac{V_a}{V_i}} = \frac{.655}{\frac{30.24}{31.08}} = .673$$

<u>Answer (d)</u>

20. The head loss through the orifice is:
    a. .667 ft lb/lb
    b. .799 ft lb/lb
    c. .845 ft lb/lb
    d. 1.62 ft lb/lb
    e. 2.08 ft lb/lb

$$\text{Loss} = H(1 - C_v^2) = H\left(1 - \left(\frac{V_a}{V_i}\right)^2\right)$$

$$= 15\left(1 - \frac{30.24^2}{31.08^2}\right) = .799 \frac{\text{ft lb}}{\text{lb}}$$

<u>Answer (b)</u>

21. The hydraulic radius for 48 in diameter reinforced concrete pipe having flow 3 ft deep is:

    a. $2p + 3$ ft   b. $2/3p$ ft
    c. $2p + 3p$ ft   d. $6p$ ft
    e. $1 + [(3\sqrt{3})/8\pi]$

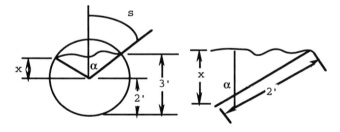

$\cos a = x/2 = 1/2 = .5$

$$\text{Hydraulic Radius} = \frac{\text{Area of flow}}{\text{Wetted Perimeter}}$$

Area of flow = Area of circular pipe − 2 (sector of arc S) + 2 (area of small triangle)

$$\text{Area of flow} = \frac{\pi(4^2)}{4} - 2\left(\frac{60}{360}\right)4\pi +$$

$$(2)\left(\frac{1}{2}\right)(1)(\sqrt{3}) = 8\left(\frac{\pi}{3}\right) + \sqrt{3}$$

$$\frac{240}{360} D \pi = \frac{8}{3} \pi$$

Hydraulic Radius = $1 + [(3\sqrt{3})/8\pi]$

Answer (e)

22. A 100% efficient pump can draw a suction of:

   a. 34 ft H$_2$O   b. 29.92 in Hg
   c. 14.696 psi   d. 1 atm
   e. all of the above

A 100% efficient pump can draw a suction of 1 atmosphere. All answers are the equivalent of 1 atmosphere. <u>Answer (e)</u>

23. A pressure of 3.34 psi is measured 11.2 ft below the surface of a liquid. The specific gravity of the liquid is?

   a. .298   b. .688
   c. 1.234   d. .798
   e. .573

$P = \rho h$: $(3.34)(144) = \rho (11.2)$: $\rho = 42.94$

$$sg = \frac{42.94}{62.4} = .688 \quad \underline{Answer\ (b)}$$

24. In an open channel of small depth and great width, the hydraulic radius is most likely equal to:

   a. 1/3 depth   b. 1.2 depth
   c. 2/3 depth   c. 3/4 depth
   e. Full depth

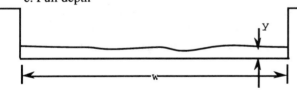

$$R = \frac{A}{WP} = \frac{wy}{w + 2y} = \frac{wy}{w} = y$$

Answer (e)

25. A cylindrical buoy 3 ft diameter weighing 150 lbs floats vertically in sea water. The anchor is one yd^3 of concrete. The mooring cable has no slack. What rise in tide will cause the anchor to lift from the bottom? Specific weights: Sea Water = 64 pcf: Concrete 150 pcf.

The anchor displaces 27 ft^3 of water x 64pcf = 1728 lbs water. In air the anchor weighs 27 ft^3 x 150pcf = 4050 lbs. The effective weight of the anchor is 4050 - 1728 = 2322 lbs.

$$\left(\frac{\pi d^2}{4}\right) \Delta (64) = 2322$$
$$D = 5.132\ ft \quad \underline{Answer\ (c)}$$

The weight of the buoy has no effect in the problem as it is already floating.

26. The actual velocity at the exit of an orifice is:

   a. $\sqrt{2gh}$ b. $C_d\sqrt{2gh}$
   c. $C_v\sqrt{2gh}$ d. $C_c\sqrt{2gh}$
   e. $C_v(2gh)$

$C_v\sqrt{2gh}$ <u>Answer (c)</u>

27. A 10 hp, 100 cfm pump is used to pump water to a storage tank. The efficiency of the pump is 62.4% and total systems losses are 3ft. With the pump level as datum, to what elevation can the tank be filled?

   a. 30 ft   b. 33 ft
   c. 36 ft   d. 39 ft
   e. 32 ft

$$Hp = \frac{Q\rho h}{33,000} = \frac{(100)(62.4)(Z + 3)}{33,000} = (10)(.624)$$

$Z = 30ft$ <u>Answer (a)</u>

Problems 28 - 30

A storm drain 12 ft diameter flows half full at a section where the slope is 1ft/1000 ft. n for the pipe is 0.18. The drain empties into a canal with a 10ft wide base and side slopes of 1 vertical to 5 horizontal. The slope of the canal is 50 ft/1000 ft. Depth of flow is 5ft.

28. The hydraulic radius of the drain pipe is:

a. 1.5 ft b. 3 ft
c. 6 ft   d. 9 ft
e. 12 ft

$$\text{Area of Flow} = \left(\frac{1}{2}\right)\left(\pi \frac{D^2}{4}\right)$$

$$\text{Wetted Perimeter} = \left(\frac{1}{2}\right)(\pi D)$$

$$R = \frac{\text{Area of flow}}{\text{Wetted perimeter}} = \frac{D}{4} = 3 \text{ ft}$$

Answer (b)

29. The flow is:

a. 5.5 cfs b. 307 cfs
c. 317 cfs d. 55 cfs
e. 395 cfs

$$A = \left(\frac{1}{2}\right)\left(\pi \frac{D^2}{4}\right); \quad V = \frac{1.486}{n} R^{\frac{2}{3}} S^{\frac{1}{2}}$$

$$Q = AV = \left[\left(\frac{1}{2}\right)\left(\pi \frac{12^2}{4}\right)\right]\left[\frac{1.486}{.018} 3^{\frac{2}{3}} .001^{\frac{1}{2}}\right]$$

$$Q = 307 \text{ cfs} \qquad \text{Answer (b)}$$

30. The value of "n" for the earth canal is:

a. .022   b. .383
c. .0017 d. .417
e. .012

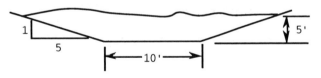

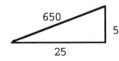

Area of flow =
$$(10)(5) + 2(.5)(25)(5) = 175$$

$$WP = 10 + 2\sqrt{650} = 60.99$$

$$R = A/WP = 175/60.99 = 2.87$$

$$S = 50/1000 = .05$$

$$n = 1.486 \frac{A}{Q} R^{\frac{2}{3}} S^{\frac{1}{2}}$$

Substituting and solving: n = .383

Answer (c)

31. A 2-in. diameter hose has a 1-in. nozzle. The pressure in the hose is 35 psi. How much force is required to hold the nozzle stationary when the flow is 1 cfs?

a. 109 lbs b. 266 lbs
c. 157 lbs d. 484 lbs
e. 593 lbs

For the 2-in. line

$$V = \frac{Q}{A} = \frac{1}{\frac{\pi}{144}} = 45.8 \text{ fps}$$

For the 1-in. nozzle

$$V = 45.8 \frac{2^2}{1^2} = 183 \text{ fps}$$

$$PA = 35 \pi \frac{2^2}{4} = 109 \text{ lb}$$

$$m\Delta V = \frac{(1)(62.4)(183 - 45.8)}{(32.2)} = 266 \text{ lb}$$

$$F = PA + mDV = 109 + 266 = 375$$

Answer (c)

Problems 32 - 34

An open top rectangular tank is 5 ft wide, 12 ft long, and 5 ft deep. The tank is filled to a depth of 3 ft. It is then accelerated forward along its 12 ft axis at 10 ft/sec².

32. How deep is the water at the forward end of the tank?

a. 0 ft   b. 1.14 ft
c. 1.86 ft d. 4.14 ft
e. 4.86 ft

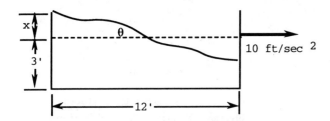

The water will increase by x ft at the aft end and decrease by the same amount in the front.

$$\tan\theta = \frac{a}{g} = \frac{x}{6}: \quad x = 1.86 \text{ ft}$$

The depth at the front will be:

$$3 - 1.86 = 1.14 \quad \underline{\text{Answer (b)}}$$

33. How deep is the water in the aft end of the tank?

a. 0 ft b. 1.14 ft
c. 1.86 ft d. 4.14 ft
e. 4.86 ft

The depth at the aft end will be:

$$3 + 1.86 = 4.86 \text{ ft} \quad \underline{\text{Answer (d)}}$$

34. What is the force on the aft end of the tank?

a. 5161 lbs b. 3870 lbs
c. 4297 lbs d. 6248 lbs
e. 3685 lbs

Average Pressure =
  r (d/2) = 62.4 (2.43) = 151.63

Area = d x w = 4.86 x 5 = 24.3

F = PA = 151.63 (24.3) = 3685
  <u>Answer (e)</u>

35. The pressure gage across a 2-in. diameter orifice in a water line indicates 20-in. Hg. The orifice coefficient is .72. The flow is?

a. 6 cfs b. 6 cfm
c. 36 cfs d. 36 cfm
e. 9.76 csm

$$Q = CA\sqrt{2gh}$$

$$h = \frac{20" \text{ Hg}}{12} \; 13.6 = 22.67 \text{ ft water}$$

$$Q = (.72)\left(\pi \frac{\left(\frac{2}{12}\right)^2}{4}\right)\sqrt{2(32.2)(22.67)}$$

$$= .6 \text{ cfs} = 36 \text{ cfm} \quad \underline{\text{Answer (d)}}$$

36. A suppressed wier:

a. has 2 end contractions
b. has 1 end contraction
c. has no end contractions
d. has 3 contractions
e. leaks around the ends

A suppressed wier has its end contractions suppressed so there are no end contractions. <u>Answer (c)</u>

37. Given the venturi as shown. $\dfrac{P_a - P_b}{w}$ is

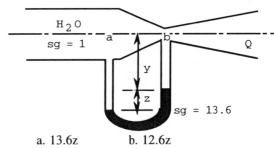

a. 13.6z b. 12.6z
c. 15.2 psi d. 18.6 psi
e. 14.7z

Let the distance from the top of z to the center line of the venturi be called y. Start at a and calculate around to b.

$$P_a + sg_{H2O}\, y + sg_{H2O}\, z +$$

$$sg_{fluid}\, y + sg_{fluid}\, z = P_b$$

$$P_a - P_b = (sg_{fluid} - sg_{H2O})\, z = 12.6z$$
  <u>Answer (b)</u>

38. A 2-in. diameter nozzle on a 4-in. diameter hose passes a stream of water. Then the nozzle is held horizontally the coordinates of a point on the stream are x = 100 ft and y = 50 ft. The nozzle coordinates are 0,0. What is the flow in the 8-in. diameter main supplying the nozzle.

a. 1.24 cfs b. 2.48 cfs
c. 4.96 cfs d. 7.44 cfs
e. 19.84 cfs

The flow rate in the main is that same as the flow rate at the 100,50 location of the nozzle.

$$y = \frac{1}{2} g t^2 : \quad t = \sqrt{\frac{2y}{g}}$$

$$x = Vt : \quad V = \frac{x}{t}$$

$$Q = AV = \left[\pi \frac{d^2}{4}\right]\left[\frac{x}{\sqrt{\frac{2y}{g}}}\right] =$$

$$\left[\pi \frac{2^2}{4}\right]\left[\frac{100}{\sqrt{\frac{2(50)}{32.2}}}\right] = 1.24 \text{ cfs}$$

<u>Answer (a)</u>

39. A fluid having a kinematic viscosity of 1 ft²/sec in the English system has a viscosity of what in the metric system?

    a. 687.9 poise   b. 929.03 stokes
    c. 11726 dynes   d. 14,849 sq units
    e. 15.298 dyne-cm/sec²

$$\left(1 \frac{ft^2}{sec}\right)\left(144 \frac{in^2}{ft^2}\right)\left(\frac{2.54 \text{ cm}^2}{in^2}\right)$$

= 929.03 Stokes   <u>Answer (b)</u>

40. Water is flowing at 10 fps and is 6 ft deep in an open channel. The water passes to a second channel where the velocity doubles and the depth of flow halves. Assuming ideal conditions, what is the difference in elevation of the channel floors?

    a. 2.84 ft   b. 1.66 ft
    c. 1.55 ft   d. 4.45 ft
    e. 6.21 ft

This is a Bernoulli problem disguised as an open channel flow problem. Let the first channel be Subscript 1 and the second 2. Let the floor of the lower channel be the datum and h be the difference in elevation between the channels. Taking points on the fluid surface:

$$\frac{P_1}{W} + Z_1 + \frac{V_1^2}{2g} = \frac{P_2}{W} + Z_2 + \frac{V_2^2}{2g}$$

Because our points were taken at the surface, the P/W terms equal and cancel.

$$(h + 6) + \frac{10^2}{2g} = 3 + \frac{20^2}{2g} = 1.66 \text{ ft}$$

<u>Answer (b)</u>

Problem 41 - 42

A rectangular wier with a crest of 10 ft is installed in a 10 ft wide channel. The lower edge of the wier is located 2 ft above the channel floor. The flow is 3,000,000 cu ft/day.

41. What is the head on the wier?

    a. 1 ft   b. 2 ft
    c. 3 ft   d. 6 in.
    e. 1.5 ft

Because the wier is 10 ft wide in a 10 ft wide channel, it is suppressed.

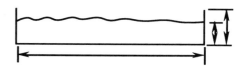

Q = 3 x 10⁶ cfd = 34.71 cfs

For $V_0 = 0$:

$$Q = 3.33 \text{ bH}^{\frac{3}{2}} = 34.71: \quad H = 1.03 \text{ ft}$$

This height is for V = 0, so we must check to see if there is a velocity contribution to the head.

$$h_v = \frac{V_0^2}{2g} = \frac{\left(\frac{Q}{A}\right)^2}{2g} = \frac{\left(\frac{34.71}{10(2+1.03)}\right)^2}{2g} = .02 \text{ ft}$$

The V term is small and 1.03 is ok.

<u>Answer (a)</u>

42. What is the velocity of the approach?

    a. 17.36 fps   b. 33.7 fps
    c. 1.14 fps   d. 22.86 fps
    e. .02 fps

$$V_0 = \left(\frac{Q}{A}\right) = \frac{34.71}{10(2+1.03)} = 1.14 \text{ fps}$$

<u>Answer (c)</u>

Problems 43 - 44

An oil spill containment scheme for protecting beaches is attached to a sea wall as shown. The oil has a sg of 0.83, is 5 ft deep and floats on top of the sea water.

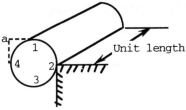

43. What is the weight of the cylinder per foot of length?

    a. 685.2 lbs   b. 758.7 lbs
    c. 739.7 lbs   d. 695.3 lbs
    e. 712.9 lbs

Force on plane 2 - 4:

$$\text{Volume 2-3-4} = \frac{1}{2} \pi r^2 \times 1$$

Wt of fluid displaced:

$$\frac{1}{2} \pi r^2 \rho$$

Head on Plane 2-4 is r and p = rρ

Force on plane 2-4 = PA
    rρ (2r x 1) = 2r²ρ

Total upward force:

$$\frac{1}{2} \pi r^2 \rho + 2r^2\rho = (\frac{1}{2} \pi r^2 + 2r^2)\,\rho$$

Unbalanced force on surface 1 - 4:
    Area 1 - a - 4
    = Area (1-a-4-0) - area (1-0-4)
    =   r² $\frac{1}{4}\pi r^2$

Weight of fluid on 1-a-4: (r² + $\frac{1}{4}\pi r^2$)ρ

Total downward force: -(r² + $\frac{1}{4}\pi r^2$)ρ

Sum of F_v = ($\frac{1}{2}\pi r^2 + 2r^2 - r^2 + \frac{1}{4}\pi r^2$)ρ
    = (r² + 3/4pr²)ρ

For r = 2 and r = (.83)(62.4):
Sum of F_v = 695 lb/ft   **Answer (d)**

44. What is the force against the sea wall?
    a. 117.8 lbs   b. 114.6 lbs
    c. 110.3 lbs   d. 106.2 lbs
    e. 103.6 lbs

Sum of F_h =
    $F_{h(1-4)} + F_{h(4-3)} + F_{h(3-2)} - F_{wall}$

But: $F_{h(4-3)} = - F_{h(3-2)}$

$F_{wall} = F_{h(1-4)} = rh_0 A$

$h_0 = r/2$ and $A = r$

$F_{wall} = (.83)(62.4)(2.2)(2) = 103.58$

                              Answer (e)

45. Q = 664 lbs/min: V = 40 ft/sec: The momentum force at a 90° bend is?

    a. 1884 lbs   b. 18.84 lbs
    c. 13.33 lbs   d. 11.31 lbs
    e. 1131 lbs

$$F = m\Delta V$$

$$m = \frac{644}{32.2 \times 60} = .333 \,\frac{\text{ft lb}}{\text{sec}^2}$$

$$\Delta V \, m = \frac{40}{\sin 45°} = .56.67 \text{ fps}$$

$$F = .333 \times 56.57 = 18.84$$

                              Answer (b)

ELECTRICITY/ELECTRONICS

Problems 1 - 2

A relay coil has a resistance of 100 Ω and operates from a 12 volt line. It is to be placed in a 120 volt system.

1. Determine the total circuit resistance required to run the 12 V system from the 120 V line.

    a. 100Ω   b. 1000Ω
    c. 900Ω   d. 925Ω
    e. 850Ω

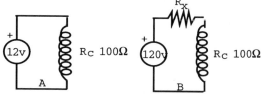

Circuit A is the original condition of 12 V, and circuit B is the new condition of 120 V.

From A:
$$I = \frac{V}{R} = \frac{12}{100} = 0.12 A$$

From B:
$$v = IR_x + IR_c$$
$$120 = .12 (R_x + 100)$$
$$R_x = 900 \Omega$$

Total circuit resistance
= 900 + 100 = 1000Ω

Answer (b)

2. The power used from the 120 V line is?

    a. 14.4 watts   b. 144 watts
    c. 28.8 watts   d. 1.44 watts
    e. 116 watts

$$P = VI = 120 (.12) = 14.4 \text{ VA}$$

Answer (a)

3. A series coil of 200 mH is connected to a parallel network made up of a 100 mH coil and a 1 μF capacitor. Determine the resonant frequency of the series-parallel circuit.

    a. 380 Hz   b. 490 Hz
    c. 616 Hz   d. 588 Hz
    e. 640 Hz

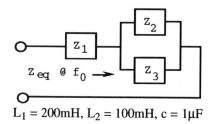

$L_1 = 200 mH, L_2 = 100 mH, c = 1 \mu F$

Using the equations:
$$W_0 = 2\pi f_0 \cdot : X_1 = W_0 L_1 : X_2 = W_0 L_2 : X_3 = \frac{1}{W_0 C}$$

$$Z_{eq} = Z_1 + \frac{Z_2 Z_3}{Z_2 + Z_3} = jX_1 + \frac{(jX_2)(-jX_3)}{j(X_2 - X_3)}$$

Substituting:
$$Z_{eq} = j\left[W_0 L_1 - \frac{\frac{L_2}{C}}{W_0 L_2 - \frac{1}{W_0 C}}\right] = 0$$

At f_0, the j term = 0 for resonance
$$(W_0 L_1)\left(W_0 L_2 - \frac{1}{W_0 C}\right) = \frac{L_2}{C}$$

$$W_0 = \sqrt{\frac{L_1 + L_2}{C(L_1 L_2)}}$$

$$= \sqrt{\frac{(200 + 100)10^{-6}}{10^{-6}[(200)(100)]10^{-6}}} = 3873$$

$$f_0 = W_0/2\pi = 616 \text{ Hz}$$

Answer (c)

4. Determine the current across coil L_1 at 1 KHz.
$R_1 = R_2 = R_3 = 100\Omega$,
$C_2 = C_3 = 1.59\mu F, L_1 = 15.9 mH$

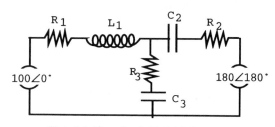

    a. .52∠-26.6°   b. 1.48∠46.6
    c. .52∠-45°   d. .95∠-18.4°
    e. .26∠45°

Identify Loops:

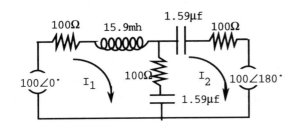

$100\angle 0° = (I_1\ 200) - (I_2\ 141\angle -45°)$

$-100\angle 180° = (I_1\ 141\angle -45°)$
$\qquad\qquad\qquad + (I_2\ 283\angle -45°)$

$$I_1 = \frac{N_1}{\Delta} = \frac{\begin{vmatrix} 100\angle 0° & -141\angle -45° \\ -100\angle 180° & 283\angle -45° \end{vmatrix}}{\begin{vmatrix} 200 & -141\angle -45° \\ -141\angle -45° & 283\angle -45° \end{vmatrix}}$$

$$= \frac{(100\angle 0°)(283\angle -45°) + (100\angle 0°)(141\angle -45°)}{(200\angle 0°)(283\angle -45°) - (141\angle -45°)(141\angle -45°)}$$

$$= \frac{(3-j3)10^4}{(4-j2)10^4} = \frac{4243\angle -45°}{4472\angle -26.6°} = .95\angle -18.4°$$

Answer (d)

5. Determine R_{A-B} if each resistor is 15Ω.

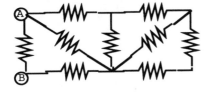

a. 7.5Ω   b. 15Ω
c. 24.6Ω   d. 18.7Ω
e. 9.28Ω

Convert Y(1,2,3) to D. D = 3Y = 45Ω

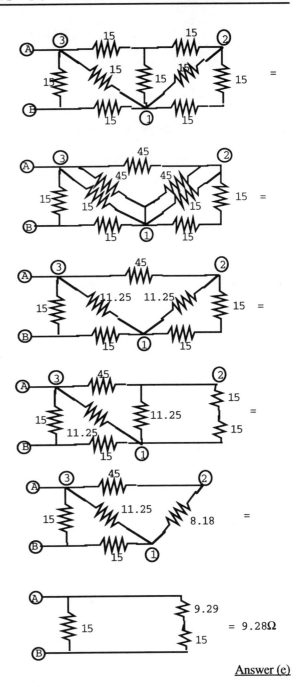

= 9.28Ω

Answer (e)

6. A Stockroom has several capacitors each rated at 10μF, 150 V. It is desired to form a capacitor from these units which will be 10μF at 300 V. What arrangement of capacitors is needed?

    a. Two 10μF in parallel, in series with two 10μF in parallel

    b. Two 10μF in series

    c. Two 10μF in parallel

d. Two 10μF in series, in parallel with two 10μF in series

e. One 10μF in series with two 10μF in parallel

The equivalent circuit that meets the problem spec is shown below:

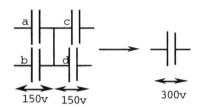

Each set of parallel capacitors (a,b) and (c,d) has:

$$C_{eq} = 10 + 10 = 20$$

These taken is series with each other will meet the voltage requirement and:

$$\frac{1}{C_{eq}} = \frac{1}{20} + \frac{1}{20}: \quad C_{eq} = 10\mu f$$

Which satisfies the capacitance requirement.

<u>Answer (a)</u>

7. An R-C high pass filter is designed so that the 3db frequency is 20kHz. What is the ratio of filter output to input amplitude at 30kHz?

    a. .707   b. .92
    c. 1.00   d. .85
    e. .833

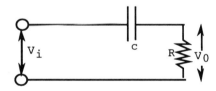

$$\frac{V_0}{V_i} = \frac{R}{r - jX_c} \quad \text{and} \quad \left|\frac{V_0}{V_i}\right| = \left|\frac{R}{Z}\right| \text{ Magnitude}$$

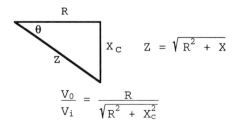

$$Z = \sqrt{R^2 + X}$$

$$\frac{V_0}{V_i} = \frac{R}{\sqrt{R^2 + X_c^2}}$$

$f_c = 20\text{kHz}$: Octave $f = 40\text{kHz}$
$\pm 1/4$ Octave $= \pm 10$ kHz

f	x	$\sqrt{R^2 + X^2}$	$\frac{V_0}{V_c}$
10 Khz	$2X_c$	$\sqrt{R^2 + 4R^2}$	.45
20 Khz	X_c	$\sqrt{R^2 + R^2}$	.707
30 Khz	$\frac{2X_c}{3}$	$\sqrt{R^2 + .44R^2}$	.833

<u>Answer (e)</u>

Problems 8 and 9

Given the following circuit:

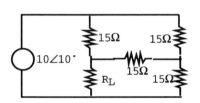

8. What value of R_L will provide maximum power transfer?

    a. 5Ω   b. 6.67Ω
    c. 9Ω   d. 2.5Ω
    e. 18Ω

Convert A,C,D (D) to a Wye:

$$R_y = \frac{R_\Delta}{3} = 5\Omega$$

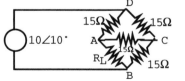

At A-B find V_{th} and Z_{th}

Temporarily short D-B:

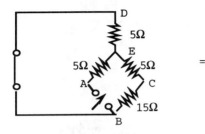

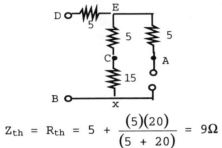

$$Z_{th} = R_{th} = 5 + \frac{(5)(20)}{(5+20)} = 9\Omega$$

Remove the short circuit and insert the voltage:

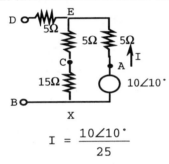

$$I = \frac{10\angle 10°}{25}$$

$$V_{EB} = V_{th} = I(20) = \left(\frac{10\angle 10°}{25}\right)(20) = 8\angle 10°\, v$$

Equivalent:

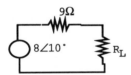

$R_L = 9\Omega$ for max power output

Answer (c)

9. What is the maximum power delivered to R_L?

a. 2.8 watts   b. 1.8 watts
c. 0.14 watts d. 1.0 watt
e. 6.67 watts

$$P_{max} = \frac{V_{th}^2}{4R} = \frac{8^2}{36} = 1.8 \text{ watts}$$

Answer (b)

10. The resistance of a coil of wire at 20°C was found to be 48.656Ω. The temperature was raised to 50°C the resistance was 49.875Ω. What is the temperature coefficient of the wire?

a. 0.000835   b. 0.000746
c. 0.000498   d. 0.000816
e. 0.000749

$$R_1 = R_1(1 + \alpha\Delta T)$$

$$R_2 = R_1(1 + \alpha(T_2 - T_1))$$

$$\alpha = \frac{R_2 - R_1}{R_1(T_2 - T_1)}$$

$$= \frac{49.875 - 48.656}{48.656(50-20)} = .000835$$

Answer (a)

11. A coil has a resistance of 50Ω and an inductance of 10mH. What time domain voltage must be applied to produce an RMS current of 20 milliamperes at a frequency of 400Hz?

a. 1.41 sin wt
b. 1.58 sin (2512t + 26.6°)
c. 1.12 sin (2512t + 26.6°)
d. 1.58 t
e. 1.24 sin (2498t + 24.3°)

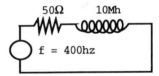

$$w = 2\pi f = 2\pi(400) = 2512$$

$$X_L = 2\pi f L = 2\pi(400)(10)(10^{-3}) = 25.12$$

$$V = IZ = (20 \times 10^{-3})(50 + j25)$$

$$= (20 \times 10^{-3})(56\angle 26.6°) = 1.12\angle 26.6°$$

$$V = 1.41(1.12\angle 26.6°) = 1.58 \text{ Peak}$$

$$V_t = 1.58 \sin(2512t + 26.6°)$$

Answer (b)

Problems 12 - 14

A 60hz, 480 V power line is terminated with a 25Ω load. The power line is equivalent to a series R-L circuit with R = 5Ω and L = 50mH.

12. What is the current at the receiving end of the transmission line?

    a. 8.7 amperes   b. 96 amperes
    c. 13.56 amperes   d. 11.32 amperes
    e. 9.6 amperes

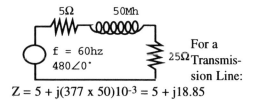

For a Transmission Line:

$Z = 5 + j(377 \times 50)10^{-3} = 5 + j18.85$

$$I = \frac{V}{Z_T} = \frac{480\angle 0°}{30 + j\,18.85} =$$

$$\frac{480\angle 0°}{35.4\angle 32°} = 13.56\angle -32°$$

Answer (c)

13. What is the voltage regulation of the line?

    a. 60%   b. 100%
    c. 56%   d. 84%
    e. 42%

$V_L = IR_L = (13.56\angle -32°)25 = 339\angle -32°$

Regulation $= \frac{480 - 339}{339} = .42$

Answer (e)

14. What is the efficiency of transmission?

    a. 83%   b. 86%
    c. 100%   d. 79%
    e. 92%

Power Delivered:
    $P_d = I^2R = 13.56^2(25) = 4597w$

Power for line loss:
    $P_{ll} = I^2R = 13.56^2(5) = 919w$

Total power $4597 + 919 = 5516w$

$\eta = \frac{4597}{5516} = .834$

Answer (a)

15. Determine the impedence at terminals A-B when the frequency is 100kHz?

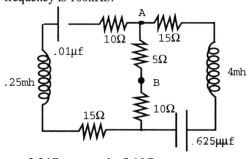

    a. 3.34Ω   b. 5.18Ω
    c. 2.34Ω   d. 4Ω
    e. 4.46Ω

Redraw the circuit:

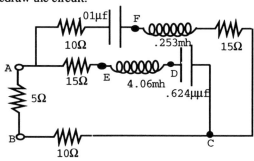

$Z_{AFGC} = 25 + j(X_L - X_C) = 25$

$X_L = (6.28)(10^5)(.253)(10^{-3}) = 159$

$X_C = \dfrac{1}{(6.28)(10^5)(.01)(10^{-6})} = 159$

$X_L - X_C @ \,0$ at 10^5 hz

$Z_{AEDC} = 15 + j(X_L - X_C) = 15$

$X_L = (6.28)(10^5)(4.06)(10^{-3}) = 2550$

$X_C = \dfrac{1}{(6.28)(10^5)(624)(10^{-12})} = 2550$

$X_L - X_C @ \,0$ at 10^5 hz

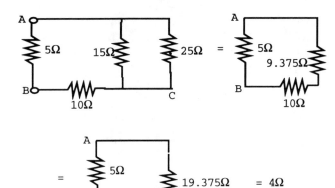

= 4Ω

Answer (d)

Problems 16 - 18

16. It is required to supply a 14a load from a 24 V battery bank. Several 20 amp hr batteries rated at 3.5 amperes, 12 V, 0.2Ω internal resistance are available. How should they be configured?

    a. Two Parallel plus one series
    b. Four parallel sets, each set two in series
    c. Two in series
    d. Four series sets, each set two in parallel
    e. Two in parallel

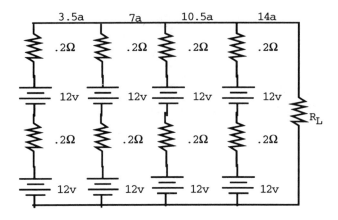

Answer (b)

17. What is the load resistance?

    a. .4Ω   b. 1.6Ω
    c. 2.85Ω   d. 3.25Ω
    e. .2Ω

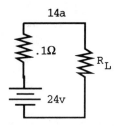

$24 = 14(.1 + R_L): R_L = 1.6Ω$

Answer (b)

18. How long will the battery operate before a decrease in voltage is apparent?

    a. 1.6 hr   b. 1.4 hr
    c. 14 hr   d. 5.7 hr
    e. 3.5 hr

For $R_L = 1.6Ω$, each battery will operate:

$$\frac{20 \text{ ampere-hr}}{3.5 \text{ ampere}} = 5.7 \text{ hr}$$

Answer (d)

19. A composite conductor has a steel core and an aluminum shell. It's outside diameter is 200 mils. The resistivity of the steel and aluminum are 80 and 24Ω circular mill per ft respectively. The ratio of steel to aluminum by volume is 2:3. What is the resistance of 1000' of the conductor?

    a. 6Ω   b. 1Ω
    c. 1.2Ω   d. .833Ω
    e. 11Ω

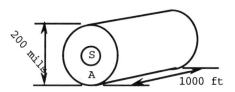

CM area = $D^2 = 200^2 = 4 \times 10^4$ cm^2
$CM_s + CM_a = 4 \times 10^4$ cm^2

$\frac{CM_s}{CM_a} = \frac{2}{3}: CM_s = \frac{2}{3} CM_a$

$\frac{2}{3} CM_a + CM_a = 4 \times 10^4$

$CM_a = 2.4 \times 10^4$ cm^2
$CM_s = 1.6 \times 10^4$ cm^2

$$R_a = \frac{\rho L}{a} = \frac{(24)(1000)}{2.4 \times 10^4} = 1\Omega$$

$$R_s = \frac{\rho L}{a} = \frac{(80)(1000)}{1.6 \times 10^4} = 5\Omega$$

$$R_{eq} = \frac{(5)(1)}{5+1} = .833\Omega$$

Answer (d)

Problems 20 - 21

A 1Ω load is fed by two batteries connected in parallel. One battery is 6.3 V with an internal resistance of .004Ω and the other is 5.4 V with an internal resistance of .008Ω.

20. What is the current flowing in the 5.4 V battery?

a. 6.0 amperes   b. 73 amperes
c. 69 amperes   d. 3.66 amperes
e. 3.23 amperes

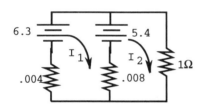

Loop 1
$$6.3 - 5.4 = .012 I_1 - .008 I_2$$
Loop 2
$$5.4 = -.008 I_1 + 1.008 I_2$$

Solve simultaneously for I_1 and I_2

$$I_1 = 73 \text{ amperes}$$
$$I_2 = 5.97 \text{ amperes}$$

Answer (b)

21. What is the current flowing in the 1Ω load?

a. 6.0 amperes   b. 73 amperes
c. 69 amperes   d. 3.66 amperes
e. 3.23 amperes

From 20
$$I_2 = 5.97a$$

Answer (a)

22. A 240 volt, 20KW single phase transformer has two secondary 120 V windings A and B. Winding A has a load which has a power factor of .6 lagging and B has a load with a power factor of .8 leading. What load impedance is seen from the 240 V winding if A and B are each 10KW?

a. 6.2   b. 10.16
c. 12   d. 16
e. 5.8

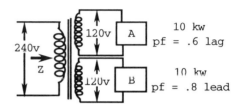

For A and B:
$$R = \frac{E^2}{P} = \frac{120^2}{10 \times 10^3} = 1.44\Omega$$

For A:
$$q = \cos^{-1} .6 = 53° \text{ lag}$$

$$X_c = 1.44 \tan 53° = 1.92\Omega$$

$$Z_A = \sqrt{1.44^2 + 1.92^2} = 2.4\angle{-53°}$$

For B:
$$q = \cos^{-1} .8 = 36.9° \text{ lead}$$

$$Z_B = 1.8\angle{36.8°}$$

Calculate Z:
$$a = \frac{N_1}{N_2} = \frac{E_1}{E_2} = \frac{240}{120} = 2$$

$$Z = a^2 Z_{eq}$$

$$\frac{1}{Z_{eq}} = \frac{1}{2.4\angle{-53°}} + \frac{1}{1.8\angle{36.9°}}$$

$$= .4167\angle{53°} + .556\angle{-36.9°}$$

$$= .25 + j33 + .44 - j.33 = .69$$

$$Z_{eq} = 1.45$$

$$Z = (4)(1.45) = 5.8\Omega$$

Answer (e)

23. A three phase three wire A-B-C system supplies 440 V to an equal impedance delta load of 10Ω. Determine the magnitude of the line current I_B in amps.

    a. 14.6 amperes   b. 76 amperes
    c. 89 amperes   d. 26 amperes
    e. 44 amperes

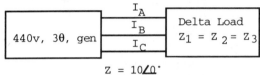

$$Z = 10\angle 0°$$

$$|I_{AB}| = |I_{BC}| = |I_{CA}| = |I_{Phase}| = \frac{|440|}{10} = 44a$$

In Delta, $I_{line} = (\sqrt{3})I_{phase} = 76.12$ amperes

$$|I_{line}| = |I_A| = |I_B| = |I_C| = |I_{Phase}| = 76.12a$$

                                 Answer (b)

24. What is the resistance in ohms looking into a 120/440 V single phase transformer connected to a load of $(30 + j30)\Omega$?

    a. 4.44Ω   b. 2.22Ω
    c. 42.2Ω   d. 11.56Ω
    e. 3.15Ω

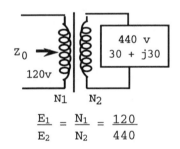

$$\frac{E_1}{E_2} = \frac{N_1}{N_2} = \frac{120}{440}$$

$$Z_0 = \left[\frac{N_1}{N_2}\right]^2 Z_L = \left(\frac{120}{440}\right)(30 + j30) =$$

$$2.22 + j2.22 \quad \text{Answer (b)}$$

25. A 20,000 ohm/volt meter is designed to measure 150 V full scale. The meter is to be redesigned to measure 100 V full scale. The change should be:

    a. Soak in water for 1 hr

    b. Place 3 megohms in parallel with 3 megohms of meter

    c. Place 1.5 megohms in parallel with 1.5 megohms of meter

    d. Place 6 megohms in parallel with 3 megohms of meter

    e. Place 3 megohms in series with 3 megohms of meter

A 20,000 ohm/volt meter at 150 V full scale needs $(150)(20,000) = 3M\Omega$.

A 20,000 ohm/volt meter at 100 V full scale needs $(100)(20,000) = 2M\Omega$

Place a resistor in parallel with the 3MΩ reistor making a 2MΩ equivalent by:

$$2M\Omega = \frac{(R_x)(3M\Omega)}{(R_x) + (3M\Omega)}: \quad R_x = 6M\Omega$$

                                   Answer (d)

26. Mechanical power developed by an armature and mechanical power delivered by a motor are:

    a. Identical
    b. Different by the stray power loss
    c. Different by armature copper loss
    d. Related by the motor slip
    e. Dependent on starting torque

Power input to an armature is equal to mechanical power by the relationship:

$$I_A^2 R_A$$

Some of the power is used to overcome stray power loss, while the bulk goes to the load.   Answer (b)

Problems 27 - 29

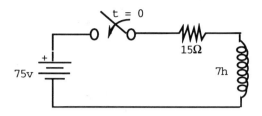

When the switch is closed at t = 0:

27. The voltage across the inductor is:

a. 0 V b. 25 V
c. 50 V d. 75 V
e. 80 V

At t = 0, current is 0, thus there is no voltage drop across the 15Ω resistor and the voltage across the inductor is 75 V.

Answer (d)

28. The charging current has a time constant of:

a. 105 msec b. .333 sec
c. .467 sec d. 105 sec
e. .750 sec

The time constant $\tau = \dfrac{L}{R} = \dfrac{7}{15} = .467$ sec

Answer (c)

29. The final energy in the inductor is:

a. 175 Joules b. 75 watts
c. 87.5 Joules d. 87.5 watts
e. 75 Joules

$$W = \frac{1}{2} LI^2 = (7)\left(\frac{75}{15}\right)^2 = 87.5 \text{ Joules}$$

Answer (c)

Problems 30 - 32

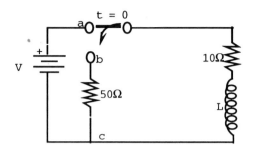

The switch has been at a for a very long time. At t = 0, it is moved to b, when a current of 8 amperes is flowing in the circuit. The voltage 2 sec later is 25 V across b-c.

30. The inductance of the coil is:

a. 33H b. 38H
c. 43H d. 45H
e. 50H

Steady state has been reached (very long time).

$$i = I_0 e^{-(t/\tau)} = I_0 e^{-t(R/L)}$$

$$V_{b-c} = iR = i \cdot 50 = 25: \; i = .5$$

$$.5 = 8 \, e^{-2\,(60/L)}$$

$$\ln 0.625 = \ln \left(e^{-2\,(60/L)} \right)$$

$$-2 \left(\frac{60}{L}\right) = \ln 0.625 = -2.77$$

$$L = \frac{120}{2.77} = 43.3 \text{ H}$$

Answer (c)

31. The maximum voltage across b-c will be?

a. 80 V b. 480 V
c. 400 V d. 444 V
e. V_{volts} DC

$V_{b-c} = (8)(50) = 400$ V   Answer (c)

32. The decaying time constant of the circuit is:

a. 0.55 sec b. 0.76 sec
c. 0.75 sec d. 0.72 sec
e. 0.80 sec

$\tau = \dfrac{L}{R} = \dfrac{43}{60} = .72$ sec   Answer (d)

Problems 33 - 35

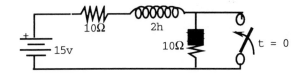

In the circuit shown, the switch has been open for a long time. At t = 0, it is closed.

33. The voltage drop across the inductor is:

   a. 15 V   b. 7.5 V
   c. 1.5 V   d. 0 V
   e. 10 V

At t<0, the inductor is a short circuit and there are zero volts across it. At t = 0, the current does not change, 0.75 amp still flows. By Kirchoff's Voltage Law:

$$15 = (0.75)(10) + V_L$$

$$V_L = 7.5 \text{ V} \qquad \text{Answer (b)}$$

34. The current through the inductor at t = 0 is:

   a. 0.75 amps   b. 0 amps
   c. 1.5 amps   d. 15 amps
   e. 1.0 amp

$$i = I_0 = .75 \text{amps} \qquad \text{Answer (a)}$$

35. The current at t = 0.1 after the switch is closed is:

   a. 1.04 amp   b. .75 amp
   c. 1.25 amp   d. 1.5 amp
   e. 1.75 amp

$$i_{(t)} = i_n + i_{ss}$$

$i_n = K\,e^{-t/\tau}$ and $\tau = \dfrac{L}{R} = \dfrac{2}{10} = .2$ sec

$i_{ss} = \dfrac{15}{10} = 1.5$ amperes

$$i(t) = K\,e^{-5t} + 1.5$$

At t = 0, i = .75

$$.75 = K + 1.5 : K = -.75$$

$$i(t) = -.75\,e^{-5t} + 1.5$$

At t = 0.1 sec:

$$i(t) = -.75\,e^{-.5} + 1.5$$

$$i(t) = 1.04 \text{ amperes} \qquad \text{Answer (a)}$$

Problems 36 - 38

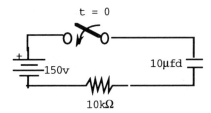

The switch in the circuit above is closed at t = 0 and opened at t = 0.03 sec.

36. The current flowing in the circuit at t = 0 is:

   a. 0 milliamperes   b. 10 milliamperes
   c. 20 milliamperes   d. 1 milliamperes
   e. 5 milliamperes

At t = 0, Vc = 0 (initially uncharged)

$$I = \dfrac{150}{10,000} = 15 \text{ma} \qquad \text{Answer (d)}$$

37. The voltage across the capacitor when the switch is opened is:

   a. 30 V   b. 36 V
   c. 39 V   d. 42 V
   e. 46 V

$v(t) = K\,e^{-t/\tau} + V_{ss}$

$t = RC = (10)(10^3)(10)(10^{-6}) = 100 \times 10^{-3}$

$V_{ss} = 150$ V

At t = 0, v(t) = 0 (uncharged)

$0 = K + 150: k = -150$

$v(t) = 150(1 - e^{-10t})$

At $t = 0.03$ sec:

$v(t) = 150(1 - e^{-10(.03)})$

$v(t) = 39$ V <u>Answer (c)</u>

38. The energy stored in the capacitor during this period is:

a. 4.5 Joules
b. 5×10^{-3} Joules
c. 7.6×10^{-3} Joules
d. 10^{-3} Joules
e. 6.6×10^{-3} Joules

$W = \frac{1}{2} CV^2 = \frac{1}{2}(10 \times 10^{-6})\, 39^2$

$= 7.6 \times 10^{-3}$ Joules <u>Answer (c)</u>

Problem 39 - 41

The rise time of an RC circuit to a step input is defined as the difference in time between the 90% and 10% points of the charging curve.

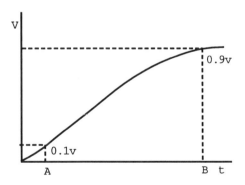

39. In terms of t (time constant) point B is:

a. $t = 3t$ b. $t = 2.5t$
c. $t = 2.3t$ d. $t = t$
e. $t = 2t$

The expression for voltage rise is:

$v(t) = (1 - e^{-t/\tau})$

At the 90% (B) point:

$e^{-t/\tau} = .1$ and $t = 2.3\tau$ <u>Answer (c)</u>

40. In terms of t (time constant) point A is:

a. $t = .1t$ b. $t = .15t$
c. $t = .2t$ d. $t = .22t$
e. $t = .25t$

At the 10% (A) point:

$e^{-t/\tau} = .9$ and $t = .1\tau$ <u>Answer (a)</u>

41. In terms of t (time constant) the rise time is:

a. $t = 2.9t$ b. $t = 2.4t$
c. $t = 1.3t$ d. $t = t$
e. $t = 2.2t$

Rise time $= B - A = 2.3t - 0.1t = 2.2t$ <u>Answer (e)</u>

42. A 100 V step is applied to an RC circuit having $R = 5\Omega$, $C = 20\,\mu Fd$, and $t = 0$. The expected rise time is:

a. .22 msec   b. .3 msec
c. .25 msec   d. .23 msec
e. .15 msec

$t_r = 2.2\, RC$
$= (2.2)(5)(20)(10^{-6}) = .22$ msec

<u>Answer (a)</u>

43. Determine the voltage transfer function ratio of the following circuit:

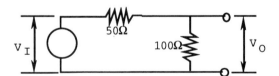

a. 3.0   b. 0.4
c. .667   d. .75
e. 1.5

Using Kirchoff's voltage law:

$V_i(t) = V_{50\Omega}(t) + V_{100\Omega}(t)$

$V_i(t) = 50\, i(t) + 100\, i(t) = 150\, i(t)$

$$i(t) = \frac{V_i(t)}{150}$$

$$V_0(t) = 100\, i(t) = 100\, \frac{V_i(t)}{150}$$

$$= .667\, V_i(t)$$

$$\frac{V_0(t)}{V_i(t)} = .667 \qquad \text{Answer (c)}$$

Problem 44 - 47

Two p-n semiconductor diodes are connected as shown below. For each diode at room temperature, $V_Z = 6.75$ V and $I_S = 1.5 \times 10^{-13}$ amps.

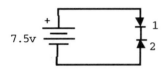

44. The state of each semiconductor diode is:

   a. Diodes 1 and 2 are forward biased

   b. Diodes 1 and 2 are reverse biased

   c. Diodes 1 is forward biased and diode 2 reverse biased.

   d. Diodes 1 is reverse biased and diode 2 forward biased.

   e. Both are reverse biased

The arrow head of diode 1 points in the direction of conventional current flow, therefore diode 1 is forward biased and diode 2 is reverse biased.

Answer (c)

45. The voltage across diode 1 is:

   a. 7.5 V   b. 0.75 V
   c. 3 V   d. 2.5 V
   e. 0.3 V

Since V (7.5) is greater than V_Z (6.75) the voltage across diode 1 is:

   7.5 - 6.75 = 0.75 V   Answer (b)

46. The voltage across diode 2 is:

   a. 7.5 V   b. 0.75 V
   c. 3 V   d. 2.5 V
   e. None of the above.

Answer (e)

47. The current through the series circuit is:

   a. 1.0 amperes   b. 1.5 amperes
   c. 1.6 amperes   d. 1.7 amperes
   e. 2.0 amperes

$$I = I_S (e^{40\,V} - 1) \text{ for diode 1}$$

$$I = 1.5 \times 10^{-13} (e^{(40)(.75)} - 1) = 1.6 \text{ amperes}$$

Answer (c)

Problems 48 - 51

A silicon transistor is used as shown:

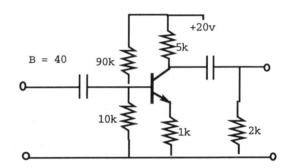

48. The quiescent value of base current I_B is:

   a. .050 milliamperes   b. .030 milliamperes
   c. .026 milliamperes   d. .025 milliamperes
   e. .020 milliamperes

$$R_B = \frac{(90)(10)}{100} = 9k$$

$$V_{BB} = \frac{(20)}{100}(10) = 2v$$

KVL: Input Loop

$$2 = 9I_B + 0.7 + I_E$$

$$1.3 = 9I_B + I_C + I_B = 10 I_B + BI_B$$

$$1.3 = 10 I_B + 40 I_B$$

$$I_B = \frac{1.3}{50} = 0.026 \text{ milliamperes}$$

<u>Answer (c)</u>

49. The quiescent value of collector current I_C is:

a. 2.0 milliamperes   b. 1.2 milliamperes
c. 1.04 milliamperes   d. 1.5 milliamperes
e. 1.5ma

$$I_C = (40)(.026) = 1.04 \text{ milliamperes}$$

<u>Answer (c)</u>

50. The quiescent value of emmiter current I_E is:

a. 2.05 milliamperes   b. 1.23 milliamperes
c. 1.066 milliamperes   d. 1.025 milliamperes
e. 1.055 milliamperes

$$I_E = 1.04 + .026 = 1.066 \text{ milliamperes}$$

<u>Answer (c)</u>

51. The quiescent value V_{CE} is:

a. 12 V   b. 13.7 V
c. 14 V   d. 15 V
e. 15.3 V

$$V_{CC} = I_C R_C + V_{CE} + I_E R_E$$

$$20 = (1.04)(5) + V_{CE} + (1.066)(1)$$

$$V_{CE} = 13.7 \text{ V}$$

<u>Answer (b)</u>

CHEMISTRY

1. The reaction below is:

$$3Mg + Cr_2(SO_4)_3 \rightarrow 3MgSO_4 + 2Cr$$

a. Synthesis
b. Analysis
c. Single Replacement
d. Double Replacement
e. Decomposition

Because the single radical SO_4 is exchanged, the reaction is single
replacement   <u>Answer (c)</u>

2. Calculate the pH of the resulting solution when 200 ml of 0.1M HCl and 300 ml of 0.05M Mg(OH)2 are reacted:

$$Mg(OH)_2 + 2HCl \rightarrow MgCl + H_2O$$

a. 7   b. 1
c. 13.3   d. 12.3
e. 10.7

There are 0.2l x 0.1M = 0.02 moles HCl and 0.3l x 0.05M = 0.15 moles Mg(OH)2 available for the reaction. By the reaction 2 moles HCl will react with 1 mole Mg(OH)2 thus 0.02 moles will react with only 0.01 moles Mg(OH)2 leaving .005 moles Mg(OH)2 excess and unreacted in a total volume of 0.5L of solution. This is 0.01M or 10^{-2}M.

$$\begin{aligned} pH &= 14.0 - pOH \\ &= 14 - \log 2 \times 10^{-2} \\ &= 12.3 \end{aligned}$$ <u>Answer (d)</u>

Problems 3 - 6

For a solution of $AlCl_3$ with a density of 1.1 g/L, and 10% by weight:

3. The molarity is:
a. 8.3M   b. .83M
c. 1M   d. 1.2M
e. .75M

Molarity is moles per liter:

$$1.1 \text{ g/ml} = 1100 \text{ g/L at } 10\% = 110 \text{ g } AlCl_3$$

$$\frac{110 \text{ g AlCl}_3}{132 \frac{\text{g AlCl}_3}{\text{Mol}}} = .83 \frac{\text{mol}}{\text{liter}}$$

Answer (b)

4. The molality of the solution is:

a. 8.4m b. .94m
c. 1m   d. .75m
e. .84m

Molality is moles solute/1000g solvent
Ratio the given solution to 1000g solvent

$$\frac{110 \text{ g AlCl}_3}{990 \text{ g H}_2\text{O}} = \frac{X \text{ AlCl}_3}{1000 \text{ g H}_2\text{O}}$$

$X = 111\text{g AlCl}_3$

$$\frac{111 \text{ g AlCl}_3}{132 \frac{\text{g AlCl}_3}{\text{mole}}} = .84 \frac{\text{mole}}{1000 \text{ g solute}}$$

Answer (e)

5. The freezing point of the solution will be:

a. 6.25°C   b. -6.25°C
c. 1.56°C   d. -1.56°C
e. 0°C

The change in freezing point will have the starting point of 0°C (H$_2$O) and be changed by the presence of the AlCl$_3$.

$$\text{Molality} = \frac{\Delta T}{\text{Constant for solute}}$$

$.84 = \frac{\Delta T}{1.86}$ $\Delta T = .84 \times 1.86 = 1.56$

Since solutes depress freezing points the correct answer is -1.56°C

Answer (d)

6. The normality of the solution is:

a. .83N   b. 84N
c. 2.5N   d. 0.4N
e. 2.5N

Normality is the no. of equivalents per liter of solution. The no. of equivalents is the gram solute/equivalent weight. Equivalent weight is molecular weight/total positive charge.

$$N = \frac{110\text{g}}{11} \times \frac{3 \text{ (pos charge)}}{132 \text{ (mw)}} = 2.5$$

Answer (e)

Problems 7 - 11

For the reaction:

$$2SO_{2(g)} + O_{2(g)} \leftrightarrow 2SO_{3(g)} + 44 \text{ Kcal}$$

What will be the effect of the stress applied to the reaction after the new equilibrium is attaintd?

1. Increase SO$_2$ 2. Increase SO$_3$
3. Increase O$_2$ 4. Decrease SO$_2$
5. Decrease SO$_3$ 6. Decrease O$_2$

7. Add SO$_2$
a. 1 only   b. 2 only
c. 3 and 4 d. 4 and 5
e. 2 and 6

Increasing a reactant will consume more of the other reactant (6) and produce more product (2) Answer (e)

8. Add a catalyst:
a. 3 only   b. 1 and 3
c. 4 and 5 d. 2 only
e. Speed up the reaction

Adding a catalyst will not change the equilibrium conditions of the reaction but will increase its rate.

Answer (e)

9. Increase the pressure:
a. 2 only   b. 1 and 3
c. 2,4 and 6 d. 7 only
e. 4 and 5

Increasing the pressure in gas phase reactions will cause the reaction to move to a higher degree of completion.

Answer (c)

10. Increase the temperature:

a. 2 only b. 1, 3 and 5
c. 2,4 and 6 d. 7 only
e. 4 and 5

When temperature is increased the reaction will proceed to the endothermic side. As this reaction is exothermic (+44 Kcal), the endothermic side contains the reactants. Thus the reactants will increase and the products will decrease moving the equilibrium to the left. Answer (b)

11. Decrease the O_2:

a. 1 only b. 1 and 3
c. 1 and 5 d. 5 only
e. 4 and 5

Decreasing the O_2 will increase the SO_2 (1) as the degree of completion will be less, and will result in less SO_3 product (5). Answer (c)

12. Find the solubility of Pb^{2+} in moles/liter:

$$PbI_{2(s)} \rightarrow Pb^{2+} + 2I^{-1}$$

$$K_{sp} = [Pb^{+2}][I^{-1}]^2 = 1.4 \times 10^{-8}$$

a. 4.66×10^{-6} b. 1.52×10^{-3}
c. 1.18×10^{-4} d. 3.54×10^{-4}
e. 0.59×10^{-4}

If the concentration of Pb^{2+} is x, then the concentration of I^{-1} must be 2x.

$$K_{sp} = [x][2x]^2 = 1.4 \times 10^{-8}$$

$$4x^3 = 1.4 \times 10^{-8}$$

$$x = 1.53 \times 10^{-3} \quad \text{Answer (b)}$$

13. If 84g of $NaHCO_3$ are decomposed, what volume of CO_2 will be produced at 47°C and 570 torrs.

$$2NaHCO_3 \rightarrow Na_2CO_3 + H_2O + CO_2$$

a. 0.5l b. 11.2l
c. 17.5l d. 1.0l
e. 22.4l

84 g $NaHCO_3$ (mW 84) is 1 mole. 1 mole $NaHCO_3$ will produce 0.5 mole CO_2.

$$V = \frac{nRT}{P} = \frac{(0.5 \text{ mole})(.0821)(320°K)}{.75 \text{ atm}} = 17.5l$$

Answer (c)

14. How many calories are needed to change 10g of ice at 0°C to steam at 100°C?

a. 100 b. 720
c. 1000 d. 7200
e. 1800

Phase change solid to liquid:

$$10g \times 80 \text{ cal/g} = 800 \text{ cal}$$

Heat to 100°C, no phase change:

Phase change liquid to vapor:

$$10g \times 540 \text{ cal/g} = 5400 \text{ cal}$$

Answer (d)

15. Which is the general formula for a saturated hydrocarbon?

a. C_nH_{2n} b. C_nH_{2n+1}
c. C_nH_{2n+2} d. C_nH_{2n-2}
e. C_nH_n

A saturated hydrocarbon is one in which all the available bonds of all the available C atoms are filled. Only answer c satisfies this condition.

Answer (c)

16. The correct name for Hg_2Cl_2 is:

a. Mercuric chloride
b. Mercurous chloride
c. Mercury chloride
d. Hydrogen chloride
e. Sodium chloride

By Definition Answer (b)

17. The correct formula for chromium III carbonate is:

   a. Cr_3CO b. $Cr_2(CO)_3$
   c. Cr_2CO_3 d. Cr_3CO_3
   e. $Cr_2(CO_3)_2$

For the formula to balance both stoichometrically and by valance

Answer (b)

18. When the equation:

$$?Cr_2O_7 + ?H_2O + ?S \rightarrow xSO_2 + yOH^- + zCr_2O_3$$

is correctly balanced, the coefficients x, y, z, will be:

   a. 2,2,3   b. 3,4,2
   c. 2.1.3   d. 2.3.4
   e. 3,3,3

$$2 \times [2Cr^{6+} + 6e^-] \rightarrow 2Cr^{3+}$$
$$3 \times [S^0 - 4e^-] \rightarrow 3S^{4+}$$

$$4Cr^{6+} + 3S^0 \rightarrow 3S^{4+} + 4Cr^{3+}$$

$$2Cr_2O_7^{-2} + 2H_2O + 3S \rightarrow 3SO_2 + 4OH^- + 2Cr_2O_3$$

Answer (b)

19. How long must a plating cell run at a current of 1.0 ampere if it must deposit 16g of copper on a plate from a $CuSO_4$ solution?

   a. 96,500 sec   b. 24,125 sec
   c. 60 min   d. 13.4 hr
   e. 604 hr

$$Cu^{2+} + 2e^- \rightarrow Cu^0$$

Since 1 mole of Cu^{2+} will produce 0.5 moles e^-, and we have 16/64 = 0.25 moles Cu^{2+} we will expect to plate 0.5 moles e^-.

   96,500 Coulombs will plate 1 mole of e^-
   48,250 Coulombs will plate 0.5 mole e^-
   and since 1 Coulomb/sec = 1 ampere

$$\text{amperes} = \frac{48,250 \text{ sec}}{3600 \frac{\text{sec}}{\text{hr}}} = 13.4 \text{ hr}$$

Answer (d)

20. Which of the following would be most effective in neutralizing the formic acid produced by a bee sting?

   a. Vinegar   b. Aspirin
   c. Lemon juice   d. Ethyl alcohol
   e. Baking soda

An acid can be neutralized only by a base. Baking soda is the only base on the list.

Answer (e)

21. The element in period 4 with chemical properties most similar to carbon is:

   a. Silicon   b. Boron
   c. Nitrogen   d. Lead
   e. Germanium

From the periodic chart, germanium (Ge)

Answer (e)

22. How many molecules of O_2 are present in a mixture of gases in a one liter vessel if it is at 2 atm and 27°C and the partial pressure of the O_2 is 0.5 atm?

   a. 0.023×10^{23} b. 0.088×10^{23}
   c. 0.496×10^{23} d. 0.120×10^{22}
   e. 0.122×10^{23}

Solve the ideal gas law, PV = nRT, for n (moles of gas).

$$\text{Total moles} = \frac{2 \times 1}{.082 \times 300} = 0.081$$

From the partial pressure of O_2 determine moles of O_2:

$$\frac{0.5}{2.0} = \frac{\text{Moles } O_2}{0.081}; \text{ Moles } O_2 = 0.02025$$

Using Avogadro's number:

$$.02025 \times 6.02 \times 10^{23} =$$
$$.122 \times 10^{23} \text{ molecules } O_2$$

Answer (e)

23. How many ml of 0.2N HCl are needed to react completely with 30 ml of a 0.1N solution of $Ba(OH)_2$?

    a. 15ml   b. 10ml
    c. 20ml   d. 25ml
    e. 30ml

equivalents of acid = equivalents of base
$N_{Acid} \times Vol_{Acid} = N_{Base} \times Vol_{Base}$
$0.2 \times Vol_{Acid} = 0.1 \times 30$
$Vol_{Acid} = 15ml$

<u>Answer (a)</u>

24. How many liters of SO_2 at 27° and 0.75atm will be produced by reacting 10.4 g of $NaHSO_3$ according to the reaction:

$$NaHSO_3 + HNO_3 \rightarrow H_2O + SO_2$$

    a. 2.24l   b. 3.284l
    c. 0.01l   d. 22.4l
    e. 0.30l

Because this is a univalent reaction, 1 mole $NaHSO_3$ will produce 1 mole SO_2.

10.4g $NaHSO_3$/104 g/gmole = 0.1 mole $NaHSO_3$ = 0.1 mole SO_2 at STP.

$$V = \frac{nRT}{P} = \frac{(0.1)(.0821)(300)}{.75} = 3.284l$$

<u>Answer (b)</u>

25. Calculate the percentage ionization of a 0.1 M acetic acid solution.

$$HAc + H_2O \rightarrow H_3O^+ + Ac^-$$

$$Ka = \frac{[H_3O^+][Ac^-]}{[HAc]} = 1.8 \times 10^{-5}$$

    a. 0.134%   b. 1.34%
    c. 13.4%   d. 98.6%
    e. 86.6%

	Concentration Before Ionization	After Ionization
HAc	0.1M	.1 - x
H_3O^+	0	x
Ac^-	0	x

$$Ka = \frac{[x][x]}{[0.1]} = 1.8 \times 10^{-5}$$
$$x = .00134 = [H_3O] = [Ac]$$

	Concentration Before Ionization	After Ionization
HAc	0.1M	.09866
H_3O^+	0	.00134
Ac^-	0	.00134

% ionization = .09886/ 0.1 x 100 = 98.66%

<u>Answer (d)</u>

26. What is the pH of the ionized solution from problem 25.

    a. 2.87   b. -2.87
    c. -1.99   d. 1.99
    e. 2.00

$pH = -\log[H_3O^+] = -\log .00135 = 2.87$

<u>Answer (a)</u>

Problems 27 - 29

A sample of nicotine was found to contain 72.73% C, 16.97% N, 10.3% H.

27. What is the empirical formula:

    a. $C_5NH_{8.5}$ b. $C_{15}N_3H_{22}$
    c. CNH   d. $C_{10}N_2H_{17}$
    e. C_6NH_{10}

Using N as a basis:

$N = 16.97/14 = 1.21$
$C = 72.73/12 = 6.06$
$H = 10.3/1 = 10.3$

Divide by 1.21, multiply by 2 to get whole numbers

$$N = 1.21 \div 1.21 \times 2 = 2$$
$$C = 6.06 \div 1.21 \times 2 = 10$$
$$H = 10.3 \div 1.21 \times 2 = 17$$

Empirical formula $C_{10}N_2H_{17}$ Answer (d)

28. The freezing point of 100g of water containing 10g nicotine was -1.127°C. What is the molecular weight?

    a. 165   b. 162
    c. 168   d. 160
    e. 170

$$\text{Molality} = \frac{\Delta T}{1.86} = \frac{(0-(-1.127))}{1.86} = .606m$$

Molality = 1 g/mole solute/1000g solvent:
10g per 100g is 100g per kg.

$$\frac{100g \text{ (Nicotine)}}{\frac{\text{MW (Nicotine)}}{1 \text{ Kg (water)}}} = \frac{.606 \frac{g}{gmole} \text{(nicotine)}}{1 \text{ Kg (water)}}$$

$$MW = 165 \quad \underline{\text{Answer (a)}}$$

29. What is the molecular formula for nicotine?

    a. $C_5NH_{8.5}$ b. $C_{15}N_3H_{22}$
    c. CNH   d. $C_{10}N_2H_{17}$
    e. C_6NH_{10}

Using the empirical formula to calculate the molecular weight of the compound gives:

C = 10 x 12 = 120
H = 17 x 1 = 17
N = 2 x 14 = 28
        MW = 165

As the empirical MW is the same as the calculated MW, the moleecular formula and the empirical formula are the same.

                                    Answer (d)

THERMODYNAMICS

1. The value of specific heat for non gases is:

    a. A function of pressure
    b. A function of volume
    c. A function of pressure and volume
    d. A function of substance only
    e. Relatively independent of process

By definition: Answer (e)

2. The Carnot efficiency for a system operating between the limits of 1000°R and 2000°F is:

    a. 22%   b. 27%
    c. 50%   e. 56%
    e. 59%

$$\eta = \frac{T_{upper} - T_{lower}}{T_{upper}} = \frac{2460 - 1000}{2460} = .593$$

3. The most general thermodynamic process is?

    a. Isentropic
    b. Polytropic
    c. Isothermal
    d. Isochoric
    e. Isobaric

The most general process of those listed is the polytropic process, $PV^n = C$.   Answer (b)

4. A reversible adiabatic process is called?

    a. Isentropic
    b. Polytropic
    c. Isobaric
    d. Isothermal
    e. Isochoric

A reversible adiabatic process (net heat transfer to the surroundings is zero) is the isentropic process. DS = 0
    Answer (a)

5. Symbolically the first law of thermodynamics can be expressed as:

    a. $DU = U_2 - U_1$ b. $DQ = C_pDT$
    c. $k = C_p/C_v$ d. $Q = DU + W$
    e. $W = \int PV\,dw$

The first law is the rule of energy balance:

Q = DU + W   Answer (d)

6. Which of the following is not true for an isentropic process?

    a. Energy gains or losses during the process can only result from work done on the process.
    b. The process is irreversible
    c. No heat is transferred
    d. P, V, and T, all vary
    e. Q = DU + W

All are true but b, an isentropic process is reversible.
    Answer (b)

7. The temperature in a 60 ft^3 tank of gas at 100°F. The molecular weight is 22 and the tank pressure is 25 psia. How many moles of an ideal gas are in the tank?

    a. 0.173   b. 0.204
    c. 0.245   d. 1.188
    e. 3.678

$$\frac{W}{n} \text{ moles} = \frac{PV}{RT} = \frac{(25)(144)(60)}{22\left(\frac{1544}{22}\right)(100+460)} = .245$$

Answer (c)

Problems 8 - 10

Two and one half pounds of air (R = 53.3) follow a polytropic process. At initial conditions the pressure is 85 psia and the temperature is 470°F. At the final conditions, volume is 16 cu ft, and temperature is 140°F.

8. What is the final pressure?

    a. 30.1 psia   b. 34.7 psia
    c. 52.1 psia   d. 57.5 psia
    e. 62.2 psia

$$P_2 = \frac{wRT_2}{V_2} = \frac{(2.5)(53.3)(600)}{(16)(144)} = 34.7 \text{ psia}$$

Answer (b)

9. What is the value of n for the process?

    a. 1.41   b. 1.68
    c. 1.82   d. 1.96
    e. 2.15

$$V_1 = \frac{wRT_1}{P_1} = \frac{(2.5)(53.3)(930)}{(85)(144)} = 10.12 \text{ ft}^3$$

$$\frac{P_1}{P_2} = \left[\frac{V_2}{V_1}\right]^n = \frac{85}{34.7} = \left[\frac{16}{10.12}\right]^n$$

n = 1.96   Answer (d)

10. The change in entropy from initial to final states is:

    a. -0.057 units   b. -1.09 units
    c. +0.057 units   d. +0.11 units
    e. -0.042 units

$$\Delta S = 2.5 \ln\left(\frac{T_2}{T_1}\right) = 2.5 \ln\left(\frac{600}{930}\right) = -1.09 \text{ units}$$

Answer (b)

Problems 11 - 13

A four cylinder, four cycle IC engine has a bore of 3.3 in and a stroke of 4.1 in. At 1800 rpm, the engine is drawing 60 cfm of entering mixture.

11. The engine displacement is:

    a. 31.7 in.^{3} b. 35.07 in.^{3}
    c. 126.7 in.^{3} d. 140.3 in.^{3}
    e. 70.14 in.^{3}

Cylinder Displacement = Area x Stroke

$$\frac{\pi D^2}{4} \times S = \frac{\pi 3.3^2}{4} \times 4.1 = 35.07 \text{ in.}^3$$

Displacement = 4 x 35.07 = 140.3 in.^{3}
    Answer (d)

12. The volume displaced per minute is:

    a. 36.53 cfm   b. 146.1 cfm
    c. 82.6 cfm   d. 46.2 cfm
    e. 73.06 cfm

For a 4 cycle engine there is one suction stroke for each two revolutions:

$$D.V. = \frac{(140.27)(0.5)(1800)}{1728} = 73.057$$

Answer (e)

13. The volumetric efficiency is:

    a. 44%   b. 58%
    c. 63%   d. 82%
    e. 90%

Volumetric efficiency is actual volume drawn divided by the piston displacement.

$$\eta = \frac{60}{73.6} = .82$$

Answer (d)

14. The enthalpy of 3.3 lb of a fluid with internal energy of 1000 Btu/lb, a pressure of 2 atm and a volume of 20 cu ft is:

    a. 1108.8 Btu   b. 3300 Btu
    c. 3408.8 Btu   d. 108.8 Btu
    e. 2208.8 Btu

$$h = u + \frac{PV}{J}$$
$$= (3.3 \times 1000) + \frac{(2 \times 14.7)(20)(144)}{778}$$
$$= 3408.8 \text{ Btu}$$

Answer (c)

15. A block of metal composed of 50% Cu and 50% Pb weighs 1 kg. The block is dropped into a tank of oil (sg = 0.82) from a height of 8 meters. The change in internal energy of the oil is:

    a. 34.52 cal   b. 18.74 cal
    c. 16.13 cal   d. 42.34 cal
    e. 20.56 cal

$$D = Q - W, \text{ but } Q = 0$$
$$W = 1 \text{ kg} \times 8 \text{ M} = 8 \text{ kg M}$$

$$\Delta U = 8 \text{ kg M} \left(\frac{\text{ft lb}}{.1383 \text{ Kg M}}\right)\left(\frac{\text{Btu}}{778 \text{ ft lb}}\right)\left(\frac{252 \text{ cal}}{\text{Btu}}\right)$$
$$= 18.74 \text{ cal}$$

Answer (b)

16. The heat of combustion of a fuel is 17,500 Btu/lb. It is used in an engine consuming the fuel at a rate of .35 lb/(hp-hr). The engine efficiency is:

    a. 17.5%   b. 25.5%
    c. 35.0%   d. 41.6%
    e. 56.2%

The strating point of this problem is the use of a conversion factor:

$$\eta = \frac{2545 \frac{\text{Btu}}{\text{Hp-Hr}}}{\left(17,500 \frac{\text{Btu}}{\text{lb}}\right)\left(.35 \frac{\text{lb}}{\text{Hp-Hr}}\right)} = .4155$$

Answer (d)

17. What volume in liters would 6.49g of CO_2 occupy at 730 mm Hg and 29°C?

    a. 3.807   b. 3.513
    c. 0.366   d. 0.387
    e. 168

The key to this problem is units:

$$V = \frac{nRT}{P} = \frac{\left(\frac{6.49}{44}\right)(0.0821)(29 + 273)}{\frac{730}{760}} = 3.8$$

Answer (a)

18. Given:
$$C + H_2O \rightarrow CO + H_2$$

How many lbs of CO are produced per ton of fuel if the fuel contains 95% Carbon?

    a. 1900 lbs   b. 2850 lbs
    c. 4433 lbs   d. 323 lbs
    e. 2000 lbs

The amount of carbon is 2000 x .95 = 1900 lbs

	C	+ H_2O	→	CO	+ H_2
MW	12	18		28	2
Per Unit	1	1.5		2.33	.17
Per Ton	1900	2850		4433	323

Answer (c)

Problems 19 - 21

The hydrocarbon C_6H_{14} is burned with zero excess air. The balanced equation for the reaction is:

$$C_6H_{14} + 9.5\ O_2 + (3.78)(9.5)\ N_2 \rightarrow 6CO_2 + 7\ H_2O + 35.9\ N_2$$

19. The % by weight of CO_2 in the product is:

a. 7.3%   b. 11.6%
c. 18.9%   d. 44.6%
e. 24.5%

Solving for CO_2:

$$CO_2 = \frac{6(12+32)}{6(44) + 7(18) + 35.9(28)} = .189$$

Answer (c)

20. If the water vapor in the product stream is condensed, what $\%CO_2$ is in the product stream:

a. 7.3%   b. 11.6%
c. 18.9%   d. 44.6%
e. 24.5%

The composition of the product stream is stoichometrically dependent, not phase dependent.

Answer (d)

21. If the water vapor in the product stream is condensed, what volume $\%CO_2$ is in the product stream:

a. 12.4%   b. 14.3%
c. 22.4%   d. 30.6%
e. 44.0%

The volume of liquid will be considered neglegable with respect to the volume of the gas phase:

$$CO_{2g} = \frac{6}{6 + 0 + 35.9} = .143$$

Answer (b)

Problems 22 - 24

500 lbs of a gas are held in a 3.5 ft diameter spherical tank made of thin wall, high strength stainless steel. The gas composition is 21% CH_4, 20% CO, 19% CO_2, 10% H_2S, 30% N_2. The tank pressure is 30 psia.

22. The average molecular weight of the gas is:

a. 15.00   b. 22.44
c. 29.12   d. 36.78
e. 38.46

Basis 100 lb-moles

	%	MW	lb
CH_4	21	16	336
CO	20	28	560
CO_2	19	44	836
H_2S	10	34	340
N_2	30	28	840
		Σ	2912

$$\text{Av MW} = \frac{2912\ \text{lbs}}{100\ \text{lb-mole}} = 29.12$$

Answer (c)

23. How many lb-moles of the gas are in the tank?

a. 17.17   b. 29.12
c. 22.40   d. 500
e. 2912

$$\frac{500\ \text{lb}}{29.12\ \frac{\text{lb}}{\text{lb-mole}}} = 17.17\ \text{lb-mole}$$

Answer (a)

24. What is the volume of the gas in the tank? (R = 1544/MW)

a. 22.4 ft³   b. 3252 ft³
c. 7.07 ft³   d. 14.14 ft³
e. 3.1416 ft³

$$V = \frac{4}{3}\pi r^3 = \frac{4}{3}\pi \left(\frac{3.5}{2}\right)^3 = 22.4\ \text{cu ft}$$

Answer (a)

25. 400 g of a metal at 212°F was dropped into an insulated tank containing 200cc of H₂O at 22°C. The temperature of the metal and water stabilized at 34°C. The specific heat of the metal is:

    a. 0.0037 cal/g   b. 0.0428 cal/g
    c. 0.0563 cal/g   d. 0.0773 cal/g
    e. 0.0909 cal/g

Heat lost by metal = heat gained by water:

$$(200cc)\left(\frac{1g}{cc}\right)\left(\frac{1\ cal}{g}\right)(34-22)$$
$$= 400g\left(C_p \frac{Cal}{g}\right)(100-34)$$

$C_p = 0.0909$ Cal/g   __Answer (e)__

Problems 26 and 27

A fluid with SG = 1.7 and C = 2.84 enters a heat exchanger at 250°F and leaves at 90°F. Air enters at 40°F, and leaves at 75°F.

26. For HE parallel, the LMTD is:

    a. 114°F   b. 100°F
    c. 74°F   d. 145°F
    e. 82.5°F

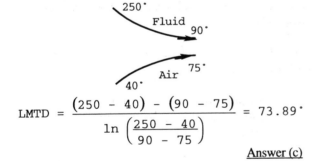

$$\text{LMTD} = \frac{(250-40)-(90-75)}{\ln\left(\frac{250-40}{90-75}\right)} = 73.89°$$

__Answer (c)__

27. For 100% heat transfer, what flow rate of the liquid is required?

    a. 0.24 lb/hr   b. 8.4 lb/hr
    c. .01849 lb/hr   d. .0845 lb/hr
    e. .0141 lb/hr

Heat lost by fluid = heat gained by air

Heat gained = $W_a C_p$ DT
    = 1 x .24 x (75 - 40) = 8.4 Btu/lb

Heat lost = $W_f C_p$ DT
    8.4 = W_f x 2.84 x (250 - 90)
    W_f = .01849 lb/hr   __Answer (c)__

28. If the actual COP is 5.3, the indicated horsepower for a compressor per ton of refrigeration is:

    a. 4.71 hp/T   b. 24.96 hp/T
    c. 23.55 hp/T   d. .889 hp/T
    e. .94 hp/T

$$\text{COP}_{actual} = \frac{4.71}{hp/T} = 5.3$$

.889 Hp/T   __Answer (d)__

29. The latent heat of evaporation of Freon 12 at 5°C is 68.2 Btu/lb. How much refrigerant must be circulated to produce one ton of refrigeration?

    a. 175.9 lb/hr   b. 152.4 lb/hr
    c. 126.3 lb/hr   d. 3.62 lb/hr
    e. 2.28 lb/hr

1 ton of refrigeration is by definition 200 Btu/Min.

$$\frac{lb_{Freon}}{hr} = \frac{200}{68.2} \times 60 = 175.9 \frac{lb}{hr}$$

__Answer (a)__

30. What is the ideal COP for a refrigerating system operating between the limits of 0°F and 500°R?

    a. .92   b. 1.15
    c. 11.5   d. 115
    e. 15.1

$$\text{COP}_{Ideal} = \frac{\text{Lower Limit}}{\text{Higher Limit} - \text{Lower Limit}}$$

$$= \frac{460}{500-460} = 11.5 \quad \underline{\text{Answer (c)}}$$

31. The saturation temperature of a water air mixture at atmospheric pressure is 535°R. How many pounds of water are there in 4.5 lbs dry air?

    a. .0189 lb   b. 0.835 lb
    c. 132 Gr wv/lb da   d. .019 lb wv/lb da
    e. .848 lb

$T_{wb} = 75°F$

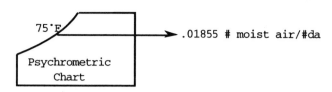

$$(4.5 \text{ lb da}) \left(\frac{.01855 \text{ lbs moisture}}{\text{lbs da}} \right) = .0835 \text{ lbs}$$

Answer (b)

For Problems 32 - 36

In cooling a liquid food, frozen water is added to the room temperature liquid while it is being stirred. Condensation appears on the outside of the container when the contents are at 52°F after being cooled from 74°F. Pressure is 14.696 psi.

32. What is the partial pressure of the dry air in the room?

    a. 14.5 psia   b. 14.6 psia
    c. 14.7 psia   d. .2 psia
    e. .198182 psia

The ambient air pressure = partial pressure of the dry air plus the partial pressure of the water vapor. The partial pressure of the water vapor is equal to the saturation pressure at the dew point which is 52°F. Table 1 of a steam table at 52°F shows P_{sat} = 0.198182 psia. The partial pressure of the dry air is:

$$\partial P_{da} = 14.696 - .198182 = 14.5 \text{ psia}$$

Answer (a)

33. What is the ratio of the weight of air to the weight of water for the ambient air in the room?

    a. 77   b. 84
    c. 93   d. 102
    e. 121.7

(MW = molecular weight = 29 for air and 18 for water). Using data from problem 7:

$$\frac{W_a}{W_w} = \frac{\partial P_a MW_a}{\partial P_a MW_a} = \frac{14.498 \times 29}{.198 \times 18} = 117.97$$

Answer (e)

34. What is the relative humidity of the air in the room?

    a. 42.3%   b. 44.5%
    c. 45.6%   d. 46.1%
    e. 47.1%

At 74°F ambient, P_{sat} = .4156
At 52°F dew point, P_{sat} = .1918

$$RH = \frac{.1918}{.4156} = .4615$$

Answer (d)

35. What is the specific humidity of the air?

    a. .0130 lb wv/lb da   b. .0119 lb wv/lb da
    c. .0108 lb wv/lb da   d. .0098 lb wv/lb da
    e. .0085 lb wv/lb da

From Problem 8, W_{da}/W_{wv} = 117.97 lb da/lb wv.

Specific humidity is the reciprocal W_{wv}/W_{da}, 1/117.97 = .008477. Answer (e)

Problems 36 - 38

A thin flat plate has a surface temperature of 150°F when placed in sunlight. The ambient air is at 72°F. The plate is well insulated on its dark side. Emissivity of the plate is 0.99, and it receives 350 Btu/hr/ft².

36. What is the heat radiated from the plate to the surrounding air?

    a. 1 Btu/hr/ft²   b. 10 Btu/hr/ft²
    c. 100 Btu/hr/ft²   d. 1000 Btu/hr/ft²
    e. 10,000 Btu/hr/ft²

$$Q_r = .173 \text{ FeA} \left[\left(\frac{T_2}{100}\right)^4 - \left(\frac{T_1}{100}\right)^4 \right]$$

$$Q_r = (.173)(.99)(1) \left[\left(\frac{610}{100}\right)^4 - \left(\frac{532}{100}\right)^4 \right]$$

$$= 99.95 \frac{\text{Btu}}{\text{hr-ft}^2}$$

Answer (c)

37. How much heat goes into the air by conduction?

   a. 2.5 Btu/Ft^2 b. 25 Btu/Ft^2
   c. 250 Btu/Ft^2 d. 2500 Btu/Ft^2
   e. .25 Btu/Ft^2

$$Q_c = Q_T - Q_r = 350 - 100 = 250 \text{ Btu/Ft}_2$$

<u>Answer (c)</u>

38. What is the unit surface conductance?

   a. 2.62 Btu/ft^2/°F
   b. 3.21 Btu/ft^2/°F
   c. 5.14 Btu/ft^2/°F
   d. 6.18 Btu/ft^2/°R
   e. 2.25 Btu/ft^2/°F

$$U = \frac{Q_c}{A \; \Delta T} = \frac{250}{(1)(150 - 72)}$$

$$= 3.205 \text{ Btu/Ft}^2/°F \quad \underline{\text{Answer (b)}}$$

39. A steam turbine receives 3500 lbs of steam per hour at 105 ft/sec and an enthalpy of 1530 Btu/lb. The exit velocity and enthalpy are 800 fps and 1250 Btu/lb respectively. The hp output of the turbine is?

   a. 123 hp   b. 283 hp
   c. 350 hp   d. 450 hp
   e. 549 hp

$$\frac{w\vec{V}_1^2}{2g} + wJh_1 = \frac{w\vec{V}_2^2}{2g} + wJh_2 + W$$

$$W = \frac{w}{g}\left(\vec{V}_1^2 - \vec{V}_2^2\right) + wJ(h_1 - h_2)$$

$$w = 3500 \frac{\text{lbs}}{\text{hr}} \times \frac{\text{hr}}{3600 \text{ Sec}} = .972 \frac{\text{lbs}}{\text{sec}}$$

$$W = .972\left[\frac{(105^2 - 800^2)}{32.2} + 778(1530 - 1250)\right]$$

$$= 192754 \frac{\text{Ft lb}}{\text{sec}} = 350.4 \text{ HP} \quad \underline{\text{Answer (c)}}$$

Problems 40 - 42

The inlet temperature to a gas turbine is 1500°F. It has an expansion pressure ratio of 6.0. The gasses leave the turbine at 995°F. $C_v = .2$ and $C_p = .26$.

40. The ideal Dh is:

   a. 142 Btu/lb   b. 173 Btu/lb
   c. 185 Btu/lb   d. 192 Btu/lb
   e. 1.3 btu/lb

The process is as follows:

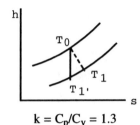

$$k = C_p/C_v = 1.3$$

$$T_1' = T_0 \left(\frac{P_1}{P_0}\right)^{\frac{k-1}{k}}$$

$$T_1' = 1960 \left(\frac{1}{6}\right)^{\frac{1.3-1}{1.3}} = 1295°R$$

$$\Delta h_I = C_p (T_0 - T_1')$$

$$.26 (1960 - 1295) = 173 \frac{\text{Btu}}{\text{lb}}$$

<u>Answer (b)</u>

41. The actual Dh is:

   a. 142 Btu/lb   b. 154 Btu/lb
   c. 166 Btu/lb   d. 173 Btu/lb
   e. 185 btu/lb

$$\Delta h_A = C_p (T_0 - T_1)$$

$$.26 (1960 - 1415) = 141.7 \frac{\text{Btu}}{\text{lb}}$$

<u>Answer (a)</u>

42. The adiabatic efficiency of the turbine is:

   a. 38%   b. 59%
   c. 64%   d. 78%
   e. 82%

$$\eta = \frac{h_A}{h_I} = \frac{141.7}{173.0} = .819$$

Answer (e)

43. A steam jet (c = .9) expands adiabatically in a nozzle from 150 psia and 150° superheat into a condenser at 2 in. Hg abs. What is the velocity of the steam jet?

a. 2812 fps   b. 3419 fps
c. 3528 fps   d. 3767 fps
e. 223.7 fps

Adiabatic is not isentropic unless the process is stated as reversible:

$$V = 223.7 \; C_v \sqrt{\Delta H}$$

From a Mollier chart:

$H_1 = 1279$ Btu/lb, and $H_{2s} = 929$ Btu/lb

$$V = (223.7)(.9)\sqrt{1279 - 929} = 3767 \text{ fps}$$

Answer (d)

44. Electric cable is used for melting snow on a driveway 8 ft wide and 45 ft long. What is the cost of melting 6 inches of snow (w = 10 lb/cu ft) at 32°F if the efficiency of operation is 50% and electricity costs $.05/kwh?

a. $7.59   b. $6.42
c. $5.83   d. 4.22
e. $3.98

Wgt of snow = 8 x 45 x .5 x 10 = 1800 lb
Latent heat of fusion = 144 Btu/lb

$$Q = 1800 \times 144 = 259{,}200 \text{ Btu}$$

$$\text{Cost} = \frac{259{,}200 \text{ Btu}}{3413 \frac{\text{kwh}}{\text{Btu}} \times .5} \times 0.05 \; \frac{\$}{\text{kwh}} = \$7.59$$

Answer (a)

45. Saturated steam at (212°F) is bubbled into a liquid bath containing 2 yds³ of a mixture of SG = 2.0 and specific heat is 0.5. The liquid bath is initially at 35°F, and must be heated to 96.3°F. For an ideal process with no losses, how much steam is required?

a. 63.4 lb   b. 1785.3 lb
c. 190.2 lb   d. 375.6 lb
e. 580.1 lb

$$W_{bath} = (2 \text{ yd}^3)\left(\frac{27 \text{ ft}^3}{\text{yd}^3}\right)\left(62.4 \; \frac{\text{lb}}{\text{ft}^3}\right)(2)$$
$$= 6739.2 \text{ lb}$$

Heat required to warm the bath is wCDT

$$Q = (6739.2)(.5)(96.3 - 35)$$
$$= 206{,}556.48 \text{ Btu}$$

The steam will give up heat to the bath when it condenses and the condensate gives up heat as it cools from 212° to 96.3°.

$$Q = w_s \; C \; DT + w_s \; h_{fg}$$
$$Q = w_s (1)(115.7) + w_s (970.3)$$
$$Q = 1086 \; w_s$$

$$w_s = \frac{206{,}556.48 \text{ Btu}}{1086 \; \frac{\text{Btu}}{\text{lb}}} = 190.2 \text{ lb}$$

Answer (c)

46. A steam throttling calorimeter receives steam at 150 psia and discharges the steam at 15 psia. The calorimeter thermometer reads 250°F. The quality of the steam is:

a. 97%   b. 93%
c. 88%   d. 76%
e. 63%

The superheated steam at 250° and 15 psia has an enthalpy of 1168.7 Btu/lb. At 150 psia $h_f = 330.6$ Btu/lb and $h_{fg} = 863.4$ Btu/lb. Since throttling is at constant enthalpy:

$$1168.7 = 330.6 + \% (863.4)$$
$$\% = .97 \quad \text{Answer (a)}$$

Problems 47 - 49

A steam turbogenerator uses 20 lb of steam per kwh produced. The steam enters at 250°F and leaves at 2 psia.
h = 934 Btu/lb, $h_f = 94$ Btu/lb

47. The Rankine cycle efficiency is:

a. 12%   b. 19%
c. 27%   d. 31%
e. 35%

At 250psi and 600°F h = 1319 Btu/lb

$$e_R = \frac{1319 - 943}{1319 - 94} = .3069$$

Answer (d)

48. Steam consumption is:

a. 11 lb/kwh   b. 10 lb/kwh
c. 9 lb/kwh   d. 8 lb/kwh
e. 5 lb/kwh

$$\text{Consumption} = \frac{3413}{1319 - 943} = 9.077 \frac{\text{lbs}}{\text{kwh}}$$

Answer (c)

Problems 49 - 50

These concepts were not covered in the notes. Static pressure and temperature are measured in a high velocity air stream. At mach 0.68 the pressure is 42 psia and the temperature is 230°F.

49. The stagnation temperature is:

a. 690 °R   b. 754 °R
c. 284 °R   d. 251 °R
e. 783 °R

$$\frac{T_0}{T} = 1 + \left[\frac{k-1}{2}\right] M^2$$

$$\frac{T_0}{230 + 460} = 1 + \left[\frac{1.4 - 1}{2}\right] .68^2$$

$$T_0 = 753.8°R \quad \text{Answer (b)}$$

50. The stagnation pressure is:

a. 42 psia   b. 57.23 psia
c. 54.7 psia   d. 65.4 psia
e. 1.22 psia

$$\frac{P_0}{P} = \left(\frac{T_0}{T}\right)^{\frac{k}{k-1}}$$

$$\frac{P_0}{42} = \left(\frac{753.8}{690}\right)^{\frac{1.4}{1.4-1}}$$

$$P_0 = 57.23 \text{ psia} \quad \text{Answer (b)}$$

MATERIAL SCIENCE

Problems 1 - 6

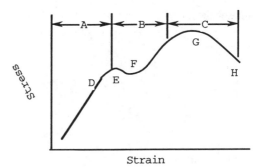

1. Range A on the stress-strain curve is:

a. Ultimate strength range
b. Plastic range
c. Elastic range
d. Ultimate strain range
e. Indeterminant range

The linear range on a stress strain curve is the elastic range.   Answer (c)

2. The proportional limit is:

a. Point D   b. Point E
c. Point F   d. Point c
e. Point H

The proportional limit would be point D. It is the point close to but below the elastic limit where the curve deviates from Hooke's Law without set.
   Answer (a)

3. The yield point is:

a. Point D   b. Point E
c. Point F   d. Point G
e. Point H

The yield point is where a permanent set would most likely occur, and above the elastic limit. Point F.
   Answer (c)

4. The elastic limit is:

a. Point D   b. Point E
c. Point F   d. Point G
e. Point H

The elastic limit is the point on the curve where when a load is released there is a permanent deformation in the material. Point E. Answer (b)

5. Failure occurs at or in:

a. B   b. F
c. C   d. G
e. E

Failure would occur at point H which is at the end of range C. Answer (c)

6. The ultimate strength is indicated by:

a. Point D   b. Point E
c. Point F   d. Point G
e. Point H

The ultimate strength point is the highest point on the curve. Point G.
Answer (d)

Problems 7 - 10

Given the following equilibrium phase diagram for materials X and Y:

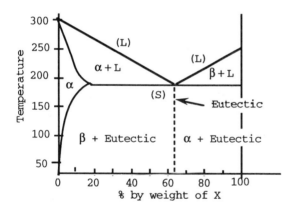

7. The eutectic composition is:

a. 18% X   b. 50% X
c. 80% X   d. 36% X
e. 62% X

The eutectic composition is the point where the liquid and solid lines intersect. 62% X. Answer (e)

8. The eutectic temperature is:

a. 50°   b. 100°
c. 180°   d. 250°
e. 325°

The eutectic temperature is the temperature of the horizontal part of the solid line. Answer (c)

9. What is the lesser composition of the solid phase of X in equilibrium at the eutectic temperature?

a. 10%   b. 18%
c. 40%   d. 62%
e. 97%

On the left, the point where the solid phase meets the eutectic temperature is the lesser composition. 18%
Answer (b)

10. For an alloy containing 70% by weight of X at 250°, what fraction exists as the a phase?

a. 10%   b. 18%
c. 40%   d. 40%
e. 0%

The intersection of 70% and 250° is in the liquid range. There is no a phase present. Answer (e)

11. A solid is produced from two components. The equilibrium condition of this solid has a and b when:

a. a and b phases have the same crystal structure
b. P and Q have the same melt points
c. The internal energy of a is greater than b
d. The internal energy of a is less than b
e. None of the above

None of the above statements are true.
Answer (e)

12. The eutectoid mixture of steel is:

a. Ferrite and cementite
b. Ferrite and austentite
c. Ferrous and cementite
d. Distentite and loostite
e. No tite at all

Answer (a)

In steel, the eutectoid mixture is a mixture of ferrite and cementite.

13. The molecular weight of vinyl chloride is 62.5. PVC has a degree of polymerization of 20,350. The molecular weight of PVC is:

    a. 325.6   b. 3.07×10^{-3}
    c. .999   d. 1.27×10^6
    e. 10,250

$$\text{Degree of polymerization} = \frac{MW_{Polymer}}{MW_{Monomer}}$$

$(20.350)(62.5) = 1.271 \times 10^6$ Answer (d)

14. What are the Miller Indices for the plane shown?

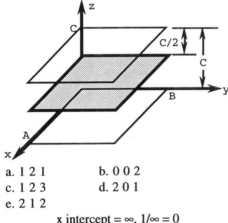

    a. 1 2 1   b. 0 0 2
    c. 1 2 3   d. 2 0 1
    e. 2 1 2

x intercept = ∞, 1/∞ = 0
y intercept = ∞, 1/∞ = 0
z intercept = C/2, 1/C/2 = 2

        0 0 2   Answer (b)

15. What are the Miller indices for the plane:

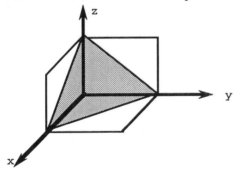

    a. 0 0 0   b. 1 1 1
    c. 2 2 2   d. 1 2 3
    e. 3 2 1

The x,y,z intercepts are all 1, thus the indices are all 1   Answer (b)

16. A unit cell is:

    a. a group of atoms in a cubic arrangement

    b. an FCC

    c. a unit cube containing the fewest number of atoms

    d. the smallest group of atoms which when regularly repeated forms the crystal

    e. a BCC

A unit cell is the smallest group of atoms which when regularly repeated forms the crystal.   Answer (d)

17. A test specimen is subjected to a stress and a crack appears. The stress is maintained at the level that caused the crack to appear.

    a. The crack will continue to propagate in its original direction

    b. The crack will stop growing

    c. The crack will propagate in a direction normal to the original

    d. The crack will continue to propagate in both normal and transverse directions.

    e. The specimen will self destruct

Once a crack has begun it will continue to propagate in its original direction
    Answer (a)

18. A force of 8N is applied to a flat tensile specimen 10 cm long, 1 cm wide, and 0.2 cm thick. The stress is:

    a. 40 N/M²   b. 8 N/M²
    c. 8 N/cm²   d. 1.6 N/M²
    e. 4 N/M²

$$\text{Stress} = \frac{F}{A} = \left(\frac{8N}{1 \times 0.2 \text{ cm}}\right)\left(\frac{100^2 \text{ cm}^2}{\text{m}^2}\right) = 4 \frac{N}{M^2}$$

    Answer (e)

19. Which of the following is not a Non-Destructive test?

    a. Rockwell hardness
    b. Magnafluxing
    c. Izod impact
    d. Radiography
    e. Acoustic testing

The question has a double negative, thus asks which is destructive. Izod Impact

<u>Answer (c)</u>

20. Molecules of a polymer are not held together by which of the following:

    a. Hydrogen bonds
    b. Primary bonds
    c. Secondary bonds
    d. Covalent bonds
    e. Tertiary bonds

Covalent bonding occurs in chemical reactions of the displacement type. The rest are internal hydrocarbon bonds.

<u>Answer (d)</u>

ENGINEERING ECONOMICS

1. A deposit of $1,500 is made in a savings account that pays 7.5% interest compounded annually. How much money will be available to the depositor at the end of 16 years?

    A. $4,438.35   D. $471.60
    B. $5,129.10   E. None of the Above
    C. $4,771.20

SOLUTION

GIVEN: P = $1,500.00, i = 7.5%, n = 16 yrs   FIND: F

    F = P(F/P, i=7.5, n=16)
        1,500 (3.1808) = $4,771.20

ANSWER IS C

2. If a deposit of $1,000 is made today in a 8% per year compounded quarterly savings account it would grow to $2,208 at the end of 10 years. If the deposit is made two years from now how much will have to be deposited so the $2,208 terminal amount will still be realized?

    A. $1,171.56   D. $1,271.56
    B. $1,071.56   E. $971.56
    C. $1,371.56

SOLUTION

GIVEN: P = $1,000, i = 8.0%/4 = 2.0%/INT. PERIOD, n = 10 yrs x 4 = 40
              PERIODS F = $2,208
FIND: P FOR F = $2,208, i=2.0%, N = 8 x 4 = 32.

    P = F(P/F, i=2, n=32)
      = 2,208 (05306)
      = 1,171.56

ANSWER IS A

3. It was planned to leave $2,500 on deposit in a savings account for 15 years at 6.5% interest. It became necessary to withdraw $750 at the end of the 5th year. How much will be on deposit at the end of the 15 year period?

    A. $5,679.50   D. $6,021.71
    B. $5,379.50   E. $5,021.71
    C. $4,679.50

SOLUTION

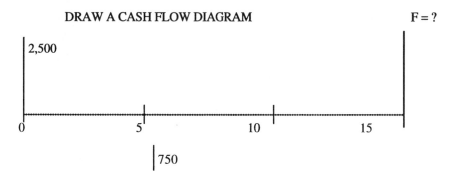

DRAW A CASH FLOW DIAGRAM

THIS SOLUTION REQUIRES FINDING THE AMOUNT OF MONEY ON DEPOSIT AT THE END OF THE 5th YEAR BEFORE THE WITHDRAWAL, THEN SUBTRACTING THE 750 WITHDRAWAL. THE NET AMOUNT ON DEPOSIT AT THE END OF THE 5th YEAR THEN RECEIVES INTEREST UNTIL THE END OF THE 15th YEAR

F_5 = 2,500(F/P, i=6.5, n=5) - 750
    = 2,500(1.3701) - 750
    = 2,675.25

F_{15} = 2,675.25(F/P, i=6.5, n=10)
    = 2,675.25(1.8771)
    = 5,021.71

ANSWER IS E

4. Fifteen years ago a deposit of $875.00 was made in a commercial bank. Today the account has a balance of $2,069.38. The bank pays interest on a semiannual basis. Although the bank interest varied over the 15-year period based on economic conditions, determine the average interest rate compounded semiannually the deposit earned.

    A. 6.27%   D. 5.84%
    B. 5.04%   E. 3.77%
    C. 4.04%

SOLUTION

GIVEN: P = 875.00, F = 2,069.38, i = ?-COMPOUNDED SEMIANNUALLY, n = 15(2) = 30 INTEREST PERIODS
  FIND: i%/YR COMPOUNDED SEMIANNUALLY

P = F(P/F, i=?, n=30)
875.00 = 2,069.38 (P/F, i=?, n=30)
P/F = 0.4228

FROM THE INTEREST TABLES

    FOR i = 3.0   P/F = 0.4120
         = ?   = 0.4228
         = 2.0   = 0.5521

$$\left(\frac{(?-2.0)}{(3.0-2.0)}\right) = \left(\frac{(0.4228-0.5521)}{(0.4120-0.5521)}\right)$$

$$? = 2.0 + (1.0)\left(\frac{(0.1293)}{(0.1401)}\right)$$

? = 2.92% / INTEREST PERIOD, ANSWER IS 2.92(2) = 5.84%/YR COMPOUNDED SEMIANNUALLY

ANSWER IS D

5. How long will it take a deposit to triple at 18% interest compounded monthly?

    A. 68 Months   D. 70 Months
    B. 72 Months   E. 76 Months
    C. 74 Months

SOLUTION

    GIVEN: P = 1.0 (UNITY), F = 3.0 (UNITY TIMES 3), i = 18%/12 = 1.5%/INT PERIOD

    FIND: F

    F = P(F/P, i=1.5%, n=?)
    3.0 = 1.0(F/P, i=1.5%, n=?)
    F/P = 3.0

FROM THE 1.5% INTEREST TABLES

    F/P = 2.9650   WHEN n = 73
        = 3.0095   n = 74

NOTE THERE IS NO NEED TO INTERPOLATE SINCE THE CHOICES OF ANSWERS ARE GIVEN AS WHOLE MONTHS, THEREFORE BY INSPECTION 3.0095 IS CLOSER TO 3.0 THAN 2.9650 AND n = 74 IS THE ANSWER.

ANSWER IS C

6. A deposit of $650 is made at the end of each year in an account paying 10%. The deposits were made for the 5th and 6th years. What amount of money is on deposit at the end of 25 years?

    A. $57,557.23   D. $55,757.23
    B. $57,755.23   E. $55,577.23
    C. $50,577.23

SOLUTION

DRAW A CASH FLOW

$A = 650$

THIS PROBLEM CAN BE VIEWED AS TWO SEPARATE PROBLEMS. 1 - WHAT IS THE FUTURE WORTH OF THE FOUR DEPOSITS MADE IN YEARS 1 THRU 4 IN YEAR 25 AND 2 - WHAT IS THE FUTURE WORTH OF THE SERIES OF DEPOSITS MADE FROM YEAR 7 THRU 25. THE SOLUTIONS TO PARTS 1 AND 2 ARE THEN ADDED FOR THE REQUIRED ANSWER. NOTE BOTH CASH AND FLOWS ARE BASED ON END OF YEAR PAYMENTS

CALCULATION 1

$F_4 = A(F/A, i=10, n=4)$
$\quad = 650(4.6410)$
$\quad = 3,016.65$

$F_{25} = F_4(F/P, i=10, n=21)$
$\quad = 3,016.65(7.4002)$
$\quad = 22,323.81$

CALCULATION 2

$F_{25} = A(F/A, i=10, n=19)$
$\quad = 650(51.1591)$
$\quad = 33,253.42$

TOTAL IN YEAR 25 = 22,323.81 + 33,253.42 = $55,577.23

ANSWER IS E

7. The U.S. sells Savings Bonds for $37.50, which are redeemable for $75.00 in 12 years. What is the annual interest rate paid by the U.S.?

  A. 5.94%   D. 7.04%
  B. 5.40%   E. None of the above
  C. 6.40%

SOLUTION

  GIVEN: P = 37.50, F = 75.00, n = 12 YEARS
  FIND: i%/YEAR

$F = P(F/P, i=?, n=12)$
$75.00 = 37.50 (F/P, i=?, n=12)$
$7.0 = (F/P)$

FROM THE INTEREST TABLES

FOR i = 6.0%   F/P = 2.0122/P, i=?, n=12)
    = ?   = 2.0
    = 5.0%   = 1.7959

$$? = 5.0 + (1.0)\left(\frac{2.0 - 179959}{2.0122 - 1.7959}\right)$$

? = 5.94%/YEAR

ANSWER IS A

8. A child receives $50,000 as a gift which is deposited in a 6% bank account compounded semiannually. If 2,500 is withdrawn at the end of each half year, how long will the money last?

   A. 21.0 Years   D. 18.0 Years
   B. 15.5 Years   E. 11.5 Years
   C. 25.0 Years

SOLUTION

   GIVEN: P = 50,000, A = 2,500, i = 6.0/2 = 3.0%/INTEREST PERIOD
   FIND: n

   P = A(P/A, i=3.0, n=?)
50,000 = 2,500(P/A, i=3.0, n=?)
    20 = (P/A)

FROM THE i = 3.0% INTEREST TABLES

   P/A = 19.6004   WHEN n = 30
   P/A = 20.0000   n = ?
   P/A = 21.4872   n = 35

$$? = 30 + (5)\left(\frac{20.0000 - 19.6004}{21.4872 - 19.6004}\right)$$

? = 31.06 PERIODS OR 15.5 YEARS, NOTE ANSWERS ARE GIVEN IN YEARS!

ANSWER IS B

9. If $10,000 is borrowed with interest at 12% compounded quarterly for 5 years, what sum should be paid at the end of the 5 years to settle the debt and accumulated interest?

   A. $18,061   D. $18,601
   B. $16,801   E. $17,816
   C. $16,108

SOLUTION

GIVEN: P = 10,000, i = 12/4 = 3%/INTEREST PERIOD, n = 5 (4) = 20 PERIODS
FIND: F

F = P(F/P, i=3, n=20)
= 10,000 (1.8061)
= 18,061

ANSWER IS A

10. A car loan for $12,500 is to be repaid in equal monthly installments over two years at a 12% interest rate. How much of the principal will have been paid after 18 payments?

    A. $8,585.55
    B. $9,585.55
    C. $8,085.55
    D. $9088.90
    E. NONE OF THE ABOVE

SOLUTION

IN ORDER TO DETERMINE HOW MUCH PRINCIPAL HAS BEEN PAID IT IS NECESSARY TO DETERMINE HOW MUCH OF THE PRINCIPAL REMAINS TO BE PAID. THE AMOUNT OF PRINCIPAL REMAINING TO BE PAID IS THE PRESENT WORTH OF THE REMAINING PAYMENTS, WHICH WOULD BE THE AMOUNT TO PAY OFF THE LOAD AFTER MAKING 18 PAYMENTS. THE AMOUNT OF PRINCIPAL PAID IS THEN THE DIFFERENCE BETWEEN THE AMOUNT BORROWED AND THE PRINCIPAL REMAINING TO BE PAID.

SINCE THE LOAN IS TO BE PAID MONTHLY, THE NUMBER OF TOTAL PAYMENTS IS 2 X 12 = 24, IF 18 PAYMENTS HAVE BEEN MADE THERE ARE 6 REMAINING.
NOTE THAT THE PROBLEM DOES NOT STATE WHAT THE MONTHLY PAYMENTS ARE.

A = P(A/P, i=1.0, n=24)
= 12,500(0.0471)
= 588.75

NOW THE REMAINING PRINCIPAL AFTER 18 PAYEMNTS CAN BE CALCULATED

P = A(P/A, i=1.0, n=6)
= 588.75(5.7955)
= 3,412.10

PRINCIPAL PAID IS THEREFORE 12,500 - 3,412.10 = 9,087.90

ANSWER IS D

11. For the loan described in Problem 10, how much would have to be paid to the bank to settle the loan after 15 years?

    A. $4,053.23
    B. $5,643.23
    C. $5,933.43
    D. $6,043.23
    E. $5,043.23

SOLUTION

FROM PROBLEM 10 A = 588.75

TO SETTLE LOAN AFTER 15 PAYMENTS = A(P/A, i=1.0, n=9)
= 588.75(8.5660)
= 5,043.23

ANSWER IS E

---

12. A debt of $300,000 due in 10 years is to be paid at maturity by means of a semiannual sinking fund. The borrower finds that only a 6% sinking fund is available at the present time, but feels that after 4 years 8% sinking fund semiannual interest will be available. Based on the borrowers assumption what semiannual equal payment into the sinking funds should be made?

    A. $13,781
    B. $10,828
    C. $15,438
    D. $9,947
    E. Problem can't be solved

SOLUTION

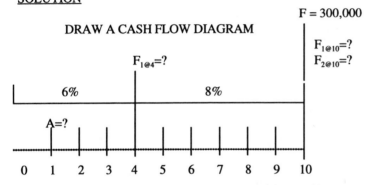

DRAW A CASH FLOW DIAGRAM

LET F_1 REPRESENT MONEY ASSOCIATED WITH THE 4% SINKING FUND
LET F_2 REPRESENT MONEY ASSOCIATED WITH THE 6% SINKING FUND

THE FUTURE VALUE OF THE TWO SINKING FUNDS MUST EQUAL 300,000 IN YEAR TEN. THE FIRST SINKING FUND LASTS FOR 4 YEARS AND RESULTS IN THE VALUE $F_{1@4}$ WHICH WILL NOW EARN INTEREST IN A SINGLE PAYMENT ACCOUNT FOR ANOTHER 6 YEARS. THE SECOND SINKING FUND LASTS FOR 6 YEARS AND ENDS IN THE 10th YEAR. THE SUM $F_{1@10}$ + $F_{2@10}$ = 300,000

REMEMBER THAT INTEREST IS SEMIANNUAL, THEREFORE i/PERIOD i/2 AND n PERIODS = 2 x YEARS

F = A(F/A, i=3, n=8) (F/P, i=3, n=12) + A(F/A, i=4, n=12)
300,000 = A(8.8923) (1.4258) + A(15.0258)
300,000 = A(12.6786) + A(15.0258)
    A = 300,000/27.7044
    A = 10,828.60

ANSWER IS B

13. A new grinding machine cost $15,000 and has a useful life of 15 years. If the asset is depreciated at 18.6%, what is its book value after 12 years based on declining balance depreciation?

    A. Can't be solved   D. $1,069
    B. $1,469   E. $1,269
    C. $984

SOLUTION

$B_x = P(1-f)^x$
$B_{12} = 15,000(1-0.186)^{12}$
     $= 1,269.36$

ANSWER IS E

For problems 14 through 16 consider a company that has purchased a piece of capital equipment for $100,000. The useful life of the asset is estimated to be 15 years with a salvage value of $12,000 at the end of the estimated life.

14. Using SLD what is the depreciation after 7 years?

    A. $6,666.67   D. $5,866.67
    B. $46,666.69   E. None of the above
    C. $41,066.69

SOLUTION

$D = (P - L)/n = (100,000 - 12,000)/15 = 5,866.67$
$D_7 = D(7) = 5,866.67(7) = 41,066.69$

ANSWER IS C

15. Using the sum of the years digits method, what is the depreciation after 7 years?

    A. $58,100   D. $60,100
    B. $63,600   E. $61,600
    C. $65,100

SOLUTION

$$D_{cum-x} = \frac{x(2n+1-x)(P-L)}{n(n+1)}$$

$$D_{cum-x} = \frac{7(2(15)+1-7)(100,000-12,000)}{15(15+1)}$$

        $= 61,600$

ANSWER IS E

16. What is the annual depreciation using the sum of the years digits depreciation method?

    A. $6,898.54
    B. $7,209.03
    C. $7,951.57
    D. $7,345.98
    E. Can't be answered

SOLUTION

THE ANSWER IS E, BECAUSE WITH SUM OF THE YEARS DIGITS EACH YEAR'S DEPRECIATION VALUE CHANGES. THE VALUE DOES NOT STAY CONSTANT.

ANSWER IS E

17. A new machine costing $144,500 will have a salvage value of $15,500 at the end of 10 years. Using the declining balance method of depreciation what will the book value be at the end of 9 years?

    A. $23,459
    B. $28,376
    C. $16,948
    D. $19,395
    E. $15,863

SOLUTION

$$f = 1 - \left|\frac{L}{P}\right|^{(1/n)} = 1 - \left|\frac{15,500}{144,500}\right|^{(1/10)}$$

$$= 0.200$$

$$B_x = P(1-f)^x$$
$$B_9 = 144,500(1 - 0.200)^9$$
$$= 19,394.46$$

ANSWER IS D

For problems 18 through 20 consider buying a new home. The selling price for the house is $375,000. You have enough money for a 15% down payment with the balance to be amortized at 12% compounded monthly for 15 years.

18. How much will the monthly payments be?

    A. $3,825
    B. $3,486
    C. $3,977
    D. $3,892
    E. $3,582

SOLUTION

IN ORDER TO FIND THE PAYMENTS THE AMOUNT FINANCED MUST BE DETERMINED

P = 375,000 - 375,000(0.15) = 318,750
i = 12%/12 = 1%/MONTH, n=15(12) = 180 MONTHLY PERIODS

A = P(A/P), i=1, n=180), NOTE THE A/P FACTOR MUST BE CALCULATED BECAUSE THE INTEREST TABLES DO NOT GO UP TO 180 INTEREST PERIODS

$$A/P = \frac{i(1+i)^N}{(1+i)^n - 1} = \frac{.01(1+.01)^{180}}{(1+.01)^{180} - 1} = 0.0120$$

A = 318,750(0.0120)
= 3,825

ANSWER IS A

19. How much interest is paid over the term of the loan?

- A. $396,570
- B. $547,639
- C. $369,750
- D. $688,500
- E. None of the above

SOLUTION

TOTAL OF PAYMENTS - PRINCIPAL = INTEREST PAID
3,825 (180) - 318,750 = 369,750

ANSWER IS C

20. If you decided to increase your down payment by borrowing $75,000 at 8% interest compounded semiannually for 15 years, how much would be saved in total interest payments?

- A. $37,000
- B. $27,000
- C. $17,000
- D. $7,000
- E. $32,000

SOLUTION

FIRST FIND THE PAYMENT ON THE 75,000 LOAN

A = P(A/P, i=4, n=30)   REMEMBER SEMIANNUAL INTEREST
= 75,000(0.0578)
= 4,335

INTEREST PAID = 4,335(30) - 75,000 = 55,050

NOW DETERMINE INTEREST PAID ON NEW MORTGAGE

P = 318,750 - 75,000 = 243,750

A = P(A/P, i=1, n=180)
= 243,750(0.0120)
= 2,925

INTEREST PAID ON MORTGAGE = 2,925(180) - 243,750 = 282,750

TOTAL INTEREST PAID FOR MORTGAGE AND LOAN = 55,000 + 282,750 = 337,800

FROM PROBLEM 19 INTEREST PAID JUST ON A MORTGAGE = 369,750

AMOUNT SAVED = 369,750 - 337,800 = 31,950 SAVINGS.

ANSWER IS E

21. The operating and maintenance cost for a mining machine are expected to be $11,000 in the first year, and increase by 800 a year during the 15-year life of the machine. What equal year end series of payments would cover these expenses over the life of the machine, if interest is 10%/year?

    A. $11,000   D. $15,223
    B. $13,423   E. $4,223
    C. $17,322

SOLUTION

DRAW A CASH FLOW DIAGRAM

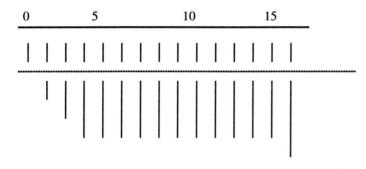

THE $800 IS ALREADY EXPRESSED AS AN ANNUAL STARTING COST, THE INCREASE IN ANNUAL COST OF $11,000/YR STARTING IN THE SECOND YEAR REPRESENTS A GRADIENT SERIES WHICH MUST BE CONVERTED TO A UNIFORM SERIES AND THEN ADDED TO THE $11,000 ANNUAL EXPENSE.

$A = G(A/G, i=10, n=15)$
    $= 800(5.2789)$
    $= 4,223.12$

TOTAL ANNUAL EXPENSE IS THEREFORE $11,000 + 4,223.12 = 15,223.12$

ANSWER IS D

Your company has been given two take over offers. Corporations A's offer is for $1,500,000 in cash upon the agreement and 10 annual payments $150,000. Corporation B's offer is for $2,000,000 now, $750,000 in one year, $500,000 in two years, and $250,000 in five years. The cost value of money to your company is 10%.

22. What is the present worth value of Corporation A's offer?

    A. $1,500,000   D. $150,008
    B. $1,140,915   E. $2,640,915
    C. $2,421,690

SOLUTION

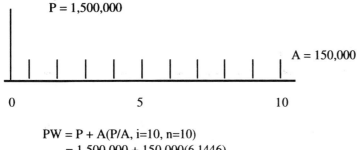

$$PW = P + A(P/A, i=10, n=10)$$
$$= 1,500,000 + 150,000(6.1446)$$
$$= 2,421,690$$

ANSWER IS C

23. What is the present worth value of Corporation B's offer?

- A. $3,250,250
- B. $1,250,250
- C. $4,250,000
- D. $2,000,000
- E. None of the above

SOLUTION

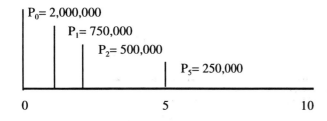

$$PW = P_0 + P_1(P/F, i=10, n=1) + p_2(P/F, i=10, n=2) = P_5(P/F, i=10, n=5)$$
$$= 2,000,000 + 750,000(0.9091) + 500,000(0.8264) + 250,000(0.6209)$$
$$= 3,250,250$$

ANSWER IS A

24. Which offer would you recommend to the owner of your company?

- A. Corporation A
- B. Corporation B
- C. Either Corporation
- D. Neither Corporation
- E. None of the above

SOLUTION

A COMPARISON OF THE ANSWERS TO PROBLEMS 2 AND 3 INDICATES THAT THE OFFER FROM CORPORATION B HAS THE HIGHER PW AND THEREFORE SHOULD BE RECOMMENDED.

ANSWER IS B

Five years ago you bought a home for $325,000. You put down 15% of the purchase price of the house and obtained a 20-year mortgage for the balance. The mortgage is at 9.5% and you make your payments annually. The closing costs on the house were $10,000 which you included in the mortgage amount. You have just made you seventh payment and have decided to purchase a new house.

If you sell your house and pay off the existing mortgage you must pay the bank a penalty of 3.5% of the current mortgage balance. In order to purchase the new house you need to put down $100,000.

25. How much are your current mortgage payments?

    A. $30,984   D. $34,289
    B. $32,489   E. $23,489
    C. $38,249

SOLUTION

$P = 325,000 - 325,000(0.15) + 10,000 = 286,250$
$A = P(A/P, i=9, n=20)$
   $= 286,250(0.1135)$
   $= 32,489.38$

ANSWER IS B

26. How much must you pay the bank to settle your existing mortgage?

    A. $236,886   D. $228,595
    B. $422,357   E. None of the above
    C. $245,178

SOLUTION

$P = A(P/A, i=9.5, n=13$
   $= 32,489.38(7.2912)$
   $= 236,886.56$
PLUS THE PENALTY $= 236,886.56(0.035) = 8,291.03$

TOTAL TO PAY OFF $= 236,886.56 + 8,291.03 = 245,177.59$

ANSWER IS C

27. How much must you sell the house for in order to make the down payment on the new house?

    A. $345,177   D. $475,177
    B. $395,177   E. NONE OF THE ABOVE
    C. $425,177

SOLUTION

SELLING PRICE $= P + P(0.035) + 100,000$
                   $= 236,886 + 8,291 + 100,000$
                   $= 345,177$

ANSWER IS A

28. How much did your house appreciate each year that you owned it?

    A. 6.2%
    B. 5.2%
    C. 7.2%
    D. 4.2%
    E. None of the above

SOLUTION

$$F = P(F/P, i=?, n=7)$$
$$345{,}177 = 325{,}000(F/P, i=?, n=7)$$
$$1.0621 = F/P$$

FROM THE INTEREST TABLES
$$@\ i = 0.75 \quad F/P = 1.0537$$
$$i = 1.00 \quad F/P = 1.0721$$

$$? = 0.75 + (0.25)\frac{1.0621 - 1.0537}{1.0721 - 1.0537}$$

$$? = 0.86\%$$

ANSWER IS E

29. An investment property cost $62,000 to purchase, another $7,500 was required to improve the property. The property produced annual return after deducting all expenses of 11,500 per year for 10 years, after which the property is worthless. What is the annual rate of return from this investment?

    A. 10.40%
    B. 10.00%
    C. 11.00%
    D. 11.40%
    E. 9.40%

SOLUTION

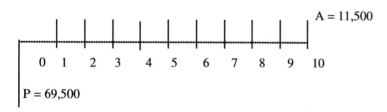

CALCULATE APPROXIMATE RATE OF RETURN

$$A = P(A/P, i=?, n=10)$$
$$11{,}500 = 69{,}500(A/P, i=?, n=10)$$
$$0.1655 = A/P$$

FROM THE INTEREST TABLES

$$@\ i = 10 \quad A/P = 0.1628$$
$$i = 11 \quad A/P = 0.1698$$
$$\text{ASSUME } i = 10\%$$

$0 = -P + A(P/A, i=10, n=10)$
$= -69{,}500 + 11{,}500(6.1446) = -69{,}500 + 70{,}662.9$
$= +1{,}162.90$

SINCE THE CALCULATION RESULTED IN A NUMBER > THEN ZERO, THE PW OF A NEEDS TO BE LOWERED BY RAISING THE INTEREST RATE, THEREFORE ASSUME i = 11% AND RECALCULATE.

$0 = -P + A(P/A, i=11, n=10)$
$= -69{,}500 + 11{,}500(5.8892) = -69{,}500 + 67{,}725.80$
$= 1{,}774.20$

BY INTERPOLATION

10%	1,162.90
?	0.00
11%	-1,774.20

$? = 10 + (1)\dfrac{0.0 - 1.162.90}{-1{,}774.20 - 1{,}162.90}$

$= 10.40\%$

ANSWER IS A

Mr. Engineer started working for the ABC Utility Company on his 25th birthday after graduating from college. Each year 5% of his salary was withheld and placed in a retirement account. The company also deposited an amount equal to 25% Mr Engineer's contribution into the account which has earned 8% interest. Mr Engineer's starting salary was $10,000 per year, with annual increases of $1,250 each year. Mr. Engineer retired on his 65th birthday.

30. How much money did Mr. Engineer deposit in his retirement account during his employment, expressed in year of hire dollars?

    A. $5,962   D. $11,840
    B. $7,878   E. $15,480
    C. $13,840

SOLUTION

THERE ARE TWO COMPONENTS TO MR. ENGINEER'S DEPOSITS, A UNIFORM SERIES AND A GRADIENT SERIES WHICH MUST BE CONVERTED TO A PRESENT WORTH IN YEAR 25.

UNIFORM SERIES

$PW = A(P/A, i=8, n=40)$
$= 10{,}000(0.05)(11.9246)$
$= 5{,}962.30$

GRADIENT SERIES

$PW = G(P/G, i=8, n=40)$
$= 1{,}250(0.05)(126.0430)$
$= 7{,}877.69$

TOTAL PW = 5,962.30 + 7,877.69 = 13,840 (ROUNDED OFF SINCE THE ANSWERS ARE EXPRESSED IN DOLLARS)

ANSWER IS C

31. How much money did the company put into Mr. Engineer's account?

A. $2,500.00
B. $3,460.00
C. $1,969.50
D. $312.50
E. $1490.50

SOLUTION

THE COMPANY DEPOSITED 25% OF THE AMOUNT MR ENGINEER DEPOSITED THEREFORE

TOTAL COMPANY DEPOSIT = 13,840(0.25) = 3,460

ANSWER IS B

32. How much money, including interest, will be in Mr. Engineer's account when he retires?

A. $75,168
B. $250,583
C. $457,279
D. $300,667
E. $375,834

SOLUTION

THE AMOUNT OF MONEY IN MR. ENGINEER'S ACCOUNT AT AGE 65 CAN BE FOUND BY TAKING THE ANSWERS TO QUESTIONS 10 AND 11 AND CALCULATING THE FUTURE WORTH OF THE SUM IN 40 YEARS.

$P_T = 13,840 + 3,460 = 17,300$
$F_{65} = P_T(F/P, i=8, n=40)$
$\quad = 17,300(21.7245)$
$\quad = 375,834$

ANSWER IS E

33. Mr. Engineer elects to receive his retirement money as a 20-year annuity with the first payment on his 65th birthday. How much will he receive each year?

A. $36,225
B. $39,124
C. $34,728
D. $38,297
E. $35,460

SOLUTION

THE SERIES OF PAYMENTS IS A BEGINNING OF YEAR SERIES WHICH CAN BE CONVERTED TO AN END OF YEAR SERIES BY MOVING THE F_{65} VALUE BACK 1 YEAR USING THE P/F FACTOR. SEE THE FOLLOWING CASH FLOW DIAGRAMS.

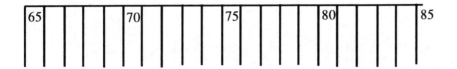

$A=?$

$F_{64} = F_{65}(P/F, i=8, n=1)$

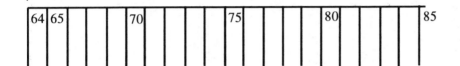

$A=?$

$F_{64} = F_{65}(P/F, i=8, n=1)$
$= 375,834(0.9259)$
$= 347,985$

$A = F_{64}(A/P, i=8, n=20)$
$= 347,985(0.1019)$
$= 35,460$

ANSWER IS E

A large manufacturing corporation pays $0.18 per mile to employees who use their personal automobiles for company business. Interest is at 9%. An engineer working for the company checked the records on his automobile and found the following data:

> First cost - $12,500.00
> Useful life - 8 years
> Trade in value - $3,000
> Cost of gas per mile travelled - $0.09
> Average cost of maintenance - $22.15 per month
> Insurance cost - $1,200/yr
> Miles per year - 18,000

34. What is the total annual owning and operating cost of the car based on straight line depreciation?

A. $3,824.24
B. $4,758.44
C. $3,404.24
D. $4,273.30
E. $5,024.24

SOLUTION

$$TAC = \frac{P-L}{n} + \left| (P-L)(i/2) \left| \frac{n+1}{n} \right| + L(i) \right| + AE$$

$$TAC = \frac{12,5000 - 3,000}{8} + \left| (12,500 - 3,000) \frac{(0.09)}{2} \frac{(8+1)}{8} + 3,000 (0.09) \right|$$

$$+ ((0.09)(18,000) + 22.15(12) + 1,200)$$

$= 1,187.50 + 750.94 + 3,085.80$
$= 5,024.24$

SOLUTION IS E

35. How many miles does the car above have to be driven, if 65% of the cars use is for the company, so the company allowance breaks even with owning the car?

    A. 38,151
    B. 18,551
    C. 28,115
    D. 43.251
    E. 22,454

SOLUTION

TO SOLVE SET UP AN EQUATION TAC = COMPANY ALLOWANCE WITH MILES PER YEAR AS THE UNKNOWN AND SOLVE

LET M = MILES/YEAR

$$\frac{P-L}{n} + I_a + AE = 0.18(M)(.65)$$

$(1,187.50 + 750.94 + (0.09)(M) + 22.15(12) + 1,200)(0.65) = 0.18(M)(0.65)$

$2,212.76 + 0.059(M) = 0.117(M)$
$M = 2,212.76/0.058$
$= 38,151$ MILES/YR

ANSWER IS A

A machine shop is considering a new piece of machinery. Three machines are under consideration. Use 14% interest and the following available data for your calculations:

	MACHINE X	MACHINE Y	MACHINE Z
First cost	$15,000	$17,000	$20,500
Annual expenses	$625	$515	$400
Five year overall	$1,000	$1,070	$1,200
Useful life - years	$15	$20	$25
Salvage value	$0	$0	$1,500

36. What is the annual cost of Machine X?

　　A. $3,067.00　　　　D. $3,218.30
　　B. $4,218.30　　　　E. $4,067.00
　　C. $2,593.30

SOLUTION

$$TAC = (P - L)(A/P, i,n) + L(i) + AE$$
$$= 15,000(A/P, i=14, n=15) + 625 + 1,000(A/F, i=14, n=5)$$
$$= 15,000(0.1628) + 625 + 1,000(0.1513)$$
$$= 2,442 + 625 + 151.3$$
$$= 3,218.30$$

ANSWER IS D

37. Which machine would you recommend for purchase?

　　A. Machine X　　　　D. Machine X & Y
　　B. Machine Y　　　　E. Machine Y & Z
　　C. Machine Z

SOLUTION

SINCE THE TAC WAS CALCULATED FOR MACHINE X, AN ANNUAL COST COMPARISON WILL THEREFORE BE USED. NOTE ANNUAL COST IS THE BEST APPROACH SINCE THE MACHINES HAVE DIFFERENT ECONOMIC LIVES.

FOR Machine Y

$$TAC = (P - L)(A/P, i,n) + L(i) + AE$$
$$= 17,000(A/P, i=14, n=20) + 515 + 1,075(A/F, i=14, n=5)$$
$$= 17,000(0.1501) + 515 + 1,075(0.1513)$$
$$= 2,551.70 + 515 + 1162.65$$
$$= 3,229.35$$

FOR Machine Z

TAC = (P - L)(A/P, i,n) + L(i) + AE
 = 19,000(A/P, i=14, n=25) + 1,5000(0.14) + 400 + 1,200(A/F, i=14, n=5)
 = 19,000(0.1455) + 210 + 400 + 1,200(0.1513)
 = 2,764.5 + 210 + 400 + 181.56
 = 3,556.06

ANSWER IS A

38. Considering unequal lives of the machines, if PW analysis was required, what period of time should the economic analysis be based on?

   A. 15 yrs
   B. 20 yrs
   C. 25 yrs
   D. 100 yrs
   E. 300 yrs

SOLUTION

THE PRESENT WORTH ANALYSIS SHOULD BE PERFORMED FOR THE LOWEST COMMON MULTIPLE, WHICH IS 300 YEARS.

ANSWER IS E

39. What is the total present worth of machine Z using a 100 year time period?

   A. $21,400
   B. $23,400
   C. $25,400
   D. $27,400
   E. $29,400

SOLUTION

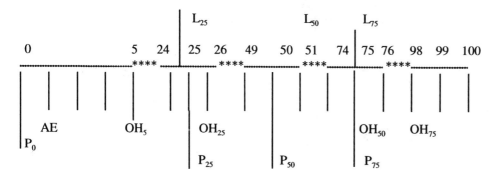

THE DETAILED PW CALCULATION DOES NOT HAVE TO BE PERFORMED IF YOU RECOGNIZE THAT THE TAC FOR MACHINE Z WAS CALCULATED AND THE PW OF THE TAC FOR 100 YEARS WILL PROVIDE THE ANSWER. IF YOU DID NOT RECOGNIZE THIS, USE THE CASH FLOW DIAGRAM TO DEVELOP THE DETAILED PW EQUATION.

PW = TAC(P/A, i=14, n=100)
 = 3,556.06(7.1428)
 = 25,400

ANSWER IS C

SOLUTIONS TO PRACTICE PROBLEMS

40. Using SFD, what is the book value of machine Y after 10 years?

    A. $18,479
    B. $12,070
    C. $16,496
    D. $7,835
    E. $13,384

SOLUTION

$B_x = P - (P - L)(A/F, i=14, n=20)(F/A, i=14, n=10)$
$= 17,000 - (17,000)(0.0101)(19.3373)$
$= 13,383.92$

ANSWER IS E

13

Simulated Exams

Eit Simulated Exam
AM Session

You have 90 minutes in which to work this scale "Simulated Exam", Morning (AM) session. You may use **only the NCEES Reference Handbook**, and battery operated silent non-print calculators. No loose papers and notes are allowed. All calculations must be done in the space allotted in the test booklet to the right of the questions.

You are to indicate all of your answers on the answer sheet provided. No credit will be given for anything written anywhere else. Be sure to blacken the chosen circle completely. In the event you erase, be sure to erase thoroughly. Only one answer will be accepted for each question. Multiple answers will result in an incorrect answer.

The questions will be grouped by subject as follows:

SUBJECT	QUESTION NUMBERS	OF QUESTIONS
Chemistry	1-4	4
Economics	5-9	4
Thermodynamics	10-16	7
Mathematics	17-24	7
Electrical Theory	25-31	7
Materials Science	32-35	4
Dynamics	36-40	5
Fluid Mechanics	41-45	5
Mechanics of Materials	46-51	5
Statics	52-56	5

1.
$$K_e = \frac{[C][D]^2}{[A]^3}$$

is the equilibrium constant for which of the reactions listed below?

a. $C + D_2 \leftrightarrow A_3$ c. $3A \leftrightarrow C + D^3$
b. $C + 3D \rightarrow A$ d. $3A \leftrightarrow C + 2D$
e. $A_3 \leftrightarrow C + D_2$

2. The hydroxyl ion concentration of a solution with a pH of 4 is:

a. 1×10 moles/L d. 4 moles/L
b. 1×10^{-10} moles/L e. 4×10^{-10} moles/L
c. 1×10^{10} moles/L

3. $2AgNO_3 + BaSO_4 \rightarrow Ag_2SO_4 + Ba(NO_3)_2$

The atomic weight for Ag = 107.88, for N = 14.0, for H = 1, for O = 16, and for S = 32. The reaction of 1 lb-mole of $AgNO_3$ will produce:

a. 311.76 lbs Ag_2SO_4 d. 73.94 lb-mole $Ba(NO_3)_2$
b. 155.88 lbs Ag_2SO_4 e. 64.11 lbs Barium
c. 1 Troy Oz Silver

4. The combination of an alkyl radical with a hydroxyl group forms:

a. and acid d. an alcohol
b. an aldehyde e. a carboxyl
c. a keytone

5. You are in charge of parts and must place the orders. You need 2000 parts per year and each part costs $25, and each time you order it costs you $50. Interest is 5%. How many times per year should you order parts?

a. 1.25 d. 7
b. 1.0 e. 3.5
c. 11.0

6. Let S be the accumulated sum, P the principal, invested, i the effective interest rate per compounding period, and n the number of compounding periods. Which of the following formulas correctly relate these quantities?

a. $S = P(1+in)$ d. $S = P(1+in)^{n-1}$
b. $S = P(1+i)^n$ e. $S = P(1+i)^{n-1}$
c. $S = P(1+n)^i$

7. Equipment is purchased for $6000. In 5 years the Salvage is expected to be $500. What is the book value of the equipment at the end of 3 years using SLD?

    a. $1400   d. $3300
    b. $2700   e. $5400
    c. $4200

8. With interest at 3.0% quarterly, $1000 will compound to how much after 5 years?

    a. $1203   d. $1806
    b. $2048   e. $1492
    c. $3698

9. A facility costs $300,000. What will be the monthly payment for principal and interest only for each $100,000 borrowed at 6.0% for 5 years?

    a. $2760   d. $1930
    b. $3800   e. $1330
    e. $8867

10. All of the following statements about work done on a process are true except:

    a. The work W depends on the process
    b. There can be an infinite number of specific heats between state points.
    c. If dQ amount of heat is transferred to W lb of a substance, the temperature change is dT.
    d. The change in internal energy is Q - W + ΔPV
    e. ΔU depends only on ΔT.

11. For any ideal gas, the volume is doubled while the absolute temperature is halved. The pressure will be:

    a. doubled
    b. quardupled
    c. quartered
    d. halved
    e. no change

12. The net entropy change for an ideal gas that undergoes no temperature change while tripling its volume will be:

    a. ΔS increases by a factor of 1.732
    b. ΔS decreases by a factor of 1.732
    c. ΔS increases by a factor of 3.0
    d. ΔS decreases by a factor of 3.0
    e. ΔS is constant

13. One of the following statements is true of the theoretical diesel cycle:

   a. It has two isentropic, one isobaric and one isochoric process.
   b. It has one polytropic, two isobaric and one adaibadic process.
   c. It is totally polytropic
   d. It has no net change on the entropy of the universe
   e. It burns gasohol.

14. A liquid has begun to freeze. A salt is added to the liquid. When the salt is dissolved:

   a. The rate of solidification increases
   b. The rate of solidification becomes zero
   c. The liquid warms up
   d. Sublimation begins
   e. The salt settles to the bottom of the tank

15. What is the enthalpy of 5 lbs of a fluid occupying 10 feet^3, with an internal energy of 1000 Btu/lb at a pressure of 1.5 atm?

   a. 5004.81 Btu   d. 2750.4 Btu
   b. 5040.81 Btu   e. 5000.28 Btu
   c. 36752 Btu

16. Air expands isentropically such that its pressure is increased by 50%. The initial temperature is 150 °F. What is the final temperature.

   a. 168.42 °F   d. 610 °R
   b. 224.91 °F   e. 672.98 °R
   c. 179 °F

17. What is the equation that must be satisfied by the coordinates of every point (x,y) that is equidistant from the origin and the point (1,1) ?

   a. $2x + 2y = 1$ d. $2x + 2y = 0$
   b. $x - y = 1$ e. $x + y = 1$
   c. $\frac{x}{2} + \frac{y}{2} = 1$

18. An Elementary game is played by rolling a die and drawing a ball from a bag containing 3 white and 7 black balls. The player wins by rolling a number less than 4 and drawing a black ball from the bag. What is the probability of winning on the 1st try?

   a. 13/20   d. 12/20
   b. 1/2   e. 7/1
   c. 7/20

19. Find the general solution to the following differential equation:

$$y\, dx - y^2\, dy = dy$$

a. $(x + y)^2 = 4y + C$ d. $2x = ye^{kt} + C$
b. $x + y = 2xy + C$ e. $y^2 = x^3 + C$
c. $2x = y^2 + \ln y^2 + C$

20. Let $\mathbf{A} = 2\mathbf{i} - 4\mathbf{j} + \mathbf{k}$, $\mathbf{B} = \mathbf{i} + \mathbf{j} - 3\mathbf{k}$, and $\mathbf{C} = -\mathbf{i} + 2\mathbf{j} + 2\mathbf{k}$. $(\mathbf{A} \times \mathbf{B}) \cdot (\mathbf{C} \times \mathbf{A})$ is ?

a. -5   d. -13
b. $11\mathbf{i} + 7\mathbf{j} + 6\mathbf{k}$ e. 15
c. 145

21. If $f(x) = \begin{vmatrix} x + 2 & x^2 \\ C x^{-2} & C \end{vmatrix}$

and C = constant

What is f'(x) ?

a. $x + C$ d. $x - C$
b. $3C$ e. 1
c. C

22. Which of the following is not true?

a. $\sin(x + y) = \sin x \cos y + \cos x \sin y$
b. $\sin x / \sin y = 2 \sin 1/2 (x + y) / \cos 1/2 (x - y)$
c. $\sin x + \sin y = 2 \sin 1/2 (x + y) \cos 1/2 (x - y)$
d. $\sin x - \sin y = 2 \cos 1/2 (x + y) \cos 1/2 (x - y)$
e. $\sin(x - y) = \sin x \cos y - \cos x \sin y$

23. If the function:

$$y = 3x^2$$

is integrated between the limits of $x = -3$ and $x = +3$, the result is:

a. 0   d. 54
b. 108   e. 27
c. 18

24. The Laplace Transform of a step function of magnitude a is:

a. $1(s + a)$ d. s/a
b. a/s e. $s + a$
c. $(s + a)/a$

25. For the circuit below, what is the voltage across the 50Ω resistor?

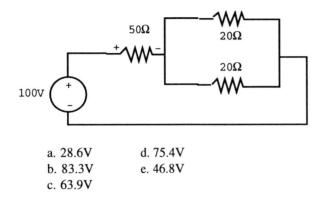

a. 28.6V
b. 83.3V
c. 63.9V
d. 75.4V
e. 46.8V

26. A parallel plate capacitor with area A, separation distance D, and separator permittivity e has a capacitance that is:

a. directly proportional to A and e, and inversely proportional to D
b. directly proportional to D and e, and inversely proportional to A
c. inversely proportional to A and D, but directly proportional to e
d. inversely proportional to e and D, but directly proportional to A
e. inversely proportional to A and e, but directly proportional to D

27. For the op amp circuit below, the voltage is:

a.
$$V_0 = -\frac{R_f}{R_i} V_I$$

b.
$$V_0 = -\frac{R_f}{R_1} V_I - \frac{R_f}{R_2} V_2$$

c.
$$V_0 = -j\frac{\omega L}{R_i} V_i$$

d.
$$V_0 = -j\frac{\omega L}{R(1 + j\omega RC)} V_i$$

e.
$$v_0 = -\frac{1}{RC}\int v_i dt$$

28. For the circuit shown below, the average power in watts dissipated in the balanced 3 phase load is:

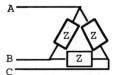

$Z = 10 + j5$
$V_{AC} = 120$ V

a. 2507 watts
b. 3456 watts
c. 5272 watts
d. 978 watts
e. 1728 watts

29. The inductance of a coil of wire is calculated using one of the formulae below:

a. $L = \dfrac{N \mu A}{l}$

b. $L = \dfrac{N \mu^2 A^2}{l}$

c. $L = \dfrac{N^2 \mu A}{l}$

d. $L = \dfrac{N \mu A^2}{l}$

e. $L = \dfrac{N^2 \mu A^2}{l}$

30. The switch in the circuit shown below has been closed for a long time. The expression to calculate L is:

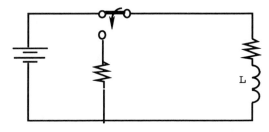

a. $i = I_0 \, e^{-RL/t}$

b. $i = I_0 \, e^{R/Lt}$

c. $i = I_0 \, e^{-t/RL}$

d. $i = I_0 \, e^{-(R/L)t}$

e. $i = I_0 \, e^{t/\tau}$

31. The power factor for the circuit below is most nearly:

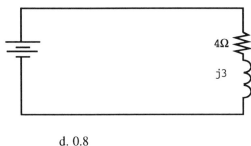

    a. 0.2   d. 0.8
    b. 0.4   e. 0.9
    c. 0.6

32. If the load on a test material is carried beyond the Ultimate Tensile Strength of the material, the material will:

    a. Return to its normal configuration
    b. Permanently maintain its deformed condition
    c. Continue bending until stress failure occurs
    d. Expand lengthwise
    e. Continue bending until strain failure occurs

33. Wrought iron coated with magnesium is left to the elements. Over enough time what will happen?

    a. The iron and the magnesium will form an irreversible chemical bond
    b. Magnesium will act as an anode and preferentially corrode
    c. Magnesium will act as a cathode and preferentially corrode
    d. Iron will act as an anode and preferentially corrode
    e. Iron will act as a cathode and preferentially corrode

34. Given the temperature/composition diagram below, at what point will the components of the mixture be totally miscible?

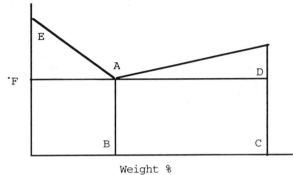

Weight %

a. A   d. D
b. B   e. E
c. C

35. Electrical and thermal conductivities for various materials differ relative magnitude and order. This phenomenon is due to:

    a. Electrons and heat travel at the same speed but encounter resistance differently.
    b. Heat and electrical energy are both wave forms that counteract each other's motion.
    c. Electrons and heat are conducted through metals by valence or free electrons and are influenced differently by lattice structure.
    d. Various materials have different densities.
    e. When heat and electrons move through a material, the material changes crystalline structure.

36. A 25000-lb truck is going up an incline of 4% and must maintain a speed of 40 mph. Neglecting friction, the engine must develop how much horsepower?

    a. 100hp   d. 120hp
    b. 109hp   e. 112hp
    c. 106hp

37. A vehicle is traveling at 50 mph on a level road and must come to a stop without its cargo sliding. The coefficient of friction between the vehicle and the cargo is 0.6. What is the minimum amount of time required for vehicle to come to a complete stop.

    a. 10 sec   d. 2.8 sec
    b. 5.2 sec   e. 3.8 sec
    c. 7.6 sec

38. Find the maximum height in meters.

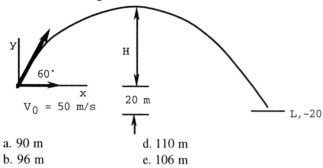

a. 90 m
b. 96 m
c. 101 m
d. 110 m
e. 106 m

39. The total distance traveled by the particle from t = 4 to t = 8 is:

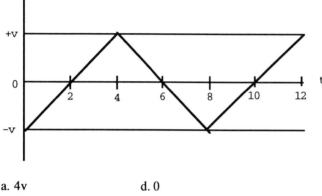

a. 4v
b. 8v
c. 10v
d. 0
e. -8v

40. For a planet, acceleration due to gravity is g, and the radius of the planet is r. If we move to a point 4r from the surface of the planet, the strength of the gravitational field will be:

a. g
b. 4g
c. g/16
d. g/4
e. 16g

41. When the load is applied as shown, the edge BC of the block moves 0.03mm to the right. Determine the shear modulus.

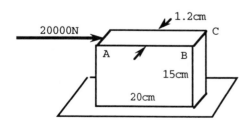

a. 50,300MPa
b. 41,700MPa
c. 38,600MPa
d. 32,500MPa
e. 26,200MPa

42. For the beam below the maximum moment occurs at:

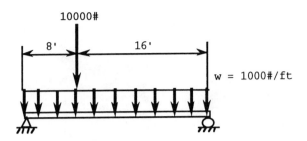

    a. 12 ft from the left
    b. 8ft 8in from left
    c. 8 ft from the left
    d. 10 ft from the left
    e. at the center of the beam

43. Determine the maximum shear stress in the shaft.

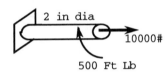

    a. 3160 psi   d. 6520 psi
    b. 4140 psi   e. 5730 psi
    c. 7340 psi

44. Brass cannot be used to reinforce concrete because:

    a. It is too dense
    b. It differs in shear capabilities
    c. It cannot carry enough load
    d. It does not stick to concrete
    e. It differs in coefficient of thermal expansion

45. What increase in termperature (°F) is necessary to cause a 1 in. dia 10 ft long steel rod with fixed ends to buckle. There is no initial stress.
    ($\alpha = 6.5 \times 10^{-6}$)
    a. 66   d. 46
    b. 26   e. 16
    c. 56

46. For a fluid the number that relates inertial force to compressibility force is?

    a. Reynolds Number   d. Drag Number
    b. Mach Number   e. Froude Number
    c. Weber Number

Problems 47 – 50
Consider the diagram of the pumping system as shown:

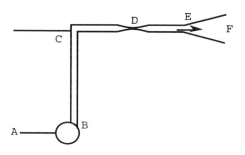

and the following equations:

a.
$$\bar{V}_a = C_v \sqrt{2gh}$$

d.
$$\bar{Q} = \left[(H + h)^{1.75} - h^{1.75}\right]^2$$

b.
$$\eta = \frac{hp\ out}{hp\ supplied}$$

e.
$$h_{xg} = \frac{k_e (V_1 - V_2)^2}{2g}$$

c.
$$\frac{P_1}{w} + H_p = \frac{P_2}{w} + Z_2$$

47. Which equation applies to the fluid as it passes through point D?

48. Which equation applies to the fluid as it passes from point B to point C?

49. Which equation applies to the fluid as it passes from point D to point E?

50. Which equation applies to the fluid as it passes from point A to point B?

51. For the system shown below, the pressure P is?

    a. 5.46psi   d. 8.92psi
    b. 6.87psi   e. 7.50psi
    e. 4.72psi

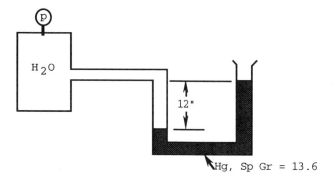

52. What is the moment of the force F about the origin O in the system below? All distances are in feet.

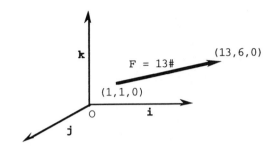

    a. -7k ft lbs   d. -17 ft lbs
    b. 7 ft lbs   e. 27k ft lbs
    c. 17k ft lbs

53. If equilibrium must be maintained, what is the maximum value that F can have?

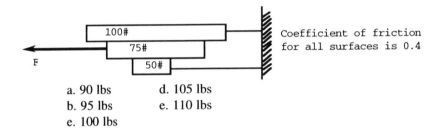

    a. 90 lbs   d. 105 lbs
    b. 95 lbs   e. 110 lbs
    e. 100 lbs

54. The magnitude of the reaction at B is:

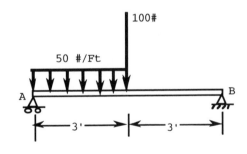

    a. 34 lbs   d. 88 lbs
    b. 44 lbs   e. 94 lbs
    c. 74 lbs

55. The y coordinate of the centroid of the area below is?

   a. 3.4   d. 4.6
   b. 2.6   e. 5.0
   c. 4.2

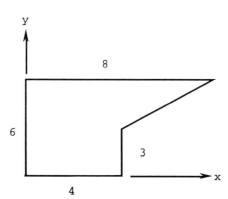

56. The force system below is best described as:

   a. Nonconcurrent, noncoplanar
   b. Concurrent
   c. Coplaner
   d. Concurrent and coplaner
   e. 2 dimensional

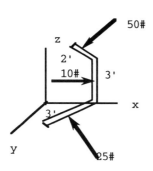

Eit Simulated Exam
AM Session
Answer Sheet

Name _____

Street _____

City _____ State _____

Zip _____

EIT Simulated Exam
PM Section

You will have 90 minutes in which to work on the simulated afternoon section of the fundamentals exam. You may use only the **NCEES Reference Handbook** and battery operated silent non printing calculators. You may not borrow or loan any of these materials during the exam. Any preliminary work must be done in the area to the right of the problems in this book.

The simulated exam consists of 26 required questions. Try to answer them all. Your score will be directly proportional to the number of correct answers. Every question has equal weight, and no points are subtracted for an incorrect answer.

You are to indicate all of your answers on the answer sheet provided. No credit will be given for anything written anywhere else. Be sure to blacken the chosen circle completely. In the event you erase, be sure to erase thoroughly. Only one answer will be accepted for each question. Multiple answers will result in an incorrect answer.

The questions are grouped as follows:

Engineering Economics	1 - 4
Electrical Circuits	5 - 8
Engineering Mechanics	9 - 15
Mathematics	16 - 22
Thermodynamics/Fluid Mechanics	23 - 26

ENGINEERING ECONOMICS

An automobile costs $5000.00 and has an estimated life of 5 years. Expenditure of extra money on certain additional maintenance items will extend this life to 10 years. Interest is 12.5%

1. How much should be spent for this additional maintenance?
   - a. $500
   - b. $450
   - c. $400
   - d. $350
   - e. $300

2. What annual payment is required in order to have the renewal cost available in 5 yrs.?
   - a. $960
   - b. $840
   - c. $780
   - d. $690
   - e. $573

3. If the interest rate was reduced to 12% per annum, compounded monthly, what is the effective interest rate?
   - a. 12.25%
   - b. 12.35%
   - c. 12.5%
   - d. 12.7%
   - e. none of the above

4. If expenditures such as gasoline, oil, tires, batteries, etc., amount to $2300 per year, insurance costs $1,000 per year, what is the Annual Cost for owning this car for the 5 year life if the salvage value is $1,000?

   - a. $43,000.00
   - b. $4,147.50
   - c. $4,548.60
   - d. $4,829.25
   - e. $3,829.25

ELECTRICAL THEORY

Problems 5-8 relate to the circuit shown below.

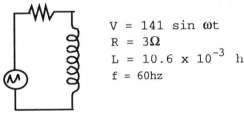

$V = 141 \sin \omega t$
$R = 3\Omega$
$L = 10.6 \times 10^{-3}$ h
$f = 60$ hz

The circuit shown is a series R-L hook with a sinusoidal input equal to V. $V = 141 \sin \omega t$; $R = 3\Omega$; $L = 10.6 \times 10^{-3}$ h; $f = 60$ Hz

5. What is the effective steady state current? (include the relative phase angle.
   a. 141 /0° b. 141 /45°
   c. 20 /-53.11° d. 100 /73.2°
   e. 5 /53.11°

6. Using the values of the components given compute the power factor.
   a. 0.7 b. 0.6
   c. 0.5 d. 0.8
   e. 0.78

7. Using the values given, what is the effective value of the voltage drop across the inductor? (include its phase angle.)
   a. 141 /89.21° b. 60 /-53.1°
   c. 80 /36.9° d. 100 /73.2°
   e. none of these

8. What is the power consumed by the load?
   a. 60 watts b. 123 watts
   c. 180 watts d. 319 watts
   e. 1200 watts

DYNAMICS

Problems 9-11 relate to the following problem statement.

The motion of a particle is defined by $a = 200 + 6t^2$ ft/sec². It has a velocity of 2 ft/sec and a distance $s = 0$, at time $= 0$ from a fixed point.

9. Determine the average velocity during the first 10 sec.
   a. 8004 fps b. 4002 fps
   c. 4440 fps d. 8040 fps
   e. 6030 fps

10. If a second particle has a motion defined by $v = 160t + 12t^2$, which equation expresses the length of time required for it reach the first particle? $S = 0$, and $t = 0$ for the second particle.
    a. $(t^3 - 8t^2 + 40t + 4$.
    b. $t(t^3 - 8t^2 + 40t + 4$.
    c. $t(t^4 - 8t^3 + 40t^2 + 4t$.
    d. $2t(t^3 - 8t^2 + 40t + 4$.
    e. $2(t^3 - 8t^2 + 40t + 4$.

11. If the weight of the particle with $v = 160t + 12t^2$ is 2 lbs, what constant force would be required to stop it if the force were applied at time $t = 2.5$ seconds. Assume that the stopping distance is 1 ft.

    a. 500 lbs   b. 5000 lbs
    c. 7000 lbs   d. 10,000 lbs
    e. 14000 lbs

MECHANICS OF MATERIALS

Questions 12 and 13 refer to the following problem statement:

A composite beam is made of wood and steel, as shown: Assume $E_{stl} = 20 \times E_{wood}$

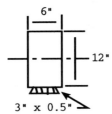

12. Determine the moment of inertia about the neutral axis.
    a. 1688 in^4 b. 3459 in^4
    c. 2586 in^4 d. 1248 in^4
    e. none of these

13. The function of the connectors in a composite unit such as this is to:
    a. Help balance the loads in the beam
    b. Transfer moment from the plates to the beam
    c. Transfer horizontal shear from the plate to the beam
    d. Transfer vertical shear from the plate to the beam
    e. It has something to do with diagonal tension

STATICS

Problems 14 and 15 refer to the beam shown below which carries a moving load of 1500 lbs. Assume the beam to be supported at B by a rod BC as shown. The supports at A and C and the joint at B permit unrestricted rotation.

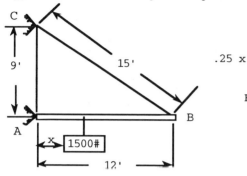

Beam Cross Section

Beam Data
Modulus of Elasticity $E = 30 \times 10^6$
Moment of Inertia $I = 20$ in^4
Cross Sectional Area $A = 3.375$ in^2
Rod Data
Modulus of Elasticity $E = 30 \times 10^6$
Cross Sectional Area $A = 0.4418$ in^2

14. The maximum unit flexural stress in the beam is most nearly
    a. 3000 psi   b. 6000 psi
    c. 8000 psi   d. 16000 psi
    e. 32000 psi

15. The maximum unit tensile stress in the rod BC is most nearly.
    a. 3000 psi   b. 6000 psi
    c. 8000 psi   d. 16000 psi
    e. 32000 psi

MATHEMATICS

16. The sum of the circumferences of two circles is 36π. The sum of the areas of the circles is 180π. The radius of one of the circles is:
    a. 10   b. 9
    c. 8   d. 7
    e. 6

17. Ground salt is poured in a pile on the ground at the rate of 4 cfm. it forms a conical pile whose height is 30% of the diameter of the base. How fast is the height of the pile increasing when the base is 5 ft. in diameter?
    a. .458 fpm   b. .687 fpm
    c. .204 fpm   d. 2.83 fpm
    e. 5.76 fpm

18. What is the general solution of the differential equation $ydx - y^2dy = dy$?
    a. $x = 3y^2 + 2y + C$
    b. $x = y^3/3 + y + C$
    c. $x = \csc y + C$
    d. $x = y^2/2 + \ln y + c$
    e. $x \sin y \cos y + \ln y + C$

19. A 3-ft long wire is cut into two pieces. One piece is used to form a square and the other piece used to form a circle. If the sum of the enclosed areas is to be a minimum, what length of the wire should be used to form the circle?
    a. 0.14 ft.   b. 0.56 ft.
    c. 1.12 ft.   d. 1.32 ft.
    e. none of the above

20. My pencil holder contains 3 red pencils, 4 green pencils and 3 blue pencils. My secretary has a pencil holder nearby which contains 4 blue pencils, 4 red pencils and 2 green pencils. What is the probability of my reaching out and grasping a blue pencil?
    a. 7/1   b. 12/1
    c. 7/10   d. 12/20
    e. 7/20

Problems 21 and 22 refer to the figure below.
Line CD is Perpendicular to line AB.

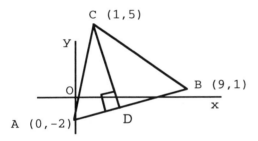

21. What is the slope of line BD?
    a. -5   b. .33333
    c. -.33333   d. -3
    e. 3

22. What is the area of triangle ABC?
    a. $5\sqrt{10}$ b. $6\sqrt{10}$
    c. 26   d. 30
    e. 60

THERMODYNAMICS

Problems 23 and 24 refer to the following problem statement

Atmospheric air inlet to an air cooler has a specific humidity of .009 lbs water vapor/lbs dry air at 80°F. Calculate the following.

23. Density of dry air

    a. .0726 lb/ft³   b. .0680 lb/ft³
    c. .0459 lb/ft³   d. .0472 lb/ft³
    e. .0858 lb/ft³

24. Enthalpy of the water vapor

    a. 41.5 Btu/lbs dry air   b. 32.5 Btu/lbs dry air
    c. 29.0 Btu/lbs dry air   d. 39.6 Btu/lbs dry air
    e. 24.8 Btu/lbs dry air

FLUID MECHANICS

Problems 25 and 26 refer to the following problem statement.

A radial gate, 5 ft wide, is installed as shown.

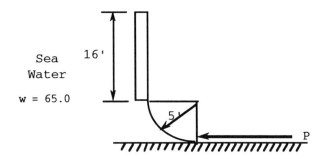

25. Neglecting the weight of the gate and any mechanical forces, such as friction, what force P is required to open the gate?
(Answers are rounded off.)

    a. 100,000 lbs   b. 50,000 lbs
    c. 25,000 lbs   d. 12,500 lbs
    e. 0 lbs

26. What force acts against the sea wall?

    a. 41600 lb.   b. 40960 lb.
    c. 79872 lb.   d. 39936 lb.
    e. none of these

EIT Simulated Exam
PM Section
Answer Sheet

1	(a)	(b)	(c)	(d)	(e)		14	(a)	(b)	(c)	(d)	(e)
2	(a)	(b)	(c)	(d)	(e)		15	(a)	(b)	(c)	(d)	(e)
3	(a)	(b)	(c)	(d)	(e)		16	(a)	(b)	(c)	(d)	(e)
4	(a)	(b)	(c)	(d)	(e)		17	(a)	(b)	(c)	(d)	(e)
5	(a)	(b)	(c)	(d)	(e)		18	(a)	(b)	(c)	(d)	(e)
6	(a)	(b)	(c)	(d)	(e)		19	(a)	(b)	(c)	(d)	(e)
7	(a)	(b)	(c)	(d)	(e)		20	(a)	(b)	(c)	(d)	(e)
8	(a)	(b)	(c)	(d)	(e)		21	(a)	(b)	(c)	(d)	(e)
9	(a)	(b)	(c)	(d)	(e)		22	(a)	(b)	(c)	(d)	(e)
10	(a)	(b)	(c)	(d)	(e)		23	(a)	(b)	(c)	(d)	(e)
11	(a)	(b)	(c)	(d)	(e)		24	(a)	(b)	(c)	(d)	(e)
12	(a)	(b)	(c)	(d)	(e)		25	(a)	(b)	(c)	(d)	(e)
13	(a)	(b)	(c)	(d)	(e)		26	(a)	(b)	(c)	(d)	(e)

Name_____

Address_____

City_____State_____

Zip_____

Solutions for the
Eit Simulated Exam
AM Session

You have 90 minutes in which to work this scale "Simulated Exam", Morning (AM) session. You may use **only the NCEES Reference Handbook,** and battery operated silent non-print calculators. No loose papers and notes are allowed. All calculations must be done in the space allotted in the test booklet to the right of the questions.

The questions will be grouped by subject as follows:

SUBJECT	QUESTION NUMBERS	OF QUESTIONS
Chemistry	1-4	4
Economics	5-9	4
Thermodynamics	10-16	7
Mathematics	17-24	7
Electrical Theory	25-31	7
Materials Science	32-35	4
Dynamics	36-40	5
Fluid Mechanics	41-45	5
Mechanics of Materials	46-51	5
Statics	52-56	5

1.
$$K_e = \frac{[C][D]^2}{[A]^3}$$

is the equilibrium constant for which of the reactions listed below?

a. $C + D_2 \leftrightarrow A_3$ c. $3A \leftrightarrow C + D^3$
b. $C + 3D \rightarrow A$ d. $3A \leftrightarrow C + 2D$
e. $A_3 \leftrightarrow C + D_2$

Answer (c)

2. The hydroxyl ion concentration of a solution with a pH of 4 is:

a. 1×10 moles/L   d. 4 moles/L
b. 1×10^{-10} moles/L   e. 4×10^{-10} moles/L
c. 1×10^{10} moles/L

Answer (b)

3. $2AgNO_3 + BaSO_4 \rightarrow Ag_2SO_4 + Ba(NO_3)_2$

The atomic weight for Ag = 107.88, for N = 14.0, for H = 1, for O = 16, and for S = 32. The reaction of 1 lb-mole of $AgNO_3$ will produce:

a. 311.76 lbs Ag_2SO_4 d. 73.94 lb-mole $Ba(NO_3)_2$
b. 155.88 lbs Ag_2SO_4 e. 64.11 lbs barium
c. 1 Troy Oz Silver

Answer (b)

The sum of the molecular weights of $AgSO_4$ is divided by 2 because 1 lb-mole $AgNO_3$ will produce 0.5 lb-mole of Ag_2SO_4.

4. The combination of an alkyl radical with a hydroxyl group forms:

a. an acid   d. an alcohol
b. an aldehyde   e. a carboxyl
c. a ketone

Answer (d)

5. You are in charge of parts and must place the orders. You need 2000 parts per year and each part costs $25, and each time you order it costs you $50. Interest is 5%. How many times per year should you order parts?

a. 1.25   d. 7
b. 1.0   e. 3.5
c. 11.0

Answer (d)

Let x be the EOQ, then the total cost is:

$$C = 2000(25) + \frac{50\,(2000)}{x} + .05\,(25x + 50)$$

Take the first derivative and equate to 0:

$$0 = -\frac{100{,}000}{x^2} + 1.25x$$

x = 282 parts per order
2000/282 = 7.06 orders per year

6. Let S be the accumulated sum, P the principal, invested, i the effective interest rate per compounding period, and n the number of compounding periods. Which of the following formulas correctly relate these quantities?

a. $S = P(1+in)$
b. $S = P(1+i)^n$
c. $S = P(1+n)^i$
d. $S = P(1+in)^{n-1}$
e. $S = P(1+i)^{n-1}$

Answer (b)

7. Equipment is purchased for $6000. In 5 years the Salvage is expected to be $500. What is the book value of the equipment at the end of 3 years using SLD?

a. $1400
b. $2700
c. $4200
d. $3300
e. $5400

$$6000 - \left[(6000 - 500)\frac{3}{5}\right] = 2700$$

Answer (b)

8. With interest at 3.0% quarterly, $1000 will compound to how much after 5 years?

a. $1203
b. $2048
c. $3698
d. $1806
e. $1492

From the Interest tables for a single payment and given
P = 1000 to find F for n = 20 and i = 3, read 1.8061.
F is 1000 x 1.8061 = 1806.

Answer (d)

9. A facility costs $300,000. What will be the monthly payment for principal and interest only for each $100,000 borrowed at 6.0% for 5 years?

a. $2760
b. $3800
e. $8867
d. $1930
e. $1330

Convert the interest rate to months: 6/12 = .5%
Calculate the periods as 5 x 12 = 60

Given P = 100,000, find A. From the interest tables the capital recovery factor A/P is .0193.

.0193 x 100000 = $1930/mo

Answer (d)

10. All of the following statements about work done on a process are true except:

    a. The work W depends on the process
    b. There can be an infinite number of specific heats between state points.
    c. If dQ amount of heat is transferred to W lb of a substance, the temperature change is dT.
    d. The change in internal energy is Q - W + ΔPV
    e. ΔU depends only on ΔT.

Answer (d)

11. For any ideal gas, the volume is doubled while the absolute temperature is halved. The pressure will be:

    a. doubled
    b. quadrupled
    c. quartered
    d. halved
    e. no change

Since $V_2 = 2V_1$

$$P_2 = \frac{P_1 V_1}{2V_1} = \frac{P_1}{2}$$

and $T_2 = T_1/2$

$$P_2 = \frac{P_1 \frac{T_1}{2}}{T_1} = \frac{P_1}{2}$$

Combining results in $P_2 = P_1/4$

Answer (c)

12. The net entropy change for an ideal gas that undergoes no temperature change while tripling its volume will be:

    a. ΔS increases by a factor of 1.732
    b. ΔS decreases by a factor of 1.732
    c. ΔS increases by a factor of 3.0
    d. ΔS decreases by a factor of 3.0
    e. ΔS is constant

Answer (c)

13. One of the following statements is true of the theoretical diesel cycle:

   a. It has two isentropic, one isobaric and one isochoric process.
   b. It has one polytropic, two isobaric and one adiabatic process.
   c. It is totally polytropic
   d. It has no net change on the entropy of the universe
   e. It burns gasohol.

<u>Answer (a)</u>

14. A liquid has begun to freeze. A salt is added to the liquid. When the salt is dissolved:

   a. The rate of solidification increases
   b. The rate of solidification becomes zero
   c. The liquid warms up
   d. Sublimation begins
   e. The salt settles to the bottom of the tank

<u>Answer (b)</u>

15. What is the enthalpy of 5 lbs of a fluid occupying 10 ft³, with an internal energy of 1000 Btu/lb at a pressure of 1.5 atm?

   a. 5004.81 Btu   d. 2750.4 Btu
   b. 5040.81 Btu   e. 5000.28 Btu
   c. 36752 Btu

$$u = 5 \text{ lbs} \times 1000 \frac{\text{Btu}}{\text{lb}} = 5000 \text{Btu}$$

$$pv = (1.5 \text{ atm})\left(14.7 \frac{\text{lbs}}{\text{atm in}^2}\right)\left(\frac{144 \text{ in}^2}{\text{ft}^2}\right)\left(\frac{\text{Btu}}{778 \text{ ft-lbs}}\right)(10 \text{ ft}^3) = 40.81 \text{Btu}$$

$$h = u + pv = 5040.81 \text{ Btu}$$

<u>Answer (b)</u>

16. Air expands isentropically such that its pressure is increased by 50%. The initial temperature is 150 °F. What is the final temperature.

   a. 168.42 °F   d. 610 °R
   b. 224.91 °F   e. 672.98 °R
   c. 179 °F

$$\frac{T_2}{T_1} = \left(\frac{P_2}{P_1}\right)^{\frac{k-1}{k}}$$

$$T_2 = T_1 \left(\frac{P_2}{P_1}\right)^{\frac{k-1}{k}}$$

$$T_2 = (150 + 460)\left(\frac{1.5}{1}\right)^{\frac{1.4-1}{1.4}} = (684.91 - 460) = 224.91 °F$$

<u>Answer (b)</u>

17. What is the equation that must be satisfied by the coordinates of every point (x,y) that is equidistant from the origin and the point (1,1) ?

    a. $2x + 2y = 1$ d. $2x + 2y = 0$
    b. $x - y = 1$ e. $x + y = 1$
    c. $\frac{x}{2} + \frac{y}{2} = 1$

The locus of points (x,y) forms the perpendicular bisector of the point drawn between (0,0) and (1,1). This line must pass through (x,y) and $\left(\frac{1}{2}, \frac{1}{2}\right)$

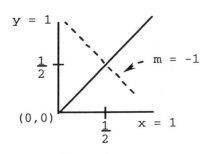

$$\frac{(y - y_1)}{(x - x_1)} = m; \quad \frac{\left(y - \frac{1}{2}\right)}{\left(x - \frac{1}{2}\right)} = -1, \quad x + y = 1$$

<u>Answer (e)</u>

18. An elementary game is played by rolling a die and drawing a ball from a bag containing 3 white and 7 black balls. The player wins by rolling a number less than 4 and drawing a black ball from the bag. What is the probability of winning on the first try?

    a. 13/20   d. 12/20
    b. 1/2   e. 7/1
    c. 7/20

The probability of rolling a number less than 4 is that of rolling a 1, 2, or 3. This is 1/2. The probability of drawing a black ball is 7/(7+3) or 7/10. The probability of the occurance of two independent events is the product of the probability of the two events or:

$$\frac{1}{2} \times \frac{7}{10} = \frac{7}{20}$$

<u>Answer (c)</u>

19. Find the general solution to the
    following differential equation:

    $$y \, dx - y^2 \, dy = dy$$

    a. $(x + y)^2 = 4y + C$ d. $2x = ye^{kt} + C$
    b. $x + y = 2xy + C$ e. $y^2 = x^3 + C$
    c. $2x = y^2 + \ln y^2 + C$

Separate variables and integrate

$$y \, dx = (1 + y^2) \, dy$$

$$\int dx = \int \frac{dy}{y} + \int y \, dy$$

$$x = \ln y + \frac{y^2}{2} + C'$$

$$2x = \ln y^2 + y^2 + C$$

Answer (c)

20. Let $\mathbf{A} = 2\mathbf{i} - 4\mathbf{j} + \mathbf{k}$, $\mathbf{B} = \mathbf{i} + \mathbf{j} - 3\mathbf{k}$, and $\mathbf{C} = -\mathbf{i} + 2\mathbf{j} + 2\mathbf{k}$. $(\mathbf{A} \times \mathbf{B}) \cdot (\mathbf{C} \times \mathbf{A})$ is ?

    a. -5   d. -13
    b. $11\mathbf{i} + 7\mathbf{j} + 6\mathbf{k}$ e. 15
    c. 145

$$\mathbf{A} \times \mathbf{B} = \begin{vmatrix} \mathbf{i} & \mathbf{j} & \mathbf{k} \\ 2 & -4 & 1 \\ 1 & 1 & -3 \end{vmatrix}$$

$$= (12\mathbf{i} + \mathbf{j} + 2\mathbf{k}) - (\mathbf{i} - 6\mathbf{j} - 4\mathbf{k})$$

$$= 11\mathbf{i} + 7\mathbf{j} + 6\mathbf{k}$$

$$\mathbf{C} \times \mathbf{A} = \begin{vmatrix} \mathbf{i} & \mathbf{j} & \mathbf{k} \\ -1 & 2 & 2 \\ 2 & -4 & 1 \end{vmatrix}$$

$$= (2\mathbf{i} + 4\mathbf{j} + 4\mathbf{k}) - (-8\mathbf{i} - \mathbf{j} + 4\mathbf{k})$$

$$= 10\mathbf{i} + 5\mathbf{j}$$

$(\mathbf{A} \times \mathbf{B}) \cdot (\mathbf{C} \times \mathbf{A}) = (11 \times 10) + (7 \times 5) + (6 \times 0) = 145$

Answer (c)

21.

$$\text{If } f(x) = \begin{vmatrix} x + 2 & x^2 \\ C\,x^{-2} & C \end{vmatrix}$$

and C = constant

What is f'(x) ?

a. x + C   d. x - C
b. 3C   e. 1
c. C

Expand the determinant:

$$f(x) = [(x + 2) \cdot (C)] - [(Cx^{-2} \cdot x^2)]$$

$$= Cx + 2C + C$$

Differentiate

$$f(x) = C$$

Answer (c)

22. Which of the following is not true?

a. sin (x + y) = sin x cos y + cos x sin y
b. sin x / sin y = 2 sin 1/2 (x + y) / cos 1/2 (x - y)
c. sin x + sin y = 2 sin 1/2 (x + y) cos 1/2 (x - y)
d. sin x - sin y = 2 cos 1/2 (x + y) cos 1/2 (x - y)
e. sin (x - y) = sin x cos y - cos x sin y

Answer (b)

23. If the function:

$$y = 3x^2$$

is integrated between the limits of x = -3 and x = +3, the result is:

a. 0   d. 54
b. 108   e. 27
c. 18

By the power rule for integration:

$$y = \left.\frac{3x^3}{6}\right|_{-3}^{3} = 27 - (-27) = 54$$

Answer (d)

24. The Laplace Transform of a step function of magnitude a is:

   a. I (s + a)   d. s/a
   b. a/s   e. s + a
   c. (s + a)/ a

Answer (b)

25. For the circuit below, what is the voltage across the 50Ω resistor?

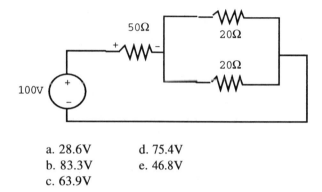

   a. 28.6V   d. 75.4V
   b. 83.3V   e. 46.8V
   c. 63.9V

The circuit simplifies to two resistors in series, the 50Ω and the 2 parallel 20Ω which are equivalent to 10Ω.

$$V = 100 = 50\,I + 10\,I: \quad I = 1.166A$$

$$V_{50} = IR_{50} = 1.166 \times 50 = 83.3V$$

Answer (b)

26. A parallel plate capacitor with area A, separation distance D, and separator permitivity e has a capacitance that is:

   a. directly proportional to A and e, and inversely proportional to D
   b. directly proportional to D and e, and inversely proportional to A
   c. inversely proportional to A and D, but directly proportional to e
   d. inversely proportional to e and D, but directly proportional to A
   e. inversely proportional to A and e, but directly proportional to D

Answer (a)

27. For the op amp circuit below, the voltage is:

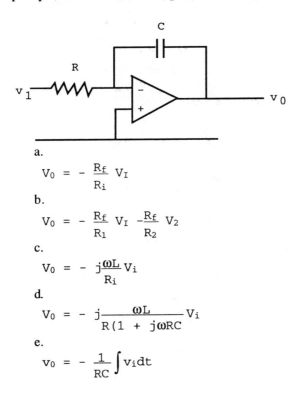

a.
$$V_0 = -\frac{R_f}{R_i} V_I$$

b.
$$V_0 = -\frac{R_f}{R_1} V_I - \frac{R_f}{R_2} V_2$$

c.
$$V_0 = -\frac{j\omega L}{R_i} V_i$$

d.
$$V_0 = -j\frac{\omega L}{R(1 + j\omega RC)} V_i$$

e.
$$v_0 = -\frac{1}{RC} \int v_i dt$$

Answer (e)

28. For the circuit shown below, the average power in watts dissipated in the balanced 3 phase load is:

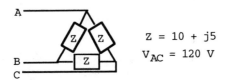

Z = 10 + j5
V_{AC} = 120 V

a. 2507 watts   d. 978 watts
b. 3456 watts   e. 1728 watts
c. 5272 watts

$$\theta_{ph} = \tan^{-1} \frac{5}{10} = 26.6°$$

$V_{ph} = 120$

$$I_{ph} = \frac{120}{|10 + j5|} = 10.73$$

$P_L = 3 V_{ph} I_{ph} \cos \theta_{ph}$
= 3 x 120 x 10.73 x .8942 = 3456w

Answer (b)

29. The inductance of a coil of wire is calculated using one of the formulae below:

a. $L = \dfrac{N \mu A}{l}$

b. $L = \dfrac{N \mu^2 A^2}{l}$

c. $L = \dfrac{N^2 \mu A}{l}$

d. $L = \dfrac{N \mu A^2}{l}$

e. $L = \dfrac{N^2 \mu A^2}{l}$

<u>Answer (c)</u>

30. The switch in the circuit shown below has been closed for a long time. The expression to calculate L is:

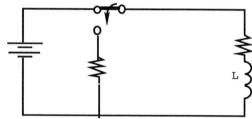

a. $i = I_0 e^{-RL/t}$

b. $i = I_0 e^{R/Lt}$

c. $i = I_0 e^{-t/RL}$

d. $i = I_0 e^{-(R/L)t}$

e. $i = I_0 e^{t/\tau}$

<u>Answer (d)</u>

31. The power factor for the circuit below is most nearly:

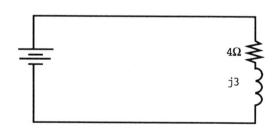

a. 0.2
b. 0.4
c. 0.6
d. 0.8
e. 0.9

$Z = 4 + j3 = 5 \angle 36.9°$
$\theta = 36.9°$
$pf = \cos \theta = \cos(36.9°) = .79968$

<u>Answer (d)</u>

32. If the load on a test material is carried beyond the ultimate tensile strength of the material, the material will:

    a. Return to its normal configuration
    b. Permanently maintain its deformed condition
    c. Continue bending until stress failure occurs
    d. Expand lengthwise.
    e. Continue bending until strain failure occurs

<u>Answer (e)</u>

33. Wrought iron coated with magnesium is left to the elements. Over enough time what will happen?

    a. The iron and the magnesium will form an irreversible chemical bond
    b. Magnesium will act as an anode and preferentially corrode
    c. Magnesium will act as a cathode and preferentially corrode
    d. Iron will act as an anode and preferentially corrode
    e. Iron will act as a cathode and preferentially corrode

<u>Answer (b)</u>

34. Given the temperature/composition diagram below, at what point will the components of the mixture be totally miscible?

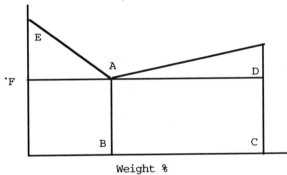

    a. A   d. D
    b. B   e. E
    c. C

Point A is the eutectic point.

<u>Answer (a)</u>

35. Electrical and thermal conductivities for various materials differ relative magnitude and order. This phenomenon is due to:

    a. Electrons and heat travel at the same speed but encounter resistance differently.
    b. Heat and electrical energy are both wave forms that counteract each other's motion.
    c. Electrons and heat are conducted through metals by valence or free electrons and are influenced differently by lattice structure.
    d. Various materials have different densities.
    e. When heat and electrons move through a material, the material changes crystalline structure.

Answer (c)

36. A 25,000-lb truck is going up an incline of 4% and must maintain a speed of 40 mph. Neglecting friction, the engine must develop how much horsepower?

    a. 100hp   d. 120hp
    b. 109hp   e. 112hp
    c. 106hp

The power required is $-\mathbf{F \cdot V}$. As friction is neglected only the weight is considered. The angle between $\mathbf{V}$ and $\mathbf{W}$ is $90° + \tan^{-1} .04 = 92.29°$.

$$-\mathbf{F} \cdot \mathbf{V} = \left(\frac{-1}{550}\right) 2500 \left(\frac{35 \times 5280}{3600}\right) \cos 92.29°$$

$$-\mathbf{F} \cdot \mathbf{V} = 106.4 \text{ hp}$$

Answer (c)

37. A vehicle is traveling at 50 mph on a level road and must come to a stop without its cargo sliding. The coefficient of friction between the vehicle and the cargo is 0.6. What is the minimum amount of time required for vehicle to come to a complete stop.

    a. 10 sec   d. 2.8 sec
    b. 5.2 sec   e. 3.8 sec
    c. 7.6 sec

To avoid sliding ma/mg must be ≤ 0.6. Thus
$$a_{max} = 0.6 \times 32.2 = 19.37 \text{ ft/sec}^2$$

$$V = \frac{50 \times 5280}{3600} = 73.33 \frac{ft}{sec}$$

$$t = \frac{73.33}{19.32} = 3.798 \text{sec}$$

Answer (e)

38. Find the maximum height in meters.

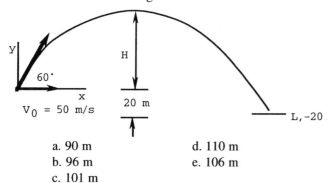

a. 90 m
b. 96 m
c. 101 m
d. 110 m
e. 106 m

$$H = \frac{V_0^2 \sin^2\theta}{2g} = \frac{50^2 \times .87^2}{2 \times 9.8} = 95.67 m$$

Answer (b)

39. The total distance traveled by the particle from t = 4 to t = 8 is:

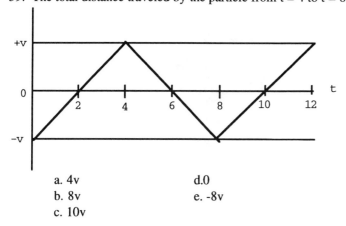

a. 4v
b. 8v
c. 10v
d. 0
e. -8v

The distance traveled is the area between the curve and the line v=0. Since the areas of the triangles above the curve between 4 and 6 is the same in magnitude but opposite in sign as the area between 6 and 8, the distance traveled is 0.

Answer (d)

40. For a planet, acceleration due to gravity is g, and the radius of the planet is r. If we move to a point 3r from the surface of the planet, the strength of the gravitational field will be:

    a. g   d. g/4
    b. 4g   e. 16g
    c. g/16

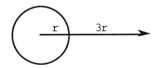

Total distance from the center of the planet is 4r

<u>Answer (c)</u>

41. When the load is applied as shown, the edge BC of the block moves 0.03mm to the right. Determine the shear modulus.

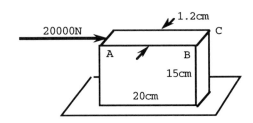

    a. 50,300 MPa   d. 32,500 MPa
    b. 41,700 MPa   e. 26,200 MPa
    c. 38,600 MPa

$$G = \frac{FL}{A \Delta L} = \frac{20000 \times .15}{.012 \times .2 \times .00003} = 41,700 \text{ MPa}$$

<u>Answer (b)</u>

42. For the beam below the maximum moment occurs at:

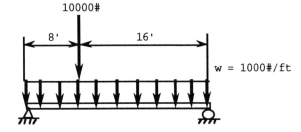

a. 12 ft from the left
b. 8 ft 8 in. from left
c. 8 ft from the left
d. 10 ft from the left
e. at the center of the beam

The maximum moment is when the point of zero shear:

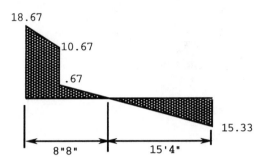

Answer (b)

43. Determine the maximum shear stress in the shaft.

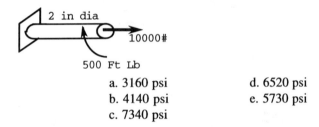

a. 3160 psi   d. 6520 psi
b. 4140 psi   e. 5730 psi
c. 7340 psi

Due to torque:
$$\tau = \frac{Tc}{J} = \frac{500 \times 12 \times 1}{\pi \times \frac{2^4}{32}} = 3820 \text{ psi}$$

Due to elongation:
$$\sigma = \frac{P}{A} = \frac{10000}{\pi \times 1^2} = 3180 \text{ psi}$$

Combined:
$$\tau_{max} = \frac{1}{2}\sqrt{3180^2 + 4 \times 3820^2} = 4137 \text{ psi}$$

Answer (b)

44. Brass cannot be used to reinforce concrete because:

a. It is too dense
b. It differs in shear capabilities
c. It cannot carry enough load
d. It does not stick to concrete
e. It differs in coefficient of thermal expansion

Answer (e)

45. What increase in temperature (°F) is necessary to cause a 1 in. dia 10 ft long steel rod eith fixed ends to buckle. There is no initial stress ($\alpha = 6.5 \times 10^{-6}$)

    a. 66   d. 46
    b. 26   e. 16
    c. 56

$$P_{cr} = \frac{4\pi^2 EI}{L^2} = \alpha \Delta T \, EA$$

$$\Delta T = \frac{4\pi^2 \times \pi \times 1 \times \frac{4}{64}}{6.5 \times 10^{-6} \times \pi \times .5^2 \times 120^2} = 26.4°F$$

<u>Answer (b)</u>

46. For a fluid the number that relates inertial force to compressibility force is?

    a. Reynolds Number   d. Drag Number
    b. Mach Number   e. Froude Number
    c. Weber Number

<u>Answer (b)</u>

Consider the diagram of the pumping system as shown:

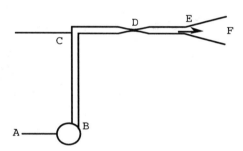

and the following equations:

a.
$$\overline{V_d} = C_v \sqrt{2gh}$$

b.
$$\eta = \frac{hp \; out}{hp \; supplied}$$

c.
$$\frac{P_1}{w} + H_p = \frac{P_2}{w} + Z_2$$

d.
$$\overline{Q} = \left[(H + h)^{1.75} - h^{1.75}\right]^2$$

e.
$$h_{xg} = \frac{k_e (V_1 - V_2)^2}{2g}$$

47. Which equation applies to the fluid as it passes through point D?

<u>Answer (a)</u>

48. Which equation applies to the fluid as it passes from point B to point C?

<u>Answer (c)</u>

49. Which equation applies to the fluid as it passes from point D to point E?

Answer (e)

50. Which equation applies to the fluid as it passes from point A to point B?

Answer (b)

51. For the system shown below, the pressure P is?

    a. 5.46psi   d. 8.92psi
    b. 6.87psi   e. 7.50psi
    e. 4.72psi

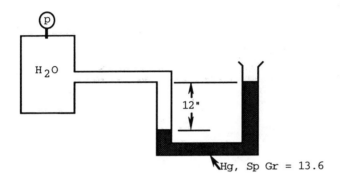

$$p + \frac{62.5 \frac{lbs}{ft^3} \times 1\ ft}{144 \frac{in^2}{ft^2}} = \frac{62.5 \frac{lbs}{ft^3} \times 1\ ft \times 13.6}{144 \frac{in^2}{ft^2}}$$

p = 5.46 psi

Answer (a)

52. What is the moment of the force F about the origin O in the system below? All distances are in feet.

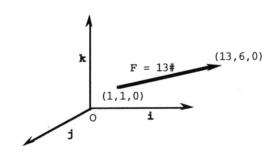

    a. -7k ft lbs   d. -17 ft lbs
    b. 7 ft lbs   e. 27**k** ft lbs
    c. 17**k** ft lbs

$$(1\mathbf{i} + 1\mathbf{j} + 0\mathbf{k})(12\mathbf{i} + 5\mathbf{j} + 0\mathbf{k}) = \begin{vmatrix} 1 & 1 & 0 \\ 12 & 5 & 0 \\ \mathbf{i} & \mathbf{j} & \mathbf{k} \end{vmatrix} = -7\mathbf{k} \text{ ft\#}$$

<u>Answer (a)</u>

53. If equilibrium must be maintained, what is the maximum value that F can have?

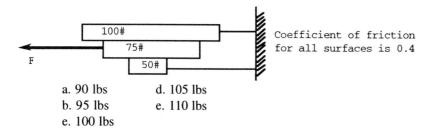

a. 90 lbs   d. 105 lbs
b. 95 lbs   e. 110 lbs
e. 100 lbs

F = (100 x .4) + (175 x .4) = 110 lbs

54. The magnitude of the reaction at B is:

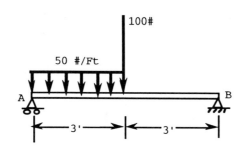

a. 34 lbs   d. 88 lbs
b. 44 lbs   e. 94 lbs
c. 74 lbs

F_B x 6 = (150 x 1.5) + (100 x 3)

F_B = 87.5 lbs   or 88

<u>Answer (d)</u>

55. The y coordinate of the centroid of the area below is?

   a. 3.4   d. 4.6
   b. 2.6   e. 5.0
   c. 4.2

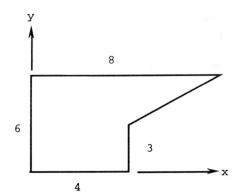

$$\bar{y} = \frac{(24 \times 3) + \left(\left(\frac{1}{2} \times 4 \times 3\right) \times \left(\left(\frac{2}{3} \times 3\right) + 3\right)\right)}{(6 \times 4) + \left(4 \times \frac{3}{2}\right)} = 3.4$$

Answer (a)

56. The force system below is best described as:

   a. Nonconcurrent, noncoplanar
   b. Concurrent
   c. Coplanar
   d. Concurrent and coplaner
   e. 2 dimensional

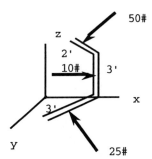

Since the forces have no common point or plane, they must be Non concurrent, non coplanar.

Answer (a)

Solutions for the EIT Simulated Exam PM Section

You will have 90 minutes in which to work on the simulated afternoon section of the fundamendals exam. This is an open book test. You may use textbooks, handbooks, and battery operated silent calculator. You may not borrow or loan any of these materials during the exam. Any preliminary work must be done in the area to the right of the problems in this book.

The simulated exam consists of 26 required questions. Try to answer them all. Your score will be directly proportional to the number of correct answers. Every question has equal weight, and no points are subtracted for an incorrect answer.

The questions are grouped as follows:

        Engineering Economics   1 - 4
        Electrical Circuits   5 - 8
        Engineering Mechanics   9 - 15
        Mathematics   16 - 22
        Thermodynamics/Fluid Mechanics   23 - 26

SOLUTIONS FOR THE EIT SIMULATED EXAM — PM SESSION

ENGINEERING ECONOMICS

An automobile costs $5000.00 and has an estimated life of 5 years. Expenditure of extra money on certain additional maintenance items will extend this life to 10 years. Interest is 12.5%

1. How much should be spent for this additional maintenance?
      a. $500   b. $450
      c. $400   d. $350
      e. $300

$A_5 = P(A/P)$ i = 12.5%, n = 5: 5000(.2809) = $1404.50
$A_{10} = P(A/P)$ i = 12.5%, n = 10: 5000(.1806) = $903
$A_5 - A_{10}$ = 1404.50 - 903 = 501.50

<u>Answer (a)</u>

2. What annual payment is required in order to have the renewal cost available in 5 yrs?
      a. $960   b. $840
      c. $780   d. $690
      e. $573

$A_5 = F(A/F)$ i = 12.5%, n = 5: 5000(.1559) = $779.50

<u>Answer (c)</u>

3. If the interest rate was reduced to 12% per annum, compounded monthly, what is the effective interest rate?

      a. 12.25%   b. 12.35%
      c. 12.5%   d. 12.7%
      e. none of the above

i = 12%/12 mo = 1%/mo
(F/P) i = 1%, n = 12 = 1.1268
Effective i = 1.1268 - 1 = .1268

<u>Answer (d)</u>

4. If expenditures such as gasoline, oil, tires, batteries, etc., amount to $2300 per year, insurance costs $1000 per year, what is the annual cost of owning this car for the 5 year life if the salvage value is $1000?

      a. $43,000.00   b. $4,147.50
      c. $4,548.60   d. $4,829.25
      e. $3,829.25

TAC = $[(P - L)(A/P) + L_i]$ + AE
    = [(5000 - 1000)(.2809) + (1000)(.125)] + 2300 + 1000
    = 4548.60

<u>Answer (c)</u>

ELECTRICAL THEORY

Problems 5-8 relate to the circuit shown below.

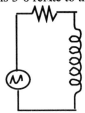

The circuit shown is a series R-L hook with a sinusoidal input equal to V. V = 141 sin ωt; R=3Ω; L=10.6 x 10⁻³h : f = 60 Hz

5. What is the effective steady state current? (include the relative phase angle.

   a) 141∠0°   b) 141∠45°
   c) 20∠-53.11° d) 100∠73.2°
   e) 5∠53.11°

Effective means RMS which equals Peak/√2 (sinusoid)

$I_{RMS} = V_{RMS}/Z$
V = 141 sin ωt = 141∠0°
V_{RMS} = 141 sin ωt/√2 = 141∠0°/√2 = 99.7∠0°
Z = R + j ωL (complex) = √(R² + (ωL)²) ∠tan⁻ (Polar)
ωL = 2π60 x (10.6 x 10⁻³) = 4
Z = √(3² + (4)²) ∠tan⁻¹ (4/3) = 5∠53.13°
I_{RMS} = 99.7∠0°/ 5∠53.13° = 20∠-53.13°

<u>Answer (c)</u>

6. Using the values of the components given compute the power factor.
        a. 0.7   b. 0.6
        c. 0.5   d. 0.8
        e. 0.78

PF = cos angle between V and I = -53.13°
     cos -53.13° = 0.6

<u>Answer (b)</u>

7. Using the values given, what is the effective value of the voltage drop across the inductor? (include its phase angle.)
        a) 141∠89.21° b) 60 ∠-53.1°
        c) 80∠36.9°   d) 100∠73.2°
        e) none of these

I_{RMS} = 20∠-53.13°, Z_L = j ωL = j4 = 4∠90°
$V_{RMS\ (Inductor)}$ = 20∠-53.13° x 4∠90°
            = 80∠36..90

<u>Answer (c)</u>

8. What is the power consumed by the load?
    a. 60 watts   b. 123 watts
    c. 180 watts d. 319 watts
    e. 1200 watts

$$P = I_{RMS}^2 R = 20^2 \times 3 = 1200 \text{ watts}$$

<u>Answer (e)</u>

DYNAMICS

Problems 9-11 relate to the following problem statement.

The motion of a particle is defined by $a = 200 + 6t^2$ ft/sec². It has a velocity of 2 ft/sec and a distance $s = 0$, at time = 0 from a fixed point.

9. Determine the average velocity during the first 10 sec.
    a. 8004 fps   b. 4002 fps
    c. 4440 fps   d. 8040 fps
    e. 6030 fps

$$s_f = s_0 \quad v_0 t + \frac{at^2}{2}$$

$$s_f = 0 + 2(10) + \frac{(200 + 6(10)^2)(10)^2}{2} = 40,020$$

$$\bar{V} = \frac{s_f - s_0}{t_f - t_0} = \frac{40,020 - 0}{10 - 0} = 4002$$

<u>Answer (b)</u>

10. If a second particle has a motion defined by $v = 160t + 12t^2$, which equation expresses the length of time required for it reach the first particle? $S = 0$, and $t = 0$ for the second particle.

    a. $(t^3 - 8t^2 + 40t + 4$.
    b. $t(t^3 - 8t^2 + 40t + 4$.
    c. $t(t^4 - 8t^3 + 40t^2 + 4t$.
    d. $2t(t^3 - 8t^2 + 40t + 4$.
    e. $2(t^3 - 8t^2 + 40t + 4$.

For particle 1 $s_1 = t^4/2 + 100t^2 + 2t + c$
    $t = 0$, $s = c$, thus $c = 0$
For particle 2 $s_2 = 4t^3 + 80t^2 + c$
    $t = 0$, $s = c$, thus $c = 0$
$s_1 = s_2$
    $4t^3 + 80t^2 = t^4/2 + 100t^2 + 2t$
Combining:
    $t(t^3 - 8t^2 + 40t + 4)$

<u>Answer (c)</u>

13-48 FUNDAMENTALS OF ENGINEERING EXAM REVIEW WORKBOOK

11. If the weight of the particle with v = 160t + 12t² is 2 lbs, what constant force would be required to stop it if the force were applied at time t = 2.5 sec. Assume that the stopping distance is 1 ft.

    a. 500 lbs   b. 5000 lbs
    c. 7000 lbs   d. 10,000 lbs
    e. 14000 lbs

$$F \times D = \frac{1}{2} M \left(V_f^2 - V_0^2\right)$$
$$V_f = 0$$
$$F \times 2 = \frac{1}{2} \frac{2}{32.2} \left(160t + 12t^2\right)^2$$

$$F = 7007 \text{ lbs}_f$$

Answer (c)

MECHANICS OF MATERIALS

Questions 12 and 13 refer to the following problem statement:

A composite beam is made of wood and steel, as shown: Assume E_{stl} = 20XE_{wood}

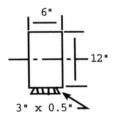

12. Determine the moment of inertia about the neutral axis.
    a. 1688 in⁴   b. 3459 in⁴
    c. 2586 in⁴   d. 1248 in⁴
    e. None of these

$$\bar{y} = \frac{(72)(6) + (3 \times 0.5 \times 20)(12.25)}{72 + 30} = 7.83"$$

$$I_G = 6 \frac{12^3}{12} + 30 (4.41)^2 + 72 (1.83)^2 = 1688 \text{ In}^4$$

Answer (a)

SOLUTIONS FOR THE EIT SIMULATED EXAM – PM SESSION   13-49

13. The function of the connectors in a composite unit such as this is to:
   a. Help balance the loads in the beam
   b. Transfer moment from the plates to the beam
   c. Transfer horizontal shear from the plate to the beam
   d. Transfer vertical shear from the plate to the beam
   e. It has something to do with diagonal tension

   <u>Answer (c)</u>

STATICS

Problems 14 and 15 refer to the beam shown below which carries a moving load of 1500 lbs. Assume the beam to be supported at B by a rod BC as shown. The supports at A and C and the joint at B permit unrestricted rotation.

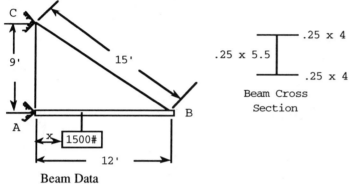

Beam Data
Modulus of Elasticity $E = 30 \times 10^6$
Moment of Inertia $I = 20$ in^4
Cross Sectional Area $A = 3.375$ in^2
   Rod Data
Modulus of Elasticity $E = 30 \times 10^6$
Cross Sectional Area $A = 0.4418$ in^2

14. The maximum unit flexural stress in the beam is most nearly
   a. 3000 psi   b. 6000 psi
   c. 8000 psi   d. 16000 psi
   e. 32000 psi

   $R = 750, M = 750 \times 6 \times 12$

   $$f = \frac{Mc}{I} = \frac{750 \times 6 \times 12 \times \frac{6}{2}}{20} = 8100 \text{ psi}$$

   <u>Answer (c)</u>

FUNDAMENTALS OF ENGINEERING EXAM REVIEW WORKBOOK

15. The maximum unit tensile stress in the rod BC is most nearly
    a. 3000 psi   b. 6000 psi
    c. 8000 psi   d. 16000 psi
    e. 32000 psi

    T = 15/9 x 1500 = 2500#
    f = T/A = 2500/.4418 = 5700#

    <u>Answer (b)</u>

MATHEMATICS

16. The sum of the circumferences of two circles is 36π. The sum of the areas of the circles is 180π. The radius of one of the circles is:
       a. 10   b. 9
       c. 8   d. 7
       e. 6

    $2\pi (R_1 + R_2) = 36\pi$, $R_1 + R_2 = 18$
    $\pi (R_1^2 + R_2^2) = 180\pi$, $R_1^2 + R_2^2 = 180$
    $R_1 = 18 - R_2$: $(18 - R_2)^2 = 180$
    $2R_2^2 - 36R_2 + 144 = 0$: $R_2 = 6$

    <u>Answer (e)</u>

17. Ground salt is poured in a pile on the ground at the rate of 4 cfm. it forms a conical pile whose height is 30% of the diameter of the base. How fast is the height of the pile increasing when the base is 5 ft in diameter?

       a. .458 fpm   b. .687 fpm
       c. .204 fpm   d. 2.83 fpm
       e. 5.76 fpm

    $V = \frac{1}{3}Bh$; $B = \frac{\pi d^2}{4}$; $h = .3d$

    $V = \frac{\pi d^3}{40}$

    $\frac{dV}{dt} = \frac{3\pi d^2}{40} \frac{dd}{dt}$

    $\frac{dd}{dt} = \frac{dV}{dt} \frac{40}{3\pi d^2} = 4 \frac{40}{3\pi 5^2} = .68$

    $\frac{dh}{dt} = .3 \frac{dd}{dt} = .3 \times .68 = .204$

    <u>Answer (c)</u>

18. What is the general solution of the differential equation $ydx - y^2dy = dy$?
    a. $x = 3y^2 + 2y + C$
    b. $x = y^3/3 + y + C$
    c. $x = \csc y + C$
    d. $x = y^2/2 + \ln y + c$
    e. $x \sin y \cos y + \ln y + C$

    $$ydx = dy + y^2 dy$$
    $$dx = \frac{dy}{y} + ydy$$
    $$x = \frac{y^2}{2} + \ln y + C$$

    <u>Answer (d)</u>

19. A 3-ft long wire is cut into two pieces. One piece is used to form a square and the other piece used to form a circle. If the sum of the enclosed areas is to be a minimum, what length of the wire should be used to form the circle?
    a. 0.14 ft.   b. 0.56 ft.
    c. 1.12 ft.   d. 1.32 ft.
    e. None of the above

    $L = 2\pi r + 4x$
    $\frac{dL}{dr} = 2\pi + 4\frac{dx}{dr} = 0; \quad \frac{dx}{dr} = -\frac{\pi}{2}$
    $A = \pi r^2 + x^2$
    $\frac{dA}{dr} = 2\pi r + 2x\frac{dx}{dr} = 0 = 2\pi r + 2x\left(\frac{-\pi}{2}\right)$
    $x = 2r$
    $L = 3 = 2\pi r + 4(2r); \; r = .21$
    $L_{circle} = 2\pi r = 2\pi(.21) = 1.32$

    <u>Answer (d)</u>

20. My pencil holder contains 3 red pencils, 4 green pencils, and 3 blue pencils. My secretary has a pencil holder nearby that contains 4 blue pencils, 4 red pencils and 2 green pencils. What is the probability of my reaching out and grasping a blue pencil?
    a. 7/1   b. 12/1
    c. 7/10   d. 12/20
    e. 7/20

    Total red $3 + 4 = 7$
    Total green $4 + 2 = 6$
    Total blue $3 + 4 = 7$
       Total $10 + 10 = 20$

    Probability of a blue pencil is 7/20.
    <u>Answer (e)</u>

Problems 21 and 22 refer to the figure below.
Line CD is Perpendicular to line AB.

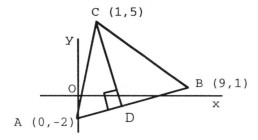

21. What is the slope of line BD?
    a. -5   b. .33333
    c. -.33333   d. -3
    e. 3

$$M = \frac{\Delta Y}{\Delta X} = \frac{1 - (-2)}{9 - 0} = \frac{3}{9} = \frac{1}{3}$$

<u>Answer (b)</u>

22. What is the area of triangle ABC?
    a. $5\sqrt{10}$ b. $6\sqrt{10}$
    c. 26   d. 30
    e. 60

We need point D

Line AB m = 1/3, B = -2 by inspection at A
$$y = \frac{1}{3} X - 2$$

Line CD perpendicular to line AB, thus m = 3
since y = mx + b: form point C
5 = -3(1) + B; b = 8
y = -3 X + 8

At point D line AB = line CD

$$\frac{1}{3} X - 2 = -3 X + 8; \quad x = 3 \text{ and } y = -1$$

$$A_{ABC} = A_{ACD} + A_{BCD} = \frac{1}{2} AD \cdot CD + \frac{1}{2} BD \cdot CD$$

$$CD = \sqrt{\Delta x^2 + \Delta y^2} = \sqrt{(3-1)^2 + (5-(-1))^2} = 2\sqrt{10}$$
$$AD = \sqrt{3^2 + 1^2} = \sqrt{10}$$
$$BD = \sqrt{6^2 + 2^2} = 2\sqrt{10}$$
$$A_{ABC} = \frac{1}{2} 2\sqrt{10} \cdot \sqrt{10} + \frac{1}{2} 2\sqrt{10} \cdot 2\sqrt{10} = 10 + (2 \times 10) = 30$$

<u>Answer (d)</u>

SOLUTIONS FOR THE EIT SIMULATED EXAM — PM SESSION   13-53

THERMODYNAMICS

Problems 23 and 24 refer to the following problem statement

Atmospheric air inlet to an air cooler has a specific humidity of .009 lbs water vapor/lbs dry air at 80°F. Calculate the following.

23. Density of dry air

    a. .0726 lb/ft³   b. .0680 lb/ft³
    c. .0459 lb/ft³   d. .0472 lb/ft³
    e. .0858 lb/ft³

See chart solution.

<u>Answer (a)</u>

24. Enthalpy of the water vapor

    a. 41.5 Btu/lbs dry air   b. 32.5 Btu/lbs dry air
    c. 29.0 Btu/lbs dry air   d. 39.6 Btu/lbs dry air
    e. 24.8 Btu/lbs dry air

See chart solution.

<u>Answer (a)</u>

FLUID MECHANICS

Problems 25 and 26 refer to the following problem statement.

A radial gate, 5 ft wide, is installed as shown.

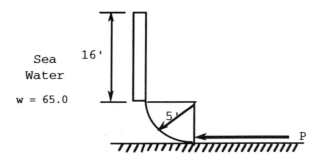

25. Neglecting the weight of the gate and any mechanical forces, such as friction, what force P is required to open the gate?
(Answers are rounded off.

    a. 100,000 lbs. b. 50,000 lbs.
    c. 25,000 lbs.   d. 12,500 lbs.
    e. 0 lbs.

Force of sea water acting down on the gate

$$\frac{\pi \times 5^2}{4} \times 5 \times 65 = 6400 \text{ lbs}$$

Force of sea water acting horizontally against the gate

$$\frac{65(16 + 21)}{2} \, 5 \times 5 = 30,060 \text{ lbs}$$

Force required to open gate is 0

<u>Answer (e)</u>

26. What force acts against the sea wall?

      a. 41600 lb.   b. 40960 lb.
      c. 79872 lb.   d. 39936 lb.
      e. none of these

$$p = \frac{wH^2}{2} = \frac{65(16)^2}{2} \times 5 = 41,600 \text{ lbs}$$

<u>Answer (a)</u>

14
Index

INDEX

Note: Section numbers are in **boldface** print

Absolute pressure, **5**-1
Absolute viscosity, **5**-41
Acceleration
   angular, **4**-40
   constant, **4**-8–13, **4**-20–25, **9**-28
   definition of, **4**-2
   gravity and, **6**-56
   maximum, **9**-34
   motion and, **4**-1, **9**-28
   reversed effective, **4**-50
   velocity and, **4**-19–20
Acetylene series, **7**-16
Acids and acidity, **7**-11–12, **7**-17, **9**-11
Acoustic emissions testing, **9**-13–14
Addition
   of complex numbers, **6**-58
   parallelogram method of, **2**-2, **2**-44
   of vectors, **1**-22, **2**-2, **2**-44, **2**-46
Admittance, electrical, **6**-21–28
After tax rate of return calculation, **10**-67–70
Aggregate, in concrete, **9**-23, **9**-24
Agitation, and corrosion, **9**-11
Air, compression of, **8**-17–18
Alcohols, **7**-17
Aldehydes, **7**-17
Algebra
   binomial formula, **1**-7–9
   block diagram, **6**-158–61
   determinants, **1**-9–12, **6**-60–62
   exponents, **1**-5–7
   fundamental laws of, **1**-5
   higher order algebraic equations, **1**-26–28
   linear algebra, **1**-12–20
   vector analysis, **1**-21–26
Aliphatic compounds, **7**-15
Alkalinity, **7**-11–12
Allotropic changes, in heat treating of metals, **9**-9
Alloys, of metals, **9**-1
Alpha decay, **7**-14
Alpha iron, **9**-9
Alternating current (AC)
   circuits, **6**-18–21, **6**-28–30
   generators, **6**-93–97, **6**-114
   measurements of, **6**-106
   motors, **6**-97–99, **6**-118
Alternatives, comparison of economic, **10**-71–75
Ampere, **7**-26
Amplifiers, operational, **6**-231–34
Analytic geometry
   conic sections, **1**-46–49
   straight lines, **1**-42–46
Angles
   of friction, **2**-57, **2**-60
   ideal bank, **4**-58
   trigonometry, **1**-33–34
   of twist, **3**-15–17
Angular acceleration, **4**-40
Angular motion, **4**-41–42
Angular speed, **4**-39–40
Angular velocity, **4**-39–42
Anions, **7**-21

Annealing, **9**-6
Annual expenses, **10**-34, **10**-37
Annuity, **10**-3
Approximate interest rate, **10**-64
Approximate rate of return, **10**-73
Archimedes Principle, **5**-14
Areas, centroids of, **2**-89–90, **2**-94
Aromatic compounds, **7**-15
Associative law, for vectors, **1**-23
Atmospheric pressure, **5**-1
Atom, definition of, **7**-2
Atomic number, **7**-1
Atomic structure
   of germanium, **6**-179–80
   of metals, **9**-1
Atomic weight, **7**-1, **7**-2
Attenuators, electrical, **6**-44–47
Austentite, **9**-10
Automatic control systems, and system engineering, **6**-149–50
Autotransformer, **6**-78
Auxiliary equation, **1**-85
Auxiliary metals, **9**-1
Avogadro's Law, **6**-56, **7**-3, **8**-28
Axial force, and beam analysis, **3**-32
Axis
   parallel axis theorem, **3**-49
   position of neutral, **3**-47–48
   three-dimensional forces and moment of force about, **2**-47–48

Band pass filter, **6**-45
Band reject filter, **6**-46
Banking of curves, **4**-57–65
Barometric pressure, **5**-1
Bazins formula, **5**-63
Beam analysis
   beam load characteristics, **3**-31
   bending moment diagram, **3**-38–45
   combined stress, **3**-65–66
   conditions of equilibrium, **3**-30, **3**-32
   flexure formula, **3**-45–46
   moment of inertia of rectangle, **3**-48–49
   moments of inertia of composite sections, **3**-49–50
   Mohr's Circle, **3**-68–73
   normal and shear stress formulas, **3**-66–67
   parallel axis theorem, **3**-49
   planes of maximum or minimum normal and shear stress, **3**-67
   planes of zero shear, **3**-67
   position of neutral axis, **3**-47–48
   reactions, **3**-33–36
   shear force diagram, **3**-36–38, **3**-43–45
   sign convention, **3**-32–33
   statically determinate beams, **3**-30
   types of beams, **3**-29
   types of loads, **3**-29
Before tax rate of return calculation, **10**-67
Bell-shaped curve, **1**-101
Belt friction, **2**-67–68
Bending moment, and beam analysis, **3**-32–33, **3**-38–45
Benefit to cost ratio method, **10**-76–80
Benzene series, **7**-16
Bernoulli's theorem, **5**-19–24

14-2 INDEX

Beta decay, **7**-14
Beta iron, **9**-9
Biconcave lens, **9**-59
Binomial expansion, **1**-8–9
Binomial formula, **1**-7–9
Binomial series, **1**-7
Bipolar junction transistor (BJT), **6**-208, **6**-219–20
Block diagram algebra, **6**-158-61
Body
    center of mass, **4**-7
    definition of, **4**-1
    rotation of rigid, **4**-65–70
Body centered cubic (BCC) structure, **9**-9
Boiler, of steam generator, **8**-50
Boiling point elevation (BPE), **7**-13
Bond calculations, **10**-21–27
Bond energy, **7**-15
Book value, of asset, **10**-28, **10**-66
Boyle's Law, **8**-1
Brake engine efficiency, **8**-39
Brake thermal efficiency, **8**-39
Branch pipe flow, **5**-50–55
Brayton cycle, **8**-15
Breakdown region, and junction field effect transistors, **6**-195
Break even cost analysis, **10**-84–87
Briggsian System, **1**-29
Brinell hardness tester, **9**-14
Brittle fracture, **9**-16
Brittle material, determining yield strength of, **3**-3
Buoyancy, and fluid mechanics, **5**-14–15

Candela (cd), **9**-62
Capacitance
    calculation of, **6**-9
    definition of, **6**-1
    equivalent, **6**-10
Capacitor AC motor, **6**-118
Capacitors, and time domain analysis, **6**-122–23
Capillary flow, **5**-44
Capital gains tax, **10**-66
Capitalized cost method, **10**-56–58
Capital recovery, **10**-35–36, **10**-40
Carbohydrates, **7**-18
Carbon
    combustion of, **8**-32
    compounds, **7**-15–18
Carbon monoxide, combustion of, **8**-32
Carburizing, **9**-7
Carnot cycle, **8**-10–11, **8**-58
Cartesian coordinate system, **1**-27
Case hardening, **9**-6
Cash flow diagrams, **10**-3, **10**-11, **10**-12, **10**-17
Cash flow table, **10**-68
Cast iron, **9**-8
Catalysts, and Law of Le Chatêlier, **7**-9
Cathodic films, **9**-11
Cations, **7**-20
Cement, **9**-23, **9**-24
Cementite, **9**-8, **9**-9
Center
    of gravity (CG), **4**-7, **5**-6
    of mass, **4**-6–7
    of percussion, **4**-72, **4**-73
    of pressure (CP), **5**-6
Centimeter, gram, second (CGS), **6**-2
Centrifugal acceleration, **4**-50
Centripetal acceleration, **4**-50
Centroidal axis, **3**-47

Centroids
    of areas, **2**-89–90, **2**-94
    definition of, **2**-87–88
    by integration, **2**-90–91
    of lines, **2**-90–91
    theorems of Pappus and Guldinus, **2**-94
Characteristic Equation, **8**-2
Charge, definition of, **6**-1
Charle's Law, **8**-1–2
Charpy test, **9**-15
Chemical reactions, **7**-6–7
Chemistry
    chemical reactions, **7**-6–7
    gas laws, **7**-3–6
    mathematics, **7**-26–27
    nuclear chemistry and radioactivity, **7**-14
    organic, **7**-15–18
    oxidation-reduction (redox) reactions, **7**-22–26
    periodic chart, **7**-1–3, **7**-34
    reaction rate and equilibrium, **7**-8–12
    solutions, **7**-12–13
    thermochemistry, **7**-14–15
    valence, **7**-19–21
Chezy Formula, **5**-68
Closed loop systems, **6**-150, **6**-162
Coefficients
    of kinetic friction, **2**-57
    of orifice, **5**-31–32
    of performance, **8**-58
    of restitution, **4**-95
    of static friction, **2**-57
    of thermal expansion, **3**-4
Color, and wavelengths, **9**-57
Combinations, and probability, **1**-95–96
Combined engine efficiency, **8**-39
Combined stress, and beam analysis, **3**-65–66, **3**-70–73
Combined thermal efficiency, **8**-39
Combustion, and thermodynamics, **8**-31–34
Common base configuration, of transistors, **6**-211–12, **6**-217–19
Common collector configuration, of transistors, **6**-219
Common emitter configuration, of transistors, **6**-212–17, **6**-226–30
Common logs, **1**-30, **1**-31
Commutators, and DC generators, **6**-82–83
Comparative economic analysis methods, **10**-34–44
Complete response, and time domain analysis, **6**-123
Complex circuits, **6**-31–44
Complex numbers, **1**-2–3, **6**-57–59
Component motion
    flight of projectiles and, **4**-46–48
    radial and tangential coordinates, **4**-48–53
Components, in vector analysis, **1**-22
Composite sections, moments of inertia of, **3**-49–50
Compound DC motor, **6**-118
Compound interest, **10**-4–5
Compound pendulum, **4**-100
Compound shafts, and torsion, **3**-17
Compression of air, **8**-17–18
Compressive strain, **3**-1
Computed rate of return, **10**-75–76
Concave lens, **9**-58, **9**-63
Concave mirrors, **9**-53–54
Concentration, and Le Chatêlier's Law, **7**-8–9
Concrete, **9**-23–25
Concurrent force systems
    parallel force systems and, **2**-2–3, **2**-6
    three-dimensional forces and, **2**-44, **2**-46
Concurrent moment vectors, **2**-49–55
Condenser, of steam generator, **8**-51

INDEX

Conduction, and heat transfer, **8**-61–65
Conductors, of electricity, **6**-4
Cone of friction, **2**-57
Conical pendulum, **4**-100
Conic sections, in analytic geometry, **1**-46–49
Conjugate, of complex numbers, **1**-3, **6**-58
Conservation of momentum, **4**-91–94
Conservative systems, dynamics of, **4**-83–84
Constants
    acceleration, **4**-8–13, **4**-20–25, **9**-28
    of chemical equilibrium, **7**-9–12
    current region, **6**-195
    force, **4**-20–25
    universal and specific gas, **7**-4–5
    velocity, **4**-8, **9**-28
Continuous function, **1**-50
Convection, and heat transfer, **8**-65
Converging lens, **9**-58
Convex lens, **9**-58, **9**-63
Convex mirrors, **9**-54
Coplanar force systems, **2**-1–43
Corporate rate of return calculations, **10**-65–66
Corrosion, of metals, **9**-10–11
Cosines
    laws of, **1**-39
    three-dimensional forces and direction, **2**-45
Coulomb's law, **6**-2
Coulomb units, **6**-1, **7**-25
Couple, in parallel force system, **2**-3–4
Covalence, **7**-19
Covalent bonding, and semiconductors, **6**-179–81
Cramer's Rule, **1**-14
Critical angle, **9**-56
Critical point, in heat treating of metals, **9**-9
Cross derivatives, **1**-61
Cross product, of vectors, **1**-24
Crystallographic planes, **9**-3
Crystal structures, **9**-1–2
Cubic equation, **1**-26
Cumulative depreciation charge, **10**-28
Cumulative law, for vectors, **1**-23
Current
    definition of, **6**-1
    density, **6**-70
Curves, banking of, **4**-57–65
Curvilinear motion, kinematics and kinetics of, **4**-43–46, **4**-53–57
Cyaniding, **9**-7

D'Alembert acceleration, **4**-50
D'Alembert's principle, **4**-25–28
Dalton's Law of Partial Pressures, **7**-5–6, **8**-30–31
Dams, and fluid mechanics, **5**-15–19
D'Arsonval meter, **6**-105
Deceleration, **4**-3
Decibel, **9**-47
Declining balance depreciation method (DBD), **10**-28–29, **10**-32
Decomposition, as chemical reaction, **7**-6
Definite integral, **1**-71–72
Deflections, **3**-1
Deformations, **3**-1
Degree
    angles, **1**-33
    differential equation, **1**-82
Delta circuit, **6**-63–65
Delta iron, **9**-9
Denseness, of metals, **9**-22
Dependent events, and probability, **1**-95
Dependent variables, **1**-49

Depletion mode MOSFET, **6**-197
Depletion region, **6**-185
Depreciation, **10**-27–34, **10**-35, **10**-68
Derivatives, in differential calculus, **1**-52–69
Destructive tests, **9**-15
Determinants, algebra of, **1**-9–12, **6**-60–62
Dew point, **8**-48, **8**-49
Diesel engine, **8**-13–15
Differentiable function, **1**-52
Differential calculus
    applications of, **1**-63–69
    derivatives, **1**-52–59
    functions, **1**-49
    higher order derivatives, **1**-59–60
    implicit differentiation, **1**-59
    limits, **1**-49–52
    partial derivatives, **1**-61–63
Differential equations
    applications of, **1**-87–88
    definitions, **1**-82
    equations with variables separable, **1**-82–84
    feedback control systems, **6**-152
    first-order linear, **1**-84–85
    Laplace transform, **1**-92–93, **6**-143
    second-order linear, **1**-85–87
Differentiation, basic rules of, **1**-53–59
Diffraction of light, **9**-62
Dimensional analysis, and fluid mechanics, **5**-73–75
Diodes, semiconductor, **6**-185–92
Direct combination, as chemical reaction, **7**-6
Direct current (DC)
    circuits, **6**-11–18, **6**-28–30
    field effect transistor (FET), **6**-200–201
    generators, **6**-81–88
    measurements of, **6**-105
    motors, **6**-88–93, **6**-118
    voltage source, **6**-14
Direction, and motion of particle, **4**-1
Direction cosines, **2**-45
Dispersion, of light, **9**-57
Displacement
    as function of time, **4**-13–15
    as quantity of motion, **4**-1
Distance, force as function of, **4**-30–32
Diverging lens, **9**-58, **9**-59
Doppler effect, **9**-47, **9**-58
Double-angle formulas, **1**-38
Double declining balance depreciation method (DDBD), **10**-29, **10**-32–33
Double displacement, and chemical reaction, **7**-7
Double integrals, **1**-75
Drain source loop, and field effect transistors, **6**-201
Dual combustion cycle, **8**-14
Ductile and malleable iron, **9**-8
Dynamic force, and momentum, **5**-69–73
Dynamic resistance, and semiconductors, **6**-193
Dynamics
    action of jets, **4**-88–90
    banking of curves, **4**-57–65
    component motion, **4**-46–53
    conservation of momentum, **4**-91–94
    conservative systems, **4**-83–84
    curvilinear motion, **4**-43–46, **4**-53–57
    definitions of terms, **4**-1–2
    elastic impact, **4**-94–99
    equations of rectilinear motion, **4**-2–3
    equivalent springs, **4**-80–83
    harmonic motion, **4**-100

    impulse and momentum, **4**-84–88
    Newton's laws of motion, **4**-4–42
    noncentroidal rotation, **4**-70–75
    rotation of rigid body, **4**-65–70
    work-energy, **4**-75–80
Dynamic viscosity, **5**-41

Economic lot size, **10**-87–90
Economics
    benefit to cost ratio, **10**-76–80
    break even cost analysis, **10**-84–87
    capitalized cost, **10**-56–58
    cash flow diagrams, **10**-3
    comparative economic analysis methods, **10**-34–44
    depreciation calculations, **10**-27–34
    financial calculations, **10**-4–27
    incremental rate of return, **10**-71–75
    minimum cost analysis/economic lot size, **10**-87–90
    present worth, **10**-46–56
    rate of return, **10**-58–76
    replacement studies, **10**-81–83
    time value of money, **10**-2–3
    total annual cost including risk, **10**-44–46
Economizer, of steam generator, **8**-51
Efficiency
    definition of, **6**-1
    of generators, **6**-95
    thermodynamics and, **8**-39–41
Elastic collision, **4**-94
Elastic impact, **4**-94–99
Elasticity, modulus of, **3**-4, **9**-17
Elastic limit, **3**-3, **9**-17
Elbows, in piping systems, **5**-36
Electrical conductivity, **9**-19–20
Electric field intensity, **6**-3–4
Electricity
    alternating current (AC), **6**-18–21, **6**-28–30, **6**-93–99, **6**-118
    complex circuits, **6**-31–44
    definitions of terms, **6**-1–2
    direct current (DC), **6**-11–18, **6**-81–93, **6**-118
    electrolysis, **6**-4–5
    electrostatics, **6**-2–4
    field effect transistors, **6**-193–97
    filters and attenuators, **6**-44–47
    impedance and admittance, **6**-21–28
    Laplace transforms, **6**-142–48
    magnetics, **6**-69–71, **6**-117
    measurements of, **6**-105–107
    metal oxide semiconductor field effect transistor (MOSFET), **6**-197–208
    resistance, capacitance, and inductance, **6**-5–11
    rotating electrical machines, **6**-79–81
    semiconductors, **6**-179–85
    systems engineering, **6**-149–78
    three-phase circuits, **6**-63–69
    three-phase induction motor, **6**-99–104
    time domain analysis, **6**-119–42
    transformers, **6**-71–79
    transistors, **6**-208–11
    unit relationships, **6**-56
    units and notation, **6**-2
Electrochemistry, **7**-25–26
Electrodynameter, **6**-106
Electrolysis, **6**-4–5, **7**-25–26, **9**-11
Electromagnetic induction, **6**-79
Electromotive force (EMF), **6**-1, **7**-24
Electron configuration, in periodic chart, **7**-1
Electron-volt constant, **6**-56

Electroplating, **7**-25, **7**-26
Electrostatics, **6**-2–4
Electrovalence, **7**-19
Empirical formula, **7**-27
Endothermic process, **7**-15
Energy, definition of, **6**-1
Engine efficiency, **8**-39
Engineering Metals, **9**-1
Engineering Units (FPS), of electricity, **6**-2
English Units (FPS), of electricity, **6**-2
Enhancement mode MOSFET, **6**-197
Enthalpy, **8**-6
Entrance losses, and fluid mechanics, **5**-32–34
Entropy, **8**-7
Equations of State, **8**-2–3
Equilibrant, **2**-1
Equilibrium
    beam analysis and, **3**-30, **3**-32
    chemical reactions and, **7**-8–12
    noncentroidal rotation and dynamic, **4**-70–75
    parallel force systems and, **2**-4–5
    three-dimensional forces and, **2**-47, **2**-49–55
    trusses in, **2**-12–13, **2**-14–21
Equivalent capacitance, **6**-10
Equivalent inductance, **6**-11
Equivalent resistance, **6**-7–8
Equivalent springs, **4**-80–83
Ericcson cycle, **8**-11–12
Esterification, **7**-18
Esters, **7**-17–18
Ethers, **7**-17
Ethylene series, **7**-16
Euler Theorem, **6**-59
Eutectic point, **9**-6
Exothermic process, **7**-15
Exponents, **1**-5–7
Extrinsic semiconductors, **6**-182

Face centered cubic (FCC) structure, **9**-9
Face value, of bond, **10**-21
Faraday's Law, **6**-4–5, **6**-79
Faraday units, **6**-1, **7**-25
Feedback control systems, **6**-152
Ferrite, **9**-9
Ferrous metals, **9**-1
Fetley and Stearns formula, **5**-63
Field effect transistors
    DC conditions and, **6**-200–201
    junction field effect transistors (JFET), **6**-193–97
    signal conditions and, **6**-204–208
Field intensity, **6**-70
Film formation, and corrosion, **9**-11
Filters, electrical, **6**-44–47
Financial calculations, **10**-4–27
Finite interest period, **10**-5
First-order linear differential equations, **1**-84–85
Fittings, in piping systems, **5**-36
Fixed percentage depreciation method, **10**-28–29, **10**-32
Fixed supports, **3**-33
Flame hardening, **9**-7
Flemming's Rules, **6**-80
Flexure formula, and beam analysis, **3**-45–46
Flight of projectiles, **4**-46–48
Flow
    measurement of, **5**-56–59
    in open channels, **5**-67–69
    trajectory of, **5**-60–66
Flow over weirs, **5**-61

FLT (force, length, time) system, **5**-73
Fluid mechanics
    Bernoulli's theorem for incompressible fluid, **5**-19–24
    branch pipe flow, **5**-50–55
    buoyancy, **5**-14–15
    dams and gates, **5**-15–19
    dimensional analysis, **5**-73–75
    dynamic force and momentum, **5**-69–73
    entrance losses, **5**-32–34
    flow measurement, **5**-56–59
    flow in open channels, **5**-67–69
    flow trajectory, **5**-60–66
    fluid pressure, **5**-1–5
    head loss due to pipe friction, **5**-24–31
    hydrostatic force, **5**-5–11
    losses in piping systems, **5**-35–39
    nozzles, **5**-59
    orifice coefficients, **5**-31–32
    pipe line supplied by pump, **5**-47–50
    sluice gate, **5**-67
    thick-walled cylinders, **5**-13–14
    thin-walled cylinders, **5**-11–13
    translation and rotation, **5**-75–76
    viscosity and Reynolds number, **5**-41–46
Fluid pressure, **5**-1–5
Force
    constant acceleration and, **4**-20–25
    distance and, **4**-30–32
    hydrostatic, **5**-5–11
    momentum and dynamic, **5**-69–73
    motion and, **4**-1
    time and, **4**-29, **4**-87–88
    variable and Newton's laws of motion, **4**-28–32
    velocity and, **4**-33–39
    work-energy method involving, **4**-77–80
Forced response, and time domain analysis, **6**-123
Force, length, time (FLT) system, **5**-73
Force-time curve, **4**-87–88
Four-stroke cycle, **8**-12
Fractions, and exponents, **1**-6
Francis formula, **5**-62
Free body diagrams, of trusses, **2**-12, **2**-22–43
Freezing point depression (FPD), **7**-13
Fretting corrosion, **9**-13
Friction
    head loss due to pipe, **5**-24–31
    Reynolds number and, **5**-77
    three-dimensional forces, **2**-55–62, **2**-65–68
Fundamental identities, trigonometric, **1**-38–39
Fusibility, of metals, **9**-22
Fusion, heat of, **7**-15
Future sum of money, **10**-3

Gage pressure, **5**-1
Galvanic action, and material science, **9**-12–13
Galvanic Series, **9**-12
Gamma iron, **9**-9
Gases
    chemistry and laws of, **7**-3–6
    mixtures of, **8**-28–29
    sound waves, **9**-44
Gas turbine cycle, **8**-15–17
Gates, and fluid mechanics, **5**-15–19
Gate source loop, and field effect transistors, **6**-201
Gaussian distribution, **1**-101–103
Gay-Lussac's Law, **7**-3
General Energy Equation, **8**-3–4
General equation, of straight line, **1**-42, **1**-43

General solution, of differential equation, **1**-82
Generators
    alternating current (AC), **6**-93–97
    direct current (DC), **6**-81–88
Geometric mean, **1**-98
Germanium, **6**-179–80
Gradient series formulas, **10**-6, **10**-7, **10**-8
Gradual enlargement, in piping systems, **5**-35
Graham's Law, **7**-5
Gram-calorie, **6**-1
Gram-equivalent weight, **7**-13
Graphical representation of motion, **4**-13–20
Gravimetric analysis, and thermodynamics, **8**-29–31
Gravity
    acceleration and, **6**-56
    center of, **4**-7, **5**-6
Gray cast iron, **9**-8
Gross revenue, **10**-35
Guldinus, theorem of, **2**-94
Gyration, radius of, **4**-70

Half-angle formulas, **1**-38
Half-life, and radioactive decay, **7**-14
Hardness testing, **9**-14
Harmonic motion, **4**-100, **9**-31–39
Hazen-Williams formula, **5**-25, **5**-27, **5**-39, **5**-40
Head loss, and fluid mechanics, **5**-24–31
Heat
    of fusion, **7**-15
    of vaporization, **7**-15
Heating value, **8**-34
Heat transfer, **8**-61–67
Heat treating, and material science, **9**-6–13
Henry units, **6**-1
Higher order algebraic equations, **1**-26–28
Higher order derivatives, **1**-59–60
Higher order determinants, **1**-10–11
High-pass filter, **6**-44
Hinge support, **3**-33
Histograms, **1**-101–103
Hooke's Law, **3**-4, **3**-15, **3**-16, **9**-17
Horsepower, definition of, **6**-1
Hybrid parameters, and transistors, **6**-222–276
Hydrocarbons
    carbon compounds and, **7**-15
    combustion of, **8**-33
Hydrogen, combustion of, **8**-31–32
Hydrostatic force, **5**-5–11
Hyperbolic functions, trigonometric, **1**-41–42
Hysteresis, **6**-117

Ideal bank angle, **4**-58
Ideal cycle efficiency, **8**-39
Ideal gas equation, **7**-4
Identical roots, **1**-86
Illuminated body, **9**-62
Imaginary numbers, **1**-2–3, **1**-6
Impact tests, **9**-15–17
Impedance
    admittance and, **6**-21–28
    transformers and, **6**-74–75
Implicit differentiation, **1**-59
Impulse, and momentum, **4**-84–88
Impurity atom, **6**-182–83
Incandescent body, **9**-62
Inclination, of slope of line, **1**-43
Income taxes, **10**-35, **10**-66
Incompressible fluid, **5**-2, **5**-19–24

Incremental rate of return, **10**-71–75
Indefinite integral, **1**-70–71
Independent events, and probability, **1**-95
Independent variables, **1**-49
Index
   of exponent, **1**-6
   of refraction, **9**-55
Inductance
   calculation of, **6**-10
   definition of, **6**-1
   electromagnetic, **6**-79
   equivalent, **6**-11
Induction hardening, **9**-7
Inductors, and time domain analysis, **6**-120–22
Inelastic collision, **4**-94
Inertia
   composite sections and, **3**-49–50
   force, **4**-25
   rectangles and, **3**-48–49
Inflation, and present worth, **10**-47
Insulators, electrical, **6**-4
Integers, **1**-1–2
Integral calculus
   applications of, **1**-75–81
   definite integral, **1**-71–72
   double integrals, **1**-75
   indefinite integral, **1**-70–71
   integration by parts, **1**-74
Integration
   by centroids, **2**-90–91
   by parts, **1**-74
Intensity of sound, **9**-47–50
Interest rate, **10**-2, **10**-4–27, **10**-35, **10**-37
Interference, and light, **9**-62
Intermediate alloy steel, **9**-8
Internal combustion (IC) engines, **8**-12
Internal energy, **8**-5–6
International System of Units (SI), **9**-62
International units (SI), of electricity, **6**-2
Intrinsic semiconductors, **6**-179, **6**-181–82
Inverse functions, trigonometric, **1**-36
Investment cost, **10**-37
Investment tax credit, **10**-66
Ionization constants, **7**-10
Iron, **9**-8, **9**-10–11
Iron oxide, **9**-10
Irrational number, **1**-1
Isotopes, **7**-2
Izod test, **9**-15

Jets, dynamic action of, **4**-88–90
Joints, and trusses, **2**-10–11
Joules and joule cycle, **6**-1, **8**-15
Junction field effect transistors (JFET), **6**-193–97

Keppler's Laws, **9**-64
Ketones, **7**-17
Kinematics
   of curvilinear motion, **4**-43–46
   definition of, **4**-1
   of rectilinear motion, **4**-8–13
Kinematic viscosity, **5**-41, **5**-42
Kinetics
   chemical reaction rate and equilibrium, **7**-8–12
   of curvilinear motion, **4**-53–57
   definition of, **4**-1
   friction, **2**-57
   fundamental equations for particle, **4**-5–6

Kirchoff's Voltage Laws, **6**-12, **6**-124, **6**-142, **6**-201

Laminar flow, **5**-44
Laplace transformation
   definition of, **1**-88–89
   differential equations, **1**-92–93
   electricity and, **6**-142–48
   systems engineering and, **6**-152–55, **6**-162
   table of, **1**-90–92
Laws
   Avogadro's, **6**-56, **7**-3, **8**-28
   of Conservation of Momentum, **4**-94
   Coulomb's, **6**-2
   Dalton's Law of Partial Pressures, **7**-5–6, **8**-30–31
   Faraday's, **6**-4–5, **6**-79
   of gases, **7**-3–6
   Gay-Lussac's, **7**-3
   Graham's, **7**-5
   Hooke's, **3**-4, **3**-15, **3**-16, **9**-17
   Keppler's, **9**-64
   Kirchoff's Voltage, **6**-12, **6**-124, **6**-142, **6**-201
   of Le Chatêlier, **7**-8–9
   Newton's laws of motion, **4**-4–32, **4**-95
   Ohm's, **6**-69, **6**-70, **6**-123, **6**-125
   Perfect Gas, **8**-2
   Raoult's, **7**-13
   of Reflection, **9**-50
   of Refraction, **9**-54–56
   Second of Thermodynamics, **8**-54
   of sines, **1**-40
   Snell's, **9**-54–56
   for triangles, **1**-39–41
   of vector summation, **1**-23
Lenses, **9**-57–63
Lensmaker's Equation, **9**-58
Lens power, **9**-59–60
Levelized cost calculation, **10**-20
L'Hospital's Rule, **1**-51
Light and optics, physics of, **9**-50–63
Limiting friction, **2**-57
Linear algebra, **1**-12–20
Linear motion, **4**-41–42
Linear speed, **4**-39–40
Linear thermal expansion, **9**-20
Line currents, **6**-65
Lines, centroids of, **2**-90
Liquid penetrant testing, **9**-14
Liquids
   sound waves and, **9**-45
   thermodynamics and, **8**-35–37
Loading, and trusses, **2**-14
Logarithms, **1**-28–33
Log Mean Temperature Difference (LMTD), **8**-63
Longitudinal waves, **9**-43–47
Low-pass filter, **6**-45
Lumen (lm), **9**-62
Luminous body, **9**-62

Magnetic particle testing, **9**-14
Magnetics, and electricity, **6**-69–71, **6**-117
Magnitude
   motion of particle and, **4**-1
   in vector analysis, **1**-22
Majority carriers, **6**-184–85
Malleable cast iron, **9**-8
Manning Formula, **5**-68
Mantissa, of logs, **1**-30, **1**-31
Martinsite, **9**-10

Mass, center of, **4**-6–7
Material science
   crystal structures, **9**-1–2
   galvanic action, **9**-12–13
   heat treating, **9**-6–13
   impact tests, **9**-15–17
   materials testing, **9**-13–15
   nonmetallic materials, **9**-23–26
   phase diagrams, **9**-5–6
   physical properties, **9**-18–22
   space lattices, **9**-2–5
Materials testing, **9**-13–15
Mathematics
   algebra, **1**-5–28
   analytic geometry, **1**-42–46
   chemistry and, **7**-26–27
   differential calculus, **1**-49–69
   differential equations, **1**-82–88
   integral calculus, **1**-70–81
   Laplace transformation, **1**-88–93
   logarithms, **1**-28–33
   numbers, **1**-1–3
   probability, **1**-94–97
   sets, **1**-3–5
   statistics, **1**-97–103
   trigonometry, **1**-33–42
Matrix, and determinant, **1**-9
Matter, structure of, **9**-1–2. *See also* Material science
Maturity period, of bond, **10**-21
Max acceleration, **9**-34
Maximum values, determination of, **1**-67–68
Max velocity, **9**-34
Mean, statistical, **1**-98
Measurements, of electricity, **6**-105–107
Mechanical efficiency, **8**-39
Mechanical force, **6**-1
Mechanical work, **6**-2
Mechanics of materials
   beam analysis, **3**-29–73
   definitions, **3**-2–9
   shearing strain, **3**-1
   statically indeterminate structures, **3**-9–13
   statics and, **3**-1
   stress-strain diagrams, **3**-2
   tensile and compressive strain, **3**-1
   torsion, **3**-13–21
Median, **1**-98
Metal oxide semiconductor field effect transistor (MOFSET), **6**-197–208
Metals and metallic elements
   corrosion, **9**-10–11
   crystalline structure of, **9**-1
   single displacement chemical reaction, **7**-7
   space lattices, **9**-2–5
   table of valences, **7**-20
Meter, kilogram, second (MKS), **6**-2
Methane series, **7**-15
Miller indices, **9**-3
Minimum attractive rate of return (MAR), **10**-74
Minimum cost analysis, **10**-87–90
Minimum values, determination of, **1**-67–68
Minority carriers, **6**-184
Mirrors, **9**-50–54
MLT (mass, length, time) system, **5**-73
Mode, statistical, **1**-98
Modulus of elasticity, **3**-4, **9**-17
Mohr's Circle, **3**-68–73
Molality, and chemical solutions, **7**-12

Molarity, and chemical solutions, **7**-12
Molecular weight, **7**-2
Molecule, definition of, **7**-2
Moles, **7**-26
Mollier Chart, **8**-38
Momentum, dynamics of
   conservation of, **4**-91–94
   fluid mechanics and, **5**-69–73
   impulse and, **4**-84–88
Moody Chart, **5**-25, **5**-39
Motion
   angular, **4**-41–42
   curvilinear, **4**-43–46, **4**-53–57
   flight of projectiles and component, **4**-46–48
   fundamental quantities of, **4**-1
   graphical representation of, **4**-13–20
   harmonic, **4**-100, **9**-31–39
   linear, **4**-41–42
   Newton's laws of, **4**-4–42
   radial and tangential coordinates of component, **4**-48–53
   rectilinear, **4**-2–3, **4**-8–13, **9**-26–30
   wave, **9**-39–47
Motors
   alternating current (AC), **6**-97–99, **6**-118
   direct current (DC), **6**-88–93, **6**-118
   three-phase induction, **6**-99–104
Multiplication, of complex numbers, **6**-58
Mutual flux, of transformers, **6**-73
Mutually exclusive events, and probability, **1**-95

Natural logs, **1**-30
Natural numbers, **1**-1
Natural response, and time domain analysis, **6**-123
Negative bending moment, **3**-33
Negative benefit, economic, **10**-76
Neutral axis, and beam analysis, **3**-47–48
Neutral current, **6**-68
Neutralization, **7**-18
Newtonian fluids, **5**-43
Newton's laws of motion
   center of gravity, **4**-7
   center of mass, **4**-6–7
   coefficient of restitution, **4**-95
   constant force-constant acceleration, **4**-20–25
   D'Alembert's principle, **4**-25–28
   equations of kinetics for particle, **4**-5–6
   graphical representation of motion, **4**-13–20
   kinematics of rectilinear motion, **4**-8–13
   rephrasing of, **4**-4–5
   space motion, **4**-7–8
   variable force, **4**-28–32
Nitriding, **9**-7
Noble metals, **9**-1
Noncentroidal rotation, **4**-70–75
Non-concurrent force systems
   parallel force systems and, **2**-3, **2**-7–8, **2**-23
   three-dimensional forces and, **2**-47
Nonferrous metals, **9**-1
Nonmetallic materials
   material science and, **9**-23–26
   single displacement chemical reaction, **7**-7
   table of valences for, **7**-21
Non-Newtonian fluids, **5**-43
Nonrepeating numbers, **1**-1
Non-saturation region, and junction field effect transistors, **6**-195
Nonterminating numbers, **1**-1
Normal distribution, **1**-101–103

Normality, and chemical solutions, **7**-12
Normalizing, and heat treatment of metals, **9**-7
Normal stress, and beam analysis, **3**-66–67
Norton's Theorem, **6**-35
Nozzles, in piping systems, **5**-36, **5**-59
N-p-n common emitter amplifier, **6**-213
N-p-n transistor, **6**-208, **6**-209
N-type semiconductors, **6**-182–83
Nuclear chemistry, **7**-14
Null set, **1**-3
Numbers
    imaginary, **1**-2–3
    real, **1**-1–2
Nusselt equation, **8**-65
Nyquist stability criterion, **6**-170

Ohm's Law, **6**-69, **6**-70, **6**-123, **6**-125
Ohm units, **6**-1
On-off controllers, **6**-150
Open channels, flow in, **5**-67–69
Open circuit test, of transformer, **6**-76
Open loop system, **6**-149, **6**-162
Operational amplifiers, **6**-231–34
Optics, **9**-50–63
Order, of differential equation, **1**-82
Ordinary differential equation, **1**-82
Organic chemistry, **7**-15–18
Orifices
    coefficients of, **5**-31–32
    losses in piping systems and, **5**-36
Otto cycle, **8**-12–13
Oxidation, and corrosion, **9**-11
Oxidation-reduction (redox) reactions, **7**-22–26

Pappus, theorem of, **2**-94
Parallel equivalent springs, **4**-80
Parallel force system
    concurrent force systems, **2**-2–3
    couple, **2**-3–4
    definition of, **2**-1
    equations of equilibrium, **2**-4–5
    non-concurrent force system, **2**-3
    planar structure with, **2**-5
    trusses, **2**-8–43
    Varignon's theorem, **2**-4
Parallelogram method, of vector addition, **2**-2, **2**-44
Parallel resonance, **6**-43–44
Partial derivatives, **1**-61–63
Partial differential equation, **1**-82
Particle, definition of, **4**-1
Particular solution, of differential equation, **1**-82, **1**-87
Pearlite, **9**-10
Pendulum, and simple harmonic motion, **4**-100
Pentavalent impurity, **6**-182–83
Percentage, and chemical solutions, **7**-12
Percussion, center of, **4**-72, **4**-73
Perfect Gas Law, **8**-2
Periodic chart, **7**-1–3, **7**-34
Permanent set, **3**-3
Permutations, and probability, **1**-95–96
Per-unit notation, and transformers, **6**-76–77
pH and pOH, values of water solutions, **7**-11–12
Phase diagrams, **9**-5–6
Physical pendulum, **4**-100
Physical properties, and material science, **9**-18–22
Physics
    harmonic motion, **9**-31–39
    Keppler's Laws, **9**-64

    light and optics, **9**-50–63
    rectilinear motion, **9**-26–30
    sound intensity, **9**-47–50
    wave motion, **9**-39–47
Pig iron, **9**-8
Pipe friction, **5**-24–31
Pipe line, supplied by pump, **5**-47–50
Pitot tube, **5**-56
Planar structure
    with concurrent force systems, **2**-6
    with non-concurrent force systems, **2**-7–8
    with parallel system of forces, **2**-5
Planets, motion of in solar system, **9**-64
P-N junction semiconductors, **6**-184–85
P-n-p transistor, **6**-208, **6**-209
Poisson's ratio, **3**-4
Polar coordinates, **1**-37–38
Polarized light, **9**-62
Poles, systems engineering and location of, **6**-163–67
Polymers, **9**-26
Polytropic processes, in thermodynamics, **8**-7–9
Populations, and statistics, **1**-97
Porosity, of metals, **9**-22
Positive bending moment, **3**-33
Power
    definition of, **6**-1
    factor correction for synchronous motors, **6**-103–104
    plants and thermodynamics, **8**-50–54
    screws, **2**-65
    transfer of in transmission lines, **6**-37
Prandtl number, **8**-65
Precious metals, **9**-1
Present sum of money, **10**-2
Present worth, **10**-21, **10**-46–56
Pressure
    center of, **5**-6
    definition of, **5**-1
    Le Chatêlier's Law, **7**-8
    standard, **7**-3
    thin-walled cylinders under, **5**-11–13
Prestressed concrete, **9**-23
Primary leakage flux, of transformers, **6**-73
Prime number, **1**-1
Prismatic volume, **5**-1
Prisms, **9**-56–57
Probability, **1**-94–97
Profit, **10**-34–35
Projectiles, flight of, **4**-46–48
Proportional limit, **3**-2, **9**-17
Psychometrics, and thermodynamics, **8**-47–50
P-type semiconductors, **6**-183
Pump
    pipe line supplied by, **5**-47–50
    steam generators and, **8**-51
Purchase cost, of asset, **10**-27
Pure imaginary numbers, **1**-2
Pythagoras' triangle, **1**-8–9
Pythagorean theorem, **1**-36–37, **1**-38

Quadratic equation, **1**-26–27
Quiescent point (Q pt), of transistor, **6**-214

Radial and tangential coordinates, of component motion, **4**-48–53
Radians, of angles, **1**-33
Radiation, and heat transfer, **8**-66
Radicals, **1**-6
Radicand, **1**-6
Radioactivity, and nuclear chemistry, **7**-14

Radiography, **9**-13
Radius of gyration, **4**-70
Range, of sample, **1**-99–100
Rankine cycle, **8**-38–39, **8**-51
Raoult's Law, **7**-13
Rate of return, **10**-37, **10**-58–76
Rational fractions, **1**-1
Reaction rate, chemical, **7**-8–12
Reactions, and beam analysis, **3**-33–36
Real numbers, **1**-1–2
Rectangle, moment of inertia of, **3**-48–49
Rectangular coordinates, of curvilinear motion, **4**-43-46
Rectangular weir, **5**-61–62
Rectilinear motion
    equations of, **4**-2–3
    kinematics of, **4**-8–13
    physics and, **9**-26–30
Reduction (redox) reactions, **7**-22–26
Reflections, **9**-50–54
Refraction, **9**-54–56
Refrigeration, **8**-57–61
Regulation, definition of, **6**-1
Reheat cycle, and thermodynamics, **8**-54–57
Reheater, of steam generator, **8**-51
Relative humidity, **8**-47
Replacement studies, **10**-81–83
Resistance
    calculation of, **6**-5–7
    definition of, **6**-1
    electric/magnetic analogies, **6**-70
    equivalent, **6**-7–8
    semiconductors and dynamic, **6**-193
Resonance, electrical
    parallel, **6**-43–44
    series, **6**-40–43
Restitution, coefficient of, **4**-95
Retardation, of velocity, **4**-3
Reversed effective acceleration, **4**-50
Reversed effective force (REF), **4**-25
Reversible reactions, chemical, **7**-7
Reynolds number, **5**-25, **5**-39, **5**-41–46, **5**-77
Rigid body, rotation of, **4**-65–70
Risk, total annual cost including, **10**-44–46
RLC series circuit, **6**-140–42
Rockwell hardness tester, **9**-14, **9**-15
Root Locus Method, and systems engineering, **6**-171
Rotating electrical machines, **6**-79–81
Rotation
    fluid mechanics and, **5**-75–76
    noncentroidal, **4**-70–75
    of rigid body, **4**-65–70
Rotational motion, **9**-32
Routh's criterion, **6**-167–69
Rule of Current Division, **6**-17
Rule of Dulong and Pettit, **7**-2–3
Rust (iron oxide), **9**-10

Salvage value, of asset, **10**-27, **10**-37
Sand, and concrete, **9**-23, **9**-24
Saponification, **7**-18
Saturation temperature, **8**-35
Saybolt viscosity, **5**-42–43
Scalars, **4**-1
Scaler, definition of, **1**-21, **6**-2
Scaler product, **1**-23–24
Scaler-vector laws, **1**-23
Screw jacks, **2**-65
Screw thread friction, **2**-65–67

Secondary leakage flux, and transformers, **6**-74
Second Law of Thermodynamics, **8**-54
Second-order derivatives, **1**-61
Second-order determinant, **1**-9
Second-order linear differential equations, **1**-85–87
Sections, trusses and method of, **2**-15–21
Self-locking wedges, **2**-64
Selling price (SP), of asset, **10**-66
Semiconductors
    covalent bonding, **6**-179–81
    diodes, **6**-185–92
    dynamic resistance, **6**-193
    electrical field intensity, **6**-4
    extrinsic, **6**-182
    intrinsic, **6**-181–82
    minority carriers, **6**-184
    N-type, **6**-182–83
    P-N junction, **6**-184–85
    P-type, **6**-183
Series, equivalent springs in, **4**-81
Series cash flow calculations, **10**-20
Series DC motor, **6**-118
Series R-C circuit, **6**-132–39
Series resistor, **6**-105
Series resonance, **6**-40–43
Series R-L circuit, **6**-124–32
Servomechanisms, **6**-150, **6**-151
Sets, **1**-3–5
Shaded pole AC motor, **6**-118
Shear force, and beam analysis, **3**-32, **3**-36–38, **3**-43–45
Shearing strain, **3**-1
Shear stress, and beam analysis, **3**-66–68
Short circuit test, of transformer, **6**-75–76
Shunt DC motor, **6**-118
Sign convention, in beam analysis, **3**-32–33
Silicon, **6**-179
Silicon rectifier diode, **9**-11
Simple filters, **6**-44–47
Simple harmonic motion, **4**-100
Simple L pad, **6**-46
Simple pendulum, **4**-100
Simultaneous linear equations, **1**-12–14
Sines, laws of, **1**-40
Single displacement, chemical reaction, **7**-7
Single payment formulas, **10**-5
Single-phase motors, **6**-98, **6**-118
Sinking-fund depreciation method (SFD), **10**-28, **10**-31, **10**-37
Slope-intercept form, of straight-line equation, **1**-42–46
Sluice gate, **5**-67
Snell's Law, **9**-54–56
Soaps, **7**-18
Solar system, **9**-64
Solids, and sound waves, **9**-44
Solutions, and chemistry, **7**-12–13
Sound
    intensity of, **9**-47–50
    testing, **9**-13
    waves, **9**-43–47
Space charge region, **6**-185
Space lattices, **9**-2–5
Space motion, **4**-7–8
Specific gravity, **9**-21
Specific heat, **8**-4–5
Spectrographic testing, **9**-14
Speed
    of DC motor, **6**-91
    definition of, **4**-2
    of light as constant, **6**-56

relation between angular and linear, **4**-39–40
Spherical lens equations, **9**-60–61
Spherical mirrors, **9**-51–54
Split-phase AC motor, **6**-118
Spring mass, **4**-100
Springs, equivalent
    in parallel, **4**-80
    in series, **4**-81
Square array, **1**-9
Square root, of negative number, **1**-2
Squirrel cage motor, **6**-98
Stability, and systems engineering, **6**-162–67, **6**-170
Stainless steel, **9**-8, **9**-11
Standard block diagram symbols, **6**-150–51
Standard deviation, **1**-99–100
Standard oxidation potentials, **7**-24
Standard pressure, **7**-3
Standard temperature, **7**-3
Standing waves, **9**-42–43
Statically determinate beams, **3**-30
Statically indeterminate structures, **3**-9–13
Statically indeterminate torsion problems, **3**-19
Static friction, **2**-57
Static head, **5**-2
Statics
    centroids, **2**-87–93
    coplanar force systems, **2**-1–43
    mechanics of materials compared to, **3**-1
    theorems of Pappus and Guldinus, **2**-94
    three-dimensional forces, **2**-44–87
Statistics, **1**-97–103
Steady-state response, and time domain analysis, **6**-123
Steam power plants, **8**-50–54
Steel, **9**-7–8, **9**-18
Stefan-Botzmann Constant, **6**-57, **8**-66
Step controllers, **6**-150
Stirling cycle, **8**-12
Stoichiometry, **7**-26–27
Straight line depreciation method (SLD), **10**-28
Straight lines, equations of, **1**-42–46
Streamline flow, **5**-44
Stress
    combined, **3**-65–66, **3**-70–73
    formulas for normal and shear, **3**-66–67
    planes of maximum or minimum normal, **3**-67
    planes of maximum or minimum shear, **3**-67–68
    stress-strain diagrams, **3**-2
Stress corrosion, **9**-13
Stress-strain diagrams, **3**-2, **9**-18
Substituted hydrocarbons, **7**-17
Subtraction
    of complex numbers, **6**-58
    of vectors, **1**-22
Sudden contraction, in piping systems, **5**-36
Sudden enlargement, in piping systems, **5**-35
Sulfur, combustion of, **8**-32
Sum of the years digits depreciation (SYD) method, **10**-30, **10**-33, **10**-37
Synchronous motors, **6**-103–104, **6**-114
Synthetic organic material, **9**-26
System engineering
    automatic control systems, **6**-149–50
    block diagram algebra, **6**-158–61
    definition of, **6**-149
    differential equations of feedback control systems, **6**-152
    examples of, **6**-171–78
    Laplace transform, **6**-152–55
    location of poles and stability, **6**-163–67

    methods of system analysis, **6**-151
    Nyquist stability criterion, **6**-170
    Root Locus Method, **6**-171
    Routh's criterion, **6**-167–69
    stability and, **6**-162–63
    standard block diagram symbols, **6**-150–51
    transfer functions, **6**-152, **6**-155–58

Tangential coordinates, of component motion, **4**-48–53
Taxable income (TI), **10**-65
Temperature
    corrosion and, **9**-11
    critical point, **9**-9
    gas laws and standard, **7**-3
    Law of Le Chatêlier and, **7**-9
    saturation temperature, **8**-35
Tempering, of metals, **9**-7
Tensile impact testing, **9**-16, **9**-17
Tensile strain, **3**-1
Terminal velocity, **4**-36
Testing, of transformers, **6**-75–76
Theorems of Pappus and Guldinus, **2**-94
Thermal conductivity, **9**-18–19, **9**-20
Thermal efficiency, **8**-39
Thermal expansion, **3**-4, **9**-20–21
Thermochemistry, **7**-14–15
Thermodynamics
    Boyle's Law, **8**-1
    Charle's Law, **8**-1–2
    combustion and, **8**-31–34
    compression of air, **8**-17–18
    cycles, **8**-10–17
    efficiencies, **8**-39–41
    enthalpy, **8**-6
    entropy, **8**-7
    Equations of State, **8**-2–3
    gas mixtures, **8**-28–29
    General Energy Equation, **8**-3–4
    gravimetric analysis, **8**-29–31
    heating value, **8**-34
    heat transfer, **8**-61–67
    internal energy, **8**-5–6
    liquids and vapors, **8**-35–37
    polytropic processes, **8**-7–9
    power plants, **8**-50–54
    psychometrics, **8**-47–50
    rankine cycle, **8**-38–39
    refrigeration, **8**-57–61
    reheat cycle, **8**-54–57
    specific heat, **8**-4–5
    throttling calorimeter, **8**-38
    volumetric analysis, **8**-29
Thevenin equivalent circuit, **6**-138
Thevenin's Theorem, **6**-34
Thick-walled cylinders, **5**-13–14
Thin-walled cylinder, **5**-11–13
Thixotropic fluids, **5**-43
Three-dimensional forces
    belt friction, **2**-67–68
    concurrent forces not mutually perpendicular, **2**-46
    concurrent forces in space, **2**-44
    concurrent moment vectors, **2**-49–55
    direction cosines, **2**-45
    equations of equilibrium and, **2**-47, **2**-49–55
    friction and, **2**-55–62
    moment of force about an axis, **2**-47–48
    nonconcurrent space forces, **2**-47
    screw thread friction, **2**-65–67

wedges and, 2-63–64
Three force members, and trusses, 2-21–22
Three-phase circuits, 6-63–69
Three-phase induction motors, 6-98, 6-99–104
Three-reservoir problem, 5-50
Throttling calorimeter, 8-38
Time
    displacement as function of, 4-13–15
    force as function of, 4-29, 4-87–88
    velocity as function of, 4-15–19
Time domain analysis
    capacitors, 6-122–23
    diagram of transient period, 6-119
    inductors, 6-120–22
    physical components of, 6-120
    RLC series circuit, 6-140–42
    series R-C circuit, 6-132–39
    series R-L circuit, 6-124–32
    total or complete response, 6-123
Time value of money, 10-2–3
Tip-to-tail method, of vector addition, 2-2, 2-44, 2-46
Top dead center (TDC), 8-14
Torsion
    angle of twist, 3-15–17
    compound shafts, 3-17
    definition of, 3-13–14
    moments of external and internal forces, 3-14–15
    statically indeterminate problems, 3-19
Torsional motion, 9-32
Torsional spring, 4-100
Torsion impact testing, 9-16
Total annual cost including risk, 10-44–46
Total annual cost (TAC) method, 10-35-44
Total internal reflection, 9-55–56
Total momentum, 4-92
Total response, and time domain analysis, 6-123
Transfer functions, and systems engineering, 6-152, 6-155–58
Transformers, electrical, 6-71–79
Transient period, diagram of, 6-119
Transistors
    bipolar junction (BJT), 6-219–20
    characteristic equations, 6-221
    common base configuration, 6-211–12, 6-217–19
    common collector configuration, 6-219
    common emitter configuration, 6-212–17, 6-226–30
    hybrid parameters, 6-222–26
    operational amplifiers, 6-231–34
    theory of operation, 6-208–11
Translation, and fluid mechanics, 5-75–76
Translational motion, 4-7, 9-32
Transverse waves, 9-40–41
Trapezoidal weir, 5-66
Triangles
    laws for, 1-39–41
    trusses, 2-9
Triangular weirs, 5-65
Trigonometry
    angles, 1-33–34
    fundamental identities, 1-38–39
    hyperbolic functions, 1-41–42
    laws for triangles, 1-39–41
    polar coordinates, 1-37–38
    trigonometric functions, 1-34–37
Trivalent impurity atom, 6-183
Trusses
    definition of, 2-8–9
    equilibrium equations, 2-12–13
    free body diagrams, 2-12, 2-22–43

    method of joints, 2-10–15
    method of sections, 2-15–21
    three force members, 2-21–22
Turbulent flow, 5-44–45
Twist, angle of, 3-15–17
Two-angle formulas, 1-39
Two-stroke cycle, 8-12

Ultimate tensile strength, 3-3, 9-18
Ultrasonic testing, 9-13
Uniform equal series of payments or receipts, 10-3
Uniform gradient series of payments or receipts, 10-3
Uniform rectilinear motion, 9-28
Uniform series formulas, 10-6, 10-7
Unity feedback, 6-162
Universal AC motor, 6-118
Universal and specific gas constant, 7-4–5
Useful life, of asset, 10-27

Valence, and chemistry, 7-19–21
Valves, in piping systems, 5-36
Vapor, and thermodynamics, 8-35–37
Vaporization, heat of, 7-15
Variable force
    Newton's laws of motion and, 4-28–32
    work-energy method involving, 4-77–80
Variables separable, equations with, 1-82–84
Variance, statistical, 1-99–100
Varignon's theorem, 2-4
Vectors
    analysis, 1-21–26
    definition of, 1-21, 6-2
    motion and, 4–1
    product of, 1-24–26
    summation laws, 1-23
Velocity
    acceleration and, 4-19–20
    angular, 4-39–42
    constant, 4–8, 9-28
    definition of, 4-2
    derivatives and, 1-63
    elastic impact, 4-95
    force and, 4-33–39
    longitudinal waves, 9-43
    maximum, 9-34
    motion and, 4-1
    terminal, 4-36
    time and, 4-15–19
    transverse wave and, 9-40–41
Vena contracta, 5-31
Venturis, and losses in piping systems, 5-36
Venturi tube, 5-56–57
Vickers hardness tester, 9-14
Viscosity, and Reynolds number, 5-41–46
Viscous flow, 5-44

Volatility, of metals, **9**-22
Voltage, and direct current source, **6**-14
Volt units, **6**-1
Volumetric analysis, and thermodynamics, **8**-29
Volumetric efficiency, **8**-39
Volumetric thermal expansion, **9**-20–21

Water
    concrete and, **9**-23, **9**-24
    freezing point depression and boiling point elevation of solutions, **7**-13
    as incompressible fluid, **5**-2
    pH and pOH values of solutions, **7**-11–12
    properties of, **5**-46
    vapor pressure, **8**-33
Wavelengths, and color, **9**-57
Wave motion, **9**-39–47
Wedges, and three-dimensional forces, **2**-63–64
Weir, definition of, **5**-61
White cast iron, **9**-8
Whole numbers, **1**-1
Wood, as structural material, **9**-25
Work-energy, dynamics of, **4**-75–80
Wound rotor, **6**-98
WYE circuit, **6**-66–67

Yield point, **3**-2
Yield strength, **3**-3, **9**-18

Zener diode, **6**-187–88
Zero shear, and beam analysis, **3**-67